DISCRETE STRUCTURES

DISCRETE STRUCTURES

For

B.Tech. and M.C.A. Courses

By

Dr SATINDER BAL GUPTA
B.Tech., (CSE), MCA, UGC-NET, Ph.D (CS)
Prof. Deptt. of Computer Science & Applications
Vaish College of Engineering, Rohtak,
Haryana.
Email:satinderbal@gmail.com

Dr C.P. GANDHI
Ph.D (Mathematics), UGC-NET
Associate Professor; Deptt. of Mathematics,
Rayat & Bahra Institute of Engineering
& Bio-Technology, Kharar,
Chandigarh.
Email:cchanderr@gmail.com

UNIVERSITY SCIENCE PRESS
(An Imprint of Laxmi Publications Pvt. Ltd.)
An ISO 9001:2015 Company

BENGALURU • CHENNAI • GUWAHATI • HYDERABAD
KOCHI • KOLKATA • LUCKNOW • RANCHI
NEW DELHI

Printed and bound in India

First edition: 2009, *Second edition:*2010, *Third Edition :* 2013, *Fourth Edition:* 2014, *Fifth Edition* : 2025

ISBN : 978-93-80386-35-5

Branches

©	Bengaluru	080-26 75 69 30
©	Chennai	044-24 34 47 26
©	Guwahati	0361-254 36 69
©	Hyderabad	040-27 55 53 83
©	Kochi	0484-405 13 03
©	Kolkata	033-40 04 77 79
©	Lucknow	0522-430 36 13
©	Ranchi	0651-224 24 64

PUBLISHED IN INDIA BY

(An Imprint of Laxmi Publications Pvt. Ltd.)

An ISO 9001:2015 Company

113, GOLDEN HOUSE, GURUDWARA ROAD, DARYAGANJ,

NEW DELHI - 110002, INDIA

Corporate Office : B-33, Sector-5, Noida,

Gautambuddha Nagar (U.P.) - 201 301, INDIA

Telephone : +91-120-44 46 900

www.laxmipublications.com

info@laxmipublications.com

Preface to the Fifth Edition

"Discrete Structures" primarily meant for B.Tech. and M.C.A. students of various Indian Universities, has been written by taking into consideration the students' capability of solving the mathematical as well as graphical problems in a systematic and logical manner. The authors have certainly left no stone unturned in presenting the subject matter in a comprehensive and lucid way.

The various topics like

- Basic Counting Principles
- Inclusion-Exclusion Principle
- Equivalence Relations and Partitions
- Recurrence Relations and their Solutions by Generating Functions
- Permutation and Symmetric Group
- Direct Products of Groups
- Dihedral Groups
- Applications of Groups in Coding Theory
- Rings
- Applications of Boolean Algebra in Logic Circuits and Switching Designs
- Applications of Graphs and Trees in Computer Science,

have been explained in an explanatory and methodological way. Further, the MCQs have also been provided at the end of the concerned chapters. This book gives an inkling of the capability of the authors and confirms that the authors must have been successful teachers.

Errors or misprints, if any, are unintentional and regretted. Any suggestion/ recommendation for improvement of the subject matter of the book, are invited.

Dr. C.P. Gandhi
Dr. Satinder Bal Gupta

CONTENTS

---

* Not meant for P.T.U. B.Tech, "Discrete Structures" (BTCS–302) Course.

* Not meant for P.T.U. B.Tech., "Discrete Structures" (BTCS–402) Course.

* Not meant for P.T.U. B.Tech., "Discrete Structures" (BTCS–402) Course.

SYLLABUS DETAILS

The book entitled "**Discrete Structures**" is strictly according to the **latest syllabus** of various Indian universities as per details given below:

Sr. N.	*Name of the University*	*Subject*	*Subject Code*
1.	Anna University, Chennai	Discrete Mathematics	(MA 035)
2.	Himachal Pradesh University, Shimla	Discrete Mathematics and Logic Design	(IT 4003)
3.	Kurukshetra University, Kurukshetra	Discrete Structures	(CSE-205 E)
4.	Maharishi Dayanand University, Rohatak	Discrete Structures	(CSE-203-F)
5.	Pondicherry University, Puducherry	Discrete Mathematics and Graph Theory	(MA T41)
6.	Punjab Technical University, Jalandhar	Computer Mathematical Foundation	(MCA-104 (N2)
7.	Punjay Technical University, Jalandhar	Discrete Structures	BTCS-402
8.	Swami Vivekanand University, Sironja, Sagar (MP)	Discrete Structures	(IT 302)
9.	University of Mumbai, Mumbai	Discrete Structures	(CSE-305)
10.	Uttar Pradesh Technical University, Lucknow	Discrete Mathematical Structures	(ECS-303)

SYLLABUS

ANNA UNIVERSITY, CHENNAI
DISCRETE MATHEMATICS (MA 035)

1. **Logic**

   Statements-Truth Tables-connectives-Normal forms-Predicate Calculus-Inference theory for statement calculus and Predicate Calculus.

2. **Combinatorics**

   Review of Permutation and combination-Mathematical Inducation-Pigeon hole principle-Principle of inclusion and exclusion-Generating function-Recurrence relations.

3. **Groups**

   Semi groups-Monoids-groups-permutation group-Consets-Lagranges theorem-Group homomorphism-Kernal-Rings and Fields (definitions and examples only).

4. **Lattices**

   Partial ordering-Posets-Hasse diagram-Lattices-Properties of Lattices-Sub Lattices-Special Lattices-Boolean Algebra.

5. **Graphs**

   Introduction to Graphs-Graph terminology-Representation of Graphs-Graph Isomorphism-Connectivity-Euler and Hamilton Paths

SYLLABUS

HIMACHAL PRADESH UNIVERSITY, SHIMLA
DISCRETE MATHEMATICS AND LOGIC DESIGN (IT 4003)

SECTION A

Mathematical Logic: Statements and Notation, Connectives; Negation; Conjunction; Disjunction; Statement Formulas and Truth Tables; Logical Capabilities of programming Languages; Conditional and Biconditional; Well formed Formulas; Tautologies; Equivalence of Formulas; Duality Law; Tautological Implications; Formulas with distinct Truth Tables; Functionally Complete sets of connectives; other connectives; Two state devices and Statement logic; Normal forms; principal disjunctive normal forms; principal conjunctive normal form; ordering and Uniqueness of Normal Forms; Completely parenthesized Infix Notation and Polish Notation; The Theory of Interference for the statement calculus; Validity using Truth Table; Rules of Inference; Consistency of Premises and Indirect Method Of Proof; Automatic Theory Proving; The Predicate Calculus; predicates; The Statement Function, variables and Quantifiers; Predicate Formulas; Free and Bound Variables; The Universe of Discourse; Inference Theory of the Predicate Calculus, Valid Formulas and Equivalences; Some Valid Formulas over Finite Universes; Special Valid Formulas Involving Quantifiers; Theory of Inference for the Predicate Calculus; Formula Involving More Than One Quantifier.

SECTION B

Permutations, Combinations, and Discrete Probability: Introduction, The Rules of Sum and product; permutations; Combinations; Generation of permutations and combinations, Discrete probability, Information and Mutual Information.

Relations and Functions: Introduction, A Relational Model for Data Bases; properties of Binary Relations; Equivalence Relations and partitions; Partial Ordering Relations and Lattices; Chains and Antichains; A Job Scheduling problem; Functions and the Pigeonhole principle.

SECTION C

Graphs and Planner Graphs: Introduction, Basic Terminology, Multigraphs and Weighted Graphs, Paths and Circuits; Shortest paths in Weighted Graphs, Eulerian paths and circuits; Hamiltonian paths and circuits, The Traveling Salesperson problem; Factors of Graph; planar Graph.

Trees and cut-sets: Trees, Rooted Trees, path, Lengths in Rooted trees; prefix codes; Binary search trees; Spanning Trees and cut-sets; Minimum Spanning Trees; Transport Networks.

SECTION D

Recurrence Relations and Recursive Algorithms: Introduction; Recurrence Relations; Linear Recurrence Relations with constant coefficients; Homogeneous Solutions; Particular Solutions; total Solutions; solution by the Method of Generating Functions; Sorting Algorithms; Matrix Multiplication Algorithms.

Groups and Rings: Introduction, Groups, Subgroups; Generators and evaluation of Powers; Cosets and Lagrange's Theorem; permutation groups and Burnside's theorem; Codes and Group codes; Isomorphisms and Automorphisms; Homomorphisms and Norma Integral Domains, and Fields; Ring Homomorphisms; Polynomial Rings and Cyclic Codes.

SYLLABUS

KURUKSHETRA UNIVERSITY, KURUKSHETRA
DISCRETE STRUCTURES (CSE-205 E)

UNIT-I

Set Theory: Introduction to set theory, Set operations, Algebra of sets, combination ofsets, Duality, Finite and Infinite sets, Classes of sets, Power Sets, Multi sets, Cartesian Product, Representation of relations, Types of relation, Binary Relations, Equivalence relations and partitions, Partial ordering relations and lattices, Mathematics Induction, Principle of Inclusion and Exclusion, Propositions. Function and its types, Composition of function and relations, Cardinality and inverse relations. Functions and Pigeon Hole principles.

UNIT-II

Propositional Calculus: Basic operations: AND, OR, NOT, Truth value of a compound statement, propositions, tautologies, contradictions. Techniques of Counting: Rules of Sum of products, Permutations with and without repetition, Combination. Recursion and Recurrence Relation: Polynomials and their evaluation, Sequences, Introduction to AP, GP and AG series, partial fractions, linear recurrence relation with constant coefficients, Homogeneous solutions, Particular solutions, Total solution of a recurrence relation using generating functions.

UNIT-III

Algebric Structures: Definition, elementary properties of algebric structures, examples of a Monoid, Submonoid, Semi-group, Groups and rings, Homomorphism, Isomorphism and Automorphism, Subgroups and Normal subgroups, Cyclic groups, Integral domain and fields, Cosets, Lagrange's theorem, Rings, Division Ring.

UNIT-IV

Graphs And Trees: Introduction to graphs, Directed and Undirected graphs, Homomorphic and Isomorphic graphs, Subgraphs, Cut points and Bridges, Multigraph and Weighted graph, Paths and circuits, Shortest path in weighted graphs, Eurelian path and circuits, Hamilton paths and circuits, Planar graphs, Euler's formula, Trees, Rooted Trees, Spanning Trees and cut-sets, Binary trees and its traversals.

SYLLABUS

MAHARISHI DAYANAND UNIVERSITY, ROHATAK
DISCRETE STRUCTURES (CSE-203-F)

SECTION A

Set Theory and Propositional Calculus: Introduction to set theory, Set operations, Algebra of sets, Duality, Finite and Infinite sets, Classes of sets, Power Sets, Multi sets, Cartesian Product, Representation of relations, Types of relation, Equivalence relations and partitions, Partial ordering relations and lattices

Function and its Types: Composition of function and relations, Cardinality and inverse relations.

Introduction to propositional Calculus: Basic operations: AND, OR, NOT, Truth value of a compound statement, propositions, tautologies, contradictions.

SECTION B

Techniques of Counting and Recursion and Recurrence Relation: Permutations with and without repetition, Combination. Polynomials and their evaluation, Sequences, Introduction to AP, GP and AG series, partial fractions, linear recurrence relation with constant coefficients, Homogeneous solutions, Particular solutions, Total solution of a recurrence relation using generating functions.

SECTION C

Algebric Structures: Definition and examples of a monoid, Semigroup, Groups and rings, Homomorphism, Isomorphism and Automorphism, Subgroups and Normal subgroups, Cyclic groups, Integral domain and fields, Cosets, Lagrange's theorem.

SECTION D

Graphs and Trees: Introduction to graphs, Directed and Undirected graphs, Homomorphic and Isomorphic graphs, Subgraphs, Cut points and Bridges, Multigraph and Weighted graph, Paths and circuits, Shortest path in weighted graphs, Eurelian path and circuits, Hamilton paths and circuits, Planar graphs, Euler's formula, Trees, Spanning trees, Binary trees and its traversals.

SYLLABUS

PONDICHERRY UNIVERSITY, PUDUCHERRY
DISCRETE MATHEMATICS AND GRAPH THEORY (MA T41)

UNIT-I

Connectives: Statement formulae, Equivalence of Statement formulae, Functionally complete set of connectives - NAND and NOR connectives, implication, Principal conjunctive and disjunctive normal forms.

UNIT-II

Inference calculus: Derivation process - Conditional proof - Indirect method of proof-Automatic theorem proving - Predicate calculus.

UNIT-III

Partial ordering: Lattices - Properties - Lattices as algebraic system - sub lattices - Direct product and homomorphism - Special lattices - Complemented and Distributive lattices.

UNIT-IV

Graphs: Applications of graphs - degree - pendant and isolated vertices - isomorphism-sub graphs - walks - paths and circuits - connected graphs - Euler graphs - operations on graphs - More on Euler graphs - Hamilton paths and circuits - complete graph.

UNIT-V

Trees: Properties of Trees - Pendant vertices in a Tree - Distance and Center in a Tree - rooted and binary trees - spanning trees - Fundamental Circuits - Distance between spanning trees shortest spanning trees - Kruskal algorithm.

SYLLABUS

PUNJAB TECHNICAL UNIVERSITY, JALANDHAR
DISCRETE STRUCTURES (BTCS-402)

PART-A

1. **Sets, Relations and Functions:** Introduction, Combination of Sets, Ordered pairs, Proofs of general identities of sets, Relations, Operations on relations, Properties of relations and Functions, Hashing Functions, Equivalence relations, Compatibility relations, Partial order relations. *[Chapters 1, 2, 3]* **[7]**
2. **Rings and Boolean Algebra:** Rings, Subrings, Morphism of rings, ideals and quotient rings, Euclidean domains, Integral domains and fields. Boolean Algebra, direct product, Boolean sub-algebra, Boolean Rings, Application of Boolean algebra (Logic Implications, Logic Gates, Karnaugh Map) *[Chapters 9, 10]* **[8]**
3. **Combinatorial Mathematics:** Basic counting principles Permutations and Combinations, Inclusion and Exclusion Principle. Recurrence relations, Generating Function, Applications. *[Chapters 5A, 5B, 6, 7]* **[7]**

PART-B

4. **Monoids and Groups:** Groups, Semigroups and Monoids. Cyclic Subgroups and Submonoids, Subgroups and Cosets, Congruence Relations on Semigroups. Morphisms. Normal Subgroups. Homomorphism, Isomorphism, Dihedral Groups, Applications. *[Chapter 8]* **[7]**
5. **Graph Theory:** Graph-Directed and Undirected, Eulerian Chains and cycles, Hamiltonian chains and cycles Trees, Chromatic number, Connectivity, Graph coloring, Planar and connected graphs, Isomorphism and Homomorphism. Applications. *[Chapters 11, 12]* **[7]**

Note. Chapters to be read for **"Discrete Structures" (BTCS 402):** Chapters No., 1, 2, 3, 5A, 5B, 6, 7, 8, 9, 10, 11, and 12 only.

SYLLABUS

PUNJAB TECHNICAL UNIVERSITY, JALANDHAR
COMPUTER MATHEMATICAL FOUNDATION (MCA-104 (N2)

SECTION A

Sets and Relations: Definition of sets, Subsets, Complement of a set, Universal set, Intersection and Union of sets, De-Morgan's laws, Cartesian products, Equivalent sets, Countable and Uncountable sets, Minset, Partitions of sets, Relations: Basic definitions, graphs of relations, Properties of relations. *[Chapters 1, 2]*

SECTION B

Algebra of Logic: Propositions and logic operations, Connectives, Tautologies and Contradiction, Equivalence and implication, Principle of Mathematical Induction, Quantifiers. *[Chapters 4, 13]*

SECTION C

Matrix Algebra: Introduction of a Matrix, its different kinds, Matrix addition and Scalar multiplication, Multiplication of matrices, Transpose etc. Square matrices, Inverse and Rank of a square matrix, Solving simultaneous equations using Gauss eliminations, Gauss Jordan and Matrix Inversion methods. *[Chapter 14]*

SECTION D

Graphs: A general introduction, Simple and Multigraphs, directed and Undirected graphs, Eulerian and Hamiltonian Graphs, Shortest path algorithms, Chromatic number, Bipartite graph, Graph coloring. *[Chapter 11]*

Note. Chapters to be read for **"Computer Mathematical Foundation (MCA 104 N2):** Chapters No., 1, 2, 4, 11, 13 and 14 only.

SYLLABUS

SWAMI VIVEKANAND UNIVERSITY, SIRONJA, SAGAR (MP)
DISCRETE STRUCTURES (IT 302)

Unit-I

Set Theory, Relation, Function, Theorem Proving Techniques : Set Theory: Definition of sets, countable and uncountable sets, Venn Diagrams, proofs of some general identities on sets Relation: Definition, types of relation, composition of relations, Pictorial representation of relation, Equivalence relation, Partial ordering relation, Job-Scheduling problem Function: Definition, type of functions, one to one, into and onto function, inverse function, composition of functions, recursively defined functions, pigeonhole principle. Theorem proving Techniques: Mathematical induction, Proof by contradiction.

Unit-II

Algebraic Structures: Definition, Properties, types: Semi Groups, Monoid, Groups, Abelian group, properties of groups, Subgroup, cyclic groups, Cosets, factor group, Permutation groups, Normal subgroup, Homomorphism and isomorphism of Groups, example and standard results, Rings and Fields: definition and standard results.

Unit-III

Propositional Logic: Proposition, First order logic, Basic logical operation, truth tables, tautologies, Contradictions, Algebra of Proposition, logical implications, logical equivalence, predicates, Normal Forms, Universal and existential quantifiers. Introduction to finite state machine Finite state machines as models of physical system equivalence machines, Finite state machines as language recognizers

Unit-IV

Graph Theory: Introduction and basic terminology of graphs, Planer graphs, Multigraphs and weighted graphs, Isomorphic graphs, Paths, Cycles and connectivity, Shortest path in weighted graph, Introduction to Eulerian paths and circuits, Hamiltonian paths and circuits, Graph coloring, chromatic number, Isomorphism and Homomorphism of graphs.

Unit-V

Posets, Hasse Diagram and Lattices: Introduction, ordered set, Hasse diagram of partially, ordered set, isomorphic ordered set, well ordered set, properties of Lattices, bounded and complemented lattices. Combinatorics: Introduction, Permutation and combination, Binomial Theorem, Multimonial Coefficients Recurrence Relation and Generating Function: Introduction to Recurrence Relation and Recursive algorithms , Linear recurrence relations with constant coefficients, Homogeneous solutions, Particular solutions, Total solutions , Generating functions , Solution by method of generating functions.

SYLLABUS

UNIVERSITY OF MUMBAI, MUMBAI
DISCRETE STRUCTURES (CSE-305)

MODULE-I

Set Theory: Sets, Venn diagrams, Operations on Sets; Laws of set theory, Power set and Products; Partitions of sets, The Principle of Inclusion and Exclusion.

MODULE-II

Logic: Propositions and logical operations, Truth tables; Equivalence, Implications; Laws of logic, Normal Forms; Predicates and Quantifiers; Mathematical Induction.

MODULE-III

Relations, Digraphs and Lattices: Relations, Paths and Digraphs; Properties and types of binary relations; Manipulation of relations, Closures, Warshall's algorithm; Equivalence and partial ordered relations; Posets and Hasse diagram; Lattice.

MODULE-IV

Functions and Pigeon Hole Principle: Definition and types of functions: Injective, Surjective and Bijective; Composition, Identity and Inverse; Pigeon-hole principle.

MODULE-V

Generating Functions and Recurrence Relations: Series and Sequences; Generating functions; Recurrence relations; Recursive Functions: Applications of recurrence relations *e.g.*, Factorial, Fibonacci, Binary search, Quick Sort etc.

MODULE-VI

Graphs and Subgraphs: Definitions, Paths and circuits: Eulerian and Hamiltonian; Planer graphs, Graph coloring; Isomorphism of graphs; Subgraphs and Subgraph isomorphism.

MODULE-VII

Trees: Trees and weighted trees; Spanning trees and minimum spanning tree; Isomorphism of trees and sub trees; Prefix codes.

MODULE-VIII

Algebraic Structures: Algebraic structures with one binary operation: semigroup, monoids and groups; Product and quotient of algebraic structures; Isomorphism, Homomorphism and Automorphism; Cyclic groups, Normal subgroups; Codes and group codes.

SYLLABUS

UTTAR PRADESH TECHNICAL UNIVERSITY, LUCKNOW
DISCRETE MATHEMATICAL STRUCTURES (ECS-303)

UNIT-I

Set Theory: Introduction, Combination of sets, Multisets, Ordered pairs. Proofs of some general identities on sets.

Relations: Definition, Operations on relations, Properties of relations, Composite Relations, Equality of relations, Recursive definition of relation, Order of relations.

Functions: Definition, Classification of functions, Operations on functions, Recursively defined functions. Growth of Functions.

Natural Numbers: Introduction, Mathematical Induction, Variants of Induction, Induction with Nonzero Base cases. Proof Methods, Proof by counter - example, Proof by contradiction.

UNIT-II

Algebraic Structures: Definition, Groups, Subgroups and order, Cyclic Groups, Cosets, Lagrange's theorem, Normal Subgroups, Permutation and Symmetric groups, Group Homomorphisms, Definition and elementary properties of Rings and Fields, Integers Modulo n.

UNIT-III

Partial order sets: Definition, Partial order sets, Combination of partial order sets, Hasse diagram.

Lattices: Definition, Properties of lattices - Bounded, Complemented, Modular and Complete lattice.

Boolean Algebra: Introduction, Axioms and Theorems of Boolean algebra, Algebraic manipulation of Boolean expressions. Simplification of Boolean Functions, Karnaugh maps, Logic gates, Digital circuits and Boolean algebra.

UNIT-IV

Propositional Logic: Proposition, well formed formula, Truth tables, Tautology, Satisfiability, Contradiction, Algebra of proposition, Theory of Inference.

Predicate Logic: First order predicate, well formed formula of predicate, quantifiers, Inference theory of predicate logic.

UNIT-V

Trees: Definition, Binary tree, Binary tree traversal, Binary search tree.

Graphs: Definition and terminology, Representation of graphs, Multigraphs, Bipartite graphs, Planar graphs, Isomorphism and Homeomorphism of graphs, Euler and Hamiltonian paths, Graph colouring.

Recurrence Relation and Generating Function: Recursive definition of functions, Recursive algorithms, Method of solving recurrences.

Combinatorics: Introduction, Counting Techniques, Pigeonhole Principle, Polya's Counting Theory.

1 SETS

1.1. INTRODUCTION

In this chapter, we will study some basic laws of algebra of sets, Venn-diagrams, representation of various sets, ordered and unordered partition of sets.

Set. A set is defined as *a collection of distinct objects of* same type or class of objects.

The objects of a set are called elements or members of the set. Objects can be numbers, alphabets, names etc.

e.g.,
$$A = \{1, 2, 3, 4, 5\}$$

A is a set of numbers containing elements 1, 2, 3, 4 and 5.

Remarks. 1. A set is usually denoted by capital letters A, B, C, D, P, Q, R, S, T etc.

2. Elements of the set are defined by p, q, r, t or p_1, q_1, r_1, t_1 etc.

1.2. SET FORMATION

The set can be formed by two ways :

(*i*) Tabular form of a set

(*ii*) Builder form of a set.

(*i*) **Tabular Form or Roster Form of a Set.** If a set is defined by actually listing its members, *e.g.,* if the set P contains elements a, b, c, d, then it is expressed as $P = \{a, b, c, d\}$.

This is called *tabular form* of a set.

(*ii*) **Builder Form of a Set.** If a set is defined by the properties which its elements must satisfy *e.g.,*

$$P = \{x : x \in N, x \text{ is a multiple of } 3\}.$$

$$R = \{x : x > 1 \text{ and } x < 10 \text{ and } x \text{ is an odd integer}\}.$$

$$T = \{x : x \text{ is even number less than } 9\}.$$

This form is called *Set builder form* of a set and the method of defining the elements is called **Property Method**.

1.3. STANDARD NOTATIONS

$x \in A$	x belongs to A or x is an element of set A.
$x \notin A$	x does not belong to set A.
ϕ	Empty set.

U	Universal set.
N	The set of all natural numbers.
I	The set of all integers.
I_0	The set of all non-zero integers.
I_+	The set of all + ve integers.
C, C_0	The set of all complex, non-zero complex numbers respectively.
Q, Q_0, Q_+	The set of rationals, non-zero rationals, +ve rational numbers respectively.
R, R_0, R_+	The set of reals, non-zero reals, +ve real numbers respectively.

1.4. KINDS OF SETS

I. Finite Set. If a set consists of specific number of different elements, then that set is called finite set. *e.g.,*

$P = \{x : x \in N, 3 < x < 9\}$.

$Q = \{2, 4, 6, 8\}$.

R = {months of year}, are finite sets.

II. Infinite Set. If a set consists of infinite number of different elements or if the counting of different elements of the set does not come to an end, the set is called infinite set.

e.g., I = {The set of all integers}.

$E = \{x : x \in N, x$ is a multiple of $2\}$, are inifinite sets.

III. (Principle of Extension) Equality of Sets. Two sets A and B are said to be equal and written as A = B if both have the same elements. Therefore, every element which belongs to A is also an element of the set B and every element which belongs to the set B is also an element of the set A.

$$A = B \quad \Leftrightarrow \quad (x \in A \quad \Leftrightarrow \quad x \in B).$$

If there is some element in set A that does not belong to set B or vice-versa then $A \neq B$ *i.e.,* A is not equal to B.

Remarks 1. A set does not change if one or more of its elements are repeated.

2. A set does not change if we change the order in which its elements are tabulated.

IV. Disjoint Sets. Two sets A and B are said to be disjoint if no element of A is in B and no element of B is in A. *e.g.,* if $R = \{a, b, c\}$, $S = \{k, p, m\}$, then R and S are disjoint sets.

V. Family of Sets. If a set A contains elements which are itself sets, then it is called family of sets or a set of sets. *e.g.,* if $A = \{\{1, 2\}, \{3, 4\}, \phi\}$, then A is set of sets.

VI. Subset of a Set. If every element of a set A is also an element of a set B, then A is called subset of B and is written as $A \subseteq B$. B is called superset of A.

e.g., $A \subseteq B = \{x : x \in A \quad \Rightarrow \quad x \in B\}$

(*a*) **Proper Subset.** If A is subset of B and $A \neq B$, then A is said to be proper subset of B. If A is a proper subset of B, then B is not subset of A *i.e.,* there is at least one element in B which is not in A. *e.g.,*

(*i*) If $A = \{2, 3, 4\}$, $B = \{2, 3, 4, 5\}$, then A is a proper subset of B.

(*ii*) The empty set ϕ is a proper subset of every set.

(*b*) **Improper Subset.** If A is a subset of B and A = B, then A is said to be an improper subset of B. *e.g.,*

(*i*) If A = {2, 3, 4}, B = {2, 3, 4}, then A is improper subset of B.

(*ii*) Every set is improper subset of itself.

VII. Null Set or Empty Set. The set that contains no element is called the null set or the empty set and is denoted by ϕ. *e.g.,*

(*a*) P = $\{x : x^2 = 4, x \text{ is odd}\}$.

(*b*) Q = $\{x : x^2 = 9, x \text{ is even}\}$.

(*c*) R = $\{x : x^2 = 9, 2x = 4\}$, are all null sets.

Remark. The set A = {0} is not a null set because 0 is the element of the set.

The set A = $\{\phi\}$ is not a null set because set ϕ is the element of the set.

VIII. Power Set

The power set of any given set A is the set of all subsets of A and is denoted by P(A). If A has n elements, then P(A) has 2^n elements.

If A = $\{r, s\}$, then its subsets are ϕ, $\{r\}$, $\{s\}$, $\{r, s\}$. The power set of A is

P(A) = $\{\phi, \{r\}, \{s\}, \{r, s\}\}$ which has $2^2 = 4$ elements.

IX. Universal Set. If all the sets under investigation are subsets of a fixed set U, then the set U is called universal set. *e.g.,* in plane geometry, the universal set consists of all the points in the plane.

X. Comparability. Two sets A and B are comparable if one of them is subset of other, *i.e.,* $A \subseteq B$ or $B \subseteq A$. If $A \subseteq B$ and $B \subseteq A$, then A = B. Set ϕ is comparable to every set. Every set is comparable to the universal set U.

Two sets A and B are said to be *incomparable* if $A \not\subseteq B$ and also $B \not\subseteq A$ *i.e.,* there is at least one element in A not in B and vice-versa.

XI. Principle of Abstraction. Given any set U and any property P, there is a set A such that the elements of A are exactly those members of U which have the property P.

Theorem I. *Prove that, the null set ϕ is a subset of every set.*

Proof. Suppose that ϕ is not a subset of A *i.e.,* $\phi \not\subset A$.

Then there exists an element $x \in \phi$ such that $x \notin A$.

But ϕ is the null set and hence for every x, $x \notin \phi$ because null set does not contain any element.

From above, $x \in \phi$ and $x \notin \phi$ which is contradiction. Hence, $\phi \not\subset A$ is wrong supposition.

Therefore, ϕ is a subset of A. Hence proved.

Theorem II. *Prove that, every set is a subset of itself i.e., $A \subseteq A$.*

Proof. Let $x \in A \Rightarrow x \in A \quad \therefore \quad A \subseteq A$

As every element belonging to set A is also an element of set A. Therefore, A is subset of A or itself. Hence proved.

Theorem III. *Prove that, if $A \subseteq B$ and $B \subseteq C$, then $A \subseteq C$.*

Proof. Let x be any element of the set A.

Since, $A \subseteq B \quad \therefore \quad x \in A \Rightarrow x \in B$

$B \subseteq C \quad \therefore \quad x \in B \Rightarrow x \in C \qquad \therefore \quad x \in A \Rightarrow x \in C$

$\therefore \quad A \subseteq C$. Hence proved.

Theorem IV. *Prove that, if $A \subseteq B$, $B \subseteq C$, $C \subseteq A$, then $A = C$.*

Proof. If $A \subseteq B$ and $B \subseteq C$, then $A \subseteq C$ | Theorem III

Also $C \subseteq A$ (Given)

$\therefore$ $A \subseteq C$ and $C \subseteq A$

$\therefore$ $A = C$. Hence proved.

Theorem V. *Prove that, if $A \subset \phi$, then $A = \phi$.*

Proof. $\phi \subset A$ (As ϕ is subset of every set).

But $A \subset \phi$ (Given)

$\therefore$ $A = \phi$. Hence proved.

Theorem VI. *The following are equivalent*

(i) $A \subseteq B$ *(ii)* $A \cap B = A$ *(iii)* $A \cup B = B$.

Proof. *(i)* ⇒ *(ii)* Let $A \subseteq B$ and $x \in A$, then, $x \in B$ and hence $x \in A \cap B \Rightarrow A \subseteq A \cap B$

Also $A \cap B \subseteq A$. Consequently $A \cap B = A$

(ii) ⇒ *(iii)* Let $A \cap B = A$ and $x \in A$. Then,

$x \in A \cap B \Rightarrow x \in A$ and $x \in B$. Hence $A \subseteq B$.

(i) ⇒ *(iii)* Let $A \subseteq B$ and $x \in A \cup B$, then,

$x \in A$ or $x \in B$.

If $x \in A$, then $x \in B$ | $\because A \subseteq B$

In either case, $x \in B \Rightarrow A \cup B \subseteq B$

Also $B \subseteq A \cup B \Rightarrow A \cup B = B$

(iii) ⇒ *(i)* Let $A \cup B = B$ and $x \in A$. Then, by definition of $A \cup B$, $x \in A \cup B = B \Rightarrow x \in B$ $\Rightarrow A \subseteq B$.

Example. *How many subsets can be formed from a set of n elements ? How many of these will be proper and how many improper ?* (P.T.U. B.Tech. May 2007)

Sol. There are ${}^nC_1$ subsets each consisting of one of the elements of the given set.

There will be ${}^nC_2$ subsets each consisting of any two of the n elements of the given set.

Also, there will be ${}^nC_3$ subsets each consisting of any three of the n elements of given set.

At last, there will be ${}^nC_n$ subsets consisting of all the n elements of the given set.

Also, there will be one set ϕ.

Hence, the total number of subsets

$$= ({}^nC_1 + {}^nC_2 + {}^nC_3 + \ldots\ldots + {}^nC_n) + 1 \text{ (For the null set)}$$

$$= {}^nC_0 + {}^nC_1 + {}^nC_2 + \ldots\ldots + {}^nC_n = 2^n \qquad (\because\ {}^nC_0 = 1)$$

Therefore, 2^n subsets are formed from n elements of a set. Out of 2^n subsets, $2^n - 1$ subsets will be proper and 1 (one) subset improper *i.e.*, the set itself.

1.5. OPERATION ON SETS

The basic set operations are :

1. Union of Sets *(P.T.U. B.Tech. May 2006 ; M.C.A. May 2007)*

Union of the sets A and B is defined to be the set of all those elements which belong to A or B or both and is denoted by $A \cup B$, *i.e.*,

$$A \cup B = \{x : x \in A \ \text{ or } \ x \in B\}$$

e.g., Let $A = \{1, 2, 3\}, B = \{3, 4, 5, 6\},$
then $A \cup B = \{1, 2, 3, 4, 5, 6\}.$

2. Intersection of Sets *(P.T.U. B.Tech. May 2006; M.C.A. May 2007)*

Intersection of two sets A and B is the set of all those elements which belong to both A and B and is denoted by $A \cap B$, *i.e.,*

$$A \cap B = \{x : x \in A \text{ and } x \in B\}$$

e.g., Let $A = \{a, b, c, d\}, B = \{a, b, l, m\},$
then $A \cap B = \{a, b\}.$

3. Difference of Sets *(P.T.U. B.Tech. May 2007)*

The difference of two sets A and B is a set of all those elements which belong to A but do not belong to B and is denoted by A – B or A/B, *i.e.,*

$$A - B = \{x : x \in A \text{ and } x \notin B\}$$

e.g., Let $A = \{a, b, c, d\}, B = \{d, l, m, n\}$
then $A - B = \{a, b, c\}.$

Note. The set A – B or A/B is also known as relative complement of B w.r.t. A.

4. Complement of a Set w.r.t. a Universal Set *(P.T.U. B.Tech. May 2007)*

The complement of a set A is a set of all those elements of the universal set which do not belong to A and is denoted by A^c,

i.e., $$A^c = U - A = \{x : x \in U \text{ and } x \notin A\} = \{x : x \notin A\}$$

e.g., Let U be the set of all natural numbers.
Let $A = \{1, 2, 3\},$
Then $A^c = \{\text{all natural numbers except } 1, 2, 3\}.$

Note. The set A^c is also known as absolute complement of the set A.

5. Symmetric Difference of Sets. The symmetric difference of two sets A and B is the set containing all the elements that are in A or in B but not in both and is denoted by $A \oplus B$. *i.e.,*

$$A \oplus B = (A \cup B) - (A \cap B)$$

1.6. ALGEBRA OF SETS

A set is obtained from a set formula by replacing the variables by definite sets. When the set variables appearing in two set formulas are replaced by any sets and both the set formulas are equal as sets, then we call that both the set formulas are equal. The equality of set formulas are called set identities. Some of the identities describe certain properties of the operations involved. These properties describe an algebra called **algebra of sets.**

Table 1
General Identities of Set Theory

(*i*) **Idempotent Laws** (*a*) $A \cup A = A$ (*b*) $A \cap A = A$ (*ii*) **Associative Laws** (*a*) $(A \cup B) \cup C = A \cup (B \cup C)$ (*b*) $(A \cap B) \cap C = A \cap (B \cap C)$ (*iii*) **Commutative Laws** (*a*) $A \cup B = B \cup A$ (*b*) $A \cap B = B \cap A$ (*iv*) **Distributive Laws** (*a*) $A \cup (B \cap C) = (A \cup B) \cap (A \cup C)$ (*b*) $A \cap (B \cup C) = (A \cap B) \cup (A \cap C)$ (*v*) **De Morgan's Laws** (*a*) $(A \cup B)^c = A^c \cap B^c$ (*b*) $(A \cap B)^c = A^c \cup B^c$	(*vi*) **Identity Laws** (*a*) $A \cup \phi = A$ (*b*) $A \cap U = A$ (*c*) $A \cup U = U$ (*d*) $A \cap \phi = \phi$ (*vii*) **Complement Laws** (*a*) $A \cup A^c = U$ (*b*) $A \cap A^c = \phi$ (*c*) $U^c = \phi$ (*d*) $\phi^c = U$ (*viii*) **Involution Law** (*a*) $(A^c)^c = A$

The **Table 1** shows general identities of sets. We now prove these identities.

Theorem I. *Prove Idempotent laws i.e.,*

(a) $A \cup A = A$ *(b)* $A \cap A = A$.

Proof. (*a*) To prove $A \cup A = A$

Since, for any set A and B, $B \subset A \cup B$, therefore $A \subset A \cup A$

Let $x \in A \cup A \Rightarrow x \in A$ or $x \in A \Rightarrow x \in A$

$\therefore$ $A \cup A \subset A$

As $A \cup A \subset A$ and $A \subset A \cup A \Rightarrow A = A \cup A$. Hence proved.

(*b*) To prove $A \cap A = A$

Since, for any set A and B, $A \cap B \subset B$, therefore $A \cap A \subset A$

Let $x \in A \Rightarrow x \in A$ and $x \in A$

$\Rightarrow x \in A \cap A$ $\therefore$ $A \subset A \cap A$

As $A \cap A \subset A$ and $A \subset A \cap A \Rightarrow A = A \cap A$. Hence proved.

Theorem II. *Prove Associative laws i.e., (a)* $(A \cup B) \cup C = A \cup (B \cup C)$

(P.T.U. M.C.A. May 2008, 2007 ; P.T.U. B.Tech Dec. 2006)

(b) $(A \cap B) \cap C = A \cap (B \cap C)$. (P.T.U. M.C.A. Dec. 2007)

Proof. (*a*) To prove $(A \cup B) \cup C = A \cup (B \cup C)$

Let $x \in (A \cup B) \cup C$

$\Rightarrow$ $(x \in A$ or $x \in B)$ or $x \in C$

$\Rightarrow$ $x \in A$ or $x \in B$ or $x \in C$

$\Rightarrow$ $x \in A$ or $(x \in B$ or $x \in C)$

$\Rightarrow$ $x \in A$ or $(x \in B \cup C)$

$\Rightarrow$ $x \in A \cup (B \cup C)$.

Similarly, if $x \in A \cup (B \cup C)$, then $x \in (A \cup B) \cup C$.

Thus, $x \in A \cup (B \cup C) \Leftrightarrow x \in (A \cup B) \cup C$.

Hence $(A \cup B) \cup C = A \cup (B \cup C)$

(*b*) To prove $(A \cap B) \cap C = A \cap (B \cap C)$

Let $x \in A \cap (B \cap C) \Rightarrow x \in A$ and $x \in B \cap C$

$\Rightarrow$ $x \in A$ and ($x \in B$ and $x \in C$) $\Rightarrow$ $x \in A$ and $x \in B$ and $x \in C$

$\Rightarrow$ ($x \in A$ and $x \in B$) and $x \in C$ $\Rightarrow$ $(x \in A \cap B)$ and $x \in C$

$\Rightarrow$ $x \in (A \cap B) \cap C$.

Similarly, if $x \in (A \cap B) \cap C$, then $x \in A \cap (B \cap C)$

Thus, any $x \in (A \cap B) \cap C \Leftrightarrow x \in A \cap (B \cap C)$

Hence $(A \cap B) \cap C = A \cap (B \cap C)$.

Theorem III. *Prove Commutative laws i.e.,*

(a) $A \cup B = B \cup A$. (P.T.U. M.C.A. May 2007)

(b) $A \cap B = B \cap A$. (P.T.U. B.Tech. May 2006)

Proof. (*a*) To prove $A \cup B = B \cup A$, we know

$$A \cup B = \{x : x \in A \text{ or } x \in B\}$$
$$= \{x : x \in B \text{ or } x \in A\} \quad (\because \text{Order is not preserved in case of sets})$$
$$= B \cup A. \text{ Hence proved.}$$

(*b*) To prove $A \cap B = B \cap A$, we know

$$A \cap B = \{x : x \in A \text{ and } x \in B\}$$
$$= \{x : x \in B \text{ and } x \in A\}$$

($\because$ Order is not preserved in case of sets)

$$= B \cap A.$$

Hence proved.

Theorem IV. *Prove Distributive Laws i.e., (a) Intersection of sets is distributive w.r.t. union of sets i.e.,*

$$A \cap (B \cup C) = (A \cap B) \cup (A \cap C)$$

(*b*) *Union of sets is distributive w.r.t. intersection of sets i.e.,*

$$A \cup (B \cap C) = (A \cup B) \cap (A \cup C).$$

Proof. (*a*) To prove $A \cap (B \cup C) = (A \cap B) \cup (A \cap C)$

Let $x \in A \cap (B \cup C) \Rightarrow x \in A$ and $x \in B \cup C$

$\Rightarrow$ ($x \in A$ and $x \in A$) and ($x \in B$ or $x \in C$)

$\Rightarrow$ ($x \in A$ and $x \in B$) or ($x \in A$ and $x \in C$)

$\Rightarrow$ $x \in A \cap B$ or $x \in A \cap C$

$\Rightarrow$ $x \in (A \cap B) \cup (A \cap C)$

Therefore, $A \cap (B \cup C) \subset (A \cap B) \cup (A \cap C)$ ...(1)

Again, let $y \in (A \cap B) \cup (A \cap C) \Rightarrow y \in A \cap B$ or $y \in A \cap C$

$\Rightarrow$ $(y \in A \text{ and } y \in B)$ or $(y \in A \text{ and } y \in C)$

$\Rightarrow$ $(y \in A \text{ or } y \in A)$ and $(y \in B \text{ or } y \in C)$

$\Rightarrow$ $y \in A$ and $(y \in B \cup C)$

$\Rightarrow$ $y \in A \cap (B \cup C)$

Therefore, $(A \cap B) \cup (A \cap C) \subset A \cap (B \cup C)$ ...(2)

Combining (1) and (2), we get $A \cap (B \cup C) = (A \cap B) \cup (A \cap C)$. Hence proved.

(*b*) To prove $A \cup (B \cap C) = (A \cup B) \cap (A \cup C)$

Let $x \in A \cup (B \cap C) \Rightarrow x \in A$ or $x \in B \cap C$

$\Rightarrow$ $(x \in A \text{ or } x \in A)$ or $(x \in B \text{ and } x \in C)$

$\Rightarrow$ $(x \in A \text{ or } x \in B)$ and $(x \in A \text{ or } x \in C)$

$\Rightarrow$ $(x \in A \cup B)$ and $(x \in A \cup C)$

$\Rightarrow$ $x \in (A \cup B) \cap (A \cup C)$

Therefore, $A \cup (B \cap C) \subset (A \cup B) \cap (A \cup C)$ ...(1)

Again, let $y \in (A \cup B) \cap (A \cup C) \Rightarrow y \in A \cup B$ and $y \in A \cup C$

$\Rightarrow$ $(y \in A \text{ or } y \in B)$ and $(y \in A \text{ or } y \in C)$

$\Rightarrow$ $(y \in A \text{ and } y \in A)$ or $(y \in B \text{ and } y \in C)$

$\Rightarrow$ $y \in A$ or $(y \in B \cap C)$

$\Rightarrow$ $y \in A \cup (B \cap C)$

Therefore, $(A \cup B) \cap (A \cup C) \subset A \cup (B \cap C)$ ...(2)

Combining (1) and (2), we get $A \cup (B \cap C) = (A \cup B) \cap (A \cup C)$. Hence proved.

Theorem V. *Prove De Morgan's Laws* (P.T.U. B.Tech. Dec. 2013)

(*a*) $(A \cup B)^c = A^c \cap B^c$ (P.T.U. M.C.A. May 2007; Dec. 2007)

(*b*) $(A \cap B)^c = A^c \cup B^c$.

(P.T.U. B.Tech. Dec. 2009, May 2008, 2007, 2006; M.C.A. Dec. 2006)

Proof. (*a*) To prove $(A \cup B)^c = A^c \cap B^c$

Let $x \in (A \cup B)^c \Rightarrow x \notin A \cup B$ $(\because a \in A \Leftrightarrow a \notin A^c)$

$\Rightarrow x \notin A$ and $x \notin B$

$\Rightarrow x \in A^c$ and $x \in B^c$

$\Rightarrow x \in A^c \cap B^c$

Therefore, $(A \cup B)^c \subset A^c \cap B^c$ ...(1)

Again, let $x \in A^c \cap B^c \Rightarrow x \in A^c$ and $x \in B^c$

$\Rightarrow x \notin A$ and $x \notin B$

$\Rightarrow x \notin A \cup B$

$\Rightarrow x \in (A \cup B)^c$

Therefore, $A^c \cap B^c \subset (A \cup B)^c$ ...(2)

Combining (1) and (2), we get $A^c \cap B^c = (A \cup B)^c$. Hence proved.

(*b*) To prove $(A \cap B)^c = A^c \cup B^c$

Let $x \in (A \cap B)^c \Rightarrow x \notin A \cap B$ $(\because a \in A \Leftrightarrow a \notin A^c)$

$\Rightarrow x \notin A$ or $x \notin B$

$\Rightarrow x \in A^c$ or $x \in B^c$

$\Rightarrow x \in A^c \cup B^c$

$\therefore$ $(A \cap B)^c \subset A^c \cup B^c$ ...(1)

Again, let $x \in A^c \cup B^c \Rightarrow x \in A^c$ or $x \in B^c$

$\Rightarrow x \notin A$ or $x \notin B$

$\Rightarrow x \notin A \cap B$

$\Rightarrow x \in (A \cap B)^c$

$\therefore$ $A^c \cup B^c \subset (A \cap B)^c$ ....(2)

Combining (1) and (2), we get $(A \cap B)^c = A^c \cup B^c$. Hence proved.

Theorem VI. *Prove Identity Laws i.e.,*

(*a*) $A \cup \phi = A$ (*b*) $A \cap \phi = \phi$ (*c*) $A \cup U = U$ (*d*) $A \cap U = A$.

Proof. (*a*) To prove $A \cup \phi = A$

Let $x \in A \cup \phi \Rightarrow x \in A$ or $x \in \phi$

$\Rightarrow x \in A$ $(\because x \notin \phi$, as ϕ is the null set)

Therefore, $x \in A \cup \phi \Rightarrow x \in A$

Hence $A \cup \phi \subset A$ ...(1)

We know that $A \subset A \cup B$ for any set B.

For $B = \phi$, we have $A \subset A \cup \phi$ ...(2)

From (2) and (1), $A \subset A \cup \phi, A \cup \phi \subset A \Rightarrow A = A \cup \phi$. Hence proved.

(*b*) To prove $A \cap \phi = \phi$

If $x \in A$, then $x \notin \phi$ $(\because \phi$ is null set)

Therefore, $x \in A, x \notin \phi \Rightarrow A \cap \phi = \phi$. Hence proved.

(*c*) To prove $A \cup U = U$

Every set is a subset of universal set

$\therefore$ $A \cup U \subseteq U$

Also, $U \subseteq A \cup U$

Therefore, $A \cup U = U$. Hence proved.

(*d*) To prove $A \cap U = A$

We know $A \cap U \subset A$ ...(1)

So we have to show that $A \subset A \cap U$

Let $x \in A \Rightarrow x \in A$ and $x \in U$ $(\because A \subset U$ so $x \in A \Rightarrow x \in U)$

$\therefore$ $x \in A \Rightarrow x \in A \cap U$

$\therefore$ $A \subset A \cap U$ ...(2)

From (1) and (2), we get $A \cap U = A$. Hence proved.

Theorem VII. *Prove Complement Laws i.e.,*

(a) $A \cup A^c = U$ *(b)* $A \cap A^c = \phi$

(c) $U^c = \phi$ *(d)* $\phi^c = U$

(e) $(A^c)^c = A$ (P.T.U. B.Tech. Dec. 2007; M.C.A. May 2007)

Proof. *(a)* To prove $A \cup A^c = U$, we know that

Every set is a subset of U.

$\therefore$ $A \cup A^c \subset U$ ...(1)

We have to show that $U \subseteq A \cup A^c$

Let $x \in U$ $\Rightarrow$ $x \in A$ or $x \notin A$

$\Rightarrow$ $x \in A$ or $x \in A^c$ $\Rightarrow$ $x \in A \cup A^c$

$\therefore$ $U \subseteq A \cup A^c$ ...(2)

From (1) and (2), we get $A \cup A^c = U$. Hence proved.

(b) To prove $A \cap A^c = \phi$, we know that

ϕ is a subset of every set

$\therefore$ $\phi \subset A \cap A^c$ ...(1)

We have to show that $A \cap A^c \subset \phi$

Let $x \in A \cap A^c$ $\Rightarrow$ $x \in A$ and $x \in A^c$

$\Rightarrow$ $x \in A$ and $x \notin A$ (A contradictory statement)

$\Rightarrow$ $x \in \phi$

$\therefore$ $A \cap A^c \subset \phi$ ...(2)

From (1) and (2), we get $A \cap A^c = \phi$. Hence proved.

(c) To prove $U^c = \phi$

Let $x \in U^c \Leftrightarrow x \notin U \Leftrightarrow x \in \phi$

$\therefore$ $U^c = \phi$. (As U is a universal set) Hence proved.

(d) To prove $\phi^c = U$

Let $x \in \phi^c \Leftrightarrow x \notin \phi \Leftrightarrow x \in U$ (As ϕ is an empty set)

$\therefore$ $\phi^c = U$. Hence proved.

(e) Let $x \in (A^c)^c \Leftrightarrow x \notin A^c \Leftrightarrow x \in A$

$\therefore$ $(A^c)^c = A$. Hence proved.

1.7. (a) PRINCIPLE OF DUALITY FOR SETS

It states that if certain axioms imply their own duals, then the dual of any theorem that is a consequence of the axioms is also a consequence of the axioms.

The dual E* of an equation E involving sets is the equation obtained by interchanging $\cup$ and $\cap$ and also U and ϕ in *E, i.e.,* by replacing each occurrence of $\cup$, $\cap$, U and ϕ in E by $\cap$, $\cup$, ϕ and U respectively.

Example. *Write the dual of the following set equation :*

(i) $(U \cap A) \cup (B \cap A) = A$ *(ii)* $(A \cap U) \cap (\phi \cup A^c) = \phi$.

Sol. *(i)* Replacing $\cup$ by $\cap$ and also U by ϕ in the given set equation, we get

$(\phi \cup A) \cap (B \cup A) = A$, as the required dual

(*ii*) Required dual is

$$(A \cup \phi) \cup (U \cap A^c) = U.$$

1.7. (b) CARDINALITY OF A SET

The total numbers of unique elements in the set is called the cardinality of the set. The cardinality of the countably infinite set is countably infinite *e.g.*, It is denoted by $n(A)$ or $| A |$ or A or card(A).

1. Let $P = \{k, l, m, n\}$. Then the cardinality of the set P is 4.
2. Let A is the set of all non-negative even integers *i.e.*, A = {0, 2, 4, 6, 8, 10, ...}

   As A is countably infinite set, hence the cardinality of this set is countably infinite.
3. The cardinality of an empty set ϕ is zero, *i.e.*, $n(\phi) = 0$

Example 1. *Find the cardinal number of each set:*

(*a*) *{Monday, Tuesday, ..., Sunday}*

(*b*) *{x : x is a letter in the word "BASEBALL"}*

(*c*) $\{x : x^2 = 9, 2x = 8\}$

(*d*) *The power set P(A) of A = {1, 5, 7, 11}*

(*e*) *Collection of functions from A = {a, b, c} into B = {1, 2, 3, 4}*

(*f*) *Set of relations on A = {a, b, c}.*

Sol. The cardinal number of a set 'A' is denoted by $n(A)$.

(*a*) Given set A = {Monday, Tuesday, ..., Sunday}

This contains 7 elements *viz.*,

Monday, Tuesday, Wednesday, Thursday, Friday, Saturday and Sunday.

$\therefore \quad n(A) = 7.$

(*b*) Let A = {*x* : *x* is a letter in the word "BASEBALL"}

The number of letters in the word "BASEBALL" is 5 (B, A, S, E, L)

$\therefore \quad n(A) = 5.$

(*c*) Let $A = \{x : x^2 = 9, 2x = 8\}$

Here $x^2 = 9 \Rightarrow x = \pm 3$. Also, $2x = 8 \Rightarrow x = 4$,

which is not possible.

$\therefore \quad n(A) = 0.$

(*d*) A = {1, 5, 7, 11} *i.e.*, A contains 4 elements

$\therefore$ P(A) will contain $2^4 = 16$ elements

$\therefore \quad n(P(A)) = \mathbf{16}.$

(*e*) A = {*a*, *b*, *c*}, B = {1, 2, 3, 4}

A contains 3 elements, B contains 4 elements

$\therefore$ Total number of functions from A to B = $4^3 = 64$.

$\therefore$ The cardinal number of the collection of functions from A to B = 64.

(*f*) Given A = {*a*, *b*, *c*} *i.e.*, A contains 3 elements.

Total number of relations on A is

$2^{3^2} = 2^9 = 512.$ | Total number of relations on a set containing n elements $= 2^{n^2}$

$\therefore$ Required cardinality = **512.**

Example 2. *Find the cardinal number :*

(*a*) *The collection X of functions from A = {a, b, c, d} into B = {1, 2, 3, 4, 5}*

(*b*) *The set of all relations on A = {a, b, c, d}.*

(*c*) *B = {n^{100} : n is a +ve integer}*

Sol. (*a*) $n(A) = 4$, $n(B) = 5$.

$\therefore$ Total number of functions from A to B = $5^4 = 625$. [If A contains '*m*' elements and B contains '*n*' elements then total number of functions from A to B = n^m]

$\therefore$ Required cardinality = 625.

(*b*) $n(A) = 4$

$\therefore$ Total number of relations on A = $2^{4^2} = 2^{16} =$ **65536.**

(*c*) Given set is B = $\{1^{100}, 2^{100}, 3^{100}, ...\}$

There is a one-one correspondence between the elements of B and the set N. Hence B is countably infinite.

Example 3. *Let A and B are any arbitrary sets. Discuss the cardinality of A ∪ B.*

Sol. We discuss the following cases :

Case I. When A and B are both finite, then A ∪ B will also be finite. Therefore,

$n(A \cup B)$ = finite.

Case II. When A is finite and B is countably infinite. Then A ∪ B will be countably infinite. Therefore, the cardinality of A ∪ B is countably infinite.

Case III. When A is countably infinite and B is finite, then A ∪ B is countably infinite. Therefore, the cardinality of A ∪ B is countably infinite.

Case IV. When A and B are both countably infinite, then A ∪ B will be countably infinite. Hence, the cardinality of A ∪ B is countably infinite.

Case V. When A and B are both uncountable, then A ∪ B is also uncountable. The cardinality of A ∪ B is uncountably infinite.

1.8. CARTESIAN PRODUCT OF TWO SETS

The cartesian product of two sets P and Q in that order is the set of all ordered pairs whose first member belongs to the set P and second member belongs to set Q and is denoted by P × Q *i.e.,*

$$P \times Q = \{(x, y) : x \in P, y \in Q\}.$$

Similarly, $A \times B \times C = \{(a, b, c) : a \in A, b \in B, c \in C\}$. It is common to denote A × A by A^2, A × A × A by A^3 etc. Also we denote (A × B) × (A × B) by $(A \times B)^2$, (A × B) × (A × B) × (A × B) by $(A \times B)^3$ etc.

Example 1. *Let P = {a, b, c} and Q = {k, l, m, n}. Determine the cartesian product of P and Q.*

Sol. The cartesian product of P and Q is

$$P \times Q = \begin{Bmatrix} (a, k), (a, l), (a, m), (a, n), \\ (b, k), (b, l), (b, m), (b, n), \\ (c, k), (c, l), (c, m), (c, n) \end{Bmatrix}.$$

Example 2. *If A = {1, 2, 4, 5}, B = {a, b, c, f}, C = {a, 5}. Compute* A ∪ C, (A ∪ C) × B.

(P.T.U. B.Tech. Dec. 2005)

Sol. Given $A = \{1, 2, 4, 5\}, B = \{a, b, c, f\}, C = \{a, 5\}$

∴ $A \cup C = \{1, 2, 4, 5, a\}$

$(A \cup C) \times B = \{1, 2, 4, 5, a\} \times \{a, b, c, f\}$
$= \{(1, a), (1, b), (1, c), (1, f), (2, a), (2, b), (2, c), (2, f), (4, a), (4, b), (4, c), (4, f), (5, a), (5, b), (5, c), (5, f), (a, a), (a, b), (a, c), (a, f)\}$

Theorem VIII. *Prove that* $(A \times B) \cap (P \times Q) = (A \cap P) \times (B \cap Q)$.

(P.T.U. M.C.A. May 2008)

Proof. Let $(x, y) \in (A \times B) \cap (P \times Q)$
⇒ $(x, y) \in (A \times B)$ and $(x, y) \in (P \times Q)$
⇒ $(x \in A$ and $y \in B)$ and $(x \in P$ and $y \in Q)$
⇒ $(x \in A$ and $x \in P)$ and $(y \in B$ and $y \in Q)$
⇒ $x \in (A \cap P)$ and $y \in (B \cap Q)$
⇒ $(x, y) \in (A \cap P) \times (B \cap Q)$

Therefore, $(A \times B) \cap (P \times Q) \subset (A \cap P) \times (B \cap Q)$ ...(1)

Now, conversely let $(x, y) \in (A \cap P) \times (B \cap Q)$
⇒ $x \in (A \cap P)$ and $y \in (B \cap Q)$
⇒ $(x \in A$ and $y \in P)$ and $(y \in B$ and $y \in Q)$
⇒ $(x \in A$ and $y \in B)$ and $(x \in P$ and $y \in Q)$
⇒ $(x, y) \in (A \times B)$ and $(x, y) \in (P \times Q)$
⇒ $(x, y) \in (A \times B) \cap (P \times Q)$

Therefore, $(A \cap P) \times (B \cap Q) \subset (A \times B) \cap (P \times Q)$ ...(2)

From (1) and (2), we have

$(A \times B) \cap (P \times Q) = (A \cap P) \times (B \cap Q)$. Hence proved.

Theorem IX. *Prove that* $A \times (B \cap C) = (A \times B) \cap (A \times C)$.

(P.T.U. B.Tech. May 2007; M.C.A. Dec. 2006)

Proof. Let $(x, y) \in A \times (B \cap C)$ ⇒ $x \in A$ and $y \in (B \cap C)$
⇒ $x \in A$ and $(y \in B$ and $y \in C)$
⇒ $(x \in A$ and $y \in B)$ and $(x \in A$ and $y \in C)$
⇒ $(x, y) \in (A \times B)$ and $(x, y) \in (A \times C)$
⇒ $(x, y) \in (A \times B) \cap (A \times C)$

Therefore, $A \times (B \cap C) \subset (A \times B) \cap (A \times C)$ ...(1)

Now, conversely let $(x, y) \in (A \times B) \cap (A \times C)$
⇒ $(x, y) \in (A \times B)$ and $(x, y) \in (A \times C)$
⇒ $(x \in A$ and $y \in B)$ and $(x \in A$ and $y \in C)$
⇒ $x \in A$ and $y \in B$ and $y \in C$
⇒ $x \in A$ and $y \in B \cap C$
⇒ $(x, y) \in A \times (B \cap C)$

Therefore, $(A \times B) \cap (A \times C) \subset A \times (B \cap C)$ ...(2)

From (1) and (2), we have $A \times (B \cap C) = (A \times B) \cap (A \times C)$. Hence proved.

Theorem X. *Prove that* $A \times (B \cup C) = (A \times B) \cup (A \times C)$.

Proof. Let $(x, y) \in A \times (B \cup C)$ $\Rightarrow$ $x \in A$ and $y \in (B \cup C)$

$\Rightarrow$ $x \in A$ and ($y \in B$ or $y \in C$) $\Rightarrow$ ($x \in A$ and $y \in B$) or ($x \in A$ and $y \in C$)

$\Rightarrow$ $(x, y) \in A \times B$ or $(x, y) \in (A \times C)$ $\Rightarrow$ $(x, y) \in (A \times B) \cup (A \times C)$

Therefore, $A \times (B \cup C) \subset (A \times B) \cup (A \times C)$ ...(1)

Now, conversely let $(x, y) \in (A \times B) \cup (A \times C)$

$\Rightarrow$ $(x, y) \in (A \times B)$ or $(x, y) \in (A \times C)$ $\Rightarrow$ ($x \in A$ and $y \in B$) or ($x \in A$ and $y \in C$)

$\Rightarrow$ $x \in A$ and $y \in B$ or $y \in C$ $\Rightarrow$ $x \in A$ and $y \in (B \cup C)$

$\Rightarrow$ $(x, y) \in A \times (B \cup C)$

Therefore, $(A \times B) \cup (A \times C) \subset A \times (B \cup C)$ ...(2)

From (1) and (2), we have $A \times (B \cup C) = (A \times B) \cup (A \times C)$. Hence proved.

Theorem XI. *Prove that if* $A \subset B$, *then* $A \times C \subset B \times C$.

Proof. Let $(x, y) \in A \times C$ $\Rightarrow$ $x \in A$ and $y \in C$

$\Rightarrow$ $x \in B$ and $y \in C$ ($\because$ $A \subset B$)

$\Rightarrow$ $(x, y) \in B \times C$

Therefore, $A \times C \subset B \times C$. Hence proved.

Theorem XII. *If S and T have n elements in common. Show that* $S \times T$ *and* $T \times S$ *have* n^2 *elements in common.*

Proof. Suppose, a set R, consisting of n common elements of S and T.

Then, $R \subset S$ and $R \subset T$

Let $(x, y) \in (R \times R)$ $\Leftrightarrow x \in R$ and $y \in R$

$\Leftrightarrow$ ($x \in R$ and $y \in R$) and ($x \in R$ and $y \in R$)

$\Leftrightarrow$ ($x \in S$ and $y \in T$) and ($x \in T$ and $y \in S$) ($\because$ $R \subset S$; $R \subset T$)

$\Leftrightarrow$ $(x, y) \in (S \times T)$ and $(x, y) \in (T \times S)$

$\Leftrightarrow$ $(x, y) \in (S \times T) \cap (T \times S)$

Therefore, $(R \times R) = (S \times T) \cap (T \times S)$.

The right hand side contains ordered pairs common to both $S \times T$ and $T \times S$.

The left hand side $R \times R$ has n^2 elements ($\because$ R has n elements)

Since, the two sets are equal, both have the same number of elements.

Hence, $S \times T$ and $T \times S$ have n^2 common elements.

ILLUSTRATIVE EXAMPLES

Example 1. *Write the following sets in builder from :*

(*a*) $A = \{2, 4, 6, 8, 10, 12, 14\}$ (*b*) $K = \{3, 6, 9, 12, 15, 18, ...\}$

(*c*) *L* = *{PUNJAB, HARYANA, DELHI, UP, ..., BIHAR}*.

Sol. (*a*) {*x* : *x* is a +ve integer divisible by 2 and less than 15}.

(*b*) {*x* : *x* is +ve integer and is multiple of 3}.

(*c*) {*x* : *x* is the state of India}.

Example 2. *Write the following sets in tabular form :*

(*a*) $A = \{x : x^2 = 9\}$ (*b*) *B* = *{x : x is a multiple of 3 and* $0 < x < 20\}$

(*c*) *C* = *{x : x is a +ve even integer}* (*d*) *D* = *{x : x is a multiple of 5}*.

Sol. (*a*) $A = \{3, -3\}$ (*b*) $B = \{3, 6, 9, 12, 15, 18\}$

(*c*) $C = \{2, 4, 6, 8, 10, ...\}$ (*d*) $D = \{5, 10, 15, 20, 25, ...\}$

Example 3. *Which of the following sets are equal :*

(*a*) $R = \{x : x^2 = 7\}$ (*b*) $S = \{x : x + 4 = 4\}$

(*c*) $T = \{x : x^2 + 2 = 8\}$.

Sol. The sets R and T are equal as they are empty sets. The set S is not an empty set, because element ($x = 0$) belongs to set S.

Example 4. *Determine the power set P(A) of the set* $A = \{1, 2, 3\}$.

Sol. $P(A) = \{\{1\}, \{2\}, \{3\}, \{1, 2\}, \{1, 3\}, \{2, 3\}, \{1, 2, 3\}, \phi\}$.

Example 5. *Let* $A = [\{a\}, \{b, c, d, e\}, \{c, d\}]$. *List the elements of A and determine whether each of the following statements are true or false.*

(*i*) $a \in A$ (*ii*) $\{a\} \in A$

(*iii*) $\{\{a\}, \{c, d\}\} \subseteq A$ (*iv*) $\{b, c, d, e\} \subseteq A$

(*v*) $\phi \subseteq A$ (*vi*) $\phi \in A$.

Sol. Given $A = \{\{a\}, \{b, c, d, e\}, \{c, d\}\}$

The elements of A are $\{a\}, \{b, c, d, e\}, \{c, d\}$

(*i*) *False*. Since the element *a* is not one of the three elements of A.

(*ii*) *True*. The set $\{a\}$ is one of the three elements of A.

(*iii*) *True*. The set $\{\{a\}, \{c, d\}\}$ is a subset of A.

(*iv*) *False*. $\{b, c, d, e\}$ is an element of A, not a subset of A.

(*v*) *True*. The empty set is a subset of every set.

(*vi*) *False*. Since, the empty set is not listed as one of the elements of A.

Example 6. *Consider* $B = \{\{1, 3, 5\}, \{2, 4, 6\}, \{0\}\}$. *List the elements of B and determine whether each of the following statements is true or false.*

(*i*) $\phi \subseteq B$ (*ii*) $3 \in B$

(*iii*) $\{1, 3, 5\} \subseteq B$ (*iv*) $\{1, 2, 3, 4, 5, 6\} \in B$

(*v*) $\{\{2, 4, 6\}, \{0\}\} \subseteq B$ (*vi*) $\{0\} \in B$.

Sol. Given $B = \{\{1, 3, 5\}, \{2, 4, 6\}, \{0\}\}$.

The elements of B are $\{1, 3, 5\}, \{2, 4, 6\}, \{0\}$

(*i*) *True*. Since the empty set is a subset of every subset.

(*ii*) *False*. Since 3 is not an element of B.

(*iii*) *False*. Since {1, 3, 5} is one of the three elements of B.

(*iv*) *False*. Since {1, 2, 3, 4, 5, 6} is not an element of B.

(*v*) *True*. Since {2, 4, 6}, {0} are elements of B, therefore {{2, 4, 6}, {0}} is a subset of B.

(*vi*) *True*. Since {0} is one of the three elements of B.

Example 7. *Let E = {{1, 2, 3}, {2, 3}, {a, b}}, F = {{a, b}, {1, 2}}. List the elements of E. Determine whether the following statements are true or not*

(*i*) $\{a, b\} \subseteq F$ (*ii*) $[\{1, 2, 3\}] \subseteq E$

(*iii*) $F \subseteq E$ (*iv*) $\phi \subseteq F$

(*v*) $1 \in E$ (*vi*) $\{2, 3\} \in E$.

Sol. Given E = {{1, 2, 3}, {2, 3}, {*a*, *b*}}

The elements of E are {1, 2, 3}, {2, 3}, {*a*, *b*}

(*i*) *False*. Since {*a*, *b*} is one of the three elements of E

(*ii*) *True*. Since {1, 2, 3} is an element from E therefore, {{1, 2, 3}} is a subset of E

(*iii*) *False*. Since $\{1, 2\} \in F$ but $\{1, 2\} \notin E$

(*iv*) *True*. The empty set is a subset of every set

(*v*) *False*. Since 1 is not an element of E

(*vi*) *True*. Since {2, 3} is one of the three elements of E.

Example 8. *If A = {x : 3x = 6}. Does A = 2 ?*

Sol. Given $A = \{x : 3x = 6\} = \{2\}$. Now $2 \in A$ but $A \neq 2$.

Example 9. *Determine which of the following sets are equal: φ, {0}, {φ}.*

Sol. The set {0} contains one element, namely, zero.

The set φ contains no element as it is an empty set.

The set {φ} also contains one element, the null set.

Hence, each given set is different from the other sets.

Example 10. *Describe a situation where the universal set U may be empty.*

Sol. Let U is the set of town councillor in a given city. It is possible that in a given year, there are no such councillors and hence U = φ.

Example 11. *Explain the difference between* $A \subseteq B$ *and* $A \subset B$.

Sol. The statement $A \subseteq B$ reads as "*A* is a subset of B"and means that every element of A also belongs to B, which includes the possibility A = B.

The statement $A \subset B$ reads as "A is a proper subset of B" and means that A is a subset of B but $A \neq B$. Hence there is at least one element in B which is not in A.

Example 12. *If A = {2, 3, 4, 5}. Is A a subset of B = {x : x ∈ N, x is even} ?*

Sol. A cannot be a subset of B if we can find at least one element in A but not in B. Now $3 \in A$ but $3 \notin B$ since B consists of even numbers. Hence A is not a subset of B.

Example 13. *Show that $A = \{2, 3, 4, 5\}$ is a proper subset of $B = \{1, 2, 3, \ldots\ldots 8, 9\}$.*

Sol. Each element of A is also an element of B. Hence $A \subseteq B$. On the other hand, $1 \in B$ but $1 \notin A$.

Hence $A \neq B$. Therefore, A is a proper subset of B.

Example 14. *Let A, B, C be non-empty sets such that $A \subseteq B$, $B \subseteq C$ and $C \subseteq A$. What can be deduced about these sets ?*

Sol. Given $\quad A \subseteq B, B \subseteq C \Rightarrow A \subseteq C$

$B \subseteq C, C \subseteq A \Rightarrow B \subseteq A$

Now $\quad A \subseteq B, B \subseteq A \Rightarrow A = B$

$C \subseteq A, A \subseteq C \Rightarrow C = A$

Hence A, B, C are equal sets.

Example 15. *If A is a subset of the null set ϕ, then $A = \phi$. Prove it.*

Sol. We know that the null set ϕ is a subset of every set. In particular, $\phi \subset A$. Also given $A \subset \phi$. Hence $A = \phi$.

Example 16. *Does every set have a proper set ?*

Sol. *No*, since the null set ϕ does not have a proper subset. Every other set does have ϕ as a proper subset.

Example 17. *Suppose*

$U = \{1, 2, 3, 4, 5, 6, 7, 8, 9\}$

$A = \{1, 4, 9\}$

$P = \{x : x \in U \text{ and } x \text{ is a perfect square}\}$

$R = \{1, 2, 3, 5, 7, 9\}$

$D = \{2, 3, 5, 7\}$

$N = \{x : x \in U \text{ and } x \text{ is a prime number}\}$

ϕ = *the empty set*

(*a*) *Determine which sets are subsets of others.*

(*b*) *Determine which sets are proper subsets of others.*

(*c*) *Determine pair of sets which are disjoint.*

(*d*) *Determine pair of sets which are comparable.*

(*e*) *Determine pair of sets which are incomparable.*

Sol. (*a*) All the sets are subsets of U since the elements of every set belong to U. Set ϕ is subset of all other sets. Set A is subset of P. Set D is subset of R and N.

(*b*) All the sets are proper subsets of U since they are not equal to U. Set D is proper subset of R. Set R is proper subset of N. Set ϕ is proper subset of all the other sets.

(*c*) The pairs A and D, A and N, P and D, and P and N are disjoint sets.

(*d*) All the sets are comparable with the set U.

Set D is comparable with R as $D \subseteq R$

Set R is comparable with N as $R \subseteq N$

All sets comparable with ϕ as $\phi \subseteq$ Every set.

(*e*) The pairs A and R, A and D, A and N, P and R, P and D, P and N, are incomparable.

Example 18. *Determine the cardinalities of the sets :*

(*a*) $P = \{n^7 : n \text{ is a positive integer}\}$ (*b*) $Q = \{n^{109} : n \text{ is a positive integer}\}$

(*c*) $P \cup Q$ (*d*) $P \cap Q$.

Sol. (*a*) The cardinality of the set P is countably infinite as the number of +ve integers are infinite.

(*b*) The cardinality of the set Q is countably infinite as the number of positive integers are infinite.

(*c*) Since sets P and Q are both infinite, their union is also infinite and hence cardinality of the set $P \cup Q$ is also infinite.

(*d*) The cardinality of $P \cap Q$ is one because for $n = 1$

$$1^7 = 1^{109}$$

But for $n = 2$; $2^7 \neq 2^{109}$ and so on.

Hence, $P \cap Q$ contains only one element.

$\therefore$ Cardinality of $P \cap Q = 0$

Example 19. (*a*) *Let N is the set of all natural numbers. Let P denotes the set of all finite subsets of N. What is the cardinality of set P ? Give reason.*

(*b*) *Find the cardinal number of the following sets*

(*a*) $A = \{a, b, c, ..., y, z\}$ (*b*) $B = \{1, -3, 5, 7, -80\}$

(*c*) $C = \{x \in N : x^2 = 7\}$ (*d*) $D = \{5, 10, 15, 20, ...\}$

(*e*) $E = \{8, 9, 10, 11, ...\}$

Sol. (*a*) The number of subsets of any set is given by 2^n, where n is number of elements in the set. As the number of subsets of set N is 2^n, hence there are 2^n number of subsets in set P. Therefore, the cardinality of set P is 2^n.

(*b*) (*a*) The cardinality of A = 26 since there are 26 letters in the English alphabet.

(*b*) The cardinality of B = 4

(*c*) The cardinality of C = 0 (zero) as there is no $x \in N$ whose square is 7.

(*d*) Countably infinite

(*e*) Countably infinite

Example 20. *Determine which of the following are true and which are false. Justify your answer.*

(*a*) $3 \in \{1, 3, 5\}$

(*b*) Let $A = \{(a, b)\}$, then

(*i*) $a \in A$ (*ii*) $A \in A$ (*iii*) $\{a, b\} \in A$ (*iv*) $n(A) = 2$

(*c*) $\{3\} \in \{1, 3, 5\}$ (*d*) $\{3\} \subset \{1, 3, 5\}$ (*e*) $\{3, 5\} \subset \{1, 3, 5\}$

(*f*) $\{1, 3, 5\} \subset \{1, 3, 5\}$ (*g*) $\phi \subseteq \phi$ (*h*) $\phi \in \phi$

(*i*) $\phi \subseteq \{\phi\}$ (*j*) $\phi \in \{\phi\}$ (*k*) $\{\phi\} \in \phi$

(*l*) $\{\phi\} \subseteq \{\phi\}$ (*m*) $\{\phi\} \in \{\phi\}$

(*n*) $\{a, b\} \subseteq \{a, b, c, \{a, b, c\}\}$ (*o*) $1 \in \{a + 2b : a, b \text{ are even integers}\}$.

Sol. (*a*) *True.* Since 3 is a member of {1, 3, 5}. Hence, $3 \in \{1, 3, 5\}$.

(*b*) (*i*) *False.* The only member of the set A is the element {*a*, *b*}. Hence, $a \notin A$.

(*ii*) *False.*

(*iii*) *True.* $\{a, b\}$ is the element of A. Therefore, $\{a, b\} \in A$.

(*iv*) *False.* Since A has only one element namely $\{a, b\}$. Therefore, $n(A) = 1$.

(*c*) *False.* Since {3} is not an element of A.

(*d*) *True.* Since {3} is a subset of {1, 3, 5} and also {3} is not equal to {1, 3, 5}. Hence, it is a proper subset of {1, 3, 5}. Therefore, $\{3\} \subset \{1, 3, 5\}$.

(*e*) *True.* Since {3, 5} is a subset of {1, 3, 5}.

(*f*) *False.* Since {1, 3, 5} is a subset of {1, 3, 5} and both are equal. Therefore, it is an improper subset of {1, 3, 5}. Hence, $\{1, 3, 5\} \subseteq \{1, 3, 5\}$.

(*g*) *True.* Both sets have no elements *i.e.*, $\phi = \phi$. Therefore, $\phi \subseteq \phi$ (Every set is a subset of itself.)

(*h*) *False.* As ϕ contains no element.

(*i*) *True.* Since empty set is a subset of every set. Hence, $\phi \subseteq \{\phi\}$.

(*j*) *True.* ϕ is an element of $\{\phi\}$.

(*k*) *False.* Since $\{\phi\}$ is a set containing the element ϕ, whereas, ϕ is an empty set containing no element.

(*l*) *True.* Every set is a subset of itself.

(*m*) *False.*

(*n*) *True.* Since $\{a, b\}$ is a subset of the set $\{a, b, c, \{a, b, c\}\}$.

(*o*) *False.* Since a and b are even integers, $a + 2b$ is also an even integer. But 1 cannot be even.

Example 21. (*i*) *If* $A = \{1, 2, \{1, 3\}, \phi\}$, *determine the following sets*

(P.T.U. B. Tech. May 2009)

(*a*) $A - \{1\}$ (*b*) $A - \phi$ (*c*) $A - \{\phi\}$ (*d*) $A - \{1, 2\}$.

(*ii*) *If* $A = \{\phi, \{\phi\}\}$, *determine whether the following statements are true or false.*

(*a*) $\phi \in P(A)$ (*b*) $\phi \subseteq P(A)$ (*c*) $\{\phi\} \subseteq P(A)$ (*d*) $\{\phi\} \subseteq A$

(*e*) $\{\phi\} \in P(A)$ (*f*) $\{\phi\} \in A$ (*g*) $\{\{\phi\}\} \subseteq P(A)$ (*h*) $\{\{\phi\}\} \subseteq A$

(*i*) $\{\{\phi\}\} \in P(A)$ (*j*) $\{\{\phi\}\} \in A$.

Sol. (*i*) Given set $A = \{1, 2, \{1, 3\}, \phi\}$

The elements of A are 1, 2, {1, 3}, ϕ.

$\therefore$ {1}, $\{\phi\}$ are subset of A

(*a*) $A - \{1\} = \{x : x \in A \text{ and } x \notin \{1\}\} = \{2, \{1, 3\}, \phi\}$.

(*b*) $A - \phi = \{x : x \in A \text{ and } x \notin \phi\} = A = \{1, 2, \{1, 3\}, \phi\}$.

(*c*) Since ϕ is an element of A, therefore, $\{\phi\}$ is a subset of A containing the element ϕ.

$\therefore$ $A - \{\phi\} = \{x : x \in A \text{ and } x \notin \{\phi\}\} = \{1, 2, \{1, 3\}\}$.

(*d*) Since 1, 2 are elements of *A* therefore, {1, 2} is a subset of A.

$\therefore$ $A - \{1, 2\} = \{x : x \in A \text{ and } x \notin \{1, 2\}\} = \{\{1, 3\}, \phi\}$.

(*ii*) Given $A = \{\phi, \{\phi\}\}$ *i.e.*, the elements of A are ϕ, $\{\phi\}$

$\therefore$ $P(A) = \{\phi, \{\phi\}, \{\{\phi\}\}, \{\phi, \{\phi\}\}\}$ *i.e.*,

The elements of P(A) are ϕ, $\{\phi\}$, $\{\{\phi\}\}$, $\{\phi, \{\phi\}\}$

(*a*) *True*, since ϕ is an element of P(A).

(*b*) *True*, since the empty set is a subset of every set

(*c*) *True*, since ϕ is an element of P(A), therefore, $\{\phi\}$ is a subset of P(A)

(*d*) *True*, since ϕ is an element of A, therefore, $\{\phi\}$ is a subset of A

(*e*) *True*, since ϕ is an element of P(A), therefore, $\{\phi\}$ is a subset of P(A)

(*f*) *True*, since $\{\phi\}$ is an element of A

(*g*) *True*, since $\{\phi\}$ is an element of P(A), therefore, $\{\{\phi\}\}$ is a subset of P(A)

(*h*) *True*, since $\{\phi\}$ is an element of A, therefore, $\{\{\phi\}\}$ is a subset of A

(*i*) *True*, since $\{\{\phi\}\}$ is an element of P(A)

(*j*) *False*, since $\{\{\phi\}\}$ is not an element of A.

Example 22. *Determine whether each of the following statements is true or false. Explain your answer.*

(*a*) $A \cup P(A) = P(A)$ (*b*) $A \cap P(A) = A$

(*c*) $\{A\} \cup P(A) = P(A)$ (*d*) $\{A\} \cap P(A) = A$

(*e*) $P(A) - \{A\} = P(A)$.

Sol. (*a*) $A \cup P(A) = P(A)$.

False. P(A) contains all subsets of A but does not contain all elements of A. Therefore, $A \cup P(A) \neq P(A)$.

(*b*) $A \cap P(A) = A$.

False. Since power set of A contains all subsets of A but no elements of A, hence there is no element common to both the sets. Therefore, $A \cap P(A) \neq A$.

(*c*) $\{A\} \cup P(A) = P(A)$

True. P(A) contains all subsets of A and {A} is also an element of P(A).

(*d*) $\{A\} \cap P(A) = A$.

False. {A} is common to both {A} and P(A) because P(A) contains all subsets of A but their intersection is not A. Therefore, $\{A\} \cap P(A) \neq \{A\}$

(*e*) $P(A) - \{A\} = P(A)$.

False. {A} is also an element of P(A). Therefore, $P(A) - \{A\} \neq P(A)$.

Example 23. *Consider the sets*

$$A = \{1, 2, 3, 4, 5, 6\},\ B = \{4, 5, 6, 7, 8, 9\},\ C = \{1, 3, 5, 7, 9\}.$$

Find (*i*) $A \oplus B$ (*ii*) $B \oplus C$

(*iii*) $A \cap (B \oplus C)$, $(A \cap B) \oplus (A \cap C)$ *and show that both are equal.*

Sol. (*i*) By definition, $A \oplus B = (A \cup B) - (A \cap B)$

But $A \cup B = \{1, 2, 3, 4, 5, 6, 7, 8, 9\}$

$A \cap B = \{4, 5, 6\}$

$\therefore$ $(A \cup B) - (A \cap B)$ = The elements which are in $A \cup B$ but not in $A \cap B$

$= \{1, 2, 3, 7, 8, 9\} = A \oplus B$

(*ii*) Proceed as in Part (*i*)

$B \oplus C = \{1, 3, 4, 6, 8\}$

(*iii*) Using part (*ii*), $A \cap (B \oplus C) = \{1, 3, 4, 6\}$ ...(1)

Also $A \cap B = \{4, 5, 6\}$, $A \cap C = \{1, 3, 5\}$

$\therefore$ $(A \cap B) \cup (A \cap C) = \{4, 5, 6, 1, 3\}$

$(A \cap B) \cap (A \cap C) = \{5\}$

$\therefore \quad (A \cap B) \oplus (A \cap C) = (A \cap B) \cup (A \cap C) - (A \cap B) \cap (A \cap C)$

$= \{1, 3, 4, 6\}$ ...(2)

From (1) and (2)

$$A \cap (B \oplus C) = (A \cap B) \oplus (A \cap C).$$

Example 24. *Determine the following sets :*

(*a*) $\phi \cup \{\phi\}$ (*b*) $\phi \cap \{\phi\}$ (*c*) $\{\phi\} \cup \{a, \phi, \{\phi\}\}$

(*d*) $\{\phi\} \cap \{a, \phi, \{\phi\}\}$ (*e*) $\phi \oplus \{a, \phi, \{\phi\}\}$ (*f*) $\{\phi\} \oplus \{a, \phi, \{\phi\}\}$.

Sol. (*a*) ϕ is an empty set and $\{\phi\}$ is a singleton set containing the element ϕ

$\therefore \quad \phi \cup \{\phi\} = \{\phi\}$

(*b*) $\phi \cap \{\phi\} = \phi$

(*c*) $\{\phi\} \cup \{a, \phi, \{\phi\}\} = \{a, \phi, \{\phi\}\}$

(*d*) $\{\phi\} \cap \{a, \phi, \{\phi\}\} = \{\phi\}$

(*e*) We know that $A \oplus B = (A - B) \cup (B - A)$

Take $A = \phi, B = \{a, \phi, \{\phi\}\}$

$\therefore$ A – B = The set of elements which are in A and not in B = ϕ

B – A = The set of elements which are in B and not in A = $\{a, \phi, \{\phi\}\}$

$\therefore \quad A \oplus B = (A - B) \cup (B - A) = \phi \cup \{a, \phi, \{\phi\}\} = \{a, \phi, \{\phi\}\}$

(*f*) Take $A = \{\phi\}, B = \{a, \phi, \{\phi\}\}$

$\therefore$ A – B = The set of elements which are in A and not in B

= $\{\phi\}$ [since ϕ is an element of A]

B – A = The set of elements which are in B and not in A = $\{a, \{\phi\}\}$

$\therefore \quad A \oplus B = (A - B) \cup (B - A) = \{\phi\} \cup \{a, \{\phi\}\} = \{a, \{\phi\}\}$.

Example 25. *Let A, B, C be sets. Under what conditions is each of the following true ?*

(*a*) $(A - B) \cup (A - C) = \phi$ (*b*) $(A - B) \cap (A - C) = \phi$

(*c*) $(A - B) \oplus (A - C) = \phi$.

Sol. (*a*) This is true if A = B = C or A, B and C are null sets.

(*b*) This is true if A = B = C or all three sets are null sets.

(*c*) This is true if $A \subseteq B$ and $A \subseteq C$.

It is also true if $A \subseteq B$ and $A \not\subseteq C$ but B = C.

Example 26. *What can you say about P and Q if*

(*a*) $P \cap Q = P$ (*b*) $P \cup Q = P$ (*c*) $P \oplus Q = P$ (*d*) $P \cap Q = P \cup Q$.

Sol. (*a*) This tells that $P \subseteq Q$. (*b*) This tells that P = Q or $Q \subseteq P$.

(*c*) $P \oplus Q = P \Rightarrow (P - Q) \cup (Q - P) = P$. This tells that Q is a null set.

(*d*) This tells that P = Q.

Example 27. *Let A, B, C be arbitrary sets :*

(*a*) *Show that* $(A - B) - C = A - (B \cup C)$.

(*b*) *Show that* $(A - B) - C = (A - C) - B$.

(*c*) *Show that* $(A - B) - C = (A - C) - (B - C)$.

Sol. (*a*) $(A - B) - C = A - (B \cup C)$

Let $x \in (A - B) - C \Rightarrow x \in (A - B)$ and $x \notin C$

$\Rightarrow x \in A$ and $x \notin B$ and $x \notin C$

$\Rightarrow x \in A$ and $x \notin$ (B or C)

$\Rightarrow x \in A$ and $x \notin B \cup C$

$\Rightarrow x \in A - (B \cup C)$.

$\therefore \quad A - (B \cup C) \subseteq (A - B) - C$ ...(*i*)

Conversely, let $x \in A - (B \cup C)$

$\Rightarrow x \in A$ and $x \notin B \cup C$

$\Rightarrow x \in A$ and $x \notin B$ and $x \notin C$

$\Rightarrow x \in A - B$ and $x \notin C$

$\Rightarrow x \in (A - B) - C$

$\therefore \quad (A - B) - C \subseteq A - (B \cup C)$ ...(*ii*)

From (*i*) and (*ii*), we get $(A - B) - C = A - (B \cup C)$.

(*b*) $(A - B) - C = (A - C) - B$

Let $x \in (A - B) - C$

$\Rightarrow x \in A - B$ and $x \notin C$ $\Rightarrow x \in A$ and $x \notin B$ and $x \notin C$

$\Rightarrow x \in A - C$ and $x \notin B$ $\Rightarrow x \in (A - C) - B$

$\therefore \quad (A - C) - B \subseteq (A - B) - C$ ...(*i*)

Conversely, let $x \in (A - C) - B$

$\Rightarrow x \in (A - C)$ and $x \notin B$ $\Rightarrow x \in A$ and $x \notin C$ and $x \notin B$

$\Rightarrow x \in A$ and $x \notin B$ and $x \notin C$ $\Rightarrow x \in (A - B)$ and $x \notin C$

$\Rightarrow x \in (A - B) - C$

$\therefore \quad (A - B) - C \subseteq (A - C) - B$ ...(*ii*)

From (*i*) and (*ii*), we get $(A - B) - C = (A - C) - B$.

(*c*) $(A - B) - C = (A - C) - (B - C)$

Let $x \in (A - B) - C$ $\Rightarrow x \in (A - B)$ and $x \notin C$

$\Rightarrow x \in A$ and $x \notin B$ and $x \notin C$ $\Rightarrow x \in A$ and $x \notin C$ and $x \notin B$ and $x \notin C$

$\Rightarrow x \in A - C$ and $x \notin B - C$ $\Rightarrow x \in (A - C) - (B - C)$

$\therefore \quad (A - C) - (B - C) \subseteq (A - B) - C$ ...(*i*)

Conversely, let $x \in (A - C) - (B - C)$

$\Rightarrow x \in (A - C)$ and $x \notin (B - C)$ $\Rightarrow x \in A$ and $x \notin C$ and $x \notin B$ and $x \notin C$

$\Rightarrow x \in A$ and $x \notin B$ and $x \notin C$ $\Rightarrow x \in A - B$ and $x \notin C$

$\Rightarrow x \in (A - B) - C$

$\therefore \quad (A - B) - C \subseteq (A - C) - (B - C)$ ...(*ii*)

From (*i*) and (*ii*), we get $(A - B) - C = (A - C) - (B - C)$.

Example 28. *Given that $P \cup Q = P \cup R$, is it necessary that $Q = R$? Justify your answer.*

Sol. This is not necessary. Since

$P \cup Q \Rightarrow$ all elements of P or Q are in this set.

$P \cup R \Rightarrow$ all elements of P or R are in this set.

If $Q \subseteq P$ and also $R \subseteq P$. Then $P \cup Q = P \cup R = P$.

Therefore, it is not necessary that $Q = R$.

e.g., Let $P = \{1, 2, 3, 4\}$, $Q = \{1, 2\}$, $R = \{3, 4\}$, then $P \cup Q = P \cup R = P$

But here $Q \neq R$.

Example 29. *Given that $P \cap Q = P \cap R$, is it necessary that $Q = R$? Justify your answer.*

Sol. This is not necessary. Since

$P \cap Q$ contains elements common to both P and Q.

$P \cap R$ contains elements common to both P and R.

But it is not necessary that $Q = R$ because set R can have elements of set Q which are elements of set P as well, but it can also have elements other than those in set Q.

e.g., Let $P = \{1, 2, 5, 6\}$, $Q = \{1, 2, 8\}$, $R = \{1, 2, 9\}$

Then $P \cap Q = P \cap R = \{1, 2\}$; but $Q \neq R$.

Example 30. *Given that $P \oplus Q = P \oplus R$, is it necessary that $Q = R$? Justify your answer.*

Sol. It is necessary that $Q = R$.

Because $P \oplus Q = (P \cup Q) - (P \cap Q)$ *i.e.,* this set contains elements which are in sets P or Q but does not contain elements common to both sets.

Similarly, $P \oplus R = (P \cup R) - (P \cap R)$ *i.e.,* this set contains elements which are in sets P or R but does not contain elements common to both sets.

Given $P \oplus Q = P \oplus R$

$\therefore \quad (P \cup Q) - (P \cap Q) = (P \cup R) - (P \cap R)$

$\Rightarrow \quad P \cup Q = P \cup R$ and $P \cap Q = P \cap R$, which is possible only when

$Q = R$

e.g., let $P = \{1, 2, 3, 4\}$, $Q = \{1, 5\}$, $R = \{1, 5\}$

$P \oplus Q = (P \cup Q) - (P \cap Q) = \{1, 2, 3, 4, 5\} - \{1\} = \{2, 3, 4, 5\}$

$P \oplus R = (P \cup R) - (P \cap R) = \{1, 2, 3, 4, 5\} - \{1\} = \{2, 3, 4, 5\}$

$\Rightarrow \quad Q = R$.

Example 31. (*a*) *Prove that $(A - B) \cap B = \phi$.*

(*b*) $A \cup (B - A) = A \cup B$

Sol. (*a*) $(A - B) \cap B = \phi$

If $x \in B$, then $x \notin A - B$ due to definition of $A - B$. Hence

$x \in B$ and $x \notin A - B \Rightarrow (A - B) \cap B = \phi$. Hence proved.

(*b*) We prove $A \cup (B - A) = A \cup B$

Let $x \in A \cup (B - A) \Leftrightarrow x \in A$ or $x \in B - A$

$\Leftrightarrow x \in A$ or $(x \in B$ and $x \notin A)$

$\Leftrightarrow$ $(x \in A \text{ or } x \in B) \cap (x \in A \text{ or } x \notin A)$

$\Leftrightarrow$ $x \in A \text{ or } x \in B$ | $A \cup (B \cap C) = (A \cup B) \cap (A \cup C)$

$\Leftrightarrow$ $x \in A \cup B$

$\Leftrightarrow$ $A \cup (B - A) \subseteq A \cup B$

Hence $A \cup (B - A) = A \cup B$

Example 32. *Prove that* $A \cup B = \phi \Rightarrow A = \phi, B = \phi$.

Sol. Let $A \cup B = \phi$

Now ϕ is a subset of every set $\Rightarrow$ $\phi \subset A, \phi \subset B$

Since, $A \subset A \cup B, B \subset A \cup B$

Hence, $A \subset \phi, B \subset \phi$ | $A \cup B = \phi$

So $A \subset \phi, \phi \subset A$ $\Rightarrow$ $A = \phi$

$B \subset \phi, \phi \subset B$ $\Rightarrow$ $B = \phi$. Hence proved.

Example 33. *Prove that* $A - B \subset B^c$.

Sol. Let $x \in A - B$ $\Rightarrow$ $x \in A$ and $x \notin B$

$\Rightarrow$ $x \in A$ and $x \in B^c$ $\Rightarrow$ $x \in A \cap B^c$ as $A \cap B^c \subset B^c$

$\Rightarrow$ $x \in B^c$

Therefore, any $x \in A - B \Rightarrow x \in B^c$

So $A - B \subset B^c$. Hence proved.

Example 34. *Given that* $(A \cap C) \subseteq (B \cap C)$, $(A \cap \overline{C}) \subseteq (B \cap \overline{C})$. *Show that* $A \subseteq B$.

Sol. For $(A \cap C) \subseteq (B \cap C)$, all elements of set A which are also elements of set C are contained in a set which contains elements common to both B and C. It implies $A \subseteq B$ for elements which are common in A and C, and B and C.

For $(A \cap \overline{C}) \subseteq (B \cap \overline{C})$, elements common to sets A and $\overline{C}$ are contained in a set containing elements common to both B and C. It implies $A \subseteq B$ for elements common in A and $\overline{C}$ and also in B and $\overline{C}$.

Hence, from both conditions it is shown that $A \subseteq B$.

Example 35. (*a*) ***If a set A has 20 elements, then how many members of P(A) are proper subsets of A.***

(*b*) ***Salad is made with combination of one or more eatables. How many different salads can be made from onion, tomato, carrot, cabbage and cucumber ?***

Sol. (*a*) If A has 20 elements, then P(A) has 2^{20} elements out of which $(2^{20} - 1)$ are proper subsets of A, as one of the member of P(A) is the set A (improper subset of A) itself.

(*b*) Let, A = {Onion, Tomato, Carrot, Cabbage, and Cucumber}.

Now the set of all salads is all possible subsets of A except ϕ (as no salad can be prepared without using at least one of the eatables)

$\therefore$ Number of salads $= n(P(A)) - 1$

$= 2^5 - 1$ | P(A) contains 2^5 elements

$= \mathbf{31}$.

TEST YOUR KNOWLEDGE 1.1

SHORT ANSWER TYPE QUESTIONS

1. Determine which of the following sets are finite. If so, find the cardinality of each set.
   (*a*) A = {seasons in a year} (*b*) B = {states in India}
   (*c*) C = {Positive integers less than one} (*d*) D = {Odd integers}
   (*e*) E = {Positive integral divisors of 12}
2. If A = {2, 3, 4, 5} and B = $\{x : x \in$ N is even}. Is A a subset of B ?
3. (*a*) Is A = {2, 3, 4, 5} a subset of B = $\{x : x \in N, x$ is even} ?
   (*b*) Is A = {2, 3, 4, 5} a proper subset of C = {1, 2, 3, ..., 8, 9}.
   (*c*) Give an example one each for sets A, B and C such that
   (*i*) $A \subseteq B$, $B \subseteq C$ and $A \not\subseteq C$
   (*ii*) $A \subseteq B$, $B \subseteq C$ and $A \subseteq C$.
4. Which of the following sets are equal ?
   (*a*) $A = \{x : x^2 - 4x + 3 = 0\}$ (*b*) $B = \{x : x^2 - 3x + 2 = 0\}$
   (*c*) $C = \{x : x \in N, x < 3\}$ (*d*) $D = \{x : x \in N, x$ is odd $< 5\}$
   (*e*) E = {1, 2} (*f*) F = {3, 1}
5. If A = {1, 2, 5, 6}, B = {2, 5, 7}, C = {1, 3, 5, 7, 9}, U = {1, 2, 3, 4, 5, 6, 7, 8, 9}, find
   (*a*) $A \cap B$ (*b*) $B \cup C$
   (*c*) A^C (*d*) $A - B$
   (*e*) $A - C$ (*f*) $A \oplus B$
   (*g*) $A \oplus C$ (*h*) $(A \cup C) - B$
   (*i*) $(A \cup B)^c$ (*j*) $(B \oplus C) - A$
6. (*a*) If A = {1, 2, 3, 4, 5}. Find the cardinal number of the power set of A.
   (*b*) If A = {1, 2, 3}. Find P(A). (*P.T.U. B.Tech. Dec. 2007*)
7. (*a*) If $A \subseteq B$, show that $A \cap B = A$, $A \cup B = B$
   (*b*) Show that we can have $A \cap B = A \cap C$ without $B = C$
   (*c*) Show that we can have $A \cup B = A \cup C$ without $B = C$
   (*d*) Let A, B, C are three sets such that $A \cup B = A \cup C$ and $A \cap B = A \cap C$. Show that $B = C$.
8. Describe the following sets in set builder form
   (*a*) {3, 5, 7, 9, ... 77, 79}.
   (*b*) The rational numbers that lie between – 1 and 1.
   (*c*) The even integers. (*P.T.U. B.Tech. Dec. 2005*)
9. Let A = {0, 2, 3}, B = {2, 3}, C = {1, 5, 9}
   Determine which of the following statements are true. Give reasons.
   (*a*) $A \subseteq B$ (*b*) $\{3\} \in A$
   (*c*) $\{3\} \subseteq A$ (*d*) $B \subseteq A$
   (*e*) $A \subseteq B$ (*f*) $\phi \subseteq C$
   (*g*) $\phi \in A$ (*h*) $B \cap A \subseteq C$

10. If $A = \{0, 2, 3\}$, $B = \{2, 3\}$, $C = \{1, 5, 9\}$ and $U = \{0, 1, 2, 3, 4, 5, 6, 7, 8, 9\}$, Determine

(*a*) $B - A$ (*b*) $A \times B$

(*c*) $A - B$ (*d*) B^2

(*e*) B^3 (*f*) $C \times \{8\}$

LONG ANSWERS TYPE QUESTIONS

11. Find the dual if each set equations

(*a*) $(U \cap A) \cup (B \cap A) = A$ (*b*) $(A \cup B \cup C)^c = (A \cup C)^c \cap (A \cup B)^c$

(*c*) $(A \cap U) \cap (\phi \cup A^c) = \phi$ (*d*) $(A \cap U)^c \cap A = \phi$

12. (*a*) Prove that $(A \cup B) - (A \cap B) = (A - B) \cup (B - A)$

(*b*) If A and B are two subsets of a universal set, prove that $A - B = A \cap B$.

(P.T.U. B.Tech. 2008)

13. (*a*) Determine the power set of $A = \{a, b, c, d\}$

(*b*) Distinguish between ϕ, $\{\phi\}$, $\{0\}$, 0. *(P.T.U. M.C.A. May 2008)*

14. Prove that

(*a*) $(A \cap B) \subseteq A \subseteq (A \cup B)$ (*b*) $(A \cap B) \subseteq B \subseteq (A \cup B)$.

15. Prove that

(*a*) $A \oplus (B \oplus C) = (A \oplus B) \oplus C$ (*b*) $A \oplus B = B \oplus A$

(*c*) If $A \oplus C = A \oplus B$, Then $C = B$ (*d*) $A \cap (B \oplus C) = (A \cap B) \oplus (A \cap C)$

where $\oplus$ denotes the symmetric difference of two sets

16. State and prove DeMorgan's law on difference of sets.

17. Two finite sets have *m* and *n* elements. The total number of subsets of the first set is 56 more than the total number of subsets of the second set. Find the value of *m* and *n*.

(P.T.U. M.C.A. Dec. 2006)

18. (*a*) If $A = \{+, -\}$ and $B = \{00, 01, 10, 11\}$. Find $A \times B$ *(P.T.U. B.Tech. May 2005)*

(*b*) How many elements of A^4 and $(A \times B)^3$ have ?

Answers

1. (*a*) $n(A) = 4$ (*b*) $n(B) = 28$

(*c*) $n(C) = 0$ (*d*) Infinite

(*e*) $n(E) = 6$.

2. No. since $3 \in A$ does not in B.

3. (*a*) No. since $3 \in A$ but $3 \notin B$ as B contains even numbers.

(*b*) Yes, since each element of A is also an element of C. Also, $1 \in C$ but $1 \notin A$. Hence, $A \neq C$. Therefore, A is a proper subset of C.

(*c*) (*i*) $A = \{a, b\}$, $B = \{\{a, b\}, a, e, d\}$, $C = \{\{a, b\}, \{a, e, d\}, g, h\}$

(*ii*) $A = \{a, b, c\}$, $B = \{\{a, b, c\}, \{d, e\}, f\}$, $C = \{\{\{a, b, c\}, \{d, e\}, f\}, \{a, b, c\}, g\}$.

4. $B = C = E$, $A = D = G$

5. (*a*) $A \cap B = \{2, 5\}$ (*b*) $B \cup C = \{1, 2, 3, 5, 7, 9\}$

(*c*) $A^c = \{3, 4, 7, 8, 9\}$ (*d*) $A - B = \{1, 6\}$

(*e*) $A - C = \{2, 6\}$ (*f*) $A \oplus B = \{1, 6, 7\}$

(*g*) $A \oplus C = \{2, 3, 6, 7, 9\}$ (*h*) $(A \cup C) - B = \{1, 3, 6, 9\}$

(*i*) $(A \cup B)^c = \{3, 4, 8, 9\}$ (*j*) $(B \oplus C) - A = \{3, 9\}$

6. (*a*) 32 (*b*) P{A} = {{1}, {2}, {3}, {1, 2}, {1, 3}, {2, 3}, {1, 2, 3}, ϕ}

8. (*a*) {2*m* + 1, *m* ∈ Z, 2 ≤ *m* ≤ 39} (*b*) {*x* : *x* ∈ Q, – 1 ≤ *x* ≤ 1}

(*c*) {2*n* : *n* ∈ Z

9. (*a*) False (*b*) False (*c*) True (*d*) True

(*e*) False (*f*) True (*g*) False (*h*) False.

10. (*a*) ϕ (*b*) {(0, 2), (0, 3), (2, 2), (2, 3), (3, 2), (3, 3)}

(*c*) {0} (*d*) {(2, 2), (2, 3), (3, 2), (3, 3)}

(*e*) {{2, 2, 2}, {2, 2, 3}, {2, 3, 2}, {2, 3, 3}, {3, 2, 2}, {3, 2, 3}, {3, 3, 2}, {3, 3, 3}}

(*f*) {{1, 8}, {5, 8}, {9, 8}}.

11. (*a*) $(\phi \cup A) \cap (B \cup A) = A$ (*b*) $(A \cap B \cap C)^c = (A \cap C)^c \cup (A \cap B)^c$

(*c*) $(A \cup \phi) \cup (U \cap A^C) = U$ (*d*) $(A \cup \phi)^c \cup A = U$

13. (*a*) P(A) = [A, {*a*, *b*, *c*}, {*a*, *b*, *d*}, {*a*, *c*, *d*}, {*b*, *c*, *d*}, {*a*, *b*}, {*a*, *c*}, {*a*, *d*}, {*b*, *c*}, {*b*, *d*}, {*c*, *d*}, {*a*}, {*b*}, {*c*}, {*d*}, ϕ]

17. $m = 6, n = 3$.

18. (*a*) A × B = { + 00, + 01, + 10, + 11, – 11, – 00, – 10, – 11}, (*b*) 16, 512.

Hints

7. (*b*) Take A = {1, 2}, B = {2, 3}, C = {2, 4}, then A ∩ B = {2}, A ∩ C = {2}, but B ≠ C

(*c*) Take A = {1, 2}, B = {1, 3}, C = {2, 3}, then A ∪ B = A ∪ C = {1, 2, 3}, but B ≠ C.

11. Interchange ∪ and ∩ and also U and ϕ in each set equations.

12. (*a*) $(A \cup B) - (A \cap B) = (A \cup B) \cap (A \cap B)^c = (A \cup B) \cap (A^c \cup B^c)$

$$= (A \cap A^c) \cup (A \cap B^c) \cup (B \cap A^c) \cup (B \cap B^c)$$

$$= \phi \cup (A \cap B^c) \cup (B \cap A^c) \cup \phi = (A \cap B^c) \cup (B \cap A^c)$$

$$= (A - B) \cup (B - A)$$

14. (*a*) Let $x \in A \cap B \Rightarrow x \in A$ and $x \in B$

Now $x \in A \Rightarrow A \cap B \subseteq A$ ...(1)

Further, if $x \in A$, then $x \in A \cup B$ | Definition of A ∪ B

$\Rightarrow A \subseteq A \cup B$ ...(2)

Combining (1) and (2), $A \cap B \subseteq A \subseteq A \cup B$.

17. Let $n(A) = m$, $n(B) = n$, Then

Total number of subsets of $A = 2^m$

Total number of subsets of $B = 2^n$

Given, $2^m = 2^n + 56 \Rightarrow 2^m - 2^n = 56$

$\Rightarrow 2^n(2^{m-n} - 1) = 56 = 8.7 = 2^3(2^3 - 1)$

$\Rightarrow n = 3, m - n = 3 \Rightarrow m = 6.$

18. (*b*) $A^4 = A \times A \times A \times A$

$\therefore n(A^4) = n(A \times A \times A \times A) = n(A) \cdot n(A) \cdot n(A) \cdot n(A)$

$$= 2 \cdot 2 \cdot 2 \cdot 2 = 16$$

Also $n(A) = 2, n(B) = 4, n(A \times B) = n(A) \,.\, n(B) = 2.4 = 8$

$\therefore n(A \times B)^3 = n(A \times B) \cdot n(A \times B) \cdot n(A \times B) = 8.8.8 = 512$

1.9. PARTITIONS OF SETS AND VENN-DIAGRAMS

Partitions of sets. Let S be a non-empty set. A partition P of S is a finite collection $\{A_i\}_{j=1}^{n}$ of non-empty subsets of S such that

(*i*) $A_i \cap A_j = \phi \ \forall \ i \neq j$ *i.e.*, the subsets A_i are mutually disjoint

(*ii*) $\bigcup_{i=1}^{n} A_i = S$

In other words, a partition P of S is a subdivision of S into disjoint non empty sets. The subsets in a partition are called **cells**.

1.10. VENN-DIAGRAMS

A Venn-diagram is a pictorial representation of sets in which sets are represented by enclosed areas in the plane. The universal set U is represented by the interior of a rectangle and the other sets are represented by circles lying with in the rectangle. For e.g.,

If A and B are two arbitrary sets such that

(*i*) $A \subseteq B$ (*ii*) $A \cap B = \phi$

The Venn-diagrams of these sets is shown in Fig. 1.1, and Fig. 1.2.

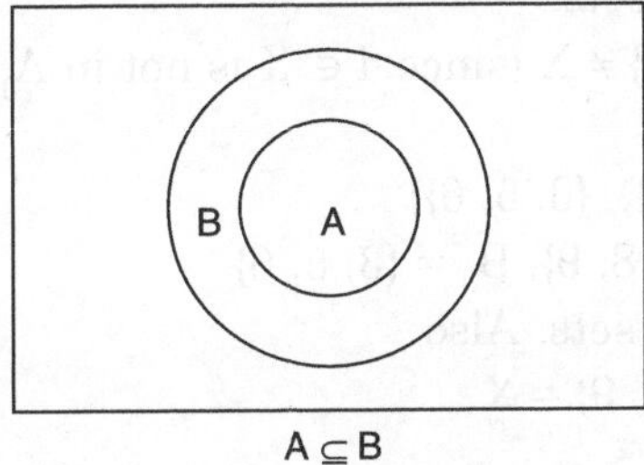

$A \subseteq B$

Fig. 1.1

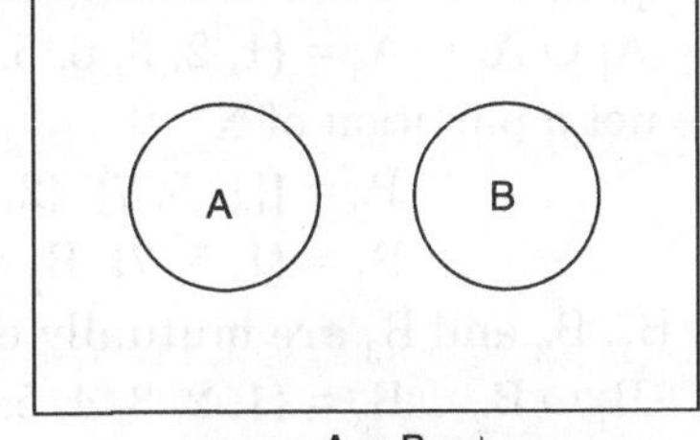

$A \cap B = \phi$

Fig. 1.2

1.11. CROSS PARTITION

If $[A_1, A_2, ..., A_m]$ and $[B_1, B_2, ... B_n]$ be partitions of a set X. Then the set $P = \{A_i \cap B_j, i = 1, 2, ... m; j = 1, 2, ... n\}$ is called cross partition.

ILLUSTRATIVE EXAMPLES

Example 1. *Let S = {red, blue, green, yellow}. Determine which of the following is a partition of S*

(*a*) P_1 = [{red}, {blue, green}] (*b*) P_2 = [{red, blue, green, yellow}]

(*c*) P_3 = [ϕ, {red, blue}, {green, yellow}] (*d*) P_4 = [{blue], [red, yellow, green}].

Sol. (*a*) No, since P_1 = [{red}, {blue, green}].

Here P_1 cannot be a partition of S since yellow does not belong to any cell in P_1.

(*b*) Yes, since P_2 = {red, blue, green, yellow}

Here P_2 is a partition of S since P_2 has one element which is S itself.

(*c*) No, since P_3 = [ϕ, {red, blue}, {green, yellow}]

Here P_3 cannot be a partition. Since the empty set ϕ cannot belong to a partition.

(*d*) Yes, since P_4 = [{blue}, {red, yellow, green}]

Here P_4 is a partition since each cell in P_4 is disjoint and their union is S.

Example 2. *Let S = {1, 2, 3, ..., 8, 9}. Determine whether or not each of the following is a partition of S*

(*a*) *[{1, 3, 5}, {2, 6}, {4, 8, 9}]*

(*b*) *{{1, 3, 5}, {2, 4, 6, 8}, {5, 7, 9}}*

(*c*) *{{1, 3, 5}, {2, 4, 6, 8}, {7, 9}}.*

(*d*) *Let X = {1, 2, 3, ... 8, 9}. Determine whether or not each of the following is a partition of X.*

(*i*) *[{1, 3, 6}, {2, 8}, {5, 7, 9}]*

(*ii*) *[{1, 5, 7}, {2, 4, 8, 9} {3, 5, 6}]*

Sol. (*a*) *No*, since 7 does not belong to any cell

(*b*) *No*, since {1, 3, 5} and {5, 7, 9} are not disjoint

(*c*) *Yes*, as each cell is disjoint and their union is S.

(*d*) (*i*) Given X = {1, 2, 3, ... 8, 9}

Let $P_1 = [\{1, 3, 6\}, \{2, 8\}, \{5, 7, 9\}]$

Take $A_1 = \{1, 3, 6\}, A_2 = \{2, 8\}, A_3 = \{5, 7, 9\}$

Clearly A_1, A_2 and A_3 are mutually disjoint sets.

Also $A_1 \cup A_2 \cup A_3 = \{1, 2, 3, 6, 5, 7, 8, 9\} \neq X$ (since $4 \in X$ is not in $A_1 \cup A_2 \cup A_3$)

$\therefore$ P_1 is not a partition of X

(*ii*) Let $P_2 = [\{1, 5, 7\}, \{2, 4, 8, 9\}, \{3, 5, 6\}]$

Take $B_1 = \{1, 5, 7\}, B_2 = \{2, 4, 8, 9\}, B_3 = \{3, 5, 6\}$

Clearly, B_1, B_2 and B_3 are mutually disjoint sets. Also

$B_1 \cup B_2 \cup B_3 = \{1, 2, 3, 4, 5, 6, 7, 8, 9\} = X$

$\therefore$ P_2 is a partition of X.

Example 3. *Find all partitions of S = {a, b, c, d}.*

Sol. Each partition of X contains either 1, 2, 3 or 4 distinct sets. The partitions are as follows:

(1) {{*a*, *b*, *c*, *d*}}

(2) {{*a*}, {*b*, *c*, *d*}}, {{*b*}, {*a*, *c*, *d*}}, {{*c*}, {*a*, *b*, *d*}}, {{*d*}, {*a*, *b*, *c*}},
{{*a*, *b*}, {*c*, *d*}}, {{*a*, *c*}, {*b*, *d*}}, {{*a*, *d*}, {*b*, *c*}}.

(3) {{*a*}, {*b*}, {*c*, *d*}}, {{*a*}, {*c*}, {*b*, *d*}}, {{*a*}, {*d*}, {*b*, *c*}},
{{*b*}, {*c*}, {*a*, *d*}}, {{*b*}, {*d*}, {*a*, *c*}}, {{*c*}, {*d*}, {*a*, *b*}}.

(4) {{*a*}, {*b*}, {*c*}, {*d*}}.

There are fifteen different partitions of X.

Example 4. *Determine whether or not each of the following is a partition of the set N of positive integers.*

(*a*) $\{\{n : n > 5\}, \{n : n < 5\}\}$ (*b*) $\{\{n : n > 5\}, \{0\}, \{1, 2, 3, 4, 5\}\}$

(*c*) $\{\{n : n^2 > 11\}, \{n : n^2 < 11\}\}$.

Sol. (*a*) *No*, since 5 does not belong to any cell

(*b*) *No*, since {0} is not a subset of N

(*c*) *Yes*, the two cells are disjoint and their union is *N*.

Example 5. *Find all the partitions of S = {1, 2, 3}.*

Sol. Each partition of S contains either, 1, 2 or 3 different cells.

The partition containing 1 cell is {1, 2, 3} = S

The partitions containing 2 different cells are {{1}, {2, 3}}, {{2}, {1, 3}}, {{3}, {1, 2}}

The partitions containing 3 different cells are {{1}, {2}, {3}}

Hence, there are five different partitions of S.

Example 6. *Consider the set Z of integers. Determine which of the following is a partition of Z ?*

(i) $P_1 = \{\{n\} : n \in Z\}$

(ii) $P_1 = \{n : n \in Z, n < 0\}$, $P_2 = \{0\}$, $P_3 = \{n : n \in Z \text{ and } n > 0\}$

(iii) $P_1 = \{n : n \in Z \text{ and } n \geq 0\}$, $P_2 = \{n : n \in Z \text{ and } n \leq 0\}$

(iv) $P_1 = \{n \in Z : | n | = k, k = -1, 0, 1, 2, ...\}$, $P_2 = \{n \in Z = | n | = -1\}$.

Sol. (i) *Yes,* since for different value of n, we have different sets and all the sets are mutually disjoint.

(ii) *Yes,* since P_1, P_2 and P_3 are mutually disjoint and

$$P_1 \cup P_2 \cup P_3 = \{n : n \in Z, n < 0\} \cup \{0\} \cup \{n : n \in Z, n > 0\} = Z$$

(iii) *No,* since P_1 and P_2 are not mutually disjoint as $0 \in$ both P_1 and P_2

(iv) *No,* since $P_2 = \{n \in Z = | n | = -1\} = \phi$, as the empty set ϕ cannot belong to any partition.

Example 7. *A student, on an exampaper, defined the term partition in the following way.*

"Let A be a set. A partition of A is any set of non-empty subsets $A_1, A_2, ...$ of A such that each element of A is in one and only one of the subsets A_1" Is this definition correct ? Why ?

Sol. From the given definition of partition, it is possible that an element of A is in one and only one of the subsets A_i. Take

$$A = \{a, b, c\}, A_1 = \{a, c\}, A_2 = \{b, c\}, A_3 = \{c, a\}$$

Clearly $A_1 \cup A_2 \cup A_3 = \{a, b, c\} = A$

But $A_1 \cap A_2 \neq \phi$. Hence the given definition is not correct.

Example 8. *Let $\{A_1, A_2, ... A_n\}$ be a partition of the set A and if B is any non-empty subset of A. Prove that $\{A_i \cap B : A_i : \cap B \neq \phi\}$ is a partition of $A \cap B$.*

Sol. Since $\{A_1, A_2, ... A_n\}$ is a partition of A, therefore, the sets A_i are non-empty and mutually disjoint and $\bigcup_{i=1}^{n} A_i = A$. Also, B is non-empty subset of A, therefore,

$$B \cap A = B \cap \bigcup_{i=1}^{n} A_i = B \cap (A_1 \cup A_2 \cup A_3 \cup ... \cup A_n)$$

$$= (B \cap A_1) \cup (B \cap A_2) \cup ... (B \cap A_n) \qquad | \text{ Distributive law}$$

$$= \bigcup_{i=1}^{n} (B \cap A_i)$$

Next, we show that all $B \cap A_i$ are mutually disjoint. For this consider, for $i \neq j$

$$(B \cap A_i) \cap (B \cap A_j) = B \cap (A_i \cap A_j) \qquad | \text{ Associative law}$$

$$= B \cap \phi \qquad | \text{ Since } A_i \text{ are mutually disjoint}$$

$$= \phi$$

Hence, the set $\{A_i \cap B : A_i \cap B \neq \phi\}$ is a partition of $A \cap B$.

Example 9. *If $A_1 = \{1, 2\}$, $A_2 = \{2, 3\}$, $A_3 = \{1, 2, 3, 6\}$. Find $\bigcup_{i=1}^{3} A_i$ and $\bigcap_{i=1}^{3} A_i$.*

(P.T.U. M.C.A. Dec. 2005)

Sol. $$\bigcup_{i=1}^{3} A_i = A_1 \cup A_2 \cup A_3 = \{1, 2\} \cup \{2, 3\} \cup \{1, 2, 3, 6\} = \{1, 2, 3, 6\}.$$

$$\bigcap_{i=1}^{3} A_i = A_1 \cap A_2 \cap A_3 = \{1, 2\} \cap \{2, 3\} \cap \{1, 2, 3, 6\} = \{2\} \cap \{1, 2, 3, 6\} = \{2\}.$$

Example 10. *Let $X = \{1, 2, 3, 4, 5, 6, 7, 8, 9\}$ and consider the following partitions of X given by $P_1 = [\{1, 3, 5, 7, 9\}, \{2, 4, 6, 8\}]$ and $P_2 = [\{1, 2, 3, 4\}, \{5, 7\}, \{6, 8, 9\}]$.*

Find the cross partition P.

Sol. Take $A_1 = \{1, 3, 5, 7, 9\}$, $A_2 = \{2, 4, 6, 8\}$

$B_1 = \{1, 2, 3, 4\}$, $B_2 = \{5, 7\}$, $B_3 = \{6, 8, 9\}$

$\therefore$ $A_1 \cap B_1 = \{1, 3\}$,

$A_1 \cap B_2 = \{5, 7\}$, $A_1 \cap B_3 = \{9\}$

$A_2 \cap B_1 = \{2, 4\}$

$A_2 \cap B_2 = \phi$, $A_2 \cap B_3 = \{6, 8\}$

$\therefore$ The required cross partition of X is given by

$$P = [\{1, 3\}, \{5, 7\}, \{9\}, \{2, 4\}, \{6, 8\}]$$

(ϕ cannot belong to the partition of a set).

Example 11. *If A and B are two sets. Represent each of the following as Venn-diagrams.*

(a) $A \cup B$ *(b) $A \cap B$* *(c) A^c.*

Sol. (*a*) $A \cup B$, $\{x : x \in A \text{ or } x \in B]$. The Venn-diagram is shown in the Fig. 1.3.

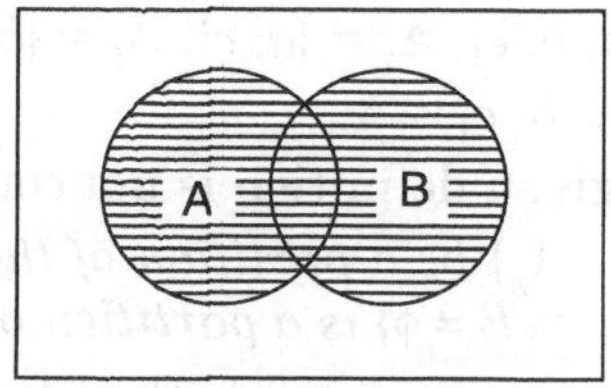

Fig. 1.3

(*b*) $A \cap B = \{x : x \in A \text{ and } x \in B]$. The Venn-diagram is shown in the Fig. 1.4.

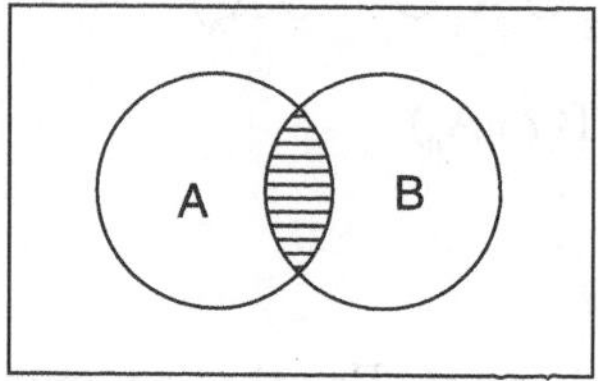

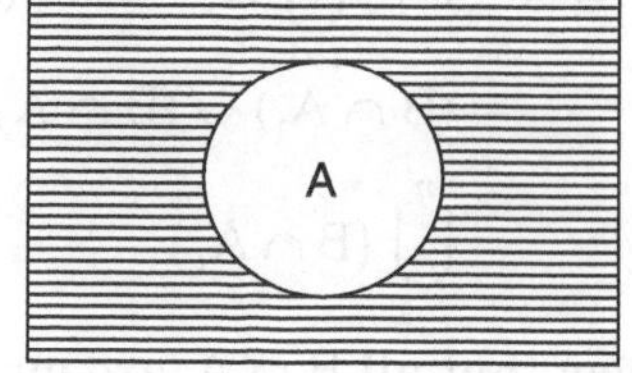

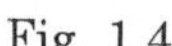

Fig. 1.4 Fig. 1.5

(*c*) $A^c = \{x : x \in U \text{ and } x \notin A]$. The Venn-diagram is shown in the Fig. 1.5.

Example 12. *Draw a Venn diagram of three arbitrary sets A, B, C which divides the universal set U into $2^3 = 8$ regions. Explain why are there eight regions ?*

Sol. The Venn diagram is shown in Fig. 1.6.

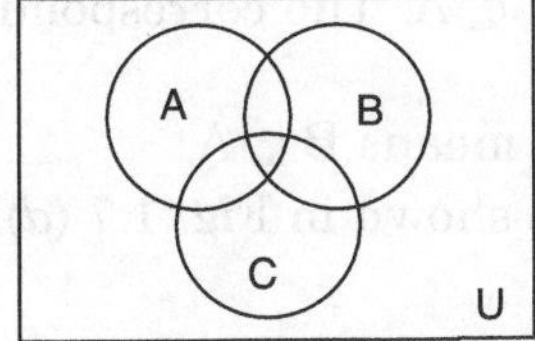

Fig. 1.6

There are eight regions since there may be elements

(*i*) in A, B and C
(*ii*) in A and B, but not in C
(*iii*) in A and C, but not in B
(*iv*) in B and C, but not in A
(*v*) in only A
(*vi*) in only B
(*vii*) in only C
(*viii*) in none of A, B, C.

Note: A Venn diagram of four sets A_1, A_2, A_3 and A_4 will contain $2^4 = 16$ regions.

Example 13. (*i*) *Let A and B be sets such that* $(A \cap B) \subseteq B$ *and* $B \not\subset A$. *Draw the Venn diagram.*

(*ii*) *Let A, B and C be sets such that* $A \subseteq B$, $A \subseteq C$, $B \cap C \subseteq A$, $A \subseteq (B \cap C)$. *Draw the Venn diagram.*

(*iii*) *Let A, B, C be sets such that* $(A \cap B \cap C) = \phi$, $A \cap B = \phi$, $B \cap C = \phi$, $A \cap C = \phi$. *Draw the Venn diagram.*

(*iv*) *Let A and B are sets such that:*

(*a*) $A \cap B = B$. *Draw its Venn diagram*

(*b*) $A \cup B = A$ *and* $A \neq B$. *Draw its Venn diagram.*

Sol. (*i*) $(A \cap B) \subseteq B$. It means every member of A is also a member of B. Hence $A \subseteq B$. Also $B \not\subset A$ means A and B are not equal.

Hence, A is a proper subset of B *i.e.*, $A \subset B$. The corresponding Venn diagram is shown in Fig. 1.7 (*a*)

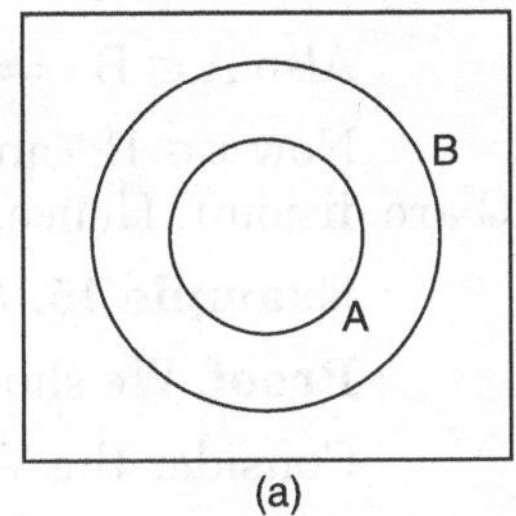

(a)

(*ii*) Given $B \cap C \subseteq A$, $A \subseteq B \cap C$. It implies $A = B \cap C$. Also $A \subseteq B$ and $A \subseteq C$. The corresponding Venn diagram is shown in Fig. 1.7 (*b*)

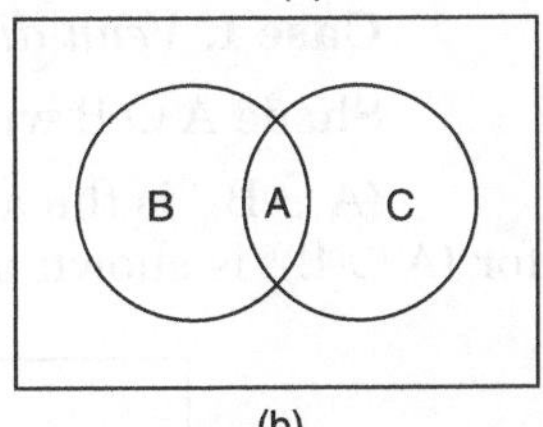

(b)

(*iii*) Given $A \cap B = \phi$, $B \cap C = \phi$, $A \cap C = \phi$. It means A, B and C are disjoint sets. The corresponding Venn diagram is shown in Fig. 1.7 (*c*)

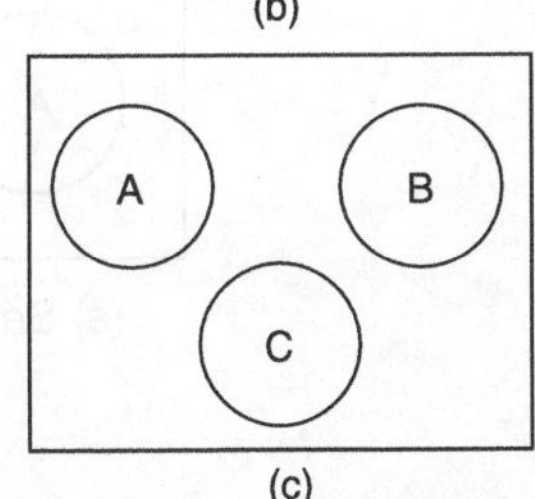

(c)

(*iv*) (*a*) $A \cap B = B$. It means $B \subseteq A$. The corresponding Venn diagram is shown in Fig. 1.7 (*d*)

(*b*) $A \cup B = A$ and $A \neq B$. This means $B \subseteq A$.

The corresponding diagram is shown in Fig. 1.7 (*d*).

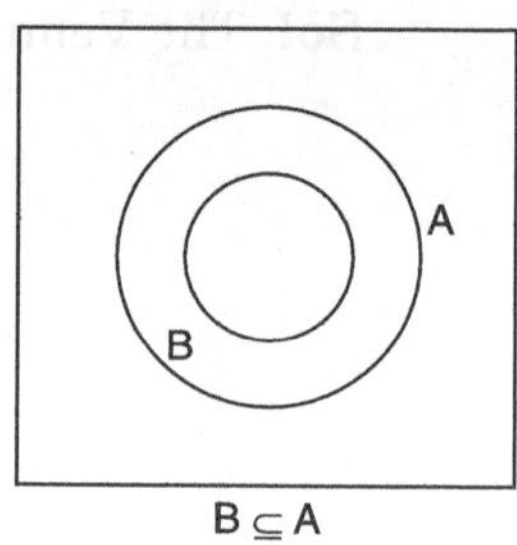

$B \subseteq A$
(d)
Fig. 1.7

Example 14. *Draw a Venn diagram of the sets A, B and C*

(*i*) *where $A \subseteq B$, A and C are disjoint, but B and C have elements in common.*

(*ii*) *where $A \subseteq B$, B and C are disjoint, but A and C have elements in common.*

Sol. (*i*) The required Venn diagram is shown in the Fig. 1.8.

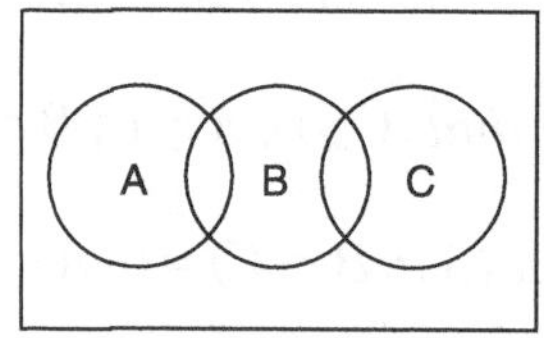

Fig. 1.8

(*ii*) If A and C have an element in common, say x, then $x \in A$ and $x \in C$

Also $A \subseteq B \Rightarrow x$ must belong to B

Now $x \in B$ and $x \in C$ and hence, B and C are not disjoint. But it is given that B and C are disjoint. Hence, No such Venn diagram exists.

Example 15. *Using Venn diagrams, prove De Morgan's laws.*

Proof. We show $(A \cup B)^c = A^c \cap B^c$

Consider the Venn diagram of two arbitrary sets A and B as shown in the Fig. 1.9(*a*)

Case I. *Venn diagram for $(A \cup B)^c$*

Shade $A \cup B$ with strokes in horizontal direction as shown in Fig. 1.9(*b*)

$(A \cup B)^c$ is the area outside of $A \cup B$ as shaded in Fig. 1.9(*b*). Hence, the Venn diagram for $(A \cup B)^c$ is shown in the shaded portion of the Fig. 1.9(*c*).

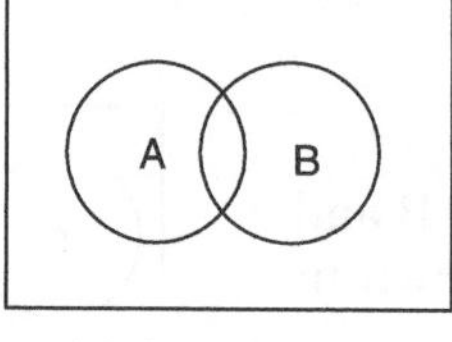

(a) Sets A and B

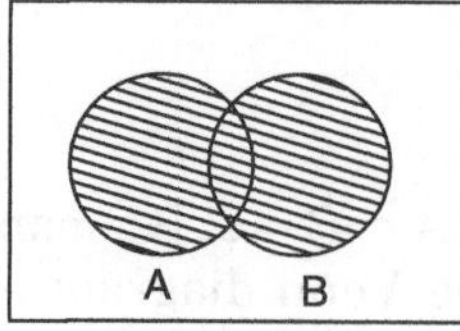

(b) $A \cup B$ is shade

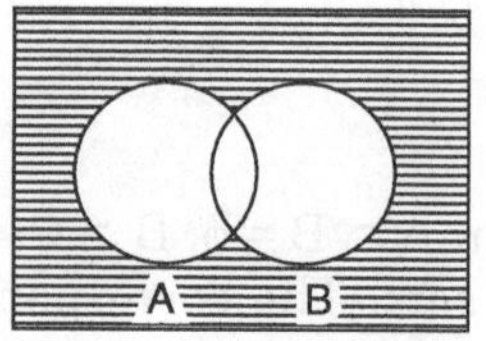

(c) $(A \cup B)^c$ is shaded

Fig. 1.9

Case II. *Venn diagram for $A^c \cap B^c$*

The area outside A is A^c. Shade A^c with strokes that slant upward to the right (⫽). Fig. 1.10(*a*)

The area outside B is B^c. Shade B^c with strokes that slant downward to the right (⧵⧵). Fig. 1.10(*a*).

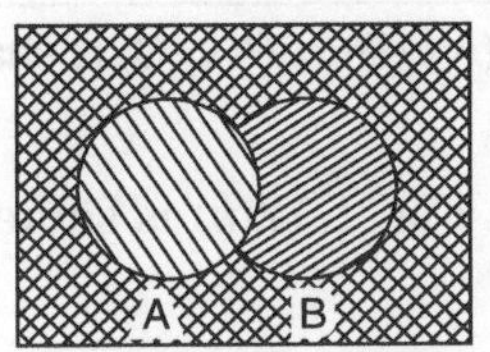

(a) A^c is shaded with ⫽ B^c is shaded with ⧵⧵

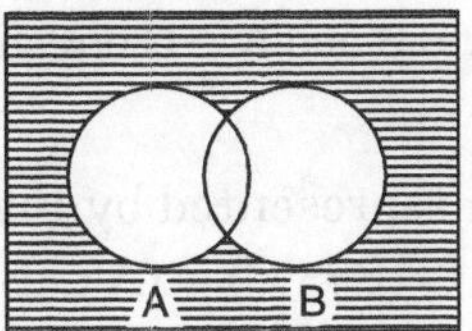

(b) $A^c \cap B^c$ is shaded

Fig. 1.10

The crosshatched area is shown shaded in Fig. 1.10 (*b*) is $A^c \cap B^c$.

Since Fig. 1.9(*c*) and Fig. 1.10(*b*) represent the same area. Hence $(A \cup B)^c = A^c \cap B^c$.

Example 16. *Using Venn diagram, show that* $A \cup (B \cap C) = (A \cup B) \cap (A \cup C)$.

Sol. Consider the Venn diagram for any three sets A, B and C as shown in Fig. 1.11(*a*)

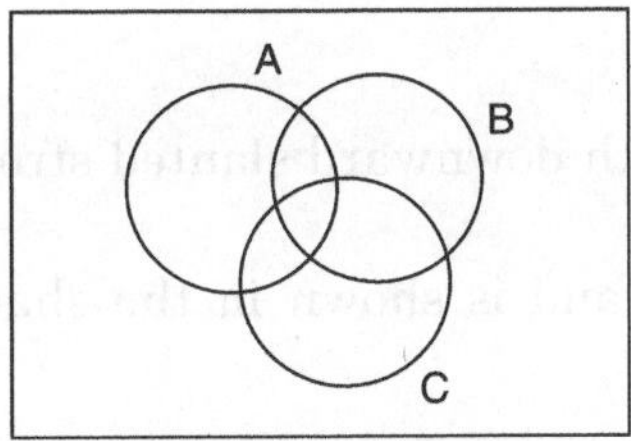

(a) Sets A, B and C

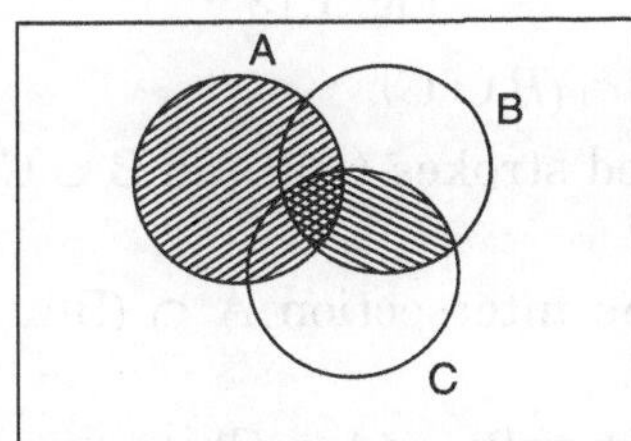

(b) A and $B \cap C$ are shaded

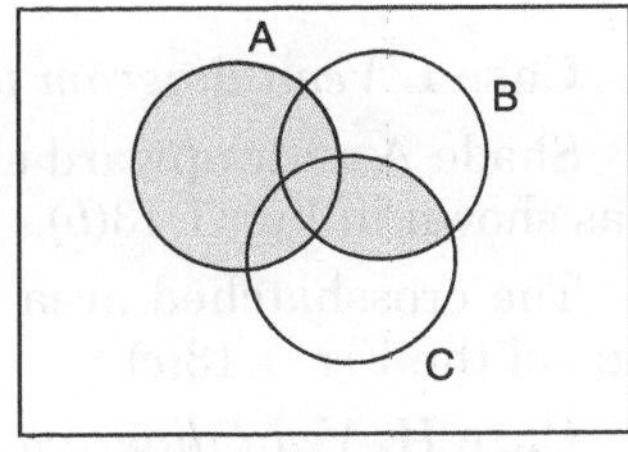

(c) $A \cup (B \cap C)$

Fig. 1.11

Case I. *Venn diagram for* $A \cup (B \cap C)$.

Shade A with upward slanted strokes (⫽) and $B \cap C$ with downward slanted strokes (⧵⧵) as shown in Fig. 1.11(*b*)

The total area is the union $A \cup (B \cap C)$ shown in the shaded portion of the Fig. 1.11(*c*)

Case II. *Venn diagram for* $(A \cup B) \cap (A \cup C)$.

Shade $A \cup B$ with upward slanted strokes (⫽) and $A \cup C$ with downward slanted strokes (⧵⧵) as shown in Fig. 1.12(*a*)

The crosshatched area is the intersection $(A \cup B) \cap (A \cup C)$ is shown in the shaded portion of the Fig. 1.12(*b*).

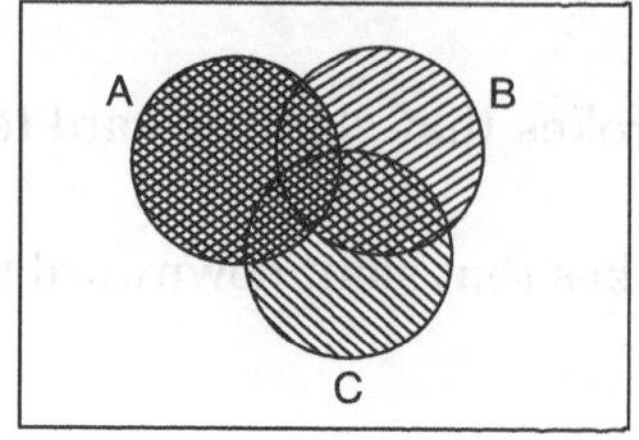

(a) A ∪ B and A ∪ C are shaded

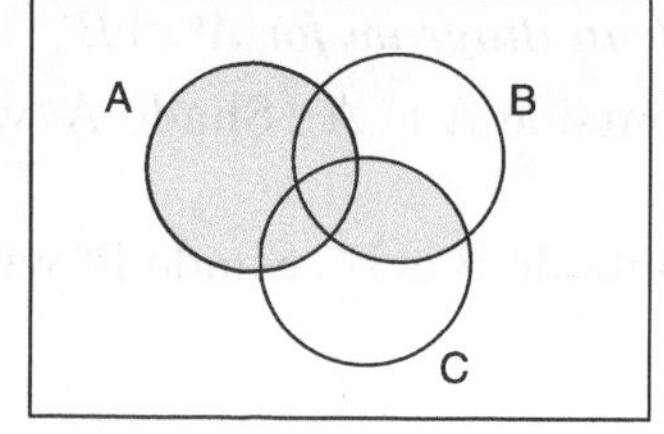

(b) (A ∪ B) ∩ (A ∪ C) is shaded

Fig. 1.12

Since the area represented by Fig. 1.11(*c*) and Fig. 1.12(*b*) is same. Hence

$$A \cup (B \cap C) = (A \cup B) \cap (A \cup C).$$

Example 17. *By using Venn diagram, Prove that $A \cap (B \cup C) = (A \cap B) \cup (A \cap C)$.*

Sol. Consider the Venn diagram for any three sets A, B, C as shown in Fig. 1.13(*a*).

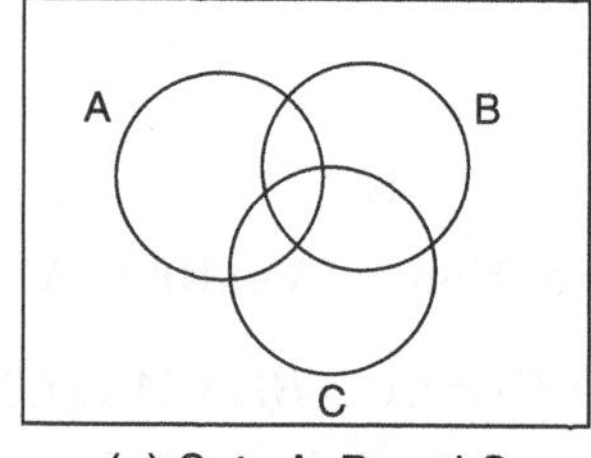

(a) Sets A, B and C

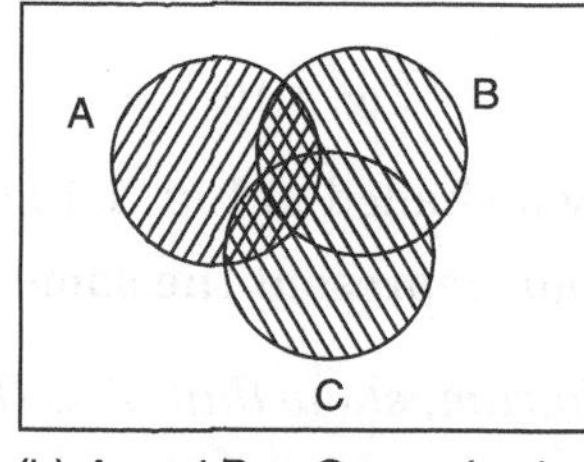

(b) A and B ∪ C are shade

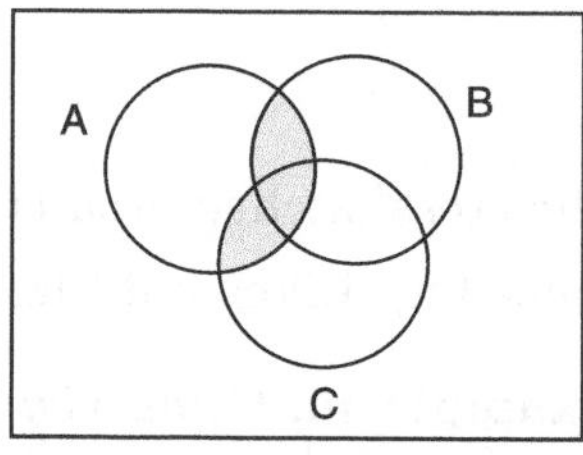

(c) A ∩ (B ∪ C) is shaded

Fig. 1.13

Case I. *Venn diagram for $A \cap (B \cup C)$.*

Shade A with upward slanted strokes (⫽) and B ∪ C with downward slanted strokes (⧵⧵) as shown in Fig. 1.13(*b*).

The crosshatched area is the intersection A ∩ (B ∪ C) and is shown in the shaded portion of the Fig. 1.13(*c*)

Case II. *Venn diagram for $(A \cap B) \cup (A \cap C)$.*

Shade A ∩ B with upward slanted strokes (⫽) and A ∩ C with downward slanted strokes (⧵⧵) as shown in Fig. 1.14(*a*)

The total area is the union (A ∩ B) ∪ (A ∩ C) as shown in the shaded portion of the Fig. 1.14(*b*).

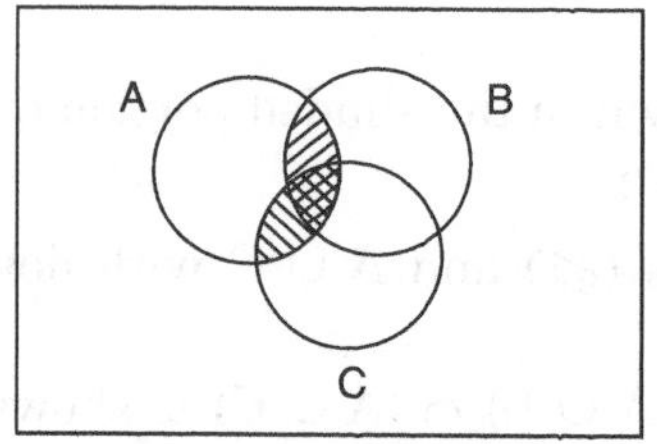

(a) A ∩ B and A ∩ C are shaded

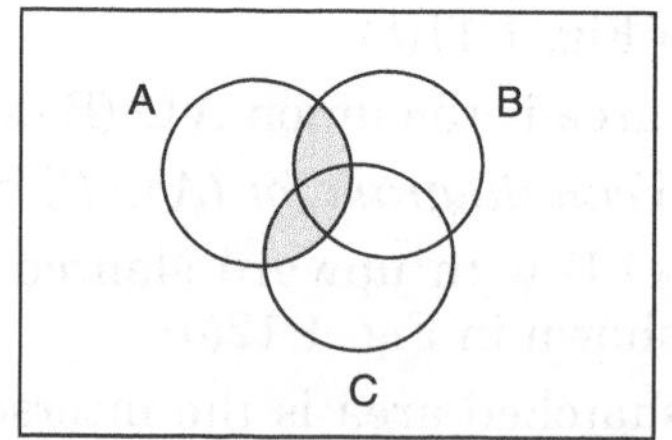

(b) (A ∩ B) ∪ (A ∩ C) is shaded

Fig. 1.14

Since the area represented by Fig. 1.13(*c*) and Fig. 1.14(*b*) is same. Hence

$$A \cap (B \cup C) = (A \cap B) \cup (A \cap C)$$

Example 18. *Describe the Venn diagram of* $A \cap B^c$.

Sol. Consider the Venn diagram for two arbitrary sets A and B as shown in Fig. 1.15(*a*) Shade A with strokes that slant upward to the right (⫽). Shade B^c, the area outside B, with strokes that slant downward to the right (⧵⧵) as shown in Fig. 1.15(*b*).

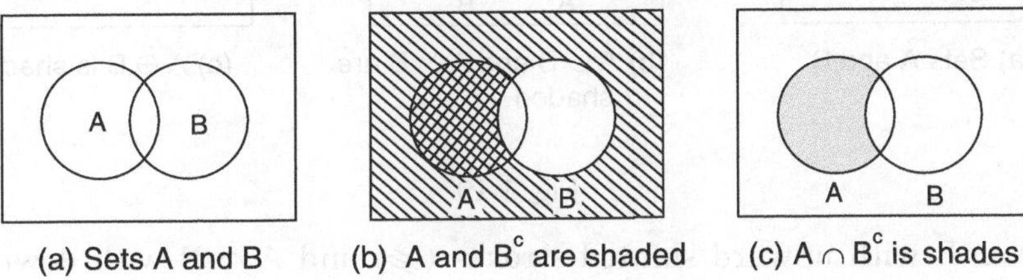

(a) Sets A and B (b) A and B^c are shaded (c) $A \cap B^c$ is shades

Fig. 1.15

The crosshatched area is the intersection $A \cap B^c$ shown in the shaded portion of Fig. 1.15(*c*)

Example 19. *If A, B are any sets. Draw the Venn-diagrams of*

(*i*) $A - B$ (P.T.U. M.C.A. May 2008)

(*ii*) $B - A$

(*iii*) $A \oplus B$, *where* $\oplus$ *denotes the symmetric difference of two sets.*

(P.T.U. M.C.A. May 2008)

Sol. (*i*) $A - B = \{x : x \in A \text{ and } x \notin B\}$. The Venn-diagram is shown in Fig. 1.16.

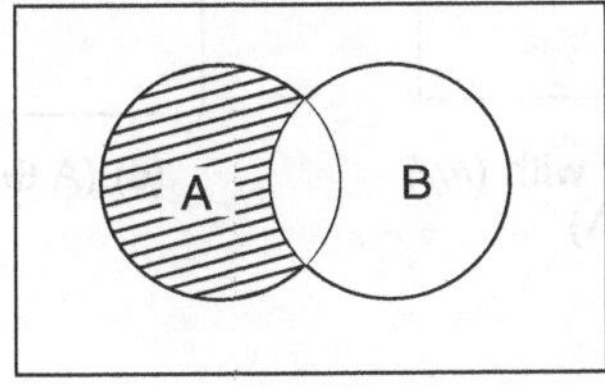

Fig. 1.16

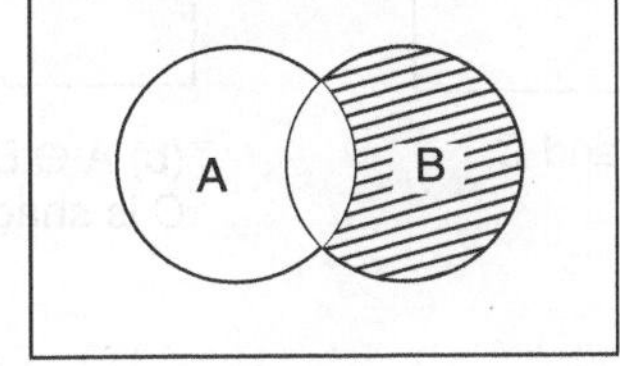

Fig. 1.17

(*ii*) $B - A = \{x : x \in B \text{ and } x \notin A\}$. The Venn-diagram is shown in the Fig. 1.17.

(*iii*) $A \oplus B = (A \cup B) - (A \cap B)$. The Venn-diagram is shown in the Fig. 1.18.

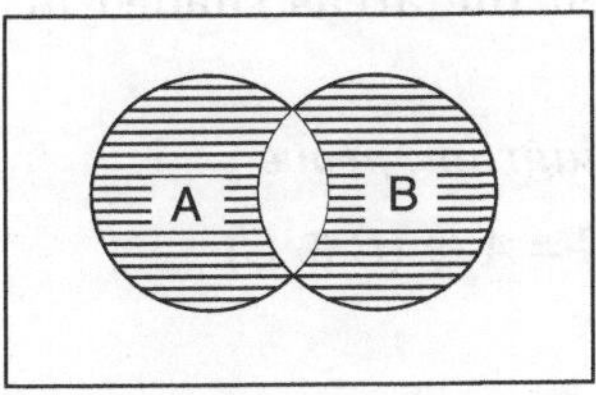

Fig. 1.18

Example 20. *Represent A ⊕ B = (A ∪ B) – (A ∩ B), using Venn diagram.*

Sol. Consider the Venn diagram for any two sets A and B as shown in the Fig. 1.19

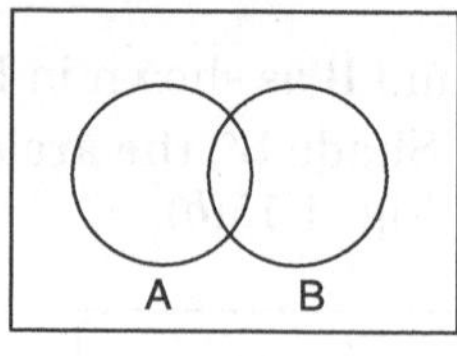

(a) Sets A and B

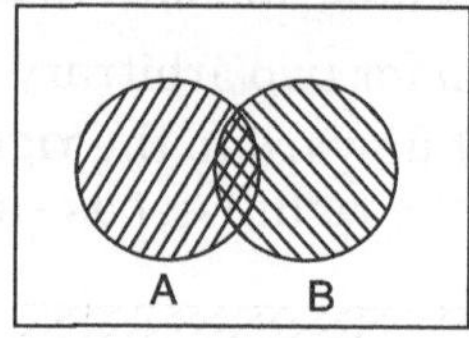

(b) A ∪ B and A ∩ B are shaded

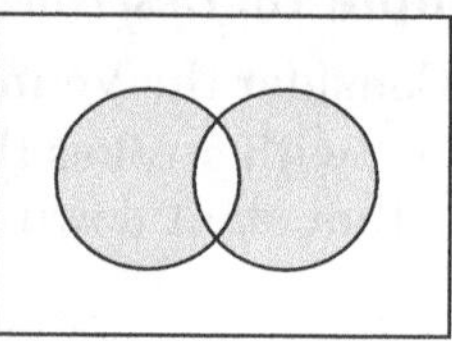

(c) A ⊕ B is shaded

Fig. 1.19

Shade A ∪ B with upward slanted strokes (////) and A ∩ B with downward slanted strokes (\\\\) as shown in Fig. 1.19(*b*)

Then, A ⊕ B consists of the area in Fig. 1.19(*b*) with strokes in one direction or another, but not both, as shown in the shaded portion of the Fig. 1.19(*c*).

Example 21. *Using Venn diagram, prove the following :*

If A ⊕ B = A ⊕ C, then B = C. *(Cancellation law)*

Sol. Consider a Venn diagram for any three sets A, B and C as shown in the Fig. 1.20(*a*)

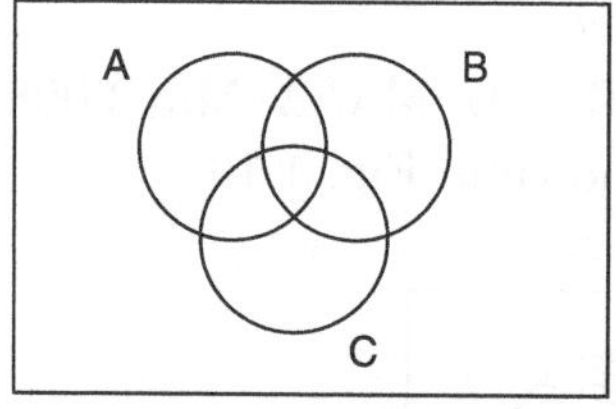

(a) Sets A, B and C

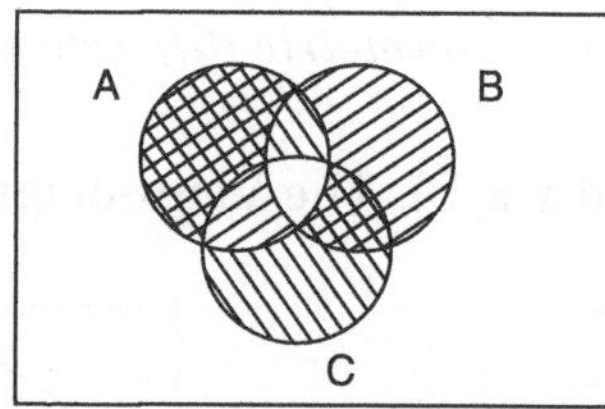

(b) A ⊕ B is shaded with (////)
C is shaded with (\\\\)

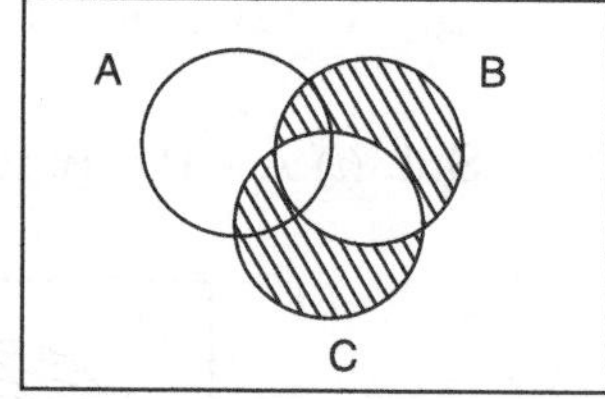

(c) (A ⊕ B) ⊕ C is shaded

Fig. 1.20

For the Venn diagram of A ⊕ B, shade A ⊕ B with strokes in one direction (////) and shade A ⊕ C with strokes in another direction (\\\\) as shown in Fig. 1.20(*b*).

Further, consider the area with strokes in one direction or another, but not in both in the following Fig. 1.20(*c*).

Now if A ⊕ B = A ⊕ C, then the areas shaded in Fig. 1.20(*c*) must be empty. Thus B = B ∩ C = C. Hence proved.

Example 22. *Using Venn diagram, prove*

$$(A \oplus B) \oplus C = A \oplus (B \oplus C). \qquad \textit{(Associative law)}$$

Sol. Consider a Venn diagram for any three sets A, B and C as shown in Fig. 1.21(*a*).

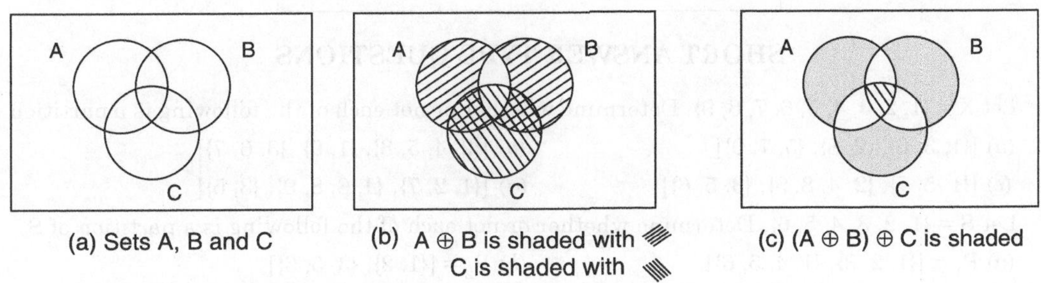

(a) Sets A, B and C (b) A ⊕ B is shaded with ⫽ C is shaded with ⧵⧵ (c) (A ⊕ B) ⊕ C is shaded

Fig. 1.21

Case I. *For the Venn diagram of (A ⊕ B) ⊕ C.*

Shade A ⊕ B with strokes in one direction (⫽) and shade C with strokes in another direction (⧵⧵) as shown in Fig. 1.21(*b*).

Then, (A ⊕ B) ⊕ C consists of the areas in Fig. 1.21(*b*) with strokes in one direction or another but not in both, as shown in the shaded portion of the Fig. 1.21(*c*).

Case II. *For the Venn diagram of A ⊕ (B ⊕ C).*

Shade A with strokes in one direction (⫽) and shade B ⊕ C with strokes in another direction (⧵⧵) as shown in the Fig. 1.22(*a*)

Then, A ⊕ (B ⊕ C) consists of the area in the Fig. 1.22(*a*) with strokes in one direction or another but not in both as shown in the shaded portion of the Fig. 1.22(*b*)

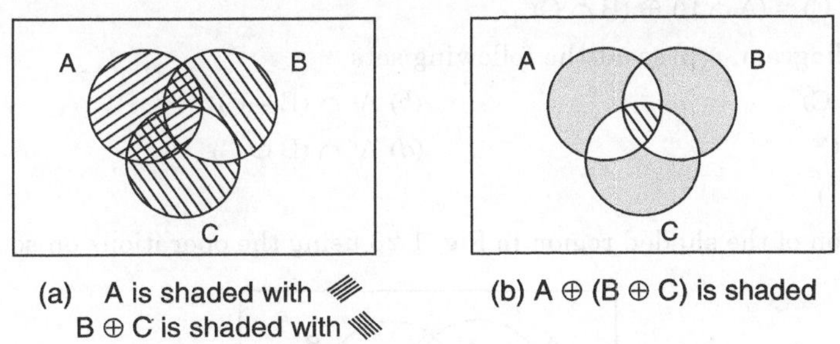

(a) A is shaded with ⫽ B ⊕ C is shaded with ⧵⧵ (b) A ⊕ (B ⊕ C) is shaded

Fig. 1.22

Since the area represented by Fig. 1.21(*c*) and Fig. 1.22(*b*) is same, therefore

(A ⊕ B) ⊕ C = A ⊕ (B ⊕ C)

TEST YOUR KNOWLEDGE 1.2

SHORT ANSWER TYPE QUESTIONS

1. Let X = {1, 2, 3, 4, 5, 6, 7, 8, 9}. Determine whether or not each of the following is a partition of X
   (*a*) [{1, 3, 6}, {2, 8}, {5, 7, 9}] (*b*) [{2, 4, 5, 8}, {1, 9}, {3, 6, 7}]
   (*c*) [{1, 5, 7}, {2, 4, 8, 9}, {3, 5, 6}] (*d*) [{1, 2, 7}, {4, 6, 8, 9}, {3, 5}]
2. Let S = {1, 2, 3, 4, 5, 6}. Determine whether or not each of the following is a partition of S
   (*a*) P_1 = [{1, 2, 3}, {1, 4, 5, 6}] (*b*) P_2 = [{1, 2}, {3, 5, 6}]
   (*c*) P_3 = [{1, 3, 5}, {2, 4}, {6}] (*d*) P_4 = [{1, 3, 5}, {2, 4, 6, 7}]
3. Determine whether or not each of the following is a partition of the set Z of positive integers.
   (*a*) $[\{n : n > 5\}, \{n : n < 5\}]$ (*b*) $[\{n : n > 5\}, \{0\}, \{1, 2, 3, 4, 5\}]$
   (*c*) $\{n : n^2 > 11\}, \{n : n^2 < 11\}$
4. Consider two sets A and B. Draw the Venn-diagrams of
   (*i*) $A \cap B^c$ (*ii*) $(B - A)^c$
   (*iii*) $B \cap A^c$ (*iv*) $A \cap (B \cup C)$
   (*v*) $(A \cap B) \cup (A \cap C)$
5. Consider the sets A, B, C. Draw the Venn-diagrams of
   (*i*) $A - (B \cup C)$ (*ii*) $A^c \cap (B \cup C)$
   (*iii*) $A^c \cap (C - B)$
6. If A and B are two sets, then
   (*i*) $A \oplus B = B \oplus A$ (Commutative law)
   (*ii*) $A \cap (B \oplus C) = (A \cap B) \oplus (B \cap C)$. (Distributive law)
7. Using Venn diagram, represent the following sets
   (*a*) $A^c \cup (B \cup C)$ (*b*) $A^c \cap (B - C)$
   (*c*) $A \cap B^c \cap C^c$ (*d*) $A^c \cap (B \cup C)$
   (*e*) $A \cup (B - C)$.
8. The description of the shaded region in Fig. 1.23 using the operations on sets, is

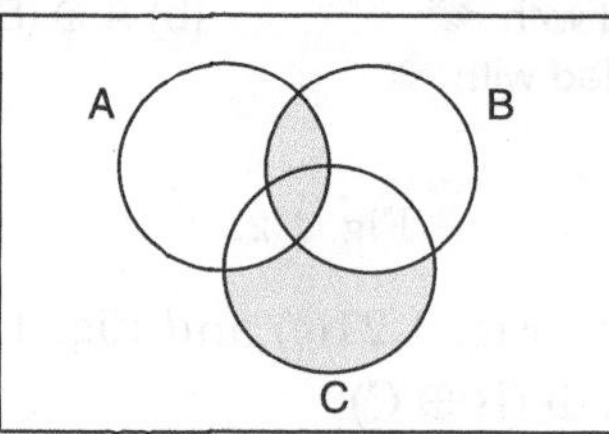

Fig. 1.23

   (*i*) $C \cup (A \cap B)$ (*ii*) $C - (A \cap C) \cup (C \cap B) \cup (A \cap B)$
   (*iii*) $C - (A \cap C) \cap (C \cap B) \cap (A \cap B)$
   (*iv*) $A \cup B \cup C - (C \cup (A \cap B))$. *(P.T.U. B.Tech. Dec. 2008)*

Answers

1. (*a*) No (*b*) Yes (*c*) No (*d*) Yes
2. (*a*) No (*b*) No (*c*) Yes (*d*) No
3. (*a*) No (*b*) No (*c*) Yes
4. (*i*)

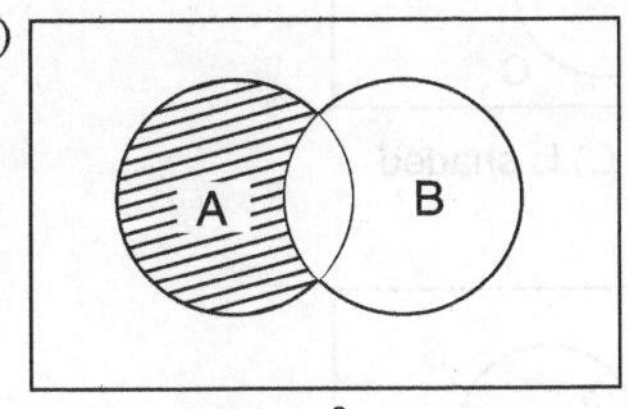

$A \cap B^c = A - B$

(*ii*)

$(B - A)^c$

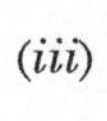

(*iii*)

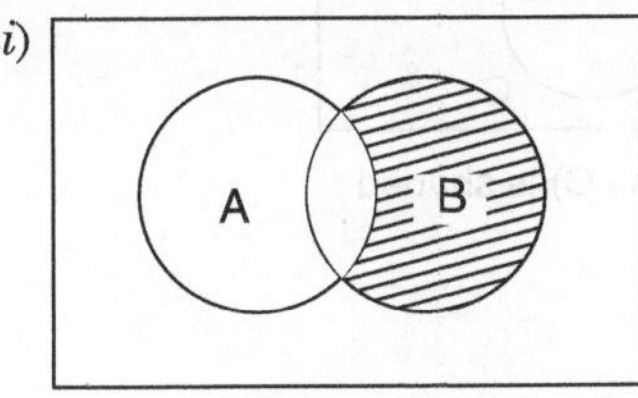

$B \cap A^c = B - A$

(*iv*), (*v*)

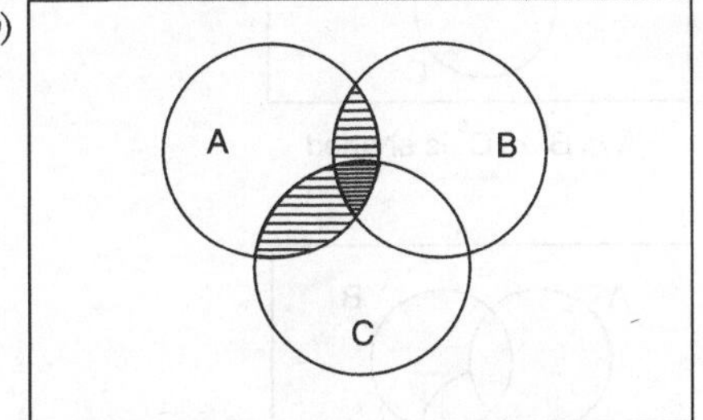

$A \cap (B \cup C) = (A \cap B) \cup (A \cap C)$

5. (*i*)

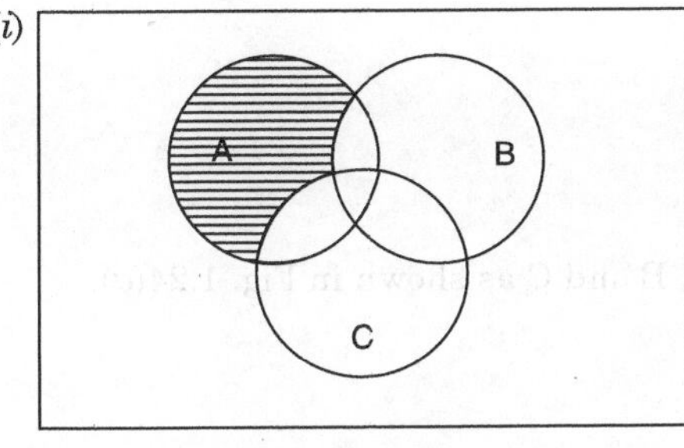

$A - (B \cup C)$

(*ii*)

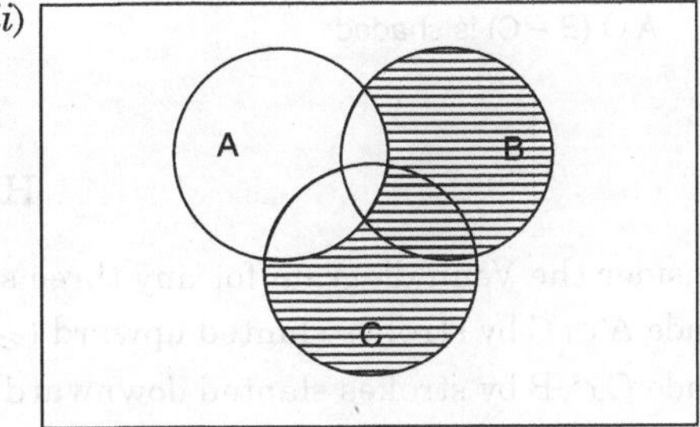

$A^c \cap (B \cup C)$

(*iii*)

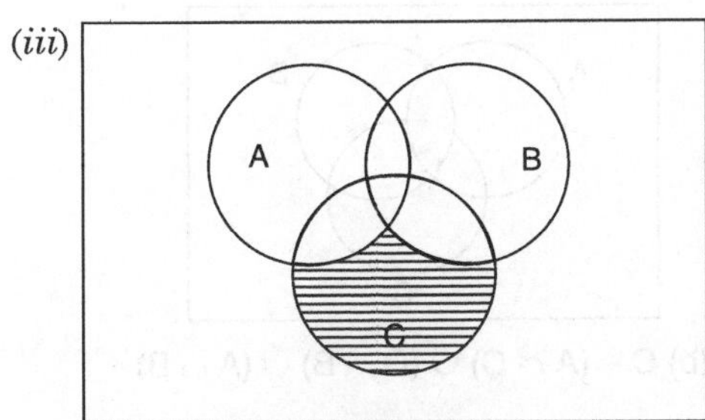

$A^c \cap (C - B)$

7. (*a*)

$A^c \cup (B \cup C)$ is shaded

(*b*)

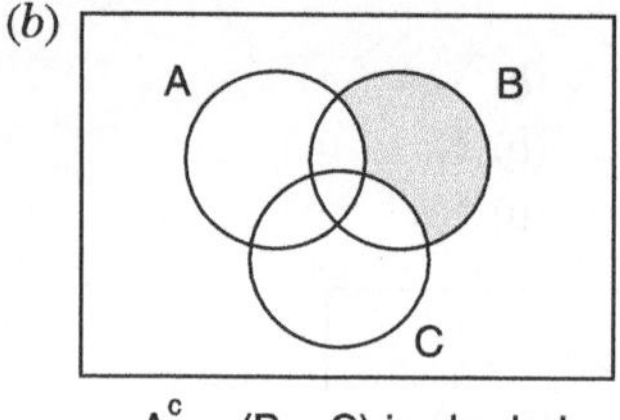

$A^c \cap (B - C)$ is shaded

(*c*)

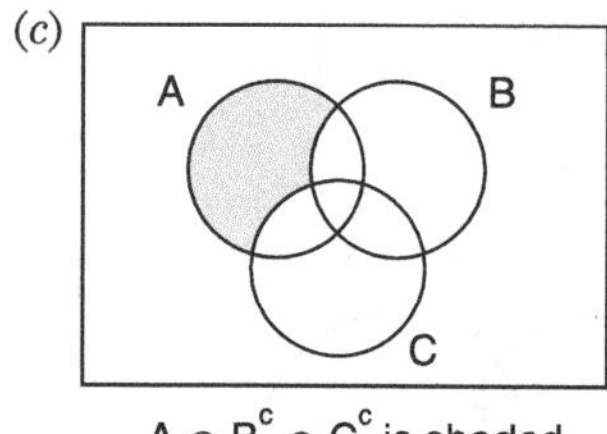

$A \cap B^c \cap C^c$ is shaded

(*d*)

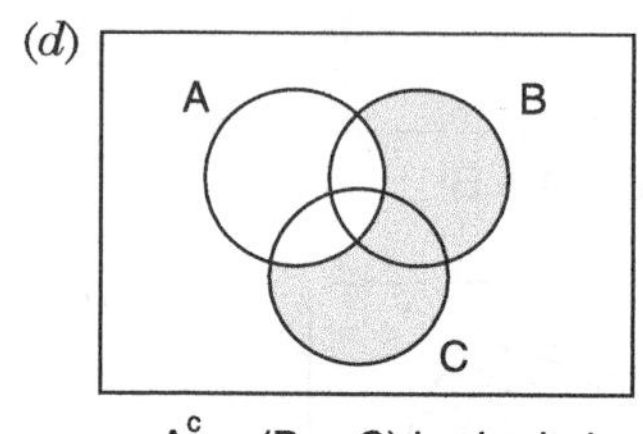

$A^c \cap (B \cup C)$ is shaded

(*e*)

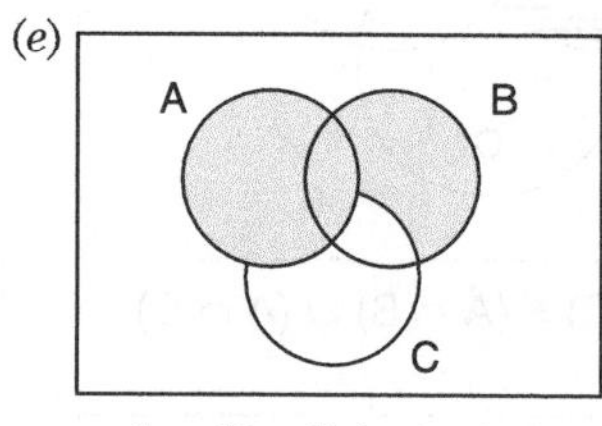

$A \cup (B - C)$ is shaded

8. (*ii*)

Hints

8. Consider the Venn diagram for any three sets A, B and C as shown in Fig. 1.24(*a*).
Shade $A \cap C$ by strokes slanted upward (⫻)
Shade $C \cap B$ by strokes slanted downward (⋛)
Shade $A \cap B$ by strokes drawn parallel (≡)
Thus, the total area is $(A \cap C) \cup (C \cap B) \cup (A \cap B)$ as shown in the shaded portion of Fig. 1.24(*a*).

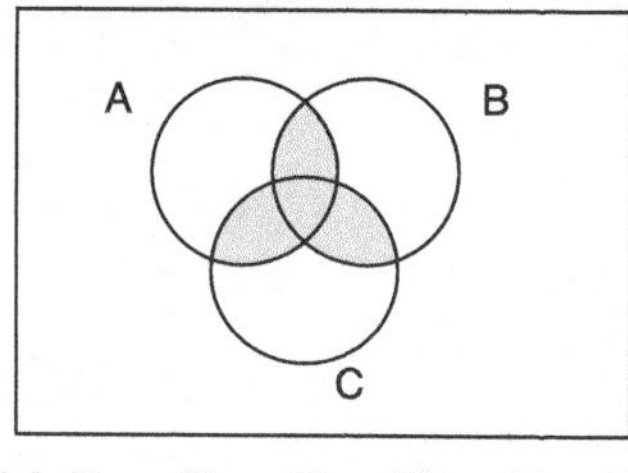

(a) $(A \cap C) \cup (C \cap B) \cup (A \cap B)$ is shaded

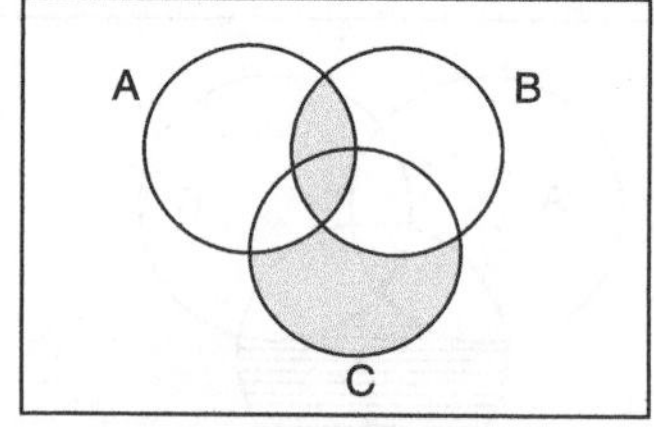

(b) $C - (A \cap C) \cup (C \cap B) \cup (A \cap B)$

Fig. 1.24

Hence $C - (A \cap C) \cup (C \cap B) \cup (A \cap B)$ is the area shown in the shaded portion of Fig. 1.24(*b*).
Hence the correct answer is (*ii*).

1.12. ENUMERABLE OR DENUMERABLE OR COUNTABLY INFINITE SET

(P.T.U., M.C.A, May 2007)

A set X is said to be *enumerable* if there exists a one-one and onto function from the set N (set of natural numbers) to the set X.

1.13. COUNTABLE SET

A set X is said to be countable if it is either finite or enumerable.

1.14. UNCOUNTABLE SET OR UNCOUNTABLY INFINITE SET

A set which is not countable is said to be uncountable.

Examples: (*i*) ϕ is countable.

(*ii*) The set {1, 5, 9, 20} is countable as it is a finite set.

(*iii*) The sets N, Z and Q are countable.

(*iv*) The set R is uncountable.

Theorem I. *Every subset of a countable set is countable.*

Proof. Let A be a countable set and B is subset of A.

Case I. If A is finite, then B being a subset of a finite set, is finite and hence countable. (Every subset of a finite set is finite).

Case II. If A is enumerable, then A can be written as

$$A = \{a_1, a_2, a_3, \ldots\ldots\}$$

If B = ϕ, then B is countable.

If B $\neq \phi$, then in the sequence of elements of A, there is first element in A is $a_{k_1} \in$ B. Take $b_1 = a_{k_1}$. The second element in A is $a_{k_2} \in$ B. Take $b_2 = a_{k_2}$. This process does or does not terminate according as B is finite or not. Since A contains all the elements of B, the possible terminating sequence of elements of B is $b_1, b_2, b_3, \ldots\ldots$

$\therefore$ B is either finite or enumerable. Hence B is countable.

Theorem II. *If A and B are countable, then A $\cap$ B is also countable or intersection of two countable sets is countable.*

Proof. Given A and B are countable sets. Also A $\cap$ B $\subseteq$ A. *i.e.,* A $\cap$ B is a subset of A and since A is countable, A $\cap$ B is also countable. (Every subset of a countable set is countable).

Theorem III. *Every superset of an uncountable set is uncountable.*

Proof. Let A is uncountable and B is such that A $\subset$ B. *i.e.,* B is a superset of A. We show B is uncountable. For if, B is countable, then, since A $\subset$ B, we can say that A is also countable, which is a contradiction as A is given uncountable. Hence B is uncountable.

1.15. FUNDAMENTAL PRODUCT

Let $A_1, A_2, \ldots A_n$ be n distinct sets. A **fundamental product** of these sets is a set of the form

$$A_1^* \cap A_2^* \cap A_3^* \cap \ldots \cap A_n^*$$

where A_i^* is either A_i or A_i^c. There are 2^n such fundamental products and any two such fundamental products are disjoint and the union of all the fundamental products is equal to the universal set.

1.15. (a) MINSETS OR MINTERMS

Let A be any set and $\{B_1, B_2, ... B_n\}$ be a set of subsets of A. Then a set of the form $D_1 \cap D_2 \cap ... \cap D_n$ where each D_i may be either B_i or B_i^c is called a minset or minterm generated by $B_1, B_2, ..., B_n$.

1.15. (b) MINSET NORMAL FORM

A set is said to be in minset normal form or canonical form if it is expressed as the union of distinct non-empty minsets or it is ϕ.

1.16. MAXSETS OR MAXTERMS

Let A be any set and $\{B_1, B_2, ... B_n\}$ be a set of subsets of A. Then a set of the form $D_1 \cup D_2 \cup ...... \cup D_n$, where each D_i may be either B_i or B_i^c is called maxset generated by $B_1, B_2, ... B_n$.

Remark 1. The set of maxsets donot necessarily form a partition of A.

Remark 2. Let A be any set and $B_1, B_2, ... B_n$ be subsets of A. Then the set of non-empty minsets generated by $B_1, B_2, ... B_n$ form a partition of A.

ILLUSTRATIVE EXAMPLES

Example 1. *Consider $A = \{1, 2, 3, 4, 5, 6\}$ and $B_1 = \{1, 3, 5\}$, $B_2 = \{1, 2, 3\}$. Obtain all possible minsets of A generated by B_1 and B_2.*

Sol. For given n sets $B_1, B_2, ... B_n$, there are 2^n minsets. The possible minsets for the given sets B_1 and B_2 are given as

$$B_1^c = \{2, 4, 6\}; \quad B_2^c = \{4, 5, 6\}$$
$$A_1 = B_1 \cap B_2 = \{1, 3\}; \quad A_2 = B_1 \cap B_2^c = \{5\}$$
$$A_3 = B_1^c \cap B_2 = \{2\}; \quad A_4 = B_1^c \cap B_2^c = \{4, 6\}$$

Also, each element of A appears exactly once in one of the four minsets $B_1 \cap B_2$, $B_1 \cap B_2^c$, $B_1^c \cap B_2$, $B_1^c \cap B_2^c$. Hence they form a partition of A.

Example 2. *Let $A = \{1, 2, 3, 4, 5, 6\}$ and let $B_1 = \{1, 3, 5\}$. $B_2 = \{1, 2, 3\}$. Find the maxsets generated by B_1 and B_2. So maxsets form a partition of A ?*

Sol. Here $B_1^c = [2, 4, 6]; \quad B_2^c = [4, 5, 6]$

The possible number of maxsets generated by n sets is 2^n. The maxsets generated by B_1 and B_2 are

$$M_1 = B_1 \cup B_2 = [1, 2, 3, 5]$$
$$M_2 = B_1^c \cup B_2 = [2, 4, 6] \cup [1, 2, 3] = [1, 2, 3, 4, 6]$$
$$M_3 = B_1 \cup B_2^c = [1, 2, 3] \cup [4, 5, 6] = [1, 2, 3, 4, 5, 6]$$
$$M_4 = B_1^c \cup B_2^c = [2, 4, 6] \cup [4, 5, 6] = [2, 4, 5, 6]$$

Now $M_1 \cap M_2 = [2, 3] \neq \phi$. Therefore M_1, M_2, M_3, M_4 cannot form a partition of A.

Example 3. *Consider $A = [1, 2, 3, 4, 5, 6, 7, 8, 9]$ and let $B_1 = [5, 6, 7]$, $B_2 = [2, 4, 5, 9]$, $B_3 = [3, 4, 5, 6, 8, 9]$. Find the minsets generated by B_1, B_2, B_3.*

(a) Do these minsets form a partition of A ?

(b) How many different subsets of A can you create using B_1, B_2 and B_3 with the standard set operations ?

Sol. *(a)* The possible number of minsets generated by *n* sets is 2^n. The minsets generated by B_1, B_2 and B_3 given below:

Here $B_1^c = [1, 2, 3, 4, 8, 9]$, $B_2^c = [1, 3, 6, 7, 8]$, $B_3^c = [1, 2, 7]$

The minsets generated by B_1, B_2 and B_3 are

$$A_1 = B_1 \cap B_2 \cap B_3 = [5]$$
$$A_2 = B_1^c \cap B_2 \cap B_3 = [4, 9]$$
$$A_3 = B_1 \cap B_2^c \cap B_3 = [6]$$
$$A_4 = B_1 \cap B_2 \cap B_3^c = \phi$$
$$A_5 = B_1^c \cap B_2^c \cap B_3 = [3, 8]$$
$$A_6 = B_1 \cap B_2^c \cap B_3^c = [7]$$
$$A_7 = B_1^c \cap B_2 \cap B_3^c = [2]$$
$$A_8 = B_1^c \cap B_2^c \cap B_3^c = [1].$$

(b) From part *(a)*, the minsets generated by B_1, B_2 and B_3 are

$$P_1 = \{5\}, P_2 = \{4, 9\}, P_3 = \{6\}, P_4 = \{3, 8\},$$
$$P_5 = \{7\}, P_6 = \{2\}, P_7 = \{1\}$$

Clearly, $P_i \cap P_j = \phi, i \neq j$

and $\bigcup_{i=1}^{7} P_i = A$, hence, they form a partition of A.

Number of subsets generated by $B_1 = 2^3 = 8$

Number of subsets generated by $B_2 = 2^4 = 16$

Number of subsets generated by $B_3 = 2^6 = 64$

∴ Total number of subsets $= 8 + 16 + 64 =$ **88.**

Example 4. *(a) Partition A = {0, 1, 2, 3, 4, 5} with the minsets generated by*

$B_1 = \{0, 2, 4\}$ and $B_2 = \{1, 5\}$

(b) How many different subsets of A can you generate from B_1 and B_2 ?

Sol. *(a)* The possible number of minsets generated by B_1 and B_2 are $2^2 = 4$.

The minsets are

$$A_1 = B_1 \cap B_2^c = \{0, 2, 4\} \cap \{0, 2, 3, 4\} = \{0, 2, 4\}$$
$$A_2 = B_1^c \cap B_2 = \{1, 3, 5\} \cap \{1, 5\} = \{1, 5\}$$
$$A_3 = B_1^c \cap B_2^c = \{1, 3, 5\} \cap \{0, 2, 3, 4\} = \{3\}$$
$$A_4 = B_1 \cap B_2 = \{0, 2, 4\} \cap \{1, 5\} = \phi$$

∴ The minsets are $\{\{3\}, \{1, 5\}, \{0, 2, 4\}\}$

(b) Number of subsets generated by $B_1 = 2^3 = 8$

Number of subsets generated by $B_2 = 2^2 = 4$

∴ Required Number of subsets $= 8 + 4 =$ **12.**

Example 5. *Let B_1, B_2 and B_3 are subsets of a universal set U.*

(a) Find all minsets generated by B_1, B_2 and B_3.

(b) Draw a Venn diagram representing all minsets obtained in part (a)

(c) Find the minset normal form of the following sets

(i) B_1^c *(ii) $B_1 \cap B_2$* *(iii) $B_2^c \cap B_3^c$.*

Sol. (*a*) The possible number of minsets generated by B_1, B_2 and B_3 is $2^3 = 8$. The minsets are

$$A_1 = B_1 \cap B_2 \cap B_3 ; \quad A_2 = B_1^c \cap B_2 \cap B_3$$
$$A_3 = B_1 \cap B_2^c \cap B_3 ; \quad A_4 = B_1 \cap B_2 \cap B_3^c$$
$$A_5 = B_1^c \cap B_2^c \cap B_3 ; \quad A_6 = B_1^c \cap B_2 \cap B_3^c$$
$$A_7 = B_1 \cap B_2^c \cap B_3^c ; \quad A_8 = B_1^c \cap B_2^c \cap B_3^c$$

(*b*) The Venn diagram for the sets A_1, A_2, ... A_8 obtained in part (*a*) is shown in Fig. 1.25

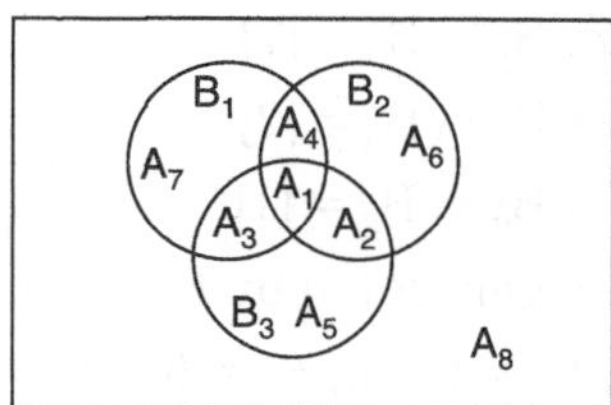

Fig. 1.25

(*c*) (*i*) The minset normal form for B_1^c (using Venn diagram) is given as

$$B_1^c = A_6 \cup A_2 \cup A_5 \cup A_8$$
$$= (B_1^c \cap B_2 \cap B_3^c) \cup (B_1^c \cap B_2 \cap B_3) \cup (B_1^c \cap B_2^c \cap B_3) \cup (B_1^c \cap B_2^c \cap B_3^c)$$

(*ii*) The minset normal form for $B_1 \cap B_2$ (using Venn diagram) is

$$B_1 \cap B_2 = A_1 \cup A_4 = (B_1 \cap B_2 \cap B_3) \cup (B_1 \cap B_2 \cap B_3^c)$$

(*iii*) The minset normal form for $B_2^c \cap B_3^c$ (using Venn diagram) is

$$B_2^c \cap B_3^c = A_5 \cup A_8 = (B_1^c \cap B_2^c \cap B_3) \cup (B_1^c \cap B_2^c \cap B_3^c)$$

Example 6. *Consider the universal set*

$U = \{1, 2, 3, 4, ... 10\}$ and the subsets

$A = \{1, 7, 8\}$, $B = \{1, 6, 9, 10\}$, $C = \{1, 9, 10\}$

(a) List the non-empty minsets generated by A, B and C. Do the minsets form a partition of U ?

(b) How many elements of U can be generated by A, B and C ?

(c) Compare the number obtained in (b) with n(P(U)).

(d) Give an example of one subset that cannot be generated by A, B and C.

Sol. (*a*) The possible number of minsets generated by *n* sets is 2^n. These minsets are given below:

$$A_1 = A \cap B \cap C^c = \{1, 7, 8\} \cap \{1, 6, 9, 10\} \cap \{2, 3, 4, 5, 6, 7, 8\} = \phi$$
$$A_2 = A \cap B^c \cap C = \{1, 7, 8\} \cap \{2, 3, 4, 5, 7, 8\} \cap \{1, 9, 10\} = \phi$$

$A_3 = A^c \cap B \cap C = \{2, 3, 4, 5, 6, 9, 10\} \cap \{1, 6, 9, 10\} \cap \{1, 9, 10\} = \{9, 10\}$

$A_4 = A \cap B^c \cap C^c = \{1, 7, 8\} \cap \{2, 3, 4, 5, 7, 8\} \cap \{2, 3, 4, 5, 6, 7, 8\} = \{7, 8\}$

$A_5 = A^c \cap B \cap C^c = \{2, 3, 4, 5, 6, 9, 10\} \cap \{1, 6, 9, 10\} \cap \{2, 3, 4, 5, 6, 7, 8\} = \{6\}$

$A_6 = A^c \cap B^c \cap C = \{2, 3, 4, 5, 6, 9, 10\} \cap \{2, 3, 4, 5, 7, 8\} \cap \{1, 9, 10\} = \phi$

$A_7 = A^c \cap B^c \cap C^c = \{2, 3, 4, 5, 6, 9, 10\} \cap \{2, 3, 4, 5, 7, 8\} \cap \{2, 3, 4, 5, 6, 7, 8\}$
$= \{2, 3, 4, 5\}$

$A_8 = A \cap B \cap C = \{1\}$.

Hence, the minsets generated by A, B and C are {9, 10}, {7, 8}, {6}, {2, 3, 4, 5}, {1}.

Clearly, $A_i \cap A_j = \phi$ for $i \neq j$ and $\bigcup_{i=1}^{8} A_i = U$. Hence, they form a partition of U.

(*b*) From part (*a*), there are five different minsets generated by A, B and C. Hence, 2^5 elements of U can be generated by A, B and C.

(*c*) As U contains 10 elements, therefore,

$$n\,(P(U)) = 2^{10}.$$

(*d*) {1, 2} cannot be generated by A, B and C since $2 \in$ A, B, and C.

Example 7. *Give an example to show that intersection of two countably infinite sets can be finite or countably infinite.*

Sol. Let A and B are two countably infinite sets.

Take A as a set of odd natural numbers and B as a set of even natural numbers *i.e.*,

$$A = \{1, 3, 5, 7, 9, ...\} \text{ and } B = \{2, 4, 6, 8, 10, ...\}$$

Then $A \cap B = \phi$, which is finite

Further, take B = N, the set of natural numbers,

Then $A \cap B = A \cap N = \{1, 3, 5, 7, 9, ...\} \cap \{1, 2, 3, 4, 5, 6, ...\}$

$= \{1, 3, 5, 7, 9, ...\} = A$, which is countably infinite.

Example 8. *Give examples to show that the intersection of two uncountably infinite sets can be:*

(i) finite *(ii) countably infinite* *(iii) uncountable.*

Sol. (*i*) Let A and B are uncountably infinite sets.

Take $A = 2^N$, B = set of negative integers, then both A and B are uncountably infinite sets. But $A \cap B = \phi$, which is finite.

(*ii*) Take $A = 2^N \cup N$, $B = (2^N \times 2^N) \cup N$

Then, $A \cap B = (2^N \cup N) \cap ((2^N \times 2^N) \cup N) = N$, which is countably infinite.

(*iii*) Take $A = 2^N$, $B = 2^N$,

Then $A \cap B = 2^N \cap 2^N = 2^N$, which is uncountable.

Example 9. *Show that at the most countably infinite number of books can be written in Punjabi.*

Sol. Assume that upto this moment, some finite number, say, N of books in Punjabi are written in Punjabi. The successor of N is N + 1.

i.e., each next book in Punjabi will be mapped with the successor of N, and so on. This means there exists 1-1 correspondence between the elements in N and elements in set of books. But N is countably infinite set, hence infinite number of books can be written is Punjabi.

Example 10. *State whether the following sets are finite, countably infinite, uncountably finite:*

(*i*) *Class of all programms that can ever be written in programming language* C^+.

(*ii*) *All movies produced by A.R. Rehman.*

(*iii*) *Number of fish in Hind Maha Sagar.*

(*iv*) *Set of all Primes.*

(*v*) *Set of real numbers in (0, 1).*

Sol. (*i*) Countably infinite

(*ii*) Finite

(*iii*) Finite

(*iv*) Countably infinite

(*v*) Uncountably infinite.

Example 11. *Let A be any set. Show that* $A \times \phi = \phi \times A = \phi$.

Sol. By definition, $\phi \times A = \{ (x, y) : x \in \phi \text{ and } y \in A\} = \phi$ as $x \in \phi$ is not true.

Similarly, $A \times \phi = \{(x, y) : x \in A \text{ and } y \in \phi\} = \phi$ as $y \in \phi$ is not true.

Example 12. *Let A and B are two sets such that* $A \times B \times \phi$. *What can you conclude ?*

Sol. Given $A \times B = \phi$ implies either A is ϕ or B is ϕ or both A and B are empty sets.

TEST YOUR KNOWLEDGE 1.3

1. Partition A = {0, 1, 2, 3, 4, 5} with the minsets generated by B_1 = {0, 2, 4}, B_2 = {1, 5}. How many different subsets of A can you generate from B_1 and B_2 ?
2. Let A = {4, 5, 6}, B_1 = {4, 5}, B_2 = {5, 6}
   (*a*) Find the minsets and maxsets generated by B_1 and B_2.
   (*b*) Do minsets form a partition of A ?
3. Let S be a set of words or string of length ≤ 2. *i.e.*, S = {0, 1, 00, 01, 10, 11} and A = {0, 00, 01}, B = {00, 01, 10, 11}.
   Find a partition of S using minsets generated by A and B.

Answers

1. {1, 5}, {0, 2, 4}, {3}. Required number of subsets = 12
2. (*a*) The minsets are A_1 = {5}, A_2 = {6}, A_3 = {4}, $A_4 = \phi$
   The maxsets are m_1 = {4, 5, 6}, m_2 = {5, 6}, m_3 = {4, 5}, m_4 = {4, 6}
   (*b*) Yes
3. The minsets are A_1 = {00, 01}, A_2 = {10, 11}, A_3 = {0}, A_4 = {1}.
   The required partition is {A_1, A_2, A_3, A_4}.

MULTIPLE CHOICE QUESTIONS

1. Let A and B be sets and A^c and B^c denote their complements. Then $(A - B) \cup (B - A) \cup (A \cup B)$ equal to

(*a*) $A \cup B$ (*b*) $A^c \cup B^c$
(*c*) $A \cap B$ (*d*) $A^c \cap B^c$.

2. The number of elements in the power set P(A) of the set $A = \{\{\phi\}, 1, \{2, 3\}\}$ is

(*a*) 2 (*b*) 4
(*c*) 8 (*d*) None.

3. Let P(A) denote the power set of A. Which of the following is true ?

(*a*) $P(P(A)) = P(A)$ (*b*) $P(A) \cap A = P(A)$
(*c*) $P(A) \cap P(P(A)) = \{\phi\}$ (*d*) $A \notin P(A)$.

4. Let A be a finite set containing n elements, then $P(A \times A)$ contains

(*a*) 2^{2^n} (*b*) 2^{n^2}
(*c*) $(2^n)^2$ (*d*) None elements.

5. Let A be an infinite set and $A_1, A_2, ..., A_n$ be n sets such that $A_1 \cup A_2 \cup A_3 \cup ...... \cup A_n = A$. Then

(*a*) At least one of the sets A_i is a finite set
(*b*) No more than one of the sets A_i can be finite
(*c*) At least one of the sets A_i is an infinite set
(*d*) None.

6. Let $A = \{3, \{1, 4\}, 5\}$, then $P(2^A)$ is

(*a*) {A, 3, 1, 4, {1, 3, 5}, {1, 4, 5}, {3, 4}, ϕ} (*b*) {A, 3, {1, 4}, 5}
(*c*) {A, {3}, {3, {1, 4}}, {3, 5}, ϕ} (*d*) None.

7. Consider the following statements :

T_1 : ∃ infinite sets A, B and C such that $A \cap (B \cup C)$ is finite
T_2 : ∃ two rational numbers x and y such that $x + y$ is rational. Then

(*a*) Only T_1 is correct (*b*) Only T_2 is correct
(*c*) Both T_1 and T_2 are correct (*d*) None of T_1 and T_2 are correct.

8. In a room containing 28 females, there are 18 females who speak English, 15 females speak French and 22 speak German, 9 females speak both English and French, 11 females speak both French and German whereas 13 speak both German and English. How many females speak all the three languages.

(*a*) 9 (*b*) 8
(*c*) 7 (*d*) 6.

9. The number of substrings of all lengths that can be formed from a character string of length n is.

(*a*) n (*b*) n^2
(*c*) $\frac{n(n-1)}{2}$ (*d*) $\frac{n(n+1)}{2}$.

10. The symmetric difference of two sets A and B is denoted by $A \oplus B$. Which of the following is true.

(*a*) $A \oplus B = (A \cup B) - (A \cap B)$ (*b*) $A \oplus B = (A - B) \cup (B - A)$

(*c*) Both (*a*) and (*b*) are true (*d*) None.

11. A set A has 10 members. Then the number of members of P(A) is

(*a*) 10 (*b*) 2^{10}

(*c*) $10^{P(A)}$ (*d*) none.

12. A set A has 5 members. The number of proper subsets of A is

(*a*) 2^5 (*b*) $2^5 - 1$

(*c*) $2^5 + 1$ (*d*) none.

Answers and Explanations

1. (*a*) $(A - B) \cup (B - A) \cup (A \cup B) = A \cup B$.

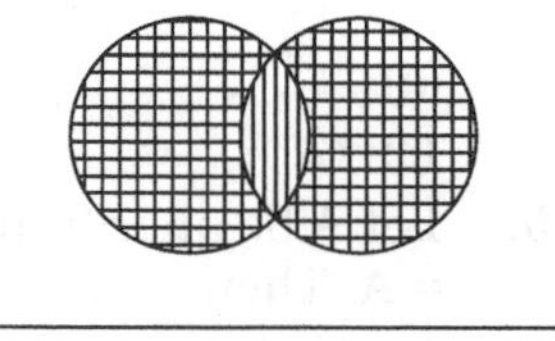

(A – B) ∪ (B – A) is shaded with ≡.
A ∪ B is shaded with III

2. (*c*) A contains 3 elements, so P(A) will contain $2^3 = 8$ elements.

3. (*c*) P(A) contains all subsets of A but no elements of A. Hence, there is no element common to A and P(A). Also, there is no element common to P(A) and P(P(A)).

$\therefore \quad P(A) \cap P(P(A)) = \phi$.

4. (*b*) $n(A) = n \quad \therefore \quad n(A \times A) = n(A) \times n(A) = n \times n = n^2$

$\therefore \quad P(A \times A)$ contains 2^{n^2} elements.

5. (*c*) 6. (*d*) 7. (*c*)

8. (*d*) $n(E) = 18$, $n(F) = 15$, $n(G) = 22$, $n(E \cap F) = 9$, $n(F \cap G) = 11$, $n(G \cap E) = 13$, $n(E \cup F \cup G) = 28$. Using

$n(E \cup F \cup G) = n(E) + n(F) + n(G) - n(E \cap F) - n(F \cap G) - n(G \cap E) + n(E \cap F \cap G)$, we have

$$28 = 18 + 15 + 22 - 9 - 11 - 13 + n(E \cap F \cap G)$$

or $\quad 28 = 22 + n(E \cap F \cap G)$

$\Rightarrow \quad n(E \cap F \cap G) = 6$.

9. (*d*) 10. (*c*) 11. (*b*)

12. (*b*) If a set A contains n elements, then P(A) contains 2^n subsets, out of which $2^n - 1$ subsets are proper and one is improper, the set A itself.

2 RELATIONS

2.1. INTRODUCTION

In the previous chapter, we have discussed various operations on sets to generate more sets from given sets. We now discuss one more property of sets which is known as cartesian products of sets which will help us in understanding the concept of relations.

2.2. ORDERED PAIR

Let A and B be any two sets. Then by an *ordered pair of elements*, we mean a pair (x, y) where $x \in A$, $y \in B$.

For example, the ordered pairs (1, 1), (2, 3), (3, 5) represent different points in a plane.

2.3. CARTESIAN PRODUCT OF SETS

Let A and B be any two non-empty sets. Then the cartesian product of the sets A and B is the set of all ordered pairs (x, y) such that $x \in A$ and $y \in B$ and it is denoted by $A \times B$. Thus

$$A \times B = \{(x, y) : x \in A \text{ and } y \in B\}.$$

For example, consider A = (1, 2), B = (3, 4, 5). We find $A \times B$, $B \times A$, $A \times A$, $B \times B$.

Here $A \times B = \{(1, 3), (1, 4), (1, 5), (2, 3), (2, 4), (2, 5)\}$

$B \times A = \{(3, 1), (3, 2), (4, 1), (4, 2), (5, 1), (5, 2)\}$

$A \times A = \{(1, 1), (1, 2), (2, 1), (2, 2)\}$

$B \times B = \{(3, 3), (3, 4), (3, 5), (4, 3), (4, 4), (4, 5), (5, 3), (5, 4), (5, 5)\}$.

2.4. RELATION (Or Binary Relation)

Let A and B be any two non-empty sets. Then a *binary relation* or simply a *relation* R from A to B is a subset of $A \times B$. Thus,

R is a relation from A to B iff $R \subseteq A \times B$.

Notation. If R is a relation from a non-empty set A to a non-empty set B and if $(x, y) \in R$, we write xRy (read as x is related to y by the relation R).

2.5. TOTAL NUMBER OF RELATIONS

Let A and B be two non-empty finite sets containing m and n elements respectively. Then the number of ordered pairs in $A \times B = mn$. Therefore, total number of subsets of $A \times B$ is 2^{mn}.

Since each subset of $A \times B$ defines a relation from A to B, so total number of relations from A to B is 2^{mn} (including the empty relation ϕ and the universal relation $A \times B$).

Similarly, if a set A contains n elements, then the total number of relations on A is 2^{n^2}.

2.6. DOMAIN AND RANGE OF A RELATION

Let R be a relation from the set A to the set B. Then, the *domain of R* is the set of all first co-ordinates of the ordered pairs belonging to R. Similarly, the *range of R* is the set of all second co-ordinates of the ordered pairs belonging to R. For *e.g.*,

Consider A = {1, 3, 5, 7}, B = {2, 4, 6, 8, 10} and R = (1, 8), (3, 6), (5, 2), (1, 4) be a relation from A to B. Then

Domain of R = {1, 3, 5}

Range of R = {2, 4, 6, 8}.

2.7. INVERSE RELATION

Let R be a relation from a set A to another set B. Then, the inverse relation is a relation from B to A. It is denoted by R^{-1}. Thus, if $R = \{(a, b) : a \in A, b \in B\}$, Then

$$R^{-1} = \{(b, a) : b \in B, a \in A\}$$

Also, Domain of R^{-1} = Range of R

Range of R^{-1} = Domain of R.

2.8. TYPES OF RELATIONS

The following are the types of relations.

(*a*) **Empty or void relation.** Let A be a set. Then $\phi \subseteq A \times A$ is true and hence it is a relation on A. This relation is called void relation or empty relation.

(*b*) **Universal relation.** Let A be any set. Then $A \times A \subseteq A \times A$ is true and hence it is a relation on A. This relation is called universal relation on the set A.

Remark. The void relation is the smallest relation on A and the universal relation is the largest relation on A.

(*c*) **Identity relation.** Let A be any set. Then the relation $R = \{x, x) : x \in A\}$ on A is called the identity relation or diagonal relation on A and is denoted by Δ_A or Δ

For example, consider $A = \{a, b, c\}$ and define relations R_1 and R_2 as follows.

$$R_1 = \{(a, a)\ (b, b), (c, c)\}$$

$$R_2 = \{(a, a), (b, b), (c, c), (a, c)\}$$

Then R_1 is an identity relation on A, but R_2 is not an identity relation on A as the element a is related to a and c.

2.9. PROPERTIES OF RELATIONS

(*a*) **Reflexive relation.** A relation R on a set A is said to be a reflexive relation if every element of A is related to itself. Thus, R is reflexive iff $(x, x) \in R$ for all $x \in A$.

A relation R on a set A is not reflexive if there is an element $x \in A$ such that $(x, x) \notin R$.

For example, consider A = (1, 2, 3). Then the relation R_1 defined by R_1 = {(1, 1), (2, 2), (3, 3), (1, 3), (2, 1)} is a reflexive relation on A.

The relation R_2 defined by R_2 = {(1, 1), (3, 3), (2, 1), (3, 2)} is not a reflexive relation on A, since $(2, 2) \notin R_2$.

The identity relation and universal relation on a non-empty set is a reflexive relation.

(*b*) **Irreflexive relation.** A relation R on a set A is said to be irreflexive if $(a, a) \notin R\ \forall\ a \in A$. In other words, a relation R on a set A is not irreflexive if there exists at least one $a \in A$ such that $(a, a) \in R$.

Example. Consider R = {(*a*, *b*) : $a \neq b$} on the set A = {*a*, *b*} *i.e.*, R is the relation of inequality on the set A, thus R is irreflexive since for every $a \in A$, $(a, a) \notin R$.

Example. Let A be a non-empty set and R be *a* relation on A = {*a* : *a* is a computer science student) defined by R = {(*a*, *b*) : a scored less than *b*}

Then for all $a \in A$, $(a, a) \notin R$ *i.e.*, R is an irreflexive relation.

Remark. Every identity relation on a non-empty set A is a reflexive relation, but not conversely. Consider A = {*a*, *b*, *c*} and define a relation R by R = {(*a*, *a*), (*b*, *b*), (*c*, *c*), (*a*, *b*)}. Then R is a reflexive relation on A but not an identity relation on A due to the element (*a*, *b*) in R.

(*c*) **Symmetric relation:** Let A be any set. Then a relation R on the set A is called symmetric iff

$$(x, y) \in R \Rightarrow (y, x) \in R \text{ for all } x, y \in A.$$

For example, the identity relation and universal relation on a non-empty set A are symmetric relations on A.

Consider A = (1, 2, 3, 4) and define relations R_1 and R_2 as follows :

$$R_1 = \{(1, 3), (1, 4), (3, 1), (2, 2), (4, 1)\}$$
$$R_2 = \{(1, 1), (2, 2), (3, 3), (1, 3)\}$$

In R_1, $(1, 3) \in R_1 \Rightarrow (3, 1) \in R_1$

$(1, 4) \in R_1 \Rightarrow (4, 1) \in R_1$

$(2, 2) \in R_1 \Rightarrow (2, 2) \in R_1$

Thus, the relation R_1 is a symmetric relation on A.

In R_2, $(1, 3) \in R_2 \nRightarrow (3, 1) \in R_2$. Thus R_2 is not a symmetric relation.

(*d*) **Transitive relation:** Let A be any set. Then a relation R is said to be transitive iff

$(x, y) \in R$ and $(y, z) \in R \Rightarrow (x, z) \in R$ for all $x, y, z \in A$.

For example, the identity relation and universal relation are transitive relation on any non-empty set.

Consider A = The set of all straight lines in a plane. Define a relation "is parallel to" on A.

Take $l_1, l_2, l_3 \in A$. Then l_1 is parallel to l_2 and l_2 is parallel to l_3. It implies that l_1 is parallel to l_3 also. Hence the relation "is parallel to" is a transitive relation.

(*e*) **Antisymmetric or Asymmetric or Non-symmetric relation:** Let A be any set. A relation R on A is said to be a antisymmetric iff

$(x, y) \in R$ and $(y, x) \in R \Rightarrow x = y$ for all $x, y \in A$.

or A relation R on a non-empty set A is **Antisymmetric** if whenever xRy and yRx, then $x = y$, *i.e.*

if whenever $(x, y), (y, x) \in R$ then $x = y$

On the other hand, a relation R on a non-empty set A is not **antisymmetric** if there exists $x, y \in A$ such that (x, y) and $(y, x) \in R$ but $x \neq y$.

Example. The relation "$\leq$" (read as less than or equal to) is antisymmetric since whenever $a \leq b$ and $b \leq a$, then $a = b$.

2.10. EQUIVALENCE RELATION (*P.T.U. B.Tech. May 2013, May 2012, May 2010, Dec. 2007*)

Let A be any set. Then a relation R on A is said to be an equivalence relation on A iff it satisfies the following :

(*i*) It is reflexive
(*ii*) It is symmetric
(*iii*) It is transitive.

Theorem I. *If R and S are two equivalence relations on a set A, then*

(*a*) *$R \cap S$ is also an equivalence relation on A.* (P.T.U. M.C.A. Dec. 2006)
(*b*) *$R \cup S$ may or may not be an equivalence relation on A.*
(P.T.U. B.Tech. Dec. 2013; P.T.U. M.C.A. Dec. 2006)

Proof. Given R is a relation on A. It means R is a subset of A × A. It implies

$$R \subseteq A \times A \quad \ldots(1)$$

Also, $$S \subseteq A \times A \quad \ldots(2)$$

(1) and (2) gives $R \cap S \subseteq A \times A$ *i.e.*, $R \cap S$ is a subset of $A \times A$. Hence $R \cap S$ is a relation on A.

We show $R \cap S$ is an equivalence relation on A.

(*i*) **Reflexivity.** Let $x \in A$ be any arbitrary element of A.

As R and S are equivalence relations on A. It implies R and S are reflexive relations on A.

If $x \in A \Rightarrow (x, x) \in R$ and $(x, x) \in S$.

$\Rightarrow \quad (x, x) \in R \cap S$ for all $x \in A$.

Hence $R \cap S$ is reflexive on A.

(*ii*) **Symmetry.** Let $x, y \in A$ such that $(x, y) \in R \cap S$. It means $(x, y) \in R$ and $(x, y) \in S$.

As R and S are symmetric relations on A.

$\therefore \quad (y, x) \in R$ and $(y, x) \in S$

$\Rightarrow \quad (y, x) \in R \cap S$.

Hence $(x, y) \in R \cap S \Rightarrow (y, x) \in R \cap S$ for all $x, y \in A$.

Therefore $R \cap S$ is symmetric on A.

(*iii*) **Transitivity.** Let $x, y, z \in A$ such that $(x, y) \in R \cap S$, $(y, z) \in R \cap S$.

We show $(x, z) \in R \cap S$.

Now, $(x, y) \in R$ and $(x, y) \in S$. Also, $(y, z) \in R$ and $(y, z) \in S$.

As R, S are transitive on A

$\therefore \quad (x, y) \in R, (y, z) \in R \Rightarrow (x, z) \in R$

Also, $(x, y) \in S, (y, z) \in S \Rightarrow (x, z) \in S$

$\Rightarrow \quad (x, z) \in R \cap S$.

Hence $R \cap S$ is an equivalence relation on A.

Further, we show $R \cup S$ may or may not be an equivalence relation.

Consider $A = \{a, b, c\}$. Define the relations R and S by

$$R = \{(a, a), (b, b), (c, c), (a, b), (b, a)\}$$
$$S = \{(a, a), (b, b), (c, c), (b, c), (c, b)\}$$

Clearly, R and S are equivalence relations on A.

But $R \cup S = \{(a, a), (b, b,), (c, c), (a, b), (b, a), (b, c), (c, b)\}$

Here $(a, b) \in R \cup S$, $(b, c) \in R \cup S$.

But $(a, c) \notin R \cup S \quad \therefore \quad R \cup S$ is not transitive.

Hence $R \cup S$ cannot be equivalence relation on A.

2.11. COMPATIBLE RELATION

A binary relation on a set which is reflexive and symmetric is called a compatible relation. Every equivalence relation is a compatible relation. But every compatible relation may not be an equivalence relation.

Let A is the set of persons and define R on A by R such that for $a, b \in A$, $(a, b) \in R$ if a is a friend of b, then R is reflexive and symmetric. Hence R is compatible.

Let $a, b, c \in A$ such that $(a, b) \in R$, $(b, c) \in R$. Here, a is a friend of b, b is a friend of c but a may or may not be a friend of c. Hence R may or may not be transitive.

Example. Let $A = \{x : x$ is a English word$\}$. Let R be a relation on A such that $R = \{(a, b) : a, b \in A$ and a and b has one or more letters common$\}$. We show R is a compatible relation.

(*i*) **Reflexive.** Every English word has letters same as itself. Therefore R is reflexive.

(*ii*) **Symmetric.** If a has one or more letters same as that of b, then b also has one or more letters same as a. Hence R is symmetric.

Therefore, R is a compatible relation. But R is not transitive.

Take a = book, b = kite, c = ten (any English word).

Here (a, b) = (book, kite) $\in$ R (as book and kite has '*k*' common)

(b, c) = (kite, ten) $\in$ R. But (a, c) = (book, ten) $\notin$ R.

$\therefore$ R is not an equivalence relation.

2.12. PARTIAL ORDER RELATION (*P.T.U. B.Tech. Dec. 2013, May 2006*)

A relation R on a set A is called a partial relation if it is

(*i*) reflexive

(*ii*) anti-symmetric

(*iii*) transitive.

For example, define a relation '$\subseteq$' on A. We show the relation '$\subseteq$' is a partial order relation on A.

(*i*) Let $B \in P(A)$. Then $B \subseteq B$ is true. Therefore, '$\subseteq$' is reflexive.

(*ii*) Let $B_1, B_2 \in P(A)$ and if $B_1 \subseteq B_2$, $B_2 \subseteq B_1$. Then $B_1 = B_2$ is true.

$\therefore$ '$\subseteq$' is anti-symmetric.

(*iii*) Let $B_1, B_2, B_3 \in P(A)$ and if $B_1 \subseteq B_2$, $B_2 \subseteq B_3$, then $B_1 \subseteq B_3$ is true.

Hence the relation '$\subseteq$' is transitive.

$\therefore$ The relation '$\subseteq$' is a partial order relation.

2.13. PRODUCT OF SETS

Let $A_1, A_2 ..., A_n$ are n sets. Then the set of all ordered n-tuples $(a_1, a_2, ..., a_n)$ where $a_1 \in A$, $a_2 \in A,..., a_n \in A$ is called the product of the sets $A_1, A_2, ... A_n$, and is denoted by $A_1 \times A_2 \times A_3 \times ... \times A_n$ or $\prod_{i=1}^{n} A_i$.

Hence, we write $A \times A = A^2$

$$A \times A \times A = A^3$$

....

....

$$A \times A \times \ldots A = A^n$$

2.14. TERNARY RELATION

Let S be any set. Then, a subset of S × S × S is called a ternary relation an S. *e.g.*, Consider A = [1, 2, 3, 4, ..., 15]. Let R be the ternary relation on A defined by the equation $x^2 + 5y = z$. We write R as a set of ordered pairs.

For $x > 3$, $x^2 > 15$. Hence, we find only solutions of the equation $x^2 + 5y = z$ for $x = 1, 2, 3$

For $x = 1$, we have $1 + 5y = z$ and the values of y, z are 1, 6; 2, 11.

For $x = 2$, we have $4 + 5y = z$ and the values of y, z are 1, 9; 2, 14

For $x = 3$, we have $9 + 5y = z$ and the values of y, z are 1, 14. Hence

R = {(1, 1, 6), (1, 2, 11), (2, 1, 9), (2, 2, 14), (3, 1, 14)}.

Example 9. *Show, how a binary operation, say, addition (+), may be viewed as a ternary relation.*

Sol. Let the binary operation '+' may be defined as a set of ordered triples as follows:

$$+ = \{(x, y, z) : x + y = z\}$$

Then, the relation '+' defined above is a ternary relation. For *e.g.*, (2, 5, 7) ∈ +, but (2, 4, 8) ∉ +.

Example. *Let A = {1, 2, 3, ... 15}. Let R be the 4-ary relation on A defined by*

$$R = \{(x, y, z, t) : 4x + 3y + z^2 = t\}. \text{ Write } R \text{ as a set of 4-tuples.}$$

Sol. Let $(x, y, z, t) \in R$ such that

$$4x + 3y + z^2 = t,$$

Then x can assume the values 1, 2 and 3 only for all $y, z, t \in A$.

Thus, R = {(1, 1, 1, 8), (1, 1, 2, 11), (1, 2, 1, 11), (1, 2, 2, 14), (1, 3, 1, 14), (2, 1, 1, 12), (2, 1, 2, 15), (2, 2, 1, 15)}.

2.15. CLOSURE PROPERTIES OF RELATIONS

Consider a relation R on a given set A. Suppose the relation R does not satisfy the desired property. After adding the least number of ordered pairs to relation R, if R satisfies the desired property, then the desired property is called closure of relation R.

(*a*) **Reflexive closure.** A relation R_R is called reflexive closure of the relation R if R_R is the smallest relation containing R having the reflexive property.

For example, consider A = {7, 8, 10}. Define a relation R by

R = {(7, 8), (7, 10), (8, 8), (10, 7)}. Here R is not reflexive. We find the reflexive closure of R.

The relation R is reflexive if $(x, x) \in R$ for all $x \in A$. ∴ We add the ordered pairs (7, 7), (10, 10) in R. The required reflexive closure of R is given by

$$R_R = \{(7, 7), (10, 10), (7, 8), (7, 10), (8, 8), (10, 7)\}$$

(*b*) **Symmetric closure.** A relation R_S is called symmetric closure of R if R_S is the smallest relation containing R having the symmetric property. The smallest symmetric relation containing R is $R_S = R \cup R^{-1}$.

For example, consider A = {4, 5, 6}. Define a relation R by R = {(4, 5), (5, 5), (5, 6)}. We find the symmetric closure of R.

Here $R^{-1} = \{(5, 4), (5, 5), (6, 5)\}$

$\therefore$ The required symmetric closure of R is given by

$$R_S = R \cup R^{-1} = \{(4, 5), (5, 5), (5, 6), (5, 4), (6, 5)\}$$

(*c*) **Transitive closure.** A relation R_T is called transitive closure of R if R_T is the smallest relation containing R having the transitive property.

For example, consider A = {4, 6, 8, 10} and define a relation R on A by

R = {(4, 4), (4, 10), (6, 6), (6, 8), (8, 10)}. We find the transitive closure of R.

Here $(6, 8) \in R$, $(8, 10) \in R$, but $(6, 10) \notin R$. So we add the ordered pair (6, 10) to R.

Hence the required transitive closure of R is given by

$$R_T = \{(4, 4), (4, 10), (6, 6), (6, 8), (8, 10)\ (6, 10)\}.$$

2.16. COMPOSITION OF RELATIONS

(P.T.U. B.Tech. May 2007)

Let R and S be two relations from sets A to B and B to C respectively. Then a relation RoS is called composite relation from A to C where $(a, c) \in$ RoS iff we can find $b \in B$ such that $(a, b) \in R$ and $(b, c) \in S$.

The relation RoS is read as *composition of R and S.*

or

Let A, B, C be any three sets. Let R be a relation from A to B and S be a relation from B to C. *i.e.,* R is a subset of A × B and S is a subset of B × C. Then R and S give rise to a relation from A to C denoted by RoS and defined by

a(RoS)c if for some $b \in B$, we have aRb and bSc. Thus,

$$\text{RoS} = \{(a, c) : \text{There exists } b \in B \text{ for which } (a, b) \in R \text{ and } (b, c) \in S\}$$

e.g., Let A = (1, 2, 3), B = (a, b, c), C = (x, y, z).

Consider the following relation: R from A to B and S from B to C, given by

$$R = \{(1, b), (2, a), (2, c)\}\ ;\ S = \{(a, y), (b, x), (c, y), (c, z)\}$$

We find the composition relation RoS

Draw the arrow diagram of R and S as shown Fig. 2.1. There is an arrow from 1 to b which is followed by an arrow from b to x. Thus 1(RoS)x or $(1, x) \in$ RoS.

Similarly, $(2, y), (2, z) \in$ RoS. No other pairs belong to RoS. Thus, RoS = {(1, x), (2, y), (2, z)}

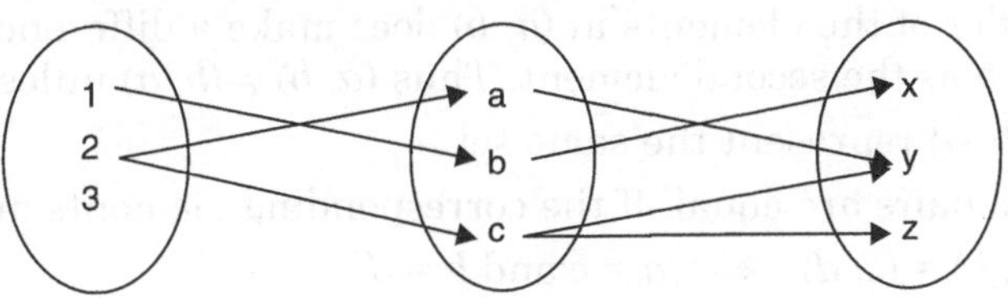

Fig. 2.1

Remark. In general RoS ≠ SoR. Also, $(\text{RoS})^{-1} = S^{-1}oR^{-1}$

Theorem II. *Let A, B and C be three sets and R be a relation from A to B, S a relation from B to C. Then* $(RoS)^{-1} = S^{-1}oR^{-1}$

Proof. Let $a \in A$, $b \in B$, $c \in C$. Then $(a, b) \in R$, $(b, c) \in S$

$\Leftrightarrow$ $(a, c) \in \text{RoS}$ $\Leftrightarrow$ $(c, a) \in (\text{RoS})^{-1}$ ...(1)

But $(b, c) \in S$ $\Leftrightarrow$ $(c, b) \in S^{-1}$,

Also $(a, b) \in R \Rightarrow (b, a) \in R^{-1} \Leftrightarrow (c, a) \in S^{-1}oR^{-1}$ ...(2)

From (1) and (2), $(RoS)^{-1} \subseteq S^{-1}oR^{-1}$

and $S^{-1}oR^{-1} \subseteq (RoS)^{-1}$

Hence $(RoS)^{-1} = S^{-1}oR^{-1}$.

2.17. DIRECTED GRAPH OR DIGRAPH OF A RELATION

Consider a relation R on A = {1, 2, 3, 4} defined by

$$R = \{(1, 2), (2, 2), (2, 4), (3, 2), (3, 4), (4, 1), (4, 3)]$$

To find the directed graph of R, write down the elements of A.

Draw an arrow from an element x to an element y whenever $(x, y) \in R$. The Fig. 2.2 so obtained is known as **directed graph** for the relation R.

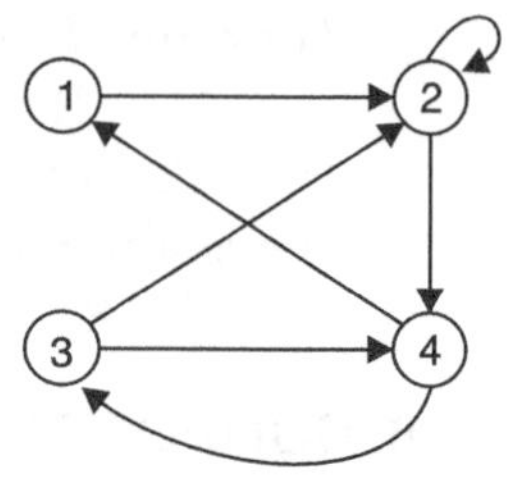

Fig. 2.2 Directed graph of R

We emphasize that a directed graph is not defined for a relation from a set to another set.

ILLUSTRATIVE EXAMPLES

Example 1. *Let R be a relation on A = {2, 3, 4, 5, 6} defined by 'x is relative prime to y'. Write R as a set of ordered pairs.*

Sol. We know that two integers x and y are said to be relative prime iff $(x, y) = 1$ *i.e.*, g.c.d of x and y is one.

∴ The required set of ordered pairs is given by

$$R = \{(2, 3), (2, 5), (3, 4), (3, 5), (4, 5), (5, 6)\}.$$

Example 2. *(a) Explain the difference between an ordered pair (a, b) and the set {a, b} with two elements.*

(b) Explain when the ordered pairs (a, b) and (c, d) are equal. When the n-tuples $(a_1, a_2, ..., a_n)$ and $(b_1, b_2, ..., b_n)$ are equal ?

(c) Let A = {1, 2}, Find A^2 and A^3

(d) Give the geometrical meaning of $R^2 = R \times R$ as points in the plane.

Sol. (*a*) The order of the elements in (a, b) does make a difference. Here, a is taken as the first element and b as the second element. Thus $(a, b) \neq (b, a)$ unless $a = b$. Whereas, $\{a, b\}$ and $\{b, a\}$ represent the same set.

(*b*) Two ordered pairs are equal iff the corresponding elements are equal. Therefore,

$$(a, b) = (c, d) \Leftrightarrow a = c \text{ and } b = d$$

Also $(a_1, a_2, ... a_n) = (b_1, b_2, ..., b_n)$

$$\Leftrightarrow \quad a_1 = b_1 \,;\, a_2 = b_2 \,;\, a_3 = b_3 \,;\, ... \, a_n = b_n$$

or $$a_i = b_i \ \forall \ i = 1, 2, ..., n.$$

(*c*) Given $A = \{1, 2\}$

Then $A^2 = A \times A = \{(1, 1), (1, 2), (2, 1), (2, 2)\}$

$A^3 = A \times A \times A = A^2 \times A$

$= \{(1, 1, 1), (1, 1, 2), (1, 2, 1), (1, 2, 2), (2, 1, 1), (2, 1, 2), (2, 2, 1), (2, 2, 2)\}$

(*d*) Let $(a, b) \in R \times R$ be any point in the plane. Then, the verticle line through P meets the x-axis at a and the horizontal line through P meets the y-axis at b. Thus,

$R^2 = R \times R$ is known as cartesian plane (Fig. 2.3).

Example 3. *(a) Show that $n(A \times B) = n(A) \,.\, n(B)$ where A and B are finite sets and n (A) denotes the number of elements in A. Generalise the result.*

(b) Let $A = \{1, 2, 3, ..., 10\}$, $B = \{a, b, c, ..., x, y, z\}$. How many elements are in $A \times B$.

(c) If $A = \{1, 2, 3, 6\}$, $B = \{8, 9, 10\}$. Determine the number of elements in

(i) $A \times B$ *(ii)* $B \times A$ *(iii)* A^2

(iv) B^4 *(v)* $A \times A \times B$ *(vi)* $B \times A \times B$.

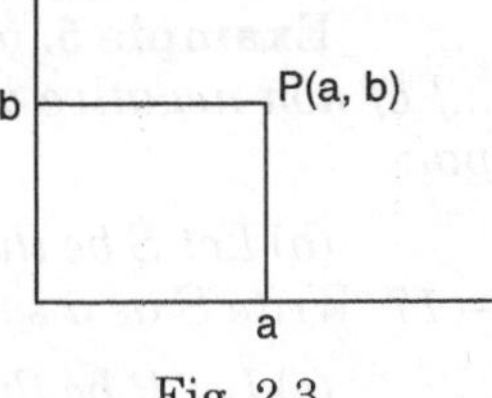

Fig. 2.3

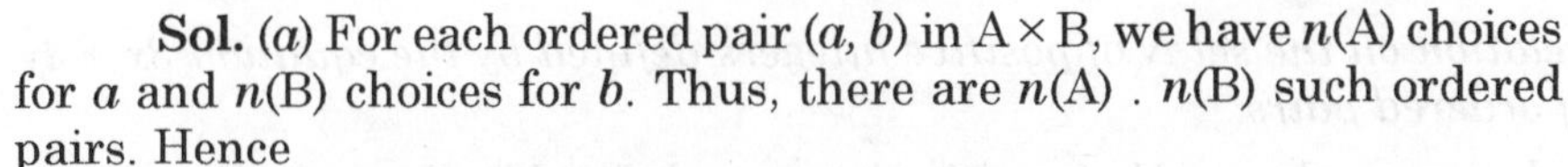

Sol. (*a*) For each ordered pair (a, b) in $A \times B$, we have $n(A)$ choices for a and $n(B)$ choices for b. Thus, there are $n(A) \,.\, n(B)$ such ordered pairs. Hence

$$n(A \times B) = n(A).\, n(B).$$

Generalisation. If $A_1, A_2, ... A_n$ are n finite sets, then

$$n(A_1 \times A_2 \times ... \times A_n) = n(A_1) \,.\, n(A_2) \,...\, n(A_n)$$

(*b*) $n(A) = 10,\ n(B) = 26$

$\therefore$ $n(A \times B) = n(A) \,.\, n(B) = 10.26 = 260$

(*c*) $n(A) = 4,\ n(B) = 3$

(*i*) $n\,(A \times B) = n(A) \,.\, n(B) = 4.3 = 12$

(*ii*) $n\,(B \times A) = n(B) \,.\, n(A) = 3.4 = 12$

(*iii*) $n(A^2) = n(A \times A) = n(A) \,.\, n(A) = 4.4 = 16$

(*iv*) $n(B^4) = n(B \times B \times B \times B) = n(B) \,.\, n(B) \,.\, n(B) \,.\, n(B) = 3.3.3.3 = 81$

(*v*) $n(A \times A \times B) = n(A) \,.\, n(A) \,.\, n(B) = 4 \,.\, 4 \,.\, 3 = 48$

(*vi*) $n(B \times A \times B) = n(B) \,.\, n(A) \,.\, n(B) = 3 \,.\, 4 \,.\, 3 = 36$

Example 4. *(a) Let $A = \{1, 2\}$, $B = \{x, y, z\}$, $C = \{3, 4\}$. Find $A \times B \times C$ and $n(A \times B \times C)$*

(b) Let $A = \{1, 2\}$, $B = \{a, b, c\}$, $C = \{c, d\}$. Find $(A \times B) \cap (A \times C)$ and hence $A \times (B \cap C)$.

(c) If $B_1 = \{1, 2\}$, $B_2 = \{3, 4\}$, $B_3 = \{5, 6\}$. Find ΠBi.

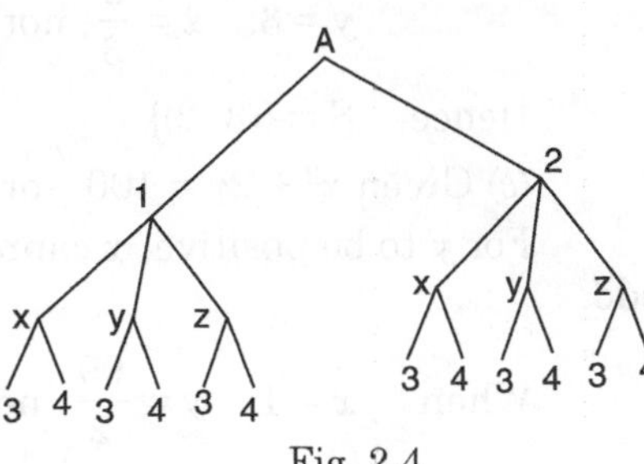

Fig. 2.4

Sol. (*a*) From the tree diagram, (Fig. 2.4)

$$A \times B \times C = \{(1, x, 3), (1, x, 4), (1, y, 3), (1, y, 4), (1, z, 3), (1, z, 4), (2, x, 3), (2, x, 4)$$
$$(2, y, 3), (2, y, 4), (2, z, 3), (2, z, 4)\}$$

Also $n(A \times B \times C) = n(A) \,.\, n(B) \,.\, n(C) = 2.\ 3.\ 2 = 12$

(*b*) Given $A = \{1, 2\}$, $B = \{a, b, c\}$, $C = \{c, d\}$

$\therefore$ $A \times B = \{(1, a), (1, b), (1, c), (2, a), (2, b), (2, c)\}$

$$A \times C = \{(1, c), (1, d), (2, c), (2, d)\}$$

$\therefore \quad (A \times B) \cap (A \times C) = \{(1, c), (2, c)\}$

Also $\quad A \times (B \cap C) = (A \times B) \cap (A \times C) = \{(1, c), (2, c)\}$

(c) Using tree diagram (Fig. 2.5)

$$\Pi B i = B_1 \times B_2 \times B_3$$
$$= \{(1, 3, 5), (1, 3, 6), (1, 4, 5), (1, 4, 6), (2, 3, 5), (2, 3, 6), (2, 4, 5), (2, 4, 6)\}$$

Fig. 2.5

Example 5. (a) *let R be the relation on the set X = {0, 1, 2, 3, ...} of non-negative integers defined by the equation $x^2 + y^2 = 25$. Write R as a set of ordered pairs.*

(b) Let S be the relation on the set N of positive integers defined by the equation $3x + 4y = 17$. Write S as a set of ordered pairs.

(c) Let R be the relation on the set N of positive integers defined by the equation $x^2 + 2y = 100$. Find the domain and range of R.

Sol. (a) Given $x^2 + y^2 = 25$. We can have

$$x = 0, \quad y = 5 \; ; \quad x = 3, \quad y = 4$$
$$x = 4, \quad y = 3 \; ; x = 5 \; ; \quad y = 0$$

Thus $\quad R = \{(0, 5), (3, 4), (4, 3), (5, 0)\}$

(b) Given $3x + 4y = 17$ or $3x = 17 - 4y$ ...(1)

From (1), we observe that no value of y can exceed 4 as x must be positive.

For $\quad y = 1, \quad x = \frac{11}{3}$, not an integer ; $y = 2, \quad x = \frac{9}{3} = 3$

$y = 3, \quad x = \frac{5}{3}$, not an integer ; $y = 4, x = -\frac{1}{3}$, not an integer

Hence $\quad S = \{(3, 2)\}$

(c) Given $x^2 + 2y = 100$ or $2y = 100 - x^2$ ...(1)

For y to be positive, x cannot exceed 9 also R.H.S. of (1) is even and hence y cannot be odd.

When $\quad x = 1, \quad y = \frac{99}{2}$, not an integer ; $x = 2, \quad y = \frac{96}{2} = 48$

$x = 3, \quad y = \frac{91}{2}$, not an integer ; $x = 4, \quad y = \frac{84}{2} = 42$

Similarly, $x \neq 5, x \neq 7, x \neq 9$

For $\quad x = 6, \quad 2y = 100 - 36 = 64 \Rightarrow y = 32$

$x = 8, \quad 2y = 100 - 64 = 36 \Rightarrow y = \frac{36}{2} = 18$

$\therefore \quad R = \{(2, 48), (4, 42), (6, 32), (8, 18)\}$

Domain $R = \{2, 4, 6, 8\}$, Range $R = \{18, 32, 42, 48\}$.

Example 6. *(a) If A is a set containing n elements. Find the number of relations from A to A.*

(b) If A = (1, 2). Find all possible relations from A to A.

Sol. *(a)* We know that if a set A has *m* elements and B has *n* elements, then total number of relations from A to B is 2^{mn}.

Therefore, the total number of relations from A to A is $2^{n \times n} = 2^{n^2}$.

(b) A contains 2 elements ∴ Total number of relations from A to A = $2^{2^2} = 4^2 = 16$.

The various relations are

ϕ, {(1, 2), (2, 1), (1, 1), (2, 2)}, {(1, 2), (2, 1)}, {(1, 2), (1, 1)}, {(1, 2), (2, 2)}, {(2, 1), (1, 1)}, {(2, 1), (2, 2)}, {(1, 1), (2, 2)}, {(1, 2), (2, 1), (1, 1)}, {(1, 2), (1, 1), (2, 2)}, {(2, 1), (2, 2), (1, 1)}, {(1, 2), (2, 1), (2, 2)}, {(1, 2), (2, 1), (1, 1), (2, 2)}, {(1, 1), (2, 2)}, {(1, 1), (2, 2), (1, 2)}, {(2, 1), (1, 1), (1, 2)}.

Example 7. *What are the properties of relations? Explain with examples.*

(P.T.U. Dec. 2005)

Sol. There are many properties of relations. These properties tell the nature and type of the relations. The following are the main properties.

(i) **Reflexive.** Consider a binary relation R on a set A. If for $x \in$ A, $(x, x) \in$ R for all $x \in$ A, then R is called reflexive relation.

For example, consider A = (1, 2) and R = {(1, 1), (2, 2)}.

Then R is reflexive relation.

(ii) **Irreflexive relation.** A relation R is called irreflexive if for every $x \in$ A, $(x, x) \notin$ R.

For example, consider A = (1, 2) and R = {(1, 2), (2, 2)}

then R is irreflexive since $(x, x) \notin$ R for every $x \in$ A.

(iii) **Symmetric relation.** Consider a binary relation R on a set A. The relation R is called symmetric if whenever $(x, y) \in R \Rightarrow (y, x) \in R$

For example, consider A = {(1, 2)} and R = {(1, 2), (2, 1), (1, 1)}

Then R is symmetric relation.

(iv) **Asymmetric relation.** A relation R is called asymmetric relation if for every

$(x, y) \in R \Rightarrow (y, x) \notin R$ for all $(x, y) \in$ A.

For example, consider A = (1, 2) and R = {(1, 1), (1, 2), (2, 2)}

This relation R is asymmetric as (1, 2) ∈ R, but (2, 1) ∉ R.

(v) **Antisymmetric relation.** A relation R on a set A is called an antisymmetric if whenever $(x, y) \in$ R, $(y, x) \in$ R then $x = y$.

For example, P(A) = The set of all subsets of A and let '⊆' is a relation on P(A).

Let A ⊆ B, B ⊆ A for A, B ∈ P(A).

Then clearly B = A. Hence '⊆' is antisymmetric relation.

(vi) **Transitive relation.** Let R be a relation on a set A. Let $(x, y) \in$ R, $(y, z) \in R \Rightarrow (x, z) \in$ R for $x, y, z \in$ R. Then R is called transitive relation.

For example, consider A = (1, 2, 3) and R = {(1, 2), (2, 1), (1, 1), (2, 2)} is a transitive relation.

Example 8. *Give an example of a relation which is*

(*a*) *Neither reflexive nor irreflexive*

(*b*) *Both symmetric and antisymmetric*

(*c*) *Both reflexive and symmetric*

(*d*) *Each reflexive, symmetric and transitive*

(*e*) *Both symmetric and transitive but not reflexive.*

Sol. (*a*) **Neither reflexive nor irreflexive**

Recall that a relation R on a set A is said to be reflexive if $a\text{R}a \ \forall \ a \in \text{A}$. Also a relation R on a set A is said to be irreflexive if $(a, a) \notin \text{R} \ \forall \ a \in \text{A}$.

Let A = {1, 2, 3} and define R = {(1, 1), (1, 2), (1, 3), (3, 3)}.

Here R is not reflexive since (2, 2) ∉ R where as 2 ∈ A.

It is also not irreflexive since (1, 1) ∈ R.

(*b*) **Both symmetric and antisymmetric**

Recall that a relation R on a set A is symmetric if $(a, b) \in \text{R} \ \Rightarrow \ (b, a) \in \text{R}$. Also a relation R is antisymmetric if $(a, b) \in \text{R}$ and $(b, a) \in \text{R} \ \Rightarrow \ a = b$.

Take A = (1, 2, 3) and R = {(1, 1), (2, 2)}

Then, R is symmetric and antisymmetric

(*c*) **Both reflexive and symmetric**

Take A = {1, 2, 3} and R = {(1, 1), (1, 2), (2, 1), (2, 2), (3, 3)}.

Here R is reflexive and symmetric relation on A.

(*d*) **Each reflexive, symmetric and transitive**

Let L be the set of lines in the Evclidean plane. Let R be the relation on L defined by "is parallel to".

If $l_1 \in \text{L}$, then $l_1 \parallel l_1 \ \Rightarrow$ R is reflexive. If $l_1, l_2 \in \text{L}$ such that $l_1 \parallel l_2$, then $l_2 \parallel l_1 \ \Rightarrow$ R is symmetric.

If $l_1, l_2, l_3 \in \text{L}$ such that $l_1 \parallel l_2$ and $l_2 \parallel l_3$, then $l_1 \parallel l_3$. Hence R is transitive also.

(*e*) **Both symmetric and transitive but not reflexive.**

The empty relation φ on any finite set A is symmetric and transitive but not reflexive.

Example 9. *Consider the following relations on the set A = (1, 2, 3, 4), defined by*

(*i*) *R* = {(1, 1), (1, 2), (1, 3), (3, 3)}

(*ii*) *S = {(1, 1), (1, 2), (2, 1), (2, 2), (3, 3)}*

(*iii*) *T = {(1, 1), (1, 2), (2, 2), (3, 3)}*

(*iv*) φ = *Empty relation*

(*v*) *U = Universal relation.*

Determine whether or not each of the above relations on A is

(*a*) *Reflexive* (*b*) *Symmetric* (*c*) *Transitive* (*d*) *Antisymmetric.*

Sol. (*a*) **Reflexive.** The universal relation on A = {1, 2, 3, 4} is reflexive. The empty relation φ on A = {1, 2, 3, 4} is not reflexive.

Also R, S and T are not reflexive since there exists an element $a \in \text{A}$ such that $(a, a) \notin$ R, S, T.

(*b*) **Symmetric**

(*i*) R = {(1, 1), (1, 2), (1, 3), (3, 3)}. Here R is not symmetric since (1, 2) ∈ R, but (2, 1) ∉ R.

(*ii*) S = {(1, 1), (1, 2), (2, 1), (2, 2), (3, 3)}. Hence $(a, b) \in \text{S} \ \Rightarrow \ (b, a) \in \text{S}$ for all $a, b \in \text{S}$

∴ S is symmetric.

(*iii*) T = {(1, 1), (1, 2), (2, 2), (2, 3)}. Here T is not symmetric since (1, 2) ∈ T $\Rightarrow$ (2, 1) ∉ T

(*iv*) The empty relation on any non-empty set is always symmetric.

(*v*) The universal relation on any non-empty set is always symmetric.

(*c*) **Transitive**

(*i*) R = {(1, 1), (1, 2), (1, 3), (3, 3)}. Here (*a*, *b*) ∈ R, (*b*, *c*) ∈ R $\Rightarrow$ (*a*, *c*) ∈ R ∀ *a*, *b*, *c* ∈ R

(*ii*) S = {(1, 1), (1, 2), (2, 1), (2, 2), (3, 3)}.

Here (*a*, *b*) ∈ S, (*b*, *c*) ∈ S $\Rightarrow$ (*a*, *c*) ∈ S ∀ *a*, *b*, *c* ∈ S. Hence S is transitive.

(*iii*) T = {(1, 1), (1, 2), (2, 2), (2, 3)}. Here (1, 2) ∈ T, (2, 3) ∈ T $\Rightarrow$ (1, 3) ∈ T

∴ T is not transitive.

(*iv*) The empty relation on A is always transitive.

(*v*) The universal relation on any non-empty set A is always transitive

(*d*) **Antisymmetric**

(*i*) R = {(1, 1), (1, 2), (1, 3), (3, 3)}. We know that a relation R is called antisymmetric relation on A iff (*a*, *b*) ∈ R, (*b*, *a*) ∈ R $\Rightarrow$ $a = b$ ∴ R is antisymmetric.

(*ii*) S = {(1, 1), (1, 2), (2, 1), (2, 2), (3, 3)}. Here (*a*, *b*) ∈ S, (*b*, *a*) ∈ S $\Rightarrow$ $a \neq b$ ∀ *a*, *b* ∈ S.

∴ S is not antisymmetric. Since (1, 2) ∈ S, (2, 1) ∈ S $\Rightarrow$ 1 ≠ 2.

(*iii*) T = {(1, 1), (1, 2), (2, 1), (1, 3), (3, 3)}. Here (1, 2) ∈ T, (2, 1) ∈ T $\Rightarrow$ 1 ≠ 2.

∴ T is not antisymmetric.

(*iv*) The empty relation on A is not antisymmetric.

(*v*) The universal relation on any non-empty set. For if, let *a*, *b* ($a \neq b$) ∈ A. If *a*R*b*, *b* R *a* and if R is antisymmetric, then, we must have $a = b$, which is a contradiction. Hence the universal relation on A is not antisymmetric.

Example 10. *Each of the following defines a relation on the set N of positive integers*

$R : x > y$

$S : x + y = 10$

$T : x + 4y = 10$ *for all x, y,* ∈ N

Determine which of the relations are

(*a*) *reflexive* (*b*) *symmetric*

(*c*) *transitive* (*d*) *antisymmetric*

Sol. (*a*) **Reflexive.** None are reflexive. For *e.g.*, (1.1) ∉ R, S and T

(*b*) **Symmetric.** Take $x = 3$, $y = 6$, then clearly $y > x$, but $x \not> y$ ∴ R is not symmetric

If (*x*, *y*) ∈ S, then $x + y = 10$ $\Rightarrow$ $y + x = 10$

$\Rightarrow$ (*y*, *x*) ∈ S. Hence S is symmetric.

Also for $x = 6, y = 1, x + 4y = 10$ holds

i.e., (6, 1) ∈ T, but (1, 6) ∉ T *i.e.*, T is not symmetric.

(*c*) **Transitive.** Let (*x*, *y*), (*y*, *z*) ∈ R $\Rightarrow$ $x > y$ and $y > z$

$\Rightarrow$ $x > z$ ∴ (*x*, *z*) ∈ R for all *x*, *z* ∈ N

∴ R is transitive

But S is not transitive.

For e.g., (3, 7) ∈ S since 3 + 7 = 10; (7, 3) ∈ S since 7 + 3 = 10

But (3, 3) ∉ S as 3 + 3 = 6 ≠ 10

Further, if $(x, y) \in T$, then $x + 4y = 10$

$(y, z) \in T$, then $y + 4z = 10$

$$\text{Consider } x + 4z = x + 10 - y = x + 10 - \left(\frac{10 - x}{4}\right)$$

$$= \frac{4x + 40 - 10 + x}{4} = \frac{5x + 30}{4} \neq 10$$

$\therefore$ $(x, z) \notin T$. Hence T is not transitive.

(*d*) **Antisymmetric.** If $x > y$ and $y > x$, then $x = y \ \forall \ x, y \in N$

$\therefore$ R is antisymmetric.

Let $(2, 8) \in S \Rightarrow (8, 2) \in S$ but $2 \neq 8$

$\therefore$ S is not antisymmetric.

If $(x, y) \in T$, then $x + 4y = 10$

$(y, x) \in T$, then $y + 4x = 10$

$\Rightarrow$ $x + 4y = y + 4x$

$\Rightarrow$ $3y = 3x \Rightarrow x = y \ \forall \ x, y \in N$

$\therefore$ T is antisymmetric.

Example 11. (*a*) *Find the numbers of relations from A = (a, b, c) to B = (1, 2).*

(*b*) *Define ternary relation and give an example.* (P.T.U., B.Tech. Dec. 2006)

(*c*) *Define anti-symmetric relation with an example.*

Sol. (*a*) We know that if a set contains m elements and B contains n elements, then total number of relations from A to B is 2^{mn}.

Here $m = 3, n = 2$ $\therefore$ required number of relations from A to B = $2^6 = 64$.

(*b*) A ternary relation is a set of ordered triples. In particular, if S is a set, then a subset of $S \times S \times S$ is called a ternary relation on S. For example, If L is a line, then "betweenness" is a ternary relation among points of L.

(*c*) A relation R is said to be antisymmetric relation if $(x, y) \in R$, $(y, x) \in R \Rightarrow y = x$

For *e.g.*, consider A = The set of natural numbers. Define R as $(x, y) \in R$ iff $x \mid y$.

Then we know that if x divides y, y divides x then $x = y$ for all $x, y \in R$

Hence the relation defined above is an antisymmetric relation.

Example 12. (*a*) *Give an example to show that a reflexive relation on a set A is not necessarily symmetric.*

(*b*) *Prove that a relation R on a set A is symmetric iff $R = R^{-1}$.*

(P.T.U. M.C.A. Dec. 2005)

Sol. (*a*) Consider A = (1, 2, 3) and define a relation R by R = {(1, 1), (2, 2), (3, 3), (1,3)}.

Here $(x, x) \in R$ for all $x \in A$. $\therefore$ The relation R is reflexive. But $(1, 3) \in R \Rightarrow (3, 1) \notin R$.

$\therefore$ The relation R is not symmetric.

(*b*) Let R is symmetric on A. We show $R = R^{-1}$

Let $(x, y) \in R$ for all $x, y \in A$

$\Rightarrow$ $(y, x) \in R$ | R is symmetric

$\Rightarrow$ $(x, y) \in R^{-1}$ | By definition of inverse relation

$\Rightarrow$ $R \subseteq R^{-1}$ ...(1)

Again let $(x, y) \in R^{-1}$ for all $x, y \in A$

$\Rightarrow$ $(y, x) \in R$ | By definition of inverse relation

$\Rightarrow$ $(x, y) \in R$ | R is symmetric

$\Rightarrow$ $R^{-1} \subseteq R$ ...(2)

From (1) and (2) $R = R^{-1}$

Converse. Let $R = R^{-1}$. We show R is symmetric

Let $(x, y) \in R$

$\Rightarrow$ $(x, y) \in R^{-1}$

$\Rightarrow$ $(y, x) \in R$ for all $x, y \in A$. | $R = R^{-1}$

i.e., R is symmetric.

Example 13. *Let R be a relation on the set of all lines in a plane defined by $(l_1, l_2) \in R \Leftrightarrow l_1$ is parallel to l_2. Show that R is an equivalence relation.*

Sol. Let A = The set of all lines in a plane and $l \in A$. Since l is parallel to l for each $l \in A$.

$\therefore$ R is reflexive.

Also let $(l_1, l_2) \in R$ for all $l_1, l_2 \in A$. Then by definition of R, l_1 is parallel to $l_2 \Rightarrow l_2$ is also parallel to l_1. It means $(l_2, l_1) \in R$.

$\therefore$ R is symmetric.

Finally, Let $(l_1, l_2) \in R$, $(l_2, l_3) \in R$ for all $l_1, l_2, l_3 \in A$.

Now, $(l_1, l_2) \in R \Rightarrow l_1$ is parallel to l_2

$(l_2, l_3) \in R \Rightarrow l_2$ is parallel to l_3.

It means l_1 is parallel to l_3. $\therefore$ $(l_1, l_3) \in R$ *i.e.,* R is transitive. Hence we can say that the relation R is an equivalence relation.

Example 14. *Show that the relation "is congruent to" on the set of all triangles in a plane is an equivalence relation.*

Sol. Let A = The set of all triangles in a plane.

R = The relation on A defined by $(T_1, T_2) \in R \Leftrightarrow T_1$ is congruent to T_2 for all $T_1, T_2 \in A$.

Reflexivity. Let $T_1 \in A$, then $T_1 \cong T_1$ is true

$\therefore$ $(T_1, T_1) \in R$ for all $T_1 \in A$. Hence R is reflexive.

Symmetry. Let $T_1, T_2 \in A$ such that $(T_1, T_2) \in R$

Now, $(T_1, T_2) \in R \Rightarrow T_1 \cong T_2 \Rightarrow T_2 \cong T_1$

$\Rightarrow$ $(T_2, T_1) \in R$ for all $T_1, T_2 \in A$

$\therefore$ R is symmetric.

Transitive. Let $T_1, T_2, T_3 \in A$ such that

$(T_1, T_2) \in R, (T_2, T_3) \in R$

Now, $(T_1, T_2) \in R \Rightarrow T_1 \cong T_2$

$(T_2, T_3) \in R \Rightarrow T_2 \cong T_3$

$\therefore$ $T_1 \cong T_3$

$\Rightarrow$ $(T_1, T_3) \in R$. $\therefore$ R is transitive.

Hence the relation R is an equivalence relation.

Example 15. *Let A be a non-empty set and R be a relation on the power set P(A) defined by $(A, B) \in R \Leftrightarrow A \subseteq B$ for all $A, B \in P(A)$. Examine whether R is an equivalence relation or not. Is R antisymmetric ?*

Sol. Let $A \in P(A)$. Since $A \subseteq A$ for all $A \in P(A)$

$\therefore$ $(A, A) \in R$. *i.e.,* R is reflexive.

Let $(A, B) \in R$ for all $A, B \in P(A)$.

Then $A \subseteq B \Rightarrow B \not\subseteq A$ *i.e.*, B may not be a subset of A. Hence R is not symmetric.

$\therefore$ R is not an equivalence relation.

But if (A, B) $\in$ R, (B, A) $\in$ R, then we have $A \subseteq B$, $B \subseteq A \Rightarrow A = B$.

Hence R is antisymmetric.

Example 16. *Consider the set Z of integers and an integer $m > 1$. We say that x is congruent to y modulo m, written as $x \equiv y \pmod m$ if $x - y$ is divisible by m or $x - y = km$, for some integer k. Show that $\equiv$ is an equivalence relation on Z.*

Sol. Reflexive. For any $x \in Z$, $x \equiv x \pmod m$

Since $x - x = 0$ is divisible by m. Hence $\equiv$ is reflexive

Symmetry. Let $x \equiv y \pmod m$ $\Rightarrow$ $x - y$ is divisible by m $\Rightarrow$ we can write $x - y = km$ for some integer k.

or $-(x - y) = -km = rm$ for some integer $r = -k$

or $y - x = rm$ or $y - x$ is divisible by m

Hence $y \equiv x \pmod m$

$\therefore$ $\equiv$ is symmetric.

Transitive. Let $x \equiv y \pmod m$, $y \equiv z \pmod m$

$x - y$ and $y - z$ are divisible by m

$\Rightarrow$ $x - y + y - z$ is also divisible by m

$\Rightarrow$ $x - z$ is divisible by m

$\Rightarrow$ $x \equiv z \pmod m$

$\therefore$ $\equiv$ is transitive

Hence $\equiv$ is an equivalence relation.

Similar Problem 1. *Let X = {1, 2, 3, 4, 5, 6, 7} and R = {(x, y) : x – y is divisible by 3}. Check whether "a is congruent to b(mod 5)" is equivalence relation or not? Justify your answer.*

(P.T.U. B.Tech. Dec. 2007)

Ans. Yes

Similar Problem 2. *Let R be a relation on the set Z of integers defined by $a \equiv b \pmod 5$ (read as "a is congruent to b(mod 5)". Then, R is an equivalence relation on Z.*

(P.T.U. B.Tech. May 2013)

Example 17. (*a*) *Let A be the set of integers and let ~ be a relation on A × A defined by*

(a, b) ~ (c, d) if $a + d = b + c$. Prove that ~ is an equivalence relation.

(P.T.U. B.Tech. May 2012, May 2010)

(*b*) *Let A be a set of non-zero integers and let ~ be the relation on A × A defined by (a, b) ~ (c, d) if ad = bc. Show that ~ is an equivalence relation.*

Sol. (*a*) **Reflexive.** $(a, b) \sim (a, b)$ if $a + b = b + a$, which is true. Hence ~ is reflexive.

Symmetric. Let $(a, b) \sim (c, d)$, then $a + d = b + c$

or $c + b = d + a$ and hence $(c, d) \sim (a, b)$

Hence ~ is symmetric.

Transitive. Let $(a, b) \sim (c, d)$. Then $a + d = b + c$

Let $(c, d) \sim (e, f)$, then $c + f = d + e$

Adding, $a + d + c + f = b + c + d + e$

$\Rightarrow$ $a + f = b + e$ $\Rightarrow$ $(a, b) \sim (e, f)$.

Hence ~ is transitive.

$\therefore$ ~ is equivalence relation.

(*b*) **Reflexive.** $(a, b) \sim (a, b)$ if $ab = ba = ab$, | $\because$ a, b are non zero integers which is true

$\therefore$ ~ is reflexive

Symmetric. Let $(a, b) \sim (c, d)$, then $ad = bc$

$\Rightarrow \quad bc = ad$

$\Rightarrow \quad cb = da$ | $\because$ a, b, c, d are non-zero integers

$\Rightarrow \quad (c, d) \sim (a, b)$

$\therefore$ ~ is symmetric

Transitive. Let $(a, b) \sim (c, d) \Rightarrow ad = bc$ ...(1)

Let $(c, d) \sim (e, f) \Rightarrow cf = de$ ...(2)

Multiplying (1) and (2), we get

$$(ad)\,(cf) = (bc)\,(de)$$

$\Rightarrow \quad cdaf = cdbe$

$\Rightarrow \quad cd(af - be) = 0$

$\Rightarrow \quad af - be = 0$ | c, d are non-zero integers $\therefore$ $cd \neq 0$

$\Rightarrow \quad af = be$

$\Rightarrow \quad (a, b) \sim (e, f)$

$\therefore$ ~ is transitive

Thus, ~ is an equivalence relation.

Example 18. (*a*) *Define partial order relation with example.*

(P.T.U. B.Tech. May 2006)

(*b*) *Let A be the set of people and R be a relation on A defined by R = {(a, b): a is a brother of b}.*

Is R equivalence relation ? Is R partial order relation ? (discard half brother, paternity brother).

(*c*) *Let R be a binary relation on A such that (a, b) ∈ R, if book a costs more and contains fewer pages than book b. Is R equivalence relation ? Is R partial order relation ?*

Sol. (*a*) **Partial order relation.** A relation R on a set A is said to be a partial order relation on A iff it is (*i*) reflexive (*ii*) antisymmetric (*iii*) transitive.

Let P(A) denotes the power set of A and define a relation R as $(A, B) \in R \Leftrightarrow A \subseteq B$.

We show R is a partial order relation.

Reflexive. Since $A \subseteq A$ for all $A \in P(A)$.

$\therefore$ $(A, A) \in R$. Hence R is reflexive.

Antisymmetric. Let $(A, B) \in R$, $(B, A) \in R$, then $A \subseteq B$, $B \subseteq A \Rightarrow A = B$ for all $A, B \in P(A)$.

$\therefore$ R is antisymmetric.

Finally, if $(A, B) \in R$, $(B, C) \in R$, then

$$A \subseteq B,\ B \subseteq C \quad \Rightarrow \quad A \subseteq C \text{ for all } A, B, C \in P(A).$$

$\therefore$ R is transitive also.

Hence the relation R is partial order relation.

(*b*) Let $a \in A$, then $(a, a) \notin R$ (if a is a female, then no female can be a brother of herself). $\therefore$ R is not reflexive.

Let $(a, b) \in R$ *i.e.*, a is a brother of b. Then (b, a) may not belong to R (If b is a female, then b cannot be brother of a). Hence R is not symmetric.

Further, if $(a, b) \in R$ and if a and b are male, then both a and b are brothers of each other. Consequently $(b, a) \in R$, but $a \neq b$.

$\therefore$ R is not antisymmetric.

Transitive. If $(a, b) \in R$ and $(b, c) \in R$. We show $(a, c) \in R$.

Case I. If $(a, b) \in R$, then a has to be male.

If $(b, c) \in R$, then b has to be male.

$\therefore$ a and b are both male. Consequently $(a, c) \in R$. Hence R is transitive.

Thus, R is neither an equivalence relation nor a partial order relation.

(c) **Reflexive.** No book costs more than itself, nor pages fewer than itself. Hence $(a, a) \notin R$, therefore R is not reflexive.

Symmetric. If $(a, b) \in R$, then the cost of the book 'a' is more than the cost of the book 'b'. Here, $(b, a) \notin R$ (since the cost of the book 'b' is less than the cost of the book 'a')

$\therefore$ R is not symmetric.

Consequently R is not an equivalence relation, not a partial order relation.

Example 19. *Let $A = (a, b, c)$. Let R be a relation on A defined by $R = \{(a, a), (a, b), (b, c), (c, a)\}$. Find the reflexive closure of R.*

Sol. Given relation $R = \{(a, a), (a, b), (b, c), (c, a)\}$ is not reflexive. Adding the ordered pairs (b, b), (c, c) in R, the required reflexive closure of R is given by

$$R_R = \{(a, a), (a, b), (b, c), (c, a), (b, b), (c, c)\}.$$

Example 20. *Let $A = (1, 2, 3, 4)$ and let R is defined by $R = \{(1, 2), (2, 3), (3, 4), (2, 1)\}$. Find the transitive closure of R.*

Sol. Given $R = \{(1, 2), (2, 3), (3, 4), (2, 1)\}$

Here $(1, 2), (2, 1) \in R$. If R is transitive. Then there should be $(1, 1) \in R$.

Similarly, $(1, 2) \in R$, $(2, 3) \in R$. If R is transitive, there should be $(1, 3) \in R$

If $(1, 3) \in R$, $(3, 4) \in R$. We should also have $(1, 4) \in R$

Also $(2, 1) \in R$, $(1, 2) \in R$. If R is transitive, there should be $(2, 2) \in R$.

If $(2, 3) \in R$, $(3, 4) \in R$, we should have $(2, 4) \in R$.

Hence the required transitive closure of R is given by

$$R_T = \{(1, 2), (2, 3), (3, 4), (2, 1), (1, 1), (1, 3), (1, 4), (2, 2), (2, 4)\}$$

Example 21. *Consider $A = \{1, 2, 3, 4\}$, and R, S be the relations defined by*

$$R = \{(1, 2), (1, 1), (1, 3), (2, 4), (3, 2)\}$$
$$S = \{(1, 4), (1, 3), (2, 3), (3, 1), (4, 1)\}.$$

Find RoS,

Sol. We compute the elements of RoS. Consider the Fig. 2.6.

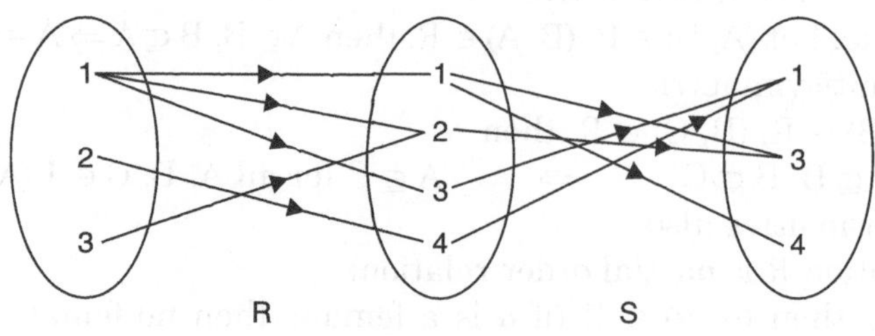

Fig. 2.6

$\therefore$ $RoS = \{(1, 3), (1, 4), (1, 1), (2, 1), (3, 3)\}$

Example 22. *Let R and S be the relations on A = (1, 2, 3, 4) A defined by*

R = {(1, 1), (3, 1), (3, 4), (4, 2), (4, 3)}

S = {(1, 3), (2, 1), (3, 1), (3, 2), (4, 4)}

(a) Find the composition relation RoS

(b) Find the composition relation SoR

(c) Find the composition relation R^2 = RoR

(d) Find the composition relation R^3 = RoRoR.

Sol. (*a*) 1R1 and 1S3 ⇒ (1, 3) ∈ RoS

Also 3R1 and 1S3 ⇒ (3, 3) ∈ RoS

3R4 and 4S4 ⇒ (3, 4) ∈ RoS

4R2 and 2S1 ⇒ (4, 1) ∈ RoS

4R3 and 3S1 ⇒ (4, 1) ∈ RoS

4R3 and 3S2 ⇒ (4, 2) ∈ RoS

Thus, RoS = {(1, 3), (3, 3), (3, 4), (4, 1), (4, 2)}

(*b*) First use S and then R.

1S3 and 3R1 ⇒ (1, 1) ∈ SoR

1S3 and 3R4 ⇒ (1, 4) ∈ SoR

2S1 and 1R1 ⇒ (2, 1) ∈ SoR

3S1 and 1R1 ⇒ (3, 1) ∈ SoR

4S4 and 4R2 ⇒ (4, 2) ∈ SoR

4S4 and 4R3 ⇒ (4, 3) ∈ SoR

Thus, SoR = {(1, 1), (1, 4), (2, 1), (3, 1), (4, 2), (4, 3)}

(*c*) Here R = {(1, 1), (3, 1), (3, 4), (4, 2), (4, 3)}

R = {(1, 1), (3, 1), (3, 4), (4, 2), (4, 3)}

(1, 1) ∈ R and (1, 1) ∈ R ⇒ (1, 1) ∈ RoR

(3, 1) ∈ R and (1, 1) ∈ R ⇒ (3, 1) ∈ RoR

(3, 4) ∈ R and (4, 2) ∈ R ⇒ (3, 2) ∈ RoR

(3, 4) ∈ R and (4, 3) ∈ R ⇒ (3, 3) ∈ RoR

(4, 3) ∈ R and (3, 4) ∈ R ⇒ (4, 4) ∈ RoR

(4, 3) ∈ R and (3, 1) ∈ R ⇒ (4, 1) ∈ RoR

∴ RoR = R^2 = {(1, 1), (3, 1), (3, 2), (3, 3), (4, 4), (4, 1)}

(*d*) Here R^3 = RoRoR = R^2oR

But R^2 = {(1, 1), (3, 1), (3, 2), (3, 3), (4, 4), (4, 1)} | Using Part (*c*)

R = {(1, 1), (3, 1), (3, 4), (4, 2), (4, 3)}

(1, 1) ∈ R^2 and (1, 1) ∈ R ⇒ (1,1) ∈ R^2oR

(3, 1) ∈ R^2 and (1, 1) ∈ R ⇒ (3, 1) ∈ R^2oR

(3, 3) ∈ R^2 and (3, 1) ∈ R ⇒ (3, 1) ∈ R^2oR

(4, 4) ∈ R^2 and (4, 2) ∈ R ⇒ (4, 2) ∈ R^2oR

(4, 1) ∈ R^2 and (1, 1) ∈ R ⇒ (4, 1) ∈ R^2oR

∴ RoRoR = R^2oR = {(1, 1), (3, 1),(4, 2), (4, 1)}.

Example 23. *Find the number of relations from A = {a, b, c} to B = {1, 2}.*

Sol. A contains 3 elements and B contains 2 elements. There are 3 × 2 = 6 ordered pairs in A × B.

Hence, total number of subsets of A × B is 2^6 = 64. Therefore,

Total number of relations from A to B = 64.

Example 24. *Let R be a relation on A = {1, 2, 3, 4} defined by "x is less than y". Write R as a set of ordered pairs. Find the inverse R^{-1} of the relation R. Can R^{-1} be described in words?*

Sol. R consists of the ordered pairs (x, y) where $x < y$. Thus

$$R = \{(1, 2), (1, 3), (1, 4), (2, 3), (2, 4), (3, 4)\}$$

Reverse the ordered pairs of R to obtain R^{-1}.

Thus, $$R^{-1} = \{(2, 1), (3, 1), (4, 1), (3, 2), (4, 2), (4, 3)\}$$

R^{-1} is the relation "x is greater than y"

Yes, R^{-1} can be described in words, *i.e.*, R^{-1} is the relation "x is greater than y".

Example 25. *Let R be a relation from A = (1, 2, 3, 4) to B = (x, y, z) defined by*

$$R = \{(1, y), (1, z), (3, y), (4, x), (4, z)\}$$

(*a*) *Find the domain and range of R*

(*b*) *Find the inverse relation R^{-1} of R.*

Sol. (*a*) The domain of R consists of the first elements of the ordered pairs of R and the range consists of the second elements. Thus,

dom (R) = {1, 3, 4} and

Range (R) = {x, y, z}

(*b*) R^{-1} is obtained by reversing the ordered pairs in R Thus,

$$R^{-1} = \{(y, 1), (z, 1), (y, 3), (x, 4), (z, 4)\}.$$

Example 26. *Let A = {1, 2, 3, 4, 6} and R be the relation "x divides y". Write R as set of ordered pairs. Find the inverse R^{-1} of the relation R. Can R^{-1} be described in words.*

Sol. We know that x/y (read as x divides y) if there exists an integer z such that $y = xz$. We find those numbers in A which are divisible by 1, 2, 3, 4 and 6.

Since 1/1, 1/2, 1/3, 1/4, 1/6, 2/2, 2/4, 2/6, 3/3, 3/6, 4/4, 4/6

Thus, R = {(1, 1), (1, 2), (1, 3), (1, 4), (1, 6), (2, 2), (2, 4), (2, 6), (3, 3), (3, 6), (4, 4), (4, 6)]

Reverse the ordered pairs of R to obtain R^{-1}. Thus,

R^{-1} = {(1, 1), (2, 1), (3, 1), (4, 1), (6, 1), (2, 2), (4, 2), (6, 2), (3, 3), (6, 3), (4, 4), (6, 4)]

Yes, R^{-1} can be described by the statement "x is a multiple of y".

Example 27. *Let R and S be the relations on A = {1, 2, 3} defined by*

$$R = \{(1, 1), (1, 2), (2, 3), (3, 1), (3, 3)\}$$
$$S = \{(1, 2), (1, 3), (2, 1), (3, 3)\}$$

(*a*) *Find $R \cap S$, $R \cup S$*

(*b*) *Find R^c.*

Sol. (*a*) $R \cap S = \{(1, 2), (3, 3)\}$

$R \cup S = \{(1, 1), (1, 2), (1, 3), (2, 1), (2, 3), (3, 1), (3, 3)\}$

(*b*) Using the fact that A × A is the universal relation on A. Hence,

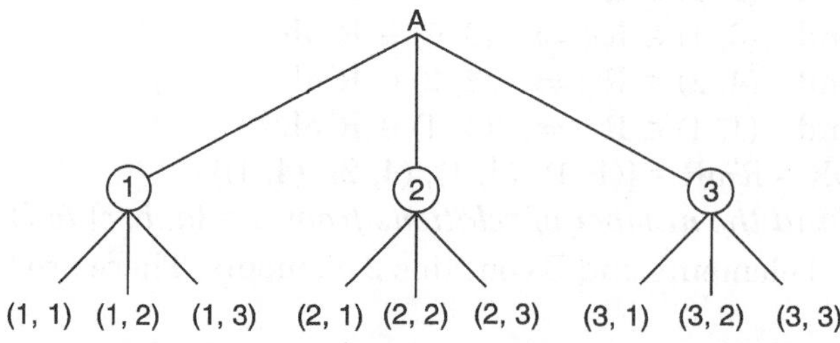

Fig. 2.7

$$A \times A = (1, 2, 3) \times (1, 2, 3)$$
$$= \{(1, 1), (1, 2), (1, 3), (2, 1), (2, 2), (2, 3), (3, 1), (3, 2), (3, 3)\}$$
$\therefore$ $R^c = A \times A - R$ = The set of ordered pairs which are in $A \times A$ but not in R
$$= \{(1, 3), (2, 1), (2, 2), (3, 2)\}.$$

Example 28. *Let* $S = \{(1, n) : n \in Z\}$, $T = \{(n, 1), n \in Z\}$ *are relations on Z. Find (i) ToS, (ii) SoT.*

Sol. (*i*) $(n, 1) \in T, (1, n) \in S \Rightarrow (n, n) \in ToS, n \in Z$

i.e., $ToS = Z \times Z$

(*ii*) To find SoT, here $(1, n) \in S, (n, 1) \in T$

$\Rightarrow$ $(1, 1) \in SoT$

$\therefore$ $SoT = \{(1, 1)\}$.

Example 29. *Consider the digraph given below Fig. 2.8. Write relation as set of ordered pairs and check for equivalence or partial ordering.*

Sol. From the Fig., $R = \{(a, b), (b, a), (a, c), (c, c), (c, d), (b, d)\}$

Reflexive. $(a, a) \notin R$ $\therefore$ R is not reflexive.

Symmetric. $(a, c) \in R$, but $(c, a) \notin R$ $\therefore$ R is not symmetric.

Antisymmetric. $(a, b) \in R$, $(b, a) \in R$, but $a \neq b$ always.

$\therefore$ R is not antisymmetric.

Transitive. $(a, c) \in R$, $(c, d) \in R$, but $(a, d) \notin R$

$\therefore$ R is not transitive.

Hence R is neither an equivalence relation nor a partial ordering relation.

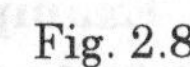

Fig. 2.8

Example 30. *Let A = (1, 2, 3, 4, 5, 6, 7, 8) and R be a relation on A defined by "(x, y) ∈ R iff y is divisible by x". Is R equivalence relation ? Is R partial order relation ? Draw its digraph.*

Sol. Here $R = \{(1, 1), (1, 2), (1, 3), (1, 4), (1, 5), (1, 6), (1, 7), (1, 8), (2, 2), (2, 4), (2, 6), (2, 8), (3, 3), (3, 6), (4, 4), (4, 8), (5, 5), (6, 6), (7, 7), (8, 8)\}$

Reflexive. $(a, a) \in R$ for all $a \in R$

$\therefore$ R is reflexive.

Antisymmetric. R is antisymmetric.

Transitive. R is transitive.

Hence, R is a partial order relation. But R is not symmetric. $(1, 2) \in R$ but $(2, 1) \notin R$.

$\therefore$ R is not an **equivalence** relation.

Digraph of R. It is shown in Fig. 2.9

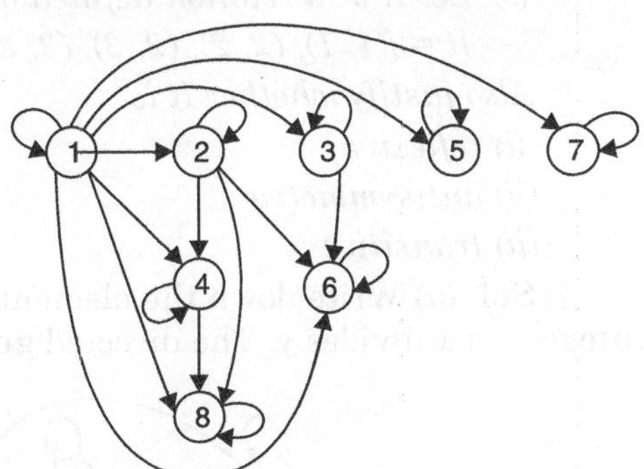

Fig. 2.9

Example 31. *Given A = {1, 2, 3, 4} and define R = {(1, 1), (2, 2), (2, 3), (3, 2), (4, 2), (4, 4)}*

(*a*) *Draw its directed graph*

(*b*) *Is R (i) reflexive (ii) symmetric (iii) transitive or (iv) antisymmetric*

(*c*) *Find* $R^2 = RoR$.

Sol. (*a*) The directed graph of *R* is shown below (Fig. 2.10).

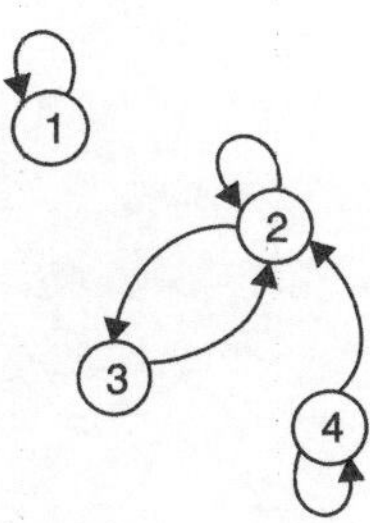

Fig. 2.10

(*b*) (*i*) $3 \in A$, but $(3, 3) \notin R$ $\therefore$ R is not reflexive.

(*ii*) $(4, 2) \in R$, but $(2, 4) \notin R$ $\therefore$ R is not symmetric.

(*iii*) $(4, 2) \in R$, $(2, 3) \in R$, but $(4, 3) \notin R$ $\therefore$ R is not transitive.

(*iv*) $(2, 3) \in R$, $(3, 2) \in R$, but $2 \neq 3$

$\therefore$ R is not antisymmetric.

(c) Here $R = \{(1, 1), (2, 2), (2, 3), (3, 2), (4, 2) (4, 4)\}$

$R = \{(1, 1), (2, 2), (2, 3), (3, 2), (4, 2), (4, 4)\}$

$(1, 1) \in R$ and $(1, 1) \in R \Rightarrow (1, 1) \in RoR$

$(2, 2) \in R$ and $(2, 2) \in R \Rightarrow (2, 2) \in RoR$

$(2, 2) \in R$ and $(2, 3) \in R \Rightarrow (2, 3) \in RoR$

$(3, 2) \in R$ and $(2, 3) \in R \Rightarrow (3, 3) \in RoR$

$(3, 2) \in R$ and $(2, 2) \in R \Rightarrow (3, 2) \in RoR$

$(4, 2) \in R$ and $(2, 3) \in R \Rightarrow (4, 3) \in RoR$

$(4, 4) \in R$ and $(4, 2) \in R \Rightarrow (4, 2) \in RoR$

$(4, 4) \in R$ and $(4, 4) \in R \Rightarrow (4, 4) \in RoR$

$\therefore$ $R^2 = RoR = \{(1, 1), (2, 2), (2, 3), (3, 2), (3, 3), (4, 2), (4, 3), (4, 4)\}$.

Example 32. *Let S be a relation on X = {a, b, c, d, e, f} defined by*

S = {(a, b), (b, b), (b, c), (c, f), (d, b), (e, a), (e, b), (e, f)}.

Draw the directed graph of S.

Sol. Write down the elements of X

Draw an arrow from the letter x to the letter y if $(x, y) \in S$.

The required directed graph of S is shown in the Fig. 2.11

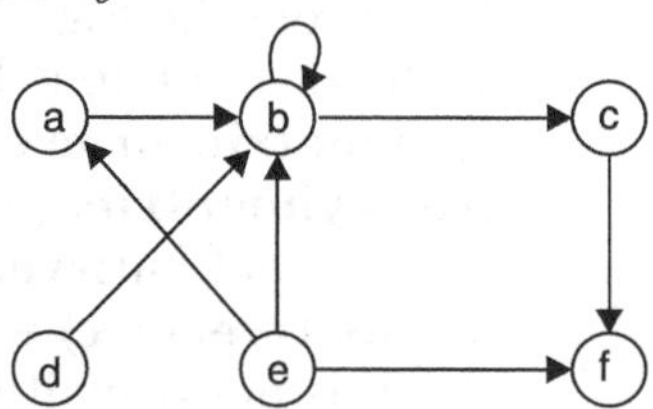

Fig. 2.11

Example 33. (*a*) *Let A = {1, 2, 3, 4, 6} and R be the relation x | y (read as x divides y) defined by*

R = {(1, 1), (1, 2), (1, 3), (1, 4), (1, 6), (2, 2), (2, 4), (2, 6), (3, 3), (3, 6), (6, 6)}.

Draw the directed graph of R

(*b*) *Let R = {(1, 1), (2, 2), (5, 5), (1, 2), (1, 3), (1, 4), (2, 1), (4, 2), (3, 5), (3, 4)} be a relation on the set A = {1, 2, 3, 4, 5}. Represent R as a directed graph* (P.T.U. B.Tech Dec 2013)

(*c*) *Let R be a relation defined on the set A given by*

R = {(1, 1), (2, 2), (2, 3), (3, 2), (4, 2), (4, 4)}. Draw its diagraph.

Also justify whether R is

(*i*) *reflexive*

(*ii*) *antisymmetric*

(*iii*) *transitive* (P.T.U. B.Tech Dec 2010)

Sol. (*a*) Write down the elements 1, 2, 3, 4, 6. Draw an arrow from the integer x to the integer y if x divides y. The directed graph is shown in Fig. 2.12(*a*)

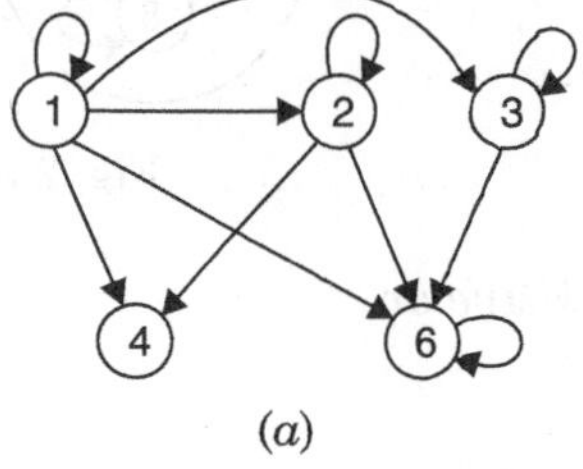

(*a*)

$$\begin{array}{c} \\ 1 \\ 2 \\ 3 \\ 4 \\ 6 \end{array} \begin{array}{c} \begin{array}{ccccc} 1 & 2 & 3 & 4 & 6 \end{array} \\ \begin{pmatrix} 1 & 1 & 1 & 1 & 1 \\ 0 & 1 & 0 & 1 & 1 \\ 0 & 0 & 1 & 0 & 1 \\ 0 & 0 & 0 & 1 & 0 \\ 0 & 0 & 0 & 0 & 1 \end{pmatrix} \end{array}$$

(*b*)

(*b*) The required diagraph is shown in Fig. 2.12(*c*)

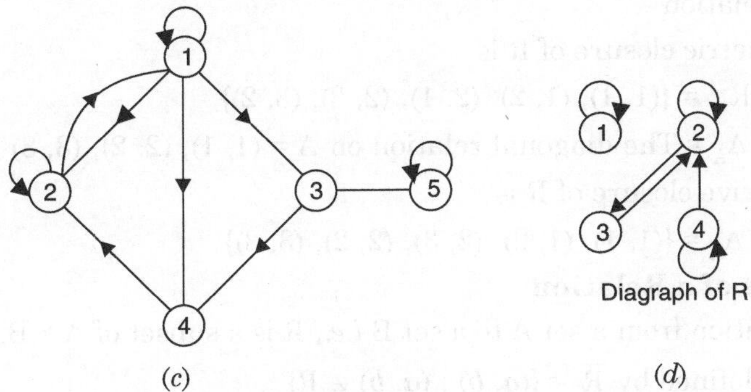

(*c*) (*d*)

Fig. 2.12

(*c*) Given R = {(1, 1), (2, 2), (2, 3), (3, 2), (4, 2), (4, 4)}. The diagraph of R is shown in Fig. 2.12(*d*):

(*i*) R is not relflexive since (3, 2) ∈ R

(*ii*) R is not antisymmetric since (4, 2) ∈ R, (2, 2) ∈ R, but 4 ≠ 2

(*iii*) R is transitive

Example 34. *Find the transitive closure R_T of the relation R on A = {1, 2, 3, 4} defined by the directed graph as shown in Fig. 2.13.*

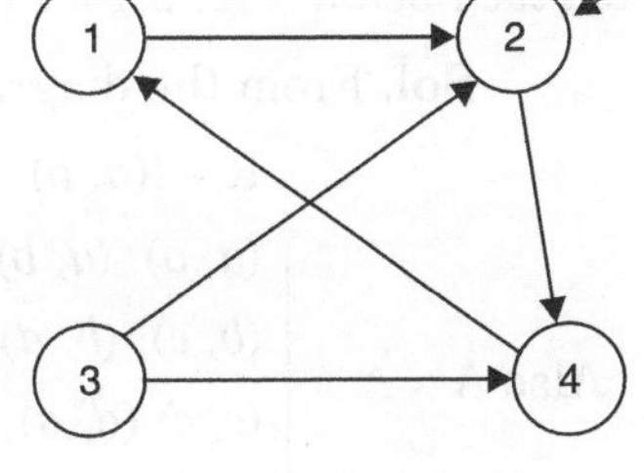

Fig. 2.13

Sol. From the directed graph,

1R2, 2R2, 2R4, 4R1, 4R3, 3R2, 3R4

∴ R = {(1, 2), (2, 2), (2, 4), (4, 1), (4, 3), (3, 2), (3, 4)}

Now (2, 4) ∈ R, (4, 1) ∈ R ⇒ (2, 1) ∉ R

∴ We add (2, 1) in R.

Also (1, 2) ∈ R, (2, 1) ∈ R ⇒ (1, 1) ∉ R

∴ We add (1, 1) in R

(1, 2) ∈ R, (2, 4) ∈ R ⇒ (1, 4) ∉ R ∴ We add (1, 4) in R

(1, 4) ∈ R, (4, 3) ∈ R ⇒ (1, 3) ∉ R ∴ We add (1, 3) in R

(2, 1) ∈ R, (1, 3) ∈ R ⇒ (2, 3) ∉ R ∴ We add (2, 3) ∈ R

(3, 2) ∈ R, (2, 1) ∈ R ⇒ (3, 1) ∉ R ∴ We add (3, 1) in R

(3, 1) ∈ R, (1, 3) ∈ R ⇒ (3, 3) ∉ R ∴ We add (3, 3) in R

(4, 1) ∈ R, (1, 2) ∈ R ⇒ (4, 2) ∉ R ∴ We add (4, 2) in R

(4, 1) ∈ R, (1, 4) ∈ R ⇒ (4, 4) ∉ R ∴ We add (4, 4) in R

∴ The transitive closure of R is given by

R_T = {(1, 2), (2, 2), (2, 4), (4, 1), (4, 3), (3, 2), (3, 4), (2, 1), (1, 1), (1, 4), (1, 3), (2, 3), (3, 1), (3, 3), (4, 2), (4, 4)}

The number of order pairs in R_T is 16

∴ $R_T = A \times A$.

Example 35. (*a*) *Let R be a relation on a set A. Give a procedure to find the symmetric and reflexive closure of R.*

(*b*) *Using the procedure of part (a), find the reflexive closure of R and symmetric closure of R where A = {1, 2, 3} and R = {(1, 1), (1, 2), (2, 3)}.*

Sol. (*a*) The symmetric closure of R is $R \cup R^{-1}$. The reflexive closure of R is $R \cup \Delta_A$ where Δ_A is the diagonal relation

(*b*) The symmetric closure of R is

$$R \cup R^{-1} = \{(1, 1), (1, 2), (2, 1), (2, 3), (3, 2)\}$$

Also Δ_A = The diagonal relation on A = (1, 1), (2, 2), (3, 3)

∴ The reflexive closure of R is

$$R \cup \Delta_A = \{(1, 1), (1, 2), (2, 3), (2, 2), (3, 3)\}$$

Complement of a Relation

Let R is a relation from a set A to a set B *i.e.*, R is a subset of A × B. The complement of R, denoted by $\overline{R}$ is defined by $\overline{R} = \{(a, b) : (a, b) \notin R\}$

Example 36. *Consider the sets A = {1, 2, 3, 4} and B {a, b, c}.*

Let $R = \{(1, a), (1, b), (2, b), (2, c), (3, b), (4, b)\}$. *Find* $\overline{R}$

Sol. A × B = {(1, *a*), (1, *b*), (1, *c*), (2, *a*), (2, *b*), (2, *c*), (3, *a*), (3, *b*), (3, *c*), (4, *a*), (4, *b*), (4, *c*)}

∴ $\overline{R}$ = A × B – R. The set of all ordered pairs in A × B but not in R

= {(1, *c*), (2, *a*), (3, *a*), (3, *c*), (4, *b*), (4, *c*)}.

Example 37. *Let A = {a, b, c, d, e} and R be a relation on A whose corresponding diagraph is given below Fig. 2.14. Find* $\overline{R}$.

Sol. From the diagraph,

R = {(*a*, *b*), (*a*, *e*), (*a*, *d*), (*b*, *c*), (*b*, *d*), (*b*, *e*), (*c*, *c*), (*d*, *d*), (*e*, *e*)}

$$\text{Also } A \times A = \left\{\begin{array}{l}(a, a), (a, b), (a, c), (a, d), (a, e), (b, a), (b, b),\\ (b, c), (b, d), (b, e), (c, a), (c, b), (c, c), (c, d),\\ (c, e), (d, a), (d, b), (d, c), (d, d), (d, e), (e, a),\\ (e, b), (e, c), (e, d), (e, e)\end{array}\right\}$$

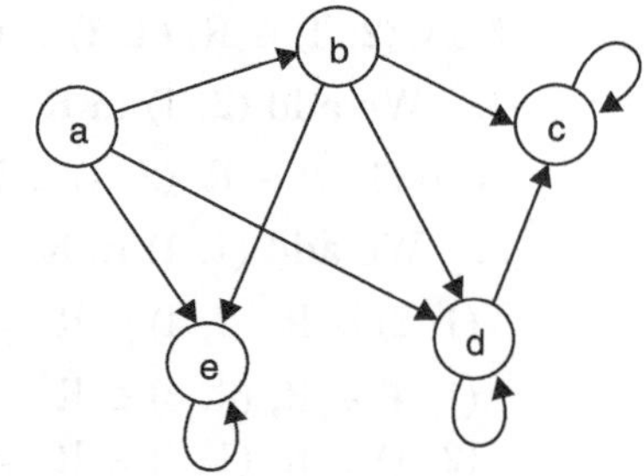

Fig. 2.14

∴ $\overline{R}$ = A × A – R. The set of all ordered pairs in A × A but not in R

= {(*a*, *a*), (*a*, *c*), (*b*, *a*), (*b*, *b*), (*c*, *a*), (*c*, *b*), (*c*, *d*), (*c*, *e*), (*d*, *a*), (*d*, *b*), (*d*, *c*), (*d*, *e*), (*e*, *a*), (*e*, *b*), (*e*, *c*), (*e*, *d*)}

TEST YOUR KNOWLEDGE 2.1

SHORT ANSWER TYPE QUESTIONS

1. What is a congruent relation ?
2. Define an equivalence relation with the help of an example.
3. What is a relation ? Give example.
4. If R is a relation on a finite set A having *n* elements. What will be the number of relations on A ?
5. (*a*) If A = (1, 2, 3, 4) and R = {(2, 2), (3, 3), (4, 4), (1, 2)} be a relation on A. Examine whether R is symmetric, transitive or reflexive.

(*b*) Let R = {(1, 1), (1, 2), (2, 1), (2, 2), (3, 3)} be a relation on the set S = {1, 2, 3}. Prove that R is an equivalence relation on S. (*P.T.U. B.Tech. Dec 2013*)

6. (*a*) Consider the relation '≤' on the set A = (2, 3, 4, 5). Find the inverse relation on A.

(*b*) Consider a set A of human beings and define a relation as "is a brother of" on A. Then which of the following is true.

(*i*) Transitive, symmetric and reflexive

(*ii*) Transitive, but neither reflexive nor symmetric

(*iii*) Transitive and reflexive but not symmetric

(*iv*) Neither reflexive, nor transitive nor symmetric.

7. (*a*) Examine whether the relation 'Divides' on the set of natural number is a partial order relation. Is it equivalence relation ? (*P.T.U., B.Tech. Dec. 2009*)

(*b*) Prove that the relation '≥' on the real numbers is not an equivalence relation. (*P.T.U., M.C.A. Dec. 2005*)

(*c*) Let A be a set of integers and R be a relation on A × A defined by (a, b) R $(c, d) \Rightarrow a + d = b + c$. Prove that R is an equivalence relation. (*P.T.U. B.Tech. May 2008, M.C.A. Dec. 2005*)

8. Which of the following collections of subsets are partitions of [1, 2, ... 6] ?

(*a*) {1, 2}, {2, 3, 4}, {4, 5, 6}
(*b*) {1}, {2, 3, 6}, {4}, {5}
(*c*) {2, 4, 6}, {1, 3, 5}
(*d*) {1, 4, 5}, {2, 6}.

9. List all partitions of A = {1, 2, 3}.

LONG ANSWER TYPE QUESTIONS

10. Consider A = {4, 5, 6, 7, 8}. Give an example of a relation which is

(*i*) Reflexive, transitive but not symmetric

(*ii*) Symmetric and antisymmetric.

(*iii*) Anti-symmetric but not reflexive

(*iv*) Neither symmetric nor anti-symmetric

(*v*) Neither symmetric, asymmetric, nor antisymmetric.

11. Consider A = (1, 2, 3, 4) and R be a relation defined by
R = {(1, 1), (1, 2), (2, 1), (2, 2), (3, 3), (3, 4), (4, 3), (4, 4)}. Determine whether or not R is

(*i*) Reflexive
(*ii*) Irreflexive
(*iii*) Symmetric
(*iv*) Asymmetric
(*v*) Antisymmetric
(*vi*) Transitive.

12. Consider A = {1, 2, 3}. Give an example of a relation on A which is

(*i*) Reflexive, symmetric, transitive and antisymmetric.

(*ii*) Neither symmetric nor antisymmetric

(*iii*) Reflexive, symmetric and not transitive

(*iv*) Not reflexive, not symmetric, not antisymmetric and not transitive.

(*v*) Symmetric and transitive but not reflexive

(*vi*) Neither reflexive nor symmetric

(*vii*) Reflexive but not symmetric.

13. Let A = (1, 2, 3, 4) and R = {(2, 1), (2, 3), (3, 2), (3, 3), (2, 2), (4, 2)}. Find

(*a*) Reflexive closure of R
(*b*) Symmetric closure of R.

14. Let R and S be relations on a set A. Determine whether each of the following statements is true or not. If not true, give counter example

(*a*) If R and S are transitive, then R ∪ S is transitive

(*b*) If R and S are transitive, then R ∩ S is transitive.

(*c*) If R is transitive, then R^{-1} is transitive.

(*d*) If R and S are reflexive, then RoS is reflexive

(*e*) If R is antisymmetric, then R^{-1} is antisymmetric.

15. If A = (*a*, *b*, *c*) and R = {(*a*, *a*), (*b*, *c*), (*a*, *b*)} be a relation on R. Find the transitive closure of R.

16. Consider a set A = {1, 3, 5}. Let R be a relation defined as *x*R*y* iff $y = x + 2$, and S be a relation defined as *x*S*y* iff $x \le y$ find

(*i*) RoS (*ii*) SoR (*iii*) Is RoS = SoR ?

17. Consider the following relations on the set N of positive integers:

R: *x* is greater than *y*, S: $x + y = 10$;

$$T = x + 4y = 10$$

(*a*) Determine which of the relations are reflexive

(*b*) Determine which of the relations are symmetric

(*c*) Determine which of the relations are antisymmetric.

(*d*) Determine which of the relations are transitive. | **Hint.** See Example 10

18. Give examples of relations R on A = {1, 2, 3} having the stated properly

(*a*) R is both symmetric and antisymmetric

(*b*) R is neither symmetric nor antisymmetric

(*c*) R is transitive but R ∪ R^{-1} is not transitive.

19. (*i*) Let A be a set of non-zero integers and let ~ be the relation on A × A defined by (*a*, *b*) ~ (*c*, *d*) if *ad* = *bc*. Prove that ~ is an equivalence relation.

(*ii*) Show that the relation ⊆ of set inclusion is not an equivalence relation.

(*iii*) Let R be the relation on the set of positive integers defined by R = {(*a*, *b*): *a* + *b* is even}. Is R an equivalence relation ?

(*iv*) Consider the relation ⊥ of perpendicularity on the set L of lines in the Euclide an plane. Is ⊥ an equivalence relation ?

20. Prove the following :

(*a*) If R and S are reflexive relations on a set A, then R ∩ S is also reflexive

(*b*) If R and S are symmetric relations on a set A, then R ∩ S is also symmetric

(*c*) If R is reflexive relation on A. Then R^{-1} and R ∪ S are reflexive for any relation S on A

(*d*) Show, by a counter example, that R and S may be transitive relations on A, but R ∪ S need not be transitive.

(*e*) If R is any relation on A, show that R ∪ R^{-1} is symmetric.

21. Let A = {*a*, *b*, *c*, *d*}, B = {1, 2, 3} and C = {*w*, *x*, *y*, *z*}. Consider the relations R from A to B and S from B to C defined by

R = {(*a*, 3), (*b*, 3), (*c*, 1), (*c*, 3), (*d*, 2)}

S = {(1, *x*), (2, *y*), (2, *z*)}

Find RoS.

22. Let A, B, C, D be any four sets. Suppose R is a relation from A to B, S is a relation from B to C and T is a relation from C to D. Then show that (RoS)oT = Ro(SoT).

23. (*a*) Show that union of two equivalence relations need not be an equivalence relation.

(P.T.U. M.C.A. Dec. 2005, 2006, May 2003)

(*b*) Consider Z, the set of integers and an integer $m > 1$. We say that x is congruent to y modulo m (written as $x \equiv y(\text{mod. } m)$. Show that this defines an equivalence relation on Z.

24. Let R be the relation on the positive integers **N** defined by the equation $x + 3y = 12$; that is, $R = \{(x, y) : x + 3y = 12\}$

(*a*) Write R as a set of ordered pairs.

(*b*) Find : (*i*) domain of R, (*ii*) range of R, and (*iii*) R^{-1}.

(*c*) Find the composition relation RoR.

25. Consider a relation whose directed graph is shown in the following (Fig. 2.15). Determine its inverse R^{-1} and its complement $\overline{R}$. Also draw the directed graphs of R^{-1} and $\overline{R}$.

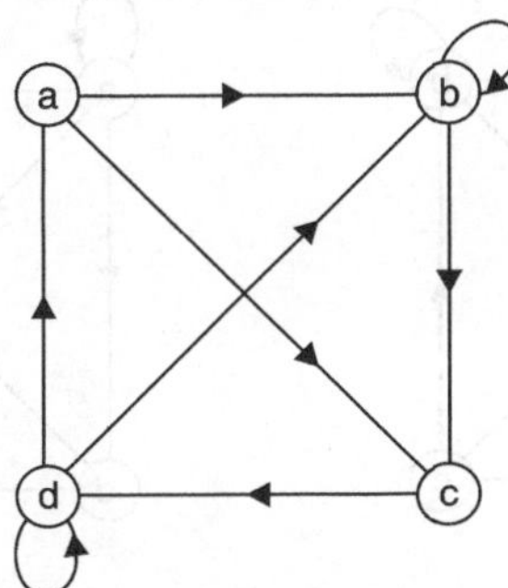

Fig. 2.15

26. Let R = (1, 2), (2, 3), (3, 3) and A = [1, 2, 3]. Find the reflexive, symmetric and transitive closure of R, using composition of relation R.

Answers

4. 2^{n^2} **5.** Reflexive and transitive.

6. (*a*) (2, 2) (3, 2), (4, 2), (5, 2), (3, 3), (4, 3), (5, 3), (4, 4), (5, 4), (5, 5).
(*b*) (*iii*). **7.** (*a*) Yes, No

8. (*b*) and (*c*) are partitions.

9. {{1}, {2}, {3}}, {{1}, {2, 3}}, {{2}, {1, 3}}, {{3}, {1, 2}}, {{1, 2, 3}}.

10. (*i*) {(4, 4), (5, 5), (6, 6), (7, 7), (8, 8), (4, 5), (5, 6), (4, 6)}
(*ii*) {(4, 4), (5, 5), (6, 6), (7, 7)} (*iii*) {(4, 4), (5, 5), (6, 6)}
(*iv*) {(4, 4), (4, 3), (3, 4), (4, 2)} (*v*) {(4, 4), (4, 5), (5, 4), (4, 6)}

11. R is reflexive, symmetric, transitive. R is not irreflexive, asymmetric, antisymmetric.

12. (*i*) R = {(1, 1), (2, 2), (3, 3)} (*ii*) R = {(1, 2), (2, 1), (2, 3)}
(*iii*) R = {(1, 1), (2, 2), (3, 3), (1, 2), (2, 3), (2, 1), (3, 2)}
(*iv*) R = {(1, 1), (2, 2), (2, 3), (3, 2)} (*v*) R = {(1, 2), (2, 1), (1, 1), (3, 2)}
(*vi*) R = {(1, 1), (2, 3), (3, 3)} (*vii*) R = {(1, 1), (2, 2), (3, 3), (2, 3)}

13. (*a*) R_R = Reflexive closure of R = {(2, 1), (2, 3), (3, 2), (3, 3), (2, 2), (4, 2), (1, 1), (4, 4)}
(*b*) R_S = Symmetric closure of R = {(2, 1), (2, 3), (3, 2), (3, 3), (2, 2), (4, 2) (1, 2), (2, 4)}

14. (*a*) Not true, take R = {(1, 2)}, S = {(2, 3)} (*b*) True (*c*) True (*d*) True (*e*) True.

15. R_T = The transitive closure of R = {(*a*, *a*), (*b*, *c*), (*a*, *b*), (*b*, *b*), (*c*, *c*), (*c*, *b*), (*b*, *a*)}

16. (*i*) RoS = {(1, 3), (1, 5), (3, 5)}
(*ii*) SoR = {(1, 3), (1, 5), (3, 5)}
(*iii*) RoS = SoR. But in general, RoS ≠ SoR

17. (*a*) None (*b*) S is symmetric
(*c*) R and T (*d*) R and T.

18. (*a*) R = {(1, 1), (2, 2)} (*b*) R = {(1, 2), (2, 1), (2, 3)}
(*c*) R = {(1, 2)}

21. RoS = {(*c*, *x*), (*d*, *y*), (*d*, *z*)}.

24. (*a*) (9, 1), (6, 2), (3, 3)
(*b*) (*i*) {9, 6, 3} (*ii*) (1, 2, 3) (*iii*) {(1, 9), (2, 6), (3, 3)}
(*c*) {3, 3}.

25. (*i*) R^{-1} = [(*b*, *a*), (*c*, *a*), (*b*, *b*), (*c*, *b*), (*d*, *c*), (*d*, *d*), (*a*, *d*), (*b*, *d*)]

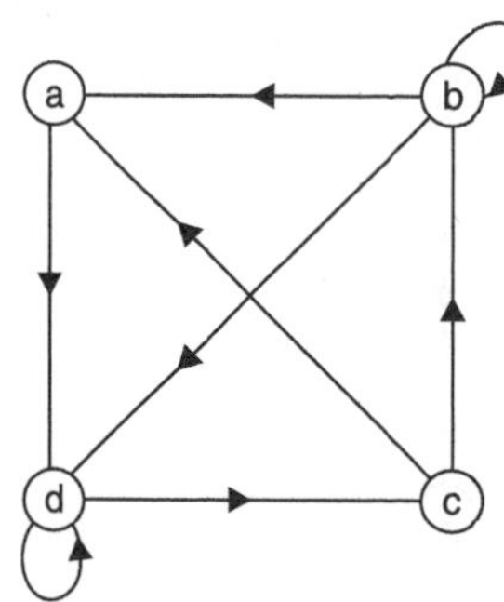

Directed graph of R^{-1}

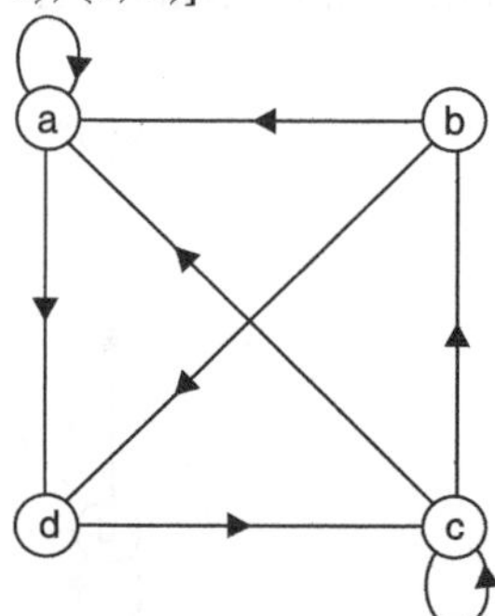

Directed graph of $\overline{R}$

(*ii*) $\overline{R}$ = [(*a*, *a*), (*a*, *d*), (*b*, *d*), (*b*, *a*), (*c*, *a*), (*c*, *b*), (*c*, *c*), (*d*, *c*)]

26. R_R = [(1, 1), (2, 2), (3, 3), (1, 2), (2, 3), (3, 1)]
R_S = [(1, 2), (2, 1), (2, 3), (3, 2), (3, 3)]
R_T = [(1, 2), (2, 3), (3, 3), (1, 3)]

Hints

19. (*i*) We must show that ~ is reflexive, symmetric, and transitive.
(*a*) *Reflexivity* : We have (*a*, *b*) ~ (*a*, *b*) since *ab* = *ba*. Hence ~ is reflective.
(*b*) *Symmetry* : Suppose (*a*, *b*) ~ (*c*, *d*). Then *ad* = *bc*. Accordingly, *cb* = *da* and hence (*c*, *d*) = (*a*, *b*). Thus, ~ is symmetric.
(*c*) *Transitivity* : Suppose (*a*, *b*) ~ (*c*, *d*) and (*c*, *d*) ~ (*e*, *f*). Then *ad* = *bc* and *cf* = *de*. Multiplying corresponding terms of the equations gives (*ad*) (*cf*) = (*bc*) (*de*). Canceling $c \neq 0$ and $d \neq 0$ from both sides of the equation yields *af* = *be*, and hence (*a*, *b*) ~ (*e*, *f*). Thus ~ is transitive. Accordingly, ~ is an equivalence relation.

20. (*a*) Let $a \in A$. Since R and S are reflexive,
$\therefore$ $(a, a) \in R$ and $(a, a) \in S$ $\Rightarrow$ $(a, a) \in R \cap S$
Hence $R \cap S$ is reflexive.
(*b*) Let $(a, b) \in R \cap S$ $\Rightarrow$ $(a, b) \in R$ and $(a, b) \in S$
But R and S are symmetric, therefore, $(b, a) \in R$ and $(b, a) \in S$ $\Rightarrow$ $(b, a) \in R \cap S$
Hence $R \cap S$ is symmetric.
(*c*) Let $a \in A$. Since R is reflexive, $\therefore$ $(a, a) \in R$
$\Rightarrow$ $(a, a) \in R^{-1}$ and $(a, a) \in R \cup S$ for any relation S on A. Hence R^{-1} and $R \cup S$ is reflexive
(*d*) Take R = (1, 2), S = (2, 3), then R and S are transitive. But
$R \cup S$ = {(1, 2), (2, 3)} is not transitive
Since $(1, 2) \in R \cup S$, $(2, 3) \in R \cup S$ $\Rightarrow$ $(1, 3) \notin R \cup S$
(*e*) Let $(a, b) \in R \cup R^{-1}$. If $(a, b) \in R$ $\Rightarrow$ $(b, a) \in R^{-1}$
Hence $(b, a) \in R \cup R^{-1}$. $\therefore$ $R \cup R^{-1}$ is symmetric.

2.18. EQUIVALENCE RELATIONS AND PARTITIONS

Let S be non-empty set. By partition P of S, we mean a finite collection $\{A_i\}_{j=1}^{n}$ of non-empty subsets of S such that,

(*i*) $A_i \cap A_J = \phi \ \forall \ i \neq j$ *i.e.*, the subsets A_i's of S are mutually disjoint

(*ii*) $\bigcup_{i=1}^{n} A_i = S$

In other words, a partition P of S is a subdivision of S into disjoint non-empty sets. Further, suppose that R be an equivalence relation on S. For each $a \in S$, let $[a]$ denotes the set of elements of S to which a is related under R. Thus,

$$[a] = \{x \in S: (a, x) \in R\} \text{ or }$$
$$= \{x \in S: aRx\}$$

The set $[a]$ is called *equivalence class* if a in S.

The collection of all equivalence classes of elements of S under an equivalence relation R is denoted by S/R. Thus,

$$S/R = \{[a] : a \in S\} \text{ where}$$

S/R (to be read as the quotient set of S by R) is called the **quotient set.**

Theorem. *Let R be an equivalence relation on a set S and let S/R denotes the quotient set of S by R. Then, the quotient set S/R is a partition of S. Also,*

(*i*) $a \in [a] \ \forall \ a \in S$

(*ii*) $[a] = [b]$ *iff* $(a, b) \in R$

(*iii*) *If* $[a] \neq [b]$, *then* $[a]$ *and* $[b]$ *are disjoint.*

Proof. (*i*) Since R is an equivalence relation on S, it must be reflexive. Thus, for every $a \in S$, $(a, a) \in R \Rightarrow a \in [a]$ | by definition of $[a]$

(*ii*) Let $(a, b) \in R$ and we show $[a] = [b]$

Let $x \in [b] \Rightarrow (b, x) \in R$. Also $(a, b) \in R$ and since R is an equivalence relation on S, it must be transitive.

Hence $(a, b) \in R, (b, x) \in R \Rightarrow (a, x) \in R$

$\Rightarrow x \in [a] \Rightarrow [b] \subseteq [a]$ ...(1)

To show $[a] \subseteq [b]$, let $(a, b) \in R$

$\Rightarrow (b, a) \in R$ | R is symmetric

$\Rightarrow a \in [b] \Rightarrow [a] \subseteq [b]$ ...(2)

Hence, combining (1) and (2), we have

$$[a] = [b]$$

Conversely, if $[a] = [b]$, then by using part (*i*),

we have $b \in [b] = [a] \Rightarrow b \in [a]$

$\Rightarrow (a, b) \in R$

(*iii*) We prove that if $[a] \cap [b] \neq \phi$, then

$$[a] = [b]$$

Since $[a] \cap [b] \neq \phi \Rightarrow \exists$ an element $x \in S$ such that

$x \in [a] \cap [b] \Rightarrow x \in [a]$ and $x \in [b]$.

Now $x \in [a] \Rightarrow (a, x) \in R$

Similarly, $(b, x) \in R$

But R is symmetric, therefore,

$$(b, x) \in R \Rightarrow (x, b) \in R$$

Again, $(a, x) \in R$, $(x, b) \in R$ and R is transitive, we must have

$$(a, b) \in R \Rightarrow [a] = [b] \qquad | \text{ Using Part } (ii)$$

ILLUSTRATIVE EXAMPLES

Example 1. *Let A = (1, 2, 3, 4, 5, 6) and R be an equivalence relation on A defined by R = {(1, 1), (1, 5), (2, 2), (2, 3), (2, 6), (3, 2), (3, 3), (3, 6), (4, 4), (5, 1), (6, 2), (6, 3), (6, 6)}.*

Find the equivalence classes of R (i.e., the partition of A induced by R). Also find A/R.

Sol. For each $a \in A$, let $[a]$ denotes the set of those elements of A to which a is related under R. Thus,

$$[a] = \{x \in A: (a, x) \in R\}$$

Here, the set $[a]$ is called *equivalence class of a in* A.

Start with an element '1' of A.

i.e., $$[1] = \{x \in A : (1, x) \in R\}$$

Clearly, (1, 1) and (1, 5) both belong to R. *i.e.,* the elements 1 and 5 are related by '1'

$$\therefore \qquad [1] = \{1, 5\} = A_1, \text{ say.}$$

Next, pick an element from A which does not belong to [1], these are 2, 3, 4 or 6.

Consider an element, '2', say, which dose not belong to [1]. We find [2].

The elements related to 2 are 2, 3 and 6 (since (2, 2), (2, 3) and (2, 6) all belong to R)

Therefore, $$[2] = \{2, 3, 6\} = A_2, \text{ say}$$

We next, pick an element which does not belong to [1] or [2]. Clearly, 4 is the only element. Now, the elements related to 4 is 4. (since $(4, 4) \in R$). Hence,

$$[4] = \{4\} = A_3, \text{ say}$$

The sets A_1, A_2, A_3 form a partition of A since A_1, A_2 and A_3 are mutually disjoint sets and $A_1 \cup A_2 \cup A_3 = A$.

Finally,

A/R = The set of all equivalence classes of elements of A under an equivalence relation R

$= [\{1\}, \{2\}, \{4\}]$

Example 2. *Consider the set W = {sheet, last, sky, wash, wind, sit}. Find W/R where R is the equivalence relation on W defined by "has the same number of letters as"*

Sol. For each $w \in W$, let $[w]$ denotes the set of all those elements of W to which w is related under R. Thus,

$$[w] = \{x \in W: (w, x) \in R\}$$

Here, the set $[w]$ is called equivalence class of w in W.

Consider the set

$$W = \{w_1, w_2, w_3, w_4, w_5, w_6\} \text{ where}$$
$$w_1 = \text{"sheet"},\ w_2 = \text{"last"},\ w_3 = \text{"sky"},$$
$$w_4 = \text{"wash"},\ w_5 = \text{"wind" and } w_6 = \text{"sit"}$$

Start with an element, say, w_1 of W. Here $[w_1]$ denotes the set of those elements of W to which 'w_1' is related under R.

Now, the element 'w_1' has 5 letters and the only element, which has same number of letters as of 'w_1' is w_1. Hence

$\{w_1\}$ = {sheet} = W_1, say,

Next, pick an element from W which does not belong to $[w_1]$, these are w_2, w_3, w_4, w_5 or w_6

Consider an element 'w_2', say, which does not belong to $[w_1]$. We find $[w_2]$.

The elements related to 'w_2' are w_2, w_4 and w_5 (since these elements have same number of letters as of 'w_2')

Therefore,

$[w_2]$ = {last, wash, wind} = W_2 say,

We next, pick an element which does not belong to $[w_1]$ or $[w_2]$. Clearly, w_3 and w_6 are the only elements.

The elements related to 'w_3' are w_3 are w_6 (since these elements have same number of letters as of 'w_3')

Thus, $[w_3]$ = {sky, sit} = W_3 say

Clearly, the sets W_1, W_2 and W_3 form a partition of W since W_1, W_2 and W_3 are mutually disjoint sets and $W_1 \cup W_2 \cup W_3 = W$. Hence

W/R = The set of all equivalence classes of elements of W under an equivalence relation R
= [{sheet}, {last, wash, wind}, {sky, sit}]

TEST YOUR KNOWLEDGE 2.2

1. Let A = {1, 2, 3, ... 9} and let '~' be the relation on A × A defined by $(a, b) \sim (c, d)$ if $a + d = b + c$
   (*a*) Prove that ~ is an equivalence relation on A
   (*b*) Find the equivalence class of (2, 5), *i.e.*, [(2, 5)]
2. Let S = {1, 2, 3, ..., 19, 20}. Let R be an equivalence relation on S defined by "$x \equiv y \pmod 5$". Find S/R, the quotient set (the partition of S induced by R).
3. Consider the set of words W = {sheet, last, sky, wash, wind, sit}
   Find W/R where R is the equivalence relation on W defined by "*begins with the same letters as*".

Answers

1. [(2, 5)] = {(1, 4), (2, 5), (3, 6), (4, 7), (5, 8), (6, 9)}
2. S/R = [{1, 6, 11, 16}, {2, 7, 12, 17}, {3, 8, 13, 18}, {4, 9, 14, 19}, {5, 10, 15, 20}]
3. W/R = [{sheet, sky, sit}, {last}, {wash, wind}]

Hints

1. (*a*) Please see example 17 page 65 in this chapter.
   (*b*) $(2, 5) \sim (c, d)$ iff $2 + d = 5 + c$ or $d - c = 5 - 2 = 3 \;\forall\; c, d \in A$
   $\therefore$ (1, 4), (2, 5), (3, 6), (4, 7), (5, 8) and (6, 9) are the only elements to which (2, 5) is related.
   $\therefore$ [(2, 5)] = {(1, 4), (2, 5), (3, 6), (4, 7), (5, 8), (6, 9)}
2. $x \equiv y \pmod 5$ (read as "x is congruent to $y \pmod 5$") means that $x - y$ is divisible by 5.
   By definition, S/R is the set of all equivalence classes of elements of S under an equivalence relation R. Therefore, we are required to find the equivalence classes of the elements of S.
   Start with an element '1' of S. The elements of S to which '1' is related under R are 1, 6, 11, 16 (since $1 \equiv 1 \pmod 5$, $1 \equiv 6 \pmod 5$, $1 \equiv 11 \pmod 5$ and $1 \equiv 6 \pmod 5$). Therefore,
   [1] = {1, 6, 11, 16} etc.
3. Please see example 2 (page 79) in this chapter.

2.19. WARSHALL'S ALGORITHM TO FIND TRANSITIVE CLOSURE

This method is to be applied for large sets and relations. This algorithm is a more efficient algorithm to compute the transitive closure.

Consider a relation R defined on a set

$A = (a_1, a_2, \ldots\ldots a_n)$. If $\{x_1, x_2, \ldots\ldots x_n\}$ is a path in R, then any vertex other than x_1 and x_n is called *interior vertex* of the path. Also for $1 < k < n$, we define a Boolean matrix W_k as follows.

The (i, j)th element of W_k is 1 iff there is a path from a_i to a_j in R whose interior vertices, if any, come from the set $\{a_1, a_2, \ldots\ldots a_n\}$. In other words, $W_n = M_R^{\infty}$.

If we define $W_0 = M_R$, then we can find a sequence $W_0, W_1, W_2, \ldots\ldots W_n$ whose first term is $W_0 = M_R$ and the last term is $W_n = M_R^{\infty}$.

Each matrix W_k can be computed from W_{k-1} using the following algorithm (**Warshall's algorithm**).

Step I. First transfer all 1's in W_{k-1} to W_k.

Step II. List the locations $p_1, p_2, \ldots$ in column k of W_{k-1}, where the entry is 1 and locations $a_1, a_2, \ldots\ldots$ in row k of w_{k-1}, where the entry is 1.

Step III. Put 1's at all the positions (p_i, q_j) of W_k (if they are not already there).

Example 1. *Using Warshall's algorithm, find the transitive closure of R defined on*

$A = \{1, 2, 3, 4\}$ *and*

$R = (1, 1), (1, 4), (2, 1), (2, 2), (3, 3), (4, 4).$

Sol. If M_R denotes the matrix representation of R, then (Take $W_0 = M_R$)

$$W_0 = M_R = \begin{pmatrix} 1 & 0 & 0 & 1 \\ 1 & 1 & 0 & 0 \\ 0 & 0 & 1 & 0 \\ 0 & 0 & 0 & 1 \end{pmatrix} \text{ and } n = 4 \qquad \text{(As } M_R \text{ is a } 4 \times 4 \text{ matrix)}$$

We compute W_4 by using **Warshall's algorithm.**

For k = 1. In column 1 of W_0, 1's are at positions 1 and 2. Hence $p_1 = 1, p_2 = 2$

In row 1 of W_0, 1's are at positions 1 and 4.

Hence $q_1 = 1, q_2 = 4$. Therefore, to obtain W_1, we put 1 at the positions :

$\{(p_1, q_1), (p_1, q_2), (p_2, q_1), (p_2, q_2) = (1, 1), (1, 4), (2, 1), (2, 4)\}$. Thus

$$W_1 = \begin{pmatrix} 1 & 0 & 0 & 1 \\ 1 & 1 & 0 & 1 \\ 0 & 0 & 1 & 0 \\ 0 & 0 & 0 & 1 \end{pmatrix}$$

For k = 2. In column 2 of W_1, 1's is at positions 2. Hence $p_1 = 2$.

In row 2 of W_1, 1's are at positions 1, 2 and 4.

Hence $\quad q_1 = 1, \quad q_2 = 2, \quad q_3 = 4.$

Therefore, to obtain W_2, we put 1s at the positions:

$\{(p_1, q_1), (p_1, q_2), (p_1, q_3) = (2, 1), (2, 2), (2, 4)\}$. Thus (using W_1)

$$W_2 = \begin{pmatrix} 1 & 0 & 0 & 1 \\ 1 & 1 & 0 & 1 \\ 0 & 0 & 1 & 0 \\ 0 & 0 & 0 & 1 \end{pmatrix}$$

For k = 3. In column 3 of W_2, '1' is at position 3.

Hence $p_1 = 3$

In row 3 of W_2, '1' is at position 3. Hence $q_1 = 3$

Thus, we put '1' at the position: $\{(p_1, q_1) = (3, 3)\}$. Thus (using W_2)

$$W_3 = \begin{pmatrix} 1 & 0 & 0 & 1 \\ 1 & 1 & 0 & 1 \\ 0 & 0 & 1 & 0 \\ 0 & 0 & 0 & 1 \end{pmatrix}$$

For k = 4. In column 4 of W_3, 1's are at positions 1, 2 and 4. Hence $p_1 = 1, p_2 = 2, p_3 = 4$. In row 4 of W_3, '1' is at position 4. Hence $q_1 = 4$

Therefore, we put 1's at the positions:

$\{(p_1, q_1), (p_2, q_1), (p_3, q_1) = (1, 4), (2, 4), (4, 4)\}$. Thus (using W_3).

$$W_4 = \begin{pmatrix} 1 & 0 & 0 & 1 \\ 1 & 1 & 0 & 1 \\ 0 & 0 & 1 & 0 \\ 0 & 0 & 0 & 1 \end{pmatrix} = M_R^{\infty}$$

Hence, from the matrix M_R^{∞}, the transitive closure of R is given by

$$R^{\infty} = \{(1, 1), (1, 4), (2, 1), (2, 2), (2, 4), (3, 3), (4, 4)\}.$$

Example 2. *Let A = {1, 2, 3, 4} and let R = {(1, 2), (2, 3), (3, 4), (2, 1)}. Find the transitive closure of R using Warshall's algorithm.*

Sol. Let M_R denotes the matrix representation of R. Take $W_0 = M_R$, we have

$$W_0 = M_R = \begin{pmatrix} 0 & 1 & 0 & 0 \\ 1 & 0 & 1 & 0 \\ 0 & 0 & 0 & 1 \\ 0 & 0 & 0 & 0 \end{pmatrix} \text{ and } n = 4 \text{ (As } M_R \text{ is a } 4 \times 4 \text{ matrix)}$$

We compute W_4 by using warshall's algorithm.

For k = 1. In column 1 of W_0, '1' is at position 2. Hence $p_1 = 2$.

In row 1 of W_0, '1' is at position 2. Hence $q_1 = 2$. Therefore, to obtain W_1, we put '1' at the position: $\{(p_1, q_1) = (2, 2)\}$. Thus

$$W_1 = \begin{pmatrix} 0 & 1 & 0 & 0 \\ 1 & 1 & 1 & 0 \\ 0 & 0 & 0 & 1 \\ 0 & 0 & 0 & 0 \end{pmatrix}$$

For k = 2. In column 2 of W_1, 1's are at positions 1, 2. Hence $p_1 = 1, p_2 = 2$.

In row 2 of W_1, 1's are at positions 1, 2 and 3. Hence $q_1 = 1, q_2 = 2, q_3 = 3$
Therefore, to obtain W_2, we put '1' at the positions :
$\{(p_1, q_1), (p_1, q_2), (p_1, q_3), (p_2, q_1), (p_2, q_2), (p_2, q_3) = (1, 1), (1, 2), (1, 3), (2, 1), (2, 2), (2, 3)\}$.
Thus (using W_1)

$$W_2 = \begin{pmatrix} 1 & 1 & 1 & 0 \\ 1 & 1 & 1 & 0 \\ 0 & 0 & 0 & 1 \\ 0 & 0 & 0 & 0 \end{pmatrix}$$

For k = 3. In column 3 of W_2, 1's are at positions 1, 2. Hence $p_1 = 1, p_2 = 2$
In row 3 of W_2, '1' is at the position 4. Hence $q_1 = 4$
Therefore, to obtain W_3 we put 1's at the positions: $\{(p_1, q_1), (p_2, q_1) = (1, 4), (2, 4)\}$.
Thus (using W_2)

$$W_3 = \begin{pmatrix} 1 & 1 & 1 & 1 \\ 1 & 1 & 1 & 1 \\ 0 & 0 & 0 & 1 \\ 0 & 0 & 0 & 0 \end{pmatrix}$$

For k = 4. In column 4 of W_3, 1's are at positions 1, 2, 3. Hence $p_1 = 1, p_2 = 2, p_3 = 3$
In row 4 of W_3, '1' is at no position, and no new 1's are added and hence $M_R^\infty = W_4 = W_3$.
Thus

$$W_4 = W_3 = \begin{pmatrix} 1 & 1 & 1 & 1 \\ 1 & 1 & 1 & 1 \\ 0 & 0 & 0 & 1 \\ 0 & 0 & 0 & 0 \end{pmatrix} = M_R^\infty$$

Thus, the transitive closure of R is given as

$$R^\infty = \{(1, 1), (1, 2), (1, 3), (1, 4), (2, 1), (2, 2), (2, 3), (2, 4), (3, 4)\}.$$

Example 3. *Let A = (1, 2, 3, 4) and let R and S be the relation on A described by*

$$M_R = \begin{pmatrix} 0 & 0 & 0 & 1 \\ 0 & 0 & 0 & 0 \\ 0 & 1 & 0 & 0 \\ 0 & 0 & 1 & 0 \end{pmatrix}, M_S = \begin{pmatrix} 1 & 1 & 0 & 0 \\ 0 & 1 & 0 & 0 \\ 0 & 0 & 1 & 0 \\ 0 & 1 & 0 & 1 \end{pmatrix}$$

Use Warshall's algorithm to compute the transitive closure of $R \cup S$.

(P.T.U. B. Tech. December 2008)

Sol. Let $M_{R \cup S}$ denotes the matrix representation or $R \cup S$, then

$$M_{R \cup S} = M_R \vee M_S = \begin{pmatrix} 0 & 0 & 0 & 1 \\ 0 & 0 & 0 & 0 \\ 0 & 1 & 0 & 0 \\ 0 & 0 & 1 & 0 \end{pmatrix} \vee \begin{pmatrix} 1 & 1 & 0 & 0 \\ 0 & 1 & 0 & 0 \\ 0 & 0 & 1 & 0 \\ 0 & 1 & 0 & 1 \end{pmatrix} = \begin{pmatrix} 1 & 1 & 0 & 1 \\ 0 & 1 & 0 & 0 \\ 0 & 1 & 1 & 0 \\ 0 & 1 & 1 & 1 \end{pmatrix}$$

We compute W_4 by using Warshall's algorithm. Here $n = 4$ (As $M_{R \cup S}$ is a 4×4 matrix) and

$$W_0 = M_{R \cup S} = \begin{pmatrix} 1 & 1 & 0 & 1 \\ 0 & 1 & 0 & 0 \\ 0 & 1 & 1 & 0 \\ 0 & 1 & 1 & 1 \end{pmatrix}$$

For k = 1. In column 1 of W_0, '1' is at position 1. Hence $p_1 = 1$.
In row 1 of W_0, 1's are at positions 1, 2, and 4.
Hence $q_1 = 1$, $q_2 = 2$, $q_3 = 4$.
Therefore, to obtain W_1, we put 1's at the positions:
$\{(p_1, q_1), (p_1, q_2), (p_1, q_3) = (1, 1), (1, 2), (1, 4)\}$. Thus (using W_0)

$$W_1 = \begin{pmatrix} 1 & 1 & 0 & 1 \\ 0 & 1 & 0 & 0 \\ 0 & 1 & 1 & 0 \\ 0 & 1 & 1 & 1 \end{pmatrix} = W_0$$

For k = 2. In column 2 of W_1, 1's are at positions 1, 2, 3 and 4. Hence $p_1 = 1, p_2 = 2, p_3 = 3, p_4 = 4$
In row 2 of W_1, '1' is at position 2. Hence $q_1 = 2$
Therefore, to obtain W_2, we put 1's at the positions:
$\{(p_1, q_1), (p_2, q_1), (p_3, q_1), (p_4, q_1) = (1, 2), (2, 2), (3, 2), (4, 2)\}$
Thus (using W_1)

$$W_2 = \begin{pmatrix} 1 & 1 & 0 & 1 \\ 0 & 1 & 0 & 0 \\ 0 & 1 & 1 & 0 \\ 0 & 1 & 1 & 1 \end{pmatrix} = W_1$$

For k = 3. In column 3 of W_2, 1's are at positions 3, 4. Hence $p_1 = 3, p_2 = 4$
In row 3 of W_2, 1's are at positions 2 and 3.
Hence $q_1 = 2$, $q_2 = 3$
Therefore, to obtain W_3, we put 1's at the positions :
$\{(p_1, q_1), (p_1, q_2), (p_2, q_1), (p_2, q_2) = (3, 2), (3, 3), (4, 2), (4, 3)\}$.
Thus (using W_2)

$$W_3 = \begin{pmatrix} 1 & 1 & 0 & 1 \\ 0 & 1 & 0 & 0 \\ 0 & 1 & 1 & 0 \\ 0 & 1 & 1 & 1 \end{pmatrix} = W_2$$

For k = 4. In column 4 of W_3, 1's are at positions 1 and 4. Hence $p_1 = 1, p_2 = 4$ in row 4 of W_3, 1's are at to obtain W_4, positions 2, 3 and 4. Hence $q_1 = 2, q_2 = 3, q_3 = 4$.
Therefore, to obtain W_4 we put 1's at the positions:
$\{(p_1, q_1), (p_1, q_2), (p_1, q_3), (p_2, q_1), (p_2, q_2), (p_2, q_3) = (1, 2), (1, 3), (1, 4), (4, 2), (4, 3), (4, 4)\}$
Thus (using W_3)

$$W_4 = \begin{pmatrix} 1 & 1 & 1 & 1 \\ 0 & 1 & 0 & 0 \\ 0 & 1 & 1 & 0 \\ 0 & 1 & 1 & 1 \end{pmatrix} = M_{(R \cup S)^{\infty}}$$

Thus, the transitive closure of $R \cup S$ is given as

$$(R \cup S)^{\infty} = \{(1, 1), (1, 2), (1, 3), (1, 4), (2, 2), (3, 2), (3, 3), (4, 2), (4, 3), (4, 4)\}.$$

Example 4. *Let A = {1, 2, 3, 4, 5} and*

$$R = \{(1, 1), (1, 2), (2, 1), (2, 2), (3, 3), (3, 4), (4, 3), (4, 4), (5, 5)\}$$

$$S = \{(1, 1), (2, 2), (3, 3), (4, 4), (4, 5), (5, 4), (5, 5)\}$$

Find the transitive closure of R ∪ S by using Warshall's algorithm.

Sol. Let M_R and M_S denote the matrix representation of R and S respectively. Then

$$M_R = \begin{pmatrix} 1 & 1 & 0 & 0 & 0 \\ 1 & 1 & 0 & 0 & 0 \\ 0 & 0 & 1 & 1 & 0 \\ 0 & 0 & 1 & 1 & 0 \\ 0 & 0 & 0 & 0 & 1 \end{pmatrix}; \quad M_S = \begin{pmatrix} 1 & 0 & 0 & 0 & 0 \\ 0 & 1 & 0 & 0 & 0 \\ 0 & 0 & 1 & 0 & 0 \\ 0 & 0 & 0 & 1 & 1 \\ 0 & 0 & 0 & 1 & 1 \end{pmatrix}$$

Also $\quad M_{R \cup S} = M_R \vee M_S$

$$= \begin{pmatrix} 1 & 1 & 0 & 0 & 0 \\ 1 & 1 & 0 & 0 & 0 \\ 0 & 0 & 1 & 1 & 0 \\ 0 & 0 & 1 & 1 & 0 \\ 0 & 0 & 0 & 0 & 1 \end{pmatrix} \vee \begin{pmatrix} 1 & 0 & 0 & 0 & 0 \\ 0 & 1 & 0 & 0 & 0 \\ 0 & 0 & 1 & 0 & 0 \\ 0 & 0 & 0 & 1 & 1 \\ 0 & 0 & 0 & 1 & 1 \end{pmatrix}$$

$$= \begin{pmatrix} 1 & 1 & 0 & 0 & 0 \\ 1 & 1 & 0 & 0 & 0 \\ 0 & 0 & 1 & 1 & 0 \\ 0 & 0 & 1 & 1 & 1 \\ 0 & 0 & 0 & 1 & 1 \end{pmatrix} \qquad \left| \begin{array}{l} 1 \vee 1 = 1 \\ 0 \vee 1 = 1 \\ 1 \vee 0 = 1 \\ 0 \vee 0 = 0 \end{array} \right. \text{etc.}$$

We now compute W_5 by Warshall's algorithm. Here $n = 5$. (As $M_{R \cup S}$ is a 5×5) matrix and

$$W_0 = M_{R \cup S} = \begin{pmatrix} 1 & 1 & 0 & 0 & 0 \\ 1 & 1 & 0 & 0 & 0 \\ 0 & 0 & 1 & 1 & 0 \\ 0 & 0 & 1 & 1 & 1 \\ 0 & 0 & 0 & 1 & 1 \end{pmatrix}$$

For k = 1. In column 1 of W_0, 1's are at positions 1, 2. Hence $p_1 = 1, p_2 = 2$.

In row 1 of W_0, 1's are at positions 1, 2. Hence $q_1 = 1, q_2 = 2$

Therefore, to obtain W_1, we put 1's at the positions:

$$\{(p_1, q_1), (p_1, q_2), (p_2, q_1), (p_2, q_2) = (1, 1), (1, 2), (2, 1), (2, 2)\}$$

Thus (using W_0)

$$W_1 = \begin{pmatrix} 1 & 1 & 0 & 0 & 0 \\ 1 & 1 & 0 & 0 & 0 \\ 0 & 0 & 1 & 1 & 0 \\ 0 & 0 & 1 & 1 & 1 \\ 0 & 0 & 0 & 1 & 1 \end{pmatrix} = W_0$$

For k = 2. In column 2 of W_1, 1's are at positions 1, 2. Hence $p_1 = 1, p_2 = 2$
In row 2 of W_1, 1's are at positions 1, 2. Hence $q_1 = 1, q_2 = 2$
Therefore, to obtain W_2, we put 1's at the positions:

$\{(p_1, q_1), (p_1, q_2), (p_2, q_1), (p_2, q_2)\} = (1, 1), (1, 2), (2, 1), (2, 2)\}$.

Thus (using W_1)

$$W_2 = \begin{pmatrix} 1 & 1 & 0 & 0 & 0 \\ 1 & 1 & 0 & 0 & 0 \\ 0 & 0 & 1 & 1 & 0 \\ 0 & 0 & 1 & 1 & 1 \\ 0 & 0 & 0 & 1 & 1 \end{pmatrix} = W_1$$

For k = 3. In column 3 of W_2, 1's are at the positions 3, 4. Hence $p_1 = 3, p_2 = 4$.
In row 3 of W_2, 1's are at the positions 3 and 4. Hence $q_1 = 3, q_2 = 4$
Therefore, to obtain W_3, we put 1's at the positions:

$\{(p_1, q_1), (p_2, q_2), (p_2, q_1), (p_2, q_2) = (3, 3), (3, 4), (4, 3), (4, 4)\}$

Thus (using W_2)

$$W_3 = \begin{pmatrix} 1 & 1 & 0 & 0 & 0 \\ 1 & 1 & 0 & 0 & 0 \\ 0 & 0 & 1 & 1 & 0 \\ 0 & 0 & 1 & 1 & 1 \\ 0 & 0 & 0 & 1 & 1 \end{pmatrix} = W_2$$

For k = 4. In column 4 of W_3, 1's are at positions 3, 4, 5. Hence $p_1 = 3, p_2 = 4, p_3 = 5$.
In row 4 of W_3, 1's are at the positions 3, 4 and 5. Hence $q_1 = 3, q_2 = 4, q_3 = 5$.
Therefore, to obtain W_4, we put 1's at the positions:
$\{(p_1, q_1), (p_1, q_2), (p_1, q_3), (p_2, q_1), (p_2, q_2), (p_2, q_3), (p_3, q_1), (p_3, q_2), (p_3, q_3)$
$=(3, 3), (3, 4), (3, 5), (4, 3), (4, 4), (4, 5), (5, 3), (5, 4), (5, 5)\}$. Thus (using W_3)

$$W_4 = \begin{pmatrix} 1 & 1 & 0 & 0 & 0 \\ 1 & 1 & 0 & 0 & 0 \\ 0 & 0 & 1 & 1 & 1 \\ 0 & 0 & 1 & 1 & 1 \\ 0 & 0 & 1 & 1 & 1 \end{pmatrix}$$

For k = 5. In column 5 of W_4, 1's are at the positions 3, 4, 5. Hence $p_1 = 3, p_2 = 4, p_3 = 5$.
In row 5 of W_4, 1's are at the positions 3, 4, 5. Hence $q_1 = 3, q_2 = 4, q_3 = 5$.

This is similar to the case for $k = 4$. Hence

$$W_5 = W_4 = \begin{pmatrix} 1 & 1 & 0 & 0 & 0 \\ 1 & 1 & 0 & 0 & 0 \\ 0 & 0 & 1 & 1 & 1 \\ 0 & 0 & 1 & 1 & 1 \\ 0 & 0 & 1 & 1 & 1 \end{pmatrix} = M_{(R \cup S)^{\infty}}$$

Thus, the transitive closure of $R \cup S$ is given as

$(R \cup S)^{\infty} = \{(1, 1), (1, 2), (2, 1), (2, 2), (3, 3), (3, 4), (3, 5), (4, 3), (4, 4), (4, 5),$
$(5, 3), (5, 4), (5, 5)\}$

TEST YOUR KNOWLEDGE 2.3

1. Let A = (1, 2, 3) and let R = {(1, 1), (1 2), (2, 3), (1, 3), (3, 1), (3, 2)}. Find the matrix $\mathbf{M}_{R^\infty}$ of the transitive closure R by using Warshall's algorithm.
2. Let A = {a_1, a_2, a_3, a_4, a_5] and R be a relation on A. If

$$M_R = \begin{pmatrix} 1 & 0 & 0 & 1 & 0 \\ 0 & 1 & 0 & 0 & 0 \\ 0 & 0 & 0 & 1 & 1 \\ 1 & 0 & 0 & 0 & 0 \\ 0 & 1 & 0 & 0 & 1 \end{pmatrix} = W_0$$

Compute W_1, W_2, W_3 as in Warshall's algorithm. Also find R^∞, the transitive closure of R.
3. Let A = (1, 2, 3, 4). For the relation R whose matrix is given, find the matrix of the transitive closure by using Warshall's algorithm.

(*i*) $M_R = \begin{pmatrix} 1 & 0 & 0 & 1 \\ 1 & 1 & 0 & 0 \\ 0 & 0 & 1 & 0 \\ 0 & 0 & 0 & 1 \end{pmatrix}$ (*ii*) $M_R = \begin{pmatrix} 1 & 1 & 0 & 0 \\ 1 & 0 & 0 & 0 \\ 0 & 0 & 0 & 0 \\ 0 & 0 & 1 & 0 \end{pmatrix}$

(*iii*) $\begin{pmatrix} 1 & 0 & 0 & 1 \\ 0 & 1 & 1 & 0 \\ 0 & 1 & 1 & 0 \\ 1 & 0 & 0 & 1 \end{pmatrix}$ (*iv*) $\begin{pmatrix} 0 & 0 & 0 & 1 \\ 1 & 0 & 0 & 1 \\ 0 & 1 & 0 & 1 \\ 0 & 0 & 1 & 0 \end{pmatrix}$.

4. Let A = {*a*, *b*, *c*, *d*, *e*] and let R and S be the relations on A described by

$$M_R = \begin{pmatrix} 1 & 0 & 1 & 0 & 1 \\ 0 & 0 & 0 & 1 & 0 \\ 1 & 0 & 0 & 0 & 0 \\ 0 & 0 & 1 & 1 & 0 \\ 1 & 0 & 1 & 0 & 0 \end{pmatrix}, M_S = \begin{pmatrix} 0 & 1 & 0 & 1 & 0 \\ 1 & 1 & 0 & 0 & 1 \\ 1 & 1 & 1 & 0 & 0 \\ 0 & 1 & 0 & 0 & 0 \\ 0 & 1 & 0 & 1 & 0 \end{pmatrix}$$

Use Warshall's algorithm to compute transitive closure of R ∪ S.

Answers

1. $\mathbf{M}_{R^\infty} = \begin{pmatrix} 1 & 1 & 1 \\ 1 & 1 & 1 \\ 1 & 1 & 1 \end{pmatrix}$, where R^∞ = {(1, 1), (1, 2), (1, 3), (2, 1), (2, 2), (2, 3), (3, 1), (3, 2), (3, 3)}

2. $W_1 = W_2 = W_3 = \begin{pmatrix} 1 & 0 & 0 & 1 & 0 \\ 0 & 1 & 0 & 0 & 0 \\ 0 & 0 & 0 & 1 & 1 \\ 1 & 0 & 0 & 1 & 0 \\ 0 & 1 & 0 & 0 & 1 \end{pmatrix}$

R^∞ = {(1 1), (1, 4), (2, 2), (3, 4), (3, 5), (4, 1), (4, 4), (5, 2), (5, 5)}

3. (*i*) $\begin{pmatrix} 1 & 0 & 0 & 1 \\ 1 & 1 & 0 & 1 \\ 0 & 0 & 1 & 0 \\ 0 & 0 & 0 & 1 \end{pmatrix}$ (*ii*) $\begin{pmatrix} 1 & 0 & 0 & 1 \\ 0 & 1 & 1 & 0 \\ 0 & 1 & 1 & 0 \\ 1 & 0 & 0 & 1 \end{pmatrix}$

(iii) $\begin{pmatrix} 1 & 0 & 0 & 1 \\ 0 & 1 & 1 & 0 \\ 0 & 1 & 1 & 0 \\ 1 & 0 & 0 & 1 \end{pmatrix}$ (iv) $\begin{pmatrix} 1 & 1 & 1 & 1 \\ 1 & 1 & 1 & 1 \\ 1 & 1 & 1 & 1 \\ 1 & 1 & 1 & 1 \end{pmatrix}$

4. $(R \cup S)^\infty = \{(a, a), (a, b), (a, c), (a, d), (a, e), (b, a), (b, b), (b, c), (b, d), (b, e), (c, a), (c, b), (c, c) (c, d), (c, e), (d, a), (d, b), (d, c), (d, d), (d, e), (e, a), (e, b), (e, c), (e, d), (e, e)\}$

MULTIPLE CHOICE QUESTIONS (MCQs)

1. The binary relation S = ϕ (empty set) on set A = {1, 2, 3} is
   (*a*) Neither reflexive nor anti-symmetric (*b*) Symmetric and reflexive
   (*c*) Transitive and reflexive (*d*) Transitive and symmetric.
2. A relation R is defined on the set of integers as xRy iff $(x + y)$ is even. Which of the following statement is TRUE ?
   (*a*) R is not an equivalence relation
   (*b*) R is an equivalence relation having one equivalence class
   (*c*) R is an equivalence relation having two equivalence classes
   (*d*) R is an equivalence relation having three equivalence classes.
3. The time complexity of computing the transitive closure of a binary relation on a set of n elements is known to be
   (*a*) $O(n)$ (*b*) $O(n \log n)$
   (*c*) $O(n^{3/2})$ (*d*) $O(n^3)$.
4. The number of equivalence relations of the set {1, 2, 3, 4} is
   (*a*) 4 (*b*) 15
   (*c*) 16 (*d*) 24.
5. Let R be non-empty relation on a collection of sets defined by ARB if and only if $A \cap B = \phi$. Then, (pick the TRUE statement)
   (*a*) R is reflexive and transitive (*b*) R is symmetric and not transitive
   (*c*) R is an equivalence relation (*d*) R is not reflexive and not transitive.
6. Let R be a symmetric and transitive relation on a set A, then
   (*a*) R is reflexive and hence an equivalence relation
   (*b*) R is reflexive and hence a partial order
   (*c*) R is not reflexive and hence not an equivalence relation
   (*d*) None of these.
7. The 'Subset' relation on a set of sets is
   (*a*) A partial ordering (*b*) An equivalence relation
   (*c*) Transitive and symmetric only (*d*) Transitive and anti-symmetric only.
8. Let R_1 and R_2 be two equivalence relations on a set. Consider the following assertions
   (*i*) $R_1 \cup R_2$ is an equivalence relation.
   (*ii*) $R_1 \cap R_2$ is an equivalence relation.
   Which of the following is correct ?
   (*a*) Both assertions are true

(*b*) Assertion (*i*) is true but assertion (*ii*) is not true
(*c*) Assertion (*ii*) is true but assertion (*i*) is not true
(*d*) Neither (*i*) nor (*ii*) is true.

9. Let R be a relation on A = {1, 2, 3, 4} defined by "x is less than y". Then R^{-1} is
(*a*) {(2, 1), (3, 1), (4, 1), (3, 2), (4, 2), (4, 3), (4, 4)}
(*b*) {(2, 1), (3, 1), (4, 1), (3, 2), (4, 2)}
(*c*) {(2, 1), (3, 1), (4, 1), (3, 2), (4, 2), (4, 4)}
(*d*) None.

10. Let R be a relation on A = {1, 2, 3} defined by R = {(1, 2), (2, 3), (3, 1)}. The transitive closure of R is
(*a*) R_T = {(1, 2), (2, 3), (1, 3), (2, 3), (3, 1) (2, 1)}
(*b*) R_T = {(1, 2), (2, 3), (3, 3), (1, 3)}
(*c*) both (*a*) and (*b*) are true
(*d*) none.

Answers and Explanations

1. (*d*) The empty relation on any non-empty set is always symmetric and transitve.

2. (*c*) R = {(x, y) : $x + y$ is even}

Reflexive. (x, x) ∈ R, since $x + x = 2x$ is even. Hence R is reflexive.

Symmetric. If (x, y) ∈ R, then $x + y$ is even ⇒ $y + x$ is also even ⇒ (y, x) ∈ R. Hence R is symmetric.

Transitive. If (x, y) ∈ R, then $x + y$ is even, say $x + y = 2k$

If (y, z) ∈ R, then $y + z$ is even, say $y + z = 2m$ where k, m are integers.

Now, $x + z = 2k - y + 2m - y = 2k + 2m - 2y = 2(k + m - y) = 2l$, say $l = k + m - y$ which is even. Hence (x, z) ∈ R

∵ R is Tansitive also.

Hence R is an equivalence relation.

To find the number of equivalence classes, we have (x, y) ∈ R ⇒ $x + y$ is even.

Also, $y + x$ is even ⇒ (y, x) ∈ R

∴ The number of equivalence classes is **2**.

3. (*d*).

4. The number of equivalence relations = $2^4 - 1 = 16 - 1 = 15$.

5. (*b*) Let A and B are two sets such that A ∩ B = ϕ. Then B ∩ A = ϕ is also true. Hence, R is symmetric.

Also, if A, B, C are sets such that A ∩ B = ϕ, B ∩ C = ϕ, then A ∩ C may or may not be ϕ. Hence, R is not transitive.

6. (*d*) Every equivalence relation is reflexive, symmetric and transitive. But the converse is not true.

7. (*a*) **8.** (*c*) **9.** (*d*) **10.** (*c*).

3 FUNCTIONS

3.1. FUNCTION

Let A and B be two non-empty sets. A function or a mapping f from a set A into a set B is a rule which associates to each element 'a' of A, a unique element 'b' of B such that $f(a) = b$.

The element 'b' is callled image of 'a' under f and 'a' is called pre-image of b.

This fact is denoted by $f : A \rightarrow B$ and read as "f is a function from A to B".

Remark. If $f : A \rightarrow B$ is a function from A to B, then

(*i*) each element of A has one and only one image in B.

(*ii*) Different elements of A can have same images in B.

Theorem I. *Let A and B be two finite sets having m and n elements respectively. Then total number of functions from A to B is* n^m.

Proof. Given, number of elements in $A = m$

Number of elements in $B = n$

Now, f is a function from A to B. It means each element of A can be associated to any one of n elements of the set B. Therefore, total number of functions from set A to set B is equal to the number of ways of doing m jobs where each job can be done in n ways *i.e.*,

Ist element of A can be associated in n ways.

2nd element of A can be associated in n ways

⋮

m^{th} element of A can be associated in n ways.

∴ By Fundamental Principle of Counting, total number of ways of associating m elements of A to any one element of B

$$= n \times n \times n \,\ldots\ldots\, m \text{ times} = n^m$$

Hence, total number of functions from A to B = n^m.

ILLUSTRATIVE EXAMPLES

Example 1. *If a set A has n elements, how many functions are there from A to A.*

Sol. If a set A has n elements, then there are n^n functions from A to A.

Example 2. *If A has m elements and B has n elements, how many functions are there from A to B and from B to A.*

Sol. A has m elements and B has n elements. So, the total number of functions from A to B are n^m and the total number of functions from B to A are m^n.

Example 3. *If A = {2, 3, 4} and B = {5, 6}. Determine all functions from A to B.*

Sol. The total number of functions from A → B are $2^3 = 8$, which are

(*i*) {(2, 5), (3, 5), (4, 5)}
(*ii*) {(2, 6), (3, 6), (4, 6)}
(*iii*) {(2, 5), (3, 5), (4, 6)}
(*iv*) {(2, 5), (3, 6), (4, 6)}
(*v*) {(2, 6), (3, 5), (4, 5)}
(*vi*) {(2, 6), (3, 6), (4, 5)}
(*vii*) {(2, 6), (3, 5), (4, 6)}
(*viii*) {(2, 5), (3, 6), (4, 5)}.

Example 4. *Let F be the set of one-one functions from A to B where A = {1, 2, 3, ..., n}, B = {1, 2, 3, ..., m} and $m \geq n \geq 1$.*

(*a*) *How many functions f are members of F ?*

(*b*) *How many functions f in F satisfy the property $f(i) = 1$ for some $1 \leq i \leq n$?*

(*c*) *How many functions f in F satisfy the property $f(i) < f(j)$ for all $1 \leq j \leq n$?*

Sol. (*a*) Given $n(A) = n$, $n(B) = m$

∴ Total number of functions which are 1-1 = $n(A) \times n(B) = nm$

Hence 'nm' functions are members of F.

(*b*) There are 'n' functions (the cardinal number of A is n) satisfying the property $f(i) = 1$.

(*c*) The number of functions satisfying

$$f(i) < f(j) = 1 + 2 + 3 + ... + n = \frac{n(n+1)}{2}.$$

3.2. DOMAIN OF A FUNCTION (*P.T.U., B.Tech. Dec. 2006*)

Let f be a function from A to B. The set A is called domain of the function f.

3.3. CO-DOMAIN OF A FUNCTION (*P.T.U., B.Tech. Dec. 2006*)

Let f be a function from A to B. The set B is called co-domain of the function f.

3.4. IMAGE OF AN ELEMENT

If the element x of A corresponds to y under function f, then y is the image of x under f and is written as $f(x) = y$.

If $f(x) = y$, then we say that x is a pre-image of y.

If $f : X \to Y$, then each element of X has unique image in Y, whereas every element in Y need not be image of some x in X.

Pre-image or **inverse image**

Let f: A → B be a function from A to B. Let T be a subset of B. Then the inverse image or pre-image of T under f, denoted by f^{-1} (T), is defined as

$$f^{-1}(T) = \{a \in A : f(a) \in T\}$$

In otherwords, f^{-1}(T) consists of the elements of A whose images belong to T.

Example 5. *Let A = {1, 2, 3, 4} and the function f : A → A as shown in Fig. 3.1. Find (i) the image of each element of A. (ii) the image f(A) of the function f.*

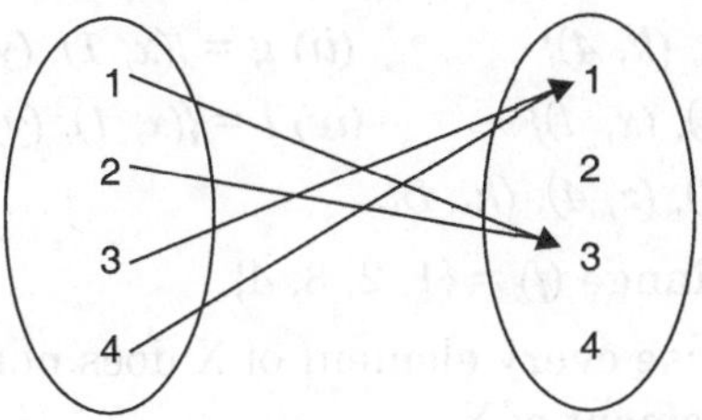

Fig. 3.1

Sol. (*i*) The image of 1 = $f(1) = 3$, $f(2) = 3$, $f(3) = 1$ and $f(4) = 1$.

(*ii*) The image of f(A) consists of all the image values.

Therefore, $f(A) = \{1, 3\}$.

3.5. RANGE OF A FUNCTION OR Im (f)

The range of a function is the set of images of its domain. In other words, we can say, it is a subset of its co-domain.

If $f : A \to B$, then $f(A) = \{f(x) : x \in A\} = \{y : y \in B \mid \exists\, x \in A, \text{ such that } f(x) = y\}$.

Example 6. *Let P = {x, y, z, u} and Q = {a, b, c, d} and f : P → Q, such that*

$$f = \{(x, a), (y, b), (z, c), (u, c)\}.$$

Find the domain, co-domain and range of function.

Sol. Domain of function f is the set P = {*x*, *y*, *z*, *u*}

Co-domain of function f is the set Q = {*a*, *b*, *c*, *d*} Range of the function f is Range (f) = {*a*, *b*, *c*}.

Example 7. *Let A = {2, 3, 4} and B = {a, b, c} and f = {(2, a), (3, b), (4, b)}*

Find domain, co-domain and range of the function.

Sol. Domain of function is Domain (f) = {2, 3, 4}

Co-domain of function is co-domain (f) = {*a*, *b*, *c*}

Range of function is Range (f) = {*a*, *b*}.

3.6. DESCRIPTION OF A FUNCTION

If $f : A \to B$ is a function such that its domain D_f consists of a finite number of elements. Then, $f(x)$ can be described by listing the values which it attains at different points of its domain.

e.g., Let A = {*a*, *b*, *c*} and B = {*x*, *y*, *z*}

and $f : A \to B$ such that $f = \{(a, x), (b, z), (c, x)\}$

Then f can be represented diagrammatically as shown in Fig.3.2.

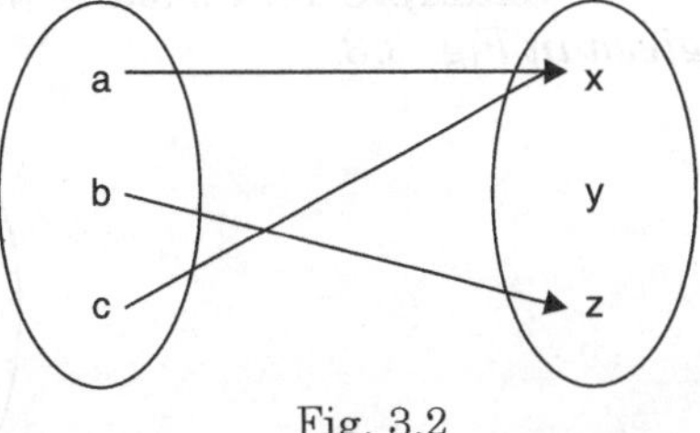

Fig. 3.2

If D_f is an infinite set, then f cannot be described by listing the values which it attains at different points of its domain. In this case, a function is generally described by a formula.

For example, $f : Z \to Z$ given by $f(x) = x^2 + 1$,

$g : R \to R$ given by $g(x) = e^x$ etc.

Example 8. *Let X = {x, y, z, k} and Y = {1, 2, 3, 4}. Determine which of the following are functions ? Give reasons if it is not. Find range if it is a function.*

(*i*) $f = \{(x, 1), (y, 2), (z, 3), (k, 4)\}$ (*ii*) $g = \{(x, 1), (y, 1), (k, 4)\}$

(*iii*) $h = \{(x, 1), (x, 2), (x, 3), (x, 4)\}$ (*iv*) $l = \{(x, 1), (y, 1), (z, 1), (k, 1)\}$

(*v*) $d = \{(x, 1), (y, 2), (y, 3), (z, 4), (k, 4)\}$.

Sol. (*i*) It is a function. Range (f) = {1, 2, 3, 4}.

(*ii*) g not a function because every element of X does not relate with some element of Y *i.e.*, z is not related with any element of Y.

(*iii*) h is not a function because element x has more than one image in set Y. *i.e.*, $x \in$ X has four images 1, 2, 3, 4.

(*iv*) It is a function. Range (l) = {1}.

(*v*) d is not a function because element y has more than one image in set Y. *i.e.*, $y \in$ X has two images 2 and 3.

3.7. (a) EVERYWHERE DEFINED FUNCTION

Consider a function f from A to B. Then the function f is everywhere defined if dom (f) = A.

Example 9. *Let A = {1, 2, 3, 4} and B = {a, b, c} and C = {α, β, γ}. Consider the following two functions f : B → C and g : A → C given as*

(*i*) $f = \{(a, \alpha), (b, \beta), (c, \gamma)\}$ (*ii*) $g = \{(1, \alpha), (3, \beta), (2, \gamma)\}$

Determine whether or not each function is everywhere defined.

Sol. (*i*) f is everywhere defined since dom (f) = {a, b, c} = B.

(*ii*) g is not everywhere defined since dom (f) = {1, 2, 3} ≠ A = {1, 2, 3, 4}.

3.7. (b) GRAPH OF A FUNCTION

Let f : A → B is a function, then the graph of f, denoted by graph f, is a subset of A × B, given by

$$\text{graph } f = \{(a, f(a)) : a \in A\}$$

This means that each $a \in$ A, there is a unique pair (a, b) in f which is equivalent to say that each **vertical** line intersects the graph in exactly one point.

Example 10. *Consider the function f from A = {a, b, c, d} into B = {x, y, z, w} defined as given in Fig. 3.3.*

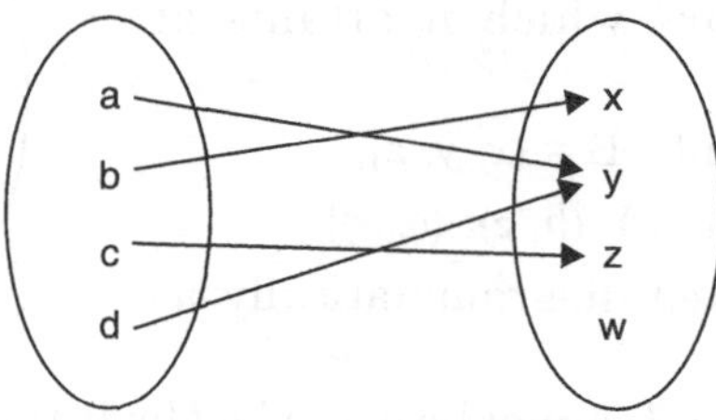

Fig. 3.3

(*a*) *Find the image of each element of A.*

(*b*) *Find the image of f*

(*c*) *Find the graph f*

(*d*) *Find f(s) where S = {a, b, d}*

(*e*) *Find f^{-1} (T) where T = {y, z}*

(*f*) *Find f^{-1} (w)*

Sol. (*a*) From the given diagram, the images of each element of A = {*a*, *b*, *c*, *d*} are given as $f(a) = y$, $f(b) = x$, $f(c) = z$, $f(d) = y$.

(*b*) $I_m(f)$ = The set of all images of the elements in A

$= \{x, y, z\}$

(*c*) Graph $f = \{(a, f(a));\ a \in A\}$

$= \{(a, y), (b, x), (c, z), (d, y)\}$

(*d*) Here S = {*a*, *b*, *d*}

$\therefore \quad f(S) = \{f(a), f(b), f(b)\} = \{y, x, y\} = \{x, y\}$

(*e*) Here, T = {*y*, *z*}

∴ From the given diagram

$f^{-1}(T) = \{a, d, c\} = \{a, c, d\}$

(*f*) $f^{-1}(w)$ does not exist since the element *w* has no its pre-image.

3.7. (c) FUNCTION AS A RELATION

Let A and B be non-empty sets. A relation from A to B is a subset of A × B. This relation is called a function from A to B if

(*i*) For each $a \in A$, there exists $b \in B$ such that $(a, b) \in f$

(*ii*) If $(a, b) \in f$, $(a, c) \in f$, then $b = c$. *i.e.*, no two ordered pairs in *f* have the same first element.

For *e.g.*, let A = (1, 2, 3), B = (2, 3, 4) and let f_1, f_2, f_3 are three subsets of A × B given by

(*i*) f_1 = (1, 2), (2, 3), (3, 4)

(*ii*) f_2 = (1, 2), (1, 3), (2, 3), (3, 4)

(*iii*) f_3 = (1, 3), (2, 4).

We examine whether or not f_1, f_2, f_3 are functions or not.

Here, f_1 is a function from A to B. As each element of A is associated to a unique element of B. (see Fig. 3.4)

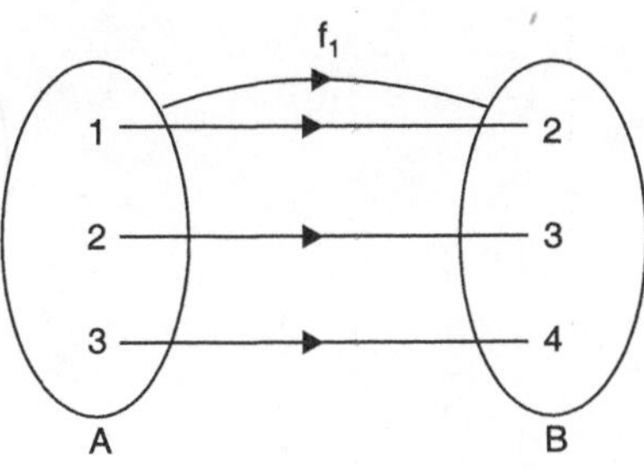

Fig. 3.4

But f_2 is not a function. Since each element of A is not associated to a unique element of B.

(1 ∈ A has two images 2 and 3) (Fig. 3.5)

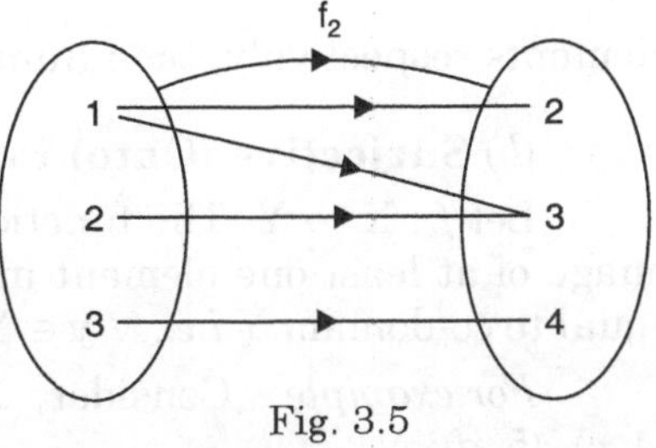

Fig. 3.5

Similarly, f_3 is not a function from A to B. ($3 \in A$ is not associated to any element of B) (Fig. 3.6)

Note. If a function f is expressed as the set of ordered pairs, then the domain of f is the set of all first components of members of f and the range of f is the set of second components of members of f.

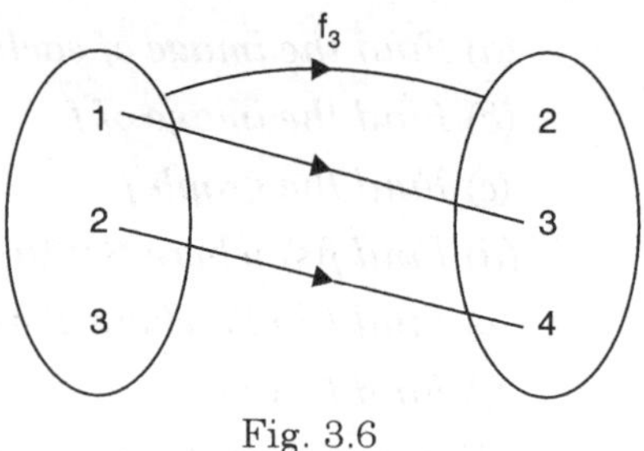

Fig. 3.6

3.8. TYPES OF FUNCTIONS

(*a*) **Injective (One-to-One) Functions.** Let $f : X \to Y$. The function f is called one-to-one or injective if different elements in X have different images in Y *i.e.*, if $f(x_1) = f(x_2)$

$$\Rightarrow \qquad x_1 = x_2 \ \forall \ x_1, x_2 \in X.$$

Another way of defining injective function is that every element of domain X has a unique image in the co-domain Y and there is no element of Y which is image of more than one element of domain X.

For example : Consider, X = {*x*, *y*, *z*, *k*} and Y = {1, 2, 3, 4} and f is function from X to Y such that

$$f = \{(x, 1), (y, 2), (z, 3), (k, 4)\}$$

The function f is injective function as every element of domain X has a unique image in the co-domain Y (Fig. 3.7).

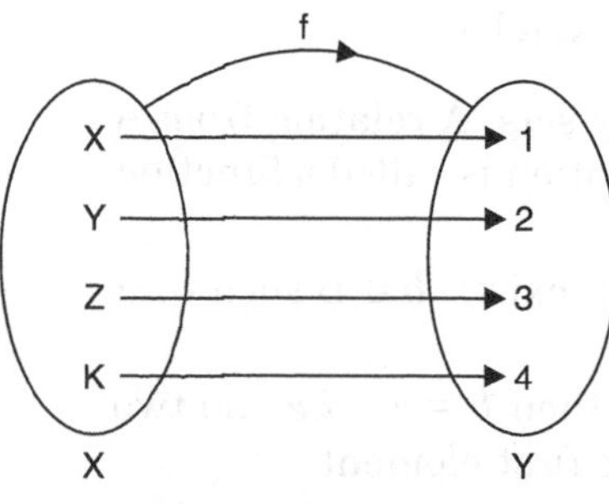

Fig. 3.7

Remark. Number of one-one functions. If A and B are finite sets having m and n elements respectively, then number of one-one functions from A to B = $\begin{cases} {}^nP_m & , \ n \geq m \\ 0 & , \ n < m \end{cases}$

(*b*) **Surjective (Onto) Functions**

Let $f : X \to Y$. The function f is called surjective function if each element in Y, is the image of at least one element in X. In other words, in surjective functions, the range of f is equal to co-domain Y *i.e.*, $\forall \ y \in Y, y = f(x)$ for some $x \in X$.

For example : Consider, X = {1, 2, 3, 4, 5}, Y = {*a*, *b*, *c*, *d*} and f = {(1, *a*), (2, *a*), (3, *b*), (4, *c*), (5, *d*)}.

It is a surjective function, as every element of Y is the image of some element of X (Fig. 3.8).

Remark: Number of onto functions. If A and B are two sets having m and n elements such that $1 \leq n \leq m$, then number of onto functions from A to B is given by $\sum_{r=1}^{n} (-1)^{n-r}\, {}^{n}C_r\, r^m$

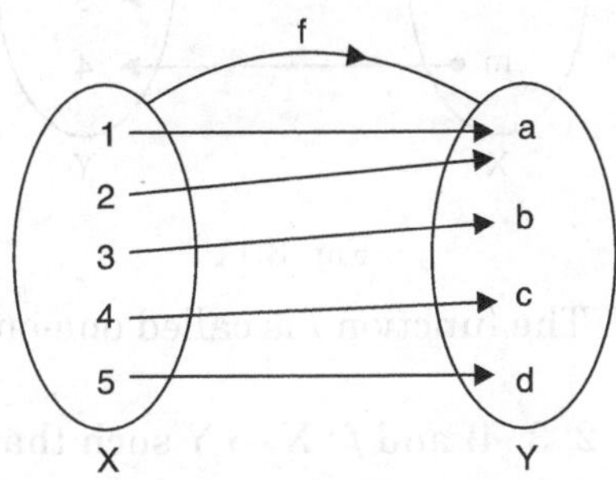

Fig. 3.8

Fig. 3.9

(*c*) **Bijective (One-to-One Onto) Functions.** A function which is both injective (one-to-one) and surjective (onto) is called a bijective (one-to-one-onto) function.

For example : Consider, P = {x, y, z}, Q = {a, b, c} and f : P → Q such that

$$f = \{(x, a), (y, b), (z, c)\}$$

The f is one-to-one and also it is onto. So it is a bijective function (Fig. 3.9).

Remark 1. To Check whether f is onto or not

Let f : A → B be the given function.

Step I. Take $y \in$ B such that $f(x) = y$

Step II. Solve $f(x) = y$ for x and obtain x in terms of y. Let $x = g(y)$.

Step III. If for all values of $y \in$ B, the values x obtained from $x = g(y)$ are in the set A, then f is onto.

If there are some $y \in$ B for which x is not in A, then f is not into.

Remark 2. Number of Bijections. If A and B are finite sets and f : A → B is a bijection, then A and B have the same number of elements. If A has n elements, then the number of bijections from A to B = Total number of arrangements of n items taken n at a time = ${}^{n}C_n = n\,!$.

(*d*) **Into Functions.** Let f : X → Y. The function f is called an into function if the range of f is not equal to the co-domain Y. Therefore, there must be an element of co-domain Y which is not the image of any element of domain X.

For example : Consider, X = {1, 2, 3}, Y = {k, l, m, n, p} and f : X → Y such that

$$f = \{(1, k), (2, n), (3, p)\}.$$

In the function f, the range *i.e.*, {k, n, p} ≠ co-domain of Y *i.e.*, {k, l, m, n, p}

Therefore, it is an into function (Fig. 3.10).

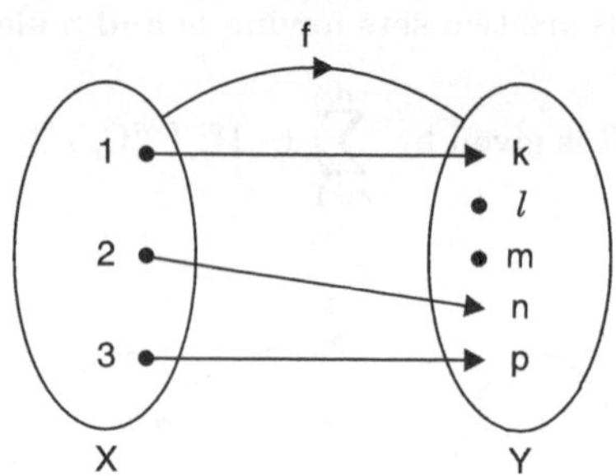

Fig. 3.10

Fig. 3.11

(*e*) **One-One Into Functions.** Let $f : X \to Y$. The function f is called one-one into function if it is one-one but not onto (See Fig. 3.11).

For example : Consider, $X = \{k, l, m\}$, $Y = \{1, 2, 3, 4\}$ and $f : X \to Y$ such that

$$f = \{(k, 1), (l, 3), (m, 4)\}$$

The function f is one-one into function (Fig. 3.11).

(*f*) **Many One Functions.** Let $f : X \to Y$. The function f is said to be many one function if there exists two or more than two different elements in X having the same image in Y.

For example : Consider $X = \{1, 2, 3, 4, 5\}$, $Y = \{x, y, z\}$ and $f : X \to Y$ such that

$$f = \{(1, x), (2, x), (3, x), (4, y), (5, z)\}$$

The function f is a many one function. (Fig. 3.12)

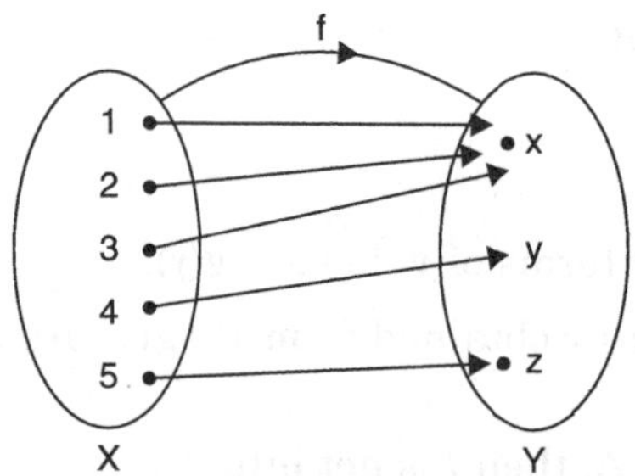

Fig. 3.12

Fig. 3.13

(*g*) **Many One Into Functions.** Let $f : X \to Y$. The function f is called many-one into function if and only if it is both many one and into function.

For example : Consider $X = \{a, b, c\}$, $Y = \{1, 2\}$ and $f : X \to Y$ such that

$$f = \{(a, 1), (b, 1), (c, 1)\}$$

As the function f is many-one and into, so it is a many-one into function (Fig. 3.13).

(*h*) **Many One Onto Functions.** Let $f : X \to Y$. The function f is called many-one-onto function if and only if it is both many one and onto.

For example : Consider, $X = \{1, 2, 3, 4\}$, $Y = \{k, l\}$ and $f : X \to Y$ such that

$$f = \{(1, k), (2, k), (3, l), (4, l)\}$$

The function f is many-one (as two elements have the same image in Y) and it is onto (as every element of Y is the image of some element X). So, it is many-one onto function (Fig. 3.14).

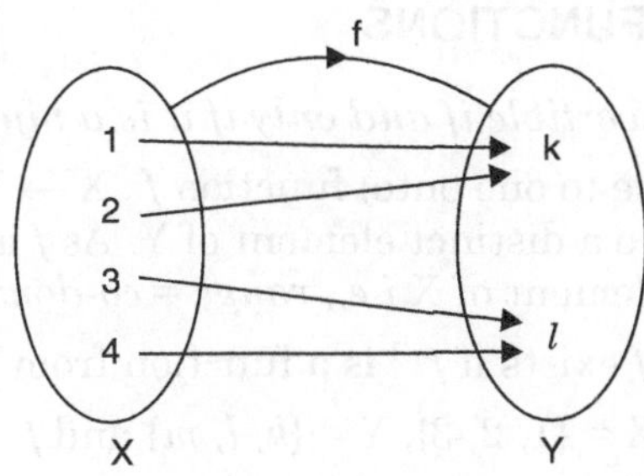

Fig. 3.14

3.9. EQUAL FUNCTIONS

Consider two functions f and g from a set X to a set Y. The functions f and g are called equal functions if and only if $f(a) = g(a)$, for every at $D = D_f \cap D_g$ or

Two functions f and g are equal iff

(i) Dom (f) = Dom (g)

(ii) Co-domain (f) = Co-domain (g)

(iii) $f(x) = g(x) \; \forall \; x \in D$ where D = Dom $(f) \cap$ Dom (g)

The functions f and g are called unequal functions if there exist at least one element $a \in X$ such that $f(a) \neq g(a)$.

3.10. (a) IDENTITY FUNCTIONS

Consider any set A. Let the function $f : A \to A$. The function f is called the identity function if each element of set A has image on itself *i.e.*, $f(a) = a \; \forall \; a \in A$.

It is denoted by I.

Example 11. Consider, A = {1, 2, 3, 4, 5} and $f : A \to A$ such that

$$f = \{(1, 1), (2, 2), (3, 3), (4, 4), (5, 5)\}.$$

The function f is an identity function as each element of A is mapped onto itself. The function f is one-one and onto (Fig. 3.15).

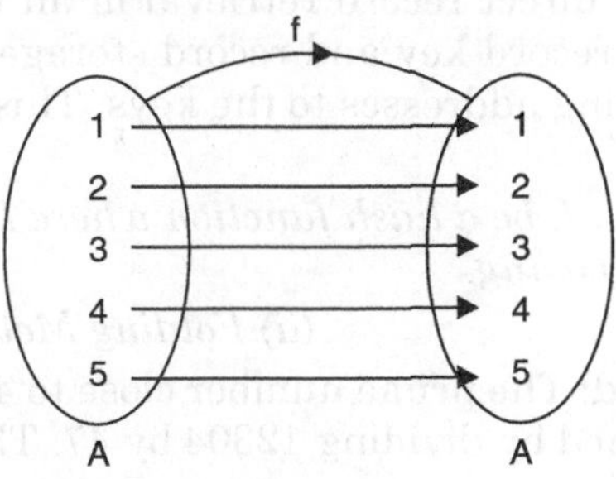

Fig. 3.15

3.10. (b) CONSTANT FUNCTION

Let f be a function with domain A. Then f is called a constant function if for every $x \in A$, $f(x) = c$, c is some constant

Example 12. *For given sets A and B, how many constant maps are there from A into B ?*

Sol. For each $x \in A$, the constant map is given by $f(x) = b$, $b \in B$

Let $n(B) = | B |$, the number of elements in B, then there are | B | constant maps.

3.11. INVERTIBLE (Inverse) FUNCTIONS

A function $f : X \to Y$ *is invertible if and only if it is a bijective function.*

Consider the bijective (one to one onto) function $f : X \to Y$. As f is one-to-one, therefore, each element of X corresponds to a distinct element of Y. As f is onto, there is no element of Y which is not the image of any element of X *i.e., range = co-domain Y.*

The inverse function for f exists if f^{-1} is a function from Y to X.

Example 13. Consider, X = {1, 2, 3}, Y = {k, l, m} and $f : X \to Y$, such that

$$f = \{(1, k), (2, m), (3, l)\} \text{ as in Fig. 3.16.}$$

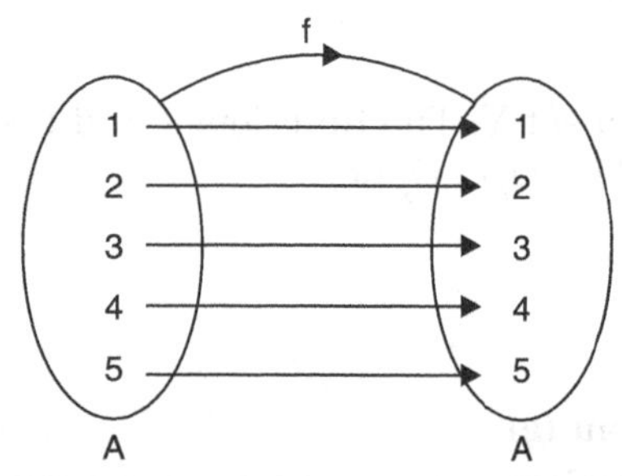

Fig. 3.16

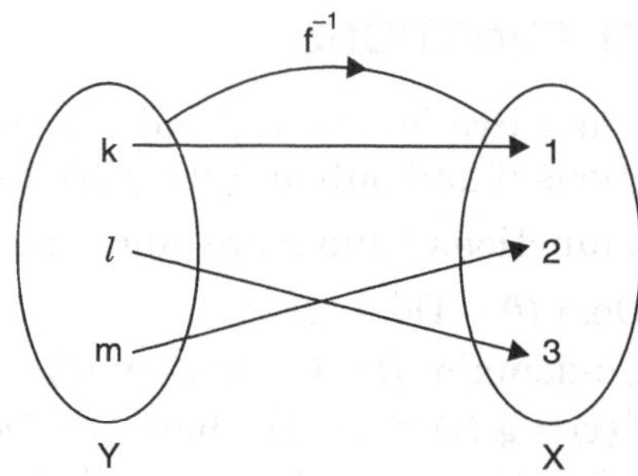

Fig. 3.17

The inverse function of f is shown in Fig. 3.17 *i.e.*, $f^{-1} = \{ (k, 1), (m, 2), (l, 3)\}$

3.12. HASHING FUNCTION

Any function defined by $f : k \to A$, where k denotes the set of keys and A is a set of physical address, is known as Hashing function.

One of the most popular Hash functions is modulus function, written as mod $f(n)$.

Uses: Hash functions are widely used in many applications such as symbol table of compilers, direct addressing in memory operations, direct access in file handling etc. In these cases, the program is supplied with a key, using this key, the program has to locate the required record of information. To make direct record retrieval in an easy and smooth way, we need some relationship between the record key and record storage position. For direct access, we require some methods of assigning addresses to the keys. This method for the same is known as Hashing function.

Example 14. *Let H : K → L be a hash function where L consists of two digit addresses 00, 01, 02,..., 49. Find H (12304) using.* (P.T.U. B.Tech May 2013)

(*i*) *Division Method* (*ii*) *Folding Method*

Sol. (*i*) Division Method: The prime number close to 49 is 47. Thus, the hash address generated by H(12304) is obtained by dividing 12304 by 47. This gives you the remainder 37, which is the required hash address.

(*ii*) Folding Method: Divide the key 12304 k_1, k_2 and k_3 into three parts and then add to obtain the hash address.

$$H(12304) = \overset{k_1}{12},\ \overset{k_2}{30},\ \overset{k_3}{4}$$

k_2 is reversed, so the required hash address is 12 + 03 + 4 = 19.

ILLUSTRATIVE EXAMPLES

Example 1. *Let A = (1, 2, 3, 4) and B = (a, b, c, d). Determine which of the following are functions. If so, find the range of each functions.*

(a) $f \subseteq A \times B$, *where* $f = \{(1, a), (2, b), (3, c), (4, d)\}$

(b) $f \subseteq A \times B$, *where* $f = \{(1, a), (2, a), (3, b), (4, d)\}$

(c) $f \subseteq A \times B$, *where* $f = \{(1, a), (2, b), (3, c)\}$

(d) $f \subseteq A \times B$, *where* $f = \{(1, a), (2, b), (2, c), (3, a), (4, a)\}$

(e) $f \subseteq A \times A$, *where* $f = \{(1, 1), (2, 1), (3, 1), (4, 1)\}$.

Sol. (a) $f = \{(1, a), (2, b), (3, c), (4, d)\}$

From Fig. 3.18, we observe that each element of A is associated to a unique element of B. ∴ f is a function. Also, range $f = (a, b, c, d)$

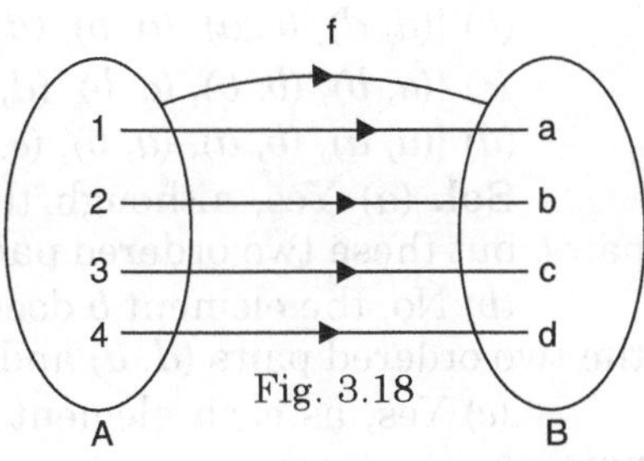

Fig. 3.18

(b) $f = \{(1, a), (2, a), (3, b), (4, d)\}$

From Fig. 3.19, we observe that each element of A is associated to a unique element of B. ∴ f is a function.

Also, range $f = (a, b, d)$

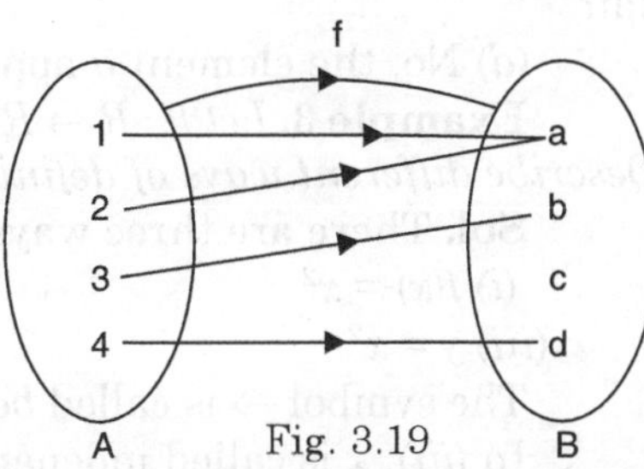

Fig. 3.19

(c) $f = \{(1, a), (2, b), (3, c)\}$

From Fig. 3.20, we observe that the element $4 \in A$ is not associated to a element of B.

∴ f is not a function.

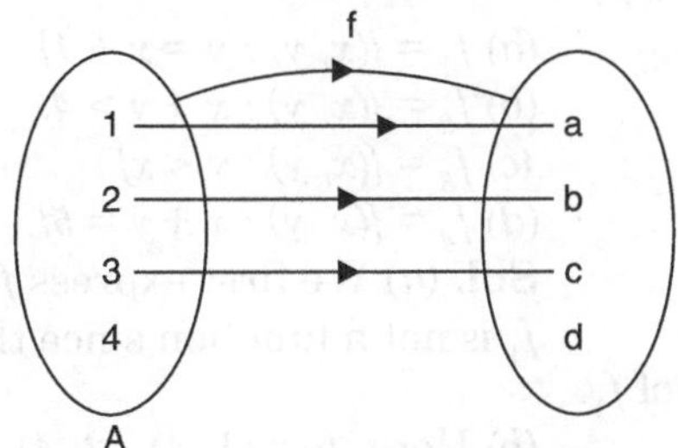

Fig. 3.20

(d) $f = \{(1, a), (2, b), (2, c), (3, a), (4, a)\}$

From Fig. 3.21, we observe that the element $2 \in A$ is not associated uniquely to elements of B.

∴ f is not a function.

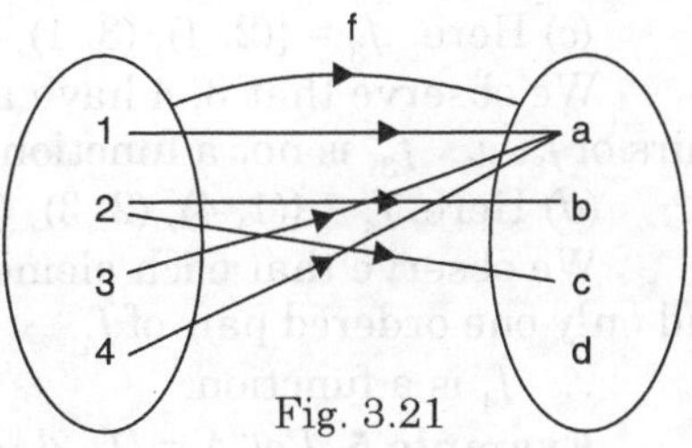

Fig. 3.21

(e) $f = \{(1, 1), (2, 1), (3, 1), (4, 1)\}$

From Fig. 3.22, we observe that each element of A is associated to a unique element of A.

$\therefore$ f is a function from A to A.

Also range $f = \{1\}$.

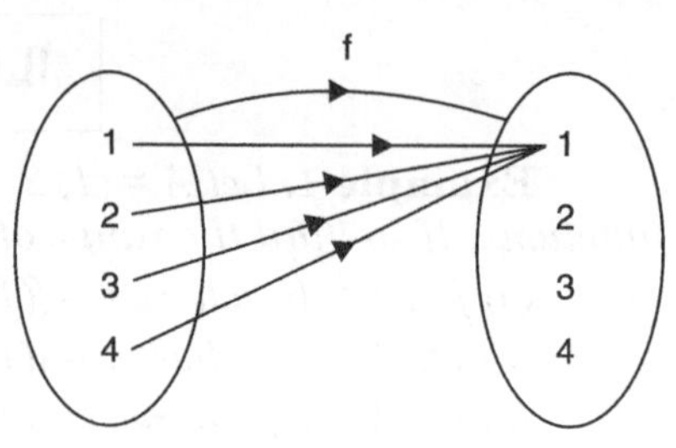

Fig. 3.22

Example 2. *Let W = {a, b, c, d}. Determine whether each of the following sets of ordered pairs is a function from W into W.*

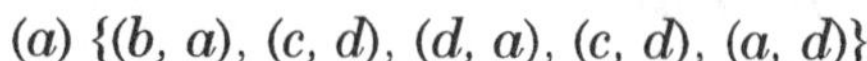

(a) $\{(b, a), (c, d), (d, a), (c, d), (a, d)\}$

(b) $\{(d, d), (c, a), (a, b), (d, b)\}$

(c) $\{(a, b), (b, b), (c, b), (d, b)\}$

(d) $\{(a, a), (b, a), (a, b), (c, d), (d, a)\}$.

Sol. (a) Yes, although the element c appears as the first coordinate in two ordered pairs, but these two ordered pairs are equal.

(b) No, the element b does not appear as the first coordinate in any ordered pair. Also, the two ordered pairs (d, d) and (d, b) have the same first element.

(c) Yes, as each element of W appears as the first coordinate exactly in one ordered pair.

(d) No, the element a appears as the first coordinate in two different ordered pairs.

Example 3. *Let $F : R \to R$ be a function which assigns to each real number x its square x^2. Describe different ways of defining f.*

Sol. There are three ways of defining f

(i) $f(x) = x^2$ (ii) $x \to x^2$

(iii) $y = x^2$

The symbol $\to$ is called *barred row* and reads as "goes into".

In (iii), x is called independent variable and y is called dependent variable.

Example 4. *If $x, y \in \{1, 2, 3, 4\}$. Then which of the following are functions in the given set ?*

(a) $f_1 = \{(x, y) : y = x + 1\}$

(b) $f_2 = \{(x, y) : x + y > 4\}$

(c) $f_3 = \{(x, y) : y < x\}$

(d) $f_4 = \{(x, y) : x + y = 5\}$.

Sol. (a) We first express f_1 as a set of ordered pairs. Here $f_1 = \{(1, 2), (2, 3), (3, 4)\}$

f_1 is not a function since the element 4 is not appeared in first place of any ordered pair of f_1.

(b) Here, $f_2 = (1, 4), (2, 4), (3, 4), (2, 3), (3, 2), (4, 1), (4, 2), (4, 3)$

We observe that 2, 3, 4 have appeared more than once as first component of the ordered pairs in f_2. $\therefore$ f_2 is not a function.

(c) Here, $f_3 = \{(2, 1), (3, 1), (4, 1), (3, 2), (4, 2), (4, 3)\}$

We observe that 3, 4 have appeared more than once as first component of the ordered pairs of f_3. $\therefore$ f_3 is not a function.

(d) Here, $f_4 = \{(1, 4), (2, 3), (3, 2), (4, 1)\}$

We observe that each element of the given set has appeared as first component in one and only one ordered pair of f_4.

$\therefore$ f_4 is a function.

Example 5. *Let A = {1, 2}, B = {3, 6} and $f : A \to B$ is given by $f(x) = x^2 + 2$, $g : A \to B$ is given by $g(x) = 3x$. Is $f = g$?*

Sol. Given $f : A \to B$ *i.e.,* f is a function from A to B.

$\therefore$ Domain $f = (1, 2)$, co-domain $g = (3, 6)$

Also $g : A \to B$ *i.e., g* is a function from A to B

$\therefore$ Domain $g = (1, 2)$, co-domain $g = (3, 6)$

Also, $f(1) = 1 + 2 = 3, g(1) = 3$

$f(2) = 4 + 2 = 6, g(2) = 6$

$\Rightarrow$ $f(x) = g(x)$ for all $x \in$ Dom. f, Dom. g. This f and g are equal functions.

Example 6. *Let f be a function with domain A and codomain B. Let the relation $k \subseteq A \times A$ is defined on A as $(x, y) \in k$ iff $f(x) = f(y)$. Show that k is an equivalence relation.*

Sol. Since $f(x) = f(x)$ for all $x \in A$

$\therefore$ $(x, x) \in k$ *i.e., k* is reflexive

Let $(x, y) \in k$ $\Rightarrow$ $f(x) = f(y)$ for all $x, y \in A$

$\Rightarrow$ $f(y) = f(x)$

$\Rightarrow$ $(y, x) \in k$ *i.e., k* is symmetric

Further, if $(x, y) \in k$, $(y, z) \in k$. Then

$f(x) = f(y), f(y) = f(z)$

$\Rightarrow$ $f(x) = f(z)$ for all $x, z \in A$

$\Rightarrow$ $(x, z) \in k$ $\therefore$ k is transitive.

Hence the relation k is an equivalence relation.

Example 7. *Determine if the following function is one-one.*

(a) To each person on the earth, assign the number which corresponds to his age.

(b) To each country in the world, assign the lattitude and longitude of its capital.

(c) To each book written by only one author, assign the author.

(d) To each country in the world, which has a prime minister, assign its prime minister.

Sol. (*a*) No, as many people in the world can have the same age.

(*b*) Yes different countries have different capitals.

(*c*) No, as we can have different books by the same author.

(*d*) Yes, as different countries in the world have different prime ministers.

Example 8. *Let A = B = {1, 2, 3, 4, 5}. Define a function $f : A \to B$ such that f is one-one and onto function.* (P.T.U. B.Tech. Dec., 2005)

Sol. Let $f(x) = x$ for all $x \in A$, then

$f(1) = 1, f(2) = 2, f(3) = 3, f(4) = 4, f(5) = 5.$

i.e., different elements of A have different images in B.

$\therefore$ f is one-one (Fig. 3.23).

Also, each element of B is the images of some element of A.

$\therefore$ f is onto also.

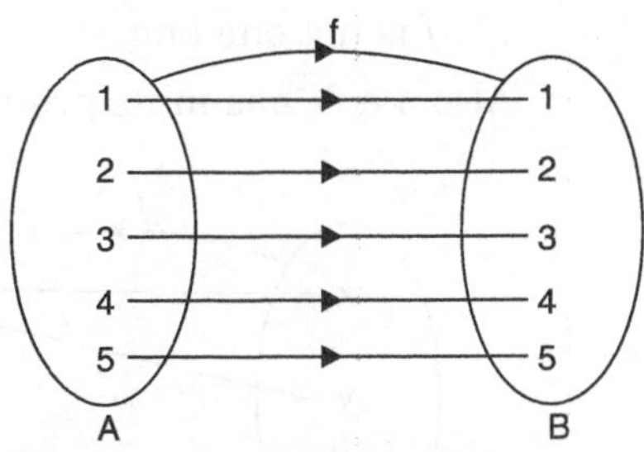

Fig. 3.23

Example 9. *Let A = B = {1, 2, 3, 4}. Define functions $f : A \to B$ (if possible) such that*

(*i*) *f is one-to-one and onto*

(*ii*) *f is neither one-to-one nor onto*

(*iii*) *f is onto but not one-to-one*

(*iv*) *f is one-to-one but not onto.*

Sol. (*i*) The function $f = \{(1, 1), (2, 4), (3, 2), (4, 3)\}$ is one-to-one and onto (Fig. 3.24).

(*ii*) The function $f = \{(1, 1), (2, 1), (3, 2), (4, 3)\}$ is neither one-to-one nor onto (Fig. 3.25).

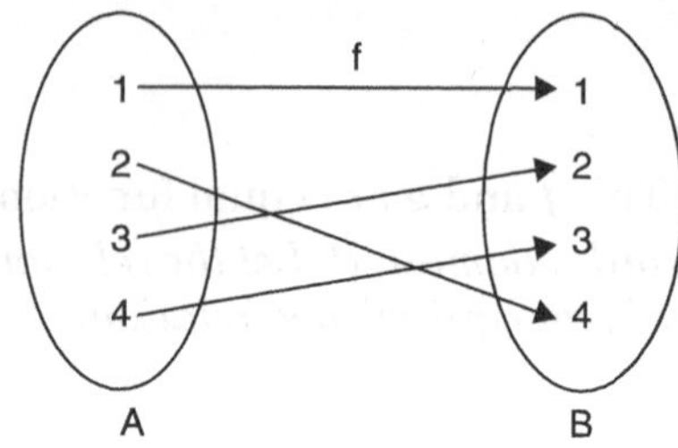

Fig. 3.24

Fig. 3.25

(*iii*) The function f which is onto but not one-to-one is not possible on the set A = B = {1, 2, 3, 4}.

(*iv*) The function f which is one-to-one but not onto is not possible on the set A = B = {1, 2, 3, 4}.

Example 10. *Consider the sets X and Y and let* $f : X \to Y$. *Determine whether the following functions are :*

(*a*) *one-to-one function* (*b*) *onto function*

(*c*) *one-to-one onto* (*d*) *neither one-to-one nor onto.*

(*i*) $X = \{x, y, z\}$; $Y = \{1, 2, 3, 4\}$, $f = \{(x, 1), (y, 1), (z, 3)\}$

(*ii*) $X = \{a, b, c, d\} = Y$, $f = \{(a, a), (b, c), (c, d), (d, b)\}$

(*iii*) $X = \left\{\frac{1}{6}, \frac{1}{9}, \frac{1}{12}\right\}$; $Y = \{a, b, c, d\}$, $f = \left\{\left(\frac{1}{6}, b\right), \left(\frac{1}{9}, d\right), \left(\frac{1}{12}, a\right)\right\}$

(*iv*) $X = \{y_1, y_2, y_3\}$; $Y = \{k, l\}$, $f = \{(y_1, l), (y_2, l), (y_3, k)\}$.

Sol. (*i*) From the Fig. 3.26, we observe that the elements $x, y \in X$ are not uniquely associated to elements of Y

$\therefore$ f is not one-one

Also $4 \in Y$ has in its pre-image in X. Hence, f is not onto.

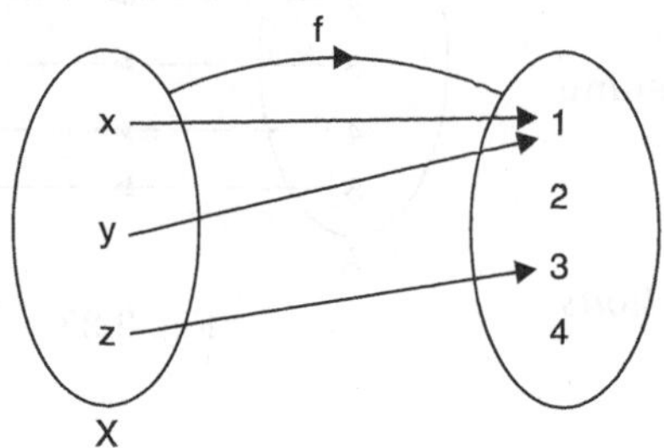

Fig. 3.26

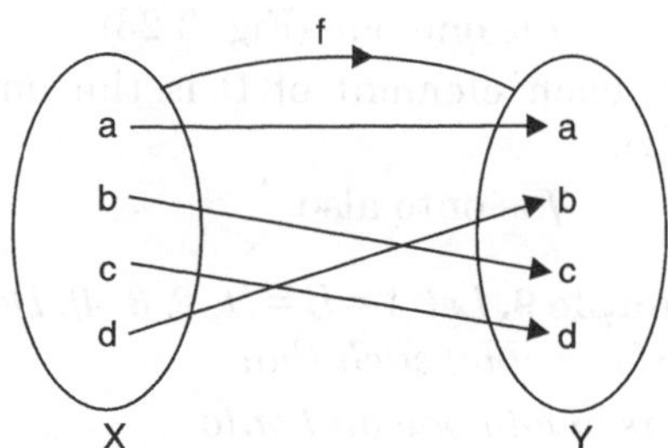

Fig. 3.27

(*ii*) From Fig. 3.27 we observe that f is one-one and onto.

(*iii*) Consider Fig. 3.28 we observe that f is

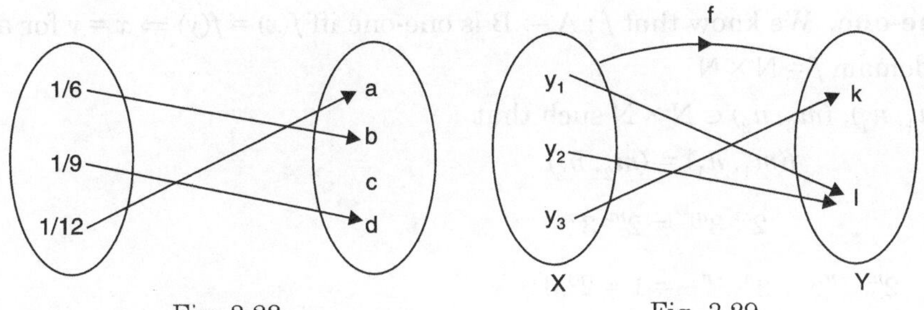

Fig. 3.28 Fig. 3.29

one-one, but not onto as the element c has not its pre-image in X.

(*iv*) From Fig. 3.29 we observe that

f is not one-one, but onto.

Example 11. *Which of the following functions are one-one, onto or both ?*

(a) $f : R \to R$ *defined by* $f(x) = x^3 - 3x$

(b) $f : Z \to Z$ *defined by* $f(x) = -x + 2$

(c) $f : N \times N \to N$ *defined by* $f(m, n) = 2^m 3^n$

(d) $f : N \to N$ *defined by* $f(n) = n^2 + n$

(e) $f : N \to N \times N$ *defined by* $f(n) = (n, |10 - n|)$.

Sol. (*a*) **Onto.** $f : \text{R} \to \text{R}$ is a function from R to R

$\therefore$ Co-domain $f = \text{R}$.

Also, range $f = \{f(x) ; x \in \text{R}\} = \{x^3 - x : x \in \text{R}\} = \text{R}$

$\therefore$ Co-domain f = range f

Hence the function $f : \text{R} \to \text{R}$ is onto.

One-one. Let $x, y \in \text{R}$ such that $f(x) = f(y)$

$\Rightarrow \quad x^3 - x = y^3 - y$

$\Rightarrow \quad x^3 - y^3 - (x - y) = 0$

$\Rightarrow \quad (x - y)(x^2 + y^2 + xy) - (x - y) = 0$

$\Rightarrow \quad (x - y)(x^2 + y^2 + xy - 1) = 0$

But $\quad x^2 + y^2 + xy - 1 \neq 0 \quad \forall\, x, y \in \text{R}$

$\therefore \quad x - y = 0 \Rightarrow x = y$

$\therefore$ f is one-one.

(*b*) **One-one.** We know that a function f is one-one if $f(x) = f(y) \Rightarrow x = y$ for all $x, y \in \text{D}_f$

Let $x, y \in \text{Z}\ (= \text{D}_f)$ and consider $f(x) = f(y)$

$\Rightarrow \quad -x + 2 = -y + 2 \Rightarrow -x = -y$

$\Rightarrow \quad x = y. \quad \therefore f : \text{Z} \to \text{Z}$ is one-one.

Onto : Let $y \in \text{Z}$ such that $f(x) = y$ for all $x \in \text{Z}$

$\Rightarrow \quad -x + 2 = y$

$\Rightarrow \quad x = 2 - y$

For each $y \in Z$, we observe that $x \in Z$

$\therefore$ f is onto.

(c) **One-one.** We know that $f : A \to B$ is one-one iff $f(x) = f(y) \Rightarrow x = y$ for all $x, y \in D_f$

Here domain $f = N \times N$

Let $(m_1, n_1), (m_2, n_2) \in N \times N$ such that

$$f(m_1, n_1) = (m_2, n_2)$$

$$\Rightarrow \quad 2^{m_1}3^{n_1} = 2^{m_2}3^{n_2}$$

$$\Rightarrow \quad 2^{m_1 - m_2} . 3^{n_1 - n_2} = 1 = 2^0.3^0$$

$$\Rightarrow \quad m_1 - m_2 = 0, n_1 - n_2 = 0$$

$$\Rightarrow \quad m_1 = m_2, n_1 = n_2$$

$$\therefore \quad (m_1, n_1) = (m_2, n_2)$$

$\therefore$ $f : N \times N \to N$ is one-one.

Onto. A function, $f : A \to B$ is said to be onto iff for $y \in B$, there exists $x \in A$ such that $f(x) = y$

Now, $1 \in N$, but there is no $(m, n) \in N \times N$ such that $f(m, n) = 1$

$\therefore$ f is not onto.

(d) **One-one.** Let $f(x) = f(y) \Rightarrow x^2 + x = y^2 + y$

$$\Rightarrow \quad x^2 - y^2 + x - y = 0 \quad \Rightarrow \quad (x - y)(x + y + 1) = 0$$

$$\Rightarrow \quad x - y = 0, \quad \Rightarrow \quad x = y \qquad | \; x + y + 1 \neq 0, \text{ for } x, y \in N$$

$f : N \to N$ is one-one.

Onto. Let $y \in N$ such that $f(x) = y \Rightarrow x^2 + x = y$

$$\Rightarrow \quad x^2 + x - y = 0 \Rightarrow \quad x = \frac{-1 \pm \sqrt{1 + 4y}}{2}$$

Now, for $y \in N$, we get some values of x which are not in N.

$$\left(\text{Take } y = 1, \text{ we get } x = \frac{-1 \pm \sqrt{5}}{2} \notin N \right)$$

$\therefore$ $f : N \to N$ is not onto.

(e) **One-one.** Let $m, n \in N$ such that $f(m) = f(n)$

$$\Rightarrow \quad (m, | 10 - m |) = (n, | 10 - n |)$$

$$\Rightarrow \quad m = n, | 10 - m | = | 10 - n |$$

$$\Rightarrow \quad m = n, 10 - m = 10 - n$$

$$\Rightarrow \quad m = n, \text{ for all } m, n \in N$$

$\therefore$ $f : N \to N \times N$ is one-one.

Onto. Given $f : N \to N \times N$ is a function from N to $N \times N$. Therefore co-domain $f = N \times N$

Also range $f = \{f(n) : n \in N\} = (n, | 10 - n |)$

$\Rightarrow$ $N \times N \neq$ co-domain

$\therefore$ $f : N \to N \times N$ is not onto function.

Example 12. *Let A = B = {1, 2, 3, 4, 5}.* Define functions *f : A → B (if possible) such that*

(*a*) *f is one-one and onto* (P.T.U. B.Tech. Dec. 2005)

(*b*) *f is neither one-one nor onto*

(*c*) *f is one-one but not onto*

(*d*) *f is onto but not one-one.*

Sol. (*a*) Define $f : A \to B$ such that

$$f(x) = x \text{ for all } x \in A = B$$

i.e., $f(1) = 1, f(2) = 2, f(3) = 3$

$f(4) = 4, f(5) = 5$

i.e., different elements of A have different images in B Fig. 3.30.

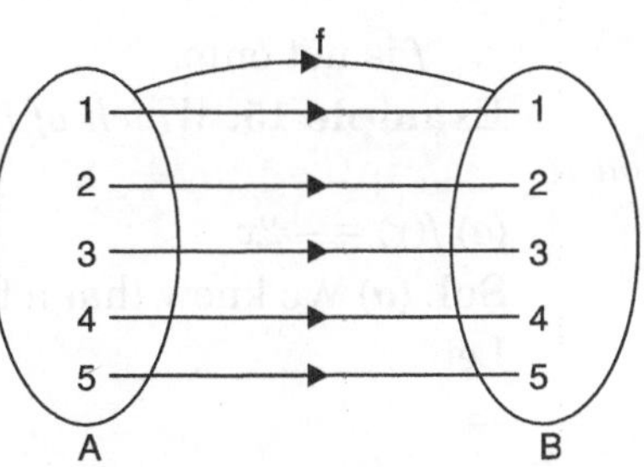

Fig. 3.30
one-one and onto

$\therefore$ $f : A \to B$ is one-one. Also, we observe that for each $y \in B$, there exists $x \in A$ such that $f(x) = y$. Hence f is onto also.

(*b*) Define $f : A \to B$ such that

$$f(x) = 1 \text{ for all } x \in A, 1 \in B$$

i.e., $f(1) = 1, f(2) = 1,$

$f(3) = 1, f(4) = 1, f(5) = 1$

i.e., different elements of A have same images in B Fig. 3.31.

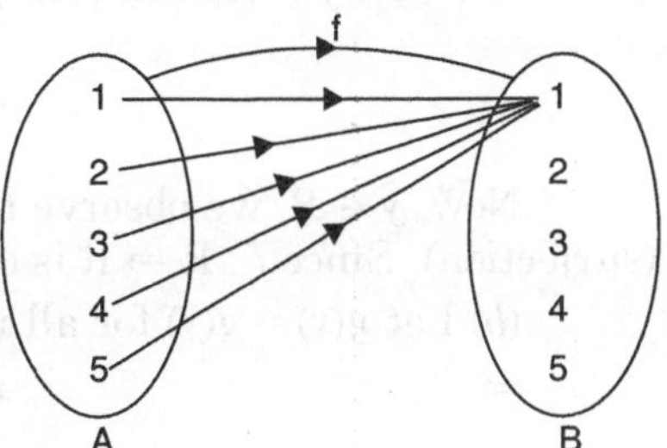

Fig. 3.31

$\therefore$ $f : A \to B$ is not one-one.

Also 2, 3, 4, 5 $\in$ B have no images in A.

$\therefore$ $f : A \to B$ is not onto

(*c*) Since A = B, $\therefore$ There is no function which is one-one but not onto.

(*d*) Again, as A = B, $\therefore$ There is no function which is onto, but not one-one.

Example 13. *Consider the function f : N × N → R defined by f(x, y) = (2x + 1) 2^y – 1, where N is the set of natural numbers including zero. Show that f is bijective.*

Sol. One-one. Let $f(x_1, y_1) = f(x_2, y_2)$ $\Rightarrow$ $(2x_1 + 1)2^{y_1} - 1 = (2x_2 + 1)2^{y_2} - 1$

$\Rightarrow$ $(2x_1 + 1)2^{y_1} = (2x_2 + 1)2^{y_2}$, which can hold if

$$2x_1 + 1 = 2x_2 + 1 \quad \text{and} \quad 2^{y_1} = 2^{y_2}$$

i.e., if $x_1 = x_2$ and $y_1 = y_2$

Hence, $(x_1, y_1) = (x_2, y_2)$.

$\therefore$ f is one-one.

Onto. For each $(x, y) \in N \times N$, $(2x + 1)\, 2^y - 1 \in R$ such that $f(x, y) = (2x + 1)\, 2^y - 1$. Hence range of f = R = co-domain f

$\therefore$ f is onto.

Hence, f is bijective.

Example 14. *Consider the function $f: N \to N$, where N is the set of natural numbers including zero defined by $f(n) = n^2 + 2$. Check whether the function f is (i) one-one (ii) onto.*

Sol. One-one. Let $m, n \in$ N such that $f(m) = f(n)$

$\Rightarrow \quad m^2 + 2 = n^2 + 2 \Rightarrow m^2 = n^2 \Rightarrow m = n \quad | \; m \neq -n$ since $m, n \in$ N

$\therefore$ f is one-one.

Onto. Given $\quad f(n) = n^2 + 2, n \in$ N

Let $\quad m = f(n) = n^2 + 2 \Rightarrow n^2 = m - 2 \Rightarrow n = \pm \sqrt{m-2}$

For $m \in$ N, $n \notin$ N (take $m = 1$, then $n = \pm\sqrt{-1} = \pm i$)

$\therefore$ f is not onto.

Example 15. *Which of the following functions are injections, surjections, or bijections on R*

(*a*) $f(x) = -2x$ (*b*) $g(x) = x^2 - 1$.

Sol. (*a*) We know that a function f : R $\to$ R is one-one iff $f(x) = f(y) \Rightarrow x = y$ for all $x, y \in$ R

Let $\quad f(x) = f(y)$

$\Rightarrow \quad -2x = -2y$

$\Rightarrow \quad x = y$ for all $x, y \in$ R.

$\therefore \quad f$: R $\to$ R is one-one (injection)

Also, let $y \in$ R such that $f(x) = y$

$$\Rightarrow \quad -2x = y \Rightarrow x = -\frac{y}{2} \quad \ldots(1)$$

Now, $y \in$ R, we observe that the values of x given by (1) are in R. $\therefore$ f : R $\to$ R is onto (surjection). Since f : R $\to$ R is one-one and onto. $\therefore$ f is bijection also.

(*b*) Let $g(x) = g(y)$ for all $x, y \in$ R

$\Rightarrow \quad x^2 - 1 = y^2 - 1 \Rightarrow x^2 = y^2$

$\Rightarrow \quad x = \pm y$

$\therefore \quad g$ = R $\to$ R is not one-one.

Also, co-domain g = R and

Range $R_g = \{g(x) : x \in R\} = \{x^2 - 1 : x \in R\}$

$= (0, \infty) \neq$ R = co-domain

$\therefore \quad g$: R $\to$ R is not **onto.**

Alternatively: Let $y = g(x) = x^2 - 1$

$\Rightarrow \quad x^2 = 1 + y \Rightarrow x = \pm\sqrt{1+y}$

For $\quad y \in$ R, $\exists\, x \notin$ R. Take

$y = -2$, then $x = \pm\sqrt{1-2} = \pm i \notin$ R

Hence g is not onto

Example 16. *Consider the function $f: N \to N$ where N is the set of natural numbers. Defined by $f(n) = n^2 + n + 1$. Show that f is one-one but not onto.* (P.T.U. B.Tech. Dec. 2008)

Sol. Given function f : N $\to$ N, where N is the set of natural numbers., is defined by

$$f(n) = n^2 + n + 1.$$

We know that a function $f : X \to Y$ is one-one if $f(x_1) = f(x_2) \Rightarrow x_1 = x_2 \; \forall \; x_1, x_2 \in x$

Also $f : x \to y$ is onto if for each $y \in y$, there exists at least one $x \in X$ such that $f(x) = y$.

One-one. Let $n_1, n_2 \in N$ such that $f(n_1) = f(n_2)$

$\Rightarrow \quad n_1^2 + n_1 + 1 = n_2^2 + n_2 + 1$

$\Rightarrow \quad n_1^2 - n_2^2 + n_1 - n_2 = 0$

$\Rightarrow \quad (n_1 - n_2)(n_1 + n_2 - 1) = 0$

$\Rightarrow$ either $\quad n_1 = n_2$ or $n_1 + n_2 - 1 = 0$

If $n_1 + n_2 - 1 = 0$, then $n_1 = 1 - n_2 \notin N$ for $n_2 \in N$

$\therefore \quad n_1 = n_2$. Hence $f : N \to N$ is one-one.

Onto. To check whether f is onto, let $m \in N$ such that $m = f(n) \; \forall \; n \in N$

or $\quad m = n^2 + n + 1 \Rightarrow n^2 + n + 1 - m = 0$

$$\Rightarrow \quad n = \frac{-1 \pm \sqrt{1 + 4(m-1)}}{2} = \frac{-1 \pm \sqrt{4m - 3}}{2}$$

Now for each $m \in N$, there exist $n \notin N$. Hence f is not onto.

Example 17. *Give an explicit formula for a function from the set of integers to the set of positive integers which is* (P.T.U. B. Tech. May 2009)

(a) one-one but not onto

(b) onto but not one-one

(c) one-one and onto

(d) neither one-one nor onto.

Sol. Recall that a function $f : X \to Y$ is called one-one if whenever

$$f(x_1) = f(x_2)$$

$$\Rightarrow \quad x_1 = x_2 \text{ for all } x_1, x_2 \in X.$$

Also a function $f : X \to Y$ is called an onto function if for each $y \in Y$, there is $x \in X$ such that $f(x) = y$.

Case I. *One-one but not onto*

Define $\quad f : Z \to Z^+$ by

$$f(x) = 2^x$$

Let $\quad f(x) = f(y) \Rightarrow 2^x = 2^y \Rightarrow x = y \; \forall \; x, y \in Z$

$\therefore \quad f$ is one-one.

If $f(x) = y$, then $2^x = y$

$\Rightarrow \quad x = \log_2 y$

Now for $y \in Z^+$, $x = \log_2 y$ is not an integer. Hence the function $f(x) : Z \to Z^+$ defined by $f(x) = 2^x$ is not an onto function.

Case II. *Not one-one, but onto*

Define $f : Z \to Z^+$ by

$$f(x) = | \, x \, |$$

Let $\quad f(x) = f(y) \Rightarrow | \, x \, | = | \, y \, | \Rightarrow x = \pm y$

$\therefore \quad f$ is not one-one.

Also $\quad f : Z \to Z^+$ is defined by $f(x) = | \, x \, |$

$\therefore \quad$ Co-domain of $f = Z^+ =$ set of positive integers range of

$$f = \{f(x) : x \in Z\} = \{| \, x \, | : x \in Z\} = Z^+$$

Since co-domain of f = range of f

$\therefore$ f is onto.

Case III. *One-one and onto*

Define $f = Z \to Z^+$ by $f(x) = x + 5$

Let $f(x) = f(y) \Rightarrow x + 5 = y + 5 \Rightarrow x = y \ \forall \ x, y \in Z.$

$\therefore$ f is one-one.

Let $f(x) = y \Rightarrow y = x + 5 \Rightarrow x = y - 5$

Now for each $y \in Z^+$, x is an integer. Hence f is onto also.

Case IV. *Not one-one, not onto*

Define $f : Z \to Z^+$ by $f(x) = x^2$

Let $f(x) = f(y) \Rightarrow x^2 = y^2 \Rightarrow x = \pm y \ \forall \ x, y \in Z$

$\therefore$ f is not one-one

Also, let $f(x) = y \Rightarrow y = x^2 \Rightarrow x = \pm \sqrt{y}$

Now for each $y \in Z^+$, x is not an integer. Hence f is not onto.

3.13. GRAPHICAL REPRESENTATION OF ONE-ONE AND ONTO FUNCTIONS

When we say that a function $f : A \to B$ is one-one, this means that there are no two distinct pairs (a_1, b) and (a_2, b) in the graph of f. Hence, each horizontal line (either below or above the x-axis) can intersect the graph of f in **atmost** one point.

When we say that a function $f : A \to B$ is onto, this means that for every $b \in B$, there must be at least one $a \in A$ such that (a, b) belongs to the graph of f. Hence, each horizontal line (either below or above the x-axis) must intersect the graph of f **in atleast one point**

Further, if each horizontal line either below or above x-axis intersects the graph of f in **exactly** one point, we say f is both one-one and onto

Example 18. *Consider the following graphs (Fig. 3.32). Determine which of these are functions from R into R.*

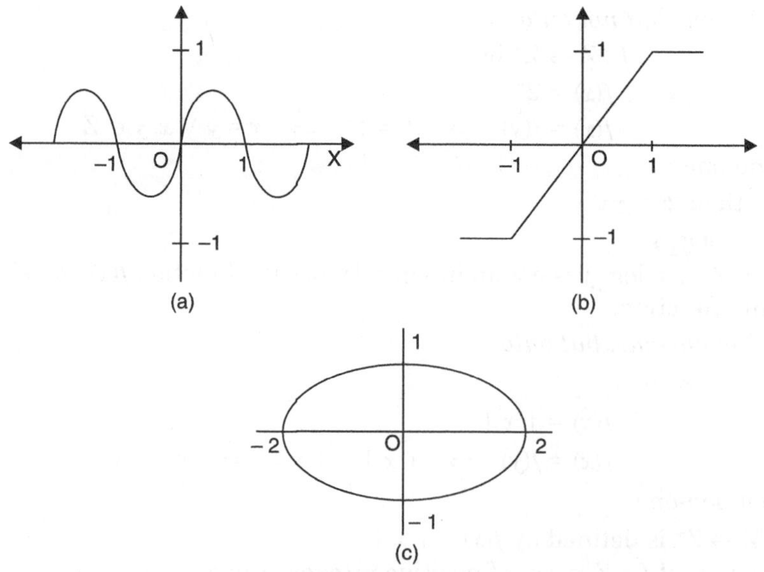

Fig. 3.32

Sol. (*a*) Recall that a set of points on a coordinate diagram is a function iff every vertical line contains exactly one point of the set. Hence the graph given in Fig. 3.32(*a*), (*b*) represents functions. But, the graph given in Fig. 3.32(*c*) does not represent a function. Since the vertical line drawn through the graph (Fig. 3.32) meets in two points.

Example 19. *Determine which of the graphs in the followings figures (Fig. 3.33) are functions from R to R.*

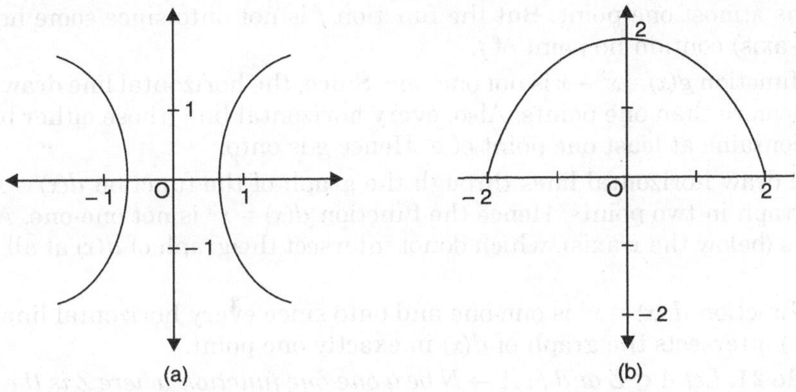

Fig. 3.33

Sol. If a vertical line is drawn, it does not contain exactly one point. Hence, both the graphs (*a*) and (*b*) do not represent any function.

Example 20. *Consider the functions defined as* $f(x) = 2^x$, $g(x) = x^3 - x$, $h(x) = x^2$, $d(x) = x^3$ *and their graphs as shown below (Fig. 3.34). Determine which of the functions are one-one. Which of them are onto ?*

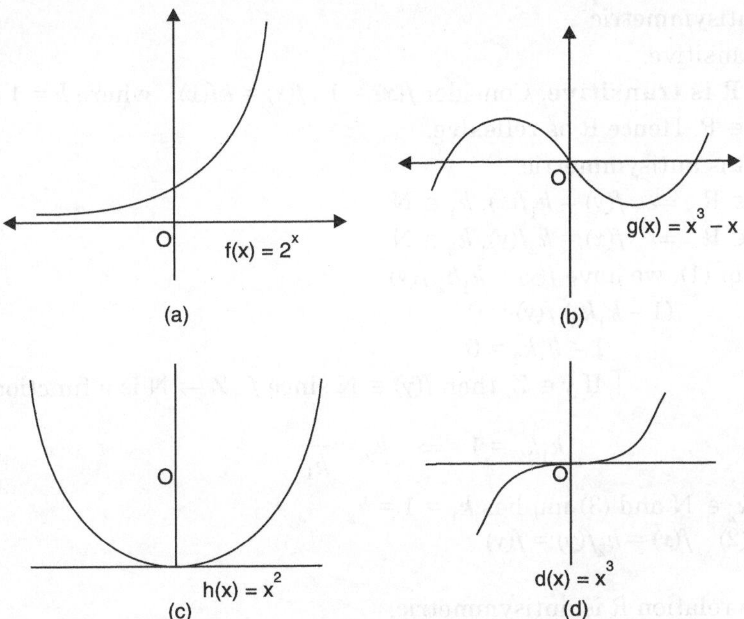

Fig. 3.34

Sol. Recall that a set of points on a coordinate diagram is a function iff every vertical line contains exactly one point of the set. Hence, the graphs shown in Fig. 3.34(*a*), (*b*), (*c*) and (*d*) represent functions.

Also, if a horizontal line is drawn through the graph and if it meets the graph in atmost one point, then the function is one-one. If each horizontal lines intersect the graph in at least one point, then the function is onto.

(*a*) The function $f(x) = 2^x$ is one-one since the horizontal line drawn through the Fig. 3.34(*a*) contains atmost one point. But the function f is not onto since some horizontal lines (those below x-axis) contain no point of f.

(*b*) The function $g(x) = x^3 - x$ is not one-one. Since, the horizontal line drawn through the graph contains more than one points. Also, every horizontal line (those either below x-axis or above x-axis) contains at least one point of g. Hence g is onto.

(*c*) If we draw horizontal lines through the graph of the function $d(x) = x^2$, these lines intersect the graph in two points. Hence the function $d(x) = x^2$ is not one-one. Also, there are horizontal lines (below the x-axis), which donot intersect the graph of $d(x)$ at all. Hence $d(x)$ is not onto.

(*d*) The function $d(x) = x^3$ is one-one and onto since every horizontal line (either above or below x-axis) intersects the graph of $d(x)$ in exactly one point.

Example 21. *Let $A \subseteq Z$ and $f : A \to N$ be a one-one function where Z is the set of integers and N is the set of natural numbers. Let R be a relation on A defined as under :*

$(x, y) \in R$ iff $f(y) = kf(x)$ for $k \in N$. Prove that R is a partial order relation.

(P.T.U. B. Tech. Dec. 2008)

Sol. Given $f : A \to N$ is a one-one function where $A \subseteq Z$, the set of integers. The relation R on A is defined as below :

$(x, y) \in R \Leftrightarrow f(y) = kf(x)$ for all $x, y \in Z$. We show R is a partial order relation *i.e.*

(*i*) R is reflexive

(*ii*) R is antisymmetric

(*iii*) R is transitive.

To show R is transitive. Consider $f(x) = 1 \,.\, f(x) = kf(x)$ where $k = 1 \in N$.

$\Rightarrow$ $(x, x) \in R$. Hence R as reflexive.

To show R is antisymmetric.

Let $(x, y) \in R \Rightarrow f(y) = k_1 f(x),\ k_1 \in N$ ...(1)

Let $(y, x) \in R \Rightarrow f(x) = k_2 f(y),\ k_2 \in N$ ...(2)

Using (2) in (1), we have $f(y) = k_1 k_2 f(y)$

$\Rightarrow \quad (1 - k_1 k_2) f(y) = 0$

$\Rightarrow \quad 1 - k_1 k_2 = 0$

| If $y \in Z$, then $f(y) \in N$ since $f : Z \to N$ is a function. Hence $f(y) \neq 0$

$$\Rightarrow \quad k_1 k_2 = 1 \Rightarrow k_2 = \frac{1}{k_1}$$

But $k_1, k_2 \in N$ and (3) implies $k_1 = 1 = k_2$

$\therefore$ from (2) $f(x) = k_2 f(y) = f(y)$

$\Rightarrow x = y$ | f is one-one

Hence the relation R is antisymmetric.

To show R is transitive.

Let $(x, y) \in R \Rightarrow f(y) = k_1 f(x)$ for $x, y \in Z, k_1 \in N$

$(y, z) \in R \Rightarrow f(z) = k_2 f(y)$ for $x, y, \in Z, k_2 \in N$

or $f(z) = k_2(k_1 f(x))$

$= k_2 k_1 f(x) = k_2 k, f(x)$

$= k f(x)$ where $k = k_2 k_1 \in N$

$\Rightarrow (x, z) \in R$

Hence R is transitive also.

Therefore, the relation R is a partial order relation.

TEST YOUR KNOWLEDGE 3.1

SHORT ANSWER TYPE QUESTIONS

1. State whether or not each diagram in the given figure, defines a function from $A = \{a, b, c\}$ into $B = \{x, y, z\}$.

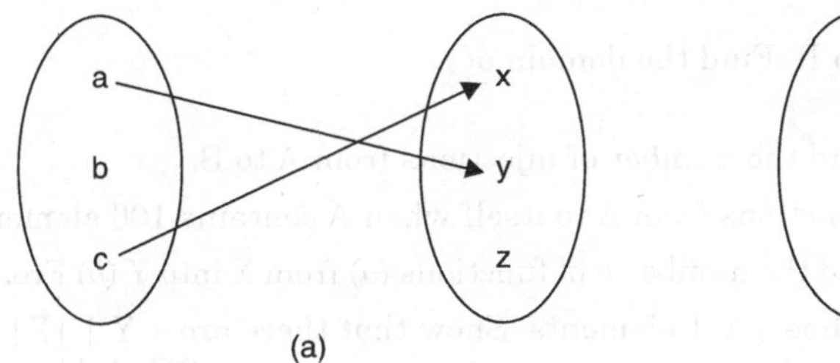

(a)

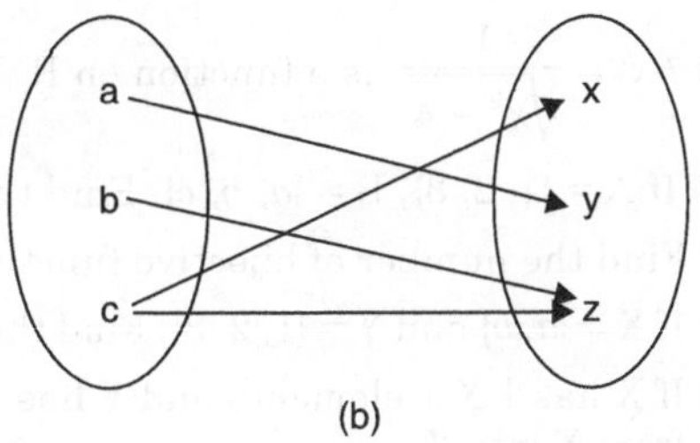

(b)

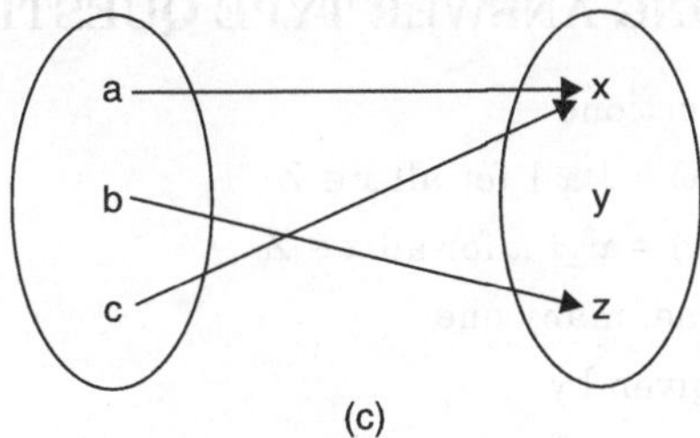

(c)

2. Consider the figure given below :

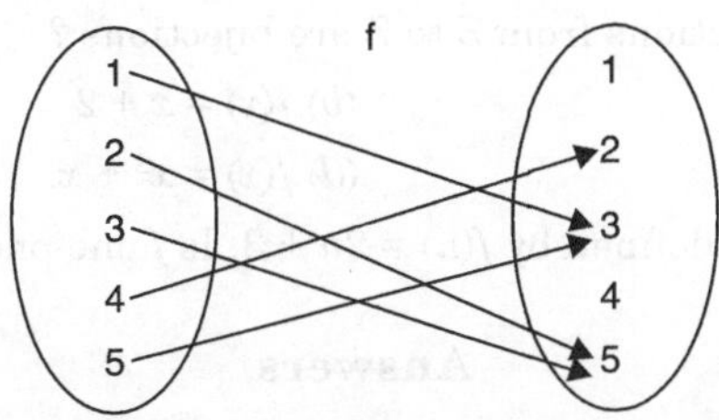

(a) Find the graph of the function and write f as a set of ordered pairs

(b) Find $f(S)$ where $S = \{1, 3, 5\}$

(c) Find $f^{-1}(T)$ where $T = \{1, 2\}$

(d) Find $f^{-1}(3)$.

3. (*a*) Let X = {1, 2, 3, 4}. Determine whether each of the following relations on X (set of ordered pairs) is a function from X into X.

(*i*) $f = \{(2, 3), (1, 4), (2, 1), (3, 2), (4, 4)\}$

(*ii*) $g = \{(3, 1), (4, 2), (1, 1)\}$

(*iii*) $h = \{(2, 1), (3, 4), (2, 1), (4, 4)\}$

(*b*) If $f(x) = x^2 - 3x - \frac{3}{x} + \frac{1}{x^2}$, then show that $f\left(\frac{1}{x}\right) = f(x)$

4. What is the range of the function $f(x)$ where $f(x) = \begin{cases} -x, & x < 0 \\ x, & 0 \le x \le 1 \\ \frac{1}{x}, & x > 1 \end{cases}$

5. Let $f : R \to R$ be a function defined by $f(x) = x^2$. Is f one-one, onto or both ?

6. Let $A = \{x : -1 < x < 1\} = B$ and $f : A \to B$ is a function defined by $f(x) = \frac{x}{2}$. Examine whether f is bijective or not.

7. Let $f(x) = \frac{1}{\sqrt{x^2 - 4}}$, is a function on R. Find the domain of f.

8. (*a*) If A = {1, 2, 3}, B = {*a*, *b*, *c*}. Find the number of injections from A to B.

(*b*) Find the number of bijective functions from A to itself when A contains 106 elements.

(*c*) If X = {*a*, *b*} and Y = {1, 2, 3}. Find the number *n* of functions (*a*) from X into Y (*b*) From Y into X

(*d*) If X has | X | elements and Y has | Y | elements. Show that there are $|Y|^{|X|}$ functions from X into Y. (**Hint.** Use theorem I)

LONG ANSWER TYPE QUESTIONS

9. Consider the following functions

(*a*) $f : Z \to Z$ defined by $f(x) = |x|$ for all $x \in Z$

(*b*) $f : Z \to Z$ defined by $f(x) = x^2 + x$ for all $x \in Z$.

Examine which are one-one, many one

10. Show that the functions given by

(*a*) $f : R \to R$ defined by $f(x) = 3x^3 + 5$

(*b*) $f : Q \to Q$ defined by $f(x) = 2x - 3$ are bijections.

11. Which of the following functions from Z to Z are bijections ?

(*a*) $f(x) = x^3$ (*b*) $f(x) = x + 2$

(*c*) $f(x) = 2x + 1$ (*d*) $f(x) = x^2 + x$.

12. Let $f : N \to N$ is a function defined by $f(n) = 2n + 3$. Is f one-one, onto or both.

Answers

1. (*a*) No, the element *b* is not associated to any element of B

(*b*) No, two elements *x* and *z* are associated to *c*

(*c*) Yes

2. (*a*) Graph $f = \{(1, 3), (2, 5), (3, 5), (4, 2), (5, 3)\}$

(*b*) $\{3, 5\}$ (*c*) $\{4\}$ (*d*) $\{1, 5\}$

3. (*a*) (*i*) No, the element 2 is repeated twice as its first coordinate.

(*ii*) No, the element 2 does not appear as the first coordinate in any ordered pair in g.

(*iii*) Yes, Although $2 \in X$ appears as the first coordinate in the two ordered pairs, but these ordered pairs , but these ordered pairs are same.

4. $(0, \infty)$

5. Neither one-one, nor onto

6. Bijective *i.e.,* one-one and onto

7. $(-\infty, -2) \cup (2, \infty)$

8. (*a*) 6 (*b*) 106 !

9. (*a*) Many-one (*b*) Many-one.

11. (*b*)

12. One-one, not onto.

3.14. COMPOSITION OF FUNCTIONS

Def. Let $f : A \to B$ and $g : B \to C$ be two functions. Then the function $gof : A \to C$, defined by

$$(gof)(x) = g(f(x)) \text{ for all } x \in A, \text{ is called composition of } f \text{ and } g.$$

Thus, to find the image of '*a*' under *gof*, we first find the image of '*a*' under *f* and then we find the image of $f(a)$ under *g*.

Further, we say that *gof* is defined only if range R_f is a subset of domain D_g. Also, *gof* is a function from A to C.

Remark. If we view *f* and *g* as relations, we use the notation *fog* for the composition of *f* and *g*. If we use *f* and *g* as functions, then, we use the notation *gof* for the composition of *f* and *g*.

For example, let $f : \{1, 2, 3\} \to \{a, b\}$ is a function defined by $f(1) = a$, $f(2) = a$, $f(3) = b$ and $g : \{a, b\} \to \{5, 6, 7\}$ is defined by $g(a) = 5$, $g(b) = 7$

We find *gof* and *fog*.

Here Range $R_f = \{a, b\}$, Domain $D_g = \{a, b\}$

Since Range $R_f \subseteq$ Domain D_g $\therefore$ *gof* is defined and $gof : \{1, 2, 3\} \to \{5, 6, 7\}$

Now, $(gof)(1) = g(f(1)) = g(a) = 5$

$(gof)(2) = g(f(2)) = g(a) = 5$

$(gof)(3) = g(f(3)) = g(b) = 7$

To find *fog*, we observe that range $R_g = \{5, 6, 7\}$ and domain $D_f = \{1, 2, 3\}$. Since range $R_g \not\subseteq$ domain D_f

$\therefore$ *fog* is not defined.

Theorem II. *Let $f : A \to B$, $g : B \to C$ and $h : C \to D$, be three functions, then*

$$fo(goh) = (fog)oh$$

Proof. Given. $f : A \to B$, $g : B \to C$, $h : C \to D$

$\Rightarrow$ $goh : B \to D$, $fog : A \to C$

$\Rightarrow$ $fo(goh) : A \to D$, $(fog)oh : A \to D$

Thus $fo(goh)$ and $(fog)oh$ are functions from A to D.

We now show $fo(goh) = (fog)oh$

Let $x \in A$ and consider

$$(fo(goh))x = f(goh)(x) = f(g(h)(x)) \quad \ldots(1)$$

Also, $$((fog)oh)(x) = (fog)(h(x)) = f((g(h(x))) \quad \ldots(2)$$

From (1) and (2), $(fo(goh))(x) = ((fog)oh)\, x$ for $x \in A$

$\Rightarrow$ $fo(goh) = (fog)oh.$

Theorem III. *Let $f : A \to B$, $g : B \to C$ are one-one (injections), Then $gof : A \to C$ is also one-one, not conversly.* (P.T.U. B.Tech. Dec. 2013)

Proof. Let $x, y \in$ A and consider $(gof)(x) = (gof)(y)$

$\Rightarrow \quad g(f(x)) = g(f(y))$

$\Rightarrow \quad f(x) = f(y)$ | g is one-one

$\Rightarrow \quad x = y$ | f is one-one

$\therefore \quad gof$ is one-one.

Theorem IV. *If $f : A \to B$, $g : B \to C$ are onto (surjections) functions, then $gof : A \to C$ is also onto (surjection) function, not conversly.*

Proof. To show gof is onto, we show that for every element in C has its pre-image in A. *i.e.,* for all $z \in$ C there exists $x \in$ A such that $(gof)(x) = z$.

Now $z \in$ C and g : B → C is onto function from B to C ∴ We can find $y \in$ B such that $g(y) = z$. Also $y \in$ B and f : A → B is onto function ∴ we can find $x \in$ A such that $f(x) = y$

$\therefore \quad (gof)(x) = g(f(x)) = g(y) = z$

$\therefore \quad gof$ is onto function from A to C.

Theorem V. *If $f : A \to B$, $g : B \to C$ are one-one and onto functions or surjections, then $gof : A \to C$ is also one-one and onto (surjection).*

Proof. Combining theorem, *II* and theorem *III,* we get the required result.

Theorem VI. *Let $f : A \to B$ and $I_B : B \to B$ be two functions, then $I_B of = f$.*

Proof. Given I_B : B → B is an identity function and f : A → B is a function from A to B. Let $x \in$ A. Consider

$$(I_B of)(x) = I_B(f(x)) = I_B(y) \qquad | \; y = f(x) \in B$$
$$= y = f(x)$$

$\Rightarrow \quad (I_B of)(x) = f(x)$

$\Rightarrow \quad I_B of = f.$

Theorem VII. *Let $I_A : A \to A$ and $f : A \to B$ be two functions, then $foI_A = f$.*

Proof. Given I_A : A → A and f : A → B ∴ $fo\,I_A$ is a function from A to B let $x \in$ A and consider

$$(foI_A)(x) = fo(I_A(x)) = f(x)$$

$\Rightarrow \quad foI_A = f.$

3.15. INVERSE FUNCTION

Let f : A → A and if there is a function g : A → A such that $gof = fog = I_A$, then g is called inverse of f and is denoted by f^{-1}. *i.e.,* $g = f^{-1}$.

If f : A → B is a one-one and onto function, then the function defined by g : B → A is called inverse of f.

Example. *Consider the functions $f : R \to R$ and $g : R \to R$ defined by $f(x) = x^4$ and $g(x) = x^{1/4}$. Then for each $x \in R$, $f(x)$ and $g(x)$ are inverses of one another.*

Sol. Here f and g are inverses of one another if

$$(fog)(x) = I_x \quad \text{and} \quad (gof)(x) = I_x$$

Now $\quad (fog)(x) = f(g(x)) = f(x^{1/4}) = (x^{1/4})^4 = x = I_x$

Also $\quad (gof)(x) = g(f(x)) = g(x^4) = (x^4)^{1/4} = x = I_x$

Hence $\quad f = g^{-1}$ and $g^{-1} = f$

3.16. METHOD TO FIND THE INVERSE OF A BIJECTION

Let $f : A \to B$ be a bijection (one-one and onto). To find f^{-1},

Step I. Take $f(x) = y$, $y \in B$, $x \in A$.

Step II. Solve $f(x) = y$ to obtain x in terms of y

Step III. Replace x by $f^{-1}(y)$ in **step II**, we get the required function.

For example, let $f : R \to R$ be a function defined by $f(x) = x^2$. Is f^{-1} exist ?
If so, find $f^{-1}(4)$, $f^{-1}(0)$, $f^{-1}(-1)$.

Here f^{-1} will exist if f is one-one and **onto.**

One-one. Let $x, y \in R$ such that $f(x) = f(y) \Rightarrow x^2 = y^2$

$\Rightarrow \quad x = \pm y.$

If $x = y$, then f is one-one.

Onto. Here $R_f = [0, \infty]$ and co-domain $f = R \neq R_f$

$\therefore$ f is not onto.

Hence $f^{-1}(4)$, $f^{-1}(0)$, $f^{-1}(-1)$ donot exist

Theorem VIII. *Let $f : A \to A$ is a function from A to A. Then f^{-1} exists iff f is one-one and onto.*

Proof. Let f^{-1} exists, we show f is one-one and onto.

Let $(a, b), (c, b) \in f$. Then, by definition of f^{-1}, (b, a) and $(b, c) \in f^{-1}$

Since f^{-1} is a function $\therefore$ $a = c$. Hence f is one-one.

Further, let $b \in A$ and since f^{-1} is a function from A to A, we can find $a \in A$ such that $f^{-1}(b) = a \Rightarrow b = f(a)$. Hence f is onto.

Converse. Let f is one-one and onto. We show f^{-1} exists.

Consider $g : A \to A$ is a function from A to A. We show $fog = I_A$ and $gof = I_A$

Let $x \in A$ and since $f : A \to A$, we can find $y \in A$ such that $f(x) = y$

Also, $g : A \to A$ is a function from A to A. $\therefore$ for $y \in A$, we can find $x \in A$ such that $g(y) = x$

Now, $\quad (gof)(x) = g(f(x)) = g(y) = x$

$\Rightarrow \quad (gof)(x) = x$ for all $x \in A$

$\Rightarrow \quad gof = I_A$

Also, $\quad (fog)(y) = f(g(y)) = f(x) = y$

$\Rightarrow \quad (fog)(y) = y$ for all $y \in A$

$\Rightarrow \quad fog = I_A$

Hence f^{-1} exists.

Theorem IX. *A function $f : A \to B$ has an inverse iff f is one-one and onto (bijective).*

Proof. Let f^{-1} exists and we show f is one-one and onto.

Let $(a, b), (c, b) \in f$ for $a, c \in A$, $b \in B$. Since f^{-1} exists. $\therefore$ by definition, $(b, a), (b, c) \in f^{-1}$. But f^{-1} is a function. $\therefore$ We must have $a = c$ $\therefore$ f is one-one.

Further, let $b \in B$ and since f^{-1} is a function from B to A, we can find $a \in A$ such that $f^{-1}(b) = a \Rightarrow b = f(a)$. Hence f is onto.

Converse. Let f is one-one and onto. We show f^{-1} exists.

Consider $g : B \to A$ is a function from B to A. We show $fog = I_B$, $gof = I_A$

Let $x \in A$ and since $f : A \to B$, there exists $y \in B$ such that $f(x) = y$

Also, $g : B \to A$. $\therefore$ we have $g(y) = x$

Now, $(gof)(x) = g(f(x)) = g(y) = x,$

$\Rightarrow$ $(gof)(x) = x$ for all $x \in A$

$\Rightarrow$ $gof = I_A$ ...(1)

Similarly, $(fog)(y) = f(g(y)) = f(x) = y$

$\Rightarrow$ $(fog)\, y = y$ for all $y \in B$

$\Rightarrow$ $fog = I_B$ ...(2)

Hence from (1) and (2), $gof = I_A$, $fog = I_B \Rightarrow f^{-1}$ exists.

Theorem X. *If $f : A \to B$ and $g : B \to C$ are two bijections, then $(gof)^{-1} = f^{-1}og^{-1}$.*

Proof. Given $f : A \to B$ is a bijection, $g : B \to C$ is a bijection. $\therefore$ $gof : A \to C$ is also a bijection (one-one and onto) (Theorem IV)

$\Rightarrow$ $(gof)^{-1} : C \to A$ exists.

Again $f : A \to B$ is a bijection $\Rightarrow f^{-1} : B \to A$ is also a bijection. Also, $g : B \to C$ is a bijection. $\Rightarrow g^{-1} : C \to B$ is also a bijection.

$\therefore$ $f^{-1}og^{-1}$ is a bijection. Hence $f^{-1}og^{-1}$ also exists.

Let $x \in A$, $y \in B$, $z \in C$ such that $f(x) = y$, $g(y) = z$

Consider $(gof)(x) = g(f(x)) = g(y) = z$

$\Rightarrow$ $(gof)^{-1}(z) = x$ ...(1)

Also, $f(x) = y \Rightarrow f^{-1}(y) = x$

$g(y) = z \Rightarrow g^{-1}(z) = y$

$\therefore$ $(f^{-1}og^{-1})(z) \Rightarrow f^{-1}(g^{-1}(z)) = f^{-1}(y) = x$ ...(2)

From (1) and (2),

$$(gof)^{-1}(z) = (f^{-1}og^{-1})(z)$$

$$(gof)^{-1} = f^{-1}og^{-1}.$$

ILLUSTRATIVE EXAMPLES

Example 1. *Let $f : \{1, 2, 3\} \to \{a, b\}$ be a function defined by $f(1) = a$, $f(2) = b$, $f(3) = b$. Let $g : \{a, b\} \to \{5, 6, 7\}$ be a function defined by $g(a) = 5$, $g(b) = 7$. Find gof, fog.*

Sol. To find *gof*. Given $f : \{1, 2, 3\} \to \{a, b\}$ and $g : \{a, b\} \to \{5, 6, 7\}$.

$\therefore$ Here Range $R_f = \{a, b\} =$ Domain D_g

$\therefore$ gof is defined (since R_f is a subset of D_g)

i.e., $gof : \{1, 2, 3\} \to \{5, 6, 7\}$

Here $(gof)(1) = g(f(1)) = g(a) = 5$

$(gof)(2) = g(f(2)) = g(b) = 7$

$(gof)(3) = g(f(3)) = g(b) = 7$

$\therefore$ $gof = \{5, 7\}$

To find *fog*. Given Range $R_g = \{5, 6, 7\}$

Domain $D_f = \{1, 2, 3\}$. $\therefore$ Since Range $R_g \not\subseteq$ Domain D_f

$\therefore$ fog is not defined.

Example 2. Let *$f : R \to R$ be defined by $f(x) = x^3$ and $g : R \to R$ be defined by $g(x) = 3x + 1$. Find $(gof)(x)$, $(fog)(x)$. Is $gof = fog$?*

Sol. Given $f : R \to R, g : R \to R$

Here Range $R_f = R$, Domain $D_g = R$

$\therefore$ *gof* is defined. (since R_f is a subset of D_g)

Also, Range $R_g = R$, Domain $D_f = R$

$\therefore$ *fog* is defined (since R_g is a subset of D_f)

Now, $(gof)(x) = g(f(x)) = g(x^3) = 3x^3 + 1$

Also, $(fog)(x) = f(g(x)) = f(3x + 1) = (3x + 1)^3$

Since $(gof)(x) = 3x^3 + 1, (fog)(x) = (3x + 1)^3$

$\therefore$ $gof \neq fog.$

Example 3. *(a) If $f(x) = x^2$, $g(x) = x^3$ find (fogofog)(x).*

(b) If $f(x) = x + 1$, $g(x) = x + 3$, find (fofofof)(x).

(c) If $f(x) = 4x - 1$, $g(x) = x^2 + 2$, find (fo(fog))(1).

(d) If $f(x) = \dfrac{x}{x+1}$, $g(x) = \dfrac{1}{x-1}$, find (fog)(x)

(e) If $f(x) = x + 5$, $g(x) = x^2$, find (gof)(x), where $f : R \to R, g : R \to R$ is given.

(f) If $f : R \to R$ and $g : R \to R$ are two functions defined by $f(x) = \sin x$ and $g(x) = x^2$. Find fog and gof. Is fog = gof? (P.T.U. B.Tech. May 2013)

Sol. *(a)* $(fogofog)(x) = f(g(f(g(x)))) = f(g(f(x^3)) = f(g(x^6))$

$= f(x^{18}) = x^{36}.$ | $f(x) = x^2$; $f(x^3) = (x^3)^2 = x^6$; $g(x) = x^3$; $g(x^6) = (x^3)^6 = x^{18}$

(b) $(fofofof)(x) = f(f(f(f(x)))) = f(f(f(x + 1)))$

$= f(f(x + 2)) = f(x + 3) = x + 4$

(c) $(fofog)(1) = f(f(g(1)) = f(f(3))$ | $g(x) = x^2 + 2, g(1) = 3$

$= f(11) = 44 - 1 = 43$ | $f(x) = 4x - 1, f(3) = 11$

(d) $(fog)(x) = f(g(x)) = f\left(\dfrac{1}{x-1}\right)$

$$= \frac{\frac{1}{x-1}}{\frac{1}{x-1}+1} = \frac{\frac{1}{x-1}}{\frac{1+x-1}{x-1}} = \frac{1}{x}$$ | $f(x) = \dfrac{x}{x+1}$

(e) $(gof)(x) = g(f(x)) = g(x + 5) = (x + 5)^2 = x^2 + 25 + 10x.$

(f) $(fog)(x) = f(g(x)) = f(x^2) = \sin x^2$

$(gof)(x) = g(f(x)) = g(\sin x) = (\sin x)^2 = \sin^2 x$

Thus, $gof \neq fog$

Example 4. *Consider the functions $f, g : R \to R$ defined by*

$$f(x) = x^2 + 3x + 1, g(x) = 2x - 3.$$

Find the composition functions

(i) fof *(ii) fog* *(iii) gof.*

Sol. *(i)* $(fof)(x) = f[f(x)] = f(x^2 + 3x + 1)$

$= (x^2 + 3x + 1)^2 + 3(x^2 + 3x + 1) + 1$

$= x^4 + 9x^2 + 1 + 6x^3 + 2x^2 + 6x + 3x^2 + 9x + 3 + 1$

$= x^4 + 6x^3 + 14x^2 + 15x + 5.$

(*ii*) $(gof)(x) = g[f(x)] = g(x^2 + 3x + 1)$

$= g(x^2 + 3x + 1) - 3 = 2x^2 + 6x + 2 - 3$

$= 2x^2 + 6x - 1.$

(*iii*) $(fog)(x) = f[g(x)] = f(2x - 3)$

$= (2x - 3)^2 + 3(2x - 3) + 1$

$= 4x^2 + 9 - 12x + 6x - 9 + 1$

$= 4x^2 - 6x + 1.$

Example 5. *Let f, g, h be functions from N to N, where N is the set of natural numbers so that*

$$f(n) = n + 1 \; ; \; g(n) = 2n \text{ and}$$

$$h(n) = \begin{cases} 0, & \text{when } n \text{ is even} \\ 1, & \text{when } n \text{ is odd.} \end{cases}$$

Determine fof, fog, gof, goh, hog, (fog)oh.

Sol. (*i*) $fof(n) = f\,[f(n)] = f(n + 1) = n + 2$

(*ii*) $fog(n) = f\,[g(n)] = f(2n) = 2n + 1$

(*iii*) $gof(n) = g[f(n)] = g(n + 1) = 2(n + 1) = 2n + 2$

(*iv*) **When *n* is even :**

$$goh(n) = g[h(n)] = g(0) = 0$$

when *n* is odd :

$$goh(n) = g[h(n)] = g(1) = 2$$

(*v*) $hog(n) = h[g(n)] = h(2n) = 1$ | $2n$ is even

(*vi*) **When *n* is even :**

$$(fog)\; oh(n) = (fog)\;[h(n)] = fog(0) = f\,[g(0)] = f(0) = 1$$

when *n* is odd :

$$(fog)\; oh(n) = (fog)\;[h(n)] = fog(1) = f\,[g(1)] = f(2) = 2 + 1 = 3$$

So, $(fog)oh = 1$, when n is even and $(fog)oh = 3$, when n is odd.

Example 6. *Consider $A = B = C = R$ and let $f : A \to B$ and $g : B \to C$ be defined by $f(x) = x + 9$ and $g(y) = y^2 + 3$.*

Find the following composition functions :

(*i*) $(fof)(a)$ (*ii*) $(gog)(a)$

(*iii*) $(fog)(b)$ (*iv*) $(gof)(b)$

(*v*) $(gof)(4)$ (*vi*) $(fog)(-4)$.

Sol. (*i*) $(fof)(a) = f[f(a)] = f(a + 9) = (a + 9) + 9 = a + 18.$

(*ii*) $(gog)(a) = g[g(a)] = g(a^2 + 3) = (a^2 + 3)^2 + 3 = a^4 + 6a^2 + 12.$

(*iii*) $(fog)(b) = f[g(b)] = f(b^2 + 3) = (b^2 + 3) + 9 = b^2 + 12.$

(*iv*) $(gof)(b) = g[f(b)] = g(b + 9) = (b + 9)^2 + 3 = b^2 + 18b + 84$.

(*v*) $(gof)(4) = g[f(4)] = g(13) = (13)^2 + 3 = 172$.

(*vi*) $(fog)(-4) = f[g(-4)] = f(19) = 19 + 9 = 28$.

Example 7. *Let* $X = \{a, b, c\}$. *Define* $f : X \to X$ *such that*

$$f = \{(a, b), (b, a), (c, c)\}$$

Find (*i*) f^{-1} (*ii*) f^2 (*iii*) f^3 (*iv*) f^4.

Sol. (*i*) The inverse function $f^{-1} = \{(b, a), (a, b), (c, c)\}$ (Fig. 3.35).

(*ii*) The f^2 is fof (Fig. 3.36).

$(fof)(a) = f[f(a)] = f(b) = a$, $(fof)(b) = f[f(b)] = f(a) = b$

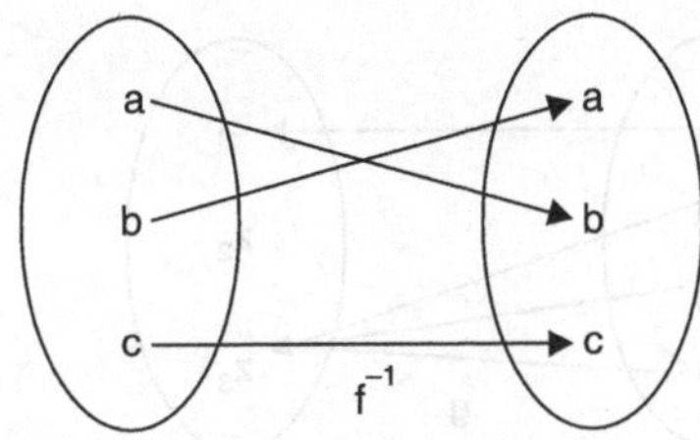

Fig. 3.35

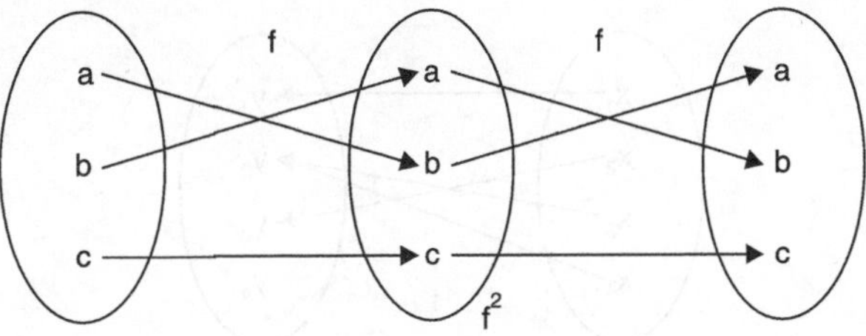

Fig. 3.36

$(fof)(c) = f[f(c)] = f(c) = c$,

So, $f^2 = \{(a, a), (b, b), (c, c)\}$.

(*iii*) The f^3 is $fofof$ *i.e.*, fof^2 (Fig. 3.37).

$(fof)^2\ (a) = f[f^2(a)] = f(a) = b$, $(fof^2)(b) = f[f^2\ (b)] = f(b) = a$

$(fof^2)(c) = f[f^2(c)] = f(c) = c$

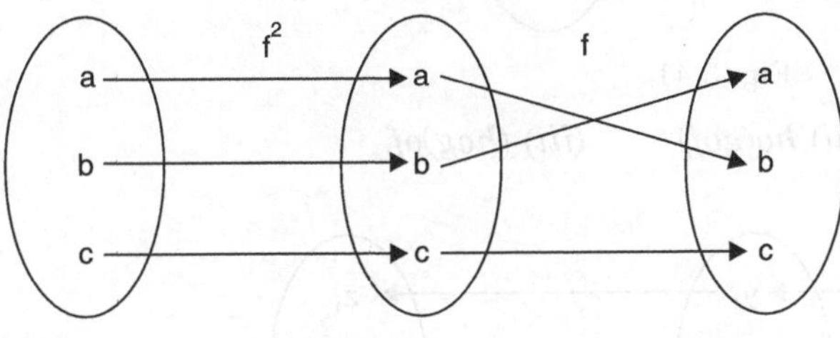

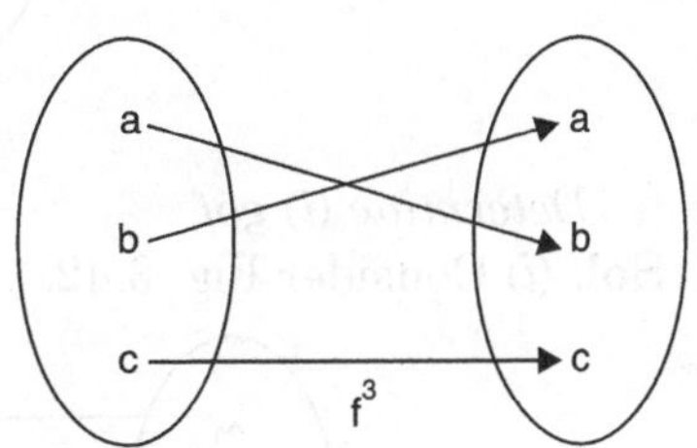

Fig. 3.37

So, $f^3 = \{(a, b), (b, a), (c, c)\}$.

(*iv*) The f^4 is $fofofof$ *i.e.*, fof^3 (Fig. 3.38).

$(fof^3)\ (a) = f[f^3\ (a)] = f(b) = a$, $(fof^3)\ (b) = f[f^3\ (b)] = f(a) = b$

$(fof^3)\ (c) = f[f^3\ (c)] = f(c) = c$

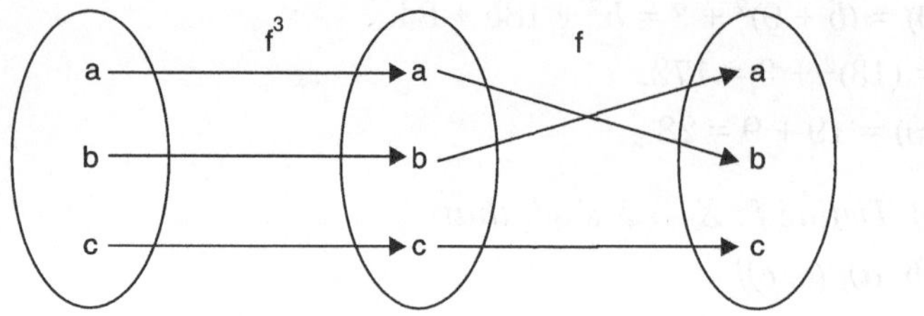

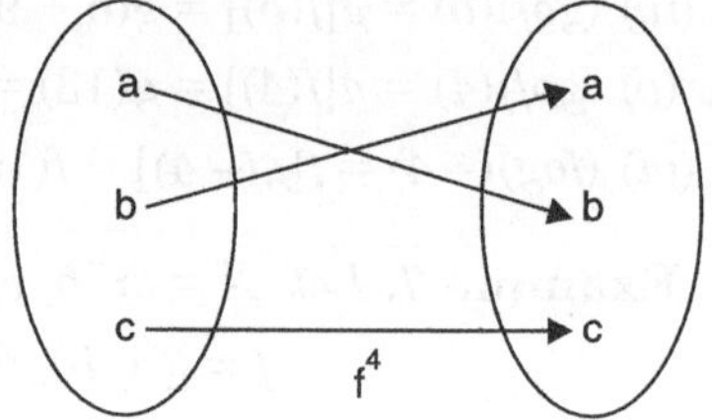

Fig. 3.38

So, $\qquad f^4 = \{(a, a), (b, b), (c, c)\}.$

Example 8. *Consider the functions f, g and h as in Figs. 3.39, 3.40 and 3.41.*

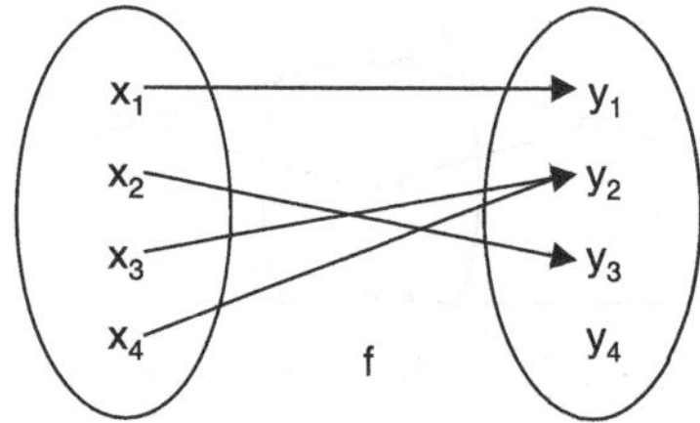

Fig. 3.39

y₁ y₂ y₃ y₄ z₁ z₂ z₃ g

Fig. 3.40

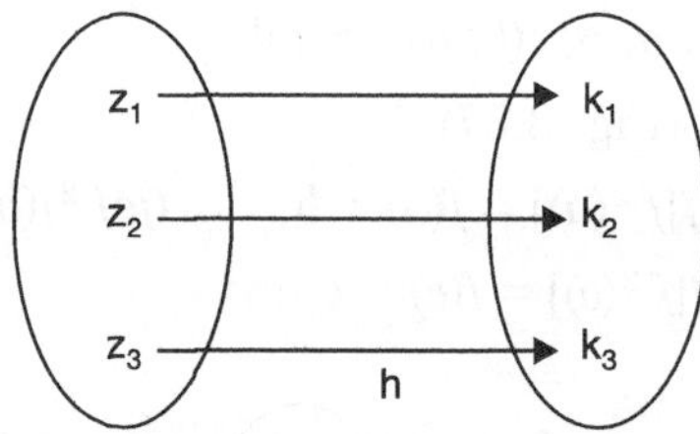

Fig. 3.41

Determine (i) gof (ii) ho(gof) (iii) (hog)of.

Sol. (*i*) Consider Fig. 3.42.

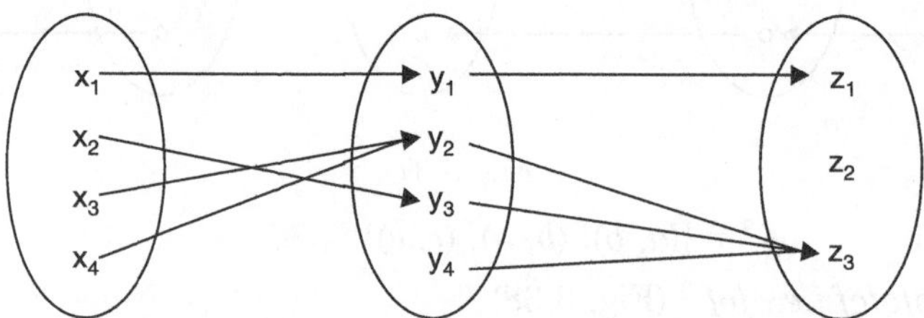

Fig. 3.42. *gof.*

$$(gof)(x_1) = g[f(x_1)] = g(y_1) = z_1, \quad (gof)(x_2) = g[f(x_2)] = g(y_3) = z_3$$
$$(gof)(x_3) = g[f(x_3)] = g(y_2) = z_3, \quad (gof)(x_4) = g[f(x_4)] = g(y_2) = z_3$$

So, $\qquad gof = \{(x_1, z_1), (x_2, z_3), (x_3, z_3), (x_4, z_3)\}.$

(*ii*) First find the composition of f with g and then with h (Fig. 3.43).

$$ho(gof)(x_1) = ho[g(f(x_1))] = ho[g(y_1)] = h(z_1) = k_1$$

$$ho(gof)(x_2) = ho[g(f(x_2))] = ho[g(y_3)] = h(z_3) = k_3$$
$$ho(gof)(x_3) = ho[g(f(x_3))] = ho[g(y_2)] = h(z_3) = k_3$$
$$ho(gof)(x_4) = ho[g(f(x_4))] = h[g(y_2)] = h(z_3) = k_3$$

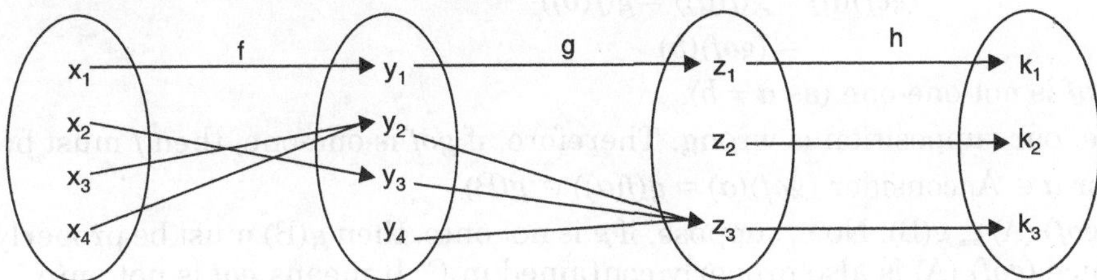

Fig. 3.43. *ho*(*gof*).

So, $ho(gof) = \{(x_1, k_1), (x_2, k_3), (x_3, k_3), (x_4, k_3)\}$.

(*iii*) First find the composition of g with h and then with f(Fig. 3.44).

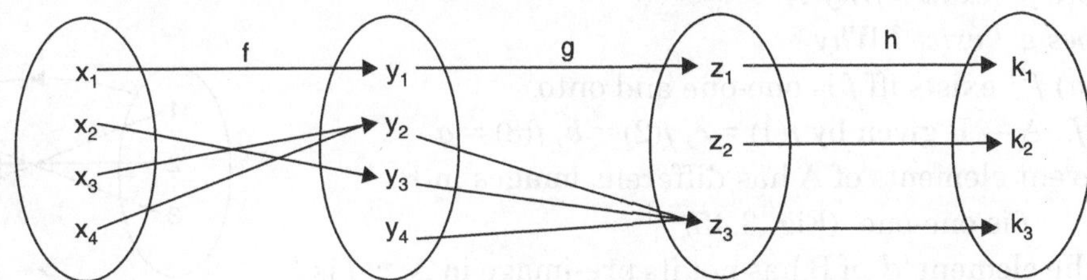

Fig. 3.44. (*hog*) *of*.

$(hog)(y_1) = h(g(y_1)) = h(z_1) = k_1$, $(hog)(y_2) = h(g(y_2)) = h(z_3) = k_3$

$(hog)(y_3) = h(g(y_3)) = h(z_3) = k_3$, $(hog)(y_4) = h(g(y_4)) = h(z_3) = k_3$

So, $hog = \{(y_1, k_1), (y_2, k_2), (y_3, k_3), (y_4, k_4)\}$

Now,

$$((hog)of)(x_1) = (hog)(f(x_1)) = hog(y_1) = k_1$$
$$((hog)of)(x_2) = (hog)(f(x_2)) = hog(y_3) = k_3$$
$$((hog)of)(x_3) = (hog)(f(x_3)) = hog(y_2) = k_3$$
$$((hog)of)(x_4) = (hog)(f(x_4)) = hog(y_2) = k_3$$

So, $(hog)of = \{(y_1, k_1), (y_2, k_3), (y_3, k_3), (y_4, k_3)\}$.

Example 9. (*a*) *If* $f : A \to B$ *and* $g : B \to C$ *are one-one functions. Show that* $gof : A \to C$ *is one-one.*

(*b*) *If* $f : A \to B$ *and* $g : B \to C$ *are onto functions. Then show that* $gof : A \to C$ *is onto*

(*c*) *In part* (*b*), *if gof is one-one, then f is one-one.*

(*d*) *In part* (*b*), *if gof is onto, then f is onto.*

Sol. (*a*) Let $(gof)(x) = (gof)(y)$

$\Rightarrow$ $g(f(x)) = g(f(y))$

$\Rightarrow$ $f(x) = f(y)$ | g is one-one

$\Rightarrow$ $x = y$ | f is one-one

Hence gof is one-one.

(*b*) As g is onto, therefore, for $c \in$ C, there exists $b \in$ B such that $g(b) = c$. As f is onto, therefore, for $b \in$ B, there exists $a \in$ A such that $f(a) = b$

$\therefore$ $(gof)\ (a) = g(f(a)) = g(b) = c$

Hence we have shown that for $a \in$ A, there exists $c \in$ C such that

$$(gof)(a) = c. \text{ Hence } gof \text{ is onto.}$$

(*c*) Suppose, f is not one-one, then, there exists $a \neq b \in$ A such that $f(a) = f(b)$

$\therefore \quad (gof)(a) = g(f(a)) = g(f(b))$

$\qquad = (gof)(b)$

$\Rightarrow$ gof is not one-one (as $a \neq b$).

Hence, our supposition is wrong. Therefore, if gof is one-one, then f must be one-one.

(*d*) For $a \in$ A, consider $(gof)(a) = g(f(a)) \in g(\text{B})$

$\Rightarrow$ (gof) (A) $\subseteq g$(B). Now, suppose, if g is not onto, then g(B) must be properly contained in C and hence (gof) (A) is also properly contained in C. It means gof is not onto.

Hence, our supposition is wrong. Therefore, if gof is onto, then g must be onto.

Example 10. *Let A = {1, 2, 3}, B = {a, b c, d}, C = {7}. Define f : A → B, by f(1) = c, f(2) = b, f(3) = a, and define g : B → C by g(x) = 7, for all x ∈ B.*

(*a*) *Does* f^{-1} *exist ? Why ?*

(*b*) *Does* g^{-1} *exist ? Why ?*

Sol. (*a*) f^{-1} exists iff f is one-one and onto.

Here f : A → B given by $f(1) = c$, $f(2) = b$, $f(3) = a$

i.e., different elements of A has different images in B.

$\therefore$ f is one-one. (Fig. 3.45)

Also the element '*d*' of B has not its pre-image in A. $\therefore$ f is not onto. $\therefore$ f^{-1} does not exist (As we know that f^{-1} exists iff it is one-one and onto).

Fig. 3.45

(*b*) Here g : B → C by $g(x) = 7$ for all $x \in$ B

i.e., $g(a) = 7, g(b) = 7, g(c) = 7, g(d) = 7$

Thus, different elements of g have different images in C

$\therefore$ g is one-one. (Fig. 3.46)

Also every element of C has its pre-image in B. $\therefore$ g is onto. Hence g^{-1} exists (we know that f^{-1} exists iff f is one-one and onto).

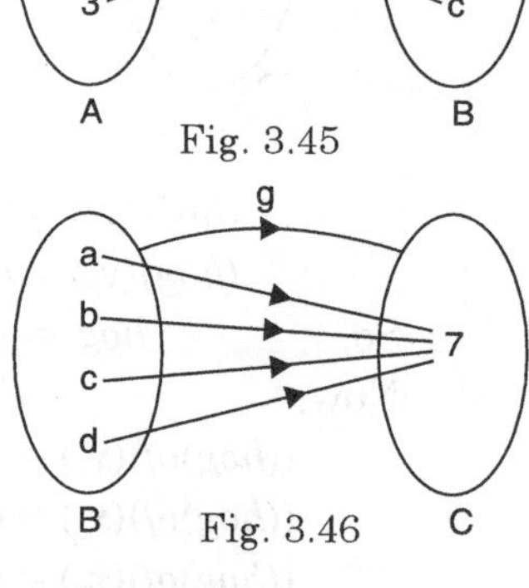

Fig. 3.46

Example 11. *Let {x} denotes the greatest integer function and f : R → Z defined by f(x) = [x]. Find* $f^{-1}(1)$.

Sol. We first define the greatest integer function. For any real number, we define the greatest integer function $[x]$ = the greatest integer less than or equal to x, $x \in$ R.

For *e.g.,* $[2.45] = 2$, $[-2.1] = -3$, $[0.32] = 0$ etc.

Here f : R → Z is defined by $f(x) = [x]$. Co-domain f = Z, Also the range R_f = Z

$\therefore$ f is onto.

Let $x, y \in$ R such that $f(x) = f(y) \Rightarrow [x] = [y]$

$\Rightarrow \quad x = y$

$\therefore$ f is one-one. Hence f^{-1} exists.

We now find $f^{-1}(1)$. Given $f(x) = [x]$, $x \in$ R

For $\quad 0 < x \leq 1, [x] = 1$

$\therefore \quad f(x) = [x] = 1, \text{ if } 0 < x \leq 1, x \in \text{R}.$

$\Rightarrow \quad f^{-1}(1) = [x : 0 < x \leq 1, x \in \text{R}].$

Example 12. *If $f : A \rightarrow B$ is a function from A to B, and if f^{-1} exists, then it is unique.*

Sol. Given f^{-1} exists $\Rightarrow$ f is one-one and onto.

Let $g : B \rightarrow A$ and $h : B \rightarrow A$ be two inverses of f. We show $g = h$.

Let $y \in B$. Since $g : B \rightarrow A$, $h : B \rightarrow A$. $\therefore$ We can find $x_1, x_2 \in A$ such that $g(y) = x_1$, $h(y) = x_2$

Now, $g(y) = x_1$

$\Rightarrow$ $y = g^{-1}(x_1) = f(x_1)$

Also, $h(y) = x_2$

$\Rightarrow$ $y = h^{-1}(x_2) = f(x_2)$

$\Rightarrow$ $f(x_1) = f(x_2)$

$\Rightarrow$ $x_1 = x_2$ | f is one-one

$\Rightarrow$ $g(y) = h(y)$ for $y \in B$

$\Rightarrow$ $g = h$.

Hence the result.

Example 13. *If $f : R \rightarrow R$ is given by*

(a) $f(x) = 3x - 5$ *(b) $f(x) = 5 - 8x$*

Find f^{-1}, if exists.

Sol. (*a*) f^{-1} exists iff $f : R \rightarrow R$ is one-one and onto.

One-one. Let $x, y \in R$ such that $f(x) = f(y)$ $\Rightarrow$ $3x - 5 = 3y - 5$

$\Rightarrow$ $3x = 3y \Rightarrow x = y$

$\therefore$ f is one-one

Onto : Let $y \in R$ such that $f(x) = y$ $\Rightarrow$ $3x - 5 = y$

$\Rightarrow$ $3x = 5 + y \Rightarrow x = \dfrac{5+y}{3}$

For each $y \in R$, we have $x = \dfrac{5+y}{3} \in R$. $\therefore$ f is onto also. Hence f^{-1} exists.

Now, $f(x) = y$ gives $x = f^{-1}(y)$

$\Rightarrow$ $f^{-1}(y) = x = \dfrac{5+y}{3}$

$\therefore$ $f^{-1}(x) = \dfrac{5+x}{3}$

(*b*) Try yourself, $f^{-1}(x) = \dfrac{5-x}{8}$.

Example 14. *Consider the function $f : N \times N \rightarrow N$ defined by $f(x, y) = (2x + 1)\, 2^y - 1$ where $N = \{0, 1, 2, 3, 4, ...\}$. Show that f is bijective.*

Sol. Let (x, y) and $(m, n) \in N \times N$ such that $f(x, y) = f(m, n)$

$\Rightarrow$ $(2x + 1)\, 2^y - 1 = (2m + 1)\, 2^n - 1$

$\Rightarrow$ $(2x + 1)\, 2^y = (2m + 1)\, 2^n$

which is possible only if

$x = m$ and $y = n$

$\Rightarrow$ $(x, y) = (m, n)$

$\therefore$ f is one-one

Let $f(a, b) = 0$ $\Rightarrow$ $(2a + 1)\, 2^b - 1 = 0$

$\Rightarrow$ $(2a + 1)\, 2^b = 1 = 2^0 = 1.2^0$ which is possible if

$\Rightarrow$ $2a + 1 = 1$ and $2^b = 2^0$ $\Rightarrow$ $a = 0$ and $b = 0$ $\Rightarrow$ $(a, b) = (0, 0)$

$\therefore$ f is onto.

Since f is 1-1 and onto, therefore f is bijective.

TEST YOUR KNOWLEDGE 3.2

SHORT ANSWER TYPE QUESTIONS

1. Let f and g are defined by $f(x) = 2x + 1$ and $g(x) = x^2 - 2$ respectively. Find

(*a*) gof (4) (*b*) $(fog)(4)$

(*c*) $(gof)(a + 2)$ (*d*) $(fog)(a + 2)$

(*e*) gof (*f*) fog

(*g*) f^2 (*h*) g^2

2. (*a*) Let $f : A \to B$ be a function from A to B. Show that

(*i*) $I_B\ of = f$ (*ii*) $foI_A = f$, where

$I_B : B \to B$ and $I_A : A \to A$ are the identity functions on B and A respectively.

(*b*) Let f, g, h are functions from Z to Z, on the set of integers, given by $f(n) = n^2$, $g(n) = n + 1$, $h(n) = n - 1$. Then find (*i*) $go(foh)$ (*ii*) $fo(goh)$ (*iii*) $ho(fog)$.

3. Let A = {1, 2, 3}. Define $f : A \to A$ by $f(1) = 2$, $f(2) = 1$, $f(3) = 3$. Find f^2 and f^{-1}.

4. Let A = B = C = R and let $f : A \to B$, $g : B \to C$ be defined by $f(a) = a + 1$, $g(b) = b^2 + 2$ find $(gof)(-2)$, $(fog)(-2)$

5. Let A = {1, 2, 3, 4}, B = {*a, b, c, d*} and $f : A \to B$ is given by

(*i*) $f = \{(1, a), (2, a), (3, c), (4, d)\}$ (*ii*) $f = \{(1, a), (2, c), (3, b), (4, d)\}$.

Determine whether f^{-1} is a function or not.

LONG ANSWER TYPE QUESTIONS

6. Let $f : A \to B$ and $g : B \to C$ be functions. Show that if gof is one-one, then f is one-one.

7. Let $f : A \to B$ and $g : B \to C$ be functions. Show that if gof is onto, then g is onto.

8. Let f ; $Z \to Z$ by $f(x) = 2x^2 + 7x$. Is f^{-1} exist ? If not, why ?

9. Find the formula for the inverse of $g(x) = x^2 - 1$.

10. Consider the functions $f(x) = 2x - 3$ and $g(x) = x^2 + 3x + 5$. Find a formula for the composition functions

(*a*) gof (*b*) fog

(*c*) f^2 (*d*) f^3

11. Find a formula for the inverse of $f(x) = \dfrac{2x - 3}{5x - 7}$

12. Let $f(x) = ax + b$, $g(x) = cx + d$ where a, b, c and d are constants. Determine for which constants a, b, c, d it is true that $fog = gof$.

Answers

1. (*a*) 79 (*b*) 29 (*c*) $4a^2 + 20a + 23$

(*d*) $2a^2 + 8a + 5$ (*e*) $4x^2 + 4x - 1$ (*f*) $2x^2 - 3$

(*g*) $4x + 3$ (*h*) $x^4 - 4x^2$

2. (*b*) (*i*) $(n - 1)^2 + 1$ (*ii*) n^2 (*iii*) $(n + 1)^2 - 1$

3. $f^2 = \{1, 2, 3\}$, $f^{-1} = f$ **4.** $(gof)(-2) = 3$, $(fog)(-2) = 5$

5. (*i*) No (*ii*) Yes

8. No, since f is not one-one. Also, f is not onto **9.** $g^{-1}(x) = \sqrt{x+1}$

10. (*a*) $4x^2 + 5$ (*b*) $2x^2 + 6x + 7$

(*c*) $4x - 9$ (*d*) $8x - 21$

11. $h^{-1}(x) = \dfrac{7x-3}{5x-7}$ **12.** $(a-1)\, d = (c-1)d$.

MULTIPLE CHOICE QUESTIONS (MCQs)

1. The number of functions from an m element set to an n element set is

(*a*) $m + n$ (*b*) m^n

(*c*) n^m (*d*) mn.

2. Let A and B be sets with cardinalities m and n respectively. The number of one-one mappings (injections) from A and B, when $m < n$, is

(*a*) m^n (*b*) ${}^nP_m$

(*c*) ${}^nC_m$ (*d*) None of these.

3. Suppose x and y are sests and $|x|$ and $|y|$ are their respective cardinalities. It is given that there are exactly 97 functions from x to y. From this one can conclude that

(*a*) $|x| = 1, \quad |y| = 97$ (*b*) $|x| = 97, \ |y| = 1$

(*c*) $|x| = 97, \ |y| = 97$ (*d*) None of these.

4. Let R denote the set of real numbers. Let $f: R \times R \to R \times R$ be a bijective function defined by $f(x, y) = (x + y, x - y)$. The inverse function of f is given by

(*a*) $f^{-1}(x, y) = \left[\dfrac{1}{x+y}, \dfrac{1}{(x-y)}\right]$ (*b*) $f^{-1}(x, y) = (x - y, x + y)$

(*c*) $f^{-1}(x, y) = \left[\dfrac{x+y}{2}, \dfrac{x-y}{2}\right]$ (*d*) $f^{-1}(x, y) = [2(x-y), 2(x+y)]$.

5. Let A and B be sets with cardinalities m and n respectively. The number of one-to-one mapping from A to b where $m < n$ is

(*a*) m^n (*b*) ${}^nP_m$

(*c*) ${}^mC_n$ (*d*) ${}^nC_m$.

6. Let A be set of integers greater than one and smaller than thousand. Let b denote set of books in library, $|B| = 999$. Let $f: A \to B$, assigning a unique number to each book f is

(*a*) one to one, onto (*b*) one to one, and not onto

(*c*) not one to one, onto (*d*) not one to one, not onto.

7. Let $f(x) = x^2 + x$ and $g(x) = x + 1$ then fog is :

(*a*) $x^2 + 3x + 1$ (*b*) $x^2 + x + 1$

(*c*) $(x+1)^2 + (x+1)$ (*d*) None of the above.

8. Each of the function 2^n and $n^{\log n}$ has growth rate ... that of any polynomial.

(*a*) Greater than (*b*) Less than

(*c*) Equal to (*d*) Proportional to.

9. The number of bijective functions from A to A when A contains 106 elements is

(*a*) 106 (*b*) 106 !

(*c*) 1 (*d*) None.

10. If $f : \mathrm{N} \to \mathrm{N}$ is a function defined by $f(n) = 2n + 3$. Then

(*a*) f is one-one, but not onto (*b*) f is onto, but not one-one

(*c*) f is neither one-one, nor onto (*d*) f is one-one and onto.

Answers and Explanations

1. (*c*) **2.** (*b*)

3. (*a*) If A is a set containing m elements and B is a set with n elements, then total number of functions from A to B is n^m. Here, $n^m = 97 = 97^1$.

$\Rightarrow \quad n = 97,\ m = 1,\ i.e.,\ |x| = 1,\ |y| = 97.$

4. (*c*) Let $f(x, y) = (x_1, y_1) \ \Rightarrow\ (x + y, x - y) = (x_1, y_1)$

$$\Rightarrow \quad x + y = x_1 \qquad \ldots(1)$$

$$x - y = y_1 \qquad \ldots(2)$$

Adding, $\quad 2x = x_1 + y_1 \ \Rightarrow\ x = \dfrac{x_1 + y_1}{2}$

Subtracting, we get $\quad 2y = x_1 - y_1 \ \Rightarrow\ y = \dfrac{x_1 - y_1}{2}$

$$\therefore \quad (x, y) = \left(\frac{x_1 + y_1}{2}, \frac{x_1 - y_1}{2}\right) = f^{-1}(x_1, y_1)$$

$$\therefore \quad f^{-1}(x, y) = \left(\frac{x + y}{2}, \frac{x - y}{2}\right).$$

5. (*b*)

6. (*b*) Consider the function $f : \mathrm{A} \to \mathrm{B}$ defined by $f(a) = a - 1$ for all $a \in \mathrm{A}$, then each element of A has different images in B. So f is one-one.

But the element 999 has not its preimage in A. Therefore, f is not onto since, if $999 \in \mathrm{B}$, then there should be 1000 in A such that

$$f(1000) = 1000 - 1 = 999$$

Hence f is one-one, but not onto.

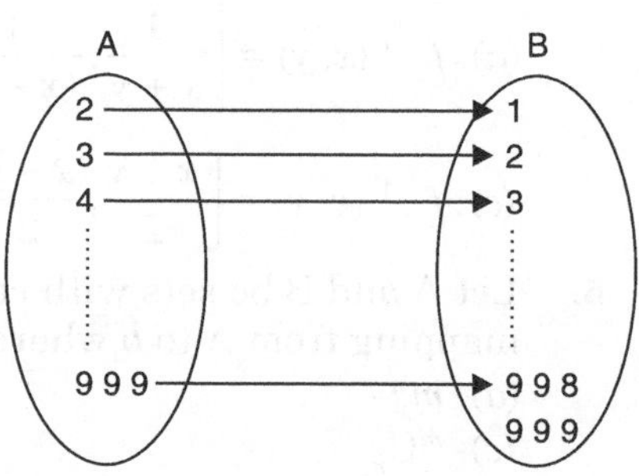

7. (*c*) $fog\,(x) = f(g(x)) = f(x + 1) = (x + 1)^2 + x + 1.$

8. (*b*)

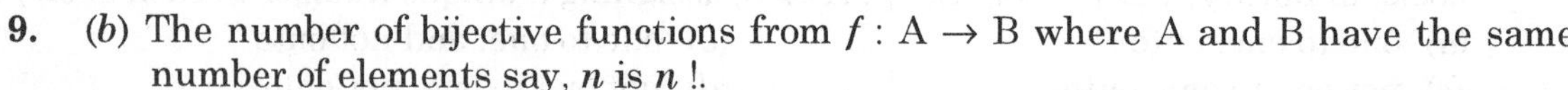

9. (*b*) The number of bijective functions from $f : \mathrm{A} \to \mathrm{B}$ where A and B have the same number of elements say, n is $n\,!$.

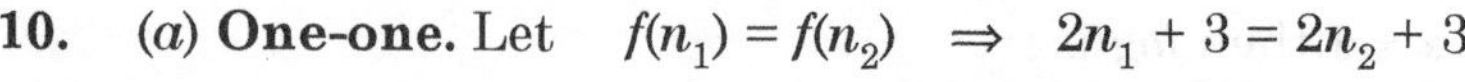

10. (*a*) **One-one.** Let $\quad f(n_1) = f(n_2) \ \Rightarrow\ 2n_1 + 3 = 2n_2 + 3$

$\Rightarrow \quad n_1 = n_2 \ \therefore\ f$ is one-one

Onto. Let $\quad f(n) = m \ \Rightarrow\ 2n + 3 = m \ \Rightarrow\ n = \dfrac{m - 3}{2}$

For each $m \in \mathrm{N}$, $\quad n = \dfrac{m - 3}{2} \notin \mathrm{N}$

$\therefore \quad f$ is not onto.

4* MATHEMATICAL INDUCTION

4.1. PRINCIPLE OF MATHEMATICAL INDUCTION

The process to establish the validity of a general result involving natural numbers is the principle of mathematical induction.

4.2. WORKING RULE

(P.T.U. M.C.A. May 2007)

Let n_0 be a fixed integer. Suppose P(n) is a statement involving the natural number n and we wish to prove that P(n) is true for all $n \geq n_0$.

1. Basis of Induction. P(n_0) is true *i.e.,* P(n) is true for $n = n_0$.

2. Induction Step. Assume that the P(k) is true for $n = k$. $(k \geq n_0)$

Then P($k + 1$) must also be true.

Then P(n) is true for all $n \geq n_0$.

4.3. PEANO'S AXIOMS

Peano, a mathematician defined the natural number in the following way. According to him, the member of set N satisfying the following properties are called natural number.

1. $1 \in \text{N}$.

2. For all $n \in \text{N}$, there exists a unique $n' \in \text{N}$, such that

(*a*) $m' = n' \iff m = n$

(*b*) There exists no element p in N such that $p' = 1$.

3. If $\text{S} \subset \text{N}$ and

(*a*) $1 \in \text{S}$

(*b*) $x \in \text{S} \Rightarrow x' \in \text{S}$ then S = N.

x' is called the successor of x and x is known as predecessor of x'.

These properties are called Peano's Axioms.

The number x' and $x + 1$.

Therefore $2 = 1'$, $3 = 2' = (1')'$, $4 = 3' = (2')' = ((1')')'$ and so on.

*Not meant for P.T.U. B.Tech., "Discrete Structures" (BTCS–302) Course.

ILLUSTRATIVE EXAMPLES

Example 1. *Prove by Mathematical Induction :*

$$1^2 + 2^2 + 3^2 + ... + n^2 = \frac{n(n+1)(2n+1)}{6}.$$

Sol. Basis of Induction. For $n = 1$,

$$P(1) = 1^2 = \frac{1(1+1)(2(1)+1)}{6} = \frac{6}{6} = 1$$

It is true for $n = 1$.

Induction Step. For $n = r$, ...(1)

$$P(r) = 1^2 + 2^2 + 3^2 + ... + r^2 = \frac{r(r+1)(2r+1)}{6} \text{ is true} \quad ...(2)$$

Adding $(r + 1)^2$ on both sides, we get

$$P(r+1) = 1^2 + 2^2 + 3^2 + ... + r^2 + (r+1)^2 = \frac{r(r+1)(2r+1)}{6} + (r+1)^2$$

| Using (2)

$$= \frac{r(r+1)(2r+1) + 6(r+1)^2}{6} = (r+1)\left[\frac{r(2r+1) + 6(r+1)}{6}\right]$$

$$= \frac{(r+1)}{6}[r(2r+1) + 6(r+1)] = \frac{(r+1)}{6}[2r^2 + 7r + 6]$$

$$= \frac{(r+1)(r+2)(2r+3)}{6} \quad ...(3)$$

As $P(r)$ is true, hence $P(r + 1)$ is true.

From (1), (2) and (3), we conclude that

$$1^2 + 2^2 + 3^2 + ... + n^2 = \frac{n(n+1)(2n+1)}{6}$$

is true for $n = 1, 2, 3, 4, 5, ...$. Hence proved.

Example 2. *Prove by Mathematical Induction :* $1^3 + 2^3 + 3^3 + ... + n^3 = \left[\frac{n(n+1)}{2}\right]^2$.

Sol. Let $P(n) = 1^3 + 2^3 + 3^3 + ... + n^3 = \left[\frac{n(n+1)}{2}\right]^2$.

Basic Step. For $n = 1$,

$$P(1) = 1^3 = \left[\frac{1(1+1)}{2}\right]^2 = (1)^2 = 1$$

Hence $P(n)$ is true for $n = 1$. ...(1)

Induction Step. For $n = r$,

$$P(r) = 1^3 + 2^3 + 3^3 + ... + r^3 = \left[\frac{r(r+1)}{2}\right]^2 \text{ is true} \quad ...(2)$$

Adding $(r + 1)^3$ on both sides

$$P(r + 1) = 1^3 + 2^3 + 3^3 + ... + r^3 + (r + 1)^3 = \left[\frac{r(r+1)}{2}\right]^2 + (r + 1)^3$$

$$= \frac{r^2(r+1)^2 + 4(r+1)(r+1)^2}{4} = \frac{(r+1)^2\{r^2 + 4(r+1)\}}{4} \quad | \text{ Using (2)}$$

$$= \frac{(r+1)^2\,[r^2 + 4r + 4]}{4} = \frac{(r+1)^2(r+2)^2}{4}$$

$$= \left[\frac{(r+1)(r+2)}{2}\right]^2 \quad ...(3)$$

As P(r) is true, hence P(r + 1) is also true.
From (1), (2) and (3), we conclude that

$$1^3 + 2^3 + 3^3 + ... + n^3 = \left[\frac{n(n+1)}{2}\right]^2$$

is true for $n = 1, 2, 3, 4, 5, ...$ Hence proved.

Example 3. *Prove by Mathematical Induction :*

$$1\,.\,2 + 2\,.\,3 + 3\,.\,4 + ... + n(n + 1) = \frac{n(n+1)(n+2)}{3}. \quad \text{(P.T.U. M.C.A. Dec. 2006)}$$

Sol. Let $\quad P(n) = 1\,.\,2 + 2\,.\,3 + 3\,.\,4 + ... + n(n + 1) = \frac{n(n+1)(n+2)}{3}.$

Basis of induction. For $n = 1$,

$$P(1) = 1\,.\,2 = \frac{1(1+1)(1+2)}{3} = 1\,.\,2$$

It is true for $n = 1$. ...(1)

Induction Step. For $n = r$,

$$P(r) = 1\,.\,2 + 2\,.\,3 + 3\,.\,4 + ... + r(r + 1) = \frac{r(r+1)(r+2)}{3} \text{ is true} \quad ...(2)$$

Adding $(r + 1)(r + 2)$ on both sides,

$$P(r + 1) = 1\,.\,2 + 2\,.\,3 + 3\,.\,4 + ... + r(r + 1) + (r + 1)(r + 2)$$

$$= \frac{r(r+1)(r+2)}{3} + (r+1)(r+2) \quad | \text{ Using (2)}$$

$$= \frac{r(r+1)(r+2) + 3(r+1)(r+2)}{3} = \frac{(r+1)(r+2)(r+3)}{3} \quad ...(3)$$

As P(r) is true, hence P(r + 1) is also true.
From, (1), (2) and (3), we conclude that

$$1\,.\,2 + 2\,.\,3 + 3\,.\,4 + ... + n(n + 1) = \frac{n(n+1)(n+2)}{3}$$

is true for $n = 1, 2, 3, 4, ...$ Hence proved.

Example 4. *Prove by Mathematical Induction :*

$$\frac{1}{1\,.\,3} + \frac{1}{3\,.\,5} + \frac{1}{5\,.\,7} + ... + \frac{1}{(2n-1)(2n+1)} = \frac{n}{2n+1}.$$

Sol. Basic Step. For $n = 1$,

$$\frac{1}{(2-1)(2+1)} = \frac{1}{3} = \frac{1}{2(1)+1} = \frac{1}{3}$$

Hence P(n) is true for $n = 1$. ...(1)

Induction Step. For $n = r$,

$$P(r) = \frac{1}{1.3} + \frac{1}{3.5} + \frac{1}{5.7} + \ldots + \frac{1}{(2r-1)(2r+1)} = \frac{r}{2r+1} \text{ is true} \qquad \ldots(2)$$

Adding $\frac{1}{(2r+1)(2r+3)}$ on both sides

$$P(r+1) = \frac{1}{1.3} + \frac{1}{3.5} + \frac{1}{5.7} + \ldots + \frac{1}{(2r-1)(2r+1)} + \frac{1}{(2r+1)(2r+3)}$$

$$= \frac{r}{2r+1} + \frac{1}{(2r+1)(2r+3)} = \frac{r(2r+3)+1}{(2r+1)(2r+3)} \qquad | \text{ Using (2)}$$

$$= \frac{2r^2+3r+1}{(2r+1)(2r+3)} = \frac{(2r+1)(r+1)}{(2r+1)(2r+3)} = \frac{(r+1)}{(2r+3)} \qquad \ldots(3)$$

As P(r) is true, hence P(r + 1) is also true.

From (1), (2) and (3), we conclude that

$$\frac{1}{1.3} + \frac{1}{3.5} + \frac{1}{5.7} + \ldots + \frac{1}{(2n-1)(2n+1)} = \frac{n}{2n+1}$$

is true for $n = 1, 2, 3, 4, \ldots$. Hence proved.

Example 5. *Prove* $1 + 3 + 5 + \ldots + (2n - 1) = n^2$ *by induction* $(n \geq 1)$.

Sol. Consider $P(n) = 1 + 3 + 5 + \ldots + (2n - 1) = n^2$

Basis of Induction. For $n = 1$,

$$P(1) = 2 \times 1 - 1 = 1 = 1^2 = 1$$

It is true for $n = 1$. ...(1)

Induction Step. For $n = r$,

$$P(r) = 1 + 3 + 5 + \ldots + (2r - 1) = r^2 \text{ is true} \qquad \ldots(2)$$

Adding $(2r + 1)$ to both sides,

$$P(r+1) = 1 + 3 + 5 + \ldots + (2r-1) + (2r+1)$$

$$= r^2 + (2r+1) = (r+1)^2 \qquad \ldots(3)$$

As P(r) is true, hence P(r + 1) is also true.

From (1), (2) and (3), we conclude that

$$1 + 3 + 5 + \ldots + (2n-1) = n^2$$

is true for $n = 1, 2, 3, \ldots$. Hence proved.

Example 6. *Show that* $1 + 2 + 2^2 + 2^3 + \ldots + 2^n = 2^{n+1} - 1$ *by induction (for* $n \geq 0$*).*

(P.T.U. M.C.A. Dec. 2005)

Sol. Consider $P(n) = 1 + 2 + 2^2 + 2^3 + \ldots + 2^n = 2^{n+1} - 1$.

Basis of Induction. For $n = 0$,

$$P(0) = 1 = 2^1 - 1 = 1$$

It is true for $n = 0$. ...(1)

Induction Step. For $n = r$,

$$P(r) = 1 + 2 + 2^2 + 2^3 + \ldots + 2^r = 2^{r+1} - 1 \text{ is true.} \qquad \ldots(2)$$

Adding 2^{r+1} to both sides,

$$P(r + 1) = 1 + 2 + 2^2 + 2^3 + \ldots + 2^r + 2^{r+1} = 2^{r+1} - 1 + 2^{r+1} \qquad | \text{ Using (2)}$$

$$= 2(2^{r+1}) - 1 = 2^{r+2} - 1 \qquad \ldots(3)$$

As P(r) is true, hence P(r + 1) is also true.

From, (1), (2) and (3), we conclude that

$$1 + 2 + 2^2 + \ldots + 2^n = 2^{n+1} - 1, \text{ is true for } n = 1, 2, 3, \ldots .$$

Example 7. *Prove by induction that for $n \geq 0$ and $a \neq 1$; $1 + a + a^2 + \ldots + a^n = \dfrac{1 - a^{n+1}}{1 - a}$.*

(P.T.U. M.C.A. May 2007)

Sol. Basis of Induction. For $n = 1$,

$$1 + a^1 = \frac{1 - a^{1+1}}{1 - a} = \frac{1 - a^2}{1 - a} = 1 + a$$

It is true for $n = 1$. ...(1)

Induction Step. For $n = r$,

$$P(r) = 1 + a + a^2 + \ldots + a^r = \frac{1 - a^{r+1}}{1 - a} \text{ is true} \qquad \ldots(2)$$

Adding a^{r+1} to both sides,

$$P(r + 1) = 1 + a + a^2 + \ldots + a^r + a^{r+1}$$

$$= \frac{1 - a^{r+1}}{1 - a} + a^{r+1} = \frac{1 - a^{r+1} + a^{r+1} - a^{r+2}}{1 - a} = \frac{1 - a^{r+2}}{1 - a} \qquad \ldots(3)$$

| Using (2)

As P(r) is true, hence P(r + 1) is also true.

From (1), (2) and (3), we conclude that

$$1 + a + a^2 + \ldots + a^n = \frac{1 - a^{n+1}}{1 - a} \text{ is true for } n \geq 0.$$

Hence proved.

Example 8. *Show that for any integer n, $11^{n+2} + 12^{2n+1}$ is divisible by 133.*

Sol. Let $\quad P(n) = 11^{n+2} + 12^{2n+1}$

Basis of Induction. For $n = 1$,

$$P(1) = 11^3 + 12^3 = 3059 = 133 \times 23$$

So, 133 divides P(1). ...(1)

Induction Step. For $n = r$,

$$P(r) = 11^{r+2} + 12^{2r+1} = 133 \times s, \text{ say,} \qquad \ldots(2)$$

Now, for $n = r + 1$,

$$P(r + 1) = 11^{r+2+1} + 12^{2(r)+3} = 11[133s - 12^{2r+1}] + 144 \,.\, 12^{2r+1} \qquad | \text{ Using (2)}$$

$$= 11 \times 133s + 12^{2r+1} \,.\, 133 = 133[11s + 12^{2r+1}] = 133 \times t, \text{ say} \qquad \ldots(3)$$

As (1), (2) and (3) all are true, hence P(n) is divisible by 133.

Example 9. *Prove by induction that the sum of the cubes of three consecutive integers is divisible by 9.*

Sol. Let $P(n) = n^3 + (n + 1)^3 + (n + 2)^3$

$P(n)$ is divisible by 9

$$P(1) = 1 + 8 + 27 = 36$$

which is divisible by 9. ...(1)

For $n = r$,

$$P(r) = r^3 + (r + 1)^3 + (r + 2)^3 = 9 \,.\, q, \text{ say,} \quad ...(2)$$

For $n = r + 1$, $P(r + 1) = (r + 1)^3 + (r + 2)^3 + (r + 3)^3$

$$= r^3 + (r + 1)^3 + (r + 2)^3 + [9 r^2 + 27r + 27] \quad | \text{ Using (2)}$$

$$= 9q + 9(r^2 + 3r + 3) = 9[q + r^2 + 3r + 3]$$

$$= 9.s, \text{ say,} \quad ...(3)$$

From (1), (2) and (3), we have the required result by induction. Hence proved.

Example 10. *Prove* $\frac{1}{n+1} + \frac{1}{n+2} + ... + \frac{1}{2n} > \frac{13}{24}$, *for* $n \geq 2$.

Sol. For $n = 2$,

$$\text{L.H.S.} = \frac{1}{2+1} + \frac{1}{2+2} = \frac{7}{12} > \frac{13}{24} = \text{R.H.S.} \quad ...(1)$$

It is true for $n = 2$. Now, for $n = r$, where $r > 2$

$$\frac{1}{r+1} + \frac{1}{r+2} + ... + \frac{1}{2r} > \frac{13}{24} \quad ...(2)$$

For $n = r + 1$,

$$\text{L.H.S.} = \frac{1}{r+2} + \frac{1}{r+3} + ... + \frac{1}{2r} + \frac{1}{2r+1} + \frac{1}{2r+2} > \frac{13}{24} + \frac{1}{2r+1} + \frac{1}{2r+2} - \frac{1}{r+1}$$

[Using (2)]

$$= \frac{13}{24} + \frac{2r+2+2r+1-2(2r+1)}{(2r+1)(2r+2)} = \frac{13}{24} + \frac{1}{(2r+1)(2r+2)} > \frac{13}{24}$$

As $n = r$ is true. Thus, the result is also true for $n = r + 1$. Hence, we have the result for $n \geq 2$ by induction. Hence proved.

Example 11. *Show that* $n^3 + 2n$ *is divisible by 3 for all* $n \geq 1$ *by induction.*

Sol. Basis of Induction. For $n = 1$,

$$P(1) = 1^3 + 2 \times 1 = 3. \text{ It is divisible by 3.}$$

Induction Step. For $n = r$,

Assume, $P(r) = r^3 + 2r = 3s$, say,

For $n = r + 1$,

$$P(r + 1) = (r + 1)^3 + 2(r + 1) = r^3 + 1 + 3r(r + 1) + 2r + 2$$

$$= r^3 + 1 + 3r^2 + 3r + 2r + 2$$

$= r^3 + 3 + 5r + 3r^2 = r^3 + 2r + 3 + 3r + 3r^2$ $(\because \quad r^3 + 2r = 3r)$

$= 3s + 3 + 3r + 3r^2 = 3r^2 + 3(r + s) + 3$

$= 3(r^2 + r + s + 1) = 3t$, say

It is divisible by 3. Hence, we have the required result by induction.

Example 12. *Show that $2^n \times 2^n - 1$ is divisible by 3 for all $n \geq 1$ by induction.*

Sol. Basis of Induction. For $n = 1$,

$2^1 \times 2^1 - 1 = 3$ divisible by 3.

It is true for $n = 1$.

Induction Step. For $n = r$,

$$P(r) = 2^r\, 2^r - 1 = 3s, \text{ say} \quad i.e., \quad 2^{r+1} - 1 = 3s \qquad ...(1)$$

For $n = r + 1$,

$$2^{r+1} \,.\, 2^{r+1} - 1 = 2^{r+1}\,(3s + 1) - 1 \qquad |\text{ Using (1)}$$

$$= 3s \,.\, 2^{r+1} + 2^{r+1} - 1 = 3s \,.\, 2^{r+1} + 3s$$

$$= 3s(2^{r+1} + 1).$$

It is divisible by 3. Hence, we have the required result by induction.

Example 13. *Show that $1^2 + 3^2 + 5^2 + ... + (2n - 1)^2 = \dfrac{n(2n-1)(2n+1)}{3}$.*

Sol. Basis of Induction. For $n = 1$,

$$(2 \times 1 - 1)^2 = 1 = \frac{1(2\times 1 - 1)(2 \times 1 + 1)}{3} = 1$$

It is true of $n = 1$.

Induction Step. For $n = r$,

$$1^2 + 3^2 + 5^2 + ... + (2r - 1)^2 = \frac{r(2r-1)(2r+1)}{3} \text{ is true.} \qquad ...(1)$$

For $n = r + 1$,

$$P(r + 1) = 1^2 + 3^2 + 5^2 + ... + (2r - 1)^2 + [2(r + 1) - 1]^2$$

$$= \frac{r(2r-1)(2r+1)}{3} + (2r + 1)^2 = \frac{r(2r-1)(2r+1) + 3(2r+1)^2}{3} \qquad |\text{ Using (1)}$$

$$= \frac{(2r+1)[2r^2 - r + 6r + 3]}{3} = \frac{(2r+1)[2r^2 + 5r + 3]}{3}$$

$$= \frac{(2r+1)[2r^2 + 3r + 2r + 3]}{3} = \frac{(2r+1)(2r+3)(r+1)}{3},$$

which proves the result for $n = r + 1$.

So, $$1^2 + 3^2 + 5^2 + ... + (2n - 1)^2 = \frac{n(2n-1)(2n+1)}{3}$$

Example 14. *Prove that $n(n + 1)(n + 2)$ is a multiple of 6.*

Sol. $P(n) = n(n + 1)(n + 2)$

Basis of Induction. For $n = 1$,

$$P(1) = 1 \,.\, (1 + 1)(1 + 2) = 1 \,.\, 2 \,.\, 3 = 6$$

which is true.

Induction Step. For $n = r$,

$$P(r) = r(r + 1)(r + 2) \text{ be a multiple of } 6$$

For $n = r + 1$,

$$P(r + 1) = (r + 1)(r + 2)(r + 3) \text{ is also a multiple of } 6.$$

Now, $P(r + 1) = (r + 1)(r + 2)(r + 3) = (r + 1)(r + 2)\, r + 3(r + 1)(r + 2)$

$$= r(r + 1)(r + 2) + 3(2r')$$

where $(r + 1)(r + 2) = \text{even} = 2r' = r(r + 1)(r + 2) + 6r'$

which is a multiple of 6.

Thus, $P(r + 1)$ is true as $P(r)$ is true.

$\therefore$ $P(n)$ is true for all natural numbers n

i.e., $n(n + 1)(n + 2)$ is a multiple of 6. Hence proved.

Example 15. *Prove that* $\forall n \in N$, $\frac{1}{5}n^5 + \frac{1}{3}n^3 + \frac{7}{15}n$ *is a natural number.*

Sol. Let $P(n) = \frac{1}{5}n^5 + \frac{1}{3}n^3 + \frac{7}{15}n$ is a natural number.

Basis of Induction. For $n = 1$,

$$\frac{1}{5} + \frac{1}{3} + \frac{7}{15} = 1, \text{ which is a natural number.}$$

It is true for $n = 1$.

Induction Step. For $n = r$,

$$\left(\frac{1}{5}r^5 + \frac{1}{3}r^3 + \frac{7}{15}r\right) \text{ is a natural number} \qquad \ldots(1)$$

For $n = r + 1$,

$$\frac{1}{5}(r + 1)^5 + \frac{1}{3}(r + 1)^3 + \frac{7}{15}(r + 1)$$

$$= \frac{1}{5}(r^5 + 5r^4 + 10r^3 + 10r^2 + 5r + 1) + \frac{1}{3}(r^3 + 3r^2 + 3r + 1) + \frac{7}{15}(r + 1)$$

$$= \left(\frac{1}{5}r^5 + \frac{1}{3}r^3 + \frac{7}{15}r\right) + (r^4 + 2r^3 + 3r^2 + 2r) + 1$$

$$= (\text{a natural number}) + (\text{a natural number}) + 1 \qquad \text{[Using (1)]}$$

$$= \text{a natural number}$$

As $(P(r))$ is true, hence $P(r + 1)$ is also true.

Hence, by induction, $P(n)$ is true $\forall n \in N$. Hence proved.

Example 16. *Prove that n(n + 1)(2n + 1) is divisible by 6.*

Sol. Let P(n) = $n(n + 1)(2n + 1)$ is divisible by 6.

Basis of Induction. For $n = 1$,

$$P(1) = 1(1 + 1)(2 + 1) = 1 . 2 . 3 = 6 \text{ which is true for } n = 1 \quad ...(1)$$

Induction Step. For $n = r$,

$$P(r) = r(r + 1)(2r + 1) \text{ be divisible by 6} \quad ...(2)$$

For $n = r + 1$,

$$P(r + 1) = (r + 1)(r + 2)(2r + 3)$$

$$\text{L.H.S.} = (r + 1)(r + 2)(2r + 3) = 2r^3 + 9r^2 + 13r + 6$$
$$= (2r^3 + 3r^2 + r) + (6r^2 + 12r + 6)$$
$$= r(r + 1)(2r + 1) + 6(r^2 + 2r + 1)$$

which is divisible by 6. [From (2)]

As (P(r)) is true, hence P(r + 1) is also true. ∴ P(n) is true for all natural numbers n

i.e., $n(n + 1)(2n + 1)$ is divisible by 6. Hence proved.

TEST YOUR KNOWLEDGE

1. If P(n) is the statement : "$n(n + 1) + 1$ is odd", then write P(3).
2. If P(n) is the statement "$n^2 + 1$ is even", then write P(4). Is it true ?
3. If P(n) is the statement "$n^3 + 2$ is a multiple of 5", then show that P(4) is not true.
4. If P(n) is the statement "$12n + 3$ is a multiple of 5", then show that P(3) is false, whereas P(6) is true.
5. Let P(n) be the statement "3" > n". If P(n) is true, show that P(n + 1) is also true.
6. If P(n) is the statement : $1^3 + 2^3 + 3^3 + ... + n^3 = \left[\frac{n(n+1)}{2}\right]^2$. Then verify that P(3), P(7) are both true.
7. If P(n) is the statement : $1^2 + 3^2 + 5^2 + ...(2n - 1)^2 = \frac{n(4n^2 - 1)}{3}$, then show that P(3), P(4), P(6) are true.
8. Let P(n) be the statement "$n^2 + n$ is even". If P(k) is true, then show that P(k + 1) is true.

Prove the following by using the principle of mathematical induction (Q. No. 9–16) :

9. $n(n + 1) + 1$ is an odd number, $n \in \mathbf{N}$.
10. $1 + 4 + 7 + ... + (3n - 2) = \frac{n(3n - 1)}{2}, n \in \mathbf{N}$.
11. $a + (a + d) + (a + 2d) + ... + [a + (n - 1)d] = \frac{n}{2}[2a + (n - 1)d], n \in \mathbf{N}$.
12. $\frac{1}{1 \times 3} + \frac{1}{3 \times 5} + \frac{1}{5 \times 7} + ... + \frac{1}{(2n - 1) \times (2n + 1)} = \frac{n}{2n + 1}, n \in \mathbf{N}$.

13. $\frac{1}{1\times 4}+\frac{1}{4\times 7}+\frac{1}{7\times 10}+......+\frac{1}{(3n-2)(3n+1)}=\frac{n}{3n+1}, n \in \mathbf{N}.$

14. $\frac{1}{2}+\frac{1}{2^2}+\frac{1}{2^3}+......+\frac{1}{2^n}=1-\frac{1}{2^n}, n \in \mathbf{N}.$

15. $a + ar + ar^2 + ... + ar^{n-1} = \frac{a(1-r^n)}{1-r}, n \in \mathbf{N}.$

16. $2 + 4 + 6 + ... + 2n = n(n + 1), n \in \mathbf{N}.$

17. Prove by mathematical induction that $5^{2n+2} - 24n - 25$ is divisible by 576.

18. Prove by mathematical induction that $(2^{3n} - 1)$ is divisible by 7 for all values of $n \in$ N.

19. Prove by Mathematical Induction that $6^{n+2} + 7^{2n+1}$ is divisible by 43 for each positive integer n.

(P.T.U. M.C.A. May. 2008)

Answers

1. 13 is odd **2.** 17 is even, No

Hint

8. Let $k^2 + k = 2\lambda$ where $\lambda \in \mathbf{N}$.

Now $(k + 1)^2 + (k + 1) = k^2 + 3k + 2 = (2\lambda - k) + 3k + 2 = 2(\lambda + k + 1)$.

5A BASIC COUNTING PRINCIPLES

5.1. INTRODUCTION

In this chapter, we will discuss some methods of counting which acts as 'building blocks' for all counting problems.

5.2. BASIC COUNTING PRINCIPLES *(P.T.U. B.Tech. Dec. 2006 ; Dec. 2005)*

There are mainly two counting principles namely

(*i*) Sum Rule (*ii*) Product Rule.

These two principles form the basis of permutations and combinations and hence known as *basic counting principles*.

5.3. SUM RULE

If there are two jobs such that they can be performed independently in m and n ways. Then number of ways in which either of the two jobs can be performed is $m + n$.

5.4. PRODUCT RULE

If there are two jobs such that one of them can be done in m ways and when it has been done, second job can be done in n ways, then the two jobs can be done in $m \times n$ ways.

ILLUSTRATIVE EXAMPLES

Example 1. *In a class there are 10 boys and 8 girls. The teacher wants to select either a boy or a girl to represent the class in a function. In how many ways the teacher can make this selection ?*

Sol. The teacher can select a boy in 10 ways and a girl in 8 ways.

$\therefore$ By sum rule, the number of ways of selecting either a boy or a girl $= 10 + 8 = 18$.

Example 2. *There are 3 students for a classical, 5 for mathematical and 4 for physical science scholarship. In how many ways, one of these scholarships be awarded ?*

Sol. Classical scholarship can be awarded in 3 ways

Mathematical scholarship can be awarded in 5 ways

Physical science scholarship can be awarded in 4 ways.

∴ Number of ways of awarding one of three scholarship

= Number of ways of awarding classical or mathematical or physical science scholarship

= 3 + 5 + 4 = 12 | By Sum Rule

Example 3. *In a class, there are 10 boys and 8 girls. The teacher wants to select a boy and a girl to represent the class in a function. In how many ways can the teacher make this selection ?*

Sol. The teacher can make this selection by

(*i*) selecting a boy among 10 boys

(*ii*) selecting a girl among 8 boys

The first job can be done in 10 ways and the second job can be done in 8 ways. Therefore, by product rule, the total number of ways of selecting a boy and a girl = $10 \times 8 = 80$.

Example 4. *A snack bar serves 5 different sandwiches and 3 different beverges. How many different lunches can a person order ?*

Sol. The person can place his order by

(*i*) asking a sandwich among 5 sandwiches

(*ii*) asking a beverge among 3 beverges

∴ By 'Product Rule', the total number of ways of placing a order for a sandwich and a beverge = $5 \times 3 = 15$.

Example 5. *A person is to complete a true-false questionaire consisting of 10 questions. How many different ways are there to answer the questionaire ?*

Sol. Each question can be answered in two ways (true or false).

Now first question can be answered in 2 ways

Second question can be answered in 2 ways

⋮

10th question can be answered in 2 ways

∴ By 'Product Rule' the total number of answering the questions

$= 2 \times 2 \times 2 \times 2 \times 2 \times 2 \times 2 \times 2 \times 2 \times 2 = 2^{10}$.

Example 6. *A questionaire contains 4 questions that have two possible answers and 3 questions with 5 possible answers. How many ways are there to answer the questionaire ?*

Sol. There are four questions each of which can be answered in two ways each.

∴ Total number of ways of answering these four questions $= 2 \times 2 \times 2 \times 2 = 16$

Also the next 3 questions each of which can be answered in 5 ways

∴ Total number of answering these 3 questions $= 5 \times 5 \times 5 = 125$

By 'Product Rule', the total number of ways of answering the questionaire

$= 16 \times 125 = 2000$.

Example 7. *A room has 6 doors. In how many ways can a man enter the room through one door and come out through a different door ?*

Sol. There are 6 ways of entering the room. Therefore, a man can enter the room in 6 ways.

After entering the room, the man can come out through any one of the remaining five doors.

∴ Number of leaving the room = 5.

∴ By 'Product rule' the required number of ways in which a man can enter a room and come out through a different room = $6 \times 5 = 30$.

Example 8. *Five persons entered the lift cabin on the ground floor of an 8 floor house. Suppose each of them can leave the cabin independently at any floor beginning with the first. Find the total number of ways in which each of the five persons can leave the cabin*

(i) At any one of the 7 floors

(ii) At different floors.

Sol. (*i*) Suppose M_1, M_2, M_3, M_4, M_5 are five persons.

Here M_1 can leave the cabin at any one of the seven floors in 7 ways

M_2 can leave the cabin at any one of the seven floors in 7 ways

Similarly, M_3, M_4, M_5 can leave the cabin at any one of the seven floors in 7 ways

∴ By 'Product Rule', Total number of ways in which five persons can leave the cabin at any one of the seven floors

$$= 7 \times 7 \times 7 \times 7 \times 7 = 7^5.$$

(*ii*) M_1 can leave the cabin at any of the seven floors in 7 ways

M_2 can leave the cabin at any one of the remaining 6 floors in 6 ways

Similarly, M_3, M_4, M_5 can leave the cabin in 5, 4 and 3 ways respectively.

∴ By Product rule, Total number of ways of leaving the cabin $= 7 \times 6 \times 5 \times 4 \times 3 = 2520.$

Example 9. *Automobile license plates in Massachusetts usually consist of 3 digits followed by 3 letters. The first digit is never zero. How many different plates of this type could be made ?*

Sol. Consider the digits 0, 1, 2, ... 9. These are 10 digits. Since the first digit of the license plate is non-zero.

∴ first place of the license plate can be filled up in 9 ways.

Similarly, the second place of the license plate can be filled up in 10 ways and the third place in 10 ways

∴ Required number of using the digits $= 9 \times 10 \times 10$

9	10	10

Again, there are 26 alphabets. The first place of the license plate can be filled up in 26 ways.

Similarly, the second place can be filled up in 26 ways and third in 26 ways

∴ Required number of using the English alphabets $= 26 \times 26 \times 26$

By 'Product Rule', total number of making the plates

$$= 9 \times 10 \times 10 \times 26 \times 26 \times 26 = 15818400.$$

Example 10. *For a set of five true or false questions, no student has written all correct answers and no students have given the same sequence of answers. Find the maximum number of students in the class for this to be possible.*

Sol. A true-false question can be answered in two ways either by making it true or false

∴ Number of ways of answering each of the five questions

$$= 2 \times 2 \times 2 \times 2 \times 2 = 32.$$

Out of these 32 ways of answering, there is only one way of answering all the five questions correctly. But no student has written all the answers correctly.

∴ Maximum number of students in the class

= Number of ways except one in which all answers are correct

$= 32 - 1 = 31.$

TEST YOUR KNOWLEDGE 5.1

1. In a class there are 8 male teachers and 5 female teachers teaching 9 mathematics class. In how many ways, can a student choose 9 mathematics professor ?
2. There are 3 students for a classical, 5 for mathematical and 4 for physical science scholarship. In how many ways can these scholarships awarded ?
3. The flag of a newly formed country is in the form □ □ □ of three blocks, each to be coloured differently. There are six different colours to be used. How many such designs are possible ?
4. In a monthly test, the teacher decides that there will be three questions, one from each chapter I, II and III of the book. If there are 12 questions in chapter I, 10 in chapter II and 6 in chapter III. In how many ways can three questions be selected ?
5. Find the total number of ways of answering 5 objective type questions, each questions having 4 choices.
6. How many three digit numbers can be formed without using the digits 0, 2, 3, 4, 5, 6 ?
7. How many numbers are there between 100 and 1000 such that 7 is in the unit place.
8. A gentleman has 6 friends to invite. In how many ways can be send invitation cards to them, if he has three servants to carry the cards.

Answers

1. 13.	**2.** $3 \times 5 \times 4 = 60$.	**3.** $6 \times 5 \times 4 = 120$
4. $12 \times 10 \times 6 = 720$	**5.** $4 \times 4 \times 4 \times 4 \times 4 = 4^5$.	**6.** $4 \times 4 \times 4 = 64$
7. $9 \times 10 \times 1 = 90$	**8.** $3 \times 3 \times 3 \times 3 \times 3 \times 3 = 3^6 = 729$.	

5B PERMUTATIONS AND COMBINATIONS

5.5. DEFINE FACTORIAL *n*

The product of first n natural number is called factorial n. It is denoted by $n\,!$ or $\underline{|n}$.
The factorial n can also be written as

$$n\,! = n(n-1)\,! = n(n-1)\,(n-2)\,!$$

$$\ldots\ldots\ldots\ldots\ldots\ldots\ldots\ldots\ldots\ldots$$

$$= n(n-1)(n-2)(n-3)\,\ldots\ldots\ldots\,1.$$

We have, $\quad 1\,! = 1 \quad$ and $\quad 0\,! = 1.$

For example : (i) Find the value of $\dfrac{10\,!}{8\,!}$.

Here, $$\frac{10\,!}{8\,!} = \frac{10 \times 9 \times 8\,!}{8\,!} = 10 \times 9 = 90.$$

(ii) Determine the value of $\dfrac{n\,!}{(n-1)\,!}$.

Here, $$\frac{n\,!}{(n-1)\,!} = \frac{n(n-1)\,!}{(n-1)\,!} = n.$$

(iii) Show that $3\,! + 4\,! \neq (3+4)\,!$

Here, $$3\,! + 4\,! = (3 \times 2 \times 1) + (4 \times 3 \times 2 \times 1)$$
$$= 6 + 24 = 30$$

and $$(3+4)\,! = 7\,! = 7 \times 6 \times 5 \times 4 \times 3 \times 2 \times 1 = 5040$$

Hence $$3\,! + 4\,! \neq (3+4)\,!$$

5.6. PERMUTATION

A permutation is an arrangement of a number of objects in some definite order taken some or all at a time. The total number of permutations of n distinct objects taken r at a time is denoted by ${}^{n}P_{r}$ or $P(n, r)$, where $1 \leq r \leq n$.

Theorem I. *Prove that the number of different permutations of n distinct objects taken r at a time, $r \le n$ is given by*

$$^nP_r = \frac{n!}{(n-r)!} = n(n-1)(n-2)\ldots(n-r+1).$$

Proof. The number of permutations of n distinct objects taken r at a time is like filling of r places with n objects.

The first place can be filled in by any one of the n objects. So, this can be done in n ways

$$\therefore \quad ^nP_1 = n.$$

The second place can be filled in by any one of the $n - 1$ objects because after filling first place. We are left with $(n - 1)$ objects. Thus, the first two places can be filled in $n(n - 1)$ ways.

$$\therefore \quad ^nP_2 = n(n-1)$$

Similarly, the third place can be filled in by any one of the remaining $(n - 2)$ objects.

Therefore, the first three places can be filled in $n(n - 1)(n - 2)$ ways.

Proceeding in this way, we have the number of permutations of n different objects taken r at a time, given by

$$^nP_r = n(n-1)(n-2)\ldots(n-\overline{r-1})$$

$$= \frac{n(n-1)(n-2)\ldots(n-r+1)(n-r)!}{(n-r)!} = \frac{n!}{(n-r)!}$$

Theorem II. (*a*) *Prove that the number of permutations of n things taken all at a time is n !.*

Proof. We know that (theorem I)

$$^nP_n = \frac{n!}{(n-n)!} = \frac{n!}{0!} = \frac{n!}{1} \qquad [\because 0! = 1]$$

$$= n!$$

Theorem II. (*b*) *Prove that* $^nP_r = n.\ ^{n-1}P_{r-1}$.

Proof. We know $\quad ^nP_r = \dfrac{n!}{(n-r)!} = \dfrac{n(n-1)!}{(n-r)!} \qquad$ | Theorem. I

$$= n \cdot \frac{(n-1)!}{(n-1-(r-1))!} = n \cdot {^{n-1}P_{r-1}}.$$

Example 1. *Determine the value of the following*

(*i*) $^4P_2$ (*ii*) $^9P_3$ (*iii*) $^{20}P_2$ (*iv*) $^{52}P_4$.

Sol. (*i*) $^4P_2 = \dfrac{4!}{(4-2)!} = \dfrac{4 \times 3 \times 2!}{2!} = 12$

(*ii*) $^9P_3 = \dfrac{9!}{(9-3)!} = \dfrac{9 \times 8 \times 7 \times 6!}{6!} = 504$

(*iii*) $^{20}P_2 = \dfrac{20!}{(20-2)!} = \dfrac{20 \times 19 \times 18!}{18!} = 380$

(*iv*) $^{52}P_4 = \dfrac{52!}{(52-4)!} = \dfrac{52 \times 51 \times 50 \times 49 \times 48!}{48!} = 6497400.$

Example 2. *Determine the value of n if*

(*i*) $4 \times {}^nP_3 = {}^{n+1}P_3$. (*ii*) $6 \times {}^nP_3 = 3 \times {}^{n+1}P_3$. (*iii*) $3 \times {}^nP_4 = 7 \times {}^{n-1}P_4$.

Sol. (*i*) $4 \times \dfrac{n!}{(n-3)!} = \dfrac{(n+1)!}{(n+1-3)!}$ $\left[\because {}^nP_r = \dfrac{n!}{(n-r)!}\right]$

$\Rightarrow \dfrac{4 \times n!}{(n-3)!} = \dfrac{(n+1) \times n!}{(n-2)(n-3)!}$

$\Rightarrow 4(n-2) = (n+1) \Rightarrow 4n - 8 = n + 1$

$\Rightarrow 3n = 9 \Rightarrow n = 3.$

(*ii*) $6 \times {}^nP_3 = 3 \times {}^{n+1}P_3$

$6 \times \dfrac{n!}{(n-3)!} = 3 \times \dfrac{(n+1)!}{(n+1-3)!}$ $\left[\because {}^nP_r = \dfrac{n!}{(n-r)!}\right]$

$\Rightarrow \dfrac{6 \times n!}{(n-3)!} = \dfrac{3(n+1)(n!)}{(n-2)(n-3)!}$

$\Rightarrow 6(n-2) = 3 \times (n+1) \Rightarrow 6n - 12 = 3n + 3$

$\Rightarrow 6n - 3n = 12 + 3 \Rightarrow 3n = 15$

$\Rightarrow n = 5.$

(*iii*) $3 \times {}^nP_4 = 7 \times {}^{n-1}P_4$

$3 \times \dfrac{n!}{(n-4)!} = 7 \times \dfrac{(n-1)!}{(n-1-4)!}$ $\left[\because {}^nP_r = \dfrac{n!}{(n-r)!}\right]$

$\Rightarrow \dfrac{3 \times n \times (n-1)!}{(n-4)(n-5)!} = \dfrac{7 \times (n-1)!}{(n-5)!}$

$\Rightarrow 3n = 7(n-4) \Rightarrow 3n = 7n - 28$

$\Rightarrow 3n - 7n = -28 \Rightarrow -4n = -28$

$\Rightarrow n = 7.$

Example 3. *How many variable names of 8 letters can be formed from the letters a, b, c, d, e, f, g, h, i if no letter is repeated ?*

Sol. There are 9 letters and 8 are to be selected.

$\therefore$ Total number of variable names of 8 letters is $= {}^9P_8 = \dfrac{9!}{(9-8)!} = \dfrac{9!}{1!} = 9!.$

Example 4. *There are 10 persons called on an interview. Each one is capable to be selected for the job. How many permutation are there to select 4 from the 10 ?*

Sol. There are 10 persons and 4 are to be selected.

$\therefore$ Total number of permutations to select 4 persons $= {}^{10}P_4$

$$= \frac{10!}{(10-4)!} = \frac{10 \times 9 \times 8 \times 7 \times 6!}{6!} = 5040.$$

Example 5. *How many 6-digit numbers can be formed from the digits 0, 1, 2, 3, 4, 5, 6, 7, if no digit is repeated.*

Sol. There are 8 numbers are 6 are to be selected.

$\therefore$ Total number of 6-digit numbers $= {}^8P_6 = \frac{8!}{(8-6)!} = \frac{8!}{2!}$

$$= \frac{8 \times 7 \times 6 \times 5 \times 4 \times 3 \times 2!}{2!} = 22560.$$

5.7. PERMUTATION WITH RESTRICTIONS

The number of permutations of n different objects taken r at a time in which p particular objects do not occur is ${}^{n-p}P_r$

The number of permutations of n different objects taken r at a time in which p particular objects are present is ${}^{n-p}P_{r-p} \times {}^rP_p$.

Example 6. *How many 6-digit numbers can be formed by using the digits 0, 1, 2, 3, 4, 5, 6, 7, 8 if every number is to start with '30' with no digit repeated.*

Sol. All the numbers begin with '30'. So, we have to choose 4-digits from the remaining 7-digits.

$\therefore$ Total number of numbers that begins with '30' is

$${}^7P_4 = \frac{7!}{(7-4)!} = \frac{7 \times 6 \times 5 \times 4 \times 3!}{3!} = 840.$$

Example 7. *In how many ways 5 different microprocessor books and 4 different digital electronics books be arranged in a shelf so that all the four digital electronics books are together ?*

Sol. Consider the four digital electronics books as one unit. Thus, we have 6 units that can be arranged in 6 ! ways.

For each of these arrangements, 4 digital electronics books can be arranged among themselves in 4 ! ways.

$\therefore$ Total number of arrangements in which all the four digital electronics books are together is

$$= 6! \times 4! = 720 \times 24 = 17280.$$

Example 8. *How many permutations can be made out of the letter of word "COMPUTER" ? How many of these*

(*i*) *begin with C ?* (*ii*) *end with R ?*

(*iii*) *begin with C and end with R ?* (*iv*) *C and R occupy the end places ?*

Sol. There are 8 letters in the word 'COMPUTER' and all are distinct.

$\therefore$ The total number of permutations of these letters is $8! = 40320$.

(*i*) Permutations which begin with C.

The first position can be filled in only one way *i.e.*, C and the remaining 7 letters can be arranged in 7 ! ways.

$\therefore$ Total number of permutations starting with C are $= 1 \times 7\,! = 5040$.

(*ii*) Permutations which end with R.

The last position can be filled in only one way *i.e.*, R and the remaining 7 letters can be arranged in 7 ! ways.

$\therefore$ The total number of permutations ending with R are $= 7\,! \times 1 = 5040$.

(*iii*) Permutations begin with C and end with R.

The first position can be filled in only one way *i.e.*, C and the last place can also be filled in only one way *i.e.*, R and the remaining 6 letters can be arranged in 6 ! ways.

$\therefore$ The total number of permutations begin with C and end with R is

$$= 1 \times 6\,! \times 1 = 720.$$

(*iv*) Permutations is which C and R occupy end places.

C and R occupy end positions in 2 ! ways *i.e.*, C, R and R, C and the remaining 6 letters can be arranged in 6 ! ways.

$\therefore$ The total number of permutations in which C and R occupy end places is

$$= 2\,! \times 6\,! = 1440.$$

5.8. PERMUTATIONS WHEN ALL OF THE OBJECTS ARE NOT DISTINCT

Theorem III. *The number of permutations of n objects, of which n_1 objects are of one kind and n_2 objects of another kind, when all are taken at a time is* $\dfrac{n\,!}{n_1\,!\,n_2\,!}$.

Proof. Let us assume that the number of required permutations be K. Now consider a single particular permutation of these K permutations, in which n_1 objects of one kind is followed by n_2 objects of other kind.

Also, assume that all n_1 object are distinct from all n_2 objects.

So, number of permutations of n_1 objects taken all at a time is $= {}^{n_1}P_{n_1} = n_1\,!$

Also, the number of permutations of n_2 objects taken all at a time is $= {}^{n_2}P_{n_2} = n_2\,!$

By the fundamental principle of counting, these K permutations will give rise to $n_1\,!\,n_2\,!$ permutations by arranging the objects of one kind within the places occupied by them.

Therefore, K permutations will give rise to K. $n_1\,!\,n_2\,!$ permutations.

For n distinct objects, the number of permutations is $= {}^nP_n = n\,!$

Therefore, $\quad K \times n_1\,!\,n_2\,! = n\,! \;\Rightarrow\; K = \dfrac{n\,!}{n_1\,!\,n_2\,!}$

This result can be generalised as follows :

If n_1 objects are of one kind, n_2 objects are of second kind, n_3 objects are of third kind, and so on upto n_1 objects are of tth type is given by

$$\frac{n\,!}{n_1\,!\,n_2\,!\,n_3\,!\ldots n_t\,!} \qquad [\text{Here } n_1 + n_2 + n_3 + \ldots + n_t = n]$$

Example 9. *Determine the number of permutations that can be made out of the letters of the word 'PROGRAMMING'.*

Sol. There are 11 letters in the word 'PROGRAMMING' out of which G's, M's and R's are two each.

$\therefore$ The total number of permutations is

$$= \frac{11!}{2! \times 2! \times 2!} = \frac{11 \times 10 \times 9 \times 8 \times 7 \times 6 \times 5 \times 4 \times 3 \times 2!}{2 \times 1 \times 2 \times 1 \times 2!} = 4989600.$$

Example 10. *There are 4 blue, 3 red and 2 black pens in a box. These are drawn one by one. Determine all the different permutations.*

Sol. There are total 9 pens in the box out of which 4 are blue, 3 are red and 2 are black.

$\therefore$ The total number of permutations

$$= \frac{9!}{4! \times 3! \times 2!} = \frac{9 \times 8 \times 7 \times 6 \times 5 \times 4!}{4! \times 3 \times 2 \times 1 \times 2 \times 1} = 1080.$$

Example 11. *How many different variable names can be formed by using the letters a, a, a, b, b, b, b, c, c, c ?*

Sol. There are total 10 letters out of which 3 are *a*'s, 4 are *b*'s and there are 3 *c*'s.

$\therefore$ Total number of permutations

$$= \frac{10!}{3! \times 4! \times 3!} = \frac{10 \times 9 \times 8 \times 7 \times 6 \times 5 \times 4!}{3 \times 2 \times 1 \times 4! \times 3 \times 2 \times 1}$$

$$= 10 \times 3 \times 4 \times 7 \times 5 = 4200.$$

Example 12. *How many 7-digits numbers can be formed using digits 1, 7, 2, 7, 6, 7, 6 ?*

Sol. There are total 7-digits out of which 3 are 7's and 2 are 6's.

$\therefore$ Total number of permutations is $= \dfrac{7!}{3! \times 2!} = 420.$

5.9. PERMUTATIONS WITH REPEATED OBJECTS

Theorem IV. *Prove that the number of different permutations of n distinct objects taken r at a time when every object is allowed to repeat any number of times is given by n^r.*

Proof. Assume that with n objects we have to fill r place when repetition of objects is allowed.

Therefore, the number of ways of filling the first place $= n$

The number of ways of filling second place $= n$

....

....

The number of ways of filling rth place $= n$

Thus, the total number of ways of filling r places with n elements is

$$= n.\ n.\ n.\ n.\ \ldots\ldots\ r \text{ times} = n^r.$$

Example 13. *How many 4-digits numbers can be formed by using the digits 2, 4, 6, 8 when repetition of digits is allowed ?*

Sol. We have 4-digits.

So, number of ways of filling unit's place = 4.
Number of ways of filling ten's place = 4.
Number of ways of filling hundred's place = 4.
Number of ways of filling thousand's place = 4.

Therefore, the total number of 4-digits numbers is $= 4 \times 4 \times 4 \times 4 = 256$.

Example 14. *How many 2-digits even numbers can be formed by using the digits 1, 3, 4, 6, 8 when repetition of digits is allowed ?*

Sol. We have three even numbers and two odd number.

Thus, number of ways of filling unit's place = 3.

Number of ways of filling ten's place = 5.

$\therefore$ Total number of two digits even numbers $= 3 \times 5 = 15$.

Example 15. *In how many ways can 5 software projects be allotted to 6 final year students when all the 5 projects are not allotted to the same student ?*

Sol. We have 5 projects and 6 students.

Each projects can be allotted in 6 ways.

Thus, the number of ways of alloting 5 projects is $= 6 \times 6 \times 6 \times 6 \times 6 = 6^5$.

Number of ways in which all projects allotted to same student = 6.

Therefore, total number of ways to allocate 5 projects to 6 students $= 6^5 - 6 = 7770$.

5.10. CIRCULAR PERMUTATIONS *(P.T.U. B.Tech. Dec. 2007)*

The circular permutations are the permutations of the objects placed in a circle. Consider the letters *k, l, m, n, o* placed along the circle as shown in Fig. 1.

If we place letters linearly, there are five different permutations *i.e.*, *k, l, m, n, o ; l, m, n, o, k ; m, n, o, k, l ; n, o, k, l, m ; o, k, l, m, n,* but there is only one circular permutation *k, l, m, n, o.* Therefore, there is no starting and ending in circular permutation. We only consider the relative positions.

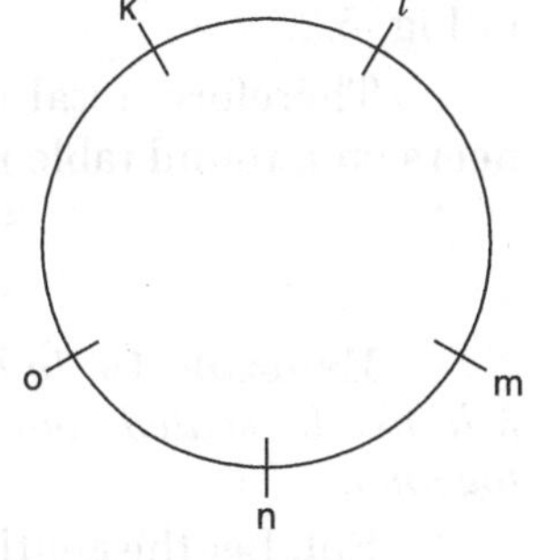

Fig. 5.1.

Theorem V. *Prove that the number of circular permutations of n different objects is (n – 1) !*

Proof. Let us consider that K be number of permutations required.

For each such circular permutation of K, there are n corresponding linear permutations. As shown earlier, we start from every object of n objects in the circular permutation. Thus, for K circular permutations, we have K. n linear permutations.

Therefore, $\quad \text{K}. n = n\,! \quad \text{or} \quad \text{K} = \dfrac{n\,!}{n}$

or $$K = \frac{n \times (n-1)!}{n} \quad \text{or} \quad K = (n-1)!$$

Hence proved.

Example 16. *In how many ways can these letters a, b, c, d, e, f be arranged in a circle ?*

Sol. There are 6 letters and hence the number of ways to arrange these 6 letters in a circle is

$$= (6 - 1) ! = 5 ! = 120.$$

Example 17. *In how many ways 10 programmers can sit on a round table to discuss the project so that project leader and a particular programmer always sit together ?*

Sol. There are total 10 programmers but project leader and a particular programmer always sit together. So, both become a single unit and hence there are (10 – 2 + 1) = 9 remains.

Thus, these 9 units can be arranged on round table in (9 – 1) ! ways.

The two programmers *i.e.*, project leader and a particular programmer can be arranged in 2 ! ways.

Therefore, the total number of ways in which 10 programmers can sit on a round table is

$$= (9 - 1) ! \times 2 ! = 8 ! \times 2 ! = 80640.$$

Example 18. *Determine the number of ways in which 5 software engineers and 6 electronics engineers can be sitted at a round table so that no two software engineers can sit together.*

Sol. There are 6 electronics engineers that can be arranged round a table in (6 – 1) ! ways. There are 5 software engineers and they are not to sit together so we have six places for software engineers and can be placed in 6 ! ways as shown in Fig. 5.2.

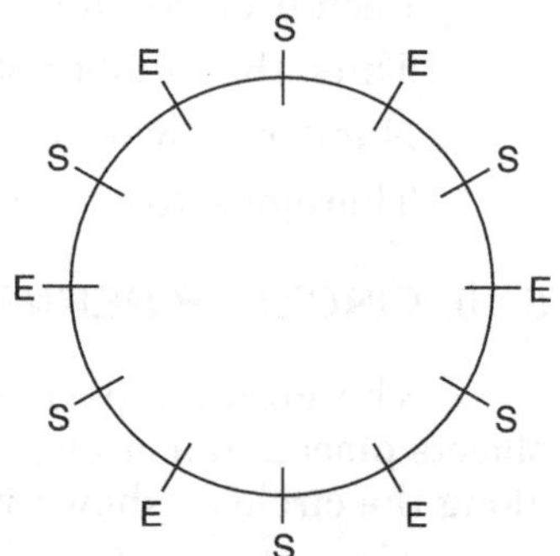

Fig. 5.2.

Therefore, total number of ways to arrange the engineers on a round table is

$$= (6 - 1) ! \times 6 ! = 5 ! \times 6 !$$
$$= 120 \times 720 = 86400.$$

Example 19. *In how many ways can 5 gentlemen and 5 ladies be seated round a table so that no two ladies are together.* (P.T.U. B. Tech. May 2009)

Sol. Let the gentlemen be seated first leaving one seat vacant in between each of gentlemen.

This can be done in (5 – 1) ! ways = 24 ways

Now 5 ladies can be seated in 5 vacant seats, as shown in Fig.5.3, in 5 P_5 ways = 5 ! ways

= 120 ways

∴ By fundamental principle of counting, required number of ways = 24 × 120 = 2880

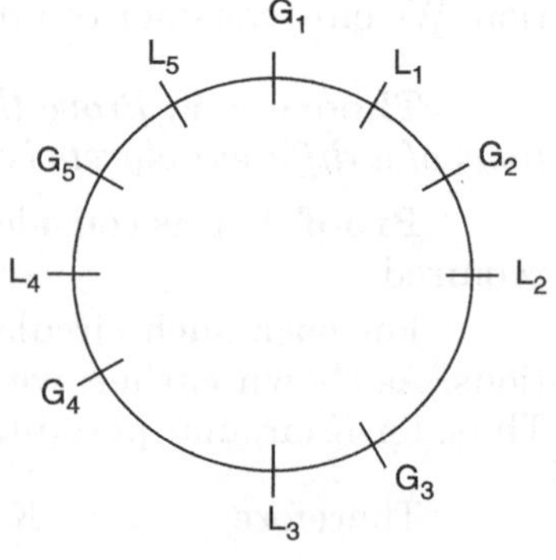

Fig. 5.3.

TEST YOUR KNOWLEDGE 5.2

SHORT ANSWER TYPE QUESTIONS

1. Compute 4 !, 5 !, 6 !, 7 !.
2. Compute (*a*) $\frac{13!}{11!}$ (*b*) $\frac{7!}{10!}$.
3. Simplify (*a*) $\frac{n!}{(n-1)!}$ (*b*) $\frac{(n+2)!}{n!}$.
4. Find *n* if (*a*) ${}^nP_2 = 72$, (*b*) ${}^nP_4 = 42\ {}^nP_2$, (*c*) ${}^{2n}P_2 = {}^{2n}P_2 - 50$.
5. Find the number of distinct permutations that can be formed all the letters of each word
   (*a*) RADAR (*b*) UNUSUAL.

LONG ANSWER TYPE QUESTIONS

6. There are four bus lines between A and B, and three bus lines between B and C. In how many ways can a man travel,
   (*a*) By bus from A to C by way of B
   (*b*) Round-trip by bus from A to C by way of B
   (*c*) Round-trip by bus from A to C by way of B if he does not want to use a bus line more than once.
7. Find the number of ways that a party of seven persons can arrange themselves
   (*a*) In a row of seven chairs (*b*) Around a circular table.
8. In how many ways can four books of mathematics, three books of history, three books of chemistry, two books of sociology be arranged on a shelf so that all books of the same subject are together ?
9. How many automobile license plates can be made if each plate contains two different letters followed by (*a*) three different digits (*b*) if the first digit is not zero.
10. There are six roads between A and B and four roads between B and C. Find the number of ways that one can drive
    (*a*) From A to C by way of B (*b*) Round-trip from A to C by way of B
    (*c*) Round-trip from A to C by way of B without using the same road more than once.
11. (*a*) Find the number of ways in which five persons can sit in a row.
    (*b*) How many ways are there if two of the persons insist on sitting next to one another.
    (*c*) Solve part (*a*) assuming they sit around a circular table.
    (*d*) Solve part (*b*) assuming they sit around a circular table.
12. Find the number of ways in which five large books, four medium size books, and three small books can be placed on a shelf so that all books of the same size are together.
13. (*a*) Find the number of permutations that can be formed from the letters of word ELEVEN.
    (*b*) How many of them begin and end with E ?
    (*c*) How many of them have the three Es together ?
    (*d*) How many begin with E and end with N ?
14. (*a*) In how many ways can three boys and 2 girls sit in a row ?
    (*b*) In how many ways can they sit in a row if the boys and girls are each to sit together ?
    (*c*) In how many ways can they sit in a row if just the girls are to sit together ?
15. (*a*) In how many ways can you arrange the letters in the word CONGRESS ?
    (*b*) In how many ways, the two S's not together ?
16. How many words can be obtained by arranging the letters of the word 'UNIVERSAL' in different ways ? In how many of them
    (*a*) E, R, S occur together ?
    (*b*) No two of the letters E, R, S occur together ? (*P.T.U. B.Tech. May 2008*)
17. Find the number of different messages that can be represented by sequences of 4 dots and 6 dashes. (*P.T.U. B.Tech May 2013*)

Answers

1. $4! = 24$, $5! = 120$, $6! = 720$, $7! = 5040$

2. (*a*) 156 (*b*) $\frac{1}{720}$ **3.** (*a*) n (*b*) $n^2 + 3n + 2$

4. (*a*) $n = 9$ (*b*) $n = 9$ (*c*) $n = 5$

5. (*a*) $\frac{5!}{2!2!} = 30$ (*b*) $\frac{7!}{3!} = 840$

6. (*a*) 12 (*b*) 144 (*c*) 72

7. (*a*) 7 ! (*b*) 6 ! **8.** 41472

9. (*a*) 468000 (*b*) 421200

10. (*a*) 24 (*b*) 576 (*c*) 360

11. (*a*) 120 (*b*) 24 (*c*) 24 (*d*) 12

12. 103680

13. (*a*) 120 (*b*) 24 (*c*) 24 (*d*) 12

14. (*a*) 120 (*b*) 24 (*c*) 48

15. (*a*) 20160 (*b*) 15120 **16.** 362880, (*a*) 30240, (*b*) 332640.

Hints

6. (*a*) There are four ways to go from A to B, and three ways to go from B to C. Hence there are $4 \times 3 = 12$ ways to go from A to C by way of B.

(*b*) There are 12 ways to go from A to C by way of B and 12 ways to return. Hence there are $12 \times 12 = 144$ ways to travel round-trip.

(*c*) The man will traval from A to B to C to B to A. $A \xrightarrow{4} B \xrightarrow{3} C \xrightarrow{2} B \xrightarrow{3} A$

Total number of ways $= 4 \times 3 \times 2 \times 3 = 72$.

8. Required number of ways $= 4! \times 4! \times 3! \times 3! \times 2!$

9. (*a*) Required no. $= 26 \times 25 \times 10 \times 9 \times 8 = 468000$

(*b*) Required no. $= 26 \times 25 \times 9 \times 9 \times 8 = 421200$

10. See Q. 6.

12. Required number of ways $= 3! \times 5! \times 4! \times 3!$

13. (*a*) ELEVEN consists 6 letters in which three Es, 2Vs, occur.

$\therefore$ Required number of ways $= \frac{6!}{3! \times 2!} = \frac{6 \times 5 \times 4}{2} = 120$.

(*b*) The first and last place are occupied with E.

Remaining places can be filled up in $4! = 24$.

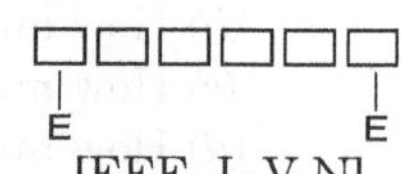

(*c*) Take three Es as one word. Total number of words are 4,

$\therefore$ Required number of words $= 4! = 24$. [EEE L,V,N]

(*d*) The first place is occupied with E and the last placed is occupied with N.

$\therefore$ Required number of ways $= \frac{4!}{2!} = 12$. [E], L, E, V, E, [N]

15. (*a*) The word CONGRESS consists of 8 words in which 2S occur.

$\therefore$ Total number of arranging the words $= \frac{8!}{2!} = 20160$.

(*b*) Consider SS as one word, then SS C, O, N, G, R, E

Total number of arranging the words in which two S's occcur together = 7 !

Hence required number of arranging the words in which no two S's occur together

= Total number of arranging the words – (number of arranging the words in which 2 S's occur together)

= 20160 – 7 ! = 15120.

17. There are total 10 symbols *i.e.* 4 dots and 6 dashes.

$\therefore$ The total number of messages that can be represented are:

$$= \frac{10!}{4! \times 6!} = \frac{10 \times 9 \times 8 \times 7 \times 6!}{4! \times 6!} = 210 \text{ messages.}$$

5.11. COMBINATION

A combination is a selection of some or all, objects from a set of given objects, where order of the objects does not matter. The number of combinations of n objects, taken r at a time is represented by ${}^nC_r$ or C(n, r).

Theorem VI. *The number of combinations of n different things, taken r at a time is given by*

$${}^nC_r = \frac{n!}{r!\,(n-r)!},\ 1 \le r \le n.$$

Proof. The number of permutations of n different things, taken r at a time is given by

$${}^nP_r = \frac{n!}{(n-r)!}$$

As there is no matter about the order of arrangement of the objects, therefore, to every combination of r things, there are r ! arrangements *i.e.*,

$${}^nP_r = r!\,{}^nC_r \quad \text{or} \quad {}^nC_r = \frac{{}^nP_r}{r!} = \frac{n!}{(n-r)!\,r!}$$

Thus, $${}^nC_r = \frac{n!}{(n-r)!\,r!},\ 1 \le r \le n.$$

Theorem VII. *Prove that the number of combinations of n things taken all at a time is one.*

Proof. We know that

$${}^nC_n = \frac{n!}{(n-n)!\,n!} = \frac{n!}{0!\,n!} = 1 \qquad [\because\ 0! = 1]$$

Theorem VIII. *Prove that the number of combinations of n things taken none at a time is one.*

Proof. We know that

$${}^nC_0 = \frac{n!}{(n-0)!\,0!} = \frac{n!}{n!\,0!} = \frac{n!}{n!} = 1 \qquad [\because\ 0! = 1]$$

Theorem IX. *Prove that* ${}^nC_{n-r} = {}^nC_r$, $1 \le r \le n$.

Proof. We know that

$${}^nC_{n-r} = \frac{n!}{n-(n-r)!\,(n-r)!} = \frac{n!}{(n-n+r)!\,(n-r)!} = \frac{n!}{r!\,(n-r)!} = {}^nC_r.$$

Remark. If ${}^nC_r = {}^nC_m$, then either $m = r$ or $m + r = n$.

Theorem X. *Prove that* $\binom{n+1}{r} = \binom{n}{r-1} + \binom{n}{r}$ *where* $1 \le r \le n$ *and r, n are natural numbers* (P.T.U. B. Tech. Dec 2010)

Proof. Here $\binom{n+1}{r} = {}^{n+1}C_r$. Consider

$$^nC_{r-1} + {}^nC_r = \frac{n!}{r!\times(n-r)!} + \frac{n!}{(r-1)!\times(n-r+1)!}$$

$$= \frac{n!}{r(r-1)!\times(n-r)!} + \frac{n!}{(r-1)!\times(n-r+1)(n-r)!}$$

$$= \frac{n!\times(n-r+1)+n!\times r}{r(r-1)!\times(n-r)!\times(n-r+1)} = \frac{n!\times(n-r+1+r)}{r!\times(n-r+1)!}$$

$$= \frac{n!\times(n+1)}{r!\times(n-r+1)!} = \frac{(n+1)!}{r!\times(n+1-r)!} = {}^{n+1}C_r$$

ILLUSTRATIVE EXAMPLES

Example 1. *Determine the value of following*

(i) $^{10}C_6$ (ii) $^{50}C_{45}$ (iii) $^{52}C_4$ (iv) $^{20}C_{10}$.

Sol. (i) $^{10}C_6 = \frac{10!}{(6)!\times(10-6)!} = \frac{10\times9\times8\times7\times6!}{6!\times4\times3\times2\times1} = 10\times3\times7 = 210.$

(ii) $^{50}C_{45} = \frac{50!}{45!\times(50-45)!} = \frac{50\times49\times48\times47\times46\times45!}{45!\times5\times4\times3\times2\times1} = 2118760.$

(iii) $^{52}C_4 = \frac{52!}{4!\times(52-4)!} = \frac{52\times51\times50\times49\times48!}{4\times3\times2\times1\times48!} = 270725.$

(iv) $^{20}C_{10} = \frac{20!}{10!\times(20-10)!} = \frac{20\times19\times18\times17\times16\times15\times14\times13\times12\times11\times10!}{10!\times10!}$

$= 184756.$

Example 2. *Determine the value of n if*

(i) $^nC_4 = {}^nC_3$ (ii) $^nC_{n-2} = 10$ (iii) $^{20}C_{n+2} = {}^{20}C_{2n-1}$.

Sol. (i) $^nC_4 = {}^nC_3$

$$\Rightarrow \quad \frac{n!}{4!\times(n-4)!} = \frac{n!}{3!(n-3)!}$$

$$\Rightarrow \quad \frac{n!}{n!} = \frac{4!\times(n-4)!}{3!\times(n-3)!} = \frac{4\times3!\times(n-4)!}{3!\times(n-3)\times(n-4)!}$$

$$\Rightarrow \quad 1 = \frac{4}{n-3} \Rightarrow n-3=4 \Rightarrow n=7.$$

(*ii*) $^nC_{n-2} = 10$

Thus $\dfrac{n!}{(n-2)!\,[n-(n-2)]!} = 10$ or $\dfrac{n!}{(n-2)!\times 2!} = 10$

or $\dfrac{n\times(n-1)\times(n-2)!}{(n-2)!\times 2!} = 10 \Rightarrow n\times(n-1) = 10\times 2! = 20$

or $n^2 - n - 20 = 0 \Rightarrow n^2 - 5n + 4n - 20 = 0$

$\Rightarrow$ $(n-5)(n+4) = 0$

$\Rightarrow$ $n = -4, 5$. Since -4 is not possible, hence $n = 5$.

(*iii*) $^{20}C_{n+2} = {}^{20}C_{2n-1}$

Therefore, we have either

$$n + 2 = 2n - 1 \quad\text{or}\quad (n+2) + (2n-1) = 25$$

| See remark below theorem IV

$\Rightarrow$ $-n = -3$ or $3n = 24$

$\Rightarrow$ $n = 3$ or $n = 8$

So $n = 3, 8$.

Example 3. *How many 16-bit strings are there containing exactly five 0's ?*

Sol. A 16-bit string having exactly five 0's is determined if we tell which bits are 0's. This can be done in $^{16}C_5$ ways.

Therefore, the total number of 16-bit strings is

$$= {}^{16}C_5 = \frac{16!}{5!\times(16-5)!}$$

$$= \frac{16\times 15\times 14\times 13\times 12\times 11!}{5\times 4\times 3\times 2\times 1\times 11!} = 4368.$$

Example 4. *How many ways can we select a software development group of 1 project leader, 5 programmers and 6 data entry operators from a group of 5 project leaders, 20 programmers and 25 data entry operators ?*

Sol. There are 5 project leaders out of which one can be selected in $^5C_1$ ways.

There are 20 programmers out of which five can be selected in $^{20}C_5$ ways.

There are 25 data entry operators out of which six can be selected in $^{25}C_6$ ways.

Therefore, the total number of ways to select the software development group is

$$= {}^5C_1 \times {}^{20}C_5 \times {}^{25}C_6 = 96101544000.$$

Example 5. *From 10 programmers in how many ways can 5 be selected when*

(a) A particular programmer is included every time.

(b) A particular programmer is not included at all.

Sol. We have to select 5 programmers from the 10 programmers. So, the number of ways to select them in $^{10}C_5$

$$= \frac{10!}{5!\times(10-5)!} = \frac{10\times 9\times 8\times 7\times 6\times 5!}{5\times 4\times 3\times 2\times 1\times 5!} = 252.$$

(*a*) When a particular programmer is included every time then the remaining = 5 – 1 = 4 programmers can be selected from the remaining = 10 – 1 = 9 programmers. This can be done in $^9C_4$ ways

$$= \frac{9!}{4!(9-4)!} = \frac{9\times8\times7\times6\times5!}{4\times3\times2\times1\times5!} = 126.$$

(*b*) When a particular programmer is not included at all, then the five programmers can be selected from the remaining = 10 – 1 = 9 programmers.

This can be done in $^9C_5$ ways

$$= \frac{9!}{5!(9-5)!} = \frac{9\times8\times7\times6\times5\times4!}{5\times4\times3\times2\times1\times4!} = 126.$$

5.12. PIGEONHOLE PRINCIPLE (*P.T.U. M.C.A. 2005*)

Theorem X. *Show that if n pigeons are assigned to m pigeonholes and m < n, then there is at least one pigeonhole that contains two or more pigeons.*

Proof. Let us label the *n* pigeons with the numbers 1 through *n* and the *m* pigeonholes with the numbers 1 through *m*. Now starting with pigeon 1 and Pigeonhole 1, assign each pigeon in order to the pigeonhole with the same number. So we can assign as many pigeons as possible to distinct pigeonholes, but as we know that pigeonholes are less than pigeons *i.e.*, $m < n$. Thus, there remains $n - m$ pigeons that have not yet been assigned to a pigeonhole. Hence, there is at least one pigeonhole that will be assigned a second pigeon.

Example 6. *Show that if any four numbers from 1 to 6 are chosen, then two of them will add to 7.*

Sol. Make three sets containing two numbers whose sum is 7.

A = {1, 6}, B = {2, 5}, C = {3, 4}. The four numbers that will be chosen assigned to the set that contains it.

As there are only three sets, two numbers that are chosen is from the same set whose sum is 7.

Example 7. *Show that at least two people must have their birthday in the same month if 13 people are assembled in a room.*

Sol. We assigned each person the month of the year on which he was born. Since there are 12 months in a year.

So, according to pigeonhole principle, there must be at least two people assigned to the same month.

Example 8. *Show that if any eight + ve integers are chosen, two of them will have same remainder when divided by 7.*

Sol. Take any eight +ve integers. When these are divided by 7 each have some remainder. Since there are eight integers and only seven distinct remainders because number 7 can generate only 7 remainders, so two +ve integers must have same remainder.

5.13. EXTENDED PIGEONHOLE PRINCIPLE

It states that if *n* pigeons are assigned to *m* pigeonholes (The number of pigeons is very large than the number of pigeonholes), then one of the pigeonholes must contain at least $[(n - 1)/m] + 1$ pigeons.

Theorem XI. *Prove extended pigeonhole principle.*

Proof. We can prove this by the method of contradiction. Assume that each pigeonhole does not contain more than $[(n - 1)/ m]$ pigeons. Then, there will be at most $m[(n - 1)/m] \leq m(n - 1)/m = n - 1$ pigeons in all. This is in contradiction to our assumptions. Hence, for given m pigeonholes, one of these must contain at least $[(n - 1)/m] + 1$ pigeons.

Example 9. *Show that if 9 colours are used to paint 100 houses, at least 12 houses will be of the same colour.*

Sol. Let us assume the colours be the pigeonholes and the houses the pigeons. Now 100 pigeons are to be assigned to 9 pigeonholes. Using the extended pigeonhole principle, $[(n - 1)/m] + 1$, where $n = 100$ and $m = 9$, we have $[(100 - 1)/9] + 1 = 12$. Thus, there are 12 houses of the same colour.

ILLUSTRATIVE EXAMPLES

Example 1. *How many different 8-bit strings are there that end with 0111 ?*

Sol. An 8-bit strings that end with 0111 can be constructed in 4 steps *i.e.,*

By selecting Ist bit, IInd bit, IIIrd bit and IVth bit and each bit can be selected in 2 ways.

Hence, the total no. of 8-bit strings that end with 0111 is

$$= 2 \,.\, 2 \,.\, 2 \,.\, 2 = 2^4.$$

Example 2. *How many 2-digits numbers greater than 40 can be formed by using the digits 1, 2, 3, 4, 6, 7*

(a) When repetition is allowed (b) When repetition is not allowed.

Sol. (*a*) When repetition is allowed

We have to find the numbers greater than 40. Therefore,

Ten's place can be filled up by 3 ways.

Unit's place can be filled up by 6 ways.

∴ The total number of 2-digits numbers greater than 40 is $= 3 \times 6 = 18$.

(*b*) When repetition is not allowed

Ten's place can be filled up by 3 ways.

Unit's place can be filled up by 5 ways.

∴ The total number of 2-digits numbers greater than 40 is $= 3 \times 5 = 15$.

Example 3. *How many words can be constructed of three English alphabets ?*

(a) When repetition of alphabets is allowed

(b) When repetition is not allowed.

Sol. There are 26 alphabets in English. Therefore,

(*a*) When repetition is allowed

First alphabet of word can be selected in 26 ways.

Second alphabet of word can be selected in 26 ways.

Third alphabet of word can be selected in 26 ways.

Hence, total number of words of three alphabets constructed is

$$= 26 \times 26 \times 26 = 17576.$$

(*b*) When repetition is not allowed

First alphabet of word can be selected in 26 ways.

Second alphabet of word can be selected in 25 ways.

Third alphabet of word can be selected in 24 ways.

Hence, the total number of words of three distinct alphabets is = 26 × 25 × 24 = 15600.

Example 4. *Show that 0 ! = 1.*

Sol. We have

$$^nP_r = \frac{n!}{(n-r)!} \qquad \text{...(1)}$$

Now, put $r = n$ in equation (1), we have

$$^nP_n = \frac{n!}{(n-n)!} \quad \Rightarrow \quad n! = \frac{n!}{0!} \quad \Rightarrow \quad 0! = \frac{n!}{n!} = 1$$

Hence 0 ! = 1.

Example 5. *There are n objects out of which r objects are to be arranged. Find the total number of permutations when*

(*a*) *four particular objects always occur.*

(*b*) *four particular objects never occur.*

Sol. (*a*) Number of ways to arrange first object = r

Number of ways to arrange second object = $r - 1$

Number of ways to arrange third object = $r - 2$

Number of ways to arrange fourth object = $r - 3$

Number of ways to arrange remaining $n - 4$ objects taking $r - 4$ at a time = $^{n-4}P_{r-4}$.

Therefore, the total number of permutation when four particular objects always occur is

$$= r(r-1)(r-2)(r-3)\ ^{n-4}P_{r-4}. \qquad \text{[Using first principle of counting]}$$

(*b*) There are four particular objects which never occur in any arrangement. Hence set aside these four particular objects.

Thus, we have to find the number of arrangements of $n - 4$ objects taking r at a time.

∴ The total number of arrangements is = $^{n-4}P_r$.

Example 6. *How many permutations can be made out of the letters of the word "Basic" ? How many of these*

(*i*) *begin with B ?* (*ii*) *end with C ?*

(*iii*) *B and C occupy the end places ?*

Sol. There are 5 letters in the word 'Basic' and all are distinct.

∴ The number of permutations of these letters

$$= 5! = 5 \times 4 \times 3 \times 2 \times 1 = 120.$$

(*i*) Permutations which begin with B

The first position can be filled in only one way *i.e.*, B and the remaining 4 letters can be arranged in 4 ! ways.

∴ Total number of permutations starting with B = 1 × 4 ! = 24.

(*ii*) Permutations which end with C

The first position can be filled in only one way *i.e.*, C and the remaining 4 letters can be arranged in 4 ! ways.

∴ Total number of permutations ending with C = 4 ! × 1 = 24.

(*iii*) Permutations in which B and C occupy end places

B and C occupy end positions in 2 ! ways *i.e.*, B, C and C, B and the remaining 3 letters can be arranged in 3 ! ways.

∴ Total number of permutations in which B and C occupy end places in $= 2! \times 3! = 12$.

Example 7. *Show that* $^nC_r + {^nC_{r-1}} = {^{n+1}C_r}$, *where* $n \geq r \geq 1$ *and n and r are natural numbers.* (P.T.U. B.Tech. Dec. 2010)

Sol. Take L.H.S. of equation *i.e.*,

$$^nC_r + {^nC_{r-1}} = \frac{n!}{r! \times (n-r)!} + \frac{n!}{(r-1)! \times (n-r+1)!}$$

$$= \frac{n!}{r(r-1)! \times (n-r)!} + \frac{n!}{(r-1)! \times (n-r+1)(n-r)!}$$

$$= \frac{n! \times (n-r+1) + n! \times r}{r(r-1)! \times (n-r)! \times (n-r+1)} = \frac{n! \times (n-r+1+r)}{r! \times (n-r+1)!}$$

$$= \frac{n! \times (n+1)}{r! \times (n-r+1)!} = {^{n+1}C_r}$$

Hence proved.

Example 8. *In the 'Discrete Structures Paper' there are 8 questions. In how many ways can an examiner select five questions in all if first question is compulsory.*

Sol. Since the first question is compulsory, the examiner has to select 4 questions from the remaining 7 questions.

Therefore, the number of ways to select 5 questions $= {^7C_4}$

$$= \frac{7!}{4! \times (7-4)!} = \frac{7 \times 6 \times 5 \times 4!}{4! \times 3 \times 2 \times 1} = 35.$$

Example 9. *Determine the number of triangles that are formed by selecting points from a set of 15 points out of which 8 are collinear.*

Sol. When we take all the 15 points, the number of triangles formed is $^{15}C_3$.

As 8 points lie on the same line, they do not form any triangle. Thus, $^8C_3$ triangles are lost.

∴ The total number of triangles produced is

$$^{15}C_3 - {^8C_3} = \frac{15!}{3 \times (15-3)!} - \frac{8!}{3!(8-3)!} = \frac{15 \times 14 \times 13 \times 12!}{3 \times 12!} - \frac{8 \times 7 \times 6 \times 5!}{3 \times 2 \times 1 \times 5!}$$

$$= \frac{15 \times 14 \times 13}{3} - \frac{8 \times 7 \times 6}{3 \times 2 \times 1} = 910 - 56 = 854.$$

Example 10. *How many lines can be drawn through 10 points on a circle ?*

Sol. As all the points on the circle are not collinear. Thus, no lines will lost.

∴ The total number of lines drawn through a circle $= {^{10}C_2}$

$$= \frac{10!}{2! \times (10-2)!} = \frac{10 \times 9 \times 8!}{2 \times 1 \times 8!} = 45.$$

Example 11. *Determine the number of diagonals that can be drawn by joining the nodes of octagon.*

Sol. The number of lines that can be formed by joining 2 out of 8 points $= {^8C_2}$

$$= \frac{8 \times 7}{2} = 28$$

Out of these 28 lines, the 8 are sides of the octagon.

$\therefore$ The number of diagonals = 28 – 8 = 20.

Example 12. *In a shipment, there are 40 floppy disks of which 5 are defective. Determine*

(*a*) *in how many ways we can select five floppy disks ?*

(*b*) *in how many ways we can select five non-defective floppy disks ?*

(*c*) *in how many ways we can select five floppy disks containing exactly three defective floppy disks ?*

(*d*) *in how many ways we can select five floppy disks containing at least 1 defective floppy disks ?*

Sol. (*a*) There are 40 floppy disks out of which we have to select 5 floppy disks. These can be done in $^{40}C_5$ ways *i.e.*,

$$= \frac{40!}{5!(40-5)!} = \frac{40 \times 39 \times 38 \times 37 \times 36 \times 35!}{5! \times 35!} = 658008.$$

(*b*) There are 40 – 5 = 35 non-defective floppy disks out of which we have to select 5. This can be done in $^{35}C_5$ ways.

$$= \frac{35!}{5!(35-5)!} = \frac{35 \times 34 \times 33 \times 32 \times 31 \times 30!}{5 \times 4 \times 3 \times 2 \times 1 \times 30!} = 324632.$$

(*c*) To select exactly three defective floppy disks out of total 5 we have $^5C_3$ ways and the remaining 2 floppy disks can be selected in $^{35}C_2$ ways.

Therefore, the total number of ways to select 5 floppy disks out of which exactly 3 are defective = $^5C_3 \times {}^{35}C_2$

$$= \frac{5!}{3!(5-3)!} \times \frac{35!}{2! \times (35-2)!} = \frac{5 \times 4 \times 3!}{3! \times 2 \times 1} \times \frac{35 \times 34 \times 33!}{2 \times 1 \times 33!}$$

$$= 5950.$$

(*d*) There are five defective floppy disks out of which at least 1 must be selected. We know that the total number of ways to select 5 floppy disks out of total 40 disks = $^{40}C_5$.

Also, the number of ways to select 5 floppy disks with number one defective

$$= {}^{35}C_5.$$

Therefore, the total number of ways to select 5 floppy disks out of which at least one is defective

$$= {}^{40}C_5 - {}^{35}C_5 = 611625.$$

Example 13. *If a finite set A has n elements. Then the power set of A i.e., P(A) has 2^n elements.*

Sol. By defination, P(A) is the set of subsets of A.

There are $^nC_1$ subsets each consisting of one of the n elements of the given set.

There are $^nC_2$ subsets each consisting of any two of the n elements of the given set

Similarly, there are $^nC_3$ subsets each consisting of any three of the n elements of the given set

⋮ ⋮ ⋮

There are ^nC_n subsets each consisting of all the n elements of the given set. Also, there will be one set $\{\phi\}$

$\therefore$ Total number of subsets $= ({}^nC_1 + ... + {}^nC_n) + 1$

$= {}^nC_0 + {}^nC_1 + ... + {}^nC_n$ $\quad | \; {}^nC_0 = 1$

$= (1 + 1)^n = 2^n.$

Remark. Number of proper subsets of A $= 2^n - 1$. There is only one improper subset *i.e.*, the set itself.

Example 14. (*a*) *How many subsets of {1, 2, 3, ... 10} contain at least 7 elements ?* (P.T.U. B Tech Dec. 2005)

(*b*) *If T = {1, 2, 3, 4, 5}. How many subsets of T have less than 4 elements ?*

(*c*) *A set contains (2n + 1) elements. If the number of subsets of this set which contains at most n elements is 8192, find n.* (P.T.U. B. Tech. December 2008)

(*d*) *A bag contains 6 white marbles and 5 red marbles. Find the number of ways in which 4 marbles can be drawn from the bag if*

(*i*) *They can be of any colour*

(*ii*) *2 must be white and 2 red.* (P.T.U. B.Tech May 2011)

Sol. (*a*) Let A = {1, 2, ... 10} *i.e.*, A has 10 elements.

There are ${}^{10}C_7$ subsets each consisting of any 7 of the 10 elements of the given set. There are ${}^{10}C_8$ subsets each consisting of any 8 of the 10 elements of the given set. There are ${}^{10}C_9$ subsets each consisting of any 9 of the 10 elements of the given set. There are ${}^{10}C_{10}$ subsets consisting of all the 10 elements of the given set.

$\therefore$ Total number of subsets containing at least 7 elements

$= {}^{10}C_7 + {}^{10}C_8 + {}^{10}C_9 + {}^{10}C_{10}$

$= {}^{10}C_3 + {}^{10}C_2 + {}^{10}C_1 + {}^{10}C_0$ $\quad | \because {}^nC_r = {}^nC_{n-r}$

$= 120 + 45 + 10 + 1 =$ **176.**

(*b*) The total number of subsets containing at least 4 elements (Proceeding as in part (*a*))

$= {}^5C_4 + {}^5C_5 = {}^5C_1 + {}^5C_0 = 5 + 1 = 6$

Also, total number of subsets $= 2^5 = 32$

$\therefore$ Required number of subsets containing less than 4 elements

= Total number of subsets – (Number of subsets containing at least 4 elements)

$= 32 - 6 = 26.$

(*c*) Let A be a given set which contains $(2n + 1)$ elements.

Number of subsets containing at most n elements = 8192

$\Rightarrow \quad 2n + {}^1C_0 + 2n + {}^1C_1 + 2n + {}^1C_2 + ... + 2n + {}^1C_n = 8192$

$\Rightarrow \quad 2^{2n+1} = 8192$ $\quad | \; {}^nC_0 + {}^nC_1 + {}^nC_2 + ... + {}^nC_n = 2^n$

$= 2^{13}$

$\Rightarrow \quad 2n + 1 = 13 \;\Rightarrow\; 2n = 12 \;\Rightarrow\; n = 6$

(*d*) Total number of balls = 11

(*i*) The number of ways in which four balls of any colour can be drawn.

$= 11 \times 11 \times 11 \times 11 = 11^4$

(*ii*) 2 white balls can be drawn in ${}^6C_2$ ways

2 Red balls can be drawn in ${}^5C_2$ ways.

$\therefore$ Required number of ways $= 6C_2 \times 5C_2 = \frac{6\times5}{1\times2}\times\frac{5\times4}{1\times2} = 150.$

Example 15. *Suppose that there are n people in a room (n ≥ 1) and they all shake hands with one another. Show that* $\frac{n(n-1)}{2}$ *hand shakes will have occured.*

(P.T.U. B. Tech. Dec. 2005)

Sol. When two persons shake hands, it is counted as one hand shake.

$\therefore$ Total number of hand shakes

= Number of ways of selecting 2 persons among n persons

$$= {}^nC_2 = \frac{n(n-1)}{1.2} = \frac{n(n-1)}{2}.$$

Example 16. *A bag contains 6 white marbles and 5 red marbles. Find the number of ways in which four marbles can be drawn from the bag if*

(a) They can be any colour

(b) Two white and 2 red

(c) They are of same colour.

Sol. Total number of marbles = 6 + 5 = 11

(*a*) Number of ways in which four marbles (of any colour) can be choosen

$$= {}^{11}C_4 = \frac{11.10.9.8.}{1.2.3.4} = 330$$

(*b*) Two white marbles can be choosen in ${}^6C_2$ ways and 2 red marbles can be choosen ${}^5C_2$ ways.

$\therefore$ Total number of ways of choosing 2 white marbles and 2 red marbles

$$= {}^6C_2 \times {}^5C_2 = \frac{6.5}{1.2}\times\frac{5.4}{1.2} = 150$$

(*c*) The four marbles drawn are of either white colour or red colour.

The four white marbles can be choosen in ${}^6C_4$

The four red marbles can be choosen in ${}^5C_4$

$\therefore$ Total number of ways of choosing either 4 white marbles or 4 red marbles

$$= {}^6C_4 + {}^5C_4 = {}^6C_2 + {}^5C_1 = 15 + 5 = 20.$$

Example 17. *How many committees of 5 with a given chair person can be selected from 12 persons ?*

Sol. The chair person can be choosen in 12 ways. Once the chair person has been selected, we have to chose 4 persons from remaining 11 persons. This can be done in ${}^{11}C_4$ ways.

$\therefore$ Required number of ways $= 12 \times {}^{11}C_4$

$$= 12 \times \frac{11\times10\times9\times8}{1\times2\times3\times4} = 3960.$$

Example 18. *Seven members of a family have total ₹ 2886 in their pockets. Show that at least one of them must have at least ₹ 416 in his pocket.*

Sol. Let us assume the members be the pigeonholes and the Rupees the pigeons. Now 2886 pigeons are to be assigned to 7 pigeonholes. Using the extended pigeonhole principle, where $n = 2886$ and $m = 7$, we have $[(2886 - 1)/7] + 1 = 416$. Hence, there are 416 Rupees in one member's pocket.

Example 19. *How many people must you have to guarantee that at least 9 of them will have birthdays in the same day of the week ?*

Sol. Let us assume the days of week the pigeonholes and the people the pigeons. Now we have 7 pigeonholes and we have to find pigeons. Using the extended pigeonhole principle, we have

$$[(n-1)/7] + 1 = 9$$
$$[(n-1)/7] = 9 - 1 = 8$$
$$n - 1 = 8 \times 7$$
$$n = 56 + 1 = 57.$$

Thus, there must be 57 people to guarantee that at least 9 of them will have birthdays in the same day of the week.

TEST YOUR KNOWLEDGE 5.3

SHORT ANSWER TYPE QUESTIONS

1. How many committees of three can be formed from eight people ?
2. A farmer buys 3 cows, 2 pigs and 4 hens from a man who has 6 cows, 5 pigs and 8 hens. How many choices does the former have ?
3. In how many ways can a committee consisting of three men and two women be choosen from seven men and five women ?
4. The congressional committees on mathematics and computer science are made up of 5 congress men each and a congressional rule is that the two committees must be disjoint. If there are 385 members of congress, how many ways could the committee be selected ?
5. How many proper subsets of {1, 2, 3, 4, 5} contain the numbers 1 and 5 ? How many of them do not contain the number 2 ?

LONG ANSWER TYPE QUESTIONS

6. In how many ways can you arrange the letters in the following words
   (*i*) COMBINE (*ii*) SUBSET (*iii*) NOON.
7. As a freshman, suppose that you had to take 2 of 4 lab science courses, 1 of 2 literature courses, 2 of 3 math courses and 1 of 7 physical education courses. Disregarding possible time conflicts, how many different schedules do you have to choose from ?
8. How many ways can a student do a 10 question true-false exam if he or she can choose not to answer any number of questions ?
9. Suppose you have a choice of fish, lamb or beef for a main course, a choice of Peas or Carrots for a vegetable and a choice of pie, cake or ice-cream for desert. If you must order one item from each catagory, how many different dinners are possible.
10. A questionaire contains 6 questions each having yes-no answers. For each yes response, there is a follow up question with 4 possible responses. In how many ways can the questionaire be answered ?
11. There are 10 points P_1, P_2, ... P_{10} on a plane but no three on the same line. Find
    (*a*) In how many ways, lines are determined by the points ?
    (*b*) How many triangles are determined by the points ?
12. A woman has 11 close friends
    (*a*) In how many ways can she invite five of them to dinner ?
    (*b*) In how many ways if two of the friends are married and will not attend separately ?
    (*c*) In how many ways if two of them are not no speaking terms and will not attend together ?

13. A woman has 11 close friends of whom six are also women

(*a*) In how many ways can she invite three or more to a party ?

(*b*) In how many ways can she invite three or more of them if she wants the same number of men as women (including herself) ?

14. A student is to answer 10 out of 13 questions in an exam.

(*a*) How many choices he has ?

(*b*) How many if he must answer the first two questions ?

(*c*) How many if he must answer the first or second question but not both ?

(*d*) How many if he must answer exactly three out of the first five questions ?

(*e*) How many if he must answer at least three of the first five questions ?

PIGEONHOLE PRINCIPLE

15. Find the minimum number of students in a class to be sure that three of them are born in the same month.

16. Suppose a laundry bag contains many red, white and blue socks. Find the minimum number of socks that one needs to choose in order to get two pairs (4 socks) of the same colour.

17. Find the minimum number of students need to gurantee that five of them belong to the same class. (Freshman, phamore, Junior, senior)

18. A student must take five classes from three areas of study. Various classes are offered in each discipline, but the student cannot take more than two classes in any given area.

(*a*) Using the Pigeonhole principle, show that the student will take at least one class in each area

(*b*) Using the inclusion-exclusion principle, show that the student will have to take at least one class in each area.

19. Use Pigeonhole Principle to prove that an injection cannot exist between a finite set A and a finite set B if the cardinality of A is greater than the cardinality of B.

Answers

1. ${}^8C_3 = 56$ **2.** ${}^6C_3 \times {}^5C_2 \times {}^8C_4 = 14000$ **3.** ${}^7C_3 \times {}^5C_2 = 350$

4. 4.41×10^{21} **5.** (*a*) $2^3 - 1 = 7$, (*b*) 4. **6.** (*i*) 7 !

(*ii*) 360 (*iii*) 6 **7.** 252 **8.** 3^{10}

9. 18 **10.** 5^6 **11.** (*a*) 45 (*b*) 120

12. (*a*) 462 (*b*) 210 (*c*) 252 **13.** (*a*) 1981

(*b*) 325 **14.** (*a*) 286 (*b*) 165 (*c*) 110

(*d*) 80 (*e*) 276 **15.** 25 **16.** 10

17. 17.

Hints

4. The first committee can be selected in ${}^{385}C_5$ ways. The second committee can be selected in ${}^{380}C_5$ ways.

∴ Required number of ways = ${}^{385}C_5 \times {}^{380}C_5$.

6. (*b*) The word SUBSET contains 6 words with 2 S's.

∴ Required number of words = $\dfrac{6!}{2!}$.

(*c*) The word NOON contains 4 words with 2 O's

∴ Required number of words = $\dfrac{4!}{2!}$.

7. The number of schedules taking lab science course = ${}^4C_2$

The number of schedules taking literature course is ${}^2C_1$, of math course is ${}^3C_2$ and of physical education course is ${}^7C_1$.

∴ Required number of schedules = ${}^4C_2 \times {}^2C_1 \times {}^3C_2 \times {}^7C_1$.

8. A question can be answered in 3 ways. Either by writing true-false or not to answer any question.

∴ Ist question can be answered in 3 ways

2nd question can be answered in 3 ways

⋮

10th question can be answered in 3 ways

∴ Total number of ways = $3 \times 3 \times 3 \ldots \times 3 = 3^{10}$

9. Required number of ways = ${}^3C_1 \times {}^2C_1 \times {}^3C_1$.

10. Each question can be answered in 5 ways.

11. (*a*) Number of straight lines formed joining the 10 points taking 2 at a time = ${}^{10}C_2 = \frac{10.9}{1.2} = 45$

(*b*) Required number of triangles = ${}^{10}C_3 = \frac{10.9.8}{1.2.3} = 120$.

12. (*b*) Two married persons out of 4 persons can be invited in ${}^4C_2 = \frac{4.3}{1.2} = 6$ ways one the two persons have been selected, the remaining 3 persons from 7 persons can be selected in ${}^7C_3 = \frac{7.6.5}{1.2.3} = 35$ ways.

∴ Total number of inviting 5 persons = 6 × 35 = 210.

13. (*a*) Total number of ways for inviting three or more friends out of 11 friends

$= {}^{11}C_3 + {}^{11}C_4 + {}^{11}C_5 + \ldots + {}^{11}C_{11}$ | ${}^nC_0 + {}^nC_1 + {}^nC_2 + \ldots + {}^nC_n = 2^n$

$= 2^{11} - {}^{11}C_0 - {}^{11}C_1 - {}^{11}C_2$

$= 2048 - 1 - 11 - 55 = 1981$

(*b*) To invite the same number of men as women including herself, the following combinations are possible.

(*i*) 2 m and 1w or (*ii*) 3 m and 2w or

(*iii*) 4 m and 3w or (*iv*) 5 m and 4w

(*i*) 2 men and 1 woman can be selecting in ${}^5C_2 \times {}^6C_1$ ways = 60

(*ii*) 3 men and 2 women can be selected in ${}^5C_3 \times {}^6C_2$ ways = 150

(*iii*) 4 men and 3 women can be selected in ${}^5C_4 \times {}^6C_3$ ways = 100

(*iv*) 5 men and 4 women can be selected in ${}^5C_5 \times {}^6C_4$ ways = 15

6 W
5 M

∴ Total number of required ways = 60 + 150 + 100 + 15 = 325.

15. There are 12 months ∴ $n = 12$ (Pigeonholes)

Also three (Pigeons) of them are to born in the same month ∴ $k + 1 = 3$ *i.e.*, $k = 2$

∴ Minimum number of required students = $kn + 1 = 25$.

16. Here $n = 3$ colours (Pigeonholes) and $k + 1 = 4$ *i.e.*, $k = 3$

∴ Required number of sucks = $kn + 1 = 9 + 1 = 10$.

17. $n = 4$ classes (Pigeonholes)

Also, $k + 1 = 5$ *i.e.*, $k = 4$

∴ Required number of students = $kn + 1 = 16 + 1 = 17$.

18. (*a*) The three areas are the Pigeonholes *i.e.*, $n = 3$ and the student must take five classes (Pigeons) *i.e.*, $k + 1 = 5 \Rightarrow k = 4$

Hence the student must take at least two classes in one area.

(*b*) Each of the three areas of study represent three disjoint sets A, B, C. Using inclusion exclusion principle, $n(A \cup B \cup C) = n(A) + n(B) + n(C)$. Since the students can take at must two classes in any area of study, the sum of classes in any two sets, say A and B must be less than or equal to four.

∴ $$n(C) = n(A \cup B \cup C) - [n(A) + n(B)] = 5 - (n(A) + n(B) \geq 1$$

Thus, the student must take at least one class in any area.

MULTIPLE CHOICE QUESTIONS (MCQs)

1. What is n, if n is minimum number of integers to be selected from I = {1, 2, 3, ..., 9} such that the sum of two of the n integers is even ?
(a) $n = 3$ (b) $n = 5$
(c) $n = 4$ (d) $n = 9$.
2. In Q. 1, the value of n such that difference of two of the n integers is 5 is
(a) 6 (b) 5
(c) 7 (d) None.
3. The minimum number of balls needed to guarantee that five of them of same colour (blue, red, green, white) is
(a) 17 (b) 5
(c) 20 (d) 6.
4. In a beauty contest, half the number of experts voted for Mr. A and two third voted for Mr. B. 10 voted for both and 6 did not vote for either. How many experts were there in all ?
(a) 18 (b) 36
(c) 84 (d) None of these.
5. The number of different permutations of the word BANANA is
(a) 720 (b) 60
(c) 120 (d) 360.
6. The number of ways in which a team of eleven players can be selected from 22 players including 2 of them and excluding 4 of them is
(a) ${}^{16}C_{11}$ (b) ${}^{16}C_5$
(c) ${}^{16}C_9$ (d) ${}^{20}C_9$.
7. Ramesh has 6 friends. In how many ways can he invite one or more of them at a dinner ?
(a) 61 (b) 62
(c) 63 (d) 64.
8. If ${}^nC_{12} = {}^nC_8$, then n is equal to
(a) 20 (b) 12
(c) 6 (d) 30.
9. The side AB, BC, CA of a triangle ABC have 3, 4 and 5 interior points respectively on them. The total number of triangles that can be constructed by using these points as vertices is
(a) 220 (b) 204
(c) 120 (d) 195.
10. Three persons enter a railway compartment. If there are 5 seats vacant, in how many ways can they take these seats ?
(a) 60 (b) 20
(c) 15 (d) 125.
11. There are three identical red balls and four identical blue balls in a bag. Three balls are drawn. The number of different colour combinations is
(a) 20 (b) 35
(c) 40 (d) 30.

Answers and Explanations

1. (*a*) The sum of two even integers or of two odd integers is always even. Consider the two subsets {1, 3, 5, 7, 9} and {2, 4, 6, 8} of I as pigeonholes. Using Pigeonhole principle, minimum number of integers = $k + 1 = 2 + 1 = 3$.

2. (*a*) Consider the following subsets (Pigeonholes) of I

{1, 6}, {2, 7}, {3, 8}, {4, 9}, {5} which are five in number. By using Pigeonhole principle, $k + 1 = 5 + 1 = 6$ will guarantee that two integers will belong to one of the subsets and their difference will be I.

3. (*a*) Number of pigeonholes $n = 4$ (blue, red, green, white)

Also, $k + 1 = 5 \Rightarrow k = 4$

$\therefore$ Minimum number of balls required = $kn + 1 = 4(4) + 1 = 17$.

4. (*c*)

5. (*b*) There are 6 letters in the word BANANA, out of which A repeats 3 times, N repeats 2 times. Therefore,

Required number of different permutations = $\frac{6!}{3!2!}$ = **60.**

6. (*c*) Take "2 players included" as one object and "4 players excluded" as another object. Therefore, we require to choose 9 players from $22 - (2 + 4) = 16$ players. Hence, required number of ways = ${}^{16}C_9$.

7. (*c*) Number of ways of inviting one friend $= {}^6C_1$

Number of ways of inviting two friends $= {}^6C_2$

⋮

Number of ways of inviting six friends $= {}^6C_6$

$\therefore$ Total number of ways $= {}^6C_1 + {}^6C_2 + {}^6C_3 + \ldots\ldots + {}^6C_6$

$= ({}^6C_0 + {}^6C_1 + {}^6C_2 + \ldots\ldots + {}^6C_6) - {}^6C_0$

$= 2^6 - 1 = 64 - 1 = \mathbf{63}$.

8. (*a*) We know that if ${}^nC_m = {}^nC_r$

Then either $m = r$ or $m + r = n$

Given ${}^nC_{12} = {}^nC_8$. Therefore, $12 + 8 = n$ or $n = 20$.

9. (*c*)

10. (*a*) seats □□□□□

First person can occupy the seat in 5 ways

Second person can occupy the seat in 4 ways

Third person can occupy the seat in 3 ways

$\therefore$ Total number of ways = $5 \times 4 \times 3 = 60$.

11. (*b*) Total number of balls = $3 + 4 = 7$

Out of these balls, 3 are red and 4 are blue.

Number of different colour combinations = $\frac{7!}{3!4!}$ = **35.**

6 INCLUSION-EXCLUSION PRINCIPLE

As we know the cardinality of the set P is the number of unique elements in set P. It is denoted as | P | and read as cardinality of set P. Some times it is also denoted by $n(P)$.

6.1. INCLUSION–EXCLUSION PRINCIPLE (*P.T.U. B.Tech. Dec. 2013, Dec. 2007, 2009*)

Theorem I. *Let P and Q be any two non-disjoint sets. Then*

$$|P \cup Q| = |P| + |Q| - |P \cap Q|.$$

Proof. Draw Venn diagram for the above as shown in Fig. 6.1.

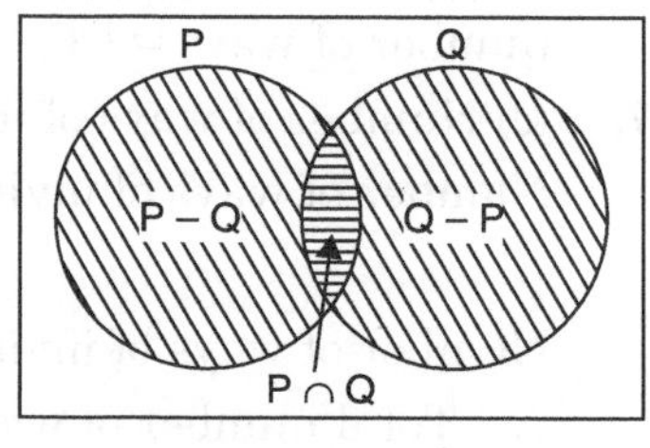

Fig. 6.1

From figure, we see that $P \cup Q$ can be seen to be the union of three disjoint sets $P - Q$, $Q - P$ and $P \cap Q$.

So $|P \cup Q| = |P - Q| + |Q - P| + |P \cap Q|$ ...(1)

Also, $|P| = |P - Q| + |P \cap Q|$ ...(2)

and $|Q| = |Q - P| + |P \cap Q|$ ...(3)

Combining (2) and (3)

$$|P| + |Q| = |P - Q| + |Q - P| + 2\,|P \cap Q| \qquad | \text{ Using (1)}$$

$$= |P \cup Q| + |P \cap Q|$$

(As $|P \cup Q| = |P - Q| + |Q - P| + |P \cap Q|$)

$\Rightarrow$ $|P \cup Q| = |P| + |Q| - |P \cap Q|$

Hence proved.

Another Statement of Inclusion–Exclusion Principle

$\therefore$ Let A and B be any finite sets, then $n(A \cup B) = n(A) + n(B) - n(A \cap B)$

In other words, to find the number of elements in the union $A \cup B$, we add $n(A)$ and $n(B)$ and then we subtract $n(A \cap B)$ *i.e.*, we include $n(A)$ and $n(B)$ and we exclude $n(A \cap B)$. This principle is known as Inclusion–Exclusion principle.

Proof. In counting the elements of $A \cup B$, we count the elements in A as well as in B. These are $n(A)$ and $n(B)$. But the elements in $n(A \cap B)$ are counted twice.

$\therefore$ $n(A \cup B) = n(A) + n(B) - n(A \cap B)$

Cor. If A and B are disjoint sets such that $A \cup B$ is finite, then

$$n(A \cup B) = n(A) + n(B)$$

Proof. In counting the elements of A∪B, we first count those elements which are in A. These are $n(A)$. The only other elements of A∪B are those elements which are in B, but not in A. Since A and B are disjoint, ∴ No element of B is in A.

Hence there are $n(B)$ elements which are in B.

$$\therefore \quad n(A \cup B) = n(A) + n(B).$$

Theorem II. *Let P, Q and R are three finite sets. Then* (P.T.U. B.Tech. Dec. 2010)

$$| P \cup Q \cup R | = | P | + | Q | + | R | - | P \cap Q | - | P \cap R | - | Q \cap R | + | P \cap Q \cap R |.$$

Proof. Using theorem I, we have

$$| P \cup (Q \cup R) | = | P | + | Q \cup R | - | P \cap (Q \cup R) |$$
$$= | P | + | Q | + | R | - | Q \cap R | - | P \cap (Q \cup R) | \quad ...(1)$$

As $\quad P \cap (Q \cup R) = (P \cap Q) \cup (P \cap R)$

So $\quad | P \cap (Q \cup R) | = | P \cap Q | + | P \cap R | - | (P \cap Q) \cap (P \cap R) |$
$$= | P \cap Q | + | P \cap R | - | P \cap Q \cap R | \quad ...(2)$$

Putting (2) in (1), we get

$$| P \cup Q \cup R | = | P | + | Q | + | R | - | P \cap Q | - | P \cap R | - | Q \cap R | + | P \cap Q \cap R |$$

Theorem III. *General Exclusion-Inclusion Principle.*

Let $A_1, A_2, ... A_m$ denote finite sets. Let s_k be the sum of the cardinalities $n(A_{i1} \cap A_{i2} \cap ... \cap A_{ik})$ of all possible k-tuple intersections of the given m sets. Then

$$n(A_1 \cup A_2 \cup A_3 ... \cup A_m) = s_1 - s_2 + s_3 ... + (-1)^{m-1} s_m$$

Remark 1. For $m = 4$, we have

$$s_1 = n(A_1) + n(A_2) + n(A_3) + n(A_4)$$
$$s_2 = n(A_1 \cap A_2) + n(A_2 \cap A_3) + n(A_3 \cap A_4) + n(A_4 \cap A_1) + n(A_1 \cap A_3) + n(A_2 \cap A_4)$$
$$s_3 = n(A_1 \cap A_2 \cap A_3) + n(A_2 \cap A_3 \cap A_4) + n(A_3 \cap A_4 \cap A_1) + n(A_1 \cap A_2 \cap A_4)$$
$$s_4 = n(A_1 \cap A_2 \cap A_3 \cap A_4)$$

Remark 2. A set of the form $A_1^* \cap A_2^* \cap A_3^* .... \cap A_n^*$ where each A_i^* is either A_i or A_i^c, is called a fundamental product of the sets A_i's.

Remark 3. Consider three sets A, B, C. Let

$P_1 = A \cap B \cap C$, $\quad P_2 = A \cap B \cap C^c$,
$P_3 = A \cap B^c \cap C$, $\quad P_4 = A \cap B^c \cap C^c$,
$P_5 = A^c \cap B \cap C$, $\quad P_6 = A^c \cap B \cap C^c$,
$P_7 = A^c \cap B^c \cap C$, $\quad P_8 = A^c \cap B^c \cap C^c$.

These eight fundamental products correspond to the eight disjoint regions as shown in the Venn-diagrams of sets A, B, C. (see Fig. 6.2)

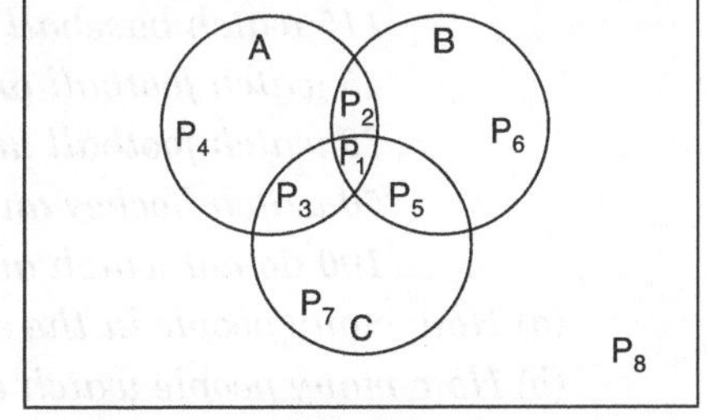

Fig. 6.2

ILLUSTRATIVE EXAMPLES

Example 1. *Out of 1200 students at a college*

582 took Economics
627 took English
543 took Mathematics

217 took both Economics and English
307 took both Economics and Mathematics
250 took both Mathematics and English
222 took all three courses.

How many took none of the three ?

Sol. Let A, B, C denote the set of students studying Economics, English, Mathematics respectively. Given

$$| A | = 582 \qquad | B | = 627$$
$$| C | = 543 \qquad | A \cap B | = 217$$
$$| A \cap C | = 307 \qquad | B \cap C | = 250$$
$$| A \cap B \cap C | = 222$$

The total number of students who took any of three subjects

$$| A \cup B \cup C | = | A | + | B | + | C | - | A \cap B | - | B \cap C | - | C \cap A | + | A \cap B \cap C |$$
$$= 582 + 627 + 543 - 217 - 307 - 250 + 222 = 1200$$

Students who took none of three subjects

= (total students in the college) – (total students who took any of three subjects)
= 1200 – 1200 = 0.

Example 2. *40 computer programmers interviewed for a job. 25 knew JAVA, 28 knew ORACLE, and 7 knew neither of language. How many knew both languages ?*

Sol. Now, $| J | = 25$

$$| O | = 28$$
$$| J \cup O | = 40 - 7 = 33$$

Computer programmers who knew both languages are

$$| J \cap O | = | J | + | O | - | J \cup O | = 25 + 28 - 33 = 20.$$

Example 3. *A survey of 550 television watchers produced the following information :*

285 watch football games
195 watch hockey games
115 watch baseball games
45 watch football and baseball games
70 watch football and hockey games
50 watch hockey and baseball games
100 do not watch any of the three games.

(*a*) *How many people in the survey watch all three games ?*

(*b*) *How many people watch exactly one of the three games ?*

Sol. (*a*) F, H, B denote the sets of watchers watching football, hockey, baseball respectively. Given

$$| F | = 285 \ ; \ | H | = 195 \ ; \ | B | = 115$$
$$| F \cap B | = 45 \ ; \ | F \cap H | = 70 \ ; \ | H \cap B | = 50$$
$$| F \cup H \cup B | = 550 - 100 = 450$$

The number of people watch all three games

$$| F \cap H \cap B | = 450 - 285 - 195 - 115 + 45 + 70 + 50 = 20.$$

(*b*) 20 watch all three games.

45 – 20 = 25 watch football and baseball but not all three.

70 – 20 = 50 watch football and hockey but not all three.

50 – 20 = 30 watch hockey and baseball but not all three.

285 – 25 – 50 – 20 = 190 watch only football.

195 – 50 – 30 – 20 = 95 watch only hockey.

115 – 25 – 30 – 20 = 40 watch only baseball.

Number of people exactly watch one of the three games

= 190 + 95 + 40 = 325.

(*b*) **Alternative.** To find the number of people watching, exactly one of the three games, we use Venn-diagrams (see Fig. 6.3).

Required number of people watching exactly one of the three games

= 190 + 95 + 40 = 325.

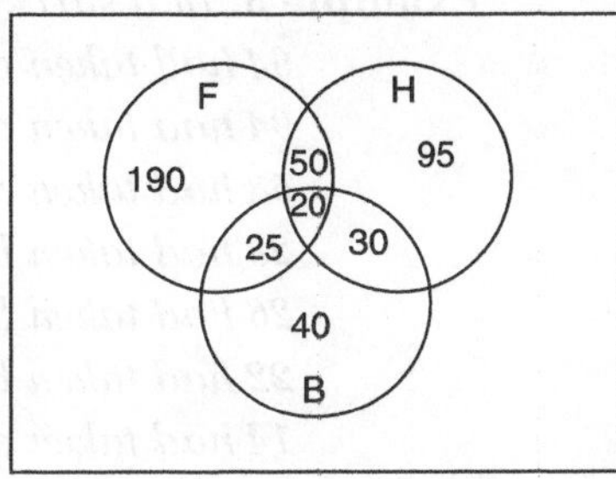

Fig. 6.3

Example 4. *Among 100 students, 32 study Mathematics, 20 study Physics, 45 study Biology, 15 study Mathematics and Biology, 7 study Mathematics and Physics, 10 study Physics and Biology and 30 do not study any of three subjects.*

(*a*) *Find the number of students studying all three subjects.*

(*b*) *Find the number of students studying exactly one of the three subjects.*

Sol. (*a*) Let M, P, B denote the sets of students studying Mathematics, Physics and Biology. Given

$$n(\mathrm{M}) = 32, \quad n(\mathrm{P}) = 20, \quad n(\mathrm{B}) = 45$$

$$n(\mathrm{M} \cap \mathrm{B}) = 15, \quad n(\mathrm{M} \cap \mathrm{P}) = 7, \quad n(\mathrm{P} \cap \mathrm{B}) = 10$$

$$n(\overline{\mathrm{M}} \cap \overline{\mathrm{B}} \cap \overline{\mathrm{P}}) = 30$$

Now, $n(\mathrm{M} \cap \mathrm{B} \cap \mathrm{P}) = 100 - n(\mathrm{M} \cup \mathrm{B} \cup \mathrm{P})$

$$= 100 - n(\overline{\mathrm{M}} \cap \overline{\mathrm{B}} \cap \overline{\mathrm{P}}) = 100 - 30 = 70$$

∴ Required number of students studying all the three subjects is given by $n(\mathrm{M} \cap \mathrm{P} \cap \mathrm{B})$

Using inclusion–exclusion principle, we have

$$n(\mathrm{M} \cap \mathrm{P} \cap \mathrm{B}) = n(\mathrm{M}) + n(\mathrm{P}) + n(\mathrm{B}) - n(\mathrm{M} \cap \mathrm{P}) - n(\mathrm{P} \cap \mathrm{B}) - n(\mathrm{B} \cap \mathrm{M}) + n(\mathrm{M} \cap \mathrm{P} \cap \mathrm{B})$$

$$\Rightarrow \quad 70 = 32 + 20 + 45 - 7 - 10 - 15 + n(\mathrm{M} \cap \mathrm{P} \cap \mathrm{B})$$

$$\Rightarrow \quad 70 = 65 + n(\mathrm{M} \cap \mathrm{P} \cap \mathrm{B})$$

$$\Rightarrow \quad n(\mathrm{M} \cap \mathrm{P} \cap \mathrm{B}) = 70 - 65 = 5$$

(*b*) To find the number of students studying exactly one of the subjects we use Venn-diagram (see Fig. 6.4)

5 study all three subjects

7 – 5 = 2 study Maths and Physics but not all three

15 – 5 = 10 study Maths and Biology but not all three

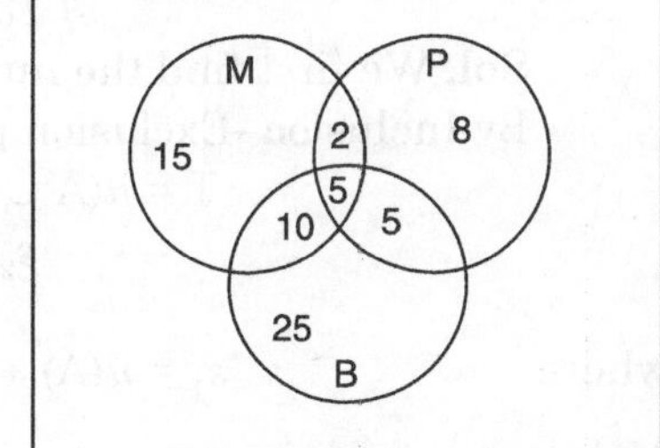

Fig. 6.4

$10 - 5 = 5$ study Biology and Physics but not all three

$32 - (10 + 2 + 5) = 15$ study Maths only

$20 - (2 + 5 + 5) = 8$ study Physics only

$45 - (10 + 5 + 5) = 25$ study only Biology only

∴ Number of students studying exactly one of three subjects = $15 + 8 + 25 = 48$.

Example 5. *In a survey of 300 students,*

64 had taken a Mathematics course
94 had taken a English course
58 had taken a Computer course
28 had taken both a Mathematics and a Computer course
26 had taken both a English and a Mathematics course
22 had taken both a English and a Computer course
14 had taken all three courses.

(*a*) *How many students were surveyed who had taken non of the three courses ?*

(*b*) *How many had taken only a Computer course ?*

Sol. Let M, E, C denote the sets of students taking mathematics, English, Computer Courses respectively. Then

$$| M | = 64 ; | E | = 94 ; | C | = 58$$

$$| M \cap C | = 28 ; | M \cap E | = 26 ; | E \cap C | = 22$$

$$| M \cap E \cap C | = 14$$

(*a*)
$$| M \cup E \cup C | = | M | + | E | + | C | - | M \cap C | - | M \cap E | - | E \cap C | + | M \cap E \cap C |$$
$$= 64 + 94 + 58 - 28 - 26 - 22 + 14 = 154$$

Students who had taken none of the courses

$$= 300 - 154 = 146.$$

(*b*) 14 had taken all three courses.

$28 - 14 = 14$ had taken both a Mathematics and a Computer but not all three

$22 - 14 = 8$ had taken both a English and a Computer courses but not all three

$58 - 14 - 8 - 14 = 22$ had taken only Computer course.

Example 6. *Let A, B, C, D denote respectively, Art, Biology, Chemistry and Drama courses. Find the number of students in a dormetory given the data.*

12 Take A,	*5 Take A and B*	*3 Take A, B, C*
20 Take B,	*7 Take A and C*	*2 Take A, B, D*
20 Take C,	*4 Take A and D*	*2 Take B, C, D*
8 Take D,	*16 Take B and C*	*3 Take A, C, D*
	4 Take B and D	*2 Take A, B, C, D*
	3 Take C and D	*71 Take none.*

Sol. We first find the number of students who take at least one course.

By Inclusion–Exclusion principle.

$$T = n(A \cup B \cup C \cup D)$$
$$= s_1 - s_2 + s_3 - s_4 \qquad \text{...(1)}$$

| General Exclusion-Inclusion Principle

where $s_1 = n(A) + n(B) + n(C) + n(D) = 12 + 20 + 20 + 8 = 60$

$s_2 = n(A \cap B) + n(B \cap C) + n(C \cap D) + n(D \cap A) + n(A \cap C) + n(B \cap D)$
$= 5 + 16 + 3 + 4 + 7 + 4 = 39$
$s_3 = n(A \cap B \cap C) + n(B \cap C \cap D) + n(A \cap C \cap D) + n(A \cap B \cap D)$
$= 3 + 2 + 3 + 2 = 10$
$s_4 = n(A \cap B \cap C \cap D) = 2$

$\therefore$ From (1) $T = 60 - 39 + 10 - 2 = 70 - 41 = 29$

Hence the required number of students

= Number of students who take at least one course + Number of students who take none course

= 29 + 71 = 100.

Example 7. *Suppose that 100 of the 120 Mathematics students at a college take at least one of the languages French, German and Russian. Also suppose*

65 study French,	*20 study French and German*
45 study German,	*25 study French and Russian*
42 study Russian,	*15 study German and Russian*

(a) Find the number of students studying all the subjects

(b) Find the number of students studying taking exactly one subject.

Sol. *(a)* Let F, G and R denote the sets of students studying French, German, Russian respectively. Given

$n(F \cup G \cup R) = 100, \quad n(F) = 65, \quad n(G) = 45,$
$n(R) = 42, \quad n(F \cap G) = 20, \quad n(F \cap R) = 25,$
$n(G \cap R) = 15$, we find $n(F \cap G \cap R)$.

Using Inclusion–Exclusion principle,

$$n(F \cup G \cup R) = n(F) + n(G) + n(R) - n(F \cap G) - n(G \cap R) - n(R \cap F) + n(F \cap G \cap R)$$

$\Rightarrow \quad 100 = 65 + 45 + 42 - 20 - 15 - 25 + n(F \cap G \cap R)$

$\Rightarrow \quad 100 = 92 + n(F \cap G \cap R)$

$\Rightarrow n(F \cap G \cap R) = 100 - 92 = 8.$

(b) To find the number of students taking exactly one subject, we use Venn-diagrams. (see Fig. 6.5)

Fig. 6.5

8 read all three subject

25 – 8 = 17 read French and Russian but not German

20 – 8 = 12 read French and German but not Russian

15 – 8 = 7 read Russian and German but not French

65 – (17 + 8 + 12) = 28 read French only

45 – (12 + 8 + 7) = 18 read German only

42 – (17 + 8 + 7) = 10 read Russian only

$\therefore$ Required number of students who study exactly one subject

= 28 + 18 + 10 = 56.

Example 8. *In a survey of 60 people, it was found that*

25 read Newsweek magazine

26 read Time, 26 read Fortune

9 read both Newsweek and Fortune

11 read both News week and Time

8 read both Time and Fortune

3 read all three magazines

(*a*) *Find the number of students who read at least one of the three magazines*

(*b*) *Find the number of students who read exactly one magazine*

(*c*) *Find the number of students who read no magazine at all.*

Sol. Let N, F, T denote the sets of students reading *Newsweek, Fortune, Time* magazines respectively.

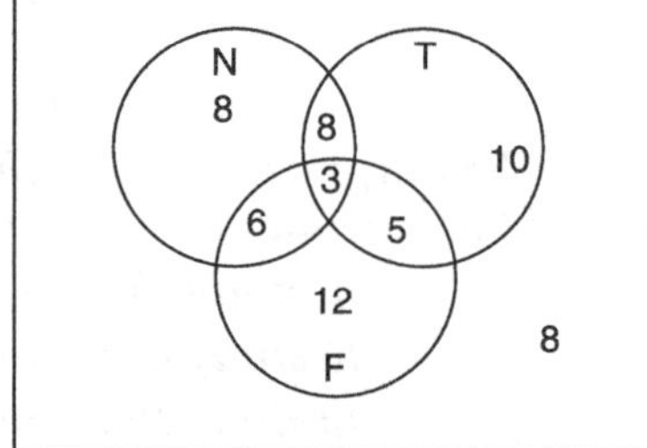

Fig. 6.6

(*a*) Required number of students

$= n(N \cup T \cup F) = n(N) + n(T) + n(F) - n(N \cap T)$
$- n(T \cap F) - n(F \cap N) + n(N \cap T \cap F)$

$= 25 + 26 + 26 - 11 - 8 - 9 + 3 = 52.$

(*b*) To find the number of students who read exactly one magazine, we use Venn-diagram. (see Fig. 6.6).

Here 3 read all three magazines

$11 - 3 = 8$ read *Newsweek* and *Time* but not all three magazines

$9 - 3 = 6$ read *Newsweek* and *Fortune* but not all three magazines

$8 - 3 = 5$ read *Time* and *Fortune* but not all three magazines

$25 - (8 + 3 + 6) = 8$ read only *Newsweek*

$26 - (8 + 3 + 5) = 10$ read only *Time*

$26 - (6 + 3 + 5) = 12$ read only *Fortune*

$\therefore$ Required number of students who read exactly one magazine

$= 8 + 10 + 12 = 30.$

(*c*) Required number of students who study no magazine

$= 60 - n(N \cup T \cup F) = 60 - 52 = 8.$

Example 9. *Among the first 500 positive integers :*

(*a*) *Determine the integers which are not divisible by 2, nor by 3, nor by 5.*

(*b*) *Determine the integers which are exactly divisible by one of them.*

Sol. Let A denotes the set of numbers of integers divisible by 2

B denotes the set of numbers of integers divisible by 3

C denotes the set of numbers of integers divisible by 5.

$$|A| = \left[\frac{500}{2}\right] = 250 \; ; \qquad |B| = \left[\frac{500}{3}\right] = 166 \; ; \qquad |C| = \left[\frac{500}{5}\right] = 100$$

$$|A \cap B| = \left[\frac{500}{2 \times 3}\right] = 83 \; ; \qquad |A \cap C| = \left[\frac{500}{2 \times 5}\right] = 50$$

$$|B \cap C| = \left[\frac{500}{3 \times 5}\right] = 33 \; ; \qquad |A \cap B \cap C| = \left[\frac{500}{3 \times 3 \times 5}\right] = 16.$$

(*a*) $|A \cup B \cup C| = 250 + 166 + 50 - 83 - 100 - 33 + 16 = 366$

The integers not divisible by 2, 3 and 5 = 500 – 366 = 134.

(*b*) The integers divisible by all the three = 16

83 – 16 = 67 integers are divisible by 2 and 3 but not all the three

50 – 16 = 34 integers are divisible by 2 and 5 but not by all the three

33 – 16 = 17 integers are divisible by 3 and 5 but not by all the three

250 – 67 – 34 – 16 = 133 integers are only divisible by 2

166 – 67 – 17 – 16 = 66 integers are only divisible by 3

100 – 34 – 17 – 16 = 33 integers are only divisible by 5

Total number of integers only divisible by 2, 3 and 5

= 133 + 33 + 66 = 232.

Example 10. *Among the first 1000 positive integers :*

(a) Determine the integers which are not divisible by 5, nor by 7, nor by 9.

(b) Determine the integers divisible by 5, but not by 7, not by 9.

(P.T.U. B.Tech. May 2012)

Sol. Let A denotes the set of numbers of integers divisible by 5

B denotes the set of number of integers divisible by 7

C denotes the set of number of integers divisible by 9.

So $|A| = \left[\frac{1000}{5}\right] = 200$; $|B| = \left[\frac{1000}{7}\right] = 142$

$|C| = \left[\frac{1000}{9}\right] = 111$; $|A \cap B| = \left[\frac{1000}{5 \times 7}\right] = 28$

$|A \cap C| = \left[\frac{1000}{5 \times 9}\right] = 22$; $|B \cap C| = \left[\frac{1000}{7 \times 9}\right] = 15$

$$|A \cap B \cap C| = \left[\frac{1000}{5 \times 7 \times 9}\right] = 3.$$

(*a*) The number of integers divisible by 5, 7 and 9

$|A \cup B \cup C| = 200 + 142 + 111 - 28 - 22 - 15 + 3$

$= 391.$

The number of integers not divisible by 5, nor by 7, nor by 9

= Total number of integers – number of integers divisible by 5, 7 and 9

= 1000 – 391 = 609.

(*b*) The integers divisible by all the three integers = 3

28 – 3 = 25 integers divisible by 5 and 7 but not by all the three

22 – 3 = 19 integers divisible by 5 and 9 but not by all the three

∴ 200 – 25 – 19 – 3 = 153 integers divisible by 5 but not by 7, not by 9.

Example 11. *How many integers between 1 and 300 (inclusive) are :*

(P.T.U. B.Tech. May 2010)

(a) divisible by at least one of 3, 5, 7 ?

(b) divisible by 3 and 5, not by 7 ?

(c) divisbile by 5 but neither by 3 nor by 7 ?

Sol. Total number of integers = 300

Let A denotes the set of numbers divisible by 3 B, the set of numbers divisible by 5, C, the set of numbers divisible by 7, then,

$$n(A) = \left[\frac{300}{3}\right] = 100, \; n(B) = \left[\frac{300}{5}\right] = 60$$

$$n(C) = \left[\frac{300}{7}\right] = 42$$

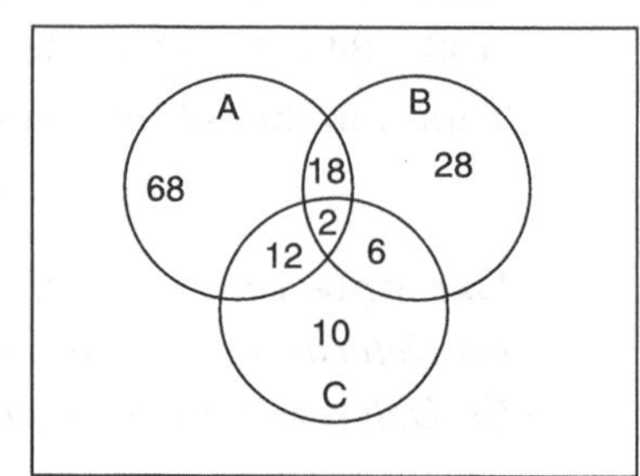

$$n(A \cap B) = \left[\frac{300}{3 \times 5}\right] = 20, \; n(B \cap C) = \left[\frac{300}{5 \times 7}\right] = 8$$

$$n(C \cap A) = \left[\frac{300}{3 \times 7}\right] = 14$$

$$n(A \cap B \cap C) = \left[\frac{300}{3 \times 5 \times 7}\right] = 2$$

Also, $n(A \cup B \cup C) = n(A) + n(B) + n(C) - n(A \cap B) - n(B \cap C) - n(C \cap A) + n(A \cap B \cap C)$

$$= 100 + 60 + 42 - 20 - 8 - 14 + 2 = 162$$

From the Venn diagram,

(*a*) number of integers divisible by at least 3, 5 or 7 = $n(A \cup B \cup C) = 162$

(*b*) the number of integers divisible by 3 and 5, not by 7 = $n(A \cap B \cap \overline{C}) = 18$

(*c*) the number of integers divisible by 5 but not by 3, 7 = $n(\overline{A} \cap B \cap \overline{C}) = 28$.

TEST YOUR KNOWLEDGE

1. (*a*) Of 32 people who save paper or bottles (or both) for recycling, 30 save paper and 14 save bottles. Find the number of people who save

(*i*) Both paper and bottles

(*ii*) Only paper

(*iii*) Only bottles

(*b*) A survey of 550 Television watchers produced the following information.

285 watch football games

195 watch hockey games

115 watch cricket

45 watch football and cricket

70 watch football and hockey

50 watch hockey and cricket

100 do not watch any of the three games.

(*i*) How many people in the survey watch all the three games ?

(*ii*) How many people watch exactly one of the three games ?

2. In a class of 300 students,
   60 study Mathematics course
   94 study English course
   58 study Discrete course
   28 study both Mathematics and Discrete course
   26 study both English and Mathematics course
   22 study both English and Discrete course
   14 study all three course.
   (*a*) Find the number of students who study none of the three courses
   (*b*) Who study only Discrete course ?
3. A survey on a sample of 25 new cars being sold at a local auto dealer was conducted to see which of three popular options, air-conditioning (A), radio (R) and Power windows (W), were already installed. The survey found
   15 had air-conditioning
   12 had radio
   11 had Power windows
   5 had air-conditioning and Power windows
   9 had air-conditioning and radio
   4 had Power windows and radio
   3 had all three options.
   Find the number of cars that had
   (*a*) Only Power windows
   (*b*) Only air-conditioning
   (*c*) Only radio
   (*d*) Radio and Power windows but not air-conditioning
   (*e*) Air-conditioning and radio but not Power windows
   (*f*) Only one of the options
   (*g*) At least one option
   (*h*) None of the options
4. Using Inclusion–Exclusion principle, prove the following.
   (*i*) $n(A \cup B) = n(A) + n(B) - n(A \cap B)$, for any finite sets A and B
   (*ii*) $n(A \cup B) = n(A) + n(B)$, where A, B are finite disjoint sets.
5. A survey was conducted among 1000 people of these 595 like Metro channel, 595 like star Movies and 550 like Zee TV. Also,
   395 of them like Metro channel and Star Movies
   350 of them like Metro channel and Zee TV
   400 of them like Star Movies and Zee TV
   250 of them like Metro channel, Star Movies and Zee TV.
   Find
   (*a*) How many of them who do not like Metro channel, do not like Star Movies and do not like Zee TV ?
   (*b*) How many of them who like Metro channel, do not like Star Movies and do not like Zee TV ?
6. In a class of 60 boys, 45 boys play cards and 30 boys play carrom. How many boys play both games ? How many play cards only and how many plays carroms only ?

(P.T.U. B.Tech. May 2008, 2010)

7. Given that

U = all students in a university

A = Day students

B = mathematics majors

C = graduate students

Also $n(U) = 16,000$, $n(A) = 9000$, $n(B) = 300$ and $n(C) = 1000$. Assume that the number of day students who are mathematics majors is 250, 50 of which are graduate students, and the total number of day graduate students is 700. Determine the number of students who are

(*a*) evening students

(*b*) non-mathematics majors

(*c*) under graduate (day or evening)

(*d*) day graduate non-mathematics majors

(*e*) evening graduate students

(*f*) evening graduate mathematics majors

(*g*) evening under graduate non-mathematics majors.

8. In a class of 80 students, 50 students know English, 55 know French and 46 know German. 37 students know English and French, 28 students know French and German, 7 students know none of the languages. Find

(*a*) How many students know all the three languages ?

(*b*) How many students know exactly two languages ?

(*c*) How many know only one language ?

9. Among integers 1 to 1000

(*a*) How many of them are not divisible by 3 nor by 5 nor by 7 ?

(*b*) How many are not divisible by 5 or 7 but divisible by 3 ?

10. It is known that in a university, 60% of teachers play tennis, 50% play bridge, 70% jog, 20% play tennis and bridge, 40% play bridge and jog and 30% play tennis and jog. It is claimed that 20% teachers jog and play tennis and bridge. Justify this claim.

Answers

1. (*a*) (*i*) 12 (*ii*) 18 (*iii*) 2 (*b*) (*i*) 450 (*ii*) 325

2. (*a*) 146 (*b*) 22 **3.** (*a*) 5 (*b*) 4 (*c*) 2
(*d*) 4 (*e*) 6 (*f*) 11 (*g*) 23 (*h*) 2.

5. (*a*) 155 (*b*) 100 **6.** 15, 30, 15.

7. (*a*) 7000 (*b*) 15700 (*c*) 15000 (*d*) 650 (*e*) 300
(*f*) and (*g*) cannot be determined with the given information.

8. (*a*) 12 (*b*) 54 (*c*) 7

9. (*a*) 457 (*b*) 229 **10.** The claim is not true.

Hints

1. (*a*) (*ii*) $n(P - B) = n(P) - n(P \cap B) = 30 - 12 = 18$

(*iii*) $n(B - P) = n(B) - n(P \cap B) = 14 - 12 = 2$

(*b*) 20 watch all the three games.

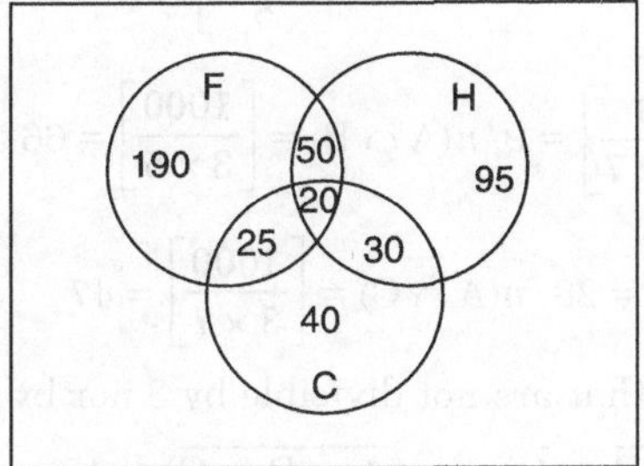

Fig. 6.7

Number of people watch exactly one of the three games = 190 + 95 + 40 = 325.

3.

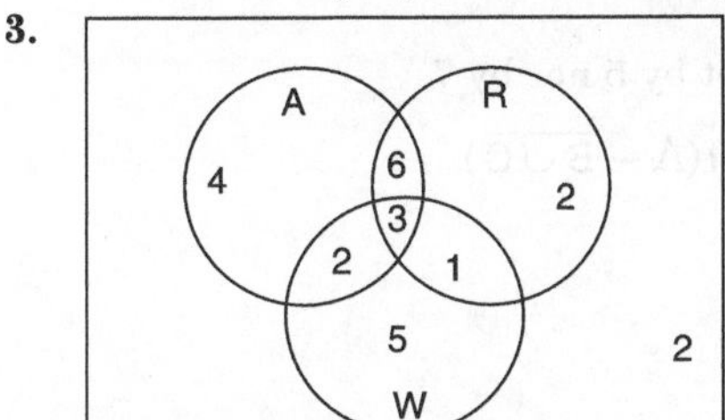

Fig. 6.8

5. (*a*) Number of People who like at least one of the channels

$$n(M \cup S \cup Z) = n(M) + n(S) + n(Z) - n(M \cap S) - n(S \cap Z) - n(Z \cap M) + n(M \cap S \cap Z)$$

$$= 595 + 595 + 550 - 395 - 350 - 400 + 250 = 845$$

$\therefore$ Required number of People who do not like Metro channel, do not like Star Movies, and do not like Zee TV = 1000 – 845 = 155.

8. Given $n(A) = 50$, $n(B) = 55$, $n(C) = 46$, $n(A \cap B) = 37$, $n(B \cap C) = 28$, $n(A \cap C) = 25$.

$n(\overline{A \cup B \cup C})$ = number of students knowing none of the language = 7

$\Rightarrow$ $n(\cup) - n(A \cup B \cup C) = 7$

$\Rightarrow$ $80 - n(A \cup B \cup C) = 7 \Rightarrow n(A \cup B \cup C) = 73$

(*a*) Number of students knowing all the three languages is given by $n(A \cap B \cap C)$

Now, $n(A \cup B \cup C) = n(A) + n(B) + n(C) - n(A \cap B) - n(B \cap C) - n(C \cap A) + n(A \cap B \cap C)$

$\Rightarrow$ $73 = 50 + 55 + 46 - 37 - 28 - 25 + n(A \cap B \cap C)$

$\Rightarrow$ $n(A \cap B \cap C) = 12$. Hence, there are 12 students who know all the three languages.

(*b*) Using Venn diagram, students knowing exactly two languages

$= n(A \cap B) + n(B \cap C) + n(A \cap C) - 3 \times n(A \cap B \cap C)$

$= 37 + 28 + 25 - 3 \times 12 = 54.$

(*c*) Using Venn diagram, number of students knowing only one language = 0 + 2 + 5 = **7.**

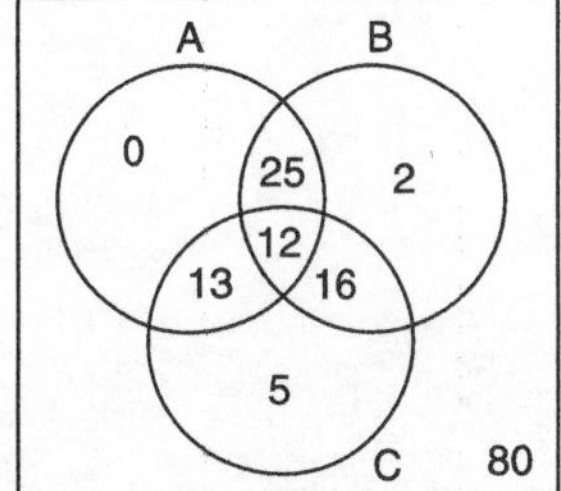

Fig. 6.9

9. Let A = The set of numbers between 1 to 1000 that are divisible by 3

B = The set of numbers between 1 to 1000 that are divisible by 5

C = The set of numbers between 1 to 1000 that are divisible by 7.

Then, $n(A) = \left[\frac{1000}{3}\right] = 333$, $n(B) = \left[\frac{1000}{5}\right] = 200$, $n(C) = \left[\frac{1000}{7}\right] = 142$

$$n(A \cap B \cap C) = \left[\frac{1000}{3 \times 5 \times 7}\right] = 9,\ n(A \cap B) = \left[\frac{1000}{3 \times 5}\right] = 66$$

$$n(B \cap C) = \left[\frac{1000}{5 \times 7}\right] = 28,\ n(A \cap C) = \left[\frac{1000}{3 \times 7}\right] = 47$$

(*a*) The number of integers that are not divisible by 3 nor by 5 nor by 7

$$= n(\overline{A} \cap \overline{B} \cap \overline{C}) = n(\overline{A \cup B \cup C}) = U - n(A \cup B \cup C) \quad \ldots(1)$$

But, $n(A \cup B \cup C) = 333 + 200 + 142 - 66 - 47 - 28 + 9 = 543$

$\therefore$ From (1), $n(\overline{A} \cap \overline{B} \cap \overline{C}) = 1000 - 543 = 457$

(*b*) Number of integers that are not divisible by 3 but not by 5 nor by 7

$$= n(A \cap \overline{B} \cap \overline{C}) = n(A \cap \overline{B \cup C}) = n(A - \overline{B \cup C})$$

$$= 333 - (57 + 9 + 38) = 229.$$

10. Let the total number of teachers = 100

$$n(T) = 60,\ n(B) = 50,\ n(J) = 70$$

$$n(T \cap B) = 20,\ n(B \cap J) = 40,\ n(T \cap J) = 30$$

Number of teachers who jog and play tennis and bridge is given by

$$n(T \cup B \cup J) = 60 + 50 + 70 - 20 - 40 - 30 + n(T \cap B \cap J)$$

$\Rightarrow$ $n(T \cap B \cap J) = 100 - (90) = 10$

i.e., 10% of Teachers jog and play tennis and bridge.

7 RECURRENCE RELATIONS AND GENERATING FUNCTIONS

7.1. INTRODUCTION

In this chapter, we will discuss the formation of recurrence relations and their solutions. The closed form expressions of recurrence relations and its solutions using generating functions is also discussed in this chapter.

7.2. RECURRENCE RELATIONS (*P.T.U. B.Tech. Dec. 2006*)

Let S be a sequence of numbers. A recurrence relation on S is a formula that relates all, but a finite number of terms of S, to previous terms of S.

For *e.g.*, the Fibonacci sequence is defined by the relation.

$$F_K = F_{K-2} + F_{K-1},\ K \geq 2, \text{ where } F_0 = 1, F_1 = 1.$$

The relation defined above is a recurrence relation and the conditions $F_0 = 1$, $F_1 = 1$ are called initial conditions.

Note. The recurrence relations are also called difference equations.

7.3. ORDER OF A RECURRENCE RELATION

The order of a recurrence relation is the difference between the highest and the lowest subscripts of S(K). For *e.g.*, consider the recurrence relation.

$$F_K = F_{K-2} + F_{K-1},\ K \geq 2$$

Here the difference between highest and lowest subscripts of F = K – (K – 2) = 2

$\therefore$ Its order is 2.

Note. The order of a recurrence relation may or may not be defined

7.4. (a) DEGREE OF RECURRENCE RELATION

It is the highest power of S(K) occurring in the recurrence relation. For example, consider the recurrence relation.

$$S^2(K + 3) + 2S^2(K + 2) + 2S(K + 1) = 0$$

Its order = (K + 3) – (K + 1) = 2

degree = 2(Highest power)

For example, consider $S^4(K) + 3S^3(K-1) + 6S^2(K-2) + 4S(K-3) = 0$

Its order $= K - (K-3) = 3$

degree $= 4$.

7.4. (b) LINEAR RECURRENCE RELATION (*P.T.U. B.Tech. Dec. 2006*)

A recurrence relation with degree one is called **linear recurrence relation.**

7.5. FORMATION OF RECURRENCE RELATIONS

We illustrate this concept in the following examples.

ILLUSTRATIVE EXAMPLES

Example 1. *Find the order of the following recurrence relations defined by*

(*i*) $T(K) = 2(T(K-1))^2 - KT(K-3)$

(*ii*) $P(K) + 2P(K-3) - K^2 = 0,$

(*iii*) $S(K) = S[K/2] + 5,\ K \geq 0$

(*iv*) $A(K) - 2A(K-1) - 2K = 0.$

Sol. (*i*) The difference between the highest and lowest subscripts $= K - (K-3) = 3$

$\therefore$ Its order $= 3$.

(*ii*) The difference between the highest and lowest subscripts $= K - (K-3) = 3$

$\therefore$ Its order $= 3$.

(*iii*) The recurrence relation is of infinite order. Since $K - \left[\frac{K}{2}\right]$ becomes larger and larger as K is made large.

(*iv*) The difference between the highest and the lowest subscripts $= K - (K-1) = 1$

$\therefore$ Its order $= 1$.

Example 2. *Obtain the linear recurrence relation from the sequence defined by* $S(K) = 5 \,.\, 2^K$.

Sol. Given $S(K) = 5 \,.\, 2^K$ ...(1)

Changing K to K – 1, we get $S(K-1) = 5 \,.\, 2^{K-1}$ ...(2)

Subtracting (1) and (2), we get

$$S(K) - S(K-1) = 5 \,.\, 2^K - 5 \,.\, 2^{K-1}$$

$$= 5 \,.\, 2^K\left(1 - \frac{1}{2}\right) = \frac{1}{2} \,.\, 5 \,.\, 2^K = \frac{S(K)}{2}$$

$$\Rightarrow \quad \left(1 - \frac{1}{2}\right) S(K) - S(K-1) = 0$$

$\Rightarrow$ $S(K) - 2S(K-1) = 0$, is the required linear recurrence relation.

Example 3. *Obtain the recurrence relation of second order from the sequence defined by*

$$S(K) = 3^{K-1} + 2^{K+1} + K.$$

Sol. Given $S(K) = 3^{K-1} + 2^{K+1} + K$ ...(1)

Changing K to K – 1 in (1), we get

$$S(K-1) = 3^{K-2} + 2^{K} + K - 1$$

$$= \frac{1}{3} \cdot 3^{K-1} + \frac{1}{2} \cdot 2^{K+1} + K - 1 \quad \text{...(2)}$$

Changing K to K – 2 in (1), we get

$$S(K-2) = 3^{K-3} + 2^{K-1} + K - 2$$

$$= \frac{1}{9} \cdot 3^{K-1} + \frac{1}{4} \cdot 2^{K+1} + K - 2 \quad \text{...(3)}$$

Eliminating 3^{K-1} from (1) and (2), we get

$$S(K) - 3S(K-1) = \left(1 - \frac{3}{2}\right) 2^{K+1} + K - 3(K-1)$$

$$= -\frac{1}{2} \cdot 2^{K+1} + 3 - 2K \quad \text{...(4)}$$

Eliminating 3^{K-1} from (2) and (3), we get

$$S(K-1) - 3S(K-2) = \left(\frac{1}{2} - \frac{3}{4}\right) 2^{K+1} + K - 1 - 3(K-2)$$

$$= -\frac{1}{4} \cdot 2^{K+1} + 5 - 2K \quad \text{...(5)}$$

Eliminating 2^{K+1} from (4) and (5), we get

$$S(K) - 3S(K-1) - 2\,[S(K-1) - 3\,S(K-2)]$$

$$= 3 - 2K - 2\,(5 - 2K) = -7 + 2K$$

or $S(K) - 5S(K-1) + 6S(K-2) = 2K - 7,$

is the required recurrence relation.

Its order = K – (K – 2) = 2.

Example 4. *Obtain the recurrence relation of second order from the sequence given by*

$$S(K) = 2 \cdot 4^{K} - 5 \cdot (-3)^{K}$$

Sol. Given $S(K) = 2 \cdot 4^{K} - 5 \cdot (-3)^{K}$ ...(1)

Changing K to K – 1 in (1), we get

$$S(K-1) = 2.4^{K-1} - 5 \cdot (-3)^{K-1} = \frac{2}{4} \cdot 4^{K} - 5. (-3)^{K} \cdot (-3)^{-1}$$

$$= \frac{1}{2} \cdot 4^{K} + \frac{5}{3} \cdot (-3)^{K} \quad \text{...(2)}$$

Changing K to K – 2 in (1), we get

$$S(K-2) = 2 \cdot 4^{K-2} - 5 \cdot (-3)^{K-2} = 2 \cdot 4^{K} \cdot 4^{-2} - 5 \cdot (-3)^{K} \cdot (-3)^{-2}$$

$$= \frac{1}{8} \cdot 4^{K} - \frac{5}{9} \cdot (-3)^{K} \quad \text{...(3)}$$

Eliminating 4^K from (1) and (2) , we get

$$S(K) - 4S(K-1) = -5 \,.\, (-3)^K - \frac{20}{3} \,.\, (-3)^K = -\frac{35}{3} \,.\, (-3)^K \qquad ...(4)$$

Eliminating 4^K from (2) and (3), we get

$$S(K-1) - 4S(K-2) = \frac{5}{3}(-3)^K + \frac{20}{9}(-3)^K = \frac{35}{9}(-3)^K \qquad ...(5)$$

Eliminating $(-3)^K$ from (4) and (5), we get

$$S(K) - 4S(K-1) + 3[S(K-1) - 4\,S(K-2)] = 0$$

or $S(K) - S(K-1) - 12S(K-2) = 0$, is the required recurrence relation of order = 2.

7.6. LINEAR RECURRENCE RELATION OF ORDER *n* WITH CONSTANT COEFFICIENTS

The general linear recurrence relation of order n with constant coefficients, is given by

$$S(K) + C_1S(K-1) + C_2S(K-2) + ... + C_nS(K-n) = f(K),\ K \geq n,$$

where $C_1, C_2, ... C_n$ are constants.

7.7. HOMOGENEOUS LINEAR RECURRENCE RELATION OF ORDER *n*

A homogeneous linear recurrence relation of order n is an equation of the form

$$S(K) + C_1S(K-1) + C_2S(K-2) + ... + C_nS(K-n) = 0.$$

7.8. CHARACTERISTIC EQUATION

Consider a linear recurrence relation of order n given by

$$S(K) + C_1S(K-1) + C_2S(K-2) + ... + C_nS(K-n) = f(K),\ K \geq n.$$

Then the equation

$a^n + C_1a^{n-1} + C_2a^{n-2} + ... + C_{n-1}\,a + C_n = 0$ is called characteristic equation.

The left-hand side of this equation is known as characteristic polynomial.

7.9. ALGORITHM FOR SOLVING HOMOGENEOUS LINEAR RECURRENCE RELATION OF ORDER *n* WITH CONSTANT COEFFICIENTS *(P.T.U. B.Tech. Dec. 2006)*

Consider the linear homogeneous recurrence relation of order n as

$$S(K) + C_1S(K-1) + C_2S(K-2) + ... + C_nS(K-n) = 0 \qquad ...(1)$$

where $C_1, C_2, ... C_n$ are constants.

Step I. Write the characteristic equation of the equation (1), which is

$$a^n + C_1a^{n-1} + C_2a^{n-2} + ... + C_{n-1}a + C_n = 0$$

Step II. Find the roots of the characteristic equation obtained in step I.

Case I. If the characteristic equation has n distinct roots, say, $m_1, m_2, ... m_n$, then the solution of (1) is given by

$$S(K) = C_1{m_1}^K + C_2{m_2}^K + ... + C_n{m_n}^K$$

Case II. If the characteristic equation has two equal roots, say $m_1 = m_2$, then the solution of (1) is given by

$$S(K) = (C_1 + C_2K)\,{m_1}^K + C_3\,{m_3}^K + ... + C_n\,{m_n}^K$$

Case III. If the characteristic equation has three equal roots, say, $m_1 = m_2 = m_3$, then the solution of (1) is given by

$$S(K) = (C_1 + C_2K + C_3K^2)\, m_1^{\,K} + C_4 m_4^{\,K} + \ldots\ldots + C_n m_n^{\,K} \text{ and so on}$$

Case IV. If the characteristic equation has imaginary roots, say, $\alpha \pm i\beta$, then solution of (1) is given by

$$S(K) = C_1\, (\alpha + i\beta)^K + C_2\, (\alpha - i\beta)^K$$

Case V. If the characteristic equation has repeated imaginary roots, say, $\alpha \pm i\beta$, $\alpha \pm i\beta$, then the solution of (1) is given by

$$S(K) = (C_1 + C_2K)\, (\alpha + i\beta)^K + (C_3 + C_4K)\, (\alpha - i\beta)^K$$

Example 5. *Solve the recurrence relation* $a_r - 3a_{r-1} + 2a_{r-2} = 0$.

Sol. The characteristic equation is given by

$$s^2 - 3s + 2 = 0 \quad \text{or} \quad (s - 1)(s - 2) = 0$$

$$\Rightarrow \qquad s = 1, 2$$

Therefore, the required solution of the given homogeneous recurrence relation is

$$a_r = C_1 + C_2 \,.\, 2^r.$$

Example 6. (*a*) *Solve the recurrence relation:* $a_r - 7a_{r-1} + 10a_{r-2} = 0$, *given that* $a_0 = 0$, $a_1 = 3$. (P.T.U. B.Tech. Dec. 2007)

(*b*) *Solve the recurrence relation:* $a_n = 3a_{n-1} + 4a_{n-2}$, $a_0 = 0$, $a_1 = 5$. (P.T.U. B.Tech Dec. 2013)

Sol. (*a*) The characteristic equation is

$$s^2 - 7s + 10 = 0 \quad \text{or} \quad (s - 2)(s - 5) = 0$$

$$\Rightarrow \qquad s = 2, 5$$

Therefore, the required solution of the given homogeneous recurrence relation is

$$a_r = c_1 \,.\, 2^r + c_2 \,.\, 5^r. \qquad \ldots(1)$$

Using $a_0 = 0$, (1) gives

$$0 = c_1 + c_2 \qquad \ldots(2)$$

Using $a_1 = 3$, (1) gives

$$3 = 2c_1 + 5c_2 \qquad \ldots(3)$$

Using (2) in (3), we get

$$3 = 2c_1 - 5c_1 = -\,3c_1$$

$$\Rightarrow \qquad c_1 = -\,1$$

From (2) $\qquad c_2 = -\,c_1 = 1$

$\therefore$ From (1), the required solution is

$$a_r = -\,2^r + 5^r.$$

(*b*) Proceed yourself as in part (*a*).

Ans. $\qquad a_n = 4^n - (-1)^n$

Example 7. *Solve the following recurrence relations*

(*a*) $t_n = 6t_{n-1} - 11\, t_{n-2} + 6t_{n-3}$, $n \geq 3$, *subject to* $t_0 = 1$, $t_1 = 5$, $t_2 = 15$.

(*b*) $t_n = -\,3\, t_{n-1} - 3\, t_{n-2} - t_{n-3}$, $n \geq 3$, *subject to* $t_0 = 1$, $t_1 = -\,2$, $t_2 = -\,1$.

(*c*) $a_n = 6a_{n-1} - 12a_{n-2} + 8a_{n-3}$, $a_0 = 3$, $a_1 = 4$, $a_2 = 12$ (*P.T.U. B.Tech Dec. 2013*)

Sol. (*a*) Given equation is $t_n - 6t_{n-1} + 11\, t_{n-2} - 6t_{n-3} = 0$

Its order $= n - (n - 3) = 3$

The charateristic equation is

$a^3 - 6a^2 + 11a - 6 = 0$ ...(1)

Here, $a = 1$ satisfies (1). So $a = 1$ is a root of (1). By Horner's method.

1	1	– 6	11	– 6
		1	– 5	6
	1	– 5	6	0 = Remainder

The quotient is $a^2 - 5a + 6 = 0 \Rightarrow a^2 - 3a - 2a + 6 = 0$

$\Rightarrow (a - 3)(a - 2) = 0 \Rightarrow a = 2, 3$

The required solution is

$t_n = C_1 + C_2 2^n + C_3 3^n$ ...(2)

Using $t_0 = 1$ in (2), $1 = C_1 + C_2 + C_3$ ...(3)

Using $t_1 = 5$ in (2), $5 = C_1 + 2C_2 + 3C_3$ ...(4)

Using $t_2 = 15$ in (2), $15 = C_1 + 4C_2 + 9C_3$ ...(5)

(3) – (4) gives $-4 = -C_2 - 2C_3$ ...(6)

(4) – (5) gives $-10 = -2C_2 - 6C_3$ ...(7)

Multiplying (6) by 2, $-8 = -2C_2 - 4C_3$ ...(8)

Subtracting (7) and (8), $-2 = -2C_3 \Rightarrow C_3 = 1$

From (7), $-10 = -2C_2 - 6 \Rightarrow C_2 = 2$

From (3), $C_1 = 1 - 2 - 1 = -2$

From (2), The required solution is

$t_n = (-2) + 2(2^n) + 3^n = -2 + 2^{n+1} + 3^n.$

(*b*) Given equation is $t_n + 3t_{n-1} + 3t_{n-2} + t_{n-3} = 0$

Its order = $n - (n - 3) = 3$

The characteristic equation is $a^3 + 3a^2 + 3a + 1 = 0$

$\Rightarrow (a + 1)^3 = 0 \Rightarrow a = -1, -1, -1$

$\therefore$ The general solution is

$t_n = (C_1 + C_2 n + C_3 n^2)(-1)^n$ ...(1)

Using $t_0 = 1$ in (1), $1 = C_1$ ...(2)

Using $t_1 = -2$ in (1), $-2 = (C_1 + C_2 + C_3)(-1)$ or $2 = C_1 + C_2 + C_3$ ...(3)

Using $t_2 = -1$ in (1), $-1 = C_1 + 2C_2 + 4C_3$ ...(4)

Put $C_1 = 1$ in (3) and (4), we get

$C_2 + C_3 = 1$ ...(5)

$2C_2 + 4C_3 = -2$ ...(6)

Solving (5) and (6), we get $C_3 = -2, C_2 = 3$

The required solution is

$t_n = (1 + 3n - 2n^2)(-1)^n.$

(*c*) Please proceed yourself as in part (*a*)

Hint. The characteristic equation is $a^3 - 6a^2 + 12a - 8 = 0$ ...(1)

Here, $a = 2$ satisfies equation (1). By Horner's method, the roots of equation (1) are 2, 2, 2

Ans. $a_n = (3 - 2n + n^2) 2^n$

Example 8. *Solve the recurrence relation for Fibonacci numbers given by*

$$f_n = f_{n-1} + f_{n-2}, \text{ subject to } f_1 = f_2 = 1.$$

Sol. Given recurrence relation is $f_n - f_{n-1} - f_{n-2} = 0$ ...(1)

Its order is $n - (n-2) = 2$

The characteristic equation is $a^2 - a - 1 = 0 \;\Rightarrow\; a = \dfrac{1 \pm \sqrt{1+4}}{2} = \dfrac{1 \pm \sqrt{5}}{2}$

Take $\alpha = \dfrac{1+\sqrt{5}}{2}$, $\beta = \dfrac{1-\sqrt{5}}{2}$. Clearly, $\alpha + \beta = 1$, $\alpha\beta = -1$

The solution of (1) is given as

$$f_n = C_1\,\alpha^n + C_2\,\beta^n \quad ...(2)$$

Given $f_1 = C_1\,\alpha + C_2\,\beta = 1$ ...(3)

$f_2 = C_1\,\alpha^2 + C_2\beta^2 = 1$ ...(4)

Solving (3) and (4), we get

$$\frac{C_1}{\beta - \beta^2} = \frac{C_2}{\alpha^2 - \alpha} = \frac{-1}{\alpha\beta^2 - \alpha^2\beta}$$

2 3 1 2

β 1 α β

β² 1 α² β²

$$\Rightarrow \quad C_1 = -\frac{\beta(1-\beta)}{\alpha\beta(\beta-\alpha)}$$

$$= -\frac{\alpha\beta}{\alpha\beta(\beta-\alpha)} \quad (\because \;\alpha+\beta = 1)$$

$$= -\frac{1}{\beta-\alpha} = -\frac{2}{1-\sqrt{5}-1-\sqrt{5}} = \frac{1}{\sqrt{5}}$$

Also,

$$C_2 = -\frac{\alpha(\alpha-1)}{\alpha\beta(\beta-\alpha)} = -\frac{-\alpha\beta}{\alpha\beta(\beta-\alpha)} \quad |\; \alpha+\beta = 1$$

$$= \frac{1}{\beta-\alpha} = -\frac{1}{\sqrt{5}}$$

$$\therefore \quad f_n = \left(\frac{1}{\sqrt{5}}\right)\left\{\left(\frac{1+\sqrt{5}}{2}\right)^n - \left(\frac{1-\sqrt{5}}{2}\right)^n\right\}, \text{ is the required solution.}$$

Example 9. *Solve the recurrence relation $t_n = 4(t_{n-1} - t_{n-2})$, subject to initial conditions $t_n = 1$ for $n = 0, 1$.*

Sol. Given equation is $t_n - 4t_{n-1} + 4t_{n-2} = 0$

Its order $= n - (n-2) = 2$

The characteristic equation is $a^2 - 4a + 4 = 0 \;\Rightarrow\; (a-2)^2 = 0 \;\Rightarrow\; a = 2, 2$

The solution is given by $t_n = (C_1 + C_2 n)2^n$

Using $t_n = 1$ for $n = 0$ in (1), we get $1 = C_1$

Using $t_n = 1$ for $n = 1$ in (1), we get $1 = (1 + C_2)2 \;\Rightarrow\; C_2 = -\dfrac{1}{2}$

$$\therefore \quad t_n = \left(1 - \frac{1}{2}n\right)2^n = (2-n)\,2^{n-1}, \text{ is the required solution.}$$

Example 10. *Given that white tiger population of Orissa is 30 at time n = 0 and 32 at time n = 1. Also the increase from time n – 1 to time n is twice the increase from time n – 2 to time n –1. Find the recurrence relation and solve it for growth rate of tiger.*

Sol. Given initial conditions are

$$t_n = \begin{cases} 30, & n = 0 \\ 32, & n = 1 \\ 3t_{n-1} - 2t_{n-2}, & n > 1 \end{cases}$$

| Increase from time $n - 1$ to time n is $t_n - t_{n-1}$.
| Increase from time $n - 2$ to time $n - 1$ is $t_{n-1} - t_{n-2}$.

According to given,

$$t_n - t_{n-1} = 2(t_{n-1} - t_{n-2})$$

$$\Rightarrow \quad t_n = 3t_{n-1} - 2t_{n-2}$$

For $n > 1$, the recurrence relation is $t_n - 3t_{n-1} + 2t_{n-2} = 0$

The characteristic equation is

$$a^2 - 3a + 2 = 0 \;\Rightarrow\; a^2 - 2a - a + 2 = 0$$

$$\Rightarrow \quad (a - 1)(a - 2) = 0 \;\Rightarrow\; a = 1, 2$$

$\therefore$ The solution is $t_n = C_1 + C_2\, 2^n$

For $n = 0$, $\quad 30 = C_1 + C_2 \quad$ ...(1)

For $n = 1$, $\quad 32 = C_1 + 2C_2 \quad$ ...(2)

Solving (1) and (2), we get $C_1 = 28$, $C_2 = 2$

$\therefore$ The required solution is $t_n = 28 + 2^{n+1}$.

Example 11. *Find the solution of the homogeneous recurrence relation*

$$y_K - y_{K-1} - y_{K-2} = 0.$$

Sol. The characteristic equation is $s^2 - s - 1 = 0$

$$s = \frac{1 \pm \sqrt{1+4}}{2} = \frac{1 \pm \sqrt{5}}{2}$$

Therefore, the required solution of the given homogeneous recurrence relation is

$$y_K = C_1\left[\frac{1+\sqrt{5}}{2}\right]^K + C_2\left[\frac{1-\sqrt{5}}{2}\right]^K.$$

Example 12. *Find the solution of the homogeneous recurrence relation* $a_{r+4} + 2a_{r+3} + 3a_{r+2} + 2a_{r+1} + a_r = 0$.

Sol. The characteristic equation is $s^4 + 2s^3 + 3s^2 + 2s + 1 = 0$

$$\Rightarrow \quad s^4 + s^2 + 1 + 2s^3 + 2s + 2s^2 = 0 \quad | \; (a + b + c)^2 = a^2 + b^2 + c^2 + 2ab + 2bc + 2ca$$

$$\Rightarrow \quad (s^2 + s + 1)^2 = 0$$

or

$$s = \frac{-1 \pm i\sqrt{3}}{2}, \frac{-1 \pm i\sqrt{3}}{2}$$

Therefore, the required solution of the given homogeneous recurrence relation is given by

$$a_r = (C_1 + C_2\, r)\left(\frac{-1 + i\sqrt{3}}{2}\right)^r + (C_3 + C_4\, r)\left(\frac{-1 - i\sqrt{3}}{2}\right)^r.$$

Example 13. *Find the solution of the homogeneous recurrence relation $y_{K+4} + 4y_{K+3} + 8y_{K+2} + 8y_{K+1} + 4y_K = 0$.*

Sol. The characteristic equation is $s^4 + 4s^3 + 8s^2 + 8s + 4 = 0$

$$\Rightarrow \quad s^4 + 4s^2 + 4 + 4s^3 + 8s + 4s^2 = 0$$

$$| \ (a + b + c)^2 = a^2 + b^2 + c^2 + 2ab + 2bc + 2ca$$

$$\Rightarrow \quad (s^2 + 2s + 2)^2 = 0$$

or $$s = -1 \pm i, -1 \pm i$$

Therefore, the required solution is given by

$$y_K = (C_1 + C_2 K)(-1 + i)^K + (C_3 + C_4 K)(-1 - i)^K.$$

Example 14. *Solve $T(K) = 7T(K-1) - 10T(K-2)$ where $T(0) = 4$, $T(1) = 17$.*

Sol. Given equation is $T(K) - 7T(K-1) + 10T(K-2) = 0$

Its order $= K - (K - 2) = 2$

$\therefore$ The characteristic equation is $a^2 - 7a + 10 = 0$

$\Rightarrow \quad a^2 - 5a - 2a + 10 = 0 \Rightarrow (a - 5)(a - 2) = 0$

$\Rightarrow \quad a = 5, 2$

$\therefore$ The required solution is

$$T(K) = C_1 5^K + C_2 2^K \quad ...(1)$$

Given $T(0) = 4 \quad \therefore \quad$ (1) gives $4 = C_1 + C_2$ ...(2)

$T(1) = 17 \quad \therefore \quad$ (1) gives $17 = 5C_1 + 2C_2$ ...(3)

Multiplying the equation (2) by 2 and subtracting from (3), we get

$$8 - 17 = 2C_1 - 5C_1 \Rightarrow -3C_1 = -9$$

$$\Rightarrow \quad C_1 = 3$$

From (2), $\quad 4 = 3 + C_2 \Rightarrow C_2 = 1$

$\therefore$ From (1), we have

$$T(K) = 3 \cdot 5^K + 2^K, \text{ is the required solution.}$$

Example 15. *Solve the recurrence relation*

(*a*) $a_r - 2a_{r-1} + a_{r-2} = 0$ *given that* $a_0 = 1$, $a_1 = 2$. (P.T.U. B.Tech. May 2006)

(*b*) $a_n = 6\,a_{n-1} - 8\,a_{n-2}$, $a_0 = 1$, $a_1 = 10$. (P.T.U. B.Tech. Dec. 2012)

(*c*) $a_n = -3\,a_{n-1} + 10\,a_{n-2}$, $n \geq 2$, *given* $a_0 = 1$, $a_1 = -4$. (P.T.U. B.Tech. Dec. 2009)

Sol. (*a*) Given recurrence relation can be written as

$$S(r) - 2S(r-1) + S(r-2) = 0 \quad ...(1)$$

where $$S(0) = 1, S(1) = 2$$

Its $\quad$ order $= r - (r - 2) = 2$

$\therefore$ The characteristic equation is $a^2 - 2a + 1 = 0$

$\Rightarrow \quad (a - 1)^2 = 0 \Rightarrow a = 1, 1$

$\therefore$ The solution to (1) is

$$S(r) = (C_1 + C_2 r) \cdot 1^r \quad ...(2)$$

Given $S(0) = 1 \quad \therefore \quad$ (2) gives $1 = C_1$

$S(1) = 2 \quad \therefore \quad$ (2) gives $2 = C_1 + 2C_2$

$\Rightarrow \quad 2C_2 = 2 - 1 = 1 \Rightarrow C_2 = \frac{1}{2}$

∴ The required solution of (1) is $S(r) = \left(1+\frac{1}{2}r\right) . 1^r$

(*b*) Given equation is $a_n - 6\,a_{n-1} + 8a_{n-2} = 0$ ...(1)

where $a_0 = 1,\ a_1 = 10$

(1) can be written as

$$S(n) - 6S(n-1) + 8S(n-2) = 0, \quad S(0) = 1,\ S(1) = 10$$

Its order $= n - (n-2) = 2$

∴ The characteristic equation is $a^2 - 6a + 8 = 0$

$\Rightarrow \quad a^2 - 4a - 2a + 8 = 0 \ \Rightarrow\ a(a-4) - 2(a-4) = 0$

$\Rightarrow \quad (a-2)(a-4) = 0 \ \Rightarrow\ a = 2, 4$

∴ The solution of (1) is given by

$$S(n) = C_1.2^n + C_2.\,4^n \quad ...(2)$$

Given $S(0) = 1$ ∴ (2) gives $1 = C_1 + C_2$ ...(3)

$S(1) = 10$ ∴ (2) gives $10 = 2C_1 + 4C_2$ ...(4)

Multiplying (3) by 2 and subtracting from (4), we get

$$-8 = -2C_2 \quad \Rightarrow \quad C_2 = 4$$

From (3), $C_1 + 4 = 1 \quad \Rightarrow \quad C_1 = -3$

∴ The required solution is

$$S(n) = -3\,.\,2^n + 4^{n+1}.$$

(*c*) Given recurrence relation is $a_n + 3a_{n-1} - 10a_{n-2} = 0$

Its order $= n - (n-2) = 2$

The characteristic equation is

$$a^2 + 3a - 10 = 0 \ \Rightarrow\ a^2 + 5a - 2a - 10 = 0$$

$\Rightarrow \quad (a+5)(a-2) = 0 \ \Rightarrow\ a = 2, -5$

The general solution is $a_n = C_1\,2^n + C_2\,(-5)^n$ ...(1)

Using $a_0 = 1$, (1) gives $1 = C_1 + C_2$ ...(2)

Using $a_1 = -4$, (1) gives $-4 = 2C_1 - 5C_2$ ...(3)

Solving (2) and (3), we get $C_2 = \frac{6}{7},\ C_1 = \frac{1}{7}$

∴ The required solution is $a_n = \left(\frac{1}{7}\right)2^n + \left(\frac{6}{7}\right)(-5)^n$.

Example 16. *Solve S(K) – 7S(K – 2) + 6S(K – 3) = 0, where S(0) = 8, S(1) = 6, S(2) = 22.*

Sol. Given equation is $S(K) - 7S(K-2) + 6S(K-3) = 0$ ...(1)

Its order $= K - (K-3) = 3$

∴ The character equation is $a^3 - 7a + 6 = 0$ ...(2)

Put $a = \pm 1, \pm 2, \pm 3, ...,$ in (2), we observe that $a = 1$ satisfies (2)

∴ $a = 1$ is root of (2).

By Horner's method

1	1	0	– 7	6	
		1	1	– 6	
	1	1	– 6	0	= Remainder

$\therefore$ The quotient is $a^2 + a - 6 = 0$

$\Rightarrow$ $a^2 + 3a - 2a - 6 = 0$

$\Rightarrow$ $a(a + 3) - 2(a + 3) = 0$

$\Rightarrow$ $(a + 3)(a - 2) = 0$

$\Rightarrow$ $a = -3, 2$

Hence the roots are 1, 2, – 3.

$\therefore$ The solution to (1) is given by

$$S(K) = C_1 . 1^K + C_2 . 2^K + C_3 . (-3)^K$$
$$= C_1 + C_2 . 2^K + C_3(-3)^K \quad ...(3)$$

Given $S(0) = 8$ $\therefore$ (3) gives $8 = C_1 + C_2 + C_3$ ...(4)

$S(1) = 6$ $\therefore$ (3) gives $6 = C_1 + 2C_2 - 3C_3$ ...(5)

$S(2) = 22$ $\therefore$ (3) gives $22 = C_1 + 4C_2 + 9C_3$ ...(6)

Multiplying (4) by 3 and adding to (5), we get

$$24 + 6 = 4C_1 + 5C_2 \quad \text{or} \quad 4C_1 + 5C_2 = 30 \quad ...(7)$$

Multiplying (5) by 3 and adding to (6), we get

$$18 + 22 = 4C_1 + 10C_2 \quad \text{or} \quad 4C_1 + 10C_2 = 40 \quad ...(8)$$

Subtracting (7) and (8), we get

$$-5C_2 = -10 \Rightarrow C_2 = 2$$

From (8), $4C_1 + 20 = 40 \Rightarrow 4C_1 = 20 \Rightarrow C_1 = 5$

From (4), $5 + 2 + C_3 = 8 \Rightarrow C_3 = 1$

$\therefore$ The required solution of (1) is given by

$$S(K) = 5 + 2 . 2^K + (-3)^K.$$

7.10. SOLUTION OF RECURRENCE RELATIONS BY THE METHOD OF SUBSTITUTION

The substitution method is also known as **iterative method** which is used to find a formula for recurrence relation. We illustrate this method by the following examples.

Example 17. *Solve the following recurrence relation by substitution method*

$$a_n = \begin{cases} a_{n-1} + 3, & n \geq 2 \\ 2, & n = 1 \end{cases}$$

Sol. For $n \geq 2$, $a_n = a_{n-1} + 3$ ...(1)

Changing n to $n - 1$ in (1), we get

$$a_{n-1} = a_{n-2} + 3 = a_{n-2} + 3(1) \quad ...(2)$$

Using (2) in (1), we get

$$a_n = (a_{n-2} + 3) + 3 = a_{n-2} + 6 = a_{n-2} + 3(2) \quad ...(3)$$

Changing n to $n - 2$ in (1), we get

$$a_{n-2} = a_{n-3} + 3 \quad ...(4)$$

Using (4) in (3), we get

$$a_n = (a_{n-3} + 3) + 6 = a_{n-3} + 9 = a_{n-3} + 3(3) \quad ...(5)$$

Proceeding in this way, we get

$$a_n = a_{n-4} + 3(4)$$
$$\vdots$$
$$a_n = a_{n-i} + 3\,(i)$$

Repeating this process for $i = n - 1$ times, we get

$$a_n = a_1 + 3(n-1)$$
$$= 2 + 3(n-1) \qquad | \; a_1 = 2$$
$$= 3n - 1, \text{ is the required solution.}$$

Example 18. *Solve the recurrence relation*

$$H_n = \begin{cases} H_{n-1} + (n-1), & n \geq 2 \\ 0, & n = 1 \end{cases}$$

by substitution (iterative) method.

Sol. For $n \geq 2$, $\quad H_n = H_{n-1} + n - 1 \qquad ...(1)$

Changing n to $(n - 1)$ in (1), we get

$$H_{n-1} = H_{n-2} + n - 2 \qquad ...(2)$$

Using (2) in (1), we get

$$H_n = (H_{n-2} + n - 2) + n - 1$$
$$= H_{n-2} + [(n-2) + (n-1)] \qquad ...(3)$$

Changing n to $n - 2$ in (1), we get

$$H_{n-2} = H_{n-3} + n - 3 \qquad ...(4)$$

Using (4) in (3), we get

$$H_n = H_{n-3} + [(n-3) + (n-2) + (n-1)]$$

Proceeding in this way, we get

$$H_n = H_{n-i} + \{(n-i) + (n - \overline{i-1}) + ... + (n-3) + (n-2) + (n-1)\}$$

For $i = n - 1$, we get

$$H_n = H_1 + [1 + 2 + 3 + ... + (n-3) + (n-2) + (n-1)]$$
$$= 0 + \Sigma(n-1) = \frac{(n-1)n}{2}, \text{ is the required solution.}$$

Example 19. *Solve the following recurrence relation by substitution method*

$$t_n = \begin{cases} 2t_{n-1} + 1, & n \geq 2 \\ 1, & n = 1 \end{cases}.$$

Sol. For $n \geq 2$, $\quad t_n = 2t_{n-1} + 1 \qquad ...(1)$

Changing n to $n - 1$ in (1), we get

$$t_{n-1} = 2t_{n-2} + 1 \qquad ...(2)$$

Using (2) in (1), we get

$$t_n = 2(2t_{n-2} + 1) + 1$$
$$= 2^2\, t_{n-2} + 2 + 1 \qquad ...(3)$$

Changing n to $n - 2$ in (1), we get

$$t_{n-2} = 2t_{n-3} + 1 \qquad ...(4)$$

Using (4) in (3), we get

$$t_n = 2^2\,(2t_{n-3} + 1) + 2 + 1 = 2^3\, t_{n-3} + 2^2 + 2 + 1$$

Proceeding in this way, we get

$$t_n = 2^i\, t_{n-i} + 2^{i-1} + 2^{i-2} + \ldots\ldots + 2 + 1$$

For $i = n - 1$, we get

$$\begin{aligned} t_n &= 2^{n-1}\, t_1 + 2^{n-2} + 2^{n-3} + \ldots\ldots + 2 + 1 \\ &= 2^{n-1} + 2^{n-2} + 2^{n-3} + \ldots\ldots + 2 + 1 \qquad | \; t_1 = 1 \\ &= 1 + 2 + 2^2 + \ldots\ldots + 2^{n-1} \\ &= 1 \cdot \frac{(1-2^n)}{1-2} = 2^n - 1, \text{ is the required solution.} \end{aligned}$$

TEST YOUR KNOWLEDGE 7.1

SHORT ANSWER TYPE QUESTIONS

1. Define recurrence relation with examples. *(P.T.U.B. Tech. Dec. 2006)*
2. Define linear homogeneous recurrence relation of order n with constant coefficients.
3. Find the order of the following recurrence relations.
   (*i*) S(K + 4) + 4S(K + 3) + 8S(K + 2) + 8S(K + 1) + 4S(K) = 0
   (*ii*) S(K) – 2S(K – 1) + 2S(K – 2) – S(K – 3) = 0
   (*iii*) S(K + 3) + S(K + 2) – 8S(K + 1) – 12S(K) = 0
   (*iv*) $S(K) = S\left[\frac{K}{2}\right] + 9,\ K \geq 0.$

LONG ANSWER TYPE QUESTIONS

4. Obtain the recurrence relations from the following closed form of the sequence given by
   (*i*) S(K) = 2K + 9 ; order 1
   (*ii*) $S(K) = K^2 - K$; order 2
   (*iii*) $S(K) = 2K^2 + 1$; order 2
   (*iv*) $S(K) = (3 + K)\, 2^K$; order 2.
5. Find the solution of the following recurrence relations.
   (*i*) S(K) – 9S(K – 1) + 18S(K – 2) = 0, where S(0) = 1, S(1) = 4
   (*ii*) S(K) – 0.25S(K – 1) = 0, where S(0) = 6
   (*iii*) S(K) – 20S(K – 1) + 100S(K – 2) = 0, where S(0) = 2, S(1) = 30
   (*iv*) S(K) – 4S(K – 1) – 11S(K – 2) + 30S(K – 3) = 0 where S(0) = 0, S(1) = – 35, S(2) = – 85.
6. Find the solution of the following recurrence relations.
   (*i*) S(K + 4) + 4S(K + 3) + 8S(K + 2) + 8S(K + 1) + 4S(K) = 0
   (*ii*) S(K) – 4S(K – 1) + 6S(K – 2) – 4S(K – 3) + S(K – 4) = 0
   (*iii*) S(K + 3) + S(K + 2) – 8S(K + 1) – 12S(K) = 0
   (*iv*) S(K + 3) + 6S(K + 2) + 12S(K + 1) + 8S(K) = 0.
7. Using iterative method or substitution method/solve the following recurrence relations.

   (*a*) $t_n = \begin{cases} 1, & n = 0 \\ t_{n-1+n}, & n \geq 1 \end{cases}$
   (*b*) $t_n = \begin{cases} 0, & n = 0 \\ t_{n-1+4n}, & n \geq 1 \end{cases}$

   (*c*) $t_n = \begin{cases} 1, & n = 0 \\ 2t_{n-1}, & n \geq 1 \end{cases}$

8. Solve the recurrence relation by iterative method

(*a*) $a_n = 2a_{n/2} + (n - 1)$ for $n \geq 2$

$a_1 = 0$, and $n = 2k$ for $k \geq 1$

(*b*) $a_n = 2a_{n-1} + 1$, and $a_1 = 7$ for $n \geq 2$

(*c*) $a_n - 7a_{n-1} + 10a_{n-2} = 6 + 8n$ with $a_0 = 1$ and $a_2 = 2$.

9. Solve the following recurrence relations by iteration (substitution) method

(*a*) $t_n = t_{n-1} + 3$ for $t_1 = 2$

(*b*) $t_n = 2t_{n-1}$ for $t_0 = 1$

(*c*) $t_n = 2t_{n-1} + 1$ for $t_1 = 1$

(*d*) $t_n = 4(t_{n-1} - t_{n-2})$, $n \geq 2$, $t_0 = t_1 = 1$

10. Solve the following recurrence relations

(*a*) $t_n = t_{n-1} + 6t_{n-2}$ for $t_0 = 32$, $t_1 = -17$

(*b*) $t_n = -8t_{n-1} - t_{n-2}$, $n \geq 2$ for $t_0 = 0$, $t_1 = 1$

(*c*) $t_{n+1} = 2t_n + 3t_{n-1}$, $n \geq 1$ for $t_0 = 0$, $t_1 = 8$

(*d*) $t_n = 2t_{n-1} + 15t_{n-2}$, $n \geq 2$ for $t_0 = 0$, $t_1 = -1$

(*e*) $t_n = -2t_{n-1} + 15t_{n-2}$, $n \geq 2$ for $t_0 = 0$, $t_1 = 1$

(*f*) $t_n = 4t_{n-1}$, $n \geq 1$ and $t_0 = 1$

(*g*) $t_n = 4t_{n-1} + 8^n$, $n \geq 1$ and $t_0 = 1$.

Answers

3. (*i*) 4 (*ii*) 3 (*iii*) 3 (*iv*) Infinite order

4. (*i*) $S(K) - S(K-2) = 2$ (*ii*) $S(K) - 2S(K-1) + S(K-2) = 2$

(*iii*) $S(K) - 2S(K-1) + S(K-2) = 4$

(*iv*) $S(K) - 4S(K-1) + 4S(K-2) = 0$

5. (*i*) $S(K) = \left(\frac{2}{3}\right) 3^K + \left(\frac{1}{3}\right) 6^K$ (*ii*) $S(K) = 6\left(\frac{1}{4}\right)^K$ (*iii*) $S(K) = (2 + K)\, 10^K$

(*iv*) $S(K) = 4\,(-3)^K + 2^K - 5^{K+1}$

6. (*i*) $S(K) = (C_1 + C_2K)\,(-1 + i)^K + (C_3 + C_4K)\,(-1 - i)^K$

(*ii*) $S(K) = (C_1 + C_2K + C_3K^2 + C_3K^3)\,.\,1^K$

(*iii*) $S(K) = C_1.3^K + (C_2 + C_3{}^K)\,(-2)^K$

(*iv*) $S(K) = (C_1 + C_2K + C_3K^2)\,(-2)^K$.

7. (*a*) $t_n = \frac{n(n+1)}{2}$, $n \geq 1$ (*b*) $t_n = 2n(n+1)$, $n \geq 1$

(*c*) $t_n = 2^n$, $n \geq 1$

8. (*a*) $a_n = 1 + n\,(\log_2 n - 1)$ (*b*) $a_n = 4.2^{n-1} - 1$

(*c*) $a_n = (-9)2^n + 2(5)^n + 8 + 2n$

9. (*a*) $t_n = 2 + 3\,(n - 1)$ (*b*) $t_n = 2^n$

(*c*) $t_n = 2^n - 1$ (*d*) $t_n = 2^n - n2^{n-1}$

10. (*a*) $t_n = 3^n$ (*b*) $t_n = \frac{1}{2\sqrt{15}}(-4+\sqrt{15})^n - \frac{1}{2\sqrt{15}}(-4-\sqrt{15})^n$

(*c*) $t_n = -2(-1)^n + 2(3)^n$ (*d*) $t_n = \frac{1}{2}(3)^n + \frac{1}{2}(-5)^n$

(*e*) $t_n = 2(3)^n + (-5)^n - 2$ (*f*) $t_n = 4^n$

(*g*) $t_n = 2(8)^n - 4^n$.

7.11. NON-HOMOGENEOUS LINEAR RECURRENCE RELATION OF ORDER *n* WITH CONSTANT COEFFICIENTS

The recurrence relation of the form

$$S(K) + C_1S(K-1) + C_2S(K-2) + \ldots\ldots + C_nS(K-n) = f(K),$$

where $C_1, C_2, \ldots\ldots C_n$ are constants and $f(K) \neq 0$, is called non-homogeneous linear recurrence relation of order n with constant coefficients.

7.12. ALGORITHM FOR SOLVING NON-HOMOGENEOUS LINEAR RECURRENCE RELATION OF ORDER *n* WITH CONSTANT COEFFICIENTS

Consider $S(K) + C_1S(K-1) + C_2S(K-2) + \ldots\ldots + C_nS(K-n) = f(K)$, where $C_1, C_2, \ldots\ldots C_n$ are constants and $f(K) \neq 0$.

Step I. Solve the corresponding homogeneous equation (*i.e.*, $f(K) = 0$) as discussed in previous section. We call this solution as homogeneous solution and denote it by $S_h(K)$.

Step II. Obtain the particular solution as explained below.

Case I. If $f(K)$ = constant, then assume the particular solution as $S_P(K)$ = constant.

Case II. If $f(K)$ = a linear equation of the form $C_0 + C_1K$, then assume the particular solution as $S_P(K) = d_0 + d_1K$.

Case III. If $f(K)$ = a quadratic polynomial of the form $C_0 + C_1K + C_2K^2$, then assume the particular solution as $S_P(K) = d_0 + d_1K + d_2K^2$

Case IV. If $f(K)$ = an exponential function of the form a_0m^K, then assume the particular solution as $S_P(K) = d_0m^K$

Case V. If $f(K)$ = an exponential function of the form a_0Km^K, then assume the particular solution as $S_P(K) = (d_0 + d_1K)m^K$

The general solution or complete solution of the given non-homogeneous linear recurrence relation of order n is given by

$$S(K) = S_h(K) + S_P(K),$$

where $S_h(K)$ = Homogeneous solution, $S_P(K)$ = Particular solution.

Imp. Note. While assuming the particular solution, it is important to note that none of the terms of $S_P(K)$, should occur in $S_h(K)$. If any term of $S_h(K)$ occurs in $S_P(K)$, then multiply the terms of $S_P(K)$, by lowest power of K. The following examples illustrate this concept more clearly.

ILLUSTRATIVE EXAMPLES

Example 1. *Solve S(K) + 5S(K – 1) = 9, S(0) = 6.*

Sol. Given equation is S(K) + 5S(K – 1) = 9 ...(1)

The characteristic equation is $a + 5 = 0 \Rightarrow a = -5$

∴ The homogeneous solution of the given recurrence relation is given by

$$S_h(K) = c_1(-5)^K$$

Now, the R.H.S. of the equation (1) is a constant, therefore assume the particular solution as

$$S_P(K) = d_0,\ d_0 \text{ is constant.}$$

Here $S_P(K-1) = d_0$

Using all these in (1), we get

$$d_0 + 5d_0 = 9 \Rightarrow d_0 = 3/2$$

Hence the general solution of (1) is given by

$$S(K) = S_h(K) + S_P(K) = c_1(-5)^K + \frac{3}{2} \quad ...(2)$$

Given S(0) = 6 ∴ (2) gives $6 = c_1 + \frac{3}{2} \Rightarrow c_1 = 6 - \frac{3}{2} = \frac{9}{2}$

∴ (2) gives $S(K) = \frac{9}{2} \cdot (-5)^K + \frac{3}{2}$, is the required solution.

Example 2. *Solve the following recurrence relation*

$$T(K) - 7T(K-1) + 10T(K-2) = 6 + 8K$$

where T(0) = 1, T(1) = 2. (P.T.U. B.Tech. Dec. 2005)

Sol. The given equation is

$$T(K) - 7T(K-1) + 10T(K-2) = 6 + 8K \quad ...(1)$$

Its order = K – (K – 2) = 2

∴ The characteristic equation is $a^2 - 7a + 10 = 0$

$$\Rightarrow a^2 - 5a - 2a + 10 = 0 \Rightarrow a(a-5) - 2(a-5) = 0$$

$$\Rightarrow (a-5)(a-2) = 0 \Rightarrow a = 5, 2$$

∴ The homogeneous solution of (1) is given by

$$T_h(K) = C_1 \cdot 5^K + C_2 \cdot 2^K$$

To find the particular solution of (1), we observe that R.H.S. of (1) is a linear equation of the form 6 + 8K. Therefore, assume the particular solution of (1) as

$$T_P(K) = d_0 + d_1K$$

Here
$$T_P(K-1) = d_0 + d_1(K-1)$$
$$T_P(K-2) = d_0 + d_1(K-2)$$

Using all these in (1), we get

$$d_0 + d_1K - 7(d_0 + d_1(K-1)) + 10(d_0 + d_1(K-2)) = 6 + 8K$$

$$\Rightarrow (d_0 - 7d_0 + 7d_1 + 10d_0 - 20d_1) + (d_1 - 7d_1 + 10d_1)K = 6 + 8K$$

$$\Rightarrow 4d_0 - 13d_1 + 4d_1K = 6 + 8K \quad ...(2)$$

Equating the coefficients of constant term, in (2), we get

$$4d_0 - 13d_1 = 6 \quad \text{...(3)}$$

Equating the coefficient of K in (2), we get

$$4d_1 = 8 \;\Rightarrow\; d_1 = 2$$

$\therefore$ From (3), $4d_0 - 26 = 6 \;\Rightarrow\; 4d_0 = 32$

$\Rightarrow$ $d_0 = 8$

$\therefore$ $T_P(K) = 8 + 2K$

Hence the general solution of (1) is given by

$$T(K) = T_h(K) + T_P(K)$$
$$= C_1 \,.\, 5^K + C_2 \,.\, 2^K + 8 + 2K \quad \text{...(4)}$$

Given $T(0) = 1$ $\therefore$ gives $1 = C_1 + C_2 + 8$ ...(5)

$T(1) = 2$ $\therefore$ (4) gives $2 = 5C_1 + 2C_2 + 10$ ...(6)

Multiplying (5) by 2 and subtracting from (6), we get

$$0 = -3C_1 + 6 \;\Rightarrow\; C_1 = 2$$

From (5), $1 = 2 + C_2 + 8 \;\Rightarrow\; C_2 = -9$

$\therefore$ (4) reduces to $T(K) = 2 \,.\, 5^K - 9 \,.\, 2^K + 8 + 2K$, is the required solution.

Example 3. *Find the general solution of the following recurrence relation*

$$S(K) - 3S(K-1) - 4S(K-2) = 4^K.$$ (P.T.U. B.Tech. May 2010)

Sol. The given equation is

$$S(K) - 3S(K-1) - 4S(K-2) = 4^K \quad \text{...(1)}$$

Its order $= K - (K-2) = 2$

$\therefore$ The characteristic equation is $a^2 - 3a - 4 = 0$

$\Rightarrow$ $a^2 - 4a + a - 4 = 0 \;\Rightarrow\; a(a-4) + (a-4) = 0$

$\Rightarrow$ $(a+1)(a-4) = 0 \;\Rightarrow\; a = -1, 4$

$\therefore$ The homogeneous solution is given by

$$S_h(K) = C_1(-1)^K + C_2 4^K \quad \text{...(2)}$$

Particular solution. Corresponding to the term 4^K, [R.H.S. of (1)], we assume the general form of the solution as $d_0\, 4^K$ but due to occurence of this term in equation (2), we multiply this by suitable power of K so that none of the term will occur in equation (2). Hence, we multiply by K. Hence the particular solution of (1) becomes as

$$S_P(K) = d_0 K\, 4^K$$

Here $S_P(K-1) = d_0(K-1)\, 4^{K-1}$

$S_P(K-2) = d_0(K-2)\, 4^{K-2}$

Using all these values in (1), we get

$$d_0 K\, 4^K - 3d_0(K-1)\, 4^{K-1} - 4d_0(K-2)\, 4^{K-2} = 4^K$$

Dividing by 4^{K-2}, we get

$$16\, Kd_0 - 12d_0(K-1) - 4d_0(K-2) = 16$$

$\Rightarrow$ $d_0(16K - 12K + 12 - 4K + 8) = 16$

$\Rightarrow$ $20d_0 = 16 \;\Rightarrow\; d_0 = 4/5$

$\therefore$ $S_P(K) = \dfrac{4}{5} K\, 4^K$

$\therefore$ The general solution of (1) is given by

$$S(K) = S_h(K) + S_P(K) = C_0(-1)^K + C_2\, 4^K + \frac{4}{5} K\, 4^K.$$

Example 4. *Find the particular solution of the recurrence relation*

$$a_{r+2} - 3a_{r+1} + 2a_r = Z^r, \quad \text{...(1)}$$

where Z is some constant.

Sol. The general form of solution is = A . Z^r

Now putting this solution on L.H.S. of equation (1), we get

$$a_{r+2} - 3a_{r+1} + 2a_r = AZ^{r+2} - 3AZ^{r+1} + 2AZ^r = (Z^2 - 3Z + 2)AZ^r \quad \text{...(2)}$$

Equating equation (2) with R.H.S. of equation (1), we get

$$(Z^2 - 3Z + 2)A = 1$$

or

$$A = \frac{1}{Z^2 - 3Z + 2} = \frac{1}{(Z-1)(Z-2)} \quad (Z \neq 1, Z \neq 2)$$

Therefore, the particular solution is $a_{r(P)} \dfrac{Z^r}{(Z-1)(Z-2)}$.

Example 5. *Find the particular solution of the recurrence relation*

$$a_{r+2} - 5a_{r+1} + 6a_r = 5^r. \quad \text{...(1)}$$

Sol. Let us assume the general form of the solution $a_{r(P)} = A \,.\, 5^r$.

Now to find the value of A, put this solution on L.H.S. of the equation (1), then this becomes

$$\begin{aligned} a_{r+2} - 5a_{r+1} + 6a_r &= A.\, 5^{r+2} - 5 \,.\, A5^{r+1} + 6 \,.\, A5^r \\ &= 25A \,.\, 5^r - 25A \,.\, 5^r + 6A \,.\, 5^r \\ &= 6A \,.\, 5^r \end{aligned} \quad \text{...(2)}$$

Equating equation (2) to R.H.S. of equation (1), we get

$$6A = 1 \;\Rightarrow\; A = \frac{1}{6}$$

Therefore, the particular solution of the difference equation $a_{r(P)} = \dfrac{1}{6} \,.\, 5^r$.

Example 6. *Find the homogeneous solution and particular solution of the recurrence relation*

$$a_{r+2} - 4a_r = r^2 + r - 1. \quad \text{...(1)}$$

Sol. The characteristic equation is given by

$$a^2 - 4 = 0$$

$$\Rightarrow \quad (a-2)(a+2) = 0$$

$$\Rightarrow \quad a = 2, -2$$

The homogeneous solution of the recurrence relation is given by

$$a_{r(h)} = C_1(2)^r + C_2(-2)^r$$

To find the particular solution, let us assume the general form of the solution is

$$a_{r(P)} = A_1 r^2 + A_2 r + A_3.$$

Putting this solution in L.H.S. of equation (1), we get

$$a_{r+2} - 4ar = A_1 (r+2)^2 + A_2 (r+2) + A_3 - 4A_1 r^2 - 4A_2 r - 4A_3$$
$$= -3A_1 r^2 + (4A_1 - 3A_2)r + (4A_1 + 2A_2 - 3A_3) \quad ...(2)$$

Equating equation (2) with R.H.S. of equation (1), we get

$$-3A_1 = 1$$
$$4A_1 - 3A_2 = 1$$
$$4A_1 + 2A_2 - 3A_3 = -1$$

After solving these three equations, we get

$$A_1 = -\frac{1}{3},\ A_2 = -\frac{7}{9},\ A_3 = -\frac{17}{27}$$

Therefore, the particular solution is $a_r(\text{P}) = -\dfrac{r^2}{3} - \dfrac{7}{9}r - \dfrac{17}{27}$.

Example 7. *Find the homogeneous solution and particular solution of the recurrence relation*

$$a_{r+2} - 2a_{r+1} + a_r = 3r + 5. \quad ...(1)$$

Sol. The characteristic equation is

$$a^2 - 2a + 1 = 0$$
$$\Rightarrow \quad (a-1)^2 = 0 \Rightarrow a = 1, 1$$

The homogeneous solution of the recurrence relation is given by

$$a_{r(h)} = C_1 + C_2 r \quad ...(2)$$

Particular solution. Corresponding to the term $3r + 5$, we assume the general form of the solution as $A_1 r + A_2$ but due to occurrence of these terms in equation (2), we multiply this by suitable power of r so that none of the term will occur in equation (2). Thus multiply by r^2.

Hence, the general form of the solution becomes

$$a_{r(\text{P})} = A_1 r^3 + A_2 r^2$$

Putting this solution in L.H.S. of equation (1), we get

$$a_{r+2} - 2a_{r+1} + a_r = A_1(r+2)^3 + A_2 (r+2)^2 - 2A_1 (r+1)^3$$
$$- 2A_2 (r+1)^2 + A_1 r^3 + A_2 r^2$$
$$= A_1 (r^3 + 8 + 6r^2 + 12r) + A_2(r^2 + 4 + 4r)$$
$$- 2A_1(r^3 + 1 + 3r^2 + 3r) - 2A_2(r^2 + 1 + 2r) + A_1 r^3 + A_2 r^2$$
$$= (12A_1 + 4A_2 - 6A_1 - 4A_2)r + (8A_1 + 4A_2 - 2A_1 - 2A_2)$$
$$= (6A_1)r + (6A_1 + 2A_2) \quad ...(3)$$

Equating equation (3) with R.H.S. of equation (1), we get

$$6A_1 = 3 \quad \therefore \quad A_1 = \frac{1}{2}$$
$$6A_1 + 2A_2 = 5 \quad \therefore \quad A_2 = 1$$

Therefore, the particular solution is $a_{r(\text{P})} = \dfrac{1}{2}r^3 + r^2$.

Example 8. *Find the particular solution of the recurrence relation*

$$a_{r+2} + a_{r+1} + a_r = r \,.\, 2^r. \qquad ...(1)$$

Sol. Let us assume the general form of the solution $a_{r(P)} = (A_0 + A_1\, r) \,.\, 2^r$

Now, put this solution in the L.H.S. of equation (1), we get

$$\begin{aligned} a_{r+2} + a_{r+1} + a_r &= 2^{r+2}\,[A_0 + A_1\,(r+2)] + 2^{r+1}\,[A_0 + A_1\,(r+1)] + 2^r\,(A_0 + A_1\,r) \\ &= 4.\, 2^r\,(A_0 + A_1 r + 2A_1) + 2 \,.\, 2^r\,(A_0 + A_1\,r + A_1) + 2^r\,(A_0 + A_1\,r) \\ &= r.\, 2^r(7A_1) + 2^r\,(7A_0 + 10A_1) \qquad ...(2) \end{aligned}$$

Equating equation (2) with R.H.S. of equation (1), we get

$$7A_1 = 1 \quad \therefore \quad A_1 = \frac{1}{7}$$

$$7A_0 + 10A_1 = 0 \quad \therefore \quad A_0 = \frac{-10}{49}$$

Therefore, the particular solution is $a_{r(P)} = 2^r \left(\frac{-10}{49} + \frac{1}{7}\,r\right)$.

Example 9. *Find the homogeneous and particular solution of the recurrence relation*

$$a_r - 4a_{r-1} + 4a_{r-2} = (r+1) \,.\, 2^r. \qquad ...(1)$$

Sol. The characteristic equation is $a^2 - 4a + 4 = 0$

$$\Rightarrow \qquad (a-2)^2 = 0 \;\Rightarrow\; a = 2, 2$$

The homogeneous solution of the difference equation is given by

$$a_{r(h)} = (C_1 + C_2 r) \,.\, 2^r \qquad ...(2)$$

To find the particular solution, let us assume the general form of the solution is $= 2^r\,(A_1 r + A_0)$, but due to occurrence of there terms in equation (2), we multiply this by suitable power of r so that none of the terms will occur in equation (2). Thus multiply by r^2.

Hence, the general form of the solution becomes $= 2^r\,(A_1 r + A_0) \,.\, r^2$

Putting this solution in L.H.S. of equation (1), we get

$$\begin{aligned} a_r - 4a_{r-1} + 4a_{r-2} &= 2^r \,.\, (A_1 r + A_0) \,.\, r^2 - 4 \,.\, 2^{r-1}\,[A_1(r-1) + A_0] \\ &\qquad .\, (r-1)^2 + 4 \,.\, 2^{r-2}\,[A_1\,(r-2) + A_0] \,.\, (r-2)^2 \\ &= 2^r \,.\, (A_1 r + A_0) \,.\, r^2 - 2(r^2 + 1 - 2r) \,.\, 2^r\,(A_1 r - A_1 + A_0) \\ &\qquad + (r^2 + 4 - 4r) \,.\, 2^r \,.\, (A_1 r - 2A_1 + A_0) \\ &= r \,.\, 2^r\,(6A_1) + 2^r\,(-6A_1 + 2A_0) \qquad ...(3) \end{aligned}$$

Equating equation (3) with R.H.S. of equation (1), we get

$$6A_1 = 1 \quad \therefore \quad A_1 = \frac{1}{6}$$

$$-6A_1 + 2A_0 = 1 \quad \therefore \quad A_0 = 1$$

Therefore, the particular solution is $a_{r(P)} = r^2.\, 2^r\left(\frac{r}{6} + 1\right)$.

Example 10. *Solve* $S(n) - 6S(n-1) + 9S(n-2) = 3^{n+1}$. (P.T.U. B. Tech. May 2010)

Sol. Given equation is $S(n) - 6S(n-1) + 9S(n-2) = 3.3^n$ ...(1)

Its order $= n - (n-2) = 2$

∴ The characteristic equation is $a^2 - 6a + 9 = 0$

⇒ $(a - 3)^2 = 0 \Rightarrow a = 3, 3$

∴ The homogeneous solution of (1) is given by

$$S_h(n) = (C_1 + C_2 n)\, 3^n \quad ...(2)$$

Corresponding to the term 3^{n+1} [R.H.S. of (1)], we assume that general form of the solution as $d_0\, 3^{n+1}$, but due to occurence of this term in equation (2), we multiply this by suitable power of n so that none of the term will occur in equation (2). Thus multiply by n^2.

Hence the particular solution of (1) becomes as

$$S_P(n) = d_0\, n^2 3^n$$

Here $S_P(n-1) = d_0(n-1)^2\, 3^{n-1}$

$$S_P(n-2) = d_0(n-2)^2\, 3^{n-2}$$

Using all these in (1), we get

$$d_0 n^2 3^n - 6d_0(n-1)^2\, 3^{n-1} + 9d_0\,(n-2)^2\, 3^{n-2} = 3 \,.\, 3^n$$

Dividing by 3^{n-2}, we get

$$9d_0 n^2 - 18\, d_0(n-1)^2 + 9d_0(n-2)^2 = 27$$

⇒ $d_0[9n^2 - 18n^2 - 18 + 36n + 9n^2 + 36 - 36n] = 27$

⇒ $d_0[54] = 27 \Rightarrow d_0 = \dfrac{1}{2}$

∴ $S_P^{(n)} = \dfrac{1}{2} n^2 \,.\, 3^n$

∴ The general solution is given by

$$S(n) = S_h(n) + S_P(n) = (C_1 + C_2 n)\, 3^n + \frac{1}{2} n^2 \,.\, 3^n.$$

Example 11. *Solve the following recurrence relation*

$$Q(J) - Q(J-1) - 12Q(J-2) = (-3)^J + 6 \,.\, 4^J$$

Sol. Given equation is

$$Q(J) - Q(J-1) - 12Q(J-2) = (-3)^J + 6 \,.\, 4^J \quad ...(1)$$

Its order = J – (J – 2) = 2

∴ The characteristic equation is $a^2 - a - 12 = 0$

⇒ $a^2 - 4a + 3a - 12 = 0 \Rightarrow a(a-4) + 3(a-4) = 0$

⇒ $(a-4)(a+3) = 0 \Rightarrow a = 4, -3$

∴ The homogeneous solution of (1) given by

$$Q_h(J) = C_1 4^J + C_2(-3)^J \quad ...(1)$$

To find the particular solution of (1), we observe that the R.H.S. of (1) is a combination of the terms $(-3)^J$ and 4^J. But these terms also occur in the homogeneous solution (1). Therefore we assume the particular solution of (1) as

$$Q_P(J) = d_0 J(-3)^J + d_1 J\, 4^J$$

Here $Q_P(J-1) = d_0(J-1)(-3)^{J-1} + d_1(J-1)\, 4^{J-1}$

$$Q_P(J-2) = d_0(J-2)(-3)^{J-2} + d_1(J-2)\, 4^{J-2}$$

Using all these values in (1), we get

$$d_0 J(-3)^J + d_1 J\, 4^J - [d_0(J-1)(-3)^{J-1} + d_1(J-1)\, 4^{J-1}] - 12\,[d_0(J-2)(-3)^{J-2} + d_1(J-2)\, 4^{J-2}]$$
$$= (-3)^J + 6 \,.\, 4^J$$

$$d_0 (-3)^{J-2} [9J + 3(J-1) - 12(J-2)] + d_1 4^{J-2} [16J - 4(J-1) - 12(J-2)] = (-3)^J + 6 . 4^J$$

or $$d_0(-3)^{J-2} (-21) + d_1 4^{J-2} (28) = (-3)^J + 6 . 4^J$$

or Equating the coefficients of $(-3)^J$, we get $-\frac{21}{9} d_0 = 1 \Rightarrow d_0 = -\frac{3}{7}$

Equating the coefficient of 4^J, we get $\frac{28}{16} d_1 = 6 \quad \Rightarrow d_1 = \frac{96}{28} = \frac{24}{7}$

$$\therefore \quad Q_P(J) = -\frac{3}{7} J(-3)^J + \frac{24}{7} J\, 4^J$$

$\therefore$ The general solution of (1) is given by

$$Q(J) = Q_h(J) + Q_P(J) = C_1 4^J + C_2(-3)^J - \frac{3}{7} J(-3)^J + \frac{24}{7} J\, 4^J.$$

Example 12. *Solve $S(K) - 4S(K-1) + 4S(K-2) = 3K + 2^K$ where $S(0) = 1$, $S(1) = 1$.*

Sol. Given equation is $S(K) - 4S(K-1) + 4S(K-2) = 3K + 2^K$ ...(1)

Its order = K – (K – 2) = 2

$\therefore$ The characteristic equation is given by $a^2 - 4a + 4 = 0$

$$\Rightarrow \quad (a-2)^2 = 0 \Rightarrow a = 2, 2$$

$\therefore$ The homogeneous solution of (1) is given by

$$S_h(K) = (C_1 + C_2K). 2^K = C_1. 2^K + C_2K . 2^K$$

To find the particular solution of (1), we observe that the R.H.S. of (1) contains the terms 3K and 2^K. But the terms 2^K and $K2^K$ also occur in $S_h(K)$. Hence we divide the particular solution in two parts.

Particular solution corresponding to the term 3K is

$$S_P^1(K) = d_0 + d_1 K$$

$$\Rightarrow \quad S_P^1(K-1) = d_0 + d_1 (K-1),$$

$$S_P^1(K-2) = d_0 + d_1 (K-2)$$

Using all these values in (1), we get

$$d_0 + d_1K - 4 (d_0 + d_1(K-1) + 4(d_0 + d_1(K-2)) = 3K$$

$$\Rightarrow \quad d_0 (1 - 4 + 4) + d_1 [K - 4(K-1) + 4(K-2)] = 3K$$

$$\Rightarrow \quad d_0 + d_1 (K - 4K + 4 + 4K - 8) = 3K$$

$$\Rightarrow \quad d_0 + d_1(K-4) = 3K$$

Equating the coefficient of constant term, we get $d_0 - 4d_1 = 0$ ...(2)

Equating the coefficient of K, we get $d_1 = 3$ ...(3)

From (2), $d_0 = 12 \quad \therefore \quad S_P^1(K) = 12 + 3K$ ...(4)

Particular solution corresponding to the term 2^K

Let $$S'_P(K) = d_2K^2 . 2^K$$

$$\therefore \quad S'_P(K-1) = d_2(K-1)^2 . 2^{K-1}$$

$$S'_P(K-2) = d_2(K-2)^2 . 2^{K-2}$$

Using all these values in (1), we get

$$d_2K^2.2^K - 4\,d_2(K-1)^2 2^{K-1} + 4\,d_2(K-2)^2\,.\,2^{K-2} = 2^K$$

$$d_2\,.\,2^{K-2}(4K^2 - 8(K-1)^2 + 4(K-2)^2) = 2^K$$

$$\Rightarrow d_2\,.\,2^{K-2}(4K^2 - 8K^2 - 8 + 16K + 4K^2 + 16 - 16K) = 2^K$$

$$\Rightarrow \quad d_2 2^{K-2}(8) = 2^K$$

$$\Rightarrow \quad \frac{8d_2}{4} = 1 \quad \Rightarrow \quad d_2 = 1/2$$

$$\therefore \quad S_P'(K) = \frac{1}{2}K^2\,.\,2^K = K^2\,.\,2^{K-1} \qquad \text{...(5)}$$

$\therefore$ The particular solution of (1) becomes

$$S_P(K) = S_P^1(K) + S_P'(K)$$

$$= 12 + 3K + K^2\,.\,2^{K-1} \qquad \text{...(6)}$$

Hence the general solution of (1) is given by

$$S(K) = S_h(K) + S_P(K) \qquad | \text{ Using (4) and (5)}$$

$$= C_1\,.\,2^K + C_2K\,.\,2^K + 12 + 3K + K^2\,.\,2^{K-1} \qquad \text{...(7)}$$

Given S(0) = 1, (7) gives

$$1 = C_1 + 12 \quad \Rightarrow \quad C_1 = -11$$

S(1) = 1 $\therefore$ (7) gives $1 = 2C_1 + 2C_2 + 12 + 3 + 1$

$$\Rightarrow \quad 1 = -22 + 2C_2 + 16 = -6 + 2C_2 \Rightarrow C_2 = \frac{7}{2}$$

$$\therefore \quad \text{From (7)}, \quad S(K) = -11\,.\,2^K + \frac{7}{2}\,.\,K\,.\,2^K + 12 + 3K + K^2\,.\,2^{K-1}$$

$$= 12 + 3K + 2^{K-1}(K^2 + 7K - 22).$$

Example 13. *Solve S(K) = rS(K – 1) + a, where S(0) = 0, where r, a > 0, r ≠ 1.*

Sol. Given equation is $S(K) - rS(K-1) = a$ ...(1)

The characteristic equation is given by $a - r = 0 \Rightarrow a = r$

$$\therefore \quad S_h(K) = C_1 r^K$$

Also, let $S_P(K) = d_0$

$$\therefore \quad S_P(K-1) = d_0$$

Using all these in (1), we get $d_0 - rd_0 = a$

$$\Rightarrow \quad d_0(1-r) = a \Rightarrow d_0 = \frac{a}{1-r}$$

$$\therefore \quad S_P(K) = \frac{a}{1-r}$$

∴ The general solution is given by

$$S(K) = S_h(K) + S_P(K) = C_1 r^K + \frac{a}{1-r} \quad \ldots(2)$$

Given S(0) = 0 ∴ (2) gives

$$0 = C_1 + \frac{a}{1-r} \Rightarrow C_1 = -\frac{a}{1-r}$$

∴ From (2), $S(K) = -\frac{a}{1-r} \cdot r^K + \frac{a}{1-r}$

$$= \frac{a(1-r^K)}{1-r}, \text{ is the required solution.}$$

Example 14. *Solve the recurrence relation* $a_r + 5a_{r-1} + 6a_{r-2} = f(r)$

where
$$f(r) = \begin{cases} 0, & r = 0, 1, 5 \\ 6, & otherwise \end{cases}$$

given that $a_0 = a_1 = 0.$ (P.T.U. B.Tech. Dec. 2006)

Sol. Given equation is

$$S(r) + 5S(r-1) + 6S(r-2) = f(r) \quad \ldots(1)$$

Its order = r – (r – 2) = 2

∴ The characteristic equation is $a^2 + 5a + 6 = 0$

$\Rightarrow$ $a^2 + 3a + 2a + 6 = 0 \Rightarrow a(a+3) + 2(a+3) = 0$

$\Rightarrow$ $(a+3)(a+2) = 0 \Rightarrow a = -3, -2$

∴ The homogeneous solution of (1) is given by

$$S_h(r) = C_1(-3)^r + C_2(-2)^r$$

To find the particular solution,

Case I. When r = 0, 1, 5, then from given, we have f(r) = 0. Consequently, $S_P(r) = 0$

Hence the general solution of (1) is given by

$$S(r) = S_h(r) + S_P(r) = C_1(-3)^r + C_2(-2)^r$$

Case II. When $r \neq 0, 1, 5$, then we are given f(r) = 6

∴ Let $S_P(r) = d_0$

$\Rightarrow$ $S_P(r-1) = d_0$

$S_P(r-2) = d_0$

Using all these in (1), we get $d_0 + 5d_0 + 6d_0 = 6$

$\Rightarrow$ $12d_0 = 6 \Rightarrow d_0 = 1/2$

∴ $S_P(r) = \frac{1}{2}$

∴ The required general solution is given by

$$S(r) = S_h(r) + S_P(r) = C_1(-3)^r + C_2(-2)^r + \frac{1}{2}.$$

Example 15. *Solve* $S(K) - 4S(K-1) + 3S(K-2) = K^2$. (P.T.U. B.Tech. May 2012)

Sol. Given equation is $S(K) - 4S(K-1) + 3S(K-2) = K^2$ ...(1)

Its order $= K - (K-2) = 2$

$\therefore$ The characteristic equation is $a^2 - 4a + 3 = 0$

$\Rightarrow \quad a^2 - 3a - a + 3 = 0 \Rightarrow a(a-3) - (a-3) = 0$

$\Rightarrow \quad (a-3)(a-1) = 0 \Rightarrow a = 1, 3$

$\therefore$ The homogeneous solution of (1) is given by

$$S_h(K) = C_1 \cdot 1^K + C_2 3^K = C_1 + C_2 3^K \quad \text{...(2)}$$

Let the particular solution of (1) is given by

$$S_P(K) = K(A_0 + A_1K + A_2K^2)$$

(We have multiplied by K since (2) contains the constant term)

$$\Rightarrow \quad S_P(K) = A_0K + A_1K^2 + A_2K^3$$

$$\therefore \quad S_P(K-1) = A_0(K-1) + A_1(K-1)^2 + A_2(K-1)^3$$

$$S_P(K-2) = A_0(K-2) + A_1(K-2)^2 + A_2(K-2)^3$$

Using all these in (1), we get

$$A_0K + A_1K^2 + A_2K^3 - 4[A_0(K-1) + A_1(K-1)^2 + A_2(K-1)^3] + 3[A_0(K-2) + A_1(K-2)^2 + A_2(K-2)^3] = K^2$$

$$\Rightarrow \quad A_0[K - 4(K-1) + 3(K-2)] + A_1[K^2 - 4K^2 + 8K - 4 + 3K^2 - 12K + 12] + A_2[K^3 - 4K^3 + 12K^2 - 12K + 4 + 3K^3 - 18K^2 + 36K - 24] = K^2$$

$$\Rightarrow \quad -2A_0 + A_1[-4K + 8] + A_2[-6K^2 + 24K - 20] = K^2$$

Equating the coefficient of powers of K, we get

Coefficient of K^2, $\quad -6A_2 = 1 \quad \Rightarrow \quad A_2 = -\dfrac{1}{6}$

Coefficient of K, $-4A_1 + 24A_2 = 0$

$$-4A_1 - 4 = 0$$

$$\Rightarrow \quad A_1 = -1$$

Coefficient of constant term, $-2A_0 + 8A_1 - 20A_2 = 0 \Rightarrow -2A_0 - 8 + \dfrac{20}{6} = 0$

$$\Rightarrow \quad 2A_0 = -\frac{48+20}{6} = -\frac{28}{6} \Rightarrow A_0 = -\frac{14}{6} = -\frac{7}{3}$$

$$\therefore \quad S_P(K) = -\frac{7}{3}K - K^2 - \frac{1}{6}K^3$$

Hence the general solution is given by

$$S(K) = S_h(K) + S_P(K) = C_1 + C_2 \cdot 3^K - \frac{7}{3}K - K^2 - \frac{K^3}{6}.$$

TEST YOUR KNOWLEDGE 7.2

1. Solve the following recurrence relations
   (*i*) $S(K) - 2S(K-1) = 3.2^K$
   (*ii*) $S(K) + 5S(K-1) + 6S(K-2) = 3K^2$ (*P.T.U. B.Tech. Dec. 2007*)
   (*iii*) $S(K) - 2S(K-1) + S(K-2) = 2$, $S(0) = 25$, $S(1) = 16$
   (*iv*) $S(K) - S(K-1) - 6S(K-2) = -30$, $S(0) = 20$, $S(1) = -5$
   (*v*) $S(K) - 5S(K-1) = 5^K$, $S(0) = 3$
   (*vi*) $S(K) - 5S(K-1) + 6\,S(K-2) = 2$, $S(0) = 1$, $S(1) = -1$.
2. Solve the following recurrence relations
   (*i*) $a_{r+2} - 2a_{r+1} + a_r = r^2 . 2^r$
   (*ii*) $a_{K+3} + a_{K+2} - 8a_{K+1} - 12a_K = 2K^2 + 5$
   (*iii*) $a_n - 4a_{n-1} = 7(5)^n,\ n \ge 1$
   (*iv*) $a_n - 4a_{n-1} = 7(4)^n,\ n \ge 1$
   (*P.T.U. B.Tech. May 2013*)

Answers

1. (*i*) $S(K) = C_1\, 2^K + 3K\, 2^K$
   (*ii*) $S(K) = C_1\,(-2)^K + C_2\,(-3)^K + \frac{115}{288} + \frac{17}{24}K + \frac{1}{4}K^2$
   (*iii*) $S(K) = K^2 - 10K + 25$
   (*iv*) $S(K) = 11(-2)^K + 4\,.\,3^K + 5$
   (*v*) $S(K) = (3 + K)5^K$
   (*vi*) $S(K) = -2\,.\,3^K + 2\,.\,2^K + 1$.
2. (*i*) $a_r = C_1 + C_2 r + 2^r(r^2 - Pr + 20)$
   (*ii*) $a_K = C_1 3^K + (C_2 + C_3 K)\,(-2)^K - \frac{K^2}{9} + \frac{1}{27}K - \frac{17}{54}$
   (*iii*) $a_n = C_1\,.\,4^n + 35(5)^n$
   (*iv*) $a_n = (C_1 + 7n)\,(4)^n$.

7.13. GENERATING FUNCTIONS OR NUMERIC FUNCTIONS

Let $<S_n>$ be a sequence defined for $n \ge 0$, then the infinite sum G(S, z) where

$$G(S, z) = S_0 + S_1 z + S_2 z^2 + S_3 z^3 + \ldots\ldots$$

$$= \sum_{n=0}^{\infty} S_n z^n,$$ is called *generating function* or *numeric function* of the sequence $<S_n>$

ILLUSTRATIVE EXAMPLES

Example 1. *Consider a sequence $<S_n>$ defined by $S(n) = 2^n$, $n \ge 0$. Obtain its generating function.* (P.T.U. B.Tech. May 2010)

Sol. By definition, the required generating function is given by

$$G(S, z) = \sum_{n=0}^{\infty} S(n)\, z^n = \sum_{n=0}^{\infty} 2^n z^n = \sum_{n=0}^{\infty} (2z)^n$$

$$= 1 + 2z + (2z)^2 + (2z)^3 + \ldots\ldots \infty$$

$$= \frac{1}{1-2z}.$$

| Sum of infinite G.P. is $S_\infty = \frac{a}{1-r}$, where a = First term, r = common ratio

Example 2. *If $S(n) = ba^n$, $n \geq 0$, obtain its generating function.*

(P.T.U. B.Tech. May 2012)

Sol. By definition, the generating function is given by

$$G(S, z) = \sum_{n=0}^{\infty} S(n)\, z^n = \sum_{n=0}^{\infty} ba^n . z^n = b \sum_{n=0}^{\infty} (az)^n$$

$$= b\,(1 + az + (az)^2 + (az)^3 + \ldots\ldots \infty)$$

$$= b . \frac{1}{1-az} = \frac{b}{1-az}. \qquad \Big|\; S_\infty = \frac{a}{1-r}$$

Example 3. *If $S(n) = n$, $n \geq 0$, obtain its generating function.*

Sol. By definition, the generating function is given by

$$G(S, z) = \sum_{n=0}^{\infty} S(n)\, z^n = \sum_{n=0}^{\infty} nz^n$$

$$= 0 + 1.z + 2z^2 + 3z^3 + 4z^4 + \ldots\ldots \infty$$

$$= z(1 + 2z + 3z^2 + 4z^3 + \ldots\ldots \infty) \qquad \big|\; (1-x)^{-2} = 1 + 2x + 3x^2 + \ldots$$

$$= z . (1-z)^{-2} = \frac{z}{(1-z)^2}.$$

Example 4. *If $S(n) = \dfrac{1}{n!}$, find its generating function.*

Sol. By definition, the generating function is given by

$$G(S, z) = \sum_{n=0}^{\infty} S(n)\, z^n = \sum_{n=0}^{\infty} \frac{z^n}{n!} = \frac{1}{0!} + \frac{z}{1!} + \frac{z^2}{2!} + \frac{z^3}{3!} + \ldots\ldots \infty$$

$$= 1 + z + \frac{z^2}{2!} + \frac{z^3}{3!} + \ldots\ldots \infty = e^z.$$

Example 5. *Obtain partial fraction decompositions for the expression*

$G(S, z) = \dfrac{6-29z}{1-11z+30z^2}$ and identify the sequence having the expression as generating function.

Sol. Consider $\quad G(S, z) = \dfrac{6-29z}{1-11z+30z^2} = \dfrac{6-29z}{30z^2 - 6z - 5z + 1}$

$$= \frac{6-29z}{6z(5z-1) - (5z-1)} = \frac{6-29z}{(6z-1)(5z-1)}$$

Let $\quad \dfrac{6-29z}{(6z-1)(5z-1)} = \dfrac{A}{6z-1} + \dfrac{B}{5z-1} \qquad \ldots(1)$

Multiplying (1) by $(6z - 1)(5z - 1)$, we get

$$6 - 29z = A(5z - 1) + B(6z - 1) \qquad ...(2)$$

Put $5z - 1 = 0$ in (2), we get $6 - \frac{29}{5} = B\left(\frac{6}{5} - 1\right) \Rightarrow B = 1$

Put $6z - 1 = 0$ in (2), we get $6 - \frac{29}{6} = A\left(\frac{5}{6} - 1\right) \Rightarrow A = -7$

$\therefore$ From (1), $\frac{6 - 29z}{(6z - 1)(5z - 1)} = \frac{-7}{6z - 1} + \frac{1}{5z - 1} = \frac{7}{1 - 6z} - \frac{1}{1 - 5z}$

$\Rightarrow$ $G(S, z) = 7.6^n - 1.5^n$

$= 7.6^n - 5^n$

If $S_n = ba^n, n \geq 0$,

Then $G(S, z) = \frac{b}{1 - az}$

7.14. GENERATING FUNCTIONS FOR SOME STANDARD SEQUENCES

Sequence *S(n)*	*Generating functions* *G(S, z)*
$S(n) = a^n, n \geq 0$	$G(S, z) = \frac{1}{1 - az}$
$S(n) = n + 1, n \geq 0$	$G(S, z) = \frac{1}{(1 - z)^2}$
$S(n) = (n + 1)\, b^n, n \geq 0$	$G(S, z) = \frac{1}{(1 - bz)^2}$
$S(n) = n, n \geq 0$	$G(S, z) = \frac{z}{(1 - z)^2}$
$S(n) = na^n, n \geq 0$	$G(S, z) = \frac{az}{(1 - az)^2}$
$S(n) = \frac{1}{n!}, n \geq 0$	$G(S, z) = e^z$
$S(n) = {}^nC_r, 0 \leq r \leq n$	$G(S, z) = (1 + z)^n$

7.15. SOLUTION OF RECURRENCE RELATION BY THE METHOD OF GENERATING FUNCTIONS

Recurrence relation can also be solved by using the generating functions. Let the recurrence relation is

$$S(n) + C_1S(n - 1) + C_2S(n - 2) + \ldots\ldots + C_kS(n - K) = f(n), n \geq K$$

Multiplying both sides by z^n and summing up from $n = K$ to ∞, where K = order of the given recurrence relation, we get

$$\sum_{n=K}^{\infty} S(n)\, z^n + C_1 \sum_{n=K}^{\infty} S(n-1)\, z^n + C_2 \sum_{n=K}^{\infty} S(n-2)\, z^n + \ldots$$

$$+ C_K \sum_{n=K}^{\infty} S(n-K)\, z^n = \sum_{n=K}^{\infty} f(n) z^n$$

Writing each term in the form $\sum_{n=0}^{\infty} S(n) z^n$, we will get the solution in terms of G(S, z).

Using the results of 7.13, we can find the required solution. The following examples illustrate the concepts more clearly.

Example 6. *Solve the recurrence relation S(K) – 3S(K – 1) – 2 = 0, K ≥ 1, where S(0) = 1, by using generating functions.*

Sol. Given equation is S(K) – 3S(K – 1) = 2 ...(1)

Multiplying (1) by z^K, we get

$$S(K)\, z^K - 3S(K-1)\, z^K = 2z^K$$

Summing up for all K ≥ 1, we get, (order = K – (K – 1) = 1)

$$\sum_{K=1}^{\infty} S(K)\, z^K - 3 \sum_{K=1}^{\infty} S(K-1)\, z^K = 2 \sum_{K=1}^{\infty} z^K \quad \ldots(2)$$

Consider $$G(S, z) = \sum_{K=0}^{\infty} S(K)\, z^K = S(0) + S(1)\, z + S(2)\, z^2 + \ldots\ldots$$

$$= S(0) + \sum_{K=1}^{\infty} S(K)\, z^K$$

$$\Rightarrow \quad \sum_{K=1}^{\infty} S(K)\, z^K = G(S, z) - S(0)$$

For the second term in (2), we have

$$\sum_{K=1}^{\infty} S(K-1)\, z^K = z \sum_{K=1}^{\infty} S(K-1)\, z^{K-1} = z \sum_{K=0}^{\infty} S(K)\, z^K = zG(S, z)$$

| Changing K to K + 1

For the third term in (2), we have

$$\sum_{K=1}^{\infty} z^K = z + z^2 + z^3 + \ldots\ldots \infty = \frac{z}{1-z}$$

Using all these in (2), we get

$$G(S, z) - S(0) - 3\, zG(S, z) = \frac{2z}{1-z}$$

$$\Rightarrow \qquad (1-3z)\,G(S, z) = \frac{2z}{1-z} + 1 \qquad | \; S(0) = 1$$

$$= \frac{2z+1-z}{1-z} = \frac{z+1}{1-z}$$

$$\Rightarrow \qquad G(S, z) = \frac{1+z}{(1-z)(1-3z)} \qquad ...(3)$$

Consider $$\frac{1+z}{(1-z)(1-3z)} = \frac{A}{1-z} + \frac{B}{1-3z} \qquad ...(4)$$

Multiplying (4) by $(1 - z)(1 - 3z)$, we get

$$1 + z = A(1 - 3z) + B(1 - z) \qquad ...(5)$$

Put $1 - 3z = 0$ in (5), we get $1 + \frac{1}{3} = B\left(1 - \frac{1}{3}\right) \Rightarrow B = 2$

Put $1 - z = 0$ in (5), we get $1 + 1 = A(1 - 3) \Rightarrow A = -1$

$\therefore$ From (3) and (4), we get $G(S, z) = \frac{-1}{1-z} + \frac{2}{1-3z}$

Hence the required solution of the given recurrence relation is

$$S(K) = -(1)^K + 2.3^K, \; K \geq 0.$$

Example 7. *Find the generating function from the recurrence relation given by*
$S(K) - 6S(K-1) + 5S(K-2) = 0$ where $S(0) = 1$, $S(1) = 2$.

(P.T.U. B.Tech. Dec. 2010)

Sol. Given recurrence relation is

$$S(K) - 6S(K-1) + 5S(K-2) = 0 \qquad ...(1)$$

Its order = K – (K – 2) = 2

Multiplying (1) both sides by z^K and summing up from K = 2 to ∞, we get

$$\sum_{K=2}^{\infty} S(K)\,z^K - 6\sum_{K=2}^{\infty} S(K-1)\,z^K + 5\sum_{K=2}^{\infty} S(K-2)\,z^K = 0$$

Consider $$G(S, z) = \sum_{K=0}^{\infty} S(K)\,z^K = S(0) + S(1)\,z + \sum_{K=2}^{\infty} S(K)\,z^K$$

$$= 1 + 2z + \sum_{K=2}^{\infty} S(K)\,z^K$$

$$\Rightarrow \qquad \sum_{K=2}^{\infty} S(K)\,z^K = G(S, z) - 1 - 2z$$

Also $$\sum_{K=2}^{\infty} S(K-1)\,z^K = z\sum_{K=2}^{\infty} S(K-1)\,z^{K-1} = z\left(\sum_{K=1}^{\infty} S(K)\,z^K\right) \qquad | \text{Changing K to K + 1}$$

$$= z\left(S(0) + \sum_{K=1}^{\infty} S(K)\,z^K - S(0)\right) = z\left(\sum_{K=0}^{\infty} S(K)\,z^K - S(0)\right)$$

$$= z(G(S, z) - 1) = zG(S, z) - z$$

Also $\sum_{K=2}^{\infty} S(K-2)\, z^K = z^2 \sum_{K=2}^{\infty} S(K-2)\, z^{K-2} = z^2 \sum_{K=0}^{\infty} S(K)\, z^K$ | Changing K to K + 2

$= z^2 G(S, z)$

Using all these in (2), we get

$[G(S, z) - 1 - 2z] - 6\,[z\, G(S, z) - z] + 5[z^2 G(S, z)] = 0$

$\Rightarrow \quad (1 - 6z + 5z^2)\, G(S, z) - 1 - 2z + 6z = 0$

$$\Rightarrow \quad G(S, z) = \frac{1-4z}{1-6z+5z^2}$$

is the required generating function.

Example 8. *Find the generating function from the recurrence relation*

$S(n-2) = S(n-1) + S(n)$ where $S(0) = 1$, $S(1) = 1$, $n \geq 0$.

(P.T.U. B.Tech. May 2012)

Sol. Given recurrence relation is

$S(n) + S(n-1) - S(n-2) = 0$

Its order $= n - (n-2) = 2$ ...(1)

Multiplying (1) both sides by z^n and summing up all the terms from $n = 2$ to ∞, we get

$$\sum_{n=2}^{\infty} S(n)\, z^n + \sum_{n=2}^{\infty} S(n-1)\, z^n - \sum_{n=2}^{\infty} S(n-2)\, z^n = 0 \qquad ...(2)$$

Consider $\quad G(S, z) = \sum_{n=0}^{\infty} S(n)\, z^n = S(0) + S(1)\, z + \sum_{n=2}^{\infty} S(n)\, z^n$

$= 1 + z + \sum_{n=2}^{\infty} S(n)\, z^n$

$\Rightarrow \quad \sum_{n=2}^{\infty} S(n)\, z^n = G(S, z) - 1 - z$

Also $\sum_{n=2}^{\infty} S(n-1)\, z^n = z \sum_{n=2}^{\infty} S(n-1)\, z^{n-1} = z \sum_{n=1}^{\infty} S(n)\, z^n$ | Changing n to $n + 1$

$= z\left[S(0) + \sum_{n=1}^{\infty} S(n)\, z^n - S(0)\right]$

$= z\left[\sum_{n=0}^{\infty} S(n)\, z^n - S(0)\right] = z[G(S, z) - 1] = zG(S, z) - z$

Also $\sum_{n=2}^{\infty} S(n-2)\, z^n = z^2 \sum_{n=2}^{\infty} S(n-2)\, z^{n-2} = z^2 \sum_{n=0}^{\infty} S(n)\, z^n$ | Changing n to $n + 2$

$= z^2\, G(S, z)$

Using all these terms in (2), we get

$[G(S, z) - 1 - z] + [z(G(S, z) - 1)] - z^2 G(S, z) = 0$

$\Rightarrow \quad (1 + z - z^2)\, G(S, z) = 1 + z + z$

$$\Rightarrow \qquad G(S, z) = \frac{2z+1}{1+z-z^2},$$

is the required generating function.

Example 9. *Use generating functions to solve the recurrence relation.*
$a_k = a_{k-1} + 2a_{k-2} + 2^k$ *where* $a_0 = 4$, $a_1 = 12$. (P.T.U. B.Tech May 2010)

Sol. Given recurrence relation can be written as

$$a_k - a_{k-1} - 2a_{k-2} = 2^k \qquad \ldots(1)$$

Multiplying both sides of (1) by z^k and summing up from $k = 2$ to $k = \infty$, we get

$$\sum_{k=2}^{\infty} a_k z^k - \sum_{k=2}^{\infty} a_{k-1} z^k - 2\sum_{k=2}^{\infty} a_{k-2} z^k = \sum_{k=2}^{\infty} 2^k z^k \qquad \ldots(2)$$

Consider $\qquad G(S, z) = \sum_{k=0}^{\infty} a_k z^k = a_0 + a_1 z + \sum_{k=2}^{\infty} a_k z^k = 4 + 12z + \sum_{k=2}^{\infty} a_k z^k$

$$\Rightarrow \qquad \sum_{k=2}^{\infty} a_k z^k = G(S, z) - 4 - 12z \qquad \ldots(3)$$

Also, $\qquad \sum_{k=2}^{\infty} a_{k-1} z^k = z\sum_{k=2}^{\infty} a_{k-1} z^{k-1} = z\sum_{k=1}^{\infty} a_k z^k \qquad$ (Changing k to $k + 1$)

$$= z\left(a_0 + \sum_{k=1}^{\infty} a_k z^k - a_0\right) = z\left(\sum_{k=0}^{\infty} a_k z^k - 4\right) = zG(S, z) - 4z \qquad \ldots(4)$$

Further $\qquad \sum_{k=2}^{\infty} a_{k-2} z^k = z^2 \sum_{k=2}^{\infty} a_{k-2} z^{k-2} = z^2 \sum_{k=0}^{\infty} a_k z^k \qquad$ (Changing k to $k + 2$)

$$= z^2 G(S, z) \qquad \ldots(5)$$

Also $\qquad \sum_{k=2}^{\infty} 2^k z^k = 1 + 2z + \sum_{k=2}^{\infty} 2^k z^k - 1 - 2z = 1 + 2z + \sum_{k=0}^{\infty} 2^k z^k$

$$= 1 + 2z + \sum_{k=0}^{\infty} (2z)^k = 1 + 2z + (1 + 2z + (2z)^2 + (2z)^3 + \ldots \infty)$$

$$= 1 + 2z + \frac{1}{1-2z} \qquad \ldots(6) \qquad \left| S_\infty = \frac{a}{1-r} \right.$$

Using (3), (4), (5) and (6) in (2), we get

$$G(S, z) - 4 - 12z - (zG(S, z) - 4z) - 2(z^2 G(S, z)) = 1 + 2z + \frac{1}{1-2z}$$

$$\Rightarrow \qquad (1 - z - 2z^2)\, G(S, z) = 1 + 2z + \frac{1}{1-2z} + 4 + 12z - 4z = 5 + 10z + \frac{1}{1-2z}$$

$$\Rightarrow \quad (2z - 1)(z + 1)\, G(S, z) = \frac{5(1+2z)(1-2z)}{(1-2z)}$$

$$\Rightarrow \qquad G(S, z) = \frac{5(1-4z^2)+1}{(2z-1)^2\,(z+1)} = \frac{6-20z^2}{(z+1)\,(2z-1)} \qquad ...(7)$$

Consider

$$\frac{6-20z^2}{(z+1)\,(2z-1)^2} = \frac{A}{z+1} + \frac{B}{2z-1} + \frac{C}{(2z-1)^2} \qquad ...(8)$$

Multiplying (8) by $(z + 1)(2z - 1)^2$, we get

$$6 - 20z^2 = A(2z - 1)^2 + B(z + 1)(2z - 1) + c(z + 1) \qquad ...(9)$$

Put $z + 1 = 0$ in (9), we get $-14 = 9A \quad \Rightarrow \quad A = -\frac{14}{9}$

Put $2z - 1 = 0$ in (9), we get $\quad \Rightarrow \quad 1 = \frac{3}{2}C \quad \Rightarrow \quad C = \frac{2}{3}$

Equating the coefficient of constant terms in (9), we get

$$6 = A - B + C = -\frac{14}{9} - B + \frac{2}{3}$$

$$\Rightarrow \qquad B = -\frac{14}{9} + \frac{2}{3} - 6 = \frac{-14+6-54}{9} = \frac{-62}{9}$$

$$\therefore \quad \text{From (7)} \quad G(S, z) = -\frac{14}{9}\cdot\frac{1}{z+1} - \frac{62}{9}\cdot\frac{1}{2z-1} + \frac{2}{3}\cdot\frac{1}{(2z-1)^2}$$

Hence, the required solution is

$$a_k = -\frac{14}{9}(-1)^k + \frac{62}{9}\,2^k + \frac{2}{3}.(k+1)\,2^k$$

| If $G(S, z) = \dfrac{1}{1-az}$, then $S(n) = a^n, n \geq 0$
| If $G(S, z) = \dfrac{1}{(1-bz)^2}$, then $S(n) = (n+1)b^n, n \geq 0$

Example 10. *Find a closed form expression (generating function) for the terms of Fibonacci sequence* (P.T.U. B.Tech. May 2005)

Sol. Consider the Fibonacci sequence F(K), given by

$$F(K) = F(K - 1) + F(K - 2), K \geq 2$$

where $\quad F(0) = 1, F(1) = 1$

Its order = K – (K – 2) = 2

Multiplying by z^K and summing up all the terms from K = 2 to ∞, we get

$$\sum_{K=2}^{\infty} F(K)\,z^K - \sum_{K=2}^{\infty} F(K-1)\,z^K - \sum_{K=2}^{\infty} F(K-2)\,z^K = 0 \qquad ...(1)$$

Consider $\quad G(F, Z) = \sum_{K=0}^{\infty} F(K)\,z^K = F(0) + F(1)\,z + \sum_{K=2}^{\infty} F(K) z^K$

$$= 1 + z + \sum_{K=2}^{\infty} F(K)\, z^K$$

$$\Rightarrow \qquad \sum_{K=2}^{\infty} F(K)\, z^K = G(F, Z) - 1 - z$$

Also $\displaystyle\sum_{K=2}^{\infty} F(K-1)\, z^K = z \sum_{K=2}^{\infty} F(K-1)\, z^{K-1} = z \sum_{K=1}^{\infty} F(K)\, z^K$ | Changing K to K + 1

$$= z\left[F(0) + \sum_{K=1}^{\infty} F(K)\, z^K - F(0)\right] = z\left[\sum_{K=0}^{\infty} F(K)\, z^K - 1\right]$$

$$= z\,(G(F, z) - 1) = zG(F, z) - z$$

$\displaystyle\sum_{K=2}^{\infty} F(K-2)\, zK = z^2 \sum_{K=2}^{\infty} F(K-2)\, z^{K-2} = z^2 \sum_{K=0}^{\infty} F(K)\, z^K$ | Changing K to K + 2.

$$= z^2 G(F, z)$$

Using all these values in (1), we get

$$[G(F, z) - 1 - z] - [zG(F, z) - z] - z^2\, G(F, z) = 0$$

$$(1 - z - z^2)\, G(F, z) - 1 - z + z = 0$$

$$\Rightarrow \qquad (1 - z - z^2)\, G(F, z) = 1$$

$$\Rightarrow \qquad G(F, z) = \frac{1}{1 - z - z^2}$$

is the required generating function (or closed form expression) for the Fibonacci sequence.

TEST YOUR KNOWLEDGE 7.3

SHORT ANSWER TYPE QUESTIONS

1. Find the generating function for the sequence $S_n = 2^n$, $n \geq 0$.
2. If $S_n = a^n$, $n \geq 0$, then find the generating function G(S, z). *(P.T.U. B.Tech. Dec. 2013)*
3. Find the generating function for the sequence $S_n = n$, $n \geq 0$.
4. Write the generating function for the sequence $U(n) = 2.3^n$, $n \geq 0$.
5. Write the generating function for the sequence $V(n) = 5$, $n \geq 0$.
6. What is the generating function for the sequence $T(n) = 7(-1)^{nP}$?

LONG ANSWER TYPE QUESTIONS

7. If S(n) is a sequence defined by $S(n) = 2.3^n + 5 + 7.(-1)^n$, find the generating function for S(n).
8. Find the generating function for the sequence
   (*i*) $V(n) = 2^{n+1} + 5^n$ (*ii*) $V(n) = 2^n\, [3 + 2(-1)^n]$.
9. If $S(n) = 2S(n-1) + 3S(n-2)$, $n \geq 2$ where $S(0) = 3$, $S(1) = 1$.
   (*a*) Find the generating function
   (*b*) Find the sequence which satisfies the given recurrence relation.

10. **If $S(n) + 3S(n-1) - 4S(n-2) = 0$, $n \geq 2$ where $S(0) = 3$, $S(1) = -2$.**

(a) Find the generating function

(b) Find the sequence which satisfies the given recurrence relation.

11. **Determine partial fraction decomposition and identify the sequence having the expression as a generating function for the following expressions**

(a) $\dfrac{3+7z}{1+3z-4z^2}$ (b) $\dfrac{3-5z}{(1-3z)(1+z)}$

(c) $\dfrac{5+2z}{1-4z^2}$ (d) $\dfrac{1}{1-5z+6z^2}$.

Answers

1. $G(S, z) = \dfrac{1}{1-2z}$ 2. $G(S, z) = \dfrac{a}{1-z}$ 3. $G(S, z) = \dfrac{z}{(1-z^2)}$

4. $G(U, z) = \dfrac{2}{1-3z}$ 5. $G(V, z) = \dfrac{5}{1-z}$ 6. $G(T, z) = \dfrac{7}{1+z}$

7. $G(S, z) = \dfrac{2}{1-3z} + \dfrac{5}{1-z} + \dfrac{7}{1+z}$

8. (i) $G(V, z) = \dfrac{2}{1-2z} + \dfrac{1}{1-5z}$ (ii) $G(V, z) = \dfrac{3}{1-2z} + \dfrac{2}{1+2z}$

9. (a) $G(S, z) = \dfrac{3-5z}{1-2z-3z^2}$ (b) $S(n) = 3^n + 2(-1)^n$.

10. (a) $G(S, z) = \dfrac{2}{1-z} + \dfrac{5}{1+4z}$ (b) $S(n) = 2 + 5(-4)^n$

11. (a) $2 + (-4)^n$ (b) $3^n + 2(-1)^n$
(c) $3 \cdot 2^n + 2(-2)^n$ (d) $-2 \cdot 2^n + 3 \cdot 3^n$

Hint

7. $S(n) = U(n) + V(n) + T(n)$ where $U(n) = 2.3^n$, $V(n) = 5$, $T(n) = 7(-1)^n$.

MULTIPLE CHOICE QUESTIONS (MCQs)

1. The recurrence $T(n) = 2T(n-1) + n$ for $n \geq 2$ and $T(1) = 1$ evaluates to

(a) $2^{n+1} - n - 2$ (b) $2^n - n$

(c) $2^n + n$ (d) $2^{n+1} - 2n - 2$.

2. **Suppose that a school principal decides to give a prize away each day. Suppose that further the principal has 3 different kinds of prizes worth ₹ 1 each and 5 different kinds of prizes worth ₹ 4. each. Find a recurrence relation for a_n = the number of different way to distribute prizes worth ₹ n.**

(a) $a_n = 3a_{n-1} + 5a_{n-4}$ (b) $a_n = a_{n-1} + a_{n-4}$

(c) $a_n = 2a_{n-1} + 4a_{n-4}$ (d) None of these.

3. **Consider a lxn chessboard. Suppose we can colour each square of the chessboard either red or white. Let a_n = the number of ways to colouring the chessboard in which no 2 red squares are adjacent. Find a recurrence relation that a_n satisfies.**

(a) $a_n = a_n + a_{n-1}$ (b) $a_n = a_{n-1} + a_{n-2}$

(c) $a_n = a_{n-1} + a_{n-2}$ (d) None of these.

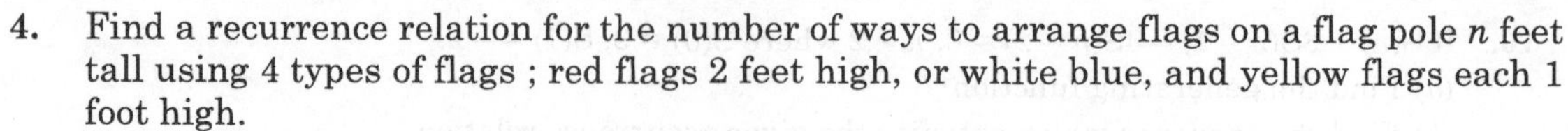

4. Find a recurrence relation for the number of ways to arrange flags on a flag pole n feet tall using 4 types of flags ; red flags 2 feet high, or white blue, and yellow flags each 1 foot high.

(*a*) $a_n = 2a_n + 3a_{n-2}$ (*b*) $a_n = a_{n-1} + a_{n-2}$

(*c*) $a_n = 3a_{n-1} + a_{n-2}$ (*d*) None of these.

5. Find a recurrence relation for the number of n-digit ternary sequences that have an even number of 0's

(*a*) $a_n = a_{n-1} + 3^{n-1}$ (*b*) $a_n = a_{n-1} + 3^n$

(*c*) $a_n = a_n + 3^n$ (*d*) None of these.

6. Suppose a coin is flipped until 2 heads appear and then the experiment stops. Find a recurrence relation for the number of experiments that end on the nth flip or sooner.

(*a*) $a_n = a_{n-1} + 2(n-1)$ (*b*) $a_n = a_{n-1} + (n-1)$

(*c*) $a_n = a_{n-1} + n$ (*d*) None of these.

7. The number of bacteria in a colony doubles every hour. If a colony begins with 5 bacteria, how many bacteria will be there after 3 hours, 6 hours? 1900 hours? after n hours? Find a recurrence relation to represent the same.

(*a*) $a_n = a_{n-1} + a_{n-2}$ (*b*) $a_n = 3a_{n-1} + 2a_{n-2}$

(*c*) $a_n = 2a_{n-1}$ (*d*) $a_n = a_{n-1} + 5$.

8. Suppose that a person deposits ₹ 10,000 in a saving account at a bank yielding 11% per year with interest compounded annually. How much amount will person have in the account after n years?

(*a*) $a_n = 1.11\, a_{n-1}$ (*b*) $a_n = 0.11\, a_{n-1}$

(*c*) $a_n = 1.10\, a_{n-1}$ (*d*) $a_n = 11.1\, a_{n-1}$.

9. Which of the following represent the sequence 1, 2, 5, 11, 26, ...

(*a*) $t_n = t_{n-1} + t_{n-2},\quad t_0 = 1, t_1 = 2$ (*b*) $t_n = t_{n-1} + 3t_{n-2},\quad t_0 = 1, t_1 = 2$

(*c*) $t_n = 2t_{n-1} + 1,\quad t_0 = 1, t_1 = 2$ (*d*) $t_n = 2t_{n-1} + 2,\quad t_0 = 1, t_1 = 2$.

10. There are n guests in a hall. Each person shakes hand with every body else exactly once. Let H_n = No. of handshakes that occur. The recurrence relation is

(*a*) $H_n = H_{n-1} + n$ (*b*) $H_n = H_{n-1} + (n-1)$

(*c*) $H_n = H_{n-1} + n - 2$ (*d*) None.

Answers and Explanations

1. (*d*).

2. (*a*) Here a_n = number of different ways of distributing prizes worth ₹ n.
a_{n-1} = number of different ways of destributing prizes worth ₹ 1 each.
a_{n-4} = number of different ways of distributing prizes worth ₹ 4 each.
∴ Required recurrence relation is $a_n = 3a_{n-1} + 5\, a_{n-4}$.

3. (*b*).

4. (*c*) Let a_n = number of ways of arrange flags on a flag pole n feet tall.
If red flags are used, then they are to be arranged on a flag pole 2 feet high
∴ number of ways to arrange such flags = a_{n-2} (As we need only 1 flag)
If white, blue or yellow flags are used, then they can be arranged on a flag pole 1 feet high.
Number of ways of arrange such flags = $3a_{n-1}$ (As we need 3 different flags)
The required recurrence relation is $a_n = 3a_{n-1} + a_{n-2}$.

5. (*a*).

6. (*b*).

7. (*c*) Initial number of bacteria = $a_0 = 5$

At the end of 1 hour, the number of bacteria doubles *i.e.*, $a_1 = 10 = 2 \times 5 = 2a_0$

At the end of 2 hours, $a_2 = 20 = 2 \times 10 = 2a_1$

At the end of 3 hours, $a_3 = 40 = 2 \times 20 = 2a_2$

$\vdots$

At the end of n hours, $a_n = 2a_{n-1}$.

8. (*a*) Initial amount A_0 = ₹ 10,000.

Now, amount after n years = Amount after (n – 1) years + Interest earned in n^{th} year

or $$A_n = A_{n-1} + (11\%)\, A_{n-1} = A_{n-1}\left(1 + \frac{11}{100}\right) = \frac{111}{100} A_{n-1} = 1.11\, A_{n-1}.$$

9. (*b*) Given sequence 1, 2, 5, 11, 26 represents

$t_0 = 1,\ t_1 = 2,\ t_2 = 5,\ t_3 = 11,\ t_4 = 26$

Now, $t_2 = 5 = 2 + 3 = t_1 + 3t_0$

$t_3 = 11 = 5 + 3 \times 2 = t_2 + 3t_1$

$t_4 = 26 = 11 + 3 \times 5 = t_3 + 3t_2$

$\vdots$

$t_n = t_{n-1} + 3t_{n-2}$.

10. (*b*) Here H_n denotes the number of handshakes that occur. Then $H_1 = 0$. For $n \geq 2$, Consider one of the guests, say, Mr. X., By definition, the number of handshakes made by remaining (n – 1) guests among themselves is H_{n-1}. Now, the guest Mr. X will shake-hands with (n – 1) guests, gives (n – 1) additional handshakes.

Hence

$$H_n = H_{n-1} + (n - 1),\ n \geq 2,\ H_1 = 0.$$

8 MONOIDS AND GROUPS

8.1. INTRODUCTION

In the present chapter, we introduce the concept of algebraic system, binary operations and groups. The study of cyclic groups, normal groups, group homomorphism etc. help us in understanding various applications of computer science. Groups play an important role in coding theory.

8.2. ALGEBRAIC STRUCTURE

If there exists a system such that it consists of a non-empty set and one or more operations on that set, then that system is called an algebraic system. It is generally denoted by (A, op_1, op_2, ..., op_n), where A is a non-empty set and op_1, op_2, ..., op_n are operations on A.

An algebraic system is also called an **algebraic structure** because the operations on the set A define a structure on the elements of A.

8.3. BINARY OPERATION

Consider a non-empty set A and a function f such that $f : A \times A \to A$, then f is called a binary operation on A whose domain is the set of ordered pairs of elements of A. If $*$ is a binary operation on A, then it may be written as $a * b$.

A binary operation can be denoted by any of the symbols $+$, $-$, $*$, $\oplus$, Δ, $\square$, $\vee$, $\wedge$ etc.

The value of the binary operation is denoted by placing the operator between the two operands.

e.g., (*i*) The operation of addition is a binary operation on the set of natural numbers.

(*ii*) The operation of subtraction is a binary operation on set of integers. But, the operation of subtraction is not a binary operation on the set of natural numbers because the subtraction of two natural numbers may or may not be a natural number.

(*iii*) The operation of multiplication is a binary operation on the set of natural numbers, set of integers and set of complex numbers.

(*iv*) The operation of set union is a binary operation on the set of subsets of a universal set. Similarly, the operation of set intersection is a binary operation on the set of subsets of a universal set.

8.4. TABLES OF OPERATION

Consider a non-empty finite set A = $\{a_1, a_2, a_3, ..., a_n\}$. A binary operation * on A can be described by means of table as shown below:

*	a_1	a_2	a_3		a_n
a_1	a_1*a_1	a_1*a_2			
a_2	a_2*a_1	a_2*a_2			
a_3			a_3*a_3		
a_n					a_n*a_n

The empty cell in the jth row and k^{th} column represent the element a_j*a_k.

ILLUSTRATIVE EXAMPLES

Example 1. *Consider the set A = {1, 2, 3} and a binary operation * on the set A defined by $a * b = 2a + 2b$. Represent operation * as a table on A.*

Sol. The table of operation is shown below: (Table 8.1)

Table 8.1

*	1	2	3
1	4	6	8
2	6	8	10
3	8	10	12

8.5. PROPERTIES OF BINARY OPERATIONS

There are many properties of the binary operations which are as follows :

1. Closure Property. Consider a non-empty set A and a binary operation * on A. Then A is closed under the operation *, if $a * b \in$ A, where a and b are elements of A.

For example, the operation of addition on the set of integers is a closed operation. *i.e.,* if $a, b \in$ Z, then $a + b \in$ Z $\forall\ a, b \in$ Z.

Example 2. *Consider the set A = {– 1, 0, 1}. Determine whether A is closed under (i) addition (ii) multiplication.*

Sol. (*i*) The sum of the elements is (– 1) + (– 1) = – 2 and 1 + 1 = 2 does not belong to A. Hence A is not closed under addition.

(*ii*) The multiplication of every two elements of the set are

$-1 * 0 = 0$; $\quad -1 * 1 = -1$; $\quad -1 * -1 = 1$

$0 * -1 = 0$; $\quad 0 * 1 = 0$; $\quad 0 * 0 = 0$

$1 * -1 = -1$; $\quad 1 * 0 = 0$; $\quad 1 * 1 = 1$

Since, each multiplication belongs to A hence A is closed under multiplication.

Example 3. *Consider the set A = {1, 3, 5, 7, 9, ...}, the set of odd +ve integers. Determine whether A is closed under (i) addition (ii) multiplication.*

Sol. (*i*) The set A is not closed under addition because the addition of two odd numbers produces an even number which does not belong to A.

(*ii*) The set A is closed under the operation multiplication because the multiplication of two odd numbers produces an odd number. So, for every a, $b \in$ A, we have $a * b \in$ A.

2. Associative Property. Consider a non-empty set A and a binary operation $*$ on A. Then the operation $*$ on A is associative, if for every a, b, c, $\in$ A, we have $(a * b) * c = a * (b * c)$.

Example 4. (*a*) *Consider the binary operation $*$ on Q, the set of rational numbers, defined by*

$$a * b = a + b - ab \ \forall \ a, b \in Q.$$

Determine whether $*$ is associative.

(*b*) *Consider the binary operation $*$ on the set N of positive integers defined by*

$$a * b = a^b$$

Determine whether $*$ is associative?

Sol. (*a*) Let us assume some elements a, b, $c \in$ Q, then by definition

$$(a * b) * c = (a + b - ab) * c = (a + b - ab) + c - (a + b - ab)c$$
$$= a + b - ab + c - ca - bc + abc = a + b + c - ab - ac - bc + abc.$$

Similarly, we have

$$a * (b * c) = a + b + c - ab - ac - bc + abc$$

Therefore, $(a * b) * c = a * (b * c)$.

Hence $*$ is associative.

(*b*) $*$ will be associative if

$$a * (b * c) = (a * b) * c \ \forall \ a, b, c \in \text{N}$$

Take $a = 2$, $b = 2$, $c = 3$ and consider

$$a * (b * c) = 2 * (2 * 3) = 2 * 2^3 = 2 * 8 = 2^8 = 256$$

and $(a * b) * c = (2 * 2) * 3 = 2^2 * 3 = 4 * 3 = 4^3 = 64$

Hence $a * (b * c) \neq (a * b) * c$

$\therefore$ $*$ is non-associative.

3. Commutative Property. Consider a non-empty set A and a binary operation $*$ on A. Then the operation $*$ on A is commutative, if for every a, $b \in$ A, we have $a * b = b * a$.

Example 5. (*a*) *Consider the binary operation $*$ on Q, the set of rational numbers, defined by*

$$a * b = a^2 + b^2 \ \forall \ a, b \in Q.$$

Determine whether $*$ is commutative.

(*b*) *Consider S = {a, b, c, d} and $*$ be a binary operation on S defined by as shown in the following table.*

Table 8.2

*	*a*	*b*	*c*	*d*
a	*a*	*b*	*c*	*d*
b	*b*	*a*	*a*	*b*
c	*c*	*b*	*a*	*a*
d	*d*	*a*	*a*	*a*

*Determine (i) whether * is associative ?*

*(ii) whether * is commutative?*

Sol. (*a*) Let us assume some elements $a, b \in Q$, then by definition

$$a * b = a^2 + b^2 = b^2 + a^2 = b * a$$

Hence * is commutative.

(*b*) (*i*) Let $a, b, c \in S$ and consider

$$b * (c * c) = b * a = b$$

and $$(b * c) * c = a * c = c$$

$\Rightarrow$ $$b * (c * c) \neq (b * c) * c$$

Thus, * is non-associative

(*ii*) $b * c = a$ and $c * b = b$

$\Rightarrow b * c * c * b$

$\therefore$ * is non-commutative

Example 6. *Consider the binary operation * and Q, the set of rational numbers defined by*

$$a * b = \frac{ab}{2} \quad \forall\, a, b \in Q.$$

*Determine whether * is (i) associative (ii) commutative.*

Sol. (*i*) Let $a, b \in Q$, then we have

$$a * b = \frac{ab}{2} = \frac{ba}{2} = b * a$$

Hence * is commutative.

(*ii*) Let $a, b, c \in Q$, then by definition we have

$$(a * b) * c = \left(\frac{ab}{2}\right) * c = \frac{\frac{ab}{2}.c}{2} = \frac{abc}{4}$$

Similarly, $$a * (b * c) = a * \left(\frac{bc}{2}\right) = \frac{\frac{abc}{2}}{2} = \frac{abc}{4}$$

Therefore, $$a * (b * c) = a * (b * c)$$

Hence, * is associative.

4. Identity. Consider a non-empty set A and a binary operation * on A. Then the operation * has an identity property if there exists an element, *e*, in A such that

$$a * e \text{ (right identity)} = e * a \text{(left identity)} = a \;\forall\, a \in A.$$

Theorem I. *Prove that $e_1' = e_1''$ where e_1' is a right identity and e_1'' is a left identity of a binary operation.*

Proof. We know that e_1'' is a right identity.

Hence, $$e_1'' * e_1' = e_1'' \qquad ...(1)$$

Also, we know that e_1'' is a left identity.

Hence, $e_1'' * e_1' = e_1'$ ...(2)

From (1) and (2), we have $e_1' = e_1''$.

Thus, we can say that if e is a right identity of a binary operation, then e is also a left identity.

Example 7. *Consider the binary operation * on I_+, the set of positive integers defined by $a * b = \frac{ab}{2}$. Determine the identity for the binary operation *, if exists.*

Sol. Let us assume that e be a +ve integer number, then

$$e * a = a, a \in I_+$$

$$\Rightarrow \quad \frac{ea}{2} = a \Rightarrow e = 2 \quad \text{...(1)}$$

Similarly,

$$a * e = a, a \in I_+$$

$$\frac{ae}{2} = a \text{ or } e = 2 \quad \text{...(2)}$$

Form (1) and (2) for $e = 2$, we have $e * a = a * e = a$

Therefore, 2 is the identity element for *.

5. Inverse. Consider a non-empty set A and a binary operation * on A. Then operation * has the inverse property if for each $a \in A$, there exists an element b in A such that

$a * b$ (right inverse) $= b * a$ (left inverse) $= e$, where b is called an inverse of a.

6. Idempotent. Consider a non-empty set A and a binary operation * on A. Then the operation * has the idempotent property, if for each $a \in A$, we have

$$a * a = a \; \forall \; a \in A.$$

7. Distributivity. Consider a non-empty set A and two binary operations * and + on A. Then the operation * distributes over +, if for every $a, b, c \in A$, we have

$$a * (b + c) = (a * b) + (a * c) \quad \text{[Left distributivity]}$$

and

$$(b + c) * a = (b * a) + (c * a) \quad \text{[Right distributivity]}$$

8. Cancellation. Consider a non-empty set A and a binary operation * on A. Then the operation * has the cancellation property, if for every $a, b, c \in A$, we have

$$a * b = a * c \Rightarrow b = c \quad \text{[Left cancellation]}$$

and

$$b * a = c * a \Rightarrow b = c \quad \text{[Right cancellation]}$$

8.6. SEMI-GROUP

(P.T.U. B.Tech. Dec. 2009, May 2008)

Let us consider, an algebraic system (A, *), where * is a binary operation on A. Then, the system (A, *) is said to be a semi-group if it satisfies the following properties :

1. The operation * is a closed operation on set A.
2. The operation * is an associative operation.

Example 8. *Consider an algebraic system (A, *), where A = {1, 3, 5, 7, 9, ...}, the set of all positive odd integers and * is a binary operation means multiplication. Determine whether (A, *) is a semi-group.*

Sol. Closure property. The operation $*$ is a closed operation because multiplication of two +ve odd integers is a +ve odd number.

Associative property. The operation $*$ is an associative operation on set A. Since for every $a, b, c \in$ A, we have

$$(a * b) * c = a * (b * c)$$

Hence, the algebraic system (A, $*$) is a semi-group.

Example 9. *Consider the algebraic system ({0, 1}, $*$), where $*$ is a multiplication operation. Determine whether ({0, 1}, $*$) is a semi-group.*

Sol. Closure property. The operation $*$ is a closed operation on the given set since

$$0 * 0 = 0 \,;\, 0 * 1 = 0 \,;\, 1 * 0 = 0 \,;\, 1 * 1 = 1.$$

Associative property. The operation $*$ is associative since we have

$$(a * b) * c = a * (b * c) \ \forall \ a, b, c$$

Since, the algebraic system is closed and associative. Hence, it is a semi-group.

Example 10. *Let S be a semi-group with an identity element e and if b and b′ are inverses of an element $a \in$ S, then $b = b'$ i.e., inverse are unique, if they exist.*

Sol. Given b is an inverse of a, therefore, we have

$$a * b = e = b * a$$

Also, b' is an inverse of a, therefore, we have

$$a * b' = e = b' * a$$

Consider $\quad b * (a * b') = b * e = b$ ...(1)

and $\quad (b * a) * b' = e * b' = b'$ ...(2)

Now, S is a semi-group, associativity holds in S *i.e.*, $b * (a * b') = (b * a) * b'$

$\Rightarrow \quad b = b'$. | Using (1) and (2)

Example 11. *Let N be the set of positive integers and let $*$ be the binary operation of least common multiple (L.C.M) on N. Find*

(a) 4 $*$ 6, 3 $*$ 5, 9 $*$ 18, 1 $*$ 6

(b) Is (N, $*$) a semi-group

(c) Is N commutative

(d) Find the identity element of N

(e) Which elements of N have inverses ?

Sol. (*a*) Let $x, y \in$ N and $x * y$ = L.C.M. of x and y

$\therefore$
4 $*$ 6 = L.C.M. of 4 and 6 = 12
3 $*$ 5 = L.C.M. of 3 and 5 = 15
9 $*$ 18 = L.C.M. of 9 and 18 = 18
1 $*$ 6 = L.C.M. of 1 and 6 = 6

(*b*) We know that the operation of L.C.M. is associative, *i.e.*,

$$a * (b * c) = (a * b) * c \quad \forall \ a, b, c \in \text{N}$$

$\therefore$ N is a semi-group under $*$.

(*c*) Also for $a, b \in$ N,

$a * b$ = L.C.M. of a and b = L.C.M. of b and $a = b * a$

$\therefore$ N is commutative also.

(*d*) For $a \in N$, consider $a * 1 = \text{L.C.M. of } a \text{ and } 1 = a$

Also, $1 * a = \text{L.C.M. of } 1 \text{ and } a = a$

$\therefore$ $a * 1 = a = 1 * a$

i.e., 1 is the identity element of N.

(*e*) Consider $a * b = 1$ *i.e.*, L.C.M. of a and b is 1, which is possible iff $a = 1$ and $b = 1$. *i.e.*, the only element which has an inverse is 1 and it is its own inverse.

Example 12. *Consider the set Q of rational numbers and let * be the operation on Q defined by* $a * b = a + b - ab$

(*a*) *Find* $3 * 4$, $2 * (-5)$, $7 * \frac{1}{2}$

(*b*) *Is* (Q, *) *a semi-group ?*

(*c*) *Is Q commutative ?*

(*d*) *Find the identity element of Q.*

(*e*) *Which elements of Q have inverses and what are they ?*

Sol. Given $a * b = a + b - ab$ for $a, b \in Q$

(*a*)
$$3 * 4 = 3 + 4 - 12 = -5$$
$$2 * (-5) = 2 + (-5) - (-10) = 2 - 5 + 10 = 7$$
$$7 * \frac{1}{2} = 7 + \frac{1}{2} - \frac{7}{2} = 4.$$

(*b*) Q will be a semi-group if it holds associativity under * for $a, b, c \in Q$.

Consider
$$a * (b * c) = a * (b + c - bc)$$
$$= a + (b + c - bc) - a(b + c - bc)$$
$$= a + b + c - bc - ab - ac + abc \quad ...(1)$$

Also,
$$(a * b) * c = (a + b - ab) * c$$
$$= a + b - ab + c - (a + b - ab)\, c$$
$$= a + b + c - ab - ac - bc + abc$$
$$= a + b + c - bc - ab - ac + abc \quad ...(2)$$

From (1) and (2),
$$a * (b * c) = (a * b) * c$$

Hence, (Q, *) is a semi-group.

(*c*) For $a, b \in Q$

Consider
$$a * b = a + b - ab = b + a - ba = b * a$$

$\therefore$ Q is commutative.

(*d*) Let e is the identity element of Q, therefore, for $a \in Q$, we have
$$a * e = a$$
$$\Rightarrow \quad a + e - ae = a$$
$$\Rightarrow \quad e - ea = 0$$
$$\Rightarrow \quad e(1 - a) = 0$$
$$\Rightarrow \quad e = 0 \text{ if } a \neq 1$$

$\therefore$ The identity of Q is 0.

(*e*) If x is the inverse of $a \in Q$, then $a * x = 0$ (identity)

$\Rightarrow \quad a + x - ax = 0$

$\Rightarrow \quad a + x(1 - a) = 0$

$\Rightarrow \quad a = x(a - 1)$

$\Rightarrow \quad x = \dfrac{a}{a-1}, a \neq 1$

Thus a has an inverse $\dfrac{a}{a-1}$.

Example 13. *Consider a non-empty set S with the operation* $a * b = a$

(*a*) *Is the operation associative ?*

(*b*) *Is the operation commutative ?*

(*c*) *Show that the right cancellation law holds.*

(*d*) *Does the left cancellation law hold ?*

Sol. (*a*) For $a, b, c \in S$,

Consider

$$a * (b * c) = a * b = a$$

and $\quad (a * b) * c = c * a = a$

$\therefore$ $*$ is associative.

(*b*) For $a \neq b \in S$,

Consider

$$a * b = a \quad \text{and} \quad b * a = b$$

$\Rightarrow \quad a * b \neq b * a$

$\therefore$ $*$ is not commutative.

(*c*) For $a, b, c \in Q$,

Consider

$$a * c = b * c$$

$\Rightarrow \quad a = b$ | Using given $a * b = a$

$\therefore$ Right cancellation laws hold.

(*d*) The left cancellation law does not hold. For example, suppose $b \neq c$, then

$$a * b = a * c$$

$\Rightarrow \quad b = c$, a contradiction

Hence, the result.

Example 14. *Let* $(A, *)$ *be semi-group. Show that for a, b, c in A, if* $a * c = c * a$ *and* $b * c = c * b$, *then* $(a * b) * c = c * (a * b)$.

Sol. Take L.H.S., we have

$(a * b) * c = a * (b * c)$ $\quad [\because\ *$ is associative$]$

$= a * (c * b)$ $\quad [\because\ b * c = c * b]$

$= (a * c) * b$ $\quad [\because\ *$ is associative$]$

$= (c * a) * b$ $\quad [\because\ a * c = c * a]$

$= c * (a * b)$ $\quad [\because\ *$ is associative$]$

Hence, $\quad (a * b) * c = c * (a * b)$.

8.7. SUBSEMI-GROUP

Let A be a non-empty subset of a semi group S. Then A is called a subsemigroup of S if A is itself a semigroup with respect of the same operation on S. Since the elements of A are also elements of S, the associative law automatically holds for the elements of A. Therefore, A is a subsemigroup of S iff A closed under the operation of S.

Example 15. *Let A and B are the sets of even and odd positive intergers. Then (A, ×) and (B, ×) are subsemigroups of (N, +). Is (A, +) subsemigroup of (N, +) ? Is (B, +) subsemigroup of (N, +) ?*

Sol. Given A = {set of even positive integers}

B = {set of odd positive integers}

Also N is a semigroup under addtion and multiplication. Let $m, n \in A$ $\Rightarrow$ $m \times n \in A$ *i.e.*, A is closed under multiplication. Therefore, A is a subsemigroup of N under multiplication.

Similarly, B is also closed under multiplication. Therefore, B is also a subsemigroup of N under multiplication.

Further, sum of two even positive integers is also a positive even integer. Therefore, A is also closed under addition. Hence A is a subsemigroup of (N, +).

But sum of two positive odd integers is not always an odd integer (*e.g.*, 5 + 3 = 8 (even))

$\therefore$ B is not closed under +. Hence B is not a subsemigroup of (N, +).

For example, Consider a semi-group (N, +), where N is the set of all natural numbers and + is an addition operation. The algebraic system (E, +) is a subsemi-group of (N, +), where E is a set of all even natural number.

8.8. FREE SEMI-GROUP

Consider a non-empty set A. A *word* w on A is a finite sequence of its elements.

For example, if we take A = $\{a, b, c\}$.

Then, the elements *abcabca*, *aabbcc*, *baccaaaa*, etc. are *words* on A.

Also, the *length* of a *word* w, denoted by $l(w)$, is the number of elements in w.

Consider the following words

$$u_1 = abcabca,\ u_2 = aabbcc,\ u_3 = baccaaaa$$

Here $l(u_1)$ = the length of the word u_1

= number of elements in u_1

= 7

Similarly, $l(u_2) = 6,\ l(u_3) = 8$

We now define *concatenation* of two or more words on the set A. By *concatenation* of words u_1 and u_2, written as $u_1 * u_2$ or u_1u_2, we mean the word obtained by writing down the elements of u_1 followed by the elements of u_2. Thus,

$$u_1 * u_2 = u_lu_2 = (abcabca)\ (aabbcc)$$
$$= abcabca^3b^2c^2$$

Here

$$l(u_1u_2) = \text{length of the word } u_1u_2$$
$$= \text{number of elements in the word } u_1u_2$$
$$= 13$$

Also, $l(u_1) = 7$ and $l(u_2) = 6$

$\Rightarrow$ $l(u_1) + l(u_2) = 7 + 6 = 13$

Thus, $l(u_1u_2) = l(u_1) + l(u_2)$

Let F denotes the collection of all words on a given set A under the operation of concatenation. Then, for $u_1, u_2 \in$ F, we have proved that $u_1u_2 \in$ F.

Also, for $u_1, u_2, u_3 \in$ F, the number of elements in the words $u_1(u_2u_3)$ and $(u_1u_2)\, u_3$ will be same. Hence, we can say that F is a semigroup under the operation of *concatenation* and this semigroup is called *free semigroup* on A and the elements of A are called *generators* of F.

8.9. CONGRUENT RELATIONS AND QUOTIENT STRUCTURES

Suppose R is an equivalence relation on a given set S and let $a \in$ S. Then,

$$[a] = \text{Equivalence class of } a \text{ in S}$$
$$= \{x \in S: (a, x) \in R\}$$

Thus, $[a]$ denotes the set of all elements of S to which a is related under R. Further, the collection of all equivalence classes of elements of S under an equivalence relation R is denoted by S/R. Thus,

$$S/R = \{[a] : a \in S\}$$

where S/R (read as the quotient space of S by R) is called the **quotient set.**

Further, suppose that the equivalence relation R on S has the following property:

If $a\text{R}a', b\text{R}b' \Rightarrow ab\text{R}a'b' \;\forall\, a, b \in S$

Then R is called a *congruence* relation on S.

We now show that S/R forms a *semigroup* under the operation $*$ on the equivalence classes, defined by

$$[a] * [b] = [a * b] \text{ or } [a]\,[b] = [ab]$$

Since S/R is closed under the operation $*$ on the equivalence classes (by definition), it remains to show S/R is associative

Consider $([a] * [b]) * [c] = [a * b] \times [c]$
$= [a * b * c]$

Also $[a] * ([b] * [c]) = [a] * [b * c]$
$= [a * b * c]$

Thus, $[a] * ([b] * [c]) = ([a] * [b]) * [c]$

Hence, associativity holds in S/R.

Thus, S/R forms a semigroup under the operation * on the equivalence classes.

This, semi-group S/R is called *quotient of S by R.*

Example 16. *Let F be a free semi-group on a set A and R be an equivalence relation on F defined by "uRu′ if l(u) = l(u′)"*

Show that R is a congruence relation on F.

Sol. Let $u\text{R}u'$, $v\text{R}v'$. then

$$l(u) = l(u') = m, \text{ say and } l(v) = l(v') = n, \text{ say}$$

Consider $l(uv) = l(u) + l(v) = m + n$

and $l(u'v') = l(u') + l(v') = m + n \Rightarrow uvRu'v'$

Thus R is a congruence relation of F.

Example 17. *Let Z be the set of integers and m > 1 be any positive integer. Consider a relation "congruence modulo m" defined by*

$a \equiv b \pmod{m}$ if $m / a - b \ \forall \ a, b \in Z$ (read as "a is congruent to b modulo m if m divides $(a - b)$").

Show that '≡' is a congruence relation on Z.

Sol. We first show that "congruence modulo m" is an equivalence relation on Z.

We know that for any integer $a \in$ Z, the difference $a - a = 0$ is divisible by m where $m > 1$ is any positive integer.

Hence $a \equiv a \pmod{m}$, which proves that '≡' is reflexive

Also, if $a \equiv b \pmod{m}$, then $m/(a - b) \ \forall \ a, b \in$ Z

$$\Rightarrow \quad m/-(a - b) = b - a \Rightarrow m/(b - a)$$

$$\Rightarrow \quad b \equiv a \pmod{m}$$

which proves that '≡' is symmetric.

Further, if $a \equiv b \pmod{m}$, $b \equiv c \pmod{m}$,

we show $a \equiv c \pmod{m} \ \forall \ a, b, c \in$ Z

Now $a \equiv b \pmod{m} \Rightarrow m/(a - b) \ \forall, a, b \in$ Z

$b \equiv c \pmod{m} \Rightarrow m/(b - c) \ \forall, b, c \in$ Z

which further implies that $m/(a - b) + (b - c) = a - c$

$$\Rightarrow \quad m/(a - c) \Rightarrow a \equiv c \pmod{m}$$

Hence the relation '≡' is an equivalence relation on Z.

Finally, we show that the equivalence relation '≡' is a congruence relation on Z.

Consider $a \equiv c \pmod{m}$ and $b \equiv d \pmod{m}$, we show $ab \equiv cd \pmod{m}$.

Now $a \equiv c \pmod{m} \Rightarrow m/(a - c) \Rightarrow m/b(a - c)$

Also, $b \equiv d \pmod{m} \Rightarrow m/(b - d) \Rightarrow m/c(b - d) \ \forall \ a, b, c \in$ Z

Hence $m/b(a - c)$ and $m/c(b - d) \ \forall \ a, b, c, d \in$ Z

$$\Rightarrow \quad m/b(a - c) + c(b - d) = ba - bc + cb - cd$$

$$\Rightarrow \quad m/ba - cd = ab - cd \Rightarrow m/ab - cd$$

$$\Rightarrow \quad ab \equiv cd \pmod{m}.$$

Hence the result.

Example 18. *Let (I, +) be a semi-group and R is an equivalence relation on I defined by*

aRb iff $a \equiv b$ (MOD 3).

Sol. If a and b yield the same remainder when divided by 3, then we have 3 divides $a - b$ *i.e.*, $3/a - b$.

Now, if $a \equiv b$ (MOD 3) and $c \equiv d$ (MOD 3), then 3 divides $a - b$ and 3 divides $c - d$.

Thus, we can write $a - b = 3m$ ...(1)

and $c - d = 3n$ ...(2)

| $a/b \Rightarrow b = at$ for some t

Here m and n are some integers of I.

Adding (1) and (2), we have

$$(a - b) + (c - d) = 3m + 3n \quad \text{or} \quad (a + c) - (b + d) = 3(m + n)$$

$\Rightarrow$ 3 divides $(a + c) - (b + d)$

or $a + c \equiv (b + d)$ (MOD 3)

Thus, the relation is a congruence relation.

Example 19. *Consider the set A = {a, b}. Let (A*, ·) is the semi-group generated by A, also let R is a relation on A defined by $\alpha R \beta$ iff α and β have the same number of a's.*

Show whether the relation R is a congruence on (A, ·).*

Sol. First of all we will show that R is an equivalence relation. So, for that we will check reflexive, symmetric and transitive properties of the relation R.

Reflexive. $\alpha R \alpha$ for any $\alpha \in A^*$ since α has same number of *a*'s as itself. Thus, R is reflexive.

Symmetric. If α and β have same number of *a*'s, then $\alpha R \beta$ or we can say $\beta R \alpha$. Thus, R is symmetric.

Transitive. If $\alpha R \beta$, it means α and β have same number of *a*'s. If $\beta R \gamma$, it means β and γ have same number of *a*'s. It implies α and γ have same number of *a*'s *i.e.*, $\alpha R \gamma$. Thus, R is transitive.

Hence, R is an equivalence relation.

To show that R is a congruence relation, let us assume that $\alpha R \alpha_1$ and $\beta R \beta_1$. It means α and α_1 have same number of *a*'s and β and β_1 have same number of *a*'s. We know that the number of *a*'s in $\alpha . \beta$ is the sum of number of *a*'s in α and the number of *a*'s in β.

From the above discussion, we can say that the number of *a*'s in $\alpha . \beta$ is same as in $\alpha_1 . \beta_1$.

Hence $(\alpha . \beta) R (\alpha_1 . \beta_1)$ which shows that R is a congruence relation.

Example 20. *Consider the semi-group (I, +), where + is an addition operation. Let $f(x) = x^2 - 2x - 3$ and also let R is a relation on I defined by aRb iff f(a) = f(b). Show whether R is a congruence relation.*

Sol. We first show that the relation R is an equivalence relation on the set I.

(*i*) $f(a) = f(a) \Rightarrow aRa$ *i.e.*, R is symmetric

(*ii*) $aRb \Rightarrow f(a) = f(b) \Rightarrow f(b) = f(a)$ *i.e.*, bRa. Hence R is symmetric

(*iii*) If aRb, bRc, then $f(b) = f(b)$ and $f(b) = f(c) \Rightarrow f(a) = f(c) \Rightarrow aRc$ *i.e.*, R is transitive

To check whether R is congruence relation or not, we will try to find two pair of numbers aRb and cRd but $(a + b) \not R (c + d)$, if possible. Then we will say R is not a congruence relation.

Thus, we have

2R0 *i.e.*, $f(2) = f(0) = -3$

and $-2R4$ *i.e.*, $f(-2) = f(4) = 5$

But $(2 + (-2)) \not R (0 + 4)$ *i.e.*, $0 \not R 4$

As $f(0) = -3$ and $f(4) = 5$

Hence, R is not a congruence relation.

Example 21. *Let (S,*) be a commutative semi-group. Show that if $x * x = x$ and $y * y = y$, then $(x * y) * (x * y) = x * y$.*

Sol. Take L.H.S. $(x * y) * (x * y)$

$= (x * y) * (y * x)$ [∵ (S, *) is a commutative semi-group]

$= x * y * y * x = x * y * x$ [∵ $y * y = y$]

$= x * x * y$ [∵ Commutative semi-group]

$= x * y$ [∵ $x * x = x$]

Hence, $= (x * y) * (x * y) = x * y$.

Example 22. *Let ({x, y}, .) be a semi-group where x . x = y show that*

(i) $x \cdot y = y \cdot x$ *(ii)* $y \cdot y = y$.

Sol. *(i)* To show that $x \cdot y = y \cdot x$

We have $x \cdot x \cdot x = x \cdot x \cdot x \Rightarrow x \cdot y = y \cdot x$ $[\because x \cdot x = y]$

Hence proved.

(ii) To show that $y \cdot y = y$

We know that the set (x, y) is closed under the operation. Therefore, we have two cases

$x \cdot y = x, \quad x \cdot y = y$

Case. I Let $x \cdot y = x$

Consider
$$\begin{aligned} y \cdot y &= y \cdot (x \cdot x) = (y \cdot x) \cdot x && [\because \text{ . is associative}] \\ &= (x \cdot y) \cdot x && [\because x \cdot y = y \cdot x] \\ &= x \cdot x && [\because x \cdot y = x] \\ &= y \end{aligned}$$

Case. II Let $x \cdot y = y$

Consider
$$\begin{aligned} y \cdot y &= (x \cdot x) \cdot y && [\because x \cdot x = y] \\ &= x \cdot (x \cdot y) && [\because \text{ . is associative}] \\ &= x \cdot y && [\because x \cdot y = y] \\ &= y \end{aligned}$$

Hence proved.

Example 23. *Let (A, ∗) be a semi-group. Further more, for every a and b in A, if* $a \neq b$ *then* $a * b \neq b * a$.

(a) Show that for every a in A, $a * a = a$

(b) Show that for every a, b in A, $a * b * a = a$

(c) Show that for every a, b, c in A, $a * b * c = a * c$.

Sol. *(a)* We know that A is a semi-group.

$$\therefore \quad (a * b) * c = a * (b * c)$$

Now putting $b = a$ and $c = a$, we have

$$(a * a) * a = a * (a * a)$$

Since A is not commutative semi-group. *i.e.*, $a * b \neq b * a$,

Hence $a * a = a$ ...(1)

(b) Let us assume that $b \in A$, then we have

$$b * b = b$$

Multiplying both sides by *a*, we get

$$a * b * b = a * b \quad \text{or} \quad (a * b) * b = a * b \qquad [\because * \text{ is associative}]$$

Hence, $a * b = a$ ...(2)

So,
$$\begin{aligned} a * b * a &= (a * b) * a && [\because * \text{ is associative}] \\ &= a * a && [\because a * b = a \text{ from (2)}] \\ &= a && [\because a * a = a \text{ from (1)}] \end{aligned}$$

(c) We know that
$$\begin{aligned} a * b * c &= (a * b) * c && [\because * \text{ is associative}] \\ &= a * c && [\because a * b = a \text{ from (2)}] \end{aligned}$$

8.10. HOMOMORPHISM OF SEMI-GROUPS

Let (S, $*$) and (S′, $*'$) be any two semi-groups. Then, a function f from S to S′ is called a *semi-group* homomorphism (or a homomorphism) if

$$f(a * b) = f(a) *' f(b) \ \forall \ a, b \in S$$

In addition, if f is one-one and onto, then f is called an isomorphism of S and S′ and we write $S \cong S'$. Also, S and S′ are said to isomorphic subgroups.

Example 24. *Consider G = {1, 3, 7, 9} under multiplication modulo 10 and f: $Z_4 \to G$ be a function defined by*

$$f(0) = 1, f(1) = 3, f(2) = 9, f(3) = 7$$

Show that f is a semi-group homomorphism. Also prove that $Z_4 \cong G$.

Sol. By definition

$$Z_4 = \{0, 1, 2, 3, +_4\}$$
$$G = \{1, 3, 7, 9, \times_{10}\}$$

We first show that $(Z_4, +_4)$ and $(G, \times_{10})$ are both semi-groups under the addition modulo 4 and multiplication modulo 10 respectively.

The addition modulo table (Table 2*a*) for Z_4 and the multiplication modulo table (Table 2*b*) for G are shown below.

Table 2

$+_4$	0	1	2	3
0	0	1	2	3
1	1	2	3	0
2	2	3	0	1
3	3	0	1	2

(*a*)

$\times_{10}$	1	3	7	9
1	1	3	7	9
3	3	9	1	7
7	7	1	9	3
9	9	7	3	1

(*b*)

From Table 2*a*, we observe that each element in the interior of the table is also on element of Z_4. It means Z_4 is closed under $+_4$.

Also, for $a, b, c \in Z_4$, we know

$$a +_4 (b +_4 c) = (a +_4 b) +_4 c \text{ is true}$$

i.e., associativity holds in Z_4.

Hence, $(Z_4, +_4)$ is a semi-group

Similarly, we can prove that $(G, \times_{10})$ is a semigroup.

To show f is a semi-group homomorphism. We need to show

$$f(a +_4 b) = f(a) \times_{10} f(b) \ \forall \ a, b \in Z_4$$

We first find $(a +_4 b)$

Now, the only elements in Z_4 are 0, 1, 2 or 3. Therefore, the value of $a +_4 b$ can be either 0 or 1 or 2 or 3 (since Z_4 is closed under $+_4$). Consider the following possibilities:

(1) If the value of $a +_4 b$ is 0

(2) If the value of $a +_4 b$ is 1

(3) If the value of $a +_4 b$ is 2

(4) If the value of $a +_4 b$ is 3

We discuss the case (1) when the value of $a +_4 b$ is 0.

From Table 2, we observe that a and b can take the following possible values.

(*i*) $a = 0, b = 0$ so that $a +_4 b = 0 \Rightarrow f(a +_4 b) = f(0) = 1$

(*ii*) $a = 1, b = 3$ so that $a +_4 b = 1 + 3 = 4 = 0 \Rightarrow f(a +_4 b) = f(0) = 1$

(*iii*) $a = 2, b = 2$ so that $a +_4 b = 2 + 2 = 4 = 0 \Rightarrow f(a +_4 b) = f(0) = 1$

(*iv*) $a = 3, b = 1$ so that $a +_4 b = 3 + 1 = 4 = 0 \Rightarrow f(a +_4 b) = f(0) = 1$

We summarise the above obtained values in the following Table 3.

Table 3. Calculation of $f(a +_4 b)$

a	b	$a +_4 b$	$f(a +_4 b)$
0	0	0	1
1	3	0	1
2	2	0	1
3	1	0	1

Similarly, the various values of $f(a) \times_{10} f(b)$ are calculated in Table 4.

Table 4. Calculation of $f(a) \times_{10} f(b)$

a	b	$f(a)$	$f(b)$	$f(a) \times_{10} f(b)$
0	0	1	1	$1 \times_{10} 1 = 1$
1	3	3	7	$3 \times_{10} 7 = 21 = 1$
2	2	9	9	$9 \times_{10} 9 = 81 = 1$
3	1	7	3	$7 \times_{10} 3 = 21 = 1$

On comparing Table 3 and Table 4, we observe that the last column of these two tables are identitcal. It means

$$f(a +_4 b) = f(a) \times_{10} f(b) \; \forall \; a, b \in Z_4$$

Hence, f is a semi-group homomrphism

To show f is 1-1 and onto: Here $f : Z_4 \rightarrow G$ such that $f(0) = 1$, $f(1) = 3$, $f(2) = 9$, $f(3) = 7$. Consider the following (Fig. 8.1)

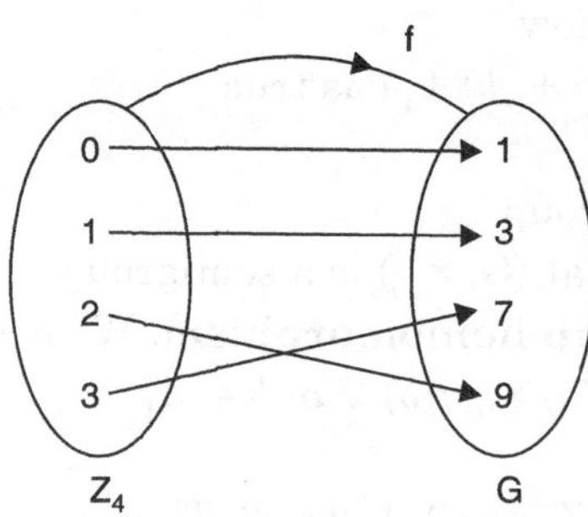

Fig. 8.1

Since different elements of Z_4 have different images in G, therefore, f is one-one.

Also, each element of G has its pre-image in Z_4. Therefore, f is onto also.

Thus, we have shown that f is homomorphism, one-one and onto. Thus, $S_4 \cong G$.

Example 25. *Let M be the set of 2 × 2 matrices over integers of the types* $\begin{pmatrix} a & b \\ c & d \end{pmatrix}$ *s.t.* $ad - bc \neq 0$.

Define $f : M \to M$ *such that* $f(A) = \det. A \ \forall\ A \in M$. *Then f is a semi-group homomorphism on (M, ×), but not on (M, +).*

Sol. We know that (M, +) and (M, ×) are both semi-groups. Let A, B ∈ M and consider

$$f(A + B) = \det. (A + B)$$
$$\neq \det . A + \det . B$$
$$\neq f(A) + f(B)$$

Thus, *f* is a not a semi-group homomorphism on (M, +).

Now, consider $f(AB) = \det. (AB)$
$$= (\det. A) (\det. B)$$
$$= f(A)\, f(B)$$

Thus, *f* is a semi-group homomorphism on (M, ×).

Example 26. *Let ~ be a congruence relation on a semi-group S. Let* $\phi : S \to S/\sim$ *be the natural mapping from "S to the quotient space of S by ~" (denoted by S/~) defined by*

$$\phi(a) = [a]$$

then ϕ *is a semi-group homomorphism.*

Sol. Let $a, b \in S$ and consider

$$\phi(ab) = [ab] = [a][b]$$
$$= \phi(a)\phi(b)$$

thus, ϕ is a semi-group homomorphism.

Remark. We know that S/R denotes the set of all equivalence classes of elements of S under an equivalence relation R. Also, if R is a congruence relation on S, then S/R forms a semi-group under the operation on the equivalence classes.

8.11 FUNDAMENTAL THEOREM OF SEMI-GROUP HOMOMORPHISM

Theorem. *Let* $f : S \to S'$ *be a semi-group homomorphism. Let ~ be a relation on S defined by "*$a \sim b$ *if* $f(a) = f(b)$*" then*

(i) ~ is a congruence relation on S

(ii) S/~ is isomorphic to f(S)

Proof. (*i*) We know that a relation R on a given set S is said to be a congruence relation on S if R is an equivalence relation on S and satisfies the following.

If aRa', bRb' ⇒ $abRa'b' \ \forall\ a, b \in S$

We first show that R is an equivalence relation on S.

(*a*) Reflexive. Since $f(a) = f(a)$ ⇒ $a \sim a \ \forall\ a \in S$

∴ ~ is reflexive

(*b*) Symmetric. Let $a \sim b$ ⇒ $f(a) = f(b)$ ⇒ $f(b) = f(a)$ ⇒ $b \sim a \ \forall\ a, b \in S$

∴ ~ is symmetric

(*c*) Transitive. Let $a \sim b$ ⇒ $f(a) = f(b) \ \forall\ a, b \in S$

Let $b \sim c$ ⇒ $f(b) = f(c) \ \forall\ b, c \in S$

It implies that $f(a) = f(b) = f(c)$

⇒ $f(a) = f(c)$ ⇒ $a \sim c \ \forall\ a, c \in S$

∴ ~ is transitive

Hence, ~ is an equivalence relation on S.

We next show that R is a congruence relation on S.

Let $a \sim a'$ and $b \sim b'$. Then

$$f(a) = f(a') \text{ and } f(b) = f(b')$$

Consider $f(ab) = f(a)\, f(b)$ | As f is a homomorphism

$$= f(a')\, f(b')$$

$$= f(a'b')$$

$\Rightarrow$ $ab \sim a'b'$

$\therefore$ ~ is a congruence relation on S

(*ii*) To show S/~ is isomorphic to f(S).

Define ϕ : S/~ $\rightarrow f$(S) by

$$\phi([a]) = f(a)$$

We show ϕ is well-defined, homomorphism, one-one and onto.

***(a)* ϕ is well-defined.** Consider $[a] = [b] \;\Rightarrow\; a \sim b$

$\Rightarrow$ $f(a) = f(b)$

$\Rightarrow$ $\phi([a]) = \phi([b])$

$\therefore$ ϕ is well-defined

***(b)* ϕ is homomorphism.** Consider $\phi([a]\,[b]) = \phi([ab])$

$$= f(ab)$$

$= f(a)\, f(b)$ | f is a semi-group homomorphism

$$= \phi([a])\, \phi([b])$$

$\therefore$ ϕ is homomorphism

***(c)* ϕ is one-one.** Let $\phi([a]) = \phi([b])$

$\Rightarrow$ $f(a) = f(b)$

$\Rightarrow$ $a \sim b$

$\Rightarrow$ $[a] = [b]$

$\therefore$ ϕ is one-one

***(d)* ϕ is onto.** Let $y \in f(S) \;\Rightarrow\; \exists\, a \in S$ such that $y = f(a)$

Now $\phi([a]) = f(a) = y$

Thus ϕ is onto

Hence, ϕ : S/~ $\rightarrow f$(S) is an isomomorphism

Example 27. *Let F be a free semi-group on A = {a, b}. Define f = F → Z such that f(u) = l(u) where l(u) denotes the length of the word 'u' in F. Let a ~ b if f(a) = f(b). Show that f is a homomorphism. Also, F/~ is isomorphic to N.*

Sol. Suppose $u_1, u_2 \in$ F and consider

$$f(u_1u_2) = l(u_1u_2) = l(u_1)\, l(u_2) = f(u_1)\, f(u_2)$$

$\therefore$ f is a homomorphism

Also, for $u \in$ F, f(F) = $l(u)$ = a natural number

$\therefore$ f(F) = N

By *fundamental theorem of semi-group homomorphism*, F/~ is isomorphic to f(F) = N

Thus F/~ is isomorphic to N

Example 28. *Let M be the set of all 2 × 2 type matrices over integers of the type* $\begin{pmatrix} a & 0 \\ 0 & 1 \end{pmatrix}$ *and let f : M → Z be a function defined by*

$$f(A) = \text{det. } A$$

Let A ~ B if f(A) = f(B). Then M/~ is isomorphic to Z.

Sol. We know that (M, ×) is a semi-group under the multiplication of matrices. Also (Z, ·) is a semi-group

Let A, B ∈ M and consider

$$f(AB) = \text{det. }(AB) = (\text{det. } A)\ (\text{det. } B)$$
$$= f(A)\ f(B)$$

Thus, f is a semi-group homomorphism from M to Z. By fundamental theorem of semi-group homomorphism, M/~ is isomorphic to f(Z).

But $$f(Z) = \{b \in Z: \text{There exists } A \in M \text{ for which } f(A) = b\}$$
$$= \{b \in Z: \text{det. } A = b\}$$
$$= Z$$

∴ M/~ is isomorphic to Z

8.12. MONOID (*P.T.U. B.Tech., Dec. 2013, May 2013, Dec. 2012, May 2008*)

Let us consider an algebraic system (A, *o*), where *o* is a binary operation on A. Then the system (A, *o*) is said to be a monoid if it satisfies the following properties.

(*i*) The operation *o* is a closed operation on set A.

(*ii*) The operation *o* is an associative operation.

(*iii*) There exists an identity element w.r.t. the operation *o*.

Examples. (N, ×), (Z, +), (Q, +) are monoids

Example 29. *Consider an algebraic system (I, +), where the set I = {0, 1, 2, 3, 4, ...} the set of natural numbers including zero and + is an addition operation. Determine whether (I, +) is a monoid.*

Sol. Closure property. The operation + is closed since sum of two natural numbers is a natural number.

Associative property. The operation + is an associative property since we have

$$(a + b) + c = a + (b + c)\ \forall\ a, b, c \in I.$$

Identity. There exists an identity element in set I w.r.t. the operation +. The element 0 is an identity element w.r.t. the operation +. Since, the operation + is a closed, associative and there exists an identity. Hence, the algebraic system (I, +) is a monoid.

Example 30. *Let S be a finite set and F(s) be the collection of all functions f : S → S under the operation of composition of functions. Show that F(s) is a semi-group. Is F(s) a monoid?*

Sol. Let *f, g, h* ∈ F(*s*), then we know that composition of functions is associative *i.e.*, *fo*(*goh*) = (*fog*)*oh* ∀ *f, g, h* ∈ F(*s*). Hence, F(*s*) is a semi-group. Also the identify function is an identify element of F(*s*).

∴ F(*s*) is a monoid.

8.13. SUBMONOID

Let us consider a monoid (M, *o*), also let S ⊆ M. Then (S, *o*) is called a submonoid of (M, *o*), if and only if it satisfies the following properties.

(*i*) S is closed under the operation *o*.

(*ii*) S is associative under the operation *o*.

(*iii*) There exists an identity element $e \in$ S.

e.g., Let us consider, a monoid (M, $*$),where $*$ is a binary operation and M is a set of all integers. Then (M_1, $*$) is a submonoid of (M, $*$), where M_1 is defined as

$$M_1 = \{a^i \mid i \text{ is from 0 to } n, \text{ a positive integer and } a \in M\}.$$

Examples 31. Since N $\subseteq$ Z and Z $\subseteq$ Q,

$\therefore$ N is a submonoid of Z and Z is a submonoid of Q.

Theorem II. *Let [M; $*$] be a monoid and K is a non-empty subset of M. Then K is a submonoid of M iff*

(*i*) $a, b \in K \Rightarrow a * b \in K$ *i.e. K is closed under* $*$

(*ii*) *For each* $a \in K$, *there exists* $e \in K$ *such that* $a * e = e * a = a$.

Proof. Let K is a submonoid of M. Then, by defination, K must be closed under $*$. Also for each $a \in$ K, there exists $e \in$ K such that $a * e = e * a$. Hence (*i*) and (*ii*) hold

Converse. Let (*i*) and (*ii*) hold

From (*i*), $a, b \in$ K $\Rightarrow$ $a * b \in$ K *i.e.*, K is closed under $*$

Also, if $a, b, c \in$ K and K $\subseteq$ M. $\therefore$ $a, b, c \in$ M. But M is a monoid

$$\therefore \quad a * (b * c) = (a * b) * c \quad \forall\, a, b, c \in K$$

i.e., the associativity holds for each element of K.

From (*ii*). K has an identity element

$\therefore$ K is a monoid. But K $\subseteq$ M

$\therefore$ K is a submonoid of M under $*$.

TEST YOUR KNOWLEDGE 8.1

1. Let $*$ be the operation on the set R of real numbers defined by $a * b = a + b + 2ab$
   (*a*) Find $2 * 3$, $3 * (-5)$, $7 * (1/2)$
   (*b*) Is (R, $*$) a semi-group ? Is it commutative ?
   (*c*) Find the identity element
   (*d*) Which elements have inverses and what are they ?
2. Let S be a semi-group with identity e and let b and b' be inverses of a. Show that $b = b'$ *i.e.*, inverses are uniques, if they exist.
3. Prove that for any commutative monoid (M ; $*$), the set of idempotent elements of M form a submonoid.
4. If a, b are elements of a monoid M and $a * b = b * a$. Show that
   $$(a * b) * (a * b) = (a * a) * (b * b)$$
5. Let S = Q $\times$ Q, the set of ordered pairs of rational numbers, with the operation $*$ defined by
   $$(a, b) * (x, y) = (ax, ay + b)$$
   (*a*) Find $(3, 4) * (1, 2)$ and $(-1, 3) * (5, 2)$
   (*b*) Is S a semi-group ? Is it commutative ?
   (*c*) Find the identity element of S
   (*d*) Which elements, if any, have inverses and what are they ?

6. Let S = N × N, the set of ordered pairs of positive integers with the operation ∗ defined by
$(a, b) * (c, d) = (ad + bc, bd)$
(*a*) Find (3, 4) ∗ (1, 5) and (2, 1) ∗ (4, 7)
(*b*) Is S a semi-group ? Is S commutative ?
7. Let A be a non-empty set with the operation ∗ defined by $a * b = a$ and assume A has more than one element. Then
(*a*) Is A a semi-groups ?
(*b*) Is A commutative ?
(*c*) Does A have an identity element ?
(*d*) Which elements, if any have inverses and what are they ?
8. Let A be a non-empty set with the operation ∗ defined by $a * b = a$, and assume A has more than one element.
(*a*) Is A a semigroup? (*b*) Is A commutative? (*c*) Does A have an identity element?
(*d*) Which elements, if any, have inverses and what are they ?
9. Let A = {*a*, *b*}. Find the number of operations on A , and exhibit one which is neither associative nor commutative.
10. Find a set A of three numbers which is closed under:
(*a*) multiplication, (*b*) addition.
11. Let S be an infinite set. Let A be the collection of finite subsets of S and let B be the collection of infinite subset of S.
(*a*) Is A closed under : (*i*) union, (*ii*) intersection, (*iii*) complements ?
(*b*) Is B closed under : (*i*) union, (*ii*) intersection, (*iii*) complements ?

Answers

1. (*a*) 17, – 32, $\frac{29}{2}$ (*b*) Yes, Yes
(*c*) zero (*d*) If $a \neq \frac{1}{2}$, then a has an inverse which is $\frac{-a}{(1+2a)}$.
5. (*a*) (3, 10), (– 5, 1) (*b*) Yes, No (*c*) (1, 0)
(*d*) The element (*a*, *b*) has an inverse if $a \neq 0$ and its inverse is $\left(\frac{1}{a}, -\frac{b}{a}\right)$
6. (*a*) (19, 20), (18, 7) (*b*) Yes, Yes
7. (*a*) Yes (*b*) No (*c*) No (*d*) It is meaningless to talk about inverses when no identity element exists.
8. (*a*) Yes (*b*) No (*c*) No (*d*) If identity element does not exist, then it is meaningless to talk about inverses.
9. 16, as there are two choices *a* or *b* for each of the four products *aa, ab, ba* and *bb* (Fig.). Further, From Fig. 8.1

$ab \neq ba$. Also

But $\begin{bmatrix} (aa)b = bb = a \\ a(ab) = aa = a \end{bmatrix}$

$\therefore \quad (aa)\, b \neq a(ab)$

∗	*a*	*b*
a	*b*	*a*
b	*b*	*a*

Fig. 8.1

10. (*a*) {– 1, 1, 0} (*b*) No such set exists
11. (*a*) (*i*) Yes (*ii*) Yes (*iii*) No (*b*) (*i*) Yes (*ii*) No (*iii*) No.

Hints

1. (*a*) Given $a * b = a + b + 2ab$
$\therefore \quad 2 * 3 = 2 + 3 + 12 = 17$
(*b*) (R, ∗) is semi-group iff we can show
$$a * (b * c) = (a * b) * c \ \forall\ a, b, c \in R$$

Consider $a * (b * c) = a * (b + c + 2bc)$

$= a + (b + c + 2bc) + 2a(b + c + 2bc)$

$= a + b + c + 2bc + 2ab + 2ac + 2abc$ ...(1)

Also $(a * b) * c = (a + b + 2ab) * c$

$= a + b + 2ab + c + 2(a + b + 2ab)\,c$

$= a + b + 2ab + c + 2ac + 2bc + 2abc$ ...(2)

From (1) and (2), we have

$$a * (b * c) = (a * b) * c$$

$\therefore$ (R, $*$) is a semi-group

Further, (R, $*$) is commutative iff we can show

$$a * b = b * a \ \forall\ a, b \in \mathrm{R}$$

Now $a * b = a + b + 2ab = b + a + 2ba = b * a$.

(*c*) Let e is an identity element of R, then for each $a \in$ R, we have

$a * e = a$

$\Rightarrow$ $a + e + 2ae = a$

$\Rightarrow$ $e(1 + 2a) = 0$

$\Rightarrow$ $e = 0$ | $1 + 2a \neq 0$

(*d*) Let $b \in$ R is the inverse of a. Then we must have

$a * b = 0$ | $e = 0$, the identity

$\Rightarrow$ $a + b + 2ab = 0$

$a + b(1 + 2a) = 0$

$\Rightarrow$ $b(1 + 2a) = -\,a$

$b = -\,a/1 + 2a,\ 1 + 2a \neq 0$

2. Since S is a semi-group with identity e and b and b' are inverses of a

$\therefore$ $b * a = e = a * b$ ...(1)

$b' * a = e = a * b'$ ...(2)

Now S is a semi-group under $*$. Therefore associativity holds among the elements of S under $*$.

i.e., $b * (a * b') = (b * a) * b'$

$\Rightarrow$ $b * e = e * b'$ | Using (1) and (2)

$\Rightarrow$ $b = b'$

3. Let S be the set of idempotent elements *i.e.,* S = $[a \in$ M: $a * a = a]$. We show S is a submonoid.

Let $a, b \in$ S, then $a * a = a,\ b \neq b = b$

Consider $(a * b) * (a * b) = a * (b * a) * b$ | Associativity

$= a * (a * b) * b$ | Commutative

$= (a * a) * (b * b)$

$= a * b$

$\therefore$ S is closed under $*$.

Also $e * e = e \ \Rightarrow\ e \in$ S *i.e.,* S has an identity element. Therefore, S is a submonoid.

4. $(a * b) * (a * b) = a * (b * (a * b))$ | Associativity

$= a * ((b * a) * b)$ | Associativity

$= a * ((a * b) * b)$ | $a * b = b * a$

$= a * (a * (b * b))$ | Associativity

$= (a * a) * (b * b)$ | Associativity

8.14. GROUP

(P.T.U. B.Tech. Dec. 2013, May 2008, 2006)

Let us consider an algebraic system (G, $*$), where $*$ is a binary operation on G. Then the system (G, $*$) is said to be a group if it satisfies the following properties:

(*i*) The operation $*$ is a closed operation.

(*ii*) The operation $*$ is an associative operation.

(*iii*) There exists an identity element w.r.t. the operation $*$.

(*iv*) For every $a \in G$, there exists an element $a^{-1} \in G$ such that $a^{-1} * a = a * a^{-1} = e$

For example, the algebraic system (I, +), where I is the set of all integers and + is an addition operation, is a group. The element 0 is the identity element w.r.t. the operation +. The inverse of every element $a \in I$ is $-a \in I$.

Examples (*i*) The sets (Q, +) (R, +) and (C, +) are groups under addition.

(*ii*) The sets R* (set of non-zero reals), Q* (set of non-zero rationals) and C* (set of non-zero complex numbers) are groups under multiplication.

ILLUSTRATIVE EXAMPLES

Example 1. *Determine whether the algebraic system (Q, +) is a group where Q is the set of all rational numbers and + is an addition operation.*

Sol. Closure Property. The set Q is closed under operation +, since the addition of two rational numbers is a rational number.

Associative Property. The operation + is associative, since $(a + b) + c = a + (b + c)\ \forall\ a, b, c \in Q$.

Identity. The element 0 is the identity element. Hence $a + 0 = 0 + a = a\ \forall\ a \in Q$.

Inverse. The inverse of every element $a \in Q$ is $-a \in Q$. Hence the inverse of every element exists.

Since, the algebraic system (Q, +) satisfies all the properties of a group, hence (Q, +) is a group.

Example 2. *Which of the following are groups under addition N, Z, Q, R, C ?*

Sol. The set of integers Z, the set of rationals Q, the set of reals R, the set of complex numbers C, are all groups under addition. (Prove yourself as in Example-1)

But N, the set of natural numbers donot form a group under addition. Since, N does not have additive identity. ($0 \notin N$).

Example 3. *Let S be the set of $n \times n$ with rational entries under the operation of matrix multiplication. Is S a group ?*

Sol. We know that matrix multiplication is associative. But inverse does not always exist. As we know that if $|A| \neq 0$, then A^{-1} exists.

Example 4. *Prove that G = {1, 2, 3, 4, 5, 6} is a finite abelian group of order 6 under multiplication modulo 7.* (P.T.U. B.Tech. May 2010, 2009)

Sol. $G = \{1, 2, 3, 4, 5, 6, \times_7\}$

Consider the multiplication modulo 7 table as shown below (Table 8.5). Recall that $a \times_7 b$ = The remainder when ab is divided by 7

$\times_7$	1	2	3	4	5	6
1	1	2	3	4	5	6
2	2	4	6	1	3	5
3	3	6	2	5	1	4
4	4	1	5	2	6	3
5	5	3	1	6	4	2
6	6	5	4	3	2	1

Table 8.5

From the table, we observe that each element inside the table is also an element of G. It means that G is closed under multiplication modulo 7.

Also for each $a, b, c \in$ G

$$a \times_7 (b \times_7 c) = (a \times_7 b) \times_7 c \quad i.e., \text{ associative law hold.}$$

From the table, we observe that the first row inside the table is identical with the top-row of the table. Therefore, 1 is the identity (multiplicative) of G.

Also, $2 \times_7 4 = 1$; $3 \times_7 5 = 1$, $4 \times_7 2 = 1$, $5 \times_7 3 = 1$, $6 \times_7 6 = 1$

Hence, each element G has an inverse, *i.e.*,

Inverse of 2 is 4 and of 4 is 2

Inverse of 3 is 5 and of 5 is 3

Inverse of 6 is 6

Hence, G is a group under the multiplication modulo **7**.

Example 5. *Consider an algebraic system (Q, *),where Q is the set of rational numbers and * is a binary operation defined by*

$$a * b = a + b - ab \ \forall \ a, b \in Q.$$

*Determine whether (Q, *) is a group.*

Sol. Closure property. Since the element $a * b \in$ Q for every $a, b \in$ Q, hence, the set Q is closed under the operation $*$.

Associative property. Let us assume $a, b, c \in$ Q, then we have

$$\begin{aligned}(a * b) * c &= (a + b - ab) * c \\ &= (a + b - ab) + c - (a + b - ab)c \\ &= a + b - ab + c - ac - bc + abc \\ &= a + b + c - ab - ac - bc + abc\end{aligned}$$

Similarly, $a * (b * c) = a + b + c - ab - ac - bc + abc.$

Therefore, $(a * b) * c = a * (b * c)$

$\therefore$ $*$ is associative.

Identity. Let e is an identity element. Then we have $a * e = a \ \forall \ a \in$ Q

$\therefore$ $a + e - ae = a$ or $e - ae = 0$

or $e(1 - a) = 0$ or $e = 0$, if $1 - a \neq 0$

Similarly, for $e * a = a \ \forall \ a \in$ Q, we have $e = 0$

Therefore, for $e = 0$, we have $a * e = e * a = a$

Thus, 0 is the identity element.

Inverse. Let us assume an element $a \in$ Q. Let a^{-1} is an inverse of a. Then we have

$a * a^{-1} = 0$ [Identity]

$\therefore$ $a + a^{-1} - aa^{-1} = 0$

or $a^{-1}(1 - a) = -a$ or $a^{-1} = \dfrac{a}{a-1}, a \neq 1$

Now, $\dfrac{a}{a-1} \in$ Q, if $a \neq 1$

Therefore, every element has inverse such that $a \neq 1$.

Since, the algebraic system (Q, *) satisfy all the properties of a group. Hence, (Q, *) is a group.

Theorem III. *Show that the identity element in a group is unique.*

Proof. Let us assume that there exists two identity elements *i.e.*, e and e' of G.
Since, $e \in G$ and e' is an identity. We have $e'e = ee' = e$
Also, $e' \in G$ and e is an identity. We have $e'e = ee' = e'$

$\therefore \quad e = e'$

Hence, identity in a group is unique.

Theorem IV. *Show that inverse of an element a in a group G is unique.*

Proof. Let us assume that $a \in G$ be an element. Also, assume that a_1^{-1} and a_2^{-1} be two inverse elements of a. Then we have,

$$a_1^{-1}\, a = aa_1^{-1} = e \quad \text{and} \quad a_2^{-1}\, a = aa_2^{-1} = e$$

Now, $\quad a_1^{-1} = a_1^{-1}\, e = a_1^{-1}(aa_2^{-1}) = (a_1^{-1}\, a)a_2^{-1} = ea_2^{-1} = a_2^{-1}$

Thus, the inverse of an element is unique.

Theorem V. *Show that $(a^{-1})^{-1} = a$ for all $a \in G$, where G is a group and a^{-1} is an inverse of a.*

Proof. Given that a^{-1} is an inverse of a. Then, we have

$$aa^{-1} = a^{-1}\, a = e$$

This implies that a is also an inverse of a^{-1}. Therefore $(a^{-1})^{-1} = a$.

Theorem VI. *Show that $(ab)^{-1} = b^{-1}\, a^{-1}$ for all $a, b \in G$.*

Proof. We have to prove that ab is inverse of $b^{-1}\, a^{-1}$. We prove

$$(ab)(b^{-1}\, a^{-1}) = (b^{-1}\, a^{-1})(ab) = e$$

Consider $\quad (ab)(b^{-1}\, a^{-1}) = [(ab)\, b^{-1}]a^{-1} = [a(bb^{-1})]\, a^{-1}$

$$= (ae)\, a^{-1} = aa^{-1} = e \quad \text{...(1)}$$

Similarly, $\quad (b^{-1}\, a^{-1})(ab) = e \quad$ ...(2)

$\therefore$ From (1) and (2), we have

$$(ab)\,(b^{-1}\, a^{-1}) = e = (b^{-1}\, a^{-1})\, ab$$ Hence proved.

Theorem VII. *Prove the left cancellation law in a group G holds i.e., $ab = ac \Rightarrow b = c$ $\forall\, a, b, c \in G$.*

Proof. Consider $\quad ab = ac.$

Then, we have $\quad b = eb = (a^{-1}\, a)b = a^{-1}\,(ab) = a^{-1}\,(ac) \quad [\because ab = ac]$

$= (a^{-1}\, a)c = ec = c \quad$ | Associativity

Hence, $\quad ab = ac \Rightarrow b = c.$

Theorem VIII. *Prove the right cancellation law in a group G holds i.e., $ba = ca \Rightarrow b = c$ $\forall\, a, b, c \in G$.*

Proof. Consider $ba = ca$.

Then, we have $\quad b = be = b(aa^{-1}) = (ba)a^{-1} = (ca)a^{-1} \quad [\because ba = ca]$

$= c(aa^{-1}) = ce = c \quad$ | Associativity

Hence, $\quad ba = ca \Rightarrow b = c.$

Theroem IX. *Let G be a group and $a, b \in G$. Then the equation $a * x = b$ has a unique solution given by $x = a^{-1} * b$.*

Proof. Given $a, b \in G$ and G is a group under $*$, therefore, a^{-1} exists in G

Hence $a^{-1} * b \in G$. | G is closed

Consider
$$a * x = b$$
$$= (a * a^{-1}) * b \qquad | \; a * a^{-1} = e$$
$$= a * (a^{-1} * b) \qquad | \text{ Associativity}$$
$$\Rightarrow \quad x = a^{-1} * b \qquad | \text{ Left cancellation law}$$

Uniqueness. Let the equation $a * x = b$ has two solutions, say, x_1 and x_2, then we have
$$a * x_1 = b \qquad ...(1)$$
$$a * x_2 = b \qquad ...(2)$$
(1) and (2) gives $a * x_1 = a * x_2$
$$\Rightarrow \quad x_1 = x_2 \qquad | \text{ Left cancellation law}$$

8.15. Z_m (THE INTEGERS MODULO m ($m \geq 1$))

The integers modulo m, denoted by Z_m, is the set given by

$Z_m = \{0, 1, 2, ... m - 1; +_m, \times_m\}$ where the operations $+_m$ (read as addition modulo m) and $\times_m$ (read as multiplication modulo m) are defined as
$$a +_m b = \text{remainder after } a + b \text{ is divided by } m$$
$$a \times_m b = \text{remainder after } a \times b \text{ is divided by } m.$$

Theroem X. *For each $m \geq 1$, $[Z_m; +_m]$ is a group known as group of integers modulo m.*

Proof. By definition, If $a, b \in Z_m$, then $a +_m b$ is remainder after $a + b$ is divided by m, which is again an element of Z_m. Hence Z_m is closed under $+_m$. Also the addition modulo m is always associative. 0 is the identity element for $+_m$ and every element of Z_m has an additive inverse. $\therefore$ Z_m is a group under addition modulo m.

8.16. FINITE AND INFINITE GROUP

A group (G, $*$) is called a finite group if G is a finite set.

A group (G, $*$) is called an infinite group if G is an infinite set.

For Example

1. The group (I, +) is an infinite group as the set I of integers is an infinite set.

2. The group G = $\{1, 2, 3, 4, 5, 6, 7, \times_8\}$ under multiplication modulo 8 is a finite group as the set G is a finite set.

3. The group G = $\{1, w, w^2\}$, is a finite group under multiplication

4. The group G = $\{\pm 1, \pm i, \pm j, \pm k\}$ where $i^2 = j^2 = k^2 = -1$ and $i \,.\, j = k, j.i = -k$ etc. is a finite group.

8.17. (a) ORDER OF GROUP

The order of the group G is the number of elements in the group G. It is denoted by $|G|$ or $o(G)$. A group of order 1 has only the identity element *i.e.,* ($\{e\}$).

8.17. (b) ORDER OF AN ELEMENT

Let G be a group under multiplication and $a \in G$. If there exists a least positive integer n such that $a^n = e$. Then a is said to be order n and written as $o(a) = n$

If G is a group under +, then a is said to be of order n if n is a least positve integer such that $na = e$. Here e is the identity element of G.

A group of order 2 has two elements *i.e.,* one identity element and one some other element.

Example 6. *Let ({e, x}, *) be a group of order 2. The table of operation is shown in (Fig. 8.3).*

*	e	x
e	e	x
x	x	e

Fig. 8.3

The group of order 3 has three elements *i.e.,* one identity element and two other elements.

Example 7. *Let ({e, x, y}, *) be a group of order 3. The table of operation is shown in (Fig. 8.4).*

*	e	x	y
e	e	x	y
x	x	y	e
y	y	e	x

Fig. 8.4

Example 8. *Consider an algebraic system ({0, 1}, +) where the operation + is defined as shown in (Fig. 8.5).*

+	0	1
0	0	1
1	1	0

Fig. 8.5

The system ({0, 1}, +) is a group. In this 0 is identity element and every element is its own inverse.

Theorem XI. *If G is a finite group of order n and $a \in$ G, then there exists a positive integer m such that $a^m = e$ and $m \le n$.*

Proof. Consider the elements of the group G as $a, a^2, a^3, \ldots a^{n+1}$. These are $n + 1$ elements. Since $|\, G \,| = n$. Therefore two of its elements, say, a^p, a^q must be equal, *i.e.,* $a^p = a^q$, $p < q$. Take $m = q - p$

$$\therefore \quad a^m = a^{q-p} = a^p \cdot a^{-p}$$
$$= a^q \cdot (a^p)^{-1} = a^q \cdot (a^q)^{-1} \qquad |\ a^p = a^q$$
$$= e$$

Further, since p, q are among $n + 1$,

$$\therefore \quad 1 \le p < q \le n + 1 \ \Rightarrow\ q - p = m \le n.$$

8.18. SUBGROUP

(*P.T.U. B.Tech. May 2013, May 2007, May 2006*)

Let us consider a group (G, *). Also, let S $\subseteq$ G ; then (S, *) is called a subgroup iff it satisfies following conditions :

(*i*) The operation * is closed operation on S.

(*ii*) The operation * is an associative operation.

(*iii*) As e is an identity element belonged to G. It must belong to the set S *i.e.,* The identity element of (G, $*$) must belongs to (S, $*$).

(*iv*) For every element $a \in$ S, a^{-1} also belongs to S.

Since Z $\subseteq$ Q, Z is a subgroup of Q under addition.

For example, let (G, +) be a group, where G is a set of all integers and (+) is an addition operation. Then (H, +) is a subgroup of the group G, where H = $\{2m : m \in$ G$\}$, the set of all even integer.

For example, let G be a group. Then the two subgroups of G are G and $G_1 = \{e\}$, e is the indentity element.

Example 9. *Let (I, +) be a group, where I is the set of all integers and (+) is an addition operation. Determine whether the following subsets of G are subgroups of G.*

(*a*) *The set (G_1, +) of all odd integers.* (*b*) *The set (G_2, +) of all positive integers.*

Sol. (*a*) The set G_1 of all odd integers is not a subgroup of G. It does not satisfy the closure property, since addition of two odd integers is always even.

(*b*) **Closure property.** The set G_2 is closed under the operation +, since addition of two even integers is always even.

Associative property. The operation + is associative since $(a + b) + c = a + (b + c)$ for every $a, b, c \in G_2$.

Identity. The element 0 is the identity element. Hence, $0 \in G_2$.

Inverse. The inverse of every element $a \in G_2$ is $-a \notin G_2$. Hence, the inverse of every element does not exists.

Since the system (G_2, +) does not satisfy all the conditions of a subgroup. Hence, (G_2, +) is not a subgroup of (I, +).

Example 10. *Consider the group Z of integers under addition. Let H be the subset of Z consisting of all multiples of a positive integer m i.e.,*

$$H = \{......, -3m, -2m, -m, 0, m, 2m, 3m, ...\}$$

Show that H is a subgroup of Z.

Sol. For $r, s \in Z$, $rm, sm \in$ H.

Consider $\quad rm + sm = (r + s)\, m \in$ H

i.e., H is closed under addition.

For $rm \in$ H, $-rm \in$ H and consider $rm + (-rm) = (r - r)\, m = 0 \in$ H

i.e., 0 is the identity of H and $-rm$ is the inverse of rm.

Hence, H is a subgroup of Z.

Theorem XII. *A subset H of a group G is a subgroup of G iff*

(*i*) *The identity element $e \in H$*

(*ii*) *H is closed under the same operation as in G*

(*iii*) *H is closed under inverses i.e., if $a \in H$, then $a^{-1} \in H$.*

Proof. Given G is a group and H is a subset of G. Let H is a subgroup of G, then, by definition, (*i*), (*ii*), (*iii*) are true.

Converse. Let (*i*), (*ii*), (*iii*) hold. We show H is a subgroup of G. We show the associativity of elements of H.

Let $a, b, c \in$ G and since H $\subseteq$ G $\therefore$ $a, b\ c \in$ H

Since elements of G are also elements of H

$\therefore$ Associativity holds for H. Hence the **Theorem.**

Another statement : A subset H of a group G is a subgroup of G iff $a * b^{-1} \in$ H. $\forall\ a, b \in$ H

Theorem XIII. *Let H_1 and H_2 be subgroup of group G, neither of which contains the other. Show that there exist an element of G belonging neither to H_1 nor H_2.*

Proof. Given H_1 and H_2 are subgroups of G. Also $H_1 \not\subset H_2$ and $H_2 \not\subset H_1$. We show that there exists an element belonging neither to H_1 nor H_2. Let, if possible, there is an element a belonging to H_1 and H_2 *i.e.,* $a \in H_1 \cap H_2$.

Now $a \in H_1$ and since H_1 is a subgroup of G $\therefore$ $a^{-1} \in H_1$ ...(1)

But $a \in H_2$ and since H_2 is a subgroup of G $\therefore$ $a^{-1} \in H_2$ ...(2)

(1) and (2) gives $H_1 \subseteq H_2$, a contradiction.

Hence the theorem.

Theorem XIV. *If H and K are two subgroups of G, then $H \cap K$ is also a subgroup of G.*

(P.T.U. B.Tech. Dec. 2012, May 2010, Dec. 2007)

Proof. We know that a subset H of a group G is a subgroup of G iff $ab^{-1} \in H \ \forall\ a, b \in H$.

Let $a, b \in H \cap K$. We show $ab^{-1} \in H \cap K$.

Now $a \in H \cap K \Rightarrow a \in H$ and $a \in K$

Also $b \in H \cap K \Rightarrow b \in H$ and $b \in K$

Since H is a subgroup of G and $a, b \in H$

$\Rightarrow$ $ab^{-1} \in H$ (Using theorem X) ...(1)

Also K is a subgroup of G and $a, b \in K$

$\Rightarrow$ $ab^{-1} \in K$ ...(2)

From (1) and (2), $ab^{-1} \in H \cap K$. Hence $H \cap K$ is a subgroup of G.

Cor. If H and K are two subgroups of a group G, then give an example to show that $H \cup K$ may not be a subgroup of G

Consider G = The group of integers under +

$H_1 = \{... -6, -4, -2, 0, 2, 4, 6 ...\}$

$H_2 = \{... -12, -9, -6, -3, 0, 3, 6, 9, 12, ...\}$ are subgroups of G under +.

But $H_1 \cup H_2 = \{... -4, -3, -2, 0, 2, 3, 4, 6, ...\}$

Since $2 \in H_1 \cup H_2$, $3 \in H_1 \cup H_2 \Rightarrow 2 + 3 = 5 \notin H_1 \cup H_2$ *i.e.,* $H_1 \cup H_2$ is not closed under +. Hence $H_1 \cup H_2$ is not a subgroup of G under +.

Theorem XV. *If H is a non-empty finite subset of a group G and H is closed under multiplication. Then H is a subgroup of G.*

Proof. We know that a non-empty subset H of a group G is a subgroups of G iff

(*i*) $a \in H, b \in H \Rightarrow ab \in H$

(*ii*) $a \in H \Rightarrow a^{-1} \in H$

The condition (*i*) is true since it is given that H is closed under multiplication.

To show (*ii*), Let $a \in H, a \in H \Rightarrow a^2 \in H$ | H is closed under multiplication

Again $a \in H, a^2 \in H \Rightarrow a^3 \in H$ and so on.

Thus the infinite collection of all the elements $a, a^2, a^3, ... a^m, ...$, belongs to H. But H is finite. $\therefore$ there must be repetetion. Let $a^r = a^s$ $r > s > 0$

$\Rightarrow$ $a^r . a^{-s} = e$

$\Rightarrow$ $a^{r-s} = e \in H$

Take $y = a^{r-s-1}$ and consider

$$ya = a^{r-s-1} . a = a^{r-s} = e$$

Similarly, $ay = e$

Hence $ya = e = ay$

$\Rightarrow$ y is the inverse of a. Hence the theorem.

Theorem XVI. Let H be a subgroup of G. Then

(*a*) $H = Ha \iff a \in H$

(*b*) $Ha = Hb \iff a\,b^{-1} \in H$

(*c*) $aH = bH \iff a^{-1}\,b \in H$

(*d*) HH = H.

Proof. (*a*) Let $Ha = H$. If $e \in H \Rightarrow e\,a \in Ha = H$

$\Rightarrow a \in H$ | $ea = a$

Conversly, Let $a \in H$. As H is a subgroup and $h \in H$, $a \in H$

$\Rightarrow h\,a \in H$ | H is closed under multiplication.

$\Rightarrow H\,a \subseteq H$ ...(1)

Again, if $h \in H$, $a \in H$ and since H is a subgroup of G, $\therefore$ $h\,a^{-1} \in H$ (Theorem X)

$\Rightarrow (ha^{-1})\,a \in H\,a$

$\Rightarrow h(a^{-1}a) \in Ha \Rightarrow h\,e \in Ha$

$\Rightarrow h \in H\,a$

$\Rightarrow H \subseteq H\,a$ ...(2)

From (1) and (2) $Ha = H$

(*b*) Let $Ha = Hb$ and we show $ab^{-1} \in H$

Now $a = e\,a \in Ha$

$\Rightarrow a \in Ha = Hb$

$\Rightarrow a \in Hb \Rightarrow a = hb,\ h \in H$

$\Rightarrow ab^{-1} = (hb)b^{-1} = h(bb^{-1}) = he = h \in H$

$\Rightarrow ab^{-1} \in H$

Conversly, Let $ab^{-1} \in H \Rightarrow ab^{-1} = h,\ h \in H$

$\Rightarrow a = hb$

$\Rightarrow Ha = Hhb = Hb$ | For $h \in H$, $Hh = H$

(*c*) Proceed yourself as in Part (*b*).

(*d*) Let $h \in H$. Then,

$H = Hh \quad \forall\ h \in H$ | Using part (*a*)

$\Leftrightarrow H \subseteq HH \subseteq H$

$\therefore HH = H$.

8.19. ABELIAN GROUP

Let us consider, an algebraic system (G, $*$), where $*$ is a binary operation on G. Then the system (G, $*$)is said to be an abelian group if it satisfies all the properties of the group plus an additional following property :

(*i*) The operation $*$ is commutative *i.e.,*

$$a * b = b * a \ \forall\ a, b \in G$$

For example, consider an algebraic system (I, +), where I is the set of all integers and + is an addition operation. The system (I, +) is an abelian group because it satisfies all the properties of a group. Also the operation + is commutative for every $a, b \in I$.

Example 11. *Consider an algebraic system (G, ∗), where G is the set of all non-zero real numbers and ∗ is a binary operation defined by* $a * b = \frac{ab}{4}$. *Show that (G, ∗) is an abelian group.*

Sol. Closure property. The set G is closed under the operation ∗. Since, $a * b = \frac{ab}{4}$ is a real number. Hence, belongs to G.

Associative property. The operation ∗ is associative. Let $a, b, c \in G$, then we have

$$(a * b) * c = \left(\frac{ab}{4}\right) * c = \frac{(ab)c}{16} = \frac{abc}{16}.$$

Similarly,
$$a * (b * c) = a * \left(\frac{bc}{4}\right) = \frac{a(bc)}{16} = \frac{abc}{16}.$$

Identity. To find the identity element, let us assume that e is a positive real number. Then for $a \in G$,

$$e * a = a \Rightarrow \frac{ea}{4} = a \text{ or } e = 4$$

Similarly,
$$a * e = a$$
$$\Rightarrow \frac{ae}{4} = a \text{ or } e = 4.$$

Thus, the identity an element in G is 4.

Inverse. Let us assume that $a \in G$. If $a^{-1} \in Q$ is an inverse of a, then $a * a^{-1} = 4$

$$\Rightarrow \frac{aa^{-1}}{4} = 4 \text{ or } a^{-1} = \frac{16}{a}$$

Similarly, $a^{-1} * a = 4$ gives

$$\Rightarrow \frac{a^{-1}a}{4} = 4 \text{ or } a^{-1} = \frac{16}{a}.$$

Thus, the inverse of an element a in G is $\frac{16}{a}$.

Commutative. The operation ∗ on G is commutative.

Since,
$$a * b = \frac{ab}{4} = \frac{ba}{4} = b * a.$$

Thus, the algebraic system (G, ∗) is closed, associative, has identity element, has inverse and commutative. Hence, the system (G, ∗) is an abelian group.

Example 12. *Let* (**Z**, ∗) *be an algebraic structure, where* Z *is the set of integers and the operation* ∗ *is defined by* $n * m$ = *maximum (n, m). Determine whether* (**Z**, ∗) *is a monoid or a group or an abelian group.*

Sol. Closure Property

We know that $n * m = \max.(n,m) \in \mathbf{Z} \quad \forall\, n, m \in \mathbf{Z}$

Hence ∗ is closed.

Associative property. Let us assume $a, b, c \in \mathbf{Z}$.

Then, we have $a * (b * c) = a * \max.\ (b, c) = \max.\ (a, \max.\ (b, c)) = \max.\ (a, b, c)$

Similarly, $(a * b) * c = \max.\ (a, b, c)$

Hence $*$ is associative.

Identity. Let e be the identity element. Then $\max.\ (a, e) = a$

Hence, the minimum element is the identity element.

Inverse. The inverse of any element does not exist. Since, the inverse does not exist, hence (**Z**, $*$) is not a group or abelian group but a monoid as it satisfies the properties of closure, associative and identity.

Example 13. *Let S = {0, 1, 2, 3, 4, 5, 6, 7} and multiplication modulo 8, that is*

$$x \otimes y = (xy) \text{ Mod } 8$$

(i) Prove that ({0, 1}, ⊗) is not a group.

(ii) Write three distinct groups (G, ⊗) where G ⊂ S and G has 2 elements.

Sol. (*i*) (*a*) **Closure property.** The set {0, 1} is closed under the operation ⊗, as shown in table of operation (Fig. 8.6).

⊗	0	1
0	0	0
1	0	1

Fig. 8.6

(*b*) **Associative property.** The operation ⊗ is associative. Let $a, b, c \in$ G, then we have

$$(a \otimes b) \otimes c = a \otimes (b \otimes c) \quad e.g., \quad (0 \otimes 1) \otimes 1 = (0) \otimes 1 = 0$$

Similarly, $0 \otimes (1 \otimes 1) = 0 \otimes (1) = 0.$

(*c*) **Identity.** The element 1 is the identity element as for every $a \in \{0, 1\}$: We have

$$1 \otimes a = a = a \otimes 1.$$

(*d*) **Inverse.** There must exist an inverse of every element $a \in \{0, 1\}$, such that

$$a \otimes a^{-1} = 1$$

But the inverse of element 0 does not exist.

Therefore, since the inverse of every element $a \in \{0, 1\}$ does not exist. Hence ({0, 1}, ⊗) is not a group.

(*ii*) The three distinct groups (G, ⊗), where G ⊂ S and G has 2 elements is as follows

(*a*) ({1, 3}, ⊗) (*b*) ({1, 5}, ⊗) (*c*) ({1, 7}, ⊗).

Example 14. *Determine whether a semi-group with more than one idempotent element can be a group.* (P.T.U. B.Tech. Dec. 2010)

Sol. Let (A, $*$) be a semi-group with two idempotent elements a and b ($a \neq b$). Then we have

$$a * a = a \quad \text{...(1)}$$

$$b * b = b. \quad \text{...(2)}$$

Now assume that A is a group with identity element e.

Then, $a * e = a$ and $b * e = b$

From (1) and (2),

$$a * e = a = a \neq a \;\Rightarrow\; e = a \qquad \text{| Left Cancellation Law}$$

Also $$b * e = b = b \neq b \;\Rightarrow\; e = b \qquad \text{| Left Cancellation Law}$$

$\Rightarrow$ $$a = e = b$$

which is a contradiction to $a \neq b$.

Hence (A, $*$) can not be group.

Example 15. *Let (G, o) be a group. Show that if (G, o) is an Abelian group then, $(a \text{ o } b)^2 = a^2 \text{ o } b^2$ for all a and b in G.* (P.T.U. B.Tech. Dec. 2010)

Sol. Let us assume G is an Abelian group, then

$$(a \text{ o } b)^2 = (a \text{ o } b) \text{ o } (a \text{ o } b) = a \text{ o } (b \text{ o } a) \text{ o } b \qquad [\because \text{ o is associative}]$$

$$= a \text{ o } (a \text{ o } b) \text{ o } b = (a \text{ o } a) \text{ o } (b \text{ o } b) = a^2 \text{ o } b^2 \qquad \text{| G is abelian}$$

Hence, $$(a \text{ o } b)^2 = a^2 \text{ o } b^2 \quad \forall\, a, b \in G.$$

Example 16. *Let G be a group of 2 × 2 matrices with rational entries and non-zero determinant. Let H be a subset of G consisting of matrices whose upper right entry is zero. Then show that H is a subgroup of G.*

Sol. Given $$G = \left[\begin{pmatrix} a & b \\ c & d \end{pmatrix} : a, b, c, d \in Q \text{ and } ad - bc \neq 0\right]$$

$$H = \left[\begin{pmatrix} a & 0 \\ c & d \end{pmatrix} : a, c, d \in Q\right]$$

H is a subgroup of G iff

(*i*) H is closed under multiplication

(*ii*) For A ∈ H, A^{-1} ∈ H

Let A, B ∈ H where $A = \begin{pmatrix} a_1 & 0 \\ c_1 & d_1 \end{pmatrix}$, $B = \begin{pmatrix} a_2 & 0 \\ c_2 & d_2 \end{pmatrix}$

Consider $$AB = \begin{pmatrix} a_1 & 0 \\ c_1 & d_1 \end{pmatrix}\begin{pmatrix} a_2 & 0 \\ c_2 & d_2 \end{pmatrix} = \begin{pmatrix} a_1a_2 & 0 \\ c_1a_2 + d_1c_2 & d_1d_2 \end{pmatrix} \in H$$

i.e., H is closed under multiplication.

Further, For A ∈ H, we have

$$A = \begin{pmatrix} a & 0 \\ c & d \end{pmatrix}, \; |A| = \begin{vmatrix} a & 0 \\ c & d \end{vmatrix} = ad$$

Also $$A_{11} = d,\; A_{12} = -c,\; A_{21} = 0,\; A_{22} = a$$

$\therefore$ $$\text{adj } A = \begin{pmatrix} A_{11} & A_{12} \\ A_{21} & A_{22} \end{pmatrix}^T = \begin{pmatrix} d & 0 \\ -c & a \end{pmatrix}$$

$\therefore$ $$A^{-1} = \frac{adjA}{|A|} = \begin{pmatrix} \dfrac{d}{ad} & 0 \\ -\dfrac{c}{ad} & \dfrac{1}{d} \end{pmatrix} \in H$$

Hence H is a subgroup of G.

Example 17. *Let G be a group of real numbers under multiplication. Let H = {– 1, 1}. Then show that H is a subgroup of G under multiplication.*

Sol. Consider the multiplication table of H under multiplication.

•	– 1	1
– 1	1	– 1
1	– 1	1

From the table, we observe that each element in the table belongs to H.

Hence H is closed under multiplication.

Also, the inverse of – 1 is – 1 and of 1 is 1. Thus each element of H has its inverse. Therefore H is a subgroup of G under multiplication.

Example 18. *Consider the group of integers Z under +. Let E = The set of even integers. Then show that E is a subgroup of Z under +.*

Sol. Given $E = \{2m : m \in Z\}$ *i.e.*, the set of even integers. Clearly E is a subset of Z.

Let $a, b \in E \Rightarrow a = 2m, m \in Z$

$b = 2n, n \in Z$

$\therefore \quad a + b = 2m + 2n = 2(m + n) \in E \qquad | \; m, n \in Z \Rightarrow m + n \in Z$

i.e., E is closed under +. Also for each $a \in E$, we have $a = 2m, m \in Z$

$\Rightarrow \quad -a = -2m = 2(-m) = 2t, \; t = -m \in Z$

$\Rightarrow \quad -a \in E$

Thus each element belonging to E has additive inverse.

$\therefore$ E is a subgroup of Z under +.

Example 19. *Let Z be a group of integers under +. Let Z^+ is the set of non-negative integers. Is Z^+ a subgroup of Z ?*

Sol. $Z^+ = \{0, 1, 2, 3, ...\}$

Clearly Z^+ is a subset of Z. But Z^+ is not a subgroup of Z. Since the elements of Z^+ do not have additive inverses. For *e.g.*, $2 \in Z^+$, but $-2 \notin Z^+$.

Example 20. *Consider Z_{12} = [0, 1, 2, ... 11], the group under addition modulo 12. Let H = [0, 3, 6, 9]. Show that H is a subgroup of Z_{12} under $+_{12}$.*

Sol. Given H = [0, 3, 6, 9]. Clearly H is a subset of Z_{12}.

Let $a, b \in H \Rightarrow a +_{12} b$ is also in H. $\therefore$ H is closed under $+_{12}$. Also we have

$3 +_{12} 9 = 0$, 0 is the identity of Z_{12}

$6 +_{12} 6 = 0$

$9 +_{12} 3 = 0$

$\therefore$ each element of H has its inverse.

$\therefore$ H is a subgroup of Z_{12} under addition modulo 12.

Example 21. *Consider the group of integers Z under +. Let 2Z and 3Z are two subgroups of [Z; +]. Is $2Z \cap 3Z$ a subgroup of Z ?*

Sol. We know that if H and K are two subgroups of a group G. Then $H \cap K$ is also a subgroup of G. Using this result, we can say that $2Z \cap 3Z$ is a subgroup of Z. (Theorem XII)

Example 22. Consider Z_{15}, the group under addition modulo 15. Let H_1 = [0, 5, 10], H_2 = [0, 4, 8, 12]. Are H_1 and H_2 subgroups of Z_{15} under $+_{15}$?

Sol. To check whether H_1 is a subgroup of Z_{15}, compute the following Table 8.6.

Table 8.6

$+_{15}$	0	5	10
0	0	5	10
5	5	10	0
10	10	0	5

From the Table 8.6, we observe that each element which is in the interior of the addition table is also in H_1. ∴ H_1 is closed under $+_{15}$. Also we have $5 +_{15} 10 = 0$, $10 +_{15} 5 = 0$, 0 is the identity ∴ each element of H, has its inverse. ∴ H_1 is a subgroup of Z_{15}.

To check, whether H_2 is a subgroup of Z_{15}, compute the following Table 8.7.

Table 8.7

$+_{15}$	0	4	8	12
0	0	4	8	12
4	4	8	12	1
8	8	12	1	5
12	12	1	5	9

From the Table 8.7, we observe that there are some elements in the interior of the addition table, which are not in H_2 (*e.g.*, $9 \notin H_2$). Hence H_2 is not closed under $+_{15}$. ∴ H_2 is not a subgroup of Z_{15}.

Example 23. *Let H and K be groups.*

(a) What do you mean by the direct product of H and K?

(b) What is the identify element of H × K ?

(c) What is the order of H × K ?

(d) Describe and write the multiplication table of the group $G = Z_2 \times Z_2$.

Sol. (*a*) Let G = H × K be the cartesian product of H and K with the operation ∗ defined by

$$(h, k) * (h', k') = (hh', kk')$$

Then G is group called the direct product of H and K.

(*b*) Let e_H is the identity element of H and e_K be the identity element of K, then (e_H, e_K) is an identity element of H × K.

(*c*) $o(G) = o(H \times K) = o(H) \times o(K)$

(*d*) As, $Z_2 = \{0, 1\}$ has two elements ∴ $o(Z_2) = 2$

∴ $o(G) = o(Z_2 \times Z_2) = o(Z_2) \times o(Z_2) = 2 \times 2 = 4$

∴ G has four elements namely $a = (1, 0)$, $b = (0, 1)$, $c = (1, 1)$, $e = (0, 0)$.

The multiplication table of $G_1 = Z_2 \times Z_2$ is shown below (Table 8.8):

Table 8.8

∗	*e*	*a*	*b*	*c*
e	*e*	*a*	*b*	*c*
a	*a*	*e*	*c*	*b*
b	*b*	*c*	*e*	*a*
c	*c*	*b*	*a*	*e*

Here, $a^2 = e$, $b^2 = e$, $c^2 = e$

Since the table is symmetric, therefore G is abelian.

TEST YOUR KNOWLEDGE 8.2

1. If a, b, c are elements of a group G and $a * b = c * a$. Then $b = c$? Explain your answer. (*P.T.U. B.Tech, Dec. 2006, May 2005*)
2. Discuss the relation between groups and monoids ? Is every monoid a group ? Is every group a monoid ?
3. Which of the following are groups ?
   (*i*) $M_{2\times3}(R)$ with matrix addition
   (*ii*) $M_{2\times2}(R)$ with matrix multiplication
   (*iii*) The positive real numbers with multiplication
   (*iv*) The non-zero real numbers with multiplication
   (*v*) The set $[-1, 1]$ with multiplication.
4. Give an example of (*i*) a finite abelian group (*ii*) an infinite non-abelian group.
5. Let $V = \{e, a, b, c\}$. Let $*$ be defined by $x * x = e$ for all $x \in V$. Write a complete table for $*$ so that $(V, *)$ is a group.
6. Which of the following subsets of the real numbers is a subgroup of $[R, +]$?
   (*a*) The rational numbers
   (*b*) The positive real numbers
   (*c*) $H = \left\{\frac{K}{2} : K \text{ is an integer}\right\}$
   (*d*) $H = \{2^K : K \text{ is an integer}\}$
   (*e*) $H = \{x : -100 \le x \le 100\}$
7. Prove that if H, K are subgroups of a group G and $H \cup K = G$. Then either $H = G$ or $K = G$.
8. Show that $[Z_n; \times_n]$, $n \ge 2$ is a monoid with identity.
9. Show that the intersection of any number of subgroups of G is a subgroup of G.
10. Let G be a group and $a, b \in G$. Then the equation $x * a = b$ has a unique solution given by $x = b * a^{-1}$.
11. Let G be a group of order p, p is prime. Find all subgroups of G.
12. Prove that congruence modulo H, $a \equiv b \pmod{H}$ is an equivalence relation in G. (*P.T.U. B.Tech. Dec. 2007*)

Answers

2. No, Yes (Every group is a monoid)
3. (*i*), (*iii*), (*iv*), (*v*) are groups
4. (*i*) $G = \{e, a, b, c\}$ is a finite abolian group under * defined by the following Table 8.9:

Table 8.9

*	*e*	*a*	*b*	*c*
e	*e*	*a*	*b*	*c*
a	*a*	*e*	*c*	*b*
b	*b*	*c*	*a*	*e*
c	*c*	*b*	*e*	*a*

(*ii*) $M_{2\times2}(R)$, the set of all 2 × 2 matrices $\begin{pmatrix} a & b \\ c & d \end{pmatrix}$; $ad - bc \neq 0$ is an infinite non-abelian group w.r.t. the matrix multiplication.

5. See Q. 4 (*i*) **6.** (*a*) and (*c*) **11.** $\{e\}$ and G itself.

Hints

1. Given $a, b, c \in G$. Consider $a * b = c * a$. We claim $b \neq c$. For, if, $b = c$, then

$a * b = a * c \neq c * a$ unless G is abelian. Hence the given statement is a false statement. However, the given statement is a true statement if G is an abelian group. In this case,

$$a * b = c * a = a * c \qquad | \text{ G is abelian}$$

$$\Rightarrow \qquad b = c \qquad | \text{ Left cancellation law}$$

2. The set N of natural numbers is a monoid under +, but not a group under +.

3. $M_{2\times2}(R)$ with matrix multiplication is not a group.

Take $A = \begin{pmatrix} 1 & 1 \\ 1 & 1 \end{pmatrix}$, $|A| = \begin{vmatrix} 1 & 1 \\ 1 & 1 \end{vmatrix} = 1 - 1 = 0$

$\Rightarrow$ $|A| = 0$ *i.e.,* the matrix is singular and hence A^{-1} does not exist.

11. By Lagrange's theorem, the order of a subgroup H of G divides the order of G. Hence O(H) = 1 or p.

If O(H) = 1, then H = $\{e\}$

If O(H) = p, then H = G.

8.20. (a) COSETS

(*P.T.U. B.Tech. Dec. 2006*)

Consider an algebraic system (G, ∗), where ∗ is a binary operation. Now, if (G, ∗) is a group and let a be an element of G and $H \subseteq G$, then

The **left coset** $a * H$ of H is the set of elements such that

$$a * H = \{a * h : h \in H\}.$$

The **right coset** $H * a$ of H is the set of elements such that

$$H * a = \{h * a : h \in H\}.$$

The subset H is itself a left and right coset since $e * H = H * e = H$.

8.20. (b) COSET REPRESENTATIVE SYSTEM FOR H IN G

A subset C of G is said to be a *coset representative system of H* if C contains exactly one element from each coset. Such an element is called a representative of the coset. The number of coset representatives is equal to [G : H], the index of H in G.

Example. *Let H be a subgroup of a finite group G. How many coset representative systems exist for the cosets of H ?*

Sol. There are n(H) ways of choosing an element from any coset and there are [G : H] distinct cosets. Hence, the desired number is $n(H)^{[G : H]}$.

ILLUSTRATIVE EXAMPLES

Example 1. *Let us consider a group (G, ∗), where G is a set having elements {0, 1} and ∗ is a binary operation. Also, let H = {1} is a subgroup of G. Determine all the left cosets of H in G.*

Sol. There are only 2 left cosets *i.e.,*

$$1 * H = H = \{1\}$$
$$0 * H = \{0\}.$$

Example 2. *Let (I, +) is a group, where I is the set of all integers and + is an addition operation and let H = {..., – 4, – 2, 0, 2, 4, 6, 8, ...} be the subgroup consisting of multiples of 2. Determine all the left cosets of H in I.*

Sol. There are two distinct left cosets of H in I.

$$0 + H = \{ ..., -6, -4, -2, 0, 2, 4, 6, ...\} = H$$
$$1 + H = \{... -5, -3, -1, 1, 3, 5, 7, ...\}$$
$$2 + H = \{... -4, -2, 0, 2, 4, ...\} = H$$
$$3 + H = \{..., -5, -3, -1, 1, 3, 5, ...\} = 1 + H$$

so on.

There is no other distinct left coset because any other left coset coincides with the cosets given above.

Example 3. *Let G = (I, +) be a group, where I is the set of integers and + is an addition operation, also let G_1 = {......– 14, – 7, 0, 7, 14, 21, ...} be a subgroup consisting of the multiples of 7. Determine the cosets of G_1 in I.*

Sol. The set I has 7 different cosets (left or right) of G_1, which are as shown below.

$$0 + H = \{...... -14, -7, 0, 7, 14, 21, ...\}$$
$$1 + H = \{...... -13, -6, 1, 8, 15, 22, ...\}$$
$$2 + H = \{...... -12, -5, 2, 9, 16, 23, ...\}$$
$$3 + H = \{...... -11, -4, 3, 10, 17, 24, ...\}$$
$$4 + H = \{...... -10, -3, 4, 11, 18, 25, ...\}$$
$$5 + H = \{...... -9, -2, 5, 12, 19, 26, ...\}$$
$$6 + H = \{...... -8, -1, 6, 13, 20, 27, ...\}$$
$$7 + H = \{...... -14, -7, 0, 7, 14, 21, ...\} = H$$

All other cosets coincides with any one of the cosets shown above, hence we will not count them.

Theorem I. *If $b \in a * H$ (left coset), then*

$a * H = b * H$. Also, if $b \in H * a$ (right coset), Then $H * a = H * b$

Proof. Case I. Let $x \in a * H$, we show $x \in b * H$

Now $x \in a * H \Rightarrow$ there exists an element $h_1 \in$ H such that $x = a * h_1$

Also $b \in a * H \Rightarrow$ There exists an element $h_2 \in$ H such that $b = a * h_2$

$\Rightarrow \quad b * h_2^{-1} = (a * h_2) * h_2^{-1} = a * (h_2 * h_2^{-1}) = a * e = a$

$\Rightarrow \quad a = b * h_2^{-1}$

$\therefore \quad x = a * h_1 = (b * h_2^{-1}) * h_1 = b * (h_2^{-1} * h_1)$

Since $h_1, h_2 \in$ H and H is subgroup of G

$\therefore \quad h_2^{-1} \in \text{H}$ and $h_2^{-1} * h_1 \in \text{H}$

Hence $\quad x = b * (h_2^{-1} * h_1) \in b * \text{H}$

$\Rightarrow \quad x \in b * \text{H}$

$\Rightarrow \quad a * \text{H} \subseteq b * \text{H}$ ...(1)

Similarly, if $x \in b * \text{H}$, we can easily show that $x \in a * \text{H}$.

$\Rightarrow \quad b * \text{H} \subseteq a * \text{H}$ ...(2)

from (1) and (2), $\quad a * \text{H} = b * \text{H}$

Case II. Let $x \in \text{H} * a$. We show $x \in \text{H} * b$. Now $x \in \text{H} * a \Rightarrow$ There exists an element $h_1 \in$ H such that $x = h_1 * a$

Also $b \in \text{H} * a \Rightarrow$ There exists an element $h_2 \in$ H such that $b = h_2 * a$

$\Rightarrow \quad h_2^{-1} * b = h_2^{-1} * (h_2 * a) = (h_2^{-1} * h_2) * a$

$= e * a = a$

$\Rightarrow \quad a = h_2^{-1} * b$

$\therefore \quad x = h_1 * a = h_1 * (h_2^{-1} * b) = (h_1 * h_2^{-1}) * b$

Since $h_1, h_2 \in$ H and H is a subgroup of G. $\therefore \quad h_1 * h_2^{-1} \in \text{H}$

Hence $\quad x = (h_1 * h_2^{-1}) * b \in \text{H} * b$

$\Rightarrow \quad x \in \text{H} * b$

$\Rightarrow \quad \text{H} * a \subseteq \text{H} * b$

Similarly, if $x \in \text{H} * b$, we can easily show that $x \in \text{H} * a$

$\Rightarrow \quad \text{H} * b \subseteq \text{H} * a$

Hence $\quad \text{H} * a = \text{H} * b$

Theorem II. *Let H be a subgroup of a group G. Then the right cosets Ha form a partition of G.*

Or

Any two right (left) cosets in a group G are either disjoint or equal.

Proof. Let Ha and Hb be two right cosets and suppose H$a \cap$ H$b \neq \phi$. We show Ha = Hb

Let $\quad x \in \text{H}a \cap \text{H}b \Rightarrow x \in \text{H}a$ and $x \in \text{H}b$

$\Rightarrow \quad x = h_1 a$ and $x = h_2 b$ for some $h_1, h_2 \in$ H

$\Rightarrow \quad h_1 a = h_2 b \in \text{H}b$

$\Rightarrow \quad h_1 a \in \text{H}b$

$\Rightarrow \quad \text{H}a \subseteq \text{H}b$

Also $\quad h_2 b = h_1 a \in \text{H}a$

$\Rightarrow \quad h_2 b \in \text{H}a$

$\Rightarrow \quad \text{H}b \subseteq \text{H}a$

Hence $\quad \text{H}a = \text{H}b$

Lagrange's Theorem

(P.T.U. B.Tech. Dec. 2007, 2006, May 2006, May 2012, May 2013)

Theorem III. *If G is a finite group and H is a subgroup of G, then o(H) | o(G).*

Proof. Since H is a subgroup of a finite group G, $\therefore$ H is also finite, say,

H = $\{h_1, h_2, \ldots h_n]$, where each h_i is distinct.

We claim H and any coset Ha have the same number of elements.

Consider Ha = $\{h_1a, h_2a, \ldots h_na\}$. We claim all h_ia's are distinct. For if,

$$h_ia = h_ja$$

$$\Rightarrow \quad h_i = h_j \quad | \text{ Right cancellation law}$$

a contradiction, since h_i's are distinct. Thus, H and Ha have same number of elements.

Now G is finite ∴ The number of distinct right cosets of H in G is also finite, say, k.

Let $$G = Ha_1 \cup Ha_2 \cup \ldots \cup Ha_k = \bigcup_{i=1}^{k} Ha_i$$

$\Rightarrow$ $o(G)$ = Number of elements in Ha_1 + number of elements in Ha_2 + ... + number of elements in Ha_K

$$= n + n + \ldots k \text{ Times} = nk$$

$$\Rightarrow \quad n \mid o(G)$$

$$\Rightarrow \quad o(H) \mid o(G)$$

Hence the Theorem.

Converse of Lagrange's theorem is however, not true *i.e., If* o(H) | o(G). Then G may not have a subgroup of order o(H).

Consider the alternating group A_4. Here $o(A_4) = \frac{4!}{2} = 12$ and $6 \mid 12$. But there is no subgroup of A_4 whose order is 6. (see the topic "Symmetric Group" in this chapter)

8.21. INDEX OF A SUBGROUP

Let G be a group and H be a subgroup of G. Then the number of right (left) cosets of H in G is called the index of H in G. The index of H in G is denoted by [G : H].

Theorem IV. *If G is a finite group and H is a subgroup of G. Then* $[G: H] = \frac{o(G)}{o(H)}$.

Proof. Proceeding in the same way as in the proof of Lagrange's theorem, we have

$o(G) = nk$, where k is the number of distinct right cosets of H in G

$$\Rightarrow \quad k = \frac{o(G)}{n} = \frac{o(G)}{o(H)}$$

$$\Rightarrow \quad [G: H] = \frac{o(G)}{o(H)}.$$

Theorem V. *Let G be a finite group of order n show that $a^n = e$ for any element $a \in$ G.*

(P.T.U. B.Tech. Dec. 2010)

Proof. Let $a \in$ G has order m, then $a^m = e$

Let H be a subgroup of G, of order m. By Lagrange's theorem,

$$o(H) \mid o(G)$$

$$\Rightarrow \quad m/n \quad \Rightarrow \quad n = mr, \text{ for some } r$$

$$\therefore \quad a^n = a^{mr} = (a^m)^r = e^r = e$$

Hence the Theorem.

8.22. NORMAL SUBGROUP

(P.T.U. B. Tech. Dec. 2007)

A subgroup H of a group G is called normal subgroup of G if for every $g \in$ G, $h \in$ H, $\Rightarrow$ $ghg^{-1} \in$ H.

or

A subgroup H of a group G is called a normal subgroup of G iff for $g \in G$, we have

$$gHg^{-1} = H \;\forall\; g \in G$$

Example 4. *Let G be the group of two by two invertible real matrices* $\begin{pmatrix} a & b \\ c & d \end{pmatrix}$; $ad - bc \neq 0$. *Let* $H = \left[\begin{pmatrix} a & 0 \\ 0 & a \end{pmatrix} : a \neq 0\right]$. *Then H is a normal subgroup of G.* (P.T.U. B.Tech. Dec. 2010)

Sol. We first show that H is a subgroup of G.

Let $h_1, h_2 \in H$ such that

$$h_1 = \begin{pmatrix} a & 0 \\ 0 & a \end{pmatrix},\; h_2 = \begin{pmatrix} a_1 & 0 \\ 0 & a_1 \end{pmatrix};\; a \neq 0,\; a_1 \neq 0$$

Now $$h_1 h_2 = \begin{pmatrix} a & 0 \\ 0 & a \end{pmatrix}\begin{pmatrix} a_1 & 0 \\ 0 & a_1 \end{pmatrix} = \begin{pmatrix} aa_1 & 0 \\ 0 & aa_1 \end{pmatrix} \in H \qquad | \; aa_1 \neq 0$$

i.e., H is closed under matrix multiplication. Further, For $A \in H$, we have

$$A = \begin{pmatrix} a & 0 \\ 0 & a \end{pmatrix},\; |A| = \begin{vmatrix} a & 0 \\ 0 & a \end{vmatrix} = a^2$$

Also $$A_{11} = a,\; A_{12} = 0,\; A_{21} = 0,\; A_{22} = a$$

$\therefore$ $$\text{adj } A = \begin{pmatrix} a & 0 \\ 0 & a \end{pmatrix}^T = \begin{pmatrix} a & 0 \\ 0 & a \end{pmatrix}$$

Hence $$A^{-1} = \frac{\text{adj} A}{|A|} = \frac{1}{a^2}\begin{pmatrix} a & 0 \\ 0 & a \end{pmatrix} = \begin{pmatrix} \frac{1}{a} & 0 \\ 0 & \frac{1}{a} \end{pmatrix} \in H,\; a \neq 0$$

Thus each element belonging to H has multiplicative inverse. Hence H is a subgroup of G.

Further, For $g = \begin{pmatrix} a & b \\ c & d \end{pmatrix} \in G$, $h = \begin{pmatrix} a & 0 \\ 0 & a \end{pmatrix} \in H$, Consider

$$ghg^{-1} = \begin{pmatrix} a & b \\ c & d \end{pmatrix}\begin{pmatrix} a & 0 \\ 0 & a \end{pmatrix}\begin{pmatrix} a & b \\ c & d \end{pmatrix}^{-1} = \begin{pmatrix} a & b \\ c & d \end{pmatrix}\begin{pmatrix} a & 0 \\ 0 & a \end{pmatrix}\begin{pmatrix} \frac{d}{ad-bc} & \frac{-b}{ad-bc} \\ \frac{-c}{ad-bc} & \frac{a}{ad-bc} \end{pmatrix}$$

$$= \begin{pmatrix} a^2 & ba \\ ca & da \end{pmatrix}\begin{pmatrix} \frac{d}{ad-bc} & \frac{-b}{ab-bc} \\ \frac{-c}{ad-bc} & \frac{a}{ad-bc} \end{pmatrix} = \begin{pmatrix} \frac{a^2d - bac}{ad-bc} & \frac{-a^2b + ba^2}{ad-bc} \\ \frac{cad - dac}{ad-bc} & \frac{-cab + da^2}{ad-bc} \end{pmatrix}$$

$$= \begin{pmatrix} \frac{a\,(ad-bc)}{ad-bc} & 0 \\ 0 & \frac{a\,(ad-bc)}{ad-bc} \end{pmatrix} = \begin{pmatrix} a & 0 \\ 0 & a \end{pmatrix} \in H$$

Hence H is a normal subgroup of G under matrix multiplication.

Example 5. *Let G be a group of two by two invertible real matrices* $\begin{pmatrix} a & b \\ c & d \end{pmatrix}$; $ad - bc \neq 0$ *under matrix multiplication. Let* $H = \left[\begin{pmatrix} a & 0 \\ 0 & b \end{pmatrix} : ab \neq 0\right]$. *Is H a normal subgroup of G ?*

Sol. We first show that H is a subgroup of G. Let A, B ∈ H such that

$$A = \begin{pmatrix} a & 0 \\ 0 & b \end{pmatrix}, ab \neq 0, \quad B = \begin{pmatrix} a_1 & 0 \\ 0 & b_1 \end{pmatrix}, a_1\, b_1 \neq 0$$

Consider $$AB = \begin{pmatrix} a & 0 \\ 0 & b \end{pmatrix}\begin{pmatrix} a_1 & 0 \\ 0 & b_1 \end{pmatrix} = \begin{pmatrix} aa_1 & 0 \\ 0 & bb_1 \end{pmatrix} \in H, \qquad | \because \; aba_1b_1 \neq 0$$

⇒ H is closed under multiplication of matrices

Further, for A ∈ H, $$A^{-1} = \begin{pmatrix} a & 0 \\ 0 & b \end{pmatrix}^{-1} = \begin{pmatrix} \frac{b}{ab} & 0 \\ 0 & \frac{a}{ab} \end{pmatrix} = \begin{pmatrix} \frac{1}{a} & 0 \\ 0 & \frac{1}{b} \end{pmatrix} \in H, \; \frac{1}{ab} \neq 0$$

Thus, every element of H has multiplicative inverse. Thus H is a subgroup of G under matrix multiplication.

Also, For $g = \begin{pmatrix} a & b \\ c & d \end{pmatrix} \in G$, $h = \begin{pmatrix} a & 0 \\ 0 & b \end{pmatrix} \in H$, Consider

$$ghg^{-1} = \begin{pmatrix} a & b \\ c & d \end{pmatrix}\begin{pmatrix} a & 0 \\ 0 & b \end{pmatrix}\begin{pmatrix} a & b \\ c & d \end{pmatrix}^{-1} = \begin{pmatrix} a & b \\ c & d \end{pmatrix}\begin{pmatrix} a & 0 \\ 0 & b \end{pmatrix}\begin{pmatrix} \frac{d}{ad-bc} & \frac{-b}{ad-bc} \\ \frac{-c}{ad-bc} & \frac{a}{ad-bc} \end{pmatrix}$$

$$= \begin{pmatrix} a^2 & b^2 \\ ca & db \end{pmatrix}\begin{pmatrix} \frac{d}{ad-bc} & \frac{-b}{ad-bc} \\ \frac{-c}{ad-bc} & \frac{a}{ad-bc} \end{pmatrix} = \begin{pmatrix} \frac{a^2d - b^2c}{ad-bc} & \frac{-a^2b + b^2a}{ad-bc} \\ \frac{cad - dbc}{ad-bc} & \frac{-cab + dab}{ad-bc} \end{pmatrix} \notin H$$

Hence H is not a normal subgroup of G under matrix multiplication.

Example 6. *Let G be the group of non-singular 2 × 2 matrices under matrix multiplication. Let H be the subset of G consisting of the lower triangular matrices i.e.; matrices of the form* $\begin{pmatrix} a & 0 \\ c & d \end{pmatrix}$ *where ad ≠ 0. Show that H is a subgroup of G, but not a normal subgroup.*

Sol. Let A, B ∈ H such that

$$A = \begin{pmatrix} a_1 & 0 \\ c_1 & d_1 \end{pmatrix}, \quad B = \begin{pmatrix} a_2 & 0 \\ c_2 & d_2 \end{pmatrix}$$

Consider $$AB = \begin{pmatrix} a_1 & 0 \\ c_1 & d_1 \end{pmatrix}\begin{pmatrix} a_2 & 0 \\ c_2 & d_2 \end{pmatrix}$$

$$= \begin{pmatrix} a_1a_2 & 0 \\ c_1a_2 + d_1c_2 & d_1d_2 \end{pmatrix} \in H$$

∴ H is closed under matrix multiplication.

Also for any M ∈ H, we have $M = \begin{pmatrix} a & 0 \\ c & d \end{pmatrix}$

⇒ $$|M| = \begin{vmatrix} a & 0 \\ c & d \end{vmatrix} = ad \neq 0 \text{ (given)}$$

$\therefore$ M^{-1} exists. Further $\begin{pmatrix} 1 & 0 \\ 0 & 1 \end{pmatrix} \in H$ is the identity of H. Hence, H is a subgroup of G. But H is not a normal subgroup of G.

Since, for example,

Take $g = \begin{pmatrix} 1 & 2 \\ 1 & 3 \end{pmatrix} \in G;\quad h = \begin{pmatrix} 1 & 0 \\ 1 & 1 \end{pmatrix} \in H$

Consider
$$ghg^{-1} = \begin{pmatrix} 1 & 2 \\ 1 & 3 \end{pmatrix}\begin{pmatrix} 1 & 0 \\ 1 & 1 \end{pmatrix}\begin{pmatrix} 1 & 2 \\ 1 & 3 \end{pmatrix}^{-1}$$
$$= \begin{pmatrix} 1 & 2 \\ 1 & 3 \end{pmatrix}\begin{pmatrix} 1 & 0 \\ 1 & 1 \end{pmatrix}\begin{pmatrix} 3 & -2 \\ -1 & 1 \end{pmatrix}$$
$$= \begin{pmatrix} 1 & 2 \\ 1 & 3 \end{pmatrix}\begin{pmatrix} 3 & -2 \\ 2 & -1 \end{pmatrix}$$
$$= \begin{pmatrix} 7 & -4 \\ 9 & -5 \end{pmatrix} \notin H.$$

Example 7. *Let G be the group of non-singular 2 × 2 matrices under matrix multiplication. Let H be a subset of G consisting of matrices with determinant 1. Show that K is a normal subgroup of G.*

Sol. We know that if $I = \begin{pmatrix} 1 & 0 \\ 0 & 1 \end{pmatrix}$, then

$$\det(I) = 1 \quad \therefore \quad I \in H. \textit{ i.e., } H \text{ has an identity.}$$

Let A, B $\in$ H $\Rightarrow$ det (A) = 1, det (B) = 1

Now, det (AB) = (det A) (det B) = 1.1 = 1

$\Rightarrow$ AB $\in$ H *i.e.,* H is closed under matrix multiplication. Let A $\in$ H $\Rightarrow$ det(A) = 1

Further, $\det(A^{-1}) = 1/\det(A) = 1/1 = 1$

$\therefore$ $A^{-1} \in H$. *i.e.,* H has an inverse.

$\therefore$ H is a subgroup of G.

Let X $\in$ G and A $\in$ H such that det A = 1

Consider $\det(XAX^{-1}) = \det(X)\det(A)\det(X^{-1})$

$$= \det(X) \,.\, 1 \,.\, \frac{1}{\det(X)} = 1$$

$\therefore$ $XAX^{-1} \in H$ for all X $\in$ G

$\therefore$ H is a normal subgroup of G.

Example 8. *Every subgroup of an abelian group is normal.*

Sol. Let H be a subgroup of a normal group G. We show H is normal. Let $h \in H$ and $g \in G$. Consider

$$ghg^{-1} = gg^{-1}h = eh = h \in H$$

$$\Rightarrow \quad ghg^{-1} \in H.$$

$\left| \begin{array}{l} h \in H \subseteq G \Rightarrow h \in G \\ \text{Also } h, g^{-1} \in G \text{ and} \\ \text{since G is abelian} \\ \therefore \ hg^{-1} = g^{-1}h \end{array} \right.$

Hence, H is a normal subgroup of G.

Theorem VI. *Let H be a subgroup and K be a normal subgroup of a group G. Show that HK is a subgroup of G.* (P.T.U. B.Tech. Dec. 2013)

Proof. We show that (*i*) HK is closed under multiplication

(*ii*) For $x \in$ HK, we should have $x^{-1} \in$ HK

(*iii*) $e \in$ HK

Let $x, y \in HK \Rightarrow x = h_1 k_1, y = h_2 k_2$, where $h_1, h_2 \in H$; $k_1, k_2 \in K$

Consider $xy = h_1 k_1 h_2 k_2 = h_1 h_2(h_2^{-1} k_1 h_2)k_2 \in HK$

$$\left[\begin{array}{l} \text{Let } h_2 \in H \Rightarrow h_2^{-1} \in H \subseteq G \Rightarrow h_2^{-1} \in G \\ \text{Since K is a normal subgroup of G, and } k_1 \in K, \\ \therefore \quad h_2^{-1} k_1 h_2 \in K \\ \Rightarrow \quad (h_2^{-1} k_1 h_2)\, k_2 \in K \\ \Rightarrow \quad h_1 h_2 (h_2^{-1} k_1 h_2)\, k_2 \in HK \end{array}\right]$$

Thus HK is closed under multiplication.

Further, For $x \in$ HK, we have

$$x^{-1} = (h_1 k_1)^{-1} = k_1^{-1} h_1^{-1} = h_1^{-1}(h_1 k_1^{-1} h_1^{-1})$$

$$\left[\begin{array}{l} \text{Let } h_1 \in H \subseteq G \Rightarrow h_1 \in G. \text{ Also } k_1 \in K \\ \Rightarrow \quad k_1^{-1} \in K \\ \text{Since K is a normal subgroup of G} \\ \therefore \quad h_1 k_1^{-1} h_1^{-1} \in K \\ \Rightarrow \quad h_1^{-1} (h_1 k_1^{-1}) h_1^{-1} \in HK \end{array}\right]$$

Thus $\quad x^{-1} \in HK.$

Finally, $\quad e \in H, e \in K \Rightarrow e \,.\, e \in HK \Rightarrow e \in HK$

Thus HK is a subgroup of G.

Theorem VII. *Let H is a subgroup of a group G. Then H is a normal subgroup of G iff*

$$aH = Ha \ \forall \ a \in G.$$

Proof. Let H is a normal subgroup of G. Then for $a \in G$, we have

$$aHa^{-1} = H$$

$$\Leftrightarrow \quad (aH\,a^{-1})a = Ha$$

$$\Leftrightarrow \quad aH(a^{-1}\,a) = Ha$$

$$\Leftrightarrow \quad aH\,e = Ha$$

$$\Leftrightarrow \quad aH = Ha$$

Theorem VIII. *The intersection of any number of normal subgroups of G is a normal subgroup of G.*

Proof. Let $H_1, H_2, H_3, \ldots$ be a collection of normal subgroups of a group G. Let

$$H = \bigcap_{i=1}^{\infty} H_i.$$ We show H is a normal subgroup of G.

Now the intersection of any collection of subgroups is a subgroup. $\therefore$ H is a subgroup of G. We show H is a normal subgroup of G. Let $g \in G$, $h \in H \Rightarrow h \in \bigcap_{i=1}^{\infty} H_i \Rightarrow h \in H_i \ \forall\ i$

But each H_i is normal in G

$$\therefore \qquad ghg^{-1} \in H_i\ \forall i \ \Rightarrow\ gh\,g^{-1} \in \bigcap_{i=1}^{\infty} = H_i$$

$$\Rightarrow \qquad ghg^{-1} \in H$$

$\Rightarrow$ H is a normal subgroup of G.

Cor. Prove that intersection of two normal subgroups is again a normal subgroup. *(P.T.U. B.Tech. May 2007, May 2006)*

Remark. Union of two normal subgroups may not be a normal subgroups

Consider G = The group of integers under +

$H_1 = \{\ldots -6, -4, -2, 0, 2, 4, 6 \ldots\}$

$H_2 = \{\ldots -12, -9, -6, -3, 0, 3, 6, 9, 12, \ldots\}$ are subgroups of G under +.

But $H_1 \cup H_2 = \{\ldots -4, -3, -2, 0, 2, 3, 4, 6, \ldots\}$

Since $2 \in H_1 \cup H_2$, $3 \in H_1 \cup H_2 \Rightarrow 2 + 3 = 5 \notin H_1 \cup H_2$ *i.e.,* $H_1 \cup H_2$ is not closed under +. Hence $H_1 \cup H_2$ is not a subgroup of G under +.

Theorem IX. *Let G is a group and H is a normal subgroup of G. Let G/H denotes the collection of right (left) cosets of H in G. Show that G/H is a group under the coset multiplication defined by*

$$aHbH = abH \quad \forall\ a, b \in G$$

Proof. (*i*) Closure Property. By definition, $G/H = \{aH : a \in G\}$

Let aH, $bH \in G/H$ and consider

$$\begin{aligned}(aH)\,(bH) &= a(Hb)H = a(bH)H \quad \mid Ha = aH \Leftrightarrow H \text{ is normal in G (Theorem VII)}\\ &= (ab)\,HH \\ &= abH \quad \mid HH = H\end{aligned}$$

Hence coset multiplication is well-defined *i.e.,* G/H is closed under coset multiplication.

(*ii*) Associativity. Let aH, bH, $cH \in G/H$ for all $a, b, c \in G$

Consider $\quad aH(bH\,cH) = aH(bcH) = abcH \quad$ | Using (*i*)

Also $\quad (aHbH)cH = (abH)cH = abcH$

$\Rightarrow \quad aH(bHcH) = (aHbH)cH$

Thus associativity holds in G/H.

(*iii*) Identity. Let $aH \in G/H$ for $a \in G$.

Consider $\quad (aH)H = a(HH) = aH \quad$ | HH = H

$\Rightarrow$ H is the identity element of G/H.

(*iv*) Inverse. Let $aH \in G/H$ and consider

$$(a^{-1}H)\,(aH) = (a^{-1}a)H = eH = H$$

i.e., $\quad a^{-1}H$ is the inverse of aH

Thus G/H is a group under coset multiplication.

8.23. QUOTIENT GROUP

Let G be a group and H be a normal subgroup of G. Let G/H denotes the set of right (left) cosets of H in G. Then G/H is a group (Proved in above theorem IX) called quotient group, or factor group under the coset multiplication defined by

$$(aH)\,(bH) = abH.$$

8.24. (a) CYCLIC SUBGROUP

Let G be any group and $a \in$ G. Then all the integral powers of a, given by

$$\ldots, a^{-3}, a^{-2}, a^{-1}, a^0 = e, a, a^2, a^3, \ldots$$

form a subgroup of G called the *cyclic group* generated by a and is denoted by $gp(a)$ or $< a >$. The element a is called generator of G.

Order of a, a ∈ G. The smallest positive integer m such that $a^m = e$, is known as order of a and is denoted by $o(a)$ or (a).

If $|a| = m$, then the cyclic subgroup $gp(a)$ has m elements given by

$$gp(a) = \{e, a, a^2, a^3, \ldots, a^{m-1}\}$$

e.g., If G = $\{\pm 1, \pm i)$, then G is a cyclic subgroup generated by i

Since $i^0 = 1, i^1 = i, i^2 = -1, i^3 = -i, i^4 = 1$

i.e., every element of G is of the form i^n, $n \in$ Z.

Also, the element i is a generator of G.

Example. The group G = $\{1, w, w^2\}$ is a cyclic group. Since

$$w^1 = w$$
$$w^2 = w^2$$
$$w^3 = 1$$

i.e., Hence each element of G is some power of w. Therefore, G is a cyclic group and w is a generator of G.

The group G = $\{\pm 1, \pm i\}$ is cyclic with 'i' as generator.

8.24. (b) CYCLIC GROUP

A group whose all elements are integral powers of one or more elements is called cyclic.

Remark. The order of a generator of the cyclic subgroup is equal to the order of the group.

e.g., $Z_{12} = [Z_{12}\,;\, +_{12}]$ is a cyclic subgroup.

Sol. $Z_{12} = \{0, 1, 2, \ldots 11, +_{12}\}$.

Consider $5 = 5$

$$5 +_{12} 5 = 10$$
$$5 +_{12} 5 +_{12} 5 = 3$$
$$5 +_{12} 5 +_{12} 5 +_{12} 5 = 8$$
$$5 +_{12} 5 +_{12} 5 +_{12} 5 +_{12} 5 = 25 = 1 \text{ etc.}$$

Thus we see that every element of Z_{12} is of the form $5n$ for some $n \in$ Z. Thus 5 is a generator of Z_{12}.

Hence $[Z_{12}, +_{12}]$ is a cyclic subgroup with 5 as generator. Since inverse of 5 is 7 ($5 +_{12} 7 = 0$), therefore, 7 is also a generator. (theorem X below)

Example 9. *The group of integers Z is cyclic under addition.*

Sol. Z = $\{0, \pm 1, \pm 2, \pm 3, \ldots\}$

Since $1 = 1$

$$1 + 1 = 2$$

$$1+1+1=3$$

$$\underbrace{1+1+1+\ldots 1}_{n \text{ times}} = n \text{ etc}$$

Thus we see that every element of Z is of the form $n(1)$. Thus Z is cyclic group. Hence $Z = <1>$. Also $Z = <-1>$.

Theorem X. *If a is a generator of a cyclic group G, show that inverse of a is also a generator.*

Proof. Let $G = <a>$ *i.e.,* G is a cyclic group and a is its generator. Let $g \in G$, then

$$g = a^r \text{ for some } r \in Z$$

Take $\quad r = -s,\ s \in Z$, we have

$$g = a^{-s} = (a^{-1})^s \text{ for some } s \in Z$$

Thus every element $g \in G$ is of the form $(a^{-1})^s$. Hence a^{-1} is a generator.

Theorem XI. *Every cyclic group is abelian.* (P.T.U. B. Tech. May 2006, May 2005)

Proof. Let G be a cyclic group with a as its generator. *i.e.,* let $G = <a>$ and $g_1 \in G$.

Then $g_1 = a^r$ for some $r \in Z$

Let $g_2 \in G$, then $g_2 = a^s$ for some $s \in Z$

Consider $\quad g_1 \cdot g_2 = a^r \,.\, a^s = a^{r+s}$

$= a^{s+r}$ $\quad\quad | \; r + s = s + r$ as Z is abelian

$= a^s \,.\, a^r = g_2 \,.\, g_1$

$\Rightarrow$ G is abelian.

Theorem XII. *Every subgroup of a cyclic group is cyclic.*

(P.T.U. B.Tech. Dec. 2009)

Proof. Let $G = <a>$ *i.e.,* G is a cyclic group with a as its generator. Let H be a subgroup of G.

Case I. If $H = \{e\}$, then $H = <e>$ *i.e.,* H is a cyclic group with e as a generator.

Case II. If $H \neq [e]$, then $o(H) \geq 2$ *i.e.,* there exists $e \neq a \in H$.

Since H is a subgroup, it must be closed under inverses and so contains positive powers of a. Let m is the smallest power of a such that $a^m \in H$. We claim $b = a^m$ is a generator of H. Let $x \in H$. But $H \subseteq G \;\therefore\; x \in G$.

Since G is a cyclic group G with a as its generator. $\therefore\; x = a^n$ for some $n \in Z$.

Dividing n by m, we get a quotient q and remainder r. *i.e.,*

$$n = mq + r,\ 0 \leq r < m$$

Now $\quad a^n = a^{mq+r} = a^{mq} \,.\, a^r = b^q \,.\, a^r$

$\Rightarrow \quad a^r = b^{-q} \,.\, a^n$

Here a^n, $b \in H$ and since H is a subgroup $\therefore\; b^{-q}\, a^n \in H$ which means $a^r \in H$.

But m was the least positive integer of a such that $a^m . \in H$ and $r < m$.

$\therefore$ We must have $r = 0$

Hence $a^n = b^q$ for some $q \in Z$

$\Rightarrow \quad x = a^n = b^q$ *i.e.,* every element $x \in H$ is of the form b^q for some $q \in Z$

$\therefore$ H is cyclic.

Theorem XIII. *If G is a cyclic group of order n and a is a generator of G. Let (n, k) = d. Then the order of the cyclic group generated by ka is $\frac{n}{d}$ where d is the greatest common divisor of n and k.*

Proof. Proof of this theorem is beyond the scope.

Example 10. *Find the order of the cyclic subgroup generated by 18 in Z_{30}.*

Sol. We know that 1 is a generator of Z_{30}. Also $18 = 18(1)$ *i.e.*, $k = 18$, $a = 1$, $n = 30$
The greatest common divisor of $(n, k) = (30, 18) = 6 = d$

$\therefore$ The order of cyclic subgroup generated by $18 = \frac{30}{6} = 5.$ (Theorem XIII)

Theorem XIV. *Every group of prime order is cyclic.*

Proof. Let G be a group of order *p*, *p* is prime. It means G must contain at least two elements. Since 2 is the least positive integer which is prime *i.e.*, if $a \in G$, then $o(a) \geq 2$.

Let $o(a) = m$ and H be a cyclic subgroup of G generated by *a*, then

$$o(H) = o(a) = m$$

| The order of a cyclic group is equal to the order of its generator

Also By Lagrange's theorem,

$$o(H) \mid o(G) \quad \Rightarrow \quad m \mid p$$

$\Rightarrow$ $\quad p = 1$ or $p = m$

But $\quad p \neq 1 \quad \therefore \quad p = m$

$\Rightarrow$ $\quad o(H) = o(G) \quad \Rightarrow \quad H = G.$

Hence G is cyclic since H is cyclic.

Theorem XV. *Let G is a cyclic group of order p* (*p* is prime). *Show that G has no proper subgroups except G and {e}.*

Proof. Let G is a cyclic group of order *p*.

Let H be any subgroup of G and $o(H) = m$.

By Lagrange theorem, $o(H) \mid o(G) \quad \Rightarrow \quad m \mid p$

$\Rightarrow$ $\quad p = 1$ or $p = m$

But $\quad p \neq 1 \quad \therefore \quad p = m$

i.e., $o(H) = m = p \quad \Rightarrow \quad$ H is a group of prime order and hence cyclic. Also $o(G) = m$

$\therefore \quad$ G = H *i.e.* G has no proper subgroups.

Remark. Cyclic subgroup generated by *a***.** Let G be any group and $a \in G$. Define $a^0 = e$; the *cyclic subgroup generated by a*, denoted by $<a>$, where $< a >$ denotes the set of all powers of *a*, is defined by $<a> = \{......, a^{-2}, a^{-1}, e, a, a^2, a^3, ...\}$

$<a>$ contains the identity element *e*, closed under group operation, contains inverses.

$\therefore$ $<a>$ is a subgroup of G and is called the *cyclic subgroup generated* by *a*.

Example 11. *Consider the group G = {1, 2, 3, 4, 5, 6} under multiplication modulo 7.*

(a) Find the multiplication table of G

(b) Find $2^{-1}, 3^{-1}, 6^{-1}$

(c) Find the orders and subgroups generated by 2 and 3

(d) Is G cyclic ? (P.T.U. B.Tech. Dec. 2013)

Sol. (*a*) By definition, $a \times_7 b$ = The remainder when *ab* is divided by 7

For *e.g.*, $5 \times_7 6 = 30 = 2$ (when 30 is divided by 7, the remainder is 2)

The multiplication table is shown below Table 8.10

Table 8.10

$\times_7$	1	2	3	4	5	6
1	1	2	3	4	5	6
2	2	4	6	1	3	5
3	3	6	2	5	1	4
4	4	1	5	2	6	3
5	5	3	1	6	4	2
6	6	5	4	3	2	1

(*b*) The identity element of G is 1. (As the first row inside the table is identical with the top most row).

∴ $2^{-1} = 4$ (In the table, the intersection of 2 and 4 is 1)

$3^{-1} = 5$

$6^{-1} = 6$

(*c*) We have $2 = 2$

$2 \times_7 2 = 4$

$2 \times_7 2 \times_7 2 = 8 = 1$

∴ $o(2) = 3$

Hence <2> = The subgroup generated by 2 = {1, 2, 4}

Also $3 = 3$

$3 \times_7 3 = 9 = 2,$

$3 \times_7 3 \times_7 3 = 27 = 6$

$3 \times_7 3 \times_7 3 \times_7 3 = 81 = 4$

$3 \times_7 3 \times_7 3 \times_7 3 \times_7 3 = 243 = 5$

$3 \times_7 3 \times_7 3 \times_7 3 \times_7 3 \times_7 3 = 729 = 1$

∴ $o(3) = 6$. ∴ The group generated by 3 is given as

<3> = {1, 2, 3, 4, 5, 6} = G

(*d*) Since $o(3) = 6 = o(G)$ ⇒ G is cyclic. Recall that a group G is cyclic if there exists an element $a \in G$ such that $o(a) = o(G)$.

Example 12. *Let G = [1, 5, 7, 11] under multiplication modulo 12.*

(a) Find the multiplication table of G

(b) Find the order of each element

(c) Is G cyclic ?

Sol. (*a*) We know $a \times_{12} b$ = The remainder when the product *ab* is divided by 12

i.e., $5 \times_{12} 7 = 35 = 11$ etc.

The multiplication table is shown below (Table 8.11)

Table 8.11

$\times_{12}$	1	5	7	11
1	1	5	7	11
5	5	1	11	7
7	7	11	1	5
11	11	7	5	1

(*b*) Order (1) = 1 (since 1 is the identity element)

To find order of 5. $5 \times_{12} 5 = 25 = 1$

∴ $o(5) = 2$

To find order of 7. $7 \times_{12} 7 = 49 = 1$

∴ $o(7) = 2$

To find order of 11. $11 \times_{12} 11 = 121 = 1$

∴ $o(11) = 2$

(*c*) We know that a group G is cyclic if there exists an element $a \in G$ such that $o(a) = o(G)$. Since $o(1) = 1$, $o(5) = 2$, $o(7) = 2$, $o(11) = 2$ *i.e.,*

There is no element of G whose order = 4

∴ G is not cyclic.

Example 13. *Consider the group G = {1, 2, 3, 4, 5, 6} under multiplication modulo 7.*

(a) Find the multiplication table of G

(b) Prove that G is a group

(c) *Find* 2^{-1}, 3^{-1}, 6^{-1}

(d) *Find the orders and subgroups generated by 2 and 3*

(e) *Is G cyclic ? Justify your answer* (P.T.U. B.Tech. Dec. 2009)

Sol. (a) and (b). Proceed yourself as in above example 12.

(c) $2^{-1} = 4$, $3^{-1} = 5$, $6^{-1} = 6$

(d) $< 2 > = \{1, 2, 4\}$, $< 3 > = \{1, 2, 3, 4, 5, 6\}$

(e) Since $o(B) = 6 = o(G) \Rightarrow$ G is cyclic.

Example 14. *Let G be a reduced residue system modulo 15 say, G = {1, 2, 4, 7, 8, 11, 13, 14}. Then G is a group under multiplication modulo 15*

(a) *Find the multiplication table of G.*

(b) *Find* 2^{-1}, 7^{-1}, 11^{-1}.

(c) *Find the orders and subgroups generated by 2, 7 and 11.*

(d) *Is G cyclic ?*

Sol. (a) The multiplication of G is shown below (Table 8.12):

Table 8.12

×	1	2	4	7	8	11	13	14
1	1	2	4	7	8	11	13	14
2	2	4	8	14	1	7	11	13
4	4	8	1	13	2	14	7	11
7	7	14	13	4	11	2	1	8
8	8	1	2	11	4	13	14	7
11	11	7	14	2	13	1	8	4
13	13	11	7	1	14	8	4	2
14	14	13	11	8	7	4	2	1

(b) From the table, $2 \times_{15} 8 = 1$ *i.e.*, 8 is the inverse of 2. Hence $2^{-1} = 8$

Also, $7 \times_{15} 13 = 1 \Rightarrow 13$ is the inverse of 7. Hence $7^{-1} = 13$

Further $11 \times_{15} 11 = 1 \Rightarrow 11$ is the inverse of 11. Hence $11^{-1} = 11$.

(c) (i) We have $2 \times_{15} 2 = 4$

$$2 \times_{15} 2 \times_{15} 2 = 8$$

$$2 \times_{15} 2 \times_{15} 2 \times_{15} 2 = 1$$

$\therefore$ order of $2 = o(2) = 4$

The group generated by 2 is given by

$$< 2 > = \{2^0, 2^1, 2^2, 2^3\} = \{1, 2, 4, 8\}$$

(ii) We have $7 \times_{15} 7 = 4$

$$7 \times_{15} 7 \times_{15} 7 = 4 \times_{15} 7 = 13$$

$$7 \times_{15} 7 \times_{15} 7 \times_{15} 7 = 13 \times_{15} 7 = 1$$

$\therefore$ order of $7 = o(7) = 4$

The group generated by 7 is given by

$$<7> = \{7^0, 7^1, 7^2, 7^3\} = \{1, 7, 4, 13\}$$

(iii) $11 \times_{15} 11 = 1 \quad \therefore$ order of $11 = o(11) = 2$

The subgroup generated by 11 is

$$<11> = \{11^0, 11^1\} = \{1, 11\}$$

(d) The group G is cyclic if $\exists$ an element whose order equals to the order of G. Here $O(G) = 8$

But we have proved that

$2 \times_{15} 2 \times_{15} 2 \times_{15} 2 = 1 \quad \therefore \quad o(2) = 4$

$4 \times_{15} 4 = 1 \quad \therefore \quad o(4) = 2$

$7 \times_{15} 7 \times_{15} 7 \times_{15} 7 = 1 \quad \therefore \quad o(7) = 4$

$8 \times_{15} 8 = 64 = 4$

$8 \times_{15} 8 \times_{15} 8 = 4 \times_{15} 8 = 2$

$8 \times_{15} 8 \times_{15} 8 \times_{15} 8 = 2 \times_{15} 8 = 1 \quad \therefore \quad o(8) = 4$

$11 \times_{15} 11 = 1 \quad \therefore \quad o(11) = 2$

Also, $13 \times_{15} 13 = 4$

$13 \times_{15} 13 \times_{15} 13 = 4 \times_{15} 13 = 7$

$13 \times_{15} 13 \times_{15} 13 \times_{15} 13 = 7 \times_{15} 13 = 1 \quad \therefore \quad o(13) = 4$

$14 \times_{15} 14 = 1 \quad \therefore \quad o(14) = 2$

Hence, there is no element $a \in$ G such that $o(a) = o(\text{G}) = 8$

$\therefore$ G is not cyclic.

8.25. MORPHISMS

The word '*morphism*' is a combination of various terms like, *homomorphism, isomorphism, actomorphism, endomorphism* etc.

8.25.1. Group Homomorphism

(*P.T.U. B. Tech. May 2007, May 2006*)

A mapping ϕ from a group (G,.). into a group ($\overline{\text{G}}$, *) is said to be a group homomorphism if

$$\phi(a \, . \, b) = \phi(a) * \phi(b) \quad \forall \; a, b \in \text{G}$$

8.25.2. Group Isomorphism (*P.T.U. B. Tech. Dec. 2007*)

A homomorphism ϕ which is one-one and onto is called **isomorphism** and the groups G and G′ are called **isomorphic**, written as G ≅ G′.

A homomorphism which is onto is called **epimorphism**

A homomorphism which is one-one is called **monomorphism.**

8.26. KERNEL f

If *f* is a homomorphism of G to $\overline{\text{G}}$, then kernel *f* is the set defined by

$$\text{Ker } f = [x \in \text{G} : f(x) = \bar{e}, \, \bar{e} \in \overline{\text{G}}]$$

8.27. IMAGE f

The image *f* is the set of the images of the elements under *f i.e.,*

$\text{Im}(f) = \{b \in \text{G}' : f(a) = b \text{ for } a \in \text{G}\}$ where *f* is a homomorphism of G to G′.

The term 'range *f*' is also used for 'image *f*'.

Example 15. *Let G be a group of real numbers under addition and let G′ be the group of positive real numbers under multiplication. Define f : G → G′ by* $f(a) = 2^a$.

Show that f is a homomorphism. Also show that G and G′ are isomorphic.

Sol. Given *f* is a mapping from (G, +) to (G′, .) defined by $f(a) = 2^a$

Let $a, b \in$ G and consider

$$f(a + b) = 2^{a+b} = 2^a \, . \, 2^b = f(a) \, . \, f(b)$$

Hence $f : \text{G} \to \text{G}'$ is homomorphism.

To check f is one-one. Let $f(a) = f(b)$

$$\Rightarrow \qquad 2^a = 2^b \Rightarrow a = b$$

$\therefore$ f is one-one.

To check f is onto : For each $a \in R$, we have 2^a is a positive real number. Thus $f(a) = 2^a$ is onto.

Hence $f : G \to G'$ is an isomorphism and the groups G and G′ are isomorphic *i.e.,* $G \cong G'$.

Example 16. *Let $f : G \to G'$ be defined by $f(Z) = | Z |$ where G = group of non-zero complex numbers under multiplication and G′ = group of non-zero real numbers under multiplication. Show that f is a group homomorphism. Also describe geometrically the Kernel K of the homomorphism f.*

Sol. Let $Z_1, Z_2 \in G$ be any two non-zero complex numbers.

Consider $\qquad f(Z_1 Z_2) = | Z_1 Z_2 | = | Z_1 | \, | Z_2 | = f(Z_1) f(Z_2)$

$\therefore$ f is a group homomorphism.

Also by definition,

$$\text{Kernel } f = \{ Z \in G : f(Z) = 1, 1 \text{ is the identity of } G' \}$$
$$= \{ Z \in G : | Z | = 1 \},$$

which is a circle with unit radius.

Example 17. *Let G be the group of two by two invertible matrices $\begin{pmatrix} a & b \\ c & d \end{pmatrix}$: $ad - bc \neq 0$.*

Define $\theta : G \to G$ by $\theta(A) = \dfrac{A}{|A|^2}$. Show that θ is a group homomorphism.

Sol. Let $A, B \in G$ such that $\theta(A) = \dfrac{A}{|A|^2}$, $\theta(B) = \dfrac{B}{|B|^2}$

Consider $$\theta(AB) = \frac{AB}{|AB|^2} = \frac{AB}{|A|^2 |B|^2}$$

$$= \frac{A}{|A|^2} \cdot \frac{B}{|B|^2} = \theta(A) \cdot \theta(B)$$

$\therefore$ θ is a group homomorphism.

Example 18. *Let G be a group and $g \in G$. Define a function $f : G \to G$ by $f(x) = gxg^{-1}$. Show that g is an isomorphism of G to G.*

Sol. Given $f : G \to G$ is a function defined by $f(x) = gxg^{-1}$ for each $x \in G$

Let $x, y \in G$ and consider

$$f(xy) = g(xy)g^{-1} = g(xg^{-1}gy)g^{-1} = (gxg^{-1})(gyg^{-1}) = f(x)f(y)$$

i.e., f is a homomorphism of G to G.

To show f is one-one : Let $f(x) = f(y)$

$$\Rightarrow \qquad gxg^{-1} = gyg^{-1}$$

$$\Rightarrow \qquad (gxg^{-1})g = (gyg^{-1})g$$

$$\Rightarrow \quad gx(g^{-1}g) = gy(g^{-1}g)$$
$$\Rightarrow \quad gxe = gye$$
$$\Rightarrow \quad gx = gy$$
$$\Rightarrow \quad g^{-1}(gx) = g^{-1}(gy)$$
$$\Rightarrow \quad (g^{-1}g)x = (g^{-1}g)y$$
$$\Rightarrow \quad ex = ey$$
$$\Rightarrow \quad x = y$$

i.e., f is one-one.

To show f is onto : Let $z \in$ G and consider

$$f(g^{-1} zg) = g(g^{-1} zg)g^{-1} = (gg^{-1})\, z(gg^{-1}) = eze = z$$

Thus for each $z \in$ G, we have $g^{-1} zg \in$ G such that $f(g^{-1} zg) = z$ *i.e.*, f is onto.

Hence G ≅ G.

Example 19. *Define* $\theta : Z_6$ *to* Z_3 *by* $\theta(n) = n(1)$ *(the sum of n ones in* Z_3*)*

Show that θ *is a homomorphism. If so, find Ker* θ *and Im(*θ*).*

Sol. By definition, $Z_6 = \{0, 1, 2, 3, 4, 5\}$ and $Z_3 = [0, 1, 2]$

Consider $\theta(0) = 0,\ \theta(1) = 1,\ \theta(2) = 1 + 1 = 2$

$\theta(3) = 1 + 1 + 1 = 3 = 0,\ \theta(4) = 1 + 1 + 1 + 1 = 4 = 1,$

$\theta(5) = 1 + 1 + 1 + 1 + 1 = 5 = 2$

Thus for each $n, m \in Z_6$, we have

$$\theta(n +_6 m) = (n + m)\,(1) = n(1) +_3 m(1) = \theta(n) +_3 \theta(m)$$

i.e., θ is a homomorphism

To find ker θ : Let $x \in$ Ker θ

$\Rightarrow \quad \theta(x) = 0$, the identity of Z_3

$x(1) = 0$

⇒ The sum of x ones which equals to zero, which means

$x = 0, 3$

i.e., Ker θ = {0, 3}

Also $\text{Im}(\theta) = \{\theta(n) : n \in Z_6\} = \{n(1) : n \in Z_6\} = [0, 1, 2] = Z_3$

Theorem XVI. *Let* $f : G \to G'$ *is a group homomorphism. Then*

(a) $f(e) = e',\ e \in G,\ e' \in G'$

(b) $f(a^{-1}) = (f(a))^{-1}\ \forall\ a \in G.$

Proof. *(a)* Given $f : G \to G'$ is a homomorphism from G to G′. For $x \in$ G, consider

$f(x)\, e' = f(x)$ | e' is identity of G′

$= f(xe) = f(x)\, f(e)$ | f is homomorphism

$\Rightarrow \quad e' = f(e)$ | Left cancellation law

$\Rightarrow \quad f(e) = e'$

(*b*) From Part (*a*), $e' = f(e) = f(aa^{-1})$

$= f(a)\, f(a^{-1})$ | f is homomorphism

$\Rightarrow \quad f(a)\, f(a^{-1}) = e'$

$\Rightarrow \quad (f(a))^{-1} f(a)\, f(a^{-1}) = (f(a))^{-1} e'$

$\Rightarrow \quad f(a^{-1}) = (f(a))^{-1}$

Theorem XVII. *If f is a homomorphism of G to $\overline{G}$ with Ker f = K. Show that K is a normal subgroup of G.* (P.T.U. B.Tech. Dec. 2007)

Proof. By definition,

$$\text{Ker } f = \{x \in G : f(x) = e',\ e' \in \overline{G}\} = K$$

We first show that Ker f is a subgroup of G

Let $\quad x, y \in \text{Ker } f \Rightarrow f(x) = e',\ f(y) = e'$

Consider $\quad f(xy^{-1}) = f(x)\, f(y^{-1})$ | Homomorphism

$= f(x)\, (f(y))^{-1} = e'\, (e')^{-1} = e'$

$\Rightarrow \quad xy^{-1} \in \text{Ker } f \Rightarrow$ Ker f is a subgroup of G.

Let $g \in G$ and $x \in \text{Ker } f$, consider

$f(gxg^{-1}) = f(g)\, f(xg^{-1})$ | f is homomorphism

$= f(g)\, f(x)\, f(g^{-1}) = f(g)\, f(x)\, (f(g))^{-1}$

$= f(g)\, e'\, (f(g))^{-1} = f(g)\, (f(g))^{-1} = e'$

$\Rightarrow \quad gxg^{-1} \in \text{Ker } f \Rightarrow$ Ker f is a normal subgroup of G.

Theorem XVIII. *Let f be a homomorphism of a group G to a group G′. Let Im(f) be the homomorphism image of G in G′. Then Im(f) is a subgroup of G′.*

Proof. By definition, $\text{Im}(f) = \{f(x) : x \in G\}$

Take $\quad e \in G \Rightarrow e' = f(e) \in \text{Im}(f)$

i.e. $\text{Im}(f) \neq \phi$, we first show that Im(f) is a subgroup of G′. Let $x', y' \in \text{Im}(f)$

$\Rightarrow$ There exists $x, y \in G$ such that $f(x) = x'$, $f(y) = y'$

Consider $\quad x'y'^{-1} = f(x)\, (f(y))^{-1}$

$= f(x)\, f(y^{-1})$ | f is a homomorphism

$= f(xy^{-1}) \in \text{Im}(f)$ | $x, y \in G$ and G is a group $\therefore\ xy^{-1} \in G$

$\Rightarrow \quad x'\, y'^{-1} \in \text{Im}(f)$

$\Rightarrow$ Im(f) is a subgroup of G′.

Theorem XIX. ***State and prove Fundamental Theorem of Group Homomorphism*** (P.T.U. B.Tech. Dec. 2012)

Statement. Let $f : G \to G'$ is a group from G to G′ homomorphism. Then $G/K \cong G'$, where K = Ker f

Proof. Given f is a group homomorphism of G to G′. Also, Ker f is a normal subgroup of G. $\therefore$ G/Ker f is defined.

Define $\theta : G/K \to G'$ by $\theta(Kx) = f(x)$, K = Ker f

We show θ is well-defined, homomorphism, one-one and onto.

θ is well-defined : Consider $Kx = Ky$

$\Rightarrow \quad xy^{-1} \in K = \text{Ker } f \quad | \ Ha = Hb \Leftrightarrow ab^{-1} \in H$

$\Rightarrow \quad f(xy^{-1}) = \bar{e}, \ \bar{e} \in G'$

$\Rightarrow \quad f(x) f(y^{-1}) = \bar{e} \quad | \text{ Homomorphism}$

$\Rightarrow \quad f(x) (f(y))^{-1} = \bar{e}$

$\Rightarrow \quad f(x) = f(y)$

$\Rightarrow \quad \theta(Kx) = \theta(Ky)$

$\Rightarrow$ θ is well-defined.

θ is homomorphism. Consider

$$\theta(KxKy) = \theta(K\,xy) = f(xy) \quad | \ HaHb = Hab$$
$$= f(x) f(y) = \theta(Kx)\, \theta(Ky)$$

$\Rightarrow$ θ is a homomorphism.

θ is one-one : Let $\theta(Kx) = \theta(Ky)$

$\Rightarrow \quad f(x) = f(y)$

$\Rightarrow \quad f(x) (f(y))^{-1} = \bar{e}$

$\Rightarrow \quad f(x) f(y^{-1}) = \bar{e}$

$\Rightarrow \quad f(xy^{-1}) = \bar{e}$

$\Rightarrow \quad xy^{-1} \in K = \text{Ker } f$

$\Rightarrow \quad Kx = Ky \quad | \ Ha = Hb \Leftrightarrow ab^{-1} \in H$

$\Rightarrow$ θ is one-one.

We lastly show that θ is onto. Let $y \in G'$. Since G' is the Image of G under f, there exists $x \in G$ such that $f(x) = y \Rightarrow \theta(Kx) = y$ *i.e.*, θ is onto. Therefore we have proved that θ is homomorphism, one-one and onto

$\therefore \quad G | K \cong G'$.

Example 20. *Let Z be the additive group of integers. Examine whether the function $f : Z \rightarrow Z$ defined by $f(x) = 2x \ \forall \ x \in Z$ is a group isomorphism?* (P.T.U. B.Tech. May 2013)

Sol. Let $x, y \in Z$ and consider

$$f(x + y) = 2(x + y)$$
$$= 2x + 2y$$
$$= f(x) + f(y)$$

$\therefore$ f is a group homomorphism

To show f is one-one:

Let $\quad f(x) = f(y)$

$\Rightarrow \quad 2x = 2y$

$\Rightarrow \quad x = y \quad \forall \ x, y \in Z$

$\therefore$ f is one-one

To show f is onto:

Let $y \in Z$ and consider $x \in Z$ such that

$$y = f(x) = 2x$$

$$\Rightarrow \quad x = \frac{y}{2}.$$

If $y = 3 \in Z$, then $x = \frac{3}{2} \notin Z$. $\therefore$ f is not onto.

Hence f is not a group isomorphism.

Example 21. (*a*) *Let G be a group of non-zero complex numbers under multiplication. Let G′ be the group of non-zero real numbers under multiplication. Consider the mapping* $f : G \to G'$ *defined by* $f(z) = | z |$. *Show that* $G \mid \text{Ker } f \cong G'$.

(*b*) *Define* $\theta : Z \to Z_{10}$ *by*

$\theta(n)$ = *The remainder when n is divided by 10. Show that θ is a homomorphism from Z to* Z_{10}. *Also find Ker θ and prove that* $Z \mid Z_{10} \cong Z_{10}$.

Sol. (*a*) Given $f : G \to G'$ defined by

$$f(z) = | z |$$

Let $z_1, z_2 \in G$ and consider

$$f(z_1 z_2) = | z_1 z_2 | = | z_1 | \, | z_2 | = f(z_1) f(z_2)$$

$\Rightarrow$ f is a homomorphism of G to G′.

By **fundamental theorem of group homomorphism**, $G \mid \text{Ker } f \cong G'$.

(*b*) Given $\theta : Z \to Z_{10}$, defined by

$$\theta(n) = \text{The remainder when } n \text{ is divided by } 10$$

Let $m, n \in Z$ and consider

$$\theta(m + n) = m +_{10} n = \theta(m) +_{10} \theta(n)$$

Thus θ is a homomorphism

Now, $\quad \text{Ker } \theta = \{n : \theta(n) = 0, \text{ The identity of } Z_{10}\}$

$= \{10n : n \in Z\} = 10Z$

By **fundamental theorem of group homomorphism.**

$$Z \mid \text{Ker } \theta \cong Z_{10}$$

$$\Rightarrow \quad Z \mid 10Z \cong Z_{10}.$$

Example 22. *Let* $S = N \times N$ *and* $*$ *be the operation on S defined by*

$$(a, b) * (a', b') = (a + a', b + b')$$

(*a*) *Show that* $*$ *is associative. Also, show that* $(S, *)$ *is a semi-group*

(*b*) *Define* $f : (S, *) \to (Z, +)$ *by* $f(a, b) = a - b$. *Show that f is a homomorphism.*

(*c*) *Find the congruence relation ~ in S determined by the homomorphism f. i.e.,* $x \sim y$ *if* $f(x) = f(y)$

(*d*) *Describe S/~. Does S/~ have an identity element? Does S/~ have an inverse element?*

Sol. (*a*) Let $x, y, z \in S$ be such that $x = (a, b)$, $y = (c, d)$, $z = (e, f)$ where a, b, c, d, e, f are elements of N. Consider.

$$\begin{aligned} x * (y * z) &= (a, b) * ((c, d) * (e, f)) \\ &= (a, b) * (c + e, d + f) \\ &= (a + (c + e), b + (d + f)) \qquad \ldots(1) \end{aligned}$$

and

$$\begin{aligned} (x * y) * z &= ((a, b) * (c, d)) * (e, f) \\ &= (a + c, b + d) * (e, f) \\ &= ((a + c) + e, (b + d) + f) \\ &= (a + (c + e), b + (d + f)) \qquad \ldots(2) \end{aligned}$$

| since a, b, c, d, e, f are elements of N, therefore associativity under addition holds in N

From (1) and (2)

$$x * (y * z) = (x * y) * z \ \forall\ x, y, z \in S$$

Also, given that S is closed under *. Therefore, (S, *) is a semigroup under *.

Consider $f(x * y) = f((a, b) * (c, d))$

$= f(a + c, b + d)$

$= a + c - (b + d)$

$= (a - b) + (c - d)$

$= f(a, b) + f(c, d)$

$= f(x) + f(y)$

$\therefore$ f is a homomorphism

(*c*) We know that a relation ~ on a non-empty set S is said to be a congruence relation on S if

(*i*) ~ is an equivalence relation

(*ii*) If $a \sim a'$, $b \sim b'$. Then $ab \sim a'b'$

Also, given $x \sim y$ if $f(x) = f(y)\ \forall\ x, y \in S$

Here $S = N \times N$ and if $x \in S$, then $\exists\ a, b \in N$ s.t. $x = (a, b)\ \forall\ a, b \in N$

Similarly, $y = (c, d)\ \forall\ c, d \in N$

Therefore, "$x \sim y$ if $f(x) = f(y)$" means $(a, b) \sim (c, d)$ if $f(a, b) = f(c, d)$

$\Rightarrow$ $a - b = c - d$

$\Rightarrow$ $a + d = b + c$

Thus, $(a, b) \sim (c, d)$ if $a + d = b + c\ \forall\ a, b, c, d \in N$ ...(1)

Here, the relation ~ defined by (1) is an equivalence relation (see example 17 (*b*), page 65 in chapter-2 "Relations")

We next show that the relation ~ is a congruence relation on S. For this, we show if $(a, b) \sim (a', b')$ and $(c, d) \sim (c', d')$. Then, we must have

$(a, b) * (c, d) \sim (a', b') * (c', d')$ ...(2)

Now $(a, b) \sim (a', b') \Rightarrow a + b' = b + a'$ ...(3)

and $(c, d) \sim (c', d') \Rightarrow c + d' = d + c'$ ...(4) | Using (1)

Therefore (2) will be true if

$(a + c, b + d) \sim (a' + c', b' + d')$ | by the definition of *

or if $a + c + b' + d' = b + d + a' + c'$

or if $a + b' + c + d' = b + a' + d + c'$ | Using (1)

or if $a + b' + c + d' = a + b' + c + d'$ | Using (3) and (4)

which is true

Hence, ~ is the required congruence relation on S.

(*d*) We know that if R is a congruence relation on a non-empty set S, then S/R, the set of all equivalence classes of elements of S, form a semi-group under the operation on the equivalence classes, defined by

$$[a]\ [b] = [ab]$$

To describe S/~. Since ~ is a congruence relation on S, S/~ is a semi-group under the operation on the equivalence classes as defined above. By using **fundamental theorem of semi-group homomorphism,** S/~ is isomorphic to f(Z) where

$f : (S, *) \to (Z, +)$ is a semi-group homomorphism defined by $f(a, b) = a - b$ where

$$S = N \times N \text{ and } a, b \in N$$

But, $f(Z) = \{m \in Z$: There exists $x \in S$ for which $f(x) = m\}$

$= \{m \in Z: f(a, b) = m \ \forall \ a, b \in N\}$

$= \{m \in Z: a - b = m \ \forall \ a, b \in N\}$

$= \{m \in Z = a = b + m \ \forall \ a, b \in N\}$

$= Z$

Hence, S/~ is isomophic to (Z, +).

It implies that S/~ and Z have same identity as well as same inverse (additive).

Since (Z, +) is an additive group under addition, (Z, +) has additive identity as well as additive inverse. Therefore, S/~ has additive identity as well as additive inverse.

Example 23. *Let S = N × N be the set of ordered pairs of positive integers with the operation* such that*

$(a, b) * (c, d) = (ad + bc, bd)$

*(a) Find (3, 4) * (1, 5) and (2, 1) * (4, 7).*

*(b) Show that * is associative. Also S is a semi-group.*

(c) Define $f: (S, *) \to (Q, +)$ by $f(a, b) = \frac{a}{b}$. Show that f is a homomorphism.

(d) Find the congruence relation ~ in S determined by the homomorphism f, i.e., x ~ y if f(x) = f(y)

(e) Describe S/~. Does S/~ have on identity elements? Does it have inverses?

Sol. (*a*) $(3, 4) * (1, 5) = (15 + 4, 20) = (19, 20)$

$(2, 1) * (4, 7) = (14 + 4, 7) = (18, 7)$

(*b*) Let $(a, b), (c, d), (e, f) \in N \times N$, then

$((a, b) * (c, d)) * (e, f) = (ad + bc, bd) * (e, f)$

$= ((a\,d + bc)\, f + bde, bdf)$

$= (adf + bcf + bde, bdf)$ ...(1)

Also, $(a, b) * ((c, d) * (e, f)) = (a, b) \times (cf + de, df)$

$= (adf + b(cf + de), bdf)$

$= (adf + bcf + bde, bdf)$ ...(2)

From (1) and (2),

$((a, b) * (c, d)) * (e, f) = (a, b) * ((c, d) * (e, f))$

∴ * is associative

Since, S is closed under * and * is associative. Hence S is a semi-group.

(*c*) Let $x = (a, b), y = (c, d)$

Then $f(x * y) = f((a, b) * (c, d)) = f(ad + bc, bd)$

$$= \frac{ad + bc}{bd} = \frac{a}{b} + \frac{c}{d}$$

$= f(a, b) + f(c, d) = f(x) + f(y)$

Thus, f is a homomorphism.

(*d*) We know that a relation ~ on a non-empty set S is said to be a congruence relation on S if

(*i*) ~ is an equivalence relation

(*ii*) If $a \sim a'$ $b \sim b'$, then $ab \sim a'b'$

Given $x \sim y$ if $f(x) = f(y)$ $\forall$ $x, y \in$ S

Here S = N × N and if $x \in$ S then, $\exists$ $a, b \in$ N such that $x = (a, b)$ $\forall$ $a, b \in$N

Similarly, $y = (c, d)$ $\forall$ $c, d \in$ N

Therefore, "$x \sim y$ if $f(x) = f(y)$" means $(a, b) \sim (c, d)$ if

$$f(a, b) = f(c, d)$$

$$\Rightarrow \quad \frac{a}{b} = \frac{c}{d}$$

$$\Rightarrow \quad ad = bc$$

Thus, $(a, b) \sim (c, d)$ if $ad = bc$ $\forall$ $a, b, c, d \in$ N ...(1)

Here, the relation ~ defined by (1) is an equivalence relation (See example 17(*b*), page 65 in chapter-2 "Relations")

We next show that the relation ~ is a congruence relation on S. For this, we show

If $(a, b) \sim (a', b')$ and $(c, d) \sim (c', d')$, then we must have

$$(a, b) * (c, d) \sim (a', b') * (c', d') \quad ...(2)$$

Now, if $(a, b) \sim (a', b')$, then

$$ab' = ba' \quad \text{[Using (1)]} \quad ...(3)$$

Again, if $(c, d) \sim (c', d')$, then

$$cd' = dc' \quad \text{[Using (1)]} \quad ...(4)$$

Also, (2) will be true if

$$(ad + bc, bd) \sim (a'd' + b'c', b'd') \quad | \text{ by the definition of } *$$

or if $(ad + bc)\, b'd' = bd(a'd' + b'c')$ | Using (1)

or if $(ad)\,(b'd') + (bc)\,(b'd') = (ba')\,(dd') + b(dc')\, b'$

or if $(bc)\,(b'd') + (bc)\,(b'd') = (ab')\,(dd') + b(cd')\, b'$ | Using (3) and (4)

or if $2bcb'd' = (ad)\, b'd' + (bc)\, b'd'$

or if $2bcb'd' = (bc)\, b'd' + (bc)\, b'd'$ | Using (1) $ad = bc$

or if $2bcb'd' = 2bcb'd'$, which is true

Hence, ~ is the requid congruence relation on S.

(*e*) We know that if R is a congruence relation on a non-empty set S, then S/R, the set of all equivalence classes of elements of S, form a semi-group under the operation on the equivalence classes, defined by

$$[a]\,[b] = [ab]$$

To describe S/~. Since ~ is a congruence relation on S, therefore, S/~ is a semi-group under the operation on the equivalence classes defined above.

Also, by using **fundamental theorem of semi-group homomorphism**, S/~ is isomorphic to f(Q) where

f : (S, *) → (Q, +) is a homomorphism defined by $f(a, b) = \dfrac{a}{b}$ where

$S = N \times N$ and $a, b \in N$

Now $f(Q) = \{q \in Q$: there exists $x \in S$ for which $f(x) = q\}$

$= \{q \in Q: f(a, b) = q \ \forall \ a, b \in N\}$

$$= \left\{q \in Q : \frac{a}{b} = q \ \forall \ a, b \in N\right\}$$

$= Q^+$, the set of +ve rational numbers

Hence, S/~ is isomorphic to $(Q^+, +)$.

Further, since S/~ and $(Q^+, +)$ are isomorphic, both S/~ and $(Q^+, +)$ have same identity as well as same inverse.

But, Q^+, the set of positive rational number has no identity (additive) as well as no inverse (additive) therefore, S/~ has no identity and no inverse.

Example 24. *Show that $G = \{1, -1, i, -i\}$ is a group under multiplication. Also $G \cong Z_4$ by giving an explicit isomorphism $f: G \to Z_4$.*

Sol. Given $G = \{1, -1, i, -i\}$ such that $i^2 = -1$. Clearly G is a group under multiplication. (prove yourself)

Also, $Z_4 = \{0, 1, 2, 3\}$

Define $f : G \to Z_4$ by

$f(1) = 0, f(-1) = 2$

$f(i) = 1, f(-i) = 3$

Then, $G \cong Z_4$.

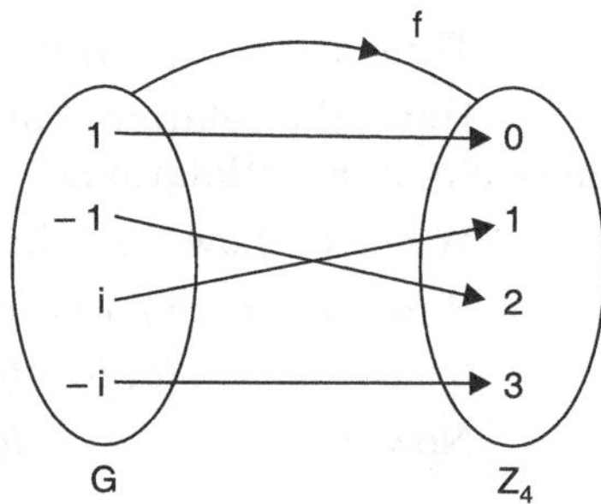

Theorem XX. *Any finite cyclic group of order n is isomorphic to Z_n.*

(P.T.U. B.Tech. Dec. 2012)

Proof. Let $G = \langle a \rangle$ be a finite cyclic group, with a as its generator and let $o(G) = n$

Define $f : Z \to G$ by $f(m) = a^m$

Let $m, r \in Z$ such that $f(m) = a^m, f(r) = a^r$

Consider $f(m + r) = a^{m+r} = a^m \cdot a^r = f(m) f(r)$

Thus f is a homomorphism of Z to G.

$\therefore$ By **fundamental theorem of group homomorphism.** $Z \mid \text{Ker } f \cong G$

But if $s \in$ Ker f, then by definition,

$f(s) = e, e \in G$

$\Leftrightarrow$ $a^s = e$

$\Leftrightarrow$ $o(a)/s$

$\Leftrightarrow$ n/s

$\Leftrightarrow$ $s = nk$, for some k

$\Leftrightarrow$ $s \in \langle n \rangle$

$\Leftrightarrow$ Ker $f = \langle n \rangle$

Hence $Z \mid \langle n \rangle \cong G$ or $G \cong Z_n$, where $Z_n = Z/\langle n \rangle$.

8.28. DIHEDRAL GROUP

Let $D_n = \{e, x, x^2, ..., x^{n-1}, y, yx, yx^2, ..., yx^{n-1}\}$ be a set of $2n$ element satisfying $x^n = e, y^2 = e, xy = yx^{-1}$, Then D_n is a group called as Dihedral group for $n \geq 3$

or

Let $D_n = \{e, y, y^2, y^3, ..., y^{n-1}, x, xy, xy^2, ..., xy^{n-1}\}$ be a set of $2n$ elements satisfying $y^n = e, x^2 = e, xy = y^{-1}x$, then D_n is a group called **Dihedral group for $n \geq 3$.**

or

Let $G = \{x^i y^j, i = 0,1; j = 0, 1, 2, ...n$, satisfying $x^2 = e, y^n = e, xy = y^{-1}x\}$.

Then G is a group, known as Dihedral group,

Remark. The order of the dihedral group D_n is given as $O(D_n) = 2n$.

Example 1. *Consider the dihedral group D_n, $n \geq 3$ defined by*

$$D_n = \{e, y, y^2, y^3, ..., y^{n-1}, x, xy, xy^2, ... xy^{n-1}\}$$

satisfying $y^n = e, x^2 = e, xy = y^{-1}x.$

Find the product (a) $(xy)(xy^2)$ (b) yxy^2 in terms of elements of D_n

Sol. (*a*) Consider $xy = y^{-1}x$ | By definition

$\Rightarrow$ $(xy)x^{-1} = (y^{-1}x)x^{-1} = y^{-1}(xx^{-1}) = y^{-1}$

$\Rightarrow$ $xyx = y^{-1}$ ...(1) | $\because$ $x^2 = e \Rightarrow x = x^{-1}$

Further, $(xy)(xy^2) = (xyx)y^2 = (y^{-1})y^2$ | Using (1)

$= y$

(*b*) Consider $xy = y^{-1}x$

$\Rightarrow$ $y(xy) = y(y^{-1}x) = (yy^{-1})x = x$

$\Rightarrow$ $(yxy)(y) = xy$

$\Rightarrow$ $yxy^2 = xy$

Example 2. *Let $H = \{y, y^2, y^3, ..., y^{n-1}, y^n = e\}$ be a subgroup of D_n, $n \geq 3$. Show that H is a normal subgroup of D_n*

Sol. By definition,

$$D_n = \{e, y, y^2, y^3, ..., y^{n-1}, x, xy, xy^2, ..., xy^{n-1}\}$$

satisfying $y^n = e, x^2 = e, xy = y^{-1}x.$

Let $g \in D_n$ and $y^j \in H$, $1 \leq j \leq n$

Case I. If $g = y^j$, then

$$gy^jg^{-1} = y^jy^jy^{-j} = y^j \in H$$

$\therefore$ H is a normal subgroup of D_n

Case II. If $g = xy^i$, $1 \leq i \leq n$, then

$gy^jg^{-1} = xy^iy^j(xy^i)^{-1} = xy^{i+j}y^{-i}x^{-1}$ | $(ab)^{-1} = b^{-1}a^{-1}$

$= xy^jx^{-1} = (xyx^{-1})^j$ | See remark below

$= (y^{-1}x\,x^{-1})^j$ | $xy = y^{-1}x$

$= (y^{-1})^j \in H$

$\therefore$ H is a normal subgroup of D_n.

Remark.

$(xyx^{-1})^j = (xyx^{-1})(xyx^{-1})(xyx^{-1}) ... (xyx^{-1})$ (j times)

$= xy(x^{-1}x)\, y\, (x^{-1}x) ... (yx^{-1})$ (j times)

$= xy^jx^{-1}$

Example 3. *Let $G = D_n$ and $H = \{y, y^2, y^3, ..., y^{n-1}, y^n = e\}$*

Then G/H is isomorphic to the Multiplicative group {1, –1}, where D_n is the dihedral group of order 2n.

Sol. Define $W = \{1, -1\}$ and $\theta : G \to W$ by

$$\theta(y^i) = 1, \quad 1 \le i \le n$$
$$\theta(xy^i) = -1, \quad 1 \le i \le n$$

We show θ is a homomorphism from G to W.

By definition,

$$D_n = \{y, y^2, y^3, ..., y^{n-1}, x, xy, xy^2, ..., xy^{n-1}\}$$

satisfying $y^n = e = x^2$, $xy = y^{-1}x$.

Let y^i, xy^k be any two elements of D_n, where $k = 0, 1, 2, ..., n - 1$

Consider

$$\theta(y^i xy^k) = \theta(y^i xy^i y^{k-i})$$
$$= \theta(y^i y^{-i} xy^{k-i}) \quad \text{| See remark below}$$
$$= \theta(xy^{k-i})$$
$$= -1 \quad ...(1)$$

Also $\theta(y^i)\,\theta(xy^k) = 1 \,.\, (-1) = -1$ ...(2)

From (1) and (2),

$$\theta(y^i xy^k) = \theta(y^i)\,\theta(xy^k)$$

Similarly $\theta(xy^k y^i) = \theta(xy^{k+i}) = -1$

and $\theta(xy^k)\,\theta(y^i) = (-1)\,(1) = -1$

∴ $\theta(xy^k y^i) = \theta(xy^k)\,\theta(y^i)$

∴ θ is a homomorphism.

Also θ is onto, therefore, By fundamental theorem of group homomorphism,

$$D_n/\ker\theta \cong W \text{ . But}$$
$$\ker\theta = \{g \in D_n : \theta(g) = 1\}$$
$$= \{y^i, 1 \le i \le n\}$$
$$= H$$

∴ D_n/H is isomorphic to W.

TEST YOUR KNOWLEDGE 8.3

1. Consider the additive group of integers under +. Let 4Z is a subgroup of Z. List the distinct cosets of 4Z in Z.
2. Is there any group with a subgroup H such that $o(H) = 6$ and $[G: H] = 6$. ? Is your answer unique ?
3. Let H be a subgroup of a group G and $a, b \in G$. Then $a \in b * H$ iff $b^{-1} * a \in H$.
4. If H is a finite subgroup of a group G. Show that H and any coset H*a* have the same number of elements.
5. Let H be a normal subgroup of a group G. Then the cosets of H in G form a group under coset multiplication defined by $(aH)(bH) = abH$
6. Let $f : G \to G'$ be a homomorphism with kernel K. Then K is a normal subgroup of G.
7. Show that any infinite cyclic group is isomorphic to additive group of integers.
8. What are the generators of the additive group *z* ?

9. Which of the following groups are cyclic ? Explain

(*i*) [Q, +] (*ii*) [R^+, .] (*iii*) [6*z*, +].

10. Determine the order of the cyclic subgroup generated by the element

(*i*) Z_{25}, 15 (*ii*) Z_{64}, 2.

11. Prove that if *o*(G) > 2 and G is cyclic, then G has atleast two generators.

12. Which of the following functions are homomorphism ?

(*i*) $f : R^* \to R^+$ defined by $f(a) = | a |$

(*ii*) $f : Z_5 \to Z_2$ defined by $f(n) = 0$ if n is even and $f(n) = 1$ if n is odd.

(*iii*) $f : Z_8 \to Z_2$ defined by $f(n) = 0$ if n is even and $f(n) = 1$ if n is odd.

13. How many homomorphism are there from ?

(*i*) Z_{12} into Z_{12} (*ii*) Z_{12} into Z_4 (*iii*) Z_9 into Z_8.

14. If G is an abelian group and let f is a function from G into G, defined by $f(x) = x^2$. Is f a homomorphism ? an isomorphism ?

15. Consider the group G = {0, 1, 2, 3, 4, 5} under addition modulo 6.

(*a*) Find the multiplication table of G.

(*b*) Prove that G is a group.

(*c*) Find 2^{-1}, 3^{-1}, 5^{-1}.

(*d*) Find the orders and subgroups generated by 2 and 3.

(*e*) Is G cyclic ? Justify your answer.

16. Let (G, ∗) be a group and $a \in$ G. Define $f : G \to G$ by $f(x) = a * x$

(*a*) Prove that f is a bijection

(*b*) On the basis of part (*a*), describe a set of bijection on the set of integers.

(*P.T.U. B.Tech. May 2005*)

17. Consider Z_{20} = {0, 1, ..., 19} under addition modulo 20. Let H be the subgroup generated by 5.

(*a*) Find the elements and order of H. (*b*) Find the cosets of H in Z_{20}.

18. Consider G = {1, 5, 7, 11, 13, 17} under multiplication modulo 18.

(*a*) Construct the multiplication table of G.

(*b*) Find 5^{-1}, 7^{-1}, and 17^{-1}.

(*c*) Find the order and subgroups generated by: (*i*) 5, (*ii*) 13.

(*d*) Is G cyclic ?

19. Consider G = {1, 5, 7, 11) under multiplication modulo 12. (*a*) Find the order of each element (*b*) Is G cyclic? (*c*) Find all subgroups.

20. Let S = **N** × **N**. Let ∗ be the operation on S defined by $(a, b) * (a', b') = (aa', bb')$.

(*a*) Show that ∗ is associative. (Hence S is a semigroup.)

(*b*) Define f : (S, ∗) → (Q, ×) by $f(a, b) = a/b$. Show that f is a homomorphism.

(*c*) Find the congruence relation ~ in S determined by the homomorphism f, *i.e.*, $x \sim y$ if $f(x) = f(y)$.

(*d*) Describe S/~. Does S/~ have an identity element? Does it have inverses?

21. Consider the set **N** of positive integers and let ∗ denote the operation of least common multiple (*l.c.m.*) on **N**.

(*a*) Find 4 ∗ 6, 3 ∗ 5, 9 ∗ 18, and 1 ∗ 6.

(*b*) Is (**N**, ∗) a semigroup ? Is it commutative ?

(*c*) Find the identity element of ∗.

(*d*) Which elements in **N**, if any, have inverse and what are they ?

22. Let Q be the set of rational numbers and let * be the relation on Q defined by $a * b = a + b - ab$

(*a*) Find $3 * 4$, $2 * (-5)$ and $7 * \frac{1}{2}$

(*b*) Is (Q, *) a semi-group ? Is it commutative ?

(*c*) Find the identity element for *.

(*d*) Do any of the elements in Q have an inverse ? why ?

Answers

1. 4Z, 1 + 4Z, 2 + 4Z, 3 + 4Z.

2. Take $G = Z_{36}$ and H = <6> *i.e.*, a cyclic group with 6 as its generator $\therefore [G : H] = \frac{o(G)}{o(H)} = \frac{36}{6} = 6$.

Also by Lagrange's theorem, $o(H) \mid o(G)$ is true. The answer is not unique.

8. 1, – 1

9. (*iii*), 6 is a generator

10. (*i*) 5, (*ii*) 32

12. (*i*) (*ii*) No (*iii*) Yes

14. Yes, Yes

15. (*a*) Table 8.13

Table 8.13

+	0	1	2	3	4	5
0	0	1	2	3	4	5
1	1	2	3	4	5	0
2	2	3	4	5	0	1
3	3	4	5	0	1	2
4	4	5	0	1	2	3
5	5	0	1	2	3	4

(*c*) $2^{-1} = 4$, $3^{-1} = 3$, $5^{-1} = 1$

(*d*) <2> = {1, 2, 4} and <3> = {1, 3}

(*e*) Yes.

16. Required bijection is $f : Z \to Z$ defined by $f(x) = x + a$ " $x \in Z$

17. (*a*) H = {0, 5, 10, 15} and o(H) = 3

(*b*)

$$H = H$$
$$H + 1 = \{1, 6, 11, 16\}$$
$$H + 2 = \{2, 7, 12, 17\}$$
$$H + 3 = \{3, 8, 13, 18\}$$
$$H + 4 = \{4, 9, 14, 19\}$$

18. (*a*)

Table 8.14

$\times_{18}$	1	5	7	11	13	17
1	1	5	7	11	13	17
5	5	7	17	1	11	13
7	7	17	13	5	1	11
11	11	1	5	13	17	7
13	13	11	1	17	7	5
17	17	13	11	7	5	1

(*b*) $5^{-1} = 11$, $7^{-1} = 13$, $17^{-1} = 17$

(*c*) (*i*) <5> = the subgroup generated by 5 = G and order (<5>) = 6 (*ii*) <13> = the subgroup generated by 13 = {1, 7, 13} and order (<13>) = 3

(*d*) Yes,

19. (*a*) o(1) = 1, o(5) = 2, o(7) = 2, o(11) = 2

(*b*) No

(*c*) G, {1}, {1, 7}, {1, 5}, {1, 11}

20. (*c*) The congruence relation ~ is given as: $(a, b) \sim (c, d)$ if $ad = bc$

(*d*) yes

21. (*a*) $4 * 6 = 12$, $3 * 5 = 15$, $9 * 18 = 18$, $1 * 6 = 6$

(*b*) Yes, since the operation of least common multiple is associative. Also $a * b = b * a$ *i.e.*, the operation of least common multiple is commutative. Hence, (N, *) is a commutative semi-group.

(*c*) 1

(*d*) The only element which has an inverse is 1 and it is its own inverse

22. (*a*) $3 * 4 = -5$, $2 * (-5) = 7$, $7 * \frac{1}{2} = 4$

(*b*) Yes, (Q, *) is a commutative semi-group

(*c*) 0 is the identity element

(*d*) Yes, if $a \neq 1 \in Q$, then a has an inverse and it is $\frac{a}{a-1}$.

Hints

4. See Proof of Lagrange's theorem.

9. (*i*), (*ii*) Let $Q = \langle g \rangle$ is a generated by g *i.e.*, $\langle g \rangle = \{nq : n \in Z\}$. But this set contains only integral multiples of q, not every element in Q. Hence Q cannot be cyclic.

10. See example 10.

12. (*ii*) Consider $f(2 +_5 4) = f(6) = f(1) = 1$

Also $f(2) +_2 f(4) = 0 + 0 = 0$.

14. $f(xy) = (xy)^2 = x^2y^2$ | Since G is abelian

$= f(x) f(y)$

15. (*d*) (*i*) $2 +_6 2 = 4$

$2 +_6 2 +_6 2 = 0 \quad \therefore \quad o(2) = 3$

$\langle 2 \rangle = \{1, 2, 4\}$

(*ii*) $3 +_6 3 = 0 \quad \therefore \quad o(3) = 2$

$\langle 3 \rangle = \{1, 3\}$

(*e*) $o(2) = 3, \quad o(3) = 2$

$4 +_6 4 = 2, \quad 4 +_6 4 +_6 4 = 2 +_6 4 = 0$

$\therefore \quad o(4) = 3$

$5 +_6 5 = 4, \quad 5 +_6 5 +_6 5 = 4 +_6 5 = 3$

$5 +_6 5 +_6 5 +_6 5 = 3 +_6 5 = 2$

$5 +_6 5 +_6 5 +_6 5 +_6 5 = 2 +_6 5 = 1$

$\therefore \quad o(5) = 5$

Also, $1 +_6 1 +_6 1 +_6 1 +_6 1 +_6 1 = 0$

$\therefore \quad o(1) = 6 = o(G)$

$\therefore$ G is cyclic as there is an element $1 \in G$ such that

$o(1) = o(G)$.

16. Given G is a group under $*$ and $a \in G$. Also $f : G \to G$ is defined by

$$f(x) = a * x \quad \forall\, x \in G$$

(*a*) We show f is a bijection. *i.e.*, f is one-one and onto.

***f* is one-one.** Let $x, y \in G$ such that $f(x) = f(y)$

$\Rightarrow \quad a * x = a * y$

$\Rightarrow \quad x = y$ | Left cancellation law

$\Rightarrow \quad f$ is one-one.

***f* is onto.** Let $a, x \in G$ and since G is a group under $*$, G must be closed under $*$ *i.e.*, $a * x \in G$.

Also given $f(x) = a * x$. Therefore for each $x \in G$, there exists an element $a * x \in G$, such that $f(x) = a * x$.

Hence f is onto. It implies f is a bijection.

(*b*) Let Z denotes the set of integers and $a \in Z$.

Define $f : Z \to Z$ such that $f(x) = x + a\ \forall\ x \in Z$. We show f is one-one and onto.

Let $\quad f(x) = f(y)$ for $x, y \in Z$

$\Rightarrow \quad x + a = y + a$ | Right cancellation law

$\Rightarrow \quad x = y.$

Hence f is one-one

Let $y \in Z$ such that $\quad f(x) = y$

$\Rightarrow \quad x + a = y$

$\Rightarrow \quad x = y - a \in Z$

Also for each $y \in Z$, we can find $x \in Z$ such that $f(x) = y$. Hence f is onto.

Since f is one-one and onto,

$\therefore \quad f$ is a bijection.

17. See examples on "cosets" and example 14, Art. 8.24 (cyclic group) in this chapter

18. See examples 11, Art. 8.24 (cyclic group) in this chapter

19. See examples 12,13 Art. 8.24 (cyclic group) in this chapter

20. (*c*) We know that a relation ~ on a non-empty set S is said to be a congruence relation on S if

(*i*) ~ is an equivalence relation on S

(*ii*) If $a \sim a'$, $b \sim b'$, then $ab \sim a'b'$

Given, $x \sim y$ if $f(x) = f(y)\ \forall\ x, y \in S$.

Here $S = N \times N$ and if $x \in S = N \times N \ \Rightarrow\ \exists\ a, b \in N$ such that $x = (a, b)$

Similarly, $y = (c, d)$ where $c, d \in N$.

Therefore, "$x \sim y$ if $f(x) = f(y)$" means $(a, b) \sim (c, d)$ if $f(a, b) = f(c, d)$

$$\Rightarrow \quad \frac{a}{b} = \frac{c}{d}$$

$$\Rightarrow \quad ad = bc$$

Thus, $(a, b) \sim (c, d)$ if $\ ad = bc$ ...(1)

Here, the relation ~ defined by (1) is an equivalence relation (example 17 (*b*) page 65 chapter 2 "Relations")

We next show that ~ is a congruence relation on S.

For this, we show if $(a, b) \sim (a', b')$ and $(c, d) \sim (c', d')$, then we must have

$$(a, b) * (c, d) \sim (a', b')\,(c', d') \qquad ...(2)$$

Now if $(a, b) \sim (a', b')$ then

$$ab' = ba' \quad \ldots(3) \text{ [Using (1)]}$$

Again, if $(c, d) \sim (c', d')$, then

$$cd' = dc' \quad \ldots(4) \text{ [Using (1)]}$$

Also (2) will be true if

$$(ac, bd) \sim (a'c', b'd') \quad | \text{ by the definition of } *$$

or if $\quad acb'd' = bda'c' \quad$ | Using (1)

or if $\quad acb'd' = ba'dc'$

or if $\quad acb'd' = ab'cd' \quad$ | Using (3) and (4)

or if $\quad acb'd' = acb'd'$

which is true

Hence, ~ is the required congruence relation on S.

(*d*) We know that if R is a congruence relation on a non-empty set S, then S/R, the set of all equivalence classes of elements of S, form a semi-group under the operation on the equivalence classes, defined by

$$[a]\,[b] = [ab]$$

To describe S/~. Since ~ is a congruence relation on S, therefore, S/~ is a semi-group under the operation on the equivalence classes as defined above.

Also, by using **fundamental theorem of semi-group homomorphism**, S/~ is isomorphic to $f(Q)$.

where $f = (S, *) \to (Q, \times)$ is a homomorphism defined by $f(a, b) = \dfrac{a}{b}$ where $S = N \times N$ and $a, b \in N$

But, $f(Q) = \{q \in Q\text{: There exists } x \in S \text{ for which } f(x) = q\}$

$$= \{q \in Q: f(a, b) = q \ \forall\ a, b \in N\}$$

$$= \left\{q \in Q: \frac{a}{b} = q \ \forall\ a, b \in N\right\}$$

$= Q^+$, the set of positive rational numbers

Hence, S/~ is isomorphic to $(Q^+, \times)$. Further, since S/~ and Q^+ are isomorphic, therefore both S/~ and Q^+ have same identities as well as same inverses.

But, since $(Q^+, \times)$ is a set of +ve rational numbers and hence it forms a group under multiplication and hence has identity as well as inverse. Therefore, S/~ has also identity (multiplicative) as well as inverse (multiplicative).

21. (*a*) Let $x, y \in N$, Then $x * y$ means *l.c.m.*, of x and y.

$4 * 6 = l.c.m.$, of 4 and 6 = 12

2	4,	6
2	2,	3
3	1,	3
	1,	1

l.c.m., of 4 and 6 = 2 × 2 × 3 = 12

(*c*) $1 * a = l.c.m.$, of 1 and $a = 1$ for any positive integer a.

Also, $a * 1 = l.c.m.$, of a and 1 = 1

∴ 1 is the identity element of N

(*d*) Let $a \in N$ and b is the inverse of a such that $a * b = 1$

⇒ l.c.m. of a and $b = 1$, which is possible iff $a = 1$, $b = 1$

∴ The only number which has an inverse is 1 and it is its own inverse.

22. (*c*) Let e is the identity element for $*$, then for every $1 \neq a \in Q$, we have

$$a * e = a \ \Rightarrow\ a + e - ae = a$$

$$\Rightarrow \quad e - ae = 0$$

$$\Rightarrow \quad e(1 - a) = 0 \ \Rightarrow\ e = 0 \quad | \ a \neq 1$$

Hence, the identity element is 0.

(*d*) Let $a, x \in Q$ are such that $a * x = 0$ (identity)

$\Rightarrow \quad a + x - ax = 0 \Rightarrow a + x(1-a) = 0$

$\Rightarrow \quad a = -x(1-a) \Rightarrow x = \dfrac{a}{a-1}, \ a \neq 1$

This $\dfrac{a}{a-1}$ is the inverse of a ($a \neq 1$).

8.29. DIRECT PRODUCT OF GROUPS

Let G_1 and G_2 be any two groups and $G = G_1 \times G_2$ denotes the cartesian product of G_1 and G_2 with the binary operation $*$ defined by

$$(g_1, g_2) * (g'_1, g'_2) = (g_1g'_1, g_2g'_2) \ \forall \ g_1, g'_1 \in G_1, g_2, g'_2 \in G_2$$

The group $G = G_1 \times G_2$, is know as *direct product* or *external direct product* of G_1 and G_2.

Theorem. *Show that the direct product of G_1 and G_2, denoted by $G = G_1 \times G_2$, is also a group under the binary composition defined by*

$$(g_1, g_2) \times (g'_1, g'_2) = (g_1 g'_1, g_2 g'_2) \ \forall \ g_1, g'_1 \in G_1, g_2, g'_2 \in G_2$$

Proof. Since G_1 and G_2 are groups, therefore,

$$g_1, g'_1 \in G_1 \text{ and } g_2, g_2 \in G_2 \Rightarrow g_1 g'_1 \in G_1 \text{ and } g_2 g'_2 \in G_2$$

$$\Rightarrow (g_1g'_1, g_2g'_2) \in G_1 \times G_2 = G$$

Thus, G is closed under $*$

Associativity. Let $g_1, g'_1, g''_1 \in G_1$ and $g_2, g'_2, g''_2 \in G_2$

Consider $(g_1, g_2) * ((g'_1, g'_2) * (g''_1, g''_2))$

$$= (g_1, g_2) \times (g'_1 g''_1, g'_2 g''_2)$$
$$= (g_1(g'_1 g''_1), g_2(g'_2 g''_2))$$
$$= ((g_1 g'_1)g''_1, (g_2 g'_2)g''_2) \qquad ...(1)$$

| As G_1 and G_2 are groups $\therefore$ associativity holds in G_1 and G_2

Also

$$((g_1, g_2) * (g'_1, g'_2)) * (g''_1, g''_2) = (g_1g'_1, g_2g'_2) * (g''_1, g''_2) = ((g_1 g'_1)g''_1, (g_2 g'_2)g''_2) \qquad ...(2)$$

From (1) and (2), we have

$$(g_1, g_2) * ((g'_1, g'_2)) * (g''_1, g''_2)) = ((g_1, g_2) * (g'_1, g'_2)) * (g''_1, g''_2)$$

Thus, associavity holds in $G = G_1 \times G_2$

Identity, let $e_1 \in G_1$ and $e_2 \in G_2$ be the identity elements of G_1 and G_2 respectively. Consider

$$(g_1, g_2) * (e_1, e_2) = (g_1 e_1, g_2 e_2) = (g_1, g_2) \qquad | \because \ g_1e_1 = g_1 \ g_2 e_2 = g_2$$

Also $\quad (e_1, e_2) * (g_1, g_2) = (e_1 g_1, e_2 g_2) = (g_1, g_2) \qquad | \because \ e_1g_1 = g_1 \ e_2 g_2 = g_2$

thus, (e_1, e_2) is the identity element of $G_1 \times G_2$.

Inverse: Consider

$$(g_1, g_2)(g_1^{-1}, g_2^{-1}) = (g_1 g_1^{-1}, g_2 g_2^{-1}) = (e_1, e_2)$$

and $\quad (g_1^{-1}, g_2^{-1})(g_1, g_2) = (g_1^{-1} g_1, g_2^{-1} g_2) = (e_1, e_2)$

| $\because$ For $g_1 \in G_1, \exists \ g_1^{-1} \in G_1$ s.t. $g_1g_1^{-1} = e_1 = g_1^{-1} g_1$ etc.

Hence $\quad (g_1, g_2)^{-1} = (g_1^{-1}, g_2^{-1})$ is the inverse element of (g_1, g_2).

Hence, $\quad G = G_1 \times G_2$ is a group under the binary composition $*$.

Generalisation. The external direct product of $G_1, G_2, \ldots G_n$, denoted by $G = G_1 \times G_2 \times G_3 \times \ldots \times G_4$ is also a group under the component wise multiplication.

ILLUSTRATIVE EXAMPLES

Example 1. *Consider $G = H \times K$ where H and K are any two groups then,*

(*a*) *What is the order of G*

(*b*) *Describe and find the multiplication table of the group $G = Z_2 \times Z_2$*

(*c*) *Is G abelian?*

(*d*) *Is G cyclic?*

(*e*) *Is $G \cong Z_4$?*

Sol. (*a*) By definition the direct product of the groups G_1 and G_2, we know $G = G_1 \times G_2$ is the cartesian product of G_1 and G_2. Therefore

$$o(G) = o(G_1 \times G_2) = o(G_1)\, o(G_2)$$

(*b*) We know $Z_2 = \{0, 1\}$

Here $o(Z_2) = 2$

Therefore, $o(G) = o(Z_2 \times Z_2) = o(Z_2) \times (Z_2) = 4$

Thus, $G = Z_2 \times Z_2$ contains 4 elements, hence

$$G = \{(0, 0), (1, 0), (0, 1), (1, 1)\} = \{e, a, b, c\}$$

where $e = (0, 0)$, $a = (1, 0)$, $b = (0, 1)$, $c = (1, 1)$.

The multiplication table of G is shown below (Table 8.15).

Table 8.15

*	*l*	*a*	*b*	*c*
l	*l*	*a*	*b*	*c*
a	*a*	*l*	*c*	*b*
b	*b*	*c*	*l*	*a*
c	*c*	*b*	*a*	*l*

Here $a * a = (1, 0) * (1, 0) = (1, 0) = a$

$b * b = (0, 1) * (0, 1) = (0, 1) = b$

$c * c = (1, 1) * (1, 1) = (1, 1) = c$

(*c*) Also, the Table 8.15 is symmetrical , therefore, G is abelian.

(*d*) Since $a * a = a$

$\Rightarrow$ $o(a) = 2$

Similarly, $o(b) = 2$, $o(c) = 2$

Thus, all the elements of G are of order 2. It means that G does not have any element whose order equals to 4, the order of G. Hence G cannot be cyclic.

(*e*) We know that a finite cyclic group of order *n* is isomorphic to Z_n, the group of integers modulo *n*. Here, G is not cyclic. Hence $G \not\cong Z_4$.

Example 2. *Let H and K be any two groups and $G = H \times K$ be the direct product of the groups H and K under the binary operation * defined by*

$$(h, k) * (h', k') = (hh', kk') \ \forall\ h, h' \in H, k, k' \in K$$

Let $H' = H \times \{e\}$. *Then show that*

(*i*) $H' \cong H$ (*ii*) $H' \Delta G$ *where* $G = H \times K$

(*iii*) $G/H' \equiv K$ *where* $G = H \times K$

Sol. (*i*) Define a mapping $\theta : H' \to H$ such that

$$\theta(h') = h \quad \forall\, h' \in H, h \in H$$

we show θ is well-defined, homomorphism, one-one and onto.

θ is well-defined. Let $h'_1, h'_2 \in H'$ such that

$$h'_1 = (h_1, e), h'_2 = (h_2, e) \;\; \forall\, h_1, h_2 \in H$$

Consider $\quad h'_1 = h'_2$

$$(h_1, e) = (h_2, e) \;\Rightarrow\; h_1 = h_2$$
$$\Rightarrow\; \theta(h'_1) = \theta(h'_2)$$

$\therefore$ θ is well-defined

Homomorphism. Consider

$$\theta(h'_1 h'_2) = (h_1 h_2, e) = (h_1, e)\,(h_2, e)$$
$$= \theta(h'_1)\,\theta(h'_2)$$

$\therefore$ θ is homomorphism

One-one. Consider

$$\theta(h'_1) = \theta(h'_2)$$
$$\Rightarrow \quad h_1 = h_2$$
$$\Rightarrow \quad (h_1, e) = (h_2, e)$$
$$\Rightarrow \quad h'_1 = h'_2$$

$\therefore$ θ is one-one

Onto. Let $h \in H$, $e \in \{e\}$, then by the definition of cartesian product of two sets.

$$(h, e) \in H \times \{e\} = H'$$

$\Rightarrow$ $\quad h' \in H'$ where $h' = (h, e)$

Also $\quad \theta(h') = (h, e)$

$\therefore$ θ is onto

Hence $\quad H' \cong H$

(*ii*) To show $H' \trianglelefteq G$ (read as H' is a normal subgroup of G), we first show H' is non-empty subset of G.

For $e \in H$, $(e, e) \in H \times \{e\} = H'$

$\Rightarrow$ $(e, e) \in H'$. Thus $H' \neq \phi$.

Also, $\quad h \in H, e \in K$

$$\Rightarrow \quad (h, e) \in H \times K = G \;\Rightarrow\; h' \in G \;\Rightarrow\; H' \subseteq G$$

Hence, H' is a non-empty subset of G.

We next show H' is a subgroup of G

Let $h'_1, h'_2 \in H' \;\Rightarrow\; h'_1 = (h_1, e), h'_2 = (h_2, e)$

Consider $\quad h'_1 h'^{-1}_2 = (h_1, e)\,(h_2, e)^{-1} = (h_1, e)\,(h_2^{-1}, e)$

$$= (h_1, h_2^{-1}, e) \;\forall\, h_1, h_2 \in H$$
$$\in H \times [e]$$

$$\Rightarrow \quad h'_1 h'^{-1}_2 \in H$$

$\therefore$ $\quad H'$ is a subgroup of G

We next show H′ is a normal subgroup of G. Let $g \in G = H \times K \Rightarrow \exists\, h_1 \in H, k_1 \in K$ such that $g = (h_1, k_1)$

Let $h' \in H' \Rightarrow h' = (h, e)$

Consider
$$gh' g^{-1} = (h_1, k_1)\,(h, e), (h_1, k_1)^{-1}$$
$$= (h_1, k_1)\,(h, e)\,(h_1^{-1}, k_1^{-1})$$
$$= (h_1, k_1)\,(hh_1^{-1}, k_1^{-1}) \qquad | \because\ e \in \{e\} \text{ is the identity}$$
$$= (h_{,1}\, hh_1^{-1}, k_1k_1^{-1})$$
$$= (h_1\, hh_1^{-1}, e) \in H \times \{e\} = H'$$

| Since $h_1, h \in H$ and H is a group, therefore $h_1^{-1} \in H \Rightarrow h_1hh_1^{-1} \in H$

$\Rightarrow \quad gh'g^{-1} \in H' \quad \forall\, h' \in H', g \in G$

Hence H′ is a normal subgroup of G.

(*iii*) Since H′ is a normal subgroup of G, therefore, G/H′ is well-defined

Define $f : H \times K \to K$ such that $f(h, k) = k\ \forall\, h \in H, k \in K$

***f* is well-defined:** Consider
$$(h_1, k_1) = (h_2, k_2)$$
$$\Rightarrow \quad k_1 = k_2$$
$$\Rightarrow \quad f(h_1, k_1) = f(h_2, k_2)$$

$\therefore$ *f* is well-defined $\qquad | \because\ f(h, k) = k\ \forall\, h \in H$

f is a group homomorphism. Let $g_1, g_2 \in G = H \times K \Rightarrow \exists\, h_1, h_2 \in H$ and $k_1, k_2 \in K$ such that
$$g_1 = (h_1, k_1) \text{ and } g_2 = (h_2, k_2)$$

By the definition of cartesian product of H and K, we have
$$g_1g_2 = (h_1, k_1)\,(h_2, k_2)$$
$$= (h_1h_2, k_1\,k_2)$$

Therefore,
$$f(g_1g_2) = f(h_1h_2, k_1k_2) = k_1k_2$$
$$= f(h_1, k_1)\, f(h_2, k_2)$$

Hence *f* is a group homomorphism.

Also,
$$\text{Ker } f = \{(h, k) \in H \times K: f(h, k) = e, e \text{ is the identity of } K$$
$$= \{(h, k) \in H \times K : k = e, e \in K$$
$$= \{(h, e) \in H \times \{e\}\} = H'$$

By **fundamental theorem of semi-group homomorphism of groups.**
$$H \times K/\text{Ker } f \cong K$$
$$\Rightarrow \quad H \times K/H' \cong K$$
$$\Rightarrow \quad G/H' \cong K$$

Hence the theorem

TEST YOUR KNOWLEDGE 8.4

1. Consider $G = Z_2 \times Z_3$. Describe and find the multiplication table of the group $G = Z_2 \times Z_3$
   (*a*) Is G abelian? (*b*) Is G cyclic?
   (*c*) Is $G \cong Z_6$?

Answer

1. $Z_2 \times Z_3 = \{(0, 0), (0, 1), (0, 2), (1, 0), (1, 1), (1, 2)\}$
   (*a*) Yes (*b*) Yes, G is cyclic group generated by (1, 1).
   (*d*) Yes, $G \cong Z_6$

Hints

(*b*) Consider

$$(1, 1) = (1, 1)$$
$$(1, 1) + (1, 1) = (1 +_2 1, 1 +_3 1) = (0, 2)$$
$$(1, 1) + (1, 1) + (1, 1) = (1 +_2 1 +_2 1, 1 +_3 1 +_3 1) = (1, 0)$$
$$(1, 1) + (1, 1) + (1, 1) + (1, 1) = (1, 0) + (1, 1)$$
$$= (1 +_2 1, 0 +_3 1) = (0, 1)$$
$$(1, 1) + (1, 1) + (1, 1) + (1, 1) + (1, 1) = (0, 1) + (1, 1) = (1, 2)$$
$$(1, 1) + (1, 1) + (1, 1) + (1, 1) + (1, 1) = (1, 2) + (1, 1) = (1 +_2 1, 2 +_3 1) = (0, 0)$$

Hence every element of G can be expressed in some powers of (1, 1).

Therefore, G is a cyclic group with (1, 1) as generator

(*c*) Since $o(1, 1) = 6 = o(Z_2 \times Z_3)$

$\therefore$ $G = Z_2 \times Z_3$ is cyclic group of order 6.

Also, any finite cyclic group of order n is isomorphic to Z_n

$\therefore$ $Z_2 \times Z_3 \cong Z_6$ or $G \cong Z_6$

PERMUTATION AND SYMMETRIC GROUP

8.30. BASIC TERMINOLOGY

Before, we define our symmetric groups, students are advised to go through the following basic terminology.

Example consider a function $f : S \to S$ defined on a nonempty set $S = \{1, 2, 3\}$ such that $f(1) = 2, f(2) = 3, f(3) = 1$

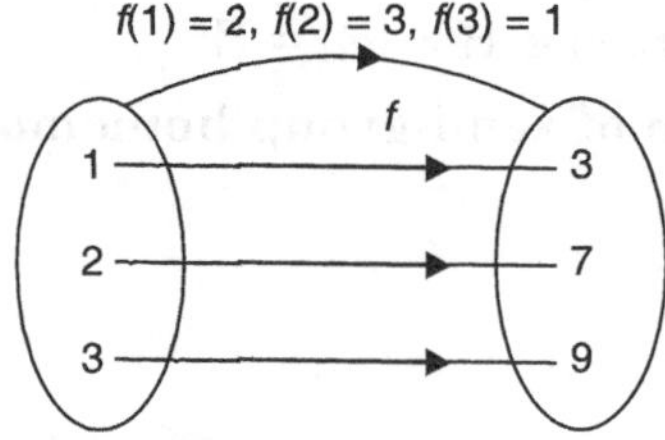

Fig. 8.6

Clearly, f is both one-one and onto (see Fig. 8.6) and we say, $f : S \to S$ is a permutation on S.

8.30. (a) PERMUTATION

Let S be any non-empty and a finite set, then a one-one and onto mapping from S to S is called a **permutation**.

Notation: The set of all permutations on a non-empty set S is denoted by A(S).

Further, if A set S has n elements, then, the total number of bijections (one-one and onto mapping) from S to S is ${}^nC_n = n!$. In the above example, the total number of one-one and onto mappings from S to S is ${}^3C_3 = 3! = 6$.

Hence, we can say that if a set S contains n elements, then A(S), the set of all permutations on S, will contain ${}^nC_n = n!$ elements.

We next introduce a notation to represent these permutations.

The function $f : S \to S$ defined on a non-empty and a finite set S = {1, 2, 3} such that $f(1) = 2$, $f(2) = 3$, $f(3) = 1$ can be expressed in **array from** as shown below:

$$f = \begin{pmatrix} 1 & 2 & 3 \\ 2 & 3 & 1 \end{pmatrix}$$

Consider another example of a permutation $g : S \to S$, defined on S = {1, 2, 3, 4, 5, 6} such that $g(1) = 5$, $g(2) = 3$, $g(3) = 1$, $g(4) = 6$, $g(5) = 2$, $g(6) = 6$, then, g can be expressed in **array form** as shown below:

$$g = \begin{pmatrix} 1 & 2 & 3 & 4 & 5 & 6 \\ 5 & 3 & 1 & 6 & 2 & 4 \end{pmatrix}$$

8.30. (b) COMPOSITION OF PERMUTATIONS IN ARRAY FORM

Consider two permutations f and g, given in **array form** as

$$f = \begin{pmatrix} 1 & 2 & 3 & 4 & 5 \\ 2 & 4 & 3 & 5 & 1 \end{pmatrix}, g = \begin{pmatrix} 1 & 2 & 3 & 4 & 5 \\ 5 & 4 & 1 & 2 & 3 \end{pmatrix}$$

then the composition of f and g (denoted by gof or sometimes, by gf), in **array form** is given as below:

$$gof = \begin{pmatrix} 1 & 2 & 3 & 4 & 5 \\ 5 & 4 & 1 & 2 & 3 \end{pmatrix} \begin{pmatrix} 1 & 2 & 3 & 4 & 5 \\ 2 & 4 & 3 & 5 & 1 \end{pmatrix}$$

Fig. 8.7

We wish to find the following elements $(gof)(1)$, $(gof)(2)$, $(gof)(3)$, $(gof)(4)$, $(gof)(5)$.

To find $(gof)(1)$: Notice that compositions of permutations expressed in array brackets, is carried out from right bracket to left bracket, by going from top to bottom (in right bracket) and then again from top to bottom (in left bracket). Hence, to find $(gof)(1)$, we start from the right bracket in the following manner:

In Fig 8.7, we notice that $1 \to 2$ (right bracket), $2 \to 4$ (left bracket), $\Rightarrow 1 \to 4$

This is equivalent to say that $(gof)(1) = g(f(1)) = g(2) = 4$, which means '1' goes to '4'.

To find $(gof)(2)$: We start from '2' in the right bracket as shown below:

$2 \to 4$ (right bracket), $4 \to 2$ (left bracket), $\Rightarrow 2 \to 2$

This is equivalent to say that $(gof)(2) = g(f)(2)) = g(4) = 2 \Rightarrow 2 \to 2$

To find $(gof)(3)$: Start from '3' (in the right bracket), $3 \to 3$, $3 \to 1 \Rightarrow 3 \to 1$.

To find $(gof)(4)$: Here $4 \to 5$, $5 \to 3 \Rightarrow 4 \to 3$

To find $(gof)(5)$: Here $5 \to 1$, $1 \to 5 \Rightarrow 5 \to 5$

Hence,
$$gof = \begin{pmatrix} 1 & 2 & 3 & 4 & 5 \\ 5 & 4 & 1 & 2 & 3 \end{pmatrix}\begin{pmatrix} 1 & 2 & 3 & 4 & 5 \\ 2 & 4 & 3 & 5 & 1 \end{pmatrix}$$

$$= \begin{pmatrix} 1 & 2 & 3 & 4 & 5 \\ 4 & 2 & 1 & 3 & 5 \end{pmatrix}$$

8.30. (c) PERMUTATION AS A SINGLE ROW

Consider $f = \begin{pmatrix} 1 & 2 & 3 & 4 & 5 & 6 \\ 2 & 3 & 5 & 6 & 1 & 4 \end{pmatrix}$

We try to express 'f' in a single row:

Here, $1 \to 2$, $2 \to 3$, $3 \to 5$, $5 \to 1$, *i.e.*, we started from '1' and ended in '1'. This fact can be represented as (1235).

Also, $4 \to 6$, $6 \to 4$ it can be expressed as (46).

Hence $f = (1235)(46)$, is in a **single row representation.**

The product of (1235) and (46)

8.30. (d) ORBIT OF A PERMUTATION

Consider a permutation f given by

$$f = \begin{pmatrix} 1 & 2 & 3 & 4 & 5 & 6 \\ 6 & 5 & 4 & 3 & 1 & 2 \end{pmatrix}$$

In cyclic notation, $f = (1625)(34)$

Here, (625) is called a **cycle** and the set {1, 6, 2, 5} is called an **orbit** of the elements 1, 6, 2 and 5.

Also {3, 4} is an orbit of the elements 3 and 4.

8.30. (e) DISJOINT PERMUTATIONS

Two permutations f and g on a non-empty set S are called *disjoint* permutations, for any $x \in$ S,

$$f(x) \neq x \;\Rightarrow\; g(x) = x \text{ and}$$
$$g(x) \neq x \;\Rightarrow\; f(x) = x \;\forall\; x \in S$$

e.g., $f = (12)$, $g = (13)$ are not disjoint since $f(1) = 2$ and $g(1) = 3 \neq 1$

But $f = (132)$ and, $g = (45)$ are disjoint permutations.

8.30. (f) CYCLIC PERMUTATIONS

Let S be a finite set containing n elements, then, a permutation f on S is called a *cyclic permutation* or a *cycle* if there exists $x_1, x_2, x_3, \ldots\ldots, x_n$ such that:

$$f(x_1) = x_2,\ f(x_2) = x_3,\ f(x_3) = x_4,\ \dots\ f(x_{n-1}) = x_n,\ f(x_n) = x_1$$

A cyclic permutation on n elements: x_1, x_2, ..., x_n is denoted by $(x_1 x_2 x_3 \dots x_n)$ and is called a *cycle of length n* or *n-recycle*. In particular, a cycle of length '2' it called a *transposition* or *2-cycle*.

Remark 1: Two permutations f and g are said to be disjoint of there is no common element in any cycles of f and g.

ILLUSTRATIVE EXAMPLES

Example 1. *Consider* $f = \begin{pmatrix} 1 & 2 & 3 & 4 & 5 & 6 & 7 \\ 2 & 4 & 6 & 1 & 7 & 3 & 5 \end{pmatrix}$. *Express f as a product of disjoint cycles.*

Sol. We wish to find the disjoint cycles of f.

Here, $1 \to 2,\ 2 \to 4,\ 4 \to 1 \Rightarrow (124)$ is a cycle of f.

Also $3 \to 6,\ 6 \to 3 \Rightarrow$ (36) is a cycle of f.

Further, $5 \to 7,\ 7 \to 5 \Rightarrow$ (57) is a cycle of f

Hence,
$$f = \begin{pmatrix} 1 & 2 & 3 & 4 & 5 & 6 & 7 \\ 2 & 4 & 6 & 1 & 7 & 3 & 5 \end{pmatrix}$$
$$= (124)(36)(57).$$

Example 2. *Consider* $f = \begin{pmatrix} 1 & 2 & 3 & 4 & 5 & 6 & 7 & 8 \\ 8 & 7 & 5 & 4 & 6 & 2 & 1 & 3 \end{pmatrix}$. *Express f as a product of disjoint cycles.*

Sol. Here, $1 \to 8,\ 8 \to 3,\ 3 \to 5,\ 5 \to 6,\ 6 \to 2,\ 2 \to 7,\ 7 \to 1$

$\therefore$ (1835527) is a cycle of f.

Also $4 \to 4$ $\therefore$ (4) is also a cycle of f

Hence f = (1835627)(4), is expressed as a product of disjoint cycles.

8.31. MULTIPLICATION OF CYCLES

We now develop a procedure for multiplication of cycles (not necessarily disjoint).

$$\text{Consider } f = \begin{pmatrix} 1 & 2 & 3 & 4 & 5 & 6 & 7 & 8 \\ 4 & 6 & 8 & 3 & 5 & 2 & 1 & 7 \end{pmatrix}$$

$$g = \begin{pmatrix} 1 & 2 & 3 & 4 & 5 & 6 & 7 & 8 \\ 2 & 4 & 6 & 8 & 1 & 3 & 5 & 7 \end{pmatrix}$$

We find *fog* in cyclic notations.

We first express f and g a product of cycles (not necessarily disjoint).

Here, $f = (14387)(26)$

$g = (124875)(36)$

Now $$fog = \overset{\sigma_1}{(14387)}\overset{\tau}{(26)}\overset{\sigma'}{(124875)}\overset{\tau'}{(36)} = \sigma\tau\sigma'\tau', \text{ say} \qquad ...(1)$$

Note that each cycle that does not contain on element fixes the element.

To find *fog*. Consider the elements 1, 2, 3, 4, 5, 6, 7, 8.

Start with '1' as shown in the below pattern.

$$1 \xrightarrow{\sigma} 4 \xrightarrow{\tau} 4 \xrightarrow{\sigma'} 8 \xrightarrow{\tau'} 8 \Rightarrow 1 \to 8$$

As '1' goes to '8', we start with '8' as shown in the below pattern.

$$8 \xrightarrow{\sigma} 7 \xrightarrow{\tau} 7 \xrightarrow{\sigma'} 5 \xrightarrow{\tau'} 5 \Rightarrow 8 \to 5$$

As '8' goes to '5', we start with '5' as shown in the below pattern

$$5 \xrightarrow{\sigma} 5 \xrightarrow{\tau} 5 \xrightarrow{\sigma'} 1 \xrightarrow{\tau'} 1 \quad \Rightarrow 5 \to 1$$

This completes the cycle (185).

Now, we are left with the remaining elements 2, 3, 4, 6, 7.

Start with '2' as shown in the below pattern.

$$2 \xrightarrow{\sigma} 2 \xrightarrow{\tau} 6 \xrightarrow{\sigma'} 6 \xrightarrow{\tau'} 3 \quad \Rightarrow 2 \to 3$$

$$3 \xrightarrow{\sigma} 8 \xrightarrow{\tau} 8 \xrightarrow{\sigma'} 7 \xrightarrow{\tau'} 7 \quad \Rightarrow 3 \to 7$$

$$7 \xrightarrow{\sigma} 1 \xrightarrow{\tau} 1 \xrightarrow{\sigma'} 2 \xrightarrow{\tau'} 2 \quad \Rightarrow 7 \to 2$$

This completes the cycle (237)

Now, we are left with the remaining elements 4 and 6 only.

Start with '4' as shown in the below patter.

$$4 \xrightarrow{\sigma} 3 \xrightarrow{\tau} 3 \xrightarrow{\sigma'} 3 \xrightarrow{\tau'} 6 \quad \Rightarrow 4 \to 6$$

$$6 \xrightarrow{\sigma} 6 \xrightarrow{\tau} 2 \xrightarrow{\sigma'} 4 \xrightarrow{\tau'} 4 \quad \Rightarrow 6 \to 4$$

This completes the cycle (46) and there are no elements left with us.

Hence, the only cycles of f are (185), (237) and (46), which are disjoint also.

Hence from (1), $fog = (14387)(26)(124875)(34)$

$= (185)(237)(46)$,

which shows that the composition of any two permutations on a finite set can be expressed as a product of disjoint cycles

Example 3. *Let $f = (13)(27)(456)(8)$ and $g = (1237)(648)(5)$*

What is the cycle form of fog? or Write fog as a product of disjoint cycles.

Sol. Let $$fog = \overset{\sigma_1}{(13)}\overset{\sigma_2}{(27)}\overset{\sigma_3}{(456)}\overset{\sigma_4}{(18)}\overset{\tau_1}{(1237)}\overset{\tau_2}{(648)}\overset{\tau_3}{(5)}, \text{ say} \qquad ...(1)$$

Consider the elements 1, 2, 3, 4, 5, 6, 7, 8.

Start with '1' as shown in the below manner.

$$1 \xrightarrow{\sigma_1} 3 \xrightarrow{\sigma_2} 3 \xrightarrow{\sigma_3} 3 \xrightarrow{\sigma_4} 3 \xrightarrow{\tau_1} 7 \xrightarrow{\tau_2} 7 \xrightarrow{\tau_3} 7 \Rightarrow 1 \to 7$$

| A cycle that does not contain an element fixes the element

Start with '7' as shown in the following pattern.

$$7 \xrightarrow{\sigma_1} 7 \xrightarrow{\sigma_2} 2 \xrightarrow{\sigma_3} 2 \xrightarrow{\sigma_4} 2 \xrightarrow{\tau_1} 3 \xrightarrow{\tau_2} 3 \xrightarrow{\tau_3} 3 \Rightarrow 7 \to 3$$

Start with '3' as shown in the below manner.

$$3 \xrightarrow{\sigma_1} 1 \xrightarrow{\sigma_2} 1 \xrightarrow{\sigma_3} 1 \xrightarrow{\sigma_4} 1 \xrightarrow{\tau_1} 2 \xrightarrow{\tau_2} 2 \xrightarrow{\tau_3} 2 \Rightarrow 3 \to 2$$

Start with '2' as shown in the below manner.

$$2 \xrightarrow{\sigma_1} 2 \xrightarrow{\sigma_2} 7 \xrightarrow{\sigma_3} 7 \xrightarrow{\sigma_4} 7 \xrightarrow{\tau_1} 1 \xrightarrow{\tau_2} 1 \xrightarrow{\tau_3} 1 \Rightarrow 2 \to 1$$

This completes the cycle (1732).
Now, we are left with the elements 4, 5, 6, 8.
Start with '4' as shown in the following manner.

$$4 \xrightarrow{\sigma_1} 4 \xrightarrow{\sigma_2} 4 \xrightarrow{\sigma_3} 5 \xrightarrow{\sigma_4} 5 \xrightarrow{\tau_1} 5 \xrightarrow{\tau_2} 5 \xrightarrow{\tau_3} 5 \Rightarrow 4 \to 5$$

Start with '5' as shown in the following manner.

$$5 \xrightarrow{\sigma_1} 5 \xrightarrow{\sigma_2} 5 \xrightarrow{\sigma_3} 6 \xrightarrow{\sigma_4} 6 \xrightarrow{\tau_1} 6 \xrightarrow{\tau_2} 4 \xrightarrow{\tau_3} 4 \Rightarrow 5 \to 4$$

This completes the cycle (45).
Now, we are left with 6 and 8.
Start with '6', as shown in the following manner.

$$6 \xrightarrow{\sigma_1} 6 \xrightarrow{\sigma_2} 6 \xrightarrow{\sigma_3} 4 \xrightarrow{\sigma_4} 4 \xrightarrow{\tau_1} 4 \xrightarrow{\tau_2} 8 \xrightarrow{\tau_3} 8 \Rightarrow 6 \to 8$$

Start with '8', as shown in the following manner.

$$8 \xrightarrow{\sigma_1} 8 \xrightarrow{\sigma_2} 8 \xrightarrow{\sigma_3} 8 \xrightarrow{\sigma_4} 8 \xrightarrow{\tau_1} 8 \xrightarrow{\tau_2} 6 \xrightarrow{\tau_3} 6 \Rightarrow 8 \to 6$$

This completes the cycle (68) and we are left with no element with us.
Thus, the only cycles of f are (1732), (45) and (68) and these cycles are disjoint also.
Hence, from (1),

$$\begin{aligned} fog &= (13)(27)(456)(8)(1237)(648)(5) \\ &= (1732)(45)(68) \end{aligned}$$

which shows that the composition of any two permutations on a finite set can be expressed as a product of disjoint cycles.

Remark 2. Every permutation on a finite set containing 'n' elements ($n > 1$) can be expressed as a product of **transpositions** (or *2-cycles*) and the decomposition of a permutation into a product of *2-cycles is not* **unique.** It means to say that

$$\begin{aligned} (a_1a_2a_3 \dots a_k) &= (a_1a_k)(a_1a_{k-1})(a_1a_{k-2}) \dots (a_1a_2) \\ &= (a_2a_k)(a_2a_{k-1})(a_2a_{k-2}) \dots (a_2a_1) \\ &= (a_3a_k)(a_3a_{k-1})(a_3a_{k-2}) \dots (a_3a_2)(a_3a_1) \end{aligned}$$

Example 4. *Express (12345) as a product of 2-cycles in atleast 4 ways.*

Sol.

$$\begin{aligned} (12345) &= (15)(14)(13)(12) \\ &= (25)(24)(23)(21) \\ &= (35)(34)(32)(31) \\ &= (45)(41)(42)(43) \end{aligned}$$

8.32. PROPERTIES OF PERMUTATIONS

We now study the following properties of permutations.

Property 1. Any two disjoint permutations on a finite set commute with each other.

Proof. Let f and g are any two disjoint permutations. We need to show $fog = gof$.

For this, let $x \in S$ and consider $f(x) \neq x$ and since f and g are disjoint, then $g(x) = x$. (By definition)

Let $f(x) = y$, then, since $f(x) \neq x \Rightarrow y \neq x$

Consider $\quad (fog)(x) = f(g(x)) = f(x) = y \quad$ ...(1)

and $\quad (gof)(x) = g(f(x)) = g(y) \quad$ ...(2)

We claim $g(y) = y$.

For if, $g(y) \neq y$ and since f and g are disjoint, then

$$f(y) = y = f(x) \quad \Rightarrow \quad y = x \text{ (since } f \text{ is a permutation), a contradiction.}$$

Hence $g(y) = y$ and therefore, from (1) and (2), we have

$$(fog)(x) = (gof)(x)$$

$$\Rightarrow \qquad fog = gof$$

Property 2. Any permutation on a finite set can be expressed as a product of transpositions (not necessarily disjoint).

Consider
$$f = \begin{pmatrix} 1 & 2 & 3 & 4 & 5 & 6 & 7 & 8 & 9 \\ 3 & 4 & 2 & 1 & 6 & 5 & 8 & 9 & 7 \end{pmatrix} = (1324)(56)(789)$$

$$= (14)(12)(13)(56)(79)(78), \text{ (See remark 2 above)}$$

Here, the transpositions (14), (12) and (13) are not disjoint (as these 2-cycles have '1' as a common element).

Property 3. Any permutation on a finite set can be expressed as a product of disjoint cycles (see examples 1, 2, 3 above)

Property 4. The nth power of a n-cycle is I.

Consider a 2-cycle (ab) containing two elements, then, we must show $(ab)^2 = \text{I}$.

Consider $\overset{\sigma_1}{(ab)}\overset{\sigma_2}{(ab)}$

Here, $\quad a \xrightarrow{\sigma_1} b \xrightarrow{\sigma_2} a \Rightarrow a \rightarrow a$

Here, $\quad b \xrightarrow{\sigma_1} a \xrightarrow{\sigma_2} b \Rightarrow b \rightarrow b$

Hence $(ab)^2 = \text{I}$ (since each element goes to itself).

Consider another example.

Let (1234) be a *4-cycle* containing 4 elements.

We show $\quad (1234)^4 = \text{I}$. For this,

Consider $\quad (1234)^2 = \overset{\sigma_1}{(1234)}\overset{\sigma_2}{(1234)}$

Here, $\quad 1 \xrightarrow{\sigma_1} 2 \xrightarrow{\sigma_2} 3 \Rightarrow 1 \rightarrow 3$

$$3 \xrightarrow{\sigma_1} 4 \xrightarrow{\sigma_2} 1 \Rightarrow 3 \rightarrow 1$$

This completes the cycle (13).

Also, $2 \xrightarrow{\sigma_1} 3 \xrightarrow{\sigma_2} 4 \Rightarrow 2 \to 4$

$4 \xrightarrow{\sigma_1} 1 \xrightarrow{\sigma_2} 2 \Rightarrow 4 \to 2$

This completes the cycle (24).

Hence, $(1234)^2 = (1234)(1234) = (13)(24)$ ...(1)

Again, $(1234)^3 = (1234)(1234)^2 = \overset{\sigma_1}{(1234)}\overset{\sigma_2}{(13)}\overset{\sigma_3}{(24)}$ | using (1)

Also, $1 \xrightarrow{\sigma_1} 2 \xrightarrow{\sigma_2} 2 \xrightarrow{\sigma_3} 4 \Rightarrow 1 \to 4$

$4 \xrightarrow{\sigma_1} 1 \xrightarrow{\sigma_2} 3 \xrightarrow{\sigma_3} 3 \Rightarrow 4 \to 3$

$3 \xrightarrow{\sigma_1} 4 \xrightarrow{\sigma_2} 4 \xrightarrow{\sigma_3} 2 \Rightarrow 3 \to 2$

$2 \xrightarrow{\sigma_1} 3 \xrightarrow{\sigma_2} 1 \xrightarrow{\sigma_3} 1 \Rightarrow 2 \to 1$

This completes the cycle (1432).

Hence, $(1234)^3 = (1432)$ ...(2)

Finally, $(1234)^4 = (1234)(1234)^3$ | using (2)

$= \overset{\sigma_1}{(1234)}\overset{\sigma_2}{(1432)}$

Here, $1 \xrightarrow{\sigma_1} 2 \xrightarrow{\sigma_2} 1 \Rightarrow 1 \to 1$

$2 \xrightarrow{\sigma_1} 3 \xrightarrow{\sigma_2} 2 \Rightarrow 2 \to 2$

$3 \xrightarrow{\sigma_1} 4 \xrightarrow{\sigma_2} 3 \Rightarrow 3 \to 3$

$4 \xrightarrow{\sigma_1} 1 \xrightarrow{\sigma_2} 4 \Rightarrow 4 \to 4$

∴ $(1234)^4 = (1234)(1432) = I$. (Since each element goes to itself).

Property 5. The product of a permutation on a finite set and its inverse is always equals to I.

Consider $f = (a_1b_1)(a_2b_2)(a_3b_3)(a_4b_4)$, then

$f^{-1} = ((a_1b_1)(a_2b_2)(a_3b_3)(a_4b_4))^{-1}$

$= (a_4b_4)(a_3b_3)(a_2b_2)(a_1b_1)$

∴ $ff^{-1} = (a_1b_1)(a_2b_2)(a_3b_3)(a_4b_4)(a_4b_4)(a_3b_3)(a_2b_2)(a_1b_1)$

$= I$ (Prove it)

8.33. EVEN AND ODD PERMUTATIONS

A permutation on a finite set is called an *even (odd)* permutation if it can be expressed as a product of even (odd) number of *transposition (2-cycles)*.

Following results can be trivially proved.

I. The product of two even permutations is an even permutation (sum of two even numbers is even)

II. The product of two odd permutations is an even permutation (sum of two odd numbers is even)

III. The product of an even and an odd permutations is an odd permutation (sum of an even and odd number is an odd number)

IV. Inverse of an even (odd) permutation is an even (odd) permutation.

V. Identity permutation is always an even permutation.

Example 5. *Consider the following permutation on S = {1, 2, 3, 4, 5, 6, 7}. Examine whether it is an odd permutation 2 or even permutation:* $\begin{pmatrix} 1 & 2 & 3 & 4 & 5 & 6 & 7 \\ 3 & 4 & 1 & 5 & 6 & 7 & 2 \end{pmatrix}$.

Sol. Let $$f = \begin{pmatrix} 1 & 2 & 3 & 4 & 5 & 6 & 7 \\ 3 & 4 & 1 & 5 & 6 & 7 & 2 \end{pmatrix}$$

We first express f as a product of *transpositions (2-cycles)*

Here $f = (13)(24567) = (13)(27)(26)(25)(24)$ (See remark 2, above)

$\therefore$ Since f is the product of 5 (odd) transpositions, hence f is an odd permutation

Theorem I. *The order of a permutation on a finite set, written as a product of disjoint cycles, is the least common multiple of the lengths of the disjoint cycles.*

Proof. This is beyond the scope of the syllabus.

Example 6. *Find the order of the following permutations*

(i) (12)(13)(14)(15)(16)

(ii) (24)(26)(28)(13)(15)(17)

Sol. (*i*) Let $f = (12)(13)(14)(15)(16) = (165432)$

$\therefore$ its order = 6 (number of distinct elements in f)

(*ii*) Let $f = (24)(26)(25)(13)(15)(17) = (2864)(1753)$

Clearly, right hand side is a product of two disjoint cycles.

$\therefore$ Required order = l.c.m. of the lengths of (2864) and (1753) = 4

Example 7. *Find the order of the following permutation.*

$$f = (123)(234)(456)(67)$$

Sol. Given permutation is not a product of disjoint cycles. Hence, we first express it as a product of disjoint cycles.

Consider $$f = \overset{\sigma_1}{(123)}\overset{\sigma_2}{(234)}\overset{\sigma_3}{(456)}\overset{\sigma_4}{(67)}, \text{ say} \quad ...(1)$$

The elements of the given permutation f are 1, 2, 3, 4, 5, 6 and 7.

Start with '1' as shown in the following manner.

$$1 \xrightarrow{\sigma_1} 2 \xrightarrow{\sigma_2} 3 \xrightarrow{\sigma_3} 3 \xrightarrow{\sigma_4} 3 \Rightarrow 1 \to 3$$

Start with '3' as shown in the following manner.

$$3 \xrightarrow{\sigma_1} 1 \xrightarrow{\sigma_2} 1 \xrightarrow{\sigma_3} 1 \xrightarrow{\sigma_4} 1 \Rightarrow 3 \to 1$$

This completes the cycle (13)

We are left with 2, 4, 5, 6 and 7

Start with '2' as shown in the following manner.

$$2 \xrightarrow{\sigma_1} 3 \xrightarrow{\sigma_2} 4 \xrightarrow{\sigma_3} 5 \xrightarrow{\sigma_4} 5 \Rightarrow 2 \to 5$$

Start with '5' as shown in the following manner.

$$5 \xrightarrow{\sigma_1} 5 \xrightarrow{\sigma_2} 5 \xrightarrow{\sigma_3} 6 \xrightarrow{\sigma_4} 7 \Rightarrow 5 \to 7$$

Start with '7' as shown in the following

$$7 \xrightarrow{\sigma_1} 7 \xrightarrow{\sigma_2} 7 \xrightarrow{\sigma_3} 7 \xrightarrow{\sigma_4} 6 \Rightarrow 7 \to 6$$

Start with '6' as shown in the following manner.

$$6 \xrightarrow{\sigma_1} 6 \xrightarrow{\sigma_2} 6 \xrightarrow{\sigma_3} 4 \xrightarrow{\sigma_4} 4 \Rightarrow 7 \to 6$$

This completes the cycle (25764) and we are left with no elements.
Hence, the only cycles of f are (13) and (25764) which are also disjoint.
∴ From (1)

$$f = (123)(234)(456)(67)$$
$$= (13)(25764)$$

Hence, order of f = l.c.m of the lengths of (13) and (25764) = l.c.m of 2 and 5 = **10**

Example 8. *Let* $f = \begin{pmatrix} 1 & 2 & 3 & 4 & 5 & 6 \\ 2 & 1 & 3 & 5 & 4 & 6 \end{pmatrix}$ *and*

$$g = \begin{pmatrix} 1 & 2 & 3 & 4 & 5 & 6 \\ 6 & 1 & 2 & 4 & 3 & 5 \end{pmatrix}$$

Compute each of the following:
(a) f^{-1} *(b) gof* *(c) fog*

Sol. *(a)* $f = \begin{pmatrix} 1 & 2 & 3 & 4 & 5 & 6 \\ 2 & 1 & 3 & 5 & 4 & 6 \end{pmatrix}$

On expressing f as a product of transpositions, we get

$$f = (12)(3)(45)(6)$$

$$f^{-1} = ((12)(3)(45)(6))^{-1} = (6)\ (45)\ (3)\ (12) \qquad ...(1)$$

We now express f^{-1} in the array form.
From (1), We notice that $6 \to 6$, $4 \to 5$ and $5 \to 4$, $3 \to 3$, $1 \to 2$ and $2 \to 1$.

$$\therefore \qquad f^{-1} = \begin{pmatrix} 1 & 2 & 3 & 4 & 5 & 6 \\ 2 & 1 & 3 & 5 & 4 & 6 \end{pmatrix}$$

(b)
$$gof = \begin{pmatrix} 1 & 2 & 3 & 4 & 5 & 6 \\ 6 & 1 & 2 & 4 & 3 & 5 \end{pmatrix} \begin{pmatrix} 1 & 2 & 3 & 4 & 5 & 6 \\ 2 & 1 & 3 & 5 & 4 & 6 \end{pmatrix}$$

$$= \begin{pmatrix} 1 & 2 & 3 & 4 & 5 & 6 \\ 1 & 6 & 2 & 3 & 4 & 5 \end{pmatrix} \text{ (see Multiplication of cycles)}$$

Example 9. *Find* a^{-1} *ba where* $a = (135)(12)$, $b = (1579)$.

Sol. $a = (135)(12) = \overset{\sigma_1}{(135)}\overset{\sigma_2}{(12)}$, say

Consider the elements 1, 2, 3, 5.

Start with '1', $1 \xrightarrow{\sigma_1} 3 \xrightarrow{\sigma_2} 3 \Rightarrow 1 \rightarrow 3$

Start with '3', $3 \xrightarrow{\sigma_1} 5 \xrightarrow{\sigma_2} 5 \Rightarrow 3 \rightarrow 5$

Start with '5', $5 \xrightarrow{\sigma_1} 1 \xrightarrow{\sigma_2} 2 \Rightarrow 5 \rightarrow 2$

Start with '2', $2 \xrightarrow{\sigma_1} 2 \xrightarrow{\sigma_2} 1 \Rightarrow 2 \rightarrow 1$

This completes the cycle (1352)

Hence, $a = (135)(12) = (1352) \Rightarrow a^{-1} = (1352)^{-1} = (2531)$

To find $a^{-1} ba$:

$$a^{-1}\, ba = (2531)(1579)(1352) = \overset{\sigma_1}{(2531)}\overset{\sigma_2}{(1579)}\overset{\sigma_3}{(1352)}, \text{ say,} \qquad ...(1)$$

Consider the elements 1, 2, 3, 5, 7, 9

Start with '1', $1 \xrightarrow{\sigma_1} 2 \xrightarrow{\sigma_2} 2 \xrightarrow{\sigma_3} 1 \Rightarrow 1 \rightarrow 1$

This completes the cycle ...(1)

Start with '2', $2 \xrightarrow{\sigma_1} 5 \xrightarrow{\sigma_2} 7 \xrightarrow{\sigma_3} 7 \Rightarrow 2 \rightarrow 7$

Start with '7', $7 \xrightarrow{\sigma_1} 7 \xrightarrow{\sigma_2} 9 \xrightarrow{\sigma_3} 9 \Rightarrow 7 \rightarrow 9$

Start with '9', $9 \xrightarrow{\sigma_1} 9 \xrightarrow{\sigma_2} 1 \xrightarrow{\sigma_3} 3 \Rightarrow 9 \rightarrow 3$

Start with '3', $3 \xrightarrow{\sigma_1} 1 \xrightarrow{\sigma_2} 5 \xrightarrow{\sigma_3} 2 \Rightarrow 3 \rightarrow 2$

This completes the cycle (2793)

We are left with the element '5'.

Therefore, (2531)(1579)(1352) = (1)(2793)(5) = (2793)

Remark 3: Let A(S) denotes the set of all permutations on a finite set S, then, either all permutations of A(S) are even or exactly half are even.

8.34. SYMMETRIC GROUP

Let S_n denotes the set of all permutations on a finite set S = (1, 2, 3, ...,n) and let $\sigma : S \rightarrow S$ is a one-one and onto mapping defined by

$$\sigma = \begin{pmatrix} 1 & 2 & 3 & \ldots & n \\ \sigma(1) & \sigma(2) & \sigma(3) & \ldots & \sigma(n) \end{pmatrix}$$

Then, S_n forms a group under the composition of mappings and called *symmetric group* of degree n.

Theorem II. *Show that S_n ($n \geq 1$) is a group under the composition of mappings.*

Proof. Let $f, g \in S_n \Rightarrow f, g$ are both one-one and onto maps on a a finite set S. Since f, g are both one-one and onto, their compostion maps *fog*, *gof* are also one-one and onto maps $\Rightarrow$ *fog*, *gof* are also permutations on S. Hence, *fog*, *gof* $\in S_n$

$\therefore$ S_n is closed under the composition of mappings.

Also, S_n has identity mapping I, each permutation on S_n has its inverse. Further, associativity under the composition of mapping also holds in S_n.

Hence, S_n forms a group under the composition of mappings.

Theorem III. Show that S_n $(n \geq 1)$ forms a group under the composition of mapping with order $o(S_n) = n!$.

Proof. By definition, elements of S_n are of the form

$$\sigma = \begin{pmatrix} 1 & 2 & 3 & \dots & n \\ \sigma(1) & \sigma(2) & \sigma(3) & \dots & \sigma(n) \end{pmatrix}$$

where σ is a one-one and onto map on the set S = (1, 2, 3, ..., n)

Now, there are 'n' choices of $\sigma(1)$. Once, $\sigma(1)$ is determined, there are $(n-1)$ choices for $\sigma(2)$ (since σ is one-one, we must have $\sigma(1) \neq \sigma(2)$). After $\sigma(2)$ is determined, there are $(n-2)$ choices for $\sigma(3)$. Continuing in this manner, we observe that there is $(n-(n-1)) = 1$ choice for $\sigma(n)$.

Hence, S_n has $n(n-1)(n-2) \dots (n-\overline{n-1}) = n!$ elements

$\therefore \; o(S_n) = n!$

Example 10. (*a*) *Find the elements and the multiplication table of* S_3.

(*b*) *Show that* S_3 *is a non-abelian group.*

Sol. (*a*) We know that S_n has $n!$ elements, therefore, S_3 has $3! = 6$ elements.

$$\therefore \qquad S_3 = \{\varepsilon, \sigma_1, \sigma_2, \sigma_3, \phi_1, \phi_2\}$$

These 6 elements are listed below:

$$\varepsilon = \begin{pmatrix} 1 & 2 & 3 \\ 1 & 2 & 3 \end{pmatrix}, \text{ the identity element of } S_3$$

$$\sigma_1 = \begin{pmatrix} 1 & 2 & 3 \\ 1 & 3 & 2 \end{pmatrix}; \; \sigma_2 = \begin{pmatrix} 1 & 2 & 3 \\ 3 & 2 & 1 \end{pmatrix}; \; \sigma_3 = \begin{pmatrix} 1 & 2 & 3 \\ 2 & 1 & 3 \end{pmatrix}$$

$$\phi_1 = \begin{pmatrix} 1 & 2 & 3 \\ 2 & 3 & 1 \end{pmatrix}; \; \phi_2 = \begin{pmatrix} 1 & 2 & 3 \\ 3 & 1 & 2 \end{pmatrix}$$

In cyclic notation,

$$S_3 = \{\varepsilon, \sigma_1, \sigma_2, \sigma_3, \phi_1, \phi_2\} = \{\varepsilon, (23), (13), (12), (132), (123)\}$$

Also, the multiplication table of S_3 is shown below

0	ε	σ_1	σ_2	σ_3	ϕ_1	ϕ_2
ε	ε	σ_1	σ_2	σ_3	ϕ_1	ϕ_2
σ_1	σ_1	ε	ϕ_1	ϕ_2	σ_2	σ_3
σ_2	σ_2	ϕ_2	ε	ϕ_1	σ_3	σ_1
σ_3	σ_3	ϕ_1	ϕ_2	ε	σ_1	σ_2
ϕ_1	ϕ_1	σ_3	σ_1	σ_2	ϕ_2	ε
ϕ_2	ϕ_2	σ_2	σ_3	σ_1	ε	ϕ_1

Table 1

Explanations: Consider

$\sigma_1 o \sigma_1 = (23)(23) = (23)^2 = I$ | The nth power of a n-cycle is I

To find $\sigma_1 o \sigma_2$: $\sigma_1 o \sigma_2 = \overset{\tau_1}{(23)}\overset{\tau_2}{(13)}$, say,

The elements are 1, 2, 3.

Start with '1', $1 \xrightarrow{\tau_1} 1 \xrightarrow{\tau_2} 3 \Rightarrow 1 \to 3$

Start with '3', $3 \xrightarrow{\tau_1} 2 \xrightarrow{\tau_2} 2 \Rightarrow 3 \to 2$

Start with '2', $2 \xrightarrow{\tau_1} 3 \xrightarrow{\tau_2} 1 \Rightarrow 2 \to 1$

This completes the cycle (132).

$\therefore$ $\sigma_1 o \sigma_2 = (23)(13) = (132) = \phi_1$

To find $\sigma_1 o \sigma_3$: $\therefore$ Here, $\sigma_1 o \sigma_3 = (23)(12) = \overset{\tau_1}{(23)}\overset{\tau_2}{(12)}$, say

The elements are 1, 2, 3.

Start with '1', $1 \xrightarrow{\tau_1} 1 \xrightarrow{\tau_2} 2 \Rightarrow 1 \to 2$

Start with '2', $2 \xrightarrow{\tau_1} 3 \xrightarrow{\tau_2} 3 \Rightarrow 2 \to 3$

Start with '3', $3 \xrightarrow{\tau_1} 2 \xrightarrow{\tau_2} 1 \Rightarrow 3 \to 1$

This completes the cycle (123).

Therefore, $\sigma_1 o \sigma_3 = (23)(12) = (123) = \phi_2$

Similarly, it can be easily proved that

$\sigma_1 o \phi_1 = \sigma_2,\ \sigma_1 o \sigma_3 = \phi_2$ etc.

(*b*) Consider $\sigma_1 o \sigma_2 = \phi_1$ and $\sigma_2 o \sigma_1 = \phi_2$

$\Rightarrow$ $\sigma_1 o \sigma_2 \neq \sigma_2 o \sigma_1$

$\therefore$ S_3 is non-abelian.

Example 11. *Consider the symmetric group $S_3 = \{\in, \sigma_1, \sigma_2, \sigma_3, \phi_1, \phi_2\}$ and $H = \{\in, \sigma_1\}$. Show that H is a subgroup of S_3. Find the right as well as left cosets of H in S_3. Is H a normal subgroup of S_3?*

Sol. Consider the multiplication table of S_3 under the composition of mappings, as shown in Table 1.

Since $\varepsilon \in H \Rightarrow H \neq \phi$. Thus, H is a non-empty subset of S_3.

We now find the cosets of H in S_3.

Consider $H = \{\varepsilon, \sigma_1\}$ where $o(H) = 2$.

Also, $o(S_3) = 3! = 6$.

$\therefore$ Number of distinct right (or left) cosets of H in $S_3 = [S_3 : H] = \dfrac{o(S_3)}{o(H)} = \dfrac{6}{2} = 3$.

Thus, there are 3 distinct right cosets as well as 3 distinct left cosets of H in S_3.

We now find these cosets.

For any $\sigma \in S_3$, all left cosets of H in S_3 are given below

$$\sigma_1 H = \{\sigma_1, \sigma_1 o \sigma_1\} = \{\sigma_1, \in\} = H$$
$$\sigma_2 H = \{\sigma_2, \sigma_2 o \sigma_1\} = \{\sigma_1, \phi_2\}$$
$$\sigma_3 H = \{\sigma_3, \sigma_3 o \sigma_1\} = \{\sigma_3, \phi_1\}$$
$$\phi_1 H = \{\phi_1, \phi_1 o \sigma_1\} = \{\phi_1, \sigma_3\} = \sigma_3 H$$
$$\phi_2 H = \{\phi_2, \phi_2 o \sigma_1\} = \{\phi_2, \sigma_2\} = \sigma_2 H$$

Thus, H, $\sigma_2 H$ and $\sigma_3 H$ are the only distinct left cosets of H in S_3.

Also, all right cosets of H in S_3 are given below.

$$H = \{\varepsilon, \sigma_1\}$$
$$H\sigma_1 = \{\sigma_1, \sigma_1 o \sigma_1\} = \{\varepsilon_1 \sigma_1\} = H = \{\sigma_1, \varepsilon\}$$
$$H\sigma_2 = \{\sigma_2, \sigma_1 o \sigma_2\} = \{\sigma_1, \phi_1\}$$
$$H\sigma_3 = \{\sigma_3, \sigma_1 o \sigma_3\} = \{\sigma_1, \phi_2\}$$
$$H\phi_1 = \{\phi_1, \sigma_1 o \phi_1\} = \{\phi_1, \sigma_2\} = H\sigma_2$$
$$H\phi_2 = \{\phi_1, \sigma_1 o \phi_2\} = \{\phi_2, \sigma_3,\} = H\sigma_3$$

Thus, H, $H\sigma_2$, $H\sigma_3$ are the only 3 distinct right cosets of H in S_3.

Further, $H\sigma_2 \neq \sigma_2 H$ and $H\sigma_3 \neq \sigma_3 H$

$\therefore$ H is not a normal subgroup of S_3 since the right cosets and the left cosets are not equal.

Example 12. *Consider the symmetric group S_3 whose multiplication table is shown in Table 1.*

(a) Find the order and the group generated by each element of S_3

(b) Find the number and all subgroups of S_3

(c) Let $A = \{\sigma_1, \sigma_2\}$, $B = \{\phi_1, \phi_2\}$. Find AB, $\sigma_3 A$, $A\sigma_3$.

(d) Let $H = gp(\sigma_1)$, $K = gp(\sigma_2)$. Show that HK is not a subgroup of S_3

(e) Is S_3 cyclic?

(f) Is S_3 abelian?

Sol. (*a*) By definition,

$$S_3 = \{\varepsilon, \sigma_1, \sigma_2, \sigma_3, \phi_1, \phi_2\} \text{ where } \varepsilon \text{ the identity element of } S_3,$$
$$\sigma_1 = (23), \sigma_2 = (13), \sigma_3 = (12), \phi_1 = (132), \phi_2 = (123)$$

We find the order of ε, σ_1, σ_2, σ_3, ϕ_1 and ϕ_2 respectively by using the multiplication Table 1.

Consider $\varepsilon^1 = \varepsilon \Rightarrow o(\varepsilon) = 1$

$$\sigma_1^2 = \sigma_1 o \sigma_1 = \varepsilon \text{ (using Table 1)}$$

$\Rightarrow$ $o(\sigma_1) = 2$

Also $\sigma_2^2 = \sigma_2 o \sigma_2 = \varepsilon \Rightarrow o(\sigma_2) = 2$

$$\sigma_3^2 = \sigma_3 o \sigma_3 = \varepsilon \Rightarrow o(\sigma_3) = 2$$

Further, $\phi_1^2 = \phi_1 o \phi_1 = \phi_2 \neq \varepsilon$

$$\phi_1^3 = \phi_1 o \phi_1 o \phi_1 = \phi_1 o \phi_2 = \varepsilon \Rightarrow o(\phi_1) = 3$$
$$\phi_2^2 = \phi_2 o \phi_2 = \phi_1 \neq \varepsilon$$

$$\phi_2^3 = \phi_2 o \phi_2 o \phi_2 = \phi_2 o \phi_1 = \varepsilon \Rightarrow o(\phi_2) = 3$$

$\therefore$ $\quad o(\varepsilon) = 1, o(\sigma_1) = o(\sigma_2) = o(\sigma_3) = 2, o(\phi_1) = o(\phi_2) = 3$

We next find the subgroups generated by each elements of S_3.

For any $x \in S_3$, if $o(x) = n$, then the subgroup generated by x, denoted by $gp(x)$ is given by

$$gp(x) = \{\varepsilon, x, x^2, x^3, \ldots, x^{n-1}\}$$

Here, $\quad gp(\varepsilon) = \{\varepsilon\}$

$gp(\sigma_1) = \{\varepsilon, \sigma_1\}$ since $o(\sigma_1) = 2$

$gp(\sigma_2) = \{\varepsilon, \sigma_2\}$ since $o(\sigma_2) = 2$

$gp(\sigma_3) = \{\varepsilon, \sigma_3\}$ since $o(\sigma_3) = 2$

$gp(\phi_1) = \{\varepsilon, \phi_1, \phi_1^2\} = \{\varepsilon, \phi_1, \phi_2\}$ since $o(\phi_1) = 3$ $\quad | \ \phi_1^2 = \phi_2$

$gp(\phi_2) = \{\varepsilon, \phi_2, \phi_2^2\} = \{\varepsilon, \phi_2, \phi_1\} = \{\varepsilon_1 \phi_1, \phi_2\}$ since $o(\phi_2) = 3$

(*b*) **By Lagrange's theorem,** if H is a subgroup of a finite group G, then order of H divides the order of G.

Therefore, any subgroup of S_3 must have order 2 or 3 since its order must divide $o(S_3) = 6$.

Using part (*a*), the only subgroups of S_3 are

$$H_1 = \{\varepsilon\}, H_2 = S_3, H_3 = \{\varepsilon, \sigma_1\}$$

$$H_4 = \{\varepsilon, \sigma_2\}, H_5 = \{\varepsilon, \sigma_3\}, H_6 = \{\varepsilon, \phi_1, \phi_2\}$$

The subgroups H_3, H_4, H_5 and H_6 are either of order 2 or 3 since 2 or 3 divide $6 = o(S_3)$.

Also, $H_1 = \{\varepsilon\}$ and $H_2 = S_3$ are trivial subgroups of S_3.

Therefore, S_3 has 6 subgroups.

(*c*) (*i*) **To find AB:** Multiplying each element of A by each element of B, we get

$$AB = \{\sigma_1, \sigma_2\}\{\phi_1, \phi_2\}$$

$$= \{\sigma_1 \phi_1, \sigma_1 \phi_2, \sigma_2 \phi_1, \sigma_2 \phi_2\}$$

But $\quad \sigma_1 \phi_1 = \sigma_2$ (from the multiplication table of S_3 (Fig. A))

$\sigma_1 \phi_2 = \sigma_3$

$\sigma_2 \phi_1 = \sigma_3 = \sigma_1 \phi_2$

$\sigma_2 \phi_2 = \sigma_1$

$\therefore$ $\quad AB = \{\sigma_2, \sigma_3, \sigma_1\} = \{\sigma_1, \sigma_2, \sigma_3\}$

(*ii*) **To find σ_3A:** Multiplying σ_3 by each element of A, we get

$\sigma_3 A = \{\sigma_3\sigma_1, \sigma_3\sigma_2\} = [\phi_1, \phi_2]$ [see multiplication table of S_3 (Table 1)].

(*iii*) **To find Aσ_3:** Multiplying each element of A by σ_3, we get

$$A_3\sigma = \{\sigma_1\sigma_3, \sigma_2\sigma_3\} = \{\phi_2, \phi_1\}$$

(*d*) Every subgroup of a finite group must divide the order of the group.

Here $\quad H = gp(\sigma_1) = \{\varepsilon, \sigma_1\}$

$K = gp(\sigma_2) = \{\varepsilon, \sigma_2\}$

$\therefore$ $\quad HK = \{\varepsilon, \sigma_1\}\{\varepsilon, \sigma_2\}$

$= \{\varepsilon, \varepsilon\sigma_2, \sigma_1\varepsilon, \sigma_1\sigma_2\} = \{\varepsilon_1, \sigma_2, \sigma_1, \phi_1\}$

$\Rightarrow$ $\quad o(HK) = 4$, which does not divide 6

$\therefore$ HK cannot be a subgroup of S_3

(*e*) A group G is cyclic if ∃ on element of $a \in G$ such that $o(a) = o(G)$.
Here $o(S_3) = 6$, but there is no element in S_3 whose order is 6.
Hence, S_3 is not cyclic.

(*f*) S_3 is non-abelian since $\sigma_1 o \sigma_2 \neq \sigma_2 o \sigma_1$.

8.35. ALTERNATING GROUP (A_n)

The set of all *even* permutations of S_n $(n \geq 2)$ is a group, under the composition of mappings, known as *alternating group of degree n*.

Theorem III. *The set A_n of all even permutations of $S_n (n \geq 2)$ is a normal subgroup of S_n and $o(A_n) = \frac{\lfloor n}{2}$ and $[S_n : A_n] = 2$.*

Proof. We know that an identity permutation on S_n is an even permutation.

Therefore, $A_n \neq \phi$ *i.e.* An is a non-empty subset of S_n.

Let $f, g \in A_n \Rightarrow f, g$ are even permutations.

Since, the inverse of an even permutation is also an even permutation, therefore, $g^{-1} \in A_n$.

Again, f and g^{-1} are even permutations and the product of two even permutation is an even permutation, therefore, $fog^{-1} \in A_n \ \forall f, g \in A_n$

$\therefore$ A_n is a subgroup of S_n.

Let $g \in S_n$ and consider $gofog^{-1}$.

The product of even permutations is an even permutation.

$$\Rightarrow \qquad gofog^{-1} \in A_n \in gf \in S_n$$

$\therefore$ A_n is a normal subgroup of S_n.

To find the order of A_n. Let σ is any odd permutation. Also, (12) being a transposition, is also an odd permutation. This implies (12) σ is an even permutation (The product of two odd permutation is an even permutation).

Thus, there are atleast as many even permutations as there are odd ones. On the other hand, for every even permutation ϕ, the permutation (12) ϕ is an odd permutation. It follows that there are atleast as many odd permutations as there are even permutations. Hence, there is an equal number of even and odd permutations. $20(A_n) = o(S_n)$

$$\Rightarrow \qquad o(A_n) = \frac{o(S_n)}{2} = \frac{n!}{2}$$

Example 13. *Consider the symmetric group $S_3 = \{\epsilon, \sigma_1, \sigma_2, \sigma_3, \phi_4, \phi_2\}$.*
Let A_3 be the alternating group defined by $A_3 = \{\epsilon, \phi_1, \phi_2\}$.
Is A_3 a normal subgroup of S_3?.

Sol. We know that A_n is a subgroup of $S_n (n \geq 2)$ (Theorem III above)

$$[S_3 : A_3] = \text{Index of } A_3 \text{ in } S_3 = \frac{o(S_3)}{o(A_3)} = \frac{3!}{\frac{3!}{2}} = 2$$

Since index of A_3 in S_3 is 2, A_3 is a normal subgroup of S_3.

TEST YOUR KNOWLEDGE 8.5

1. Find the order of each of the following permutations

   (*a*) (14) (*b*) (147)

   (*c*) (14762) (*d*) (124)(357)

   (*e*) (124)(356) (*f*) (124)(35)

   (*g*) (124)(3578)

2. What is the order of the product of a pair of disjoint cycles of lengths 4 and 6.

3. Determine which of the following permutations are even? odd?

   (*a*) (135) (*b*) (1356)

   (*c*) (13567) (*d*) (12)(134)(152)

   (*e*) (1243)(3521)

4. Show that A_n is a normal subgroup of $S_n (n \geq 2)$.

5. Let $f = \begin{pmatrix} 1 & 2 & 3 & 4 & 5 & 6 \\ 2 & 1 & 3 & 5 & 4 & 6 \end{pmatrix}$; $g = \begin{pmatrix} 1 & 2 & 3 & 4 & 5 & 6 \\ 6 & 1 & 2 & 4 & 3 & 5 \end{pmatrix}$

   Compute

   (*a*) f^{-1} (*b*) *gof* (*c*) *fog*

6. Consider the symetric group S_4.

   Let $f = \begin{pmatrix} 1 & 2 & 3 & 4 \\ 3 & 4 & 2 & 1 \end{pmatrix}$, $g = \begin{pmatrix} 1 & 2 & 3 & 4 \\ 2 & 4 & 3 & 1 \end{pmatrix}$

   (*a*) Find *fog*, *gof*, *fof*, f^{-1} (or find *fg*, *gf*, f^2, f^{-1})

   (*b*) Find the orders of *f*,*g*, *fog*

7. Let H = {∈, (12)}, K = {∈, (123), (132)} be any two subgroups of S_4. Show that HK is a subgroup of S_4 but H, k are not normal subgroups of S_4.

Answers

1. (*a*) 2 (*b*) 3

   (*c*) 5 (*d*) 3

   (*e*) 3 (*f*) 6

   (*g*) 12

2. l.c.m of 4 and 6 = 12

3. (*a*) even (*b*) odd

   (*c*) even (*b*) odd

   (*e*) even

5. (*a*) $f^{-1} = f$ (*b*) $\begin{pmatrix} 1 & 2 & 3 & 4 & 5 & 6 \\ 1 & 6 & 2 & 3 & 4 & 5 \end{pmatrix}$

   (*c*) $\begin{pmatrix} 1 & 2 & 3 & 4 & 5 & 6 \\ 6 & 2 & 1 & 5 & 3 & 4 \end{pmatrix}$

6. (*a*) $fog = \begin{pmatrix} 1 & 2 & 3 & 4 \\ 3 & 1 & 4 & 2 \end{pmatrix}$; $gof = \begin{pmatrix} 1 & 2 & 3 & 4 \\ 4 & 1 & 2 & 3 \end{pmatrix}$; $fof = \begin{pmatrix} 1 & 2 & 3 & 4 \\ 2 & 1 & 4 & 3 \end{pmatrix}$; $f^{-1} = \begin{pmatrix} 1 & 2 & 3 & 4 \\ 4 & 3 & 1 & 2 \end{pmatrix}$

   (*b*) *o*(*f*) = 4, *o*(*g*) = 3, o(*fog*) = 4

8.36. APPLICATIONS OF GROUPS IN CODING THEORY

Coding theory plays an important role in developing the techniques to provide reduant information in transmitted data. These techniques are useful in detecting, and sometimes in correcting, errors. Some of these techniques make use of group theory. To understand these techniques, we first introduce the basic terminology of coding theory.

8.37. MESSAGE

It is a basic unit of information. A message is a sequence of characters from a finite alphabete defined by the set B = {0, 1}.

8.38. WORD

To represent data, every character or symbol is first expressed in binary form. Then a 'word' is a sequenc of *m* 0's and 1's.

Let B = {0, 1} and let + be the operation defined on B as in the following table.

+	0	1
0	0	1
1	1	0

then, B is a group under +. Also every element in B is its own inverse

Generally, we define

$$B^m = B \times B \times B \times B \ldots \times B \ (m \text{ times})$$

then, B^m is also *a* group under the binary operation ⊕ defined by

$$(x_1, x_2, \ldots x_m) \oplus (y_1, y_2, \ldots, y_m) = (x_1 + y_1, x_2 + y_2, \ldots, x_m + y_m)$$

The identity of B^m is (0, 0, ..., 0) and every element is its own inverse. Now B has two elements and since $B^m = B \times B \times B \times \ldots \times B$ is a product of *m* factors, therefore B^m has 2^m elements. Thus, order of the group B^m is 2^m.

Consider the basic process of sending a word from one point to another point over a transmission channel as shown in the following figure.

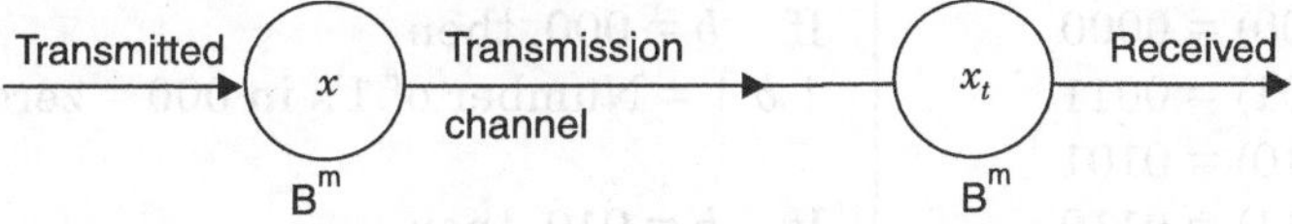

An element $x \in B^m$ is sent via a transmission channel and is received as an element $x_t \in B^m$. But in actual practice, the transmission may suffer some disturbances because of whether, electrical problems and this may cause a '0' to be received as a '1' or vice versa. Due to this, we can have a situation where the word received is different from the word that was sent. That is to say, $x \neq x_t$ where $x, x_t \in B^m$. To reduce the likelihood of receiving a word that differs from the word that was sent, we define encoding function.

8.39. ENCODING FUNCTION

For $n > m$, consider *a* function $e : B^m \to B^n$ such that *e* is a one-one function. Then the function *e* is called (*m*, *n*) encoding function. If $b \in B^m$, then *e*(*b*) is called *code word* representing *b*.

To transmit the code words by means of a transmission channel, consider the following figure.

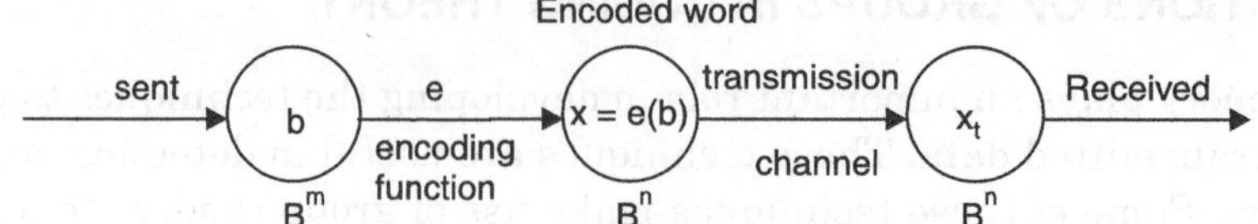

If a word $b \in B^m$ is transmitted, then each code word $x = e(b)$ is received as the word x_t in B^n. If different words in B^m will be assigned different code words in B^n. Then the encoding function $e : B^m \to B^n$ is one-one.

For the transmission channel to be *noiseless,* we should have $x = x_t$ for all $x \in B^n$.

Generally, errors in transmission do occur. We say that the *code word* $x = e(b)$ has been transmitted with k or *fewer errors* if x and x_t differ in atleast one but not more than k positions.

Let $e : B^m \to B^n$ be an (m, n) encoding function. We say that *e detects k or fewer errors.* If whenever $x = e(b)$ is transmitted with k or fewer errors; then x_t is not a code word.

8.40. WEIGHT

If $x \in B^n$, the number of 1's in x is called the weight of x and is denoted by $| x |$.

8.41. PARITY CHECK CODE

Consider the following encoding function $e : B^m \to B^{m+1}$ defined by $e(b) = b_1 b_2 .... b_m$

where $b \in B^m$ and $b_{m+1} = \begin{cases} 0, & \text{if } |b| \text{ is even} \\ 1, & \text{if } |b| \text{ is odd} \end{cases}$

i.e., b_{m+1} is zero iff the number of 1's in b is an even number and we say that **every code word *e* (*b*) has even weight.** Also b_{m+1} is 1 iff the number of 1's in b is an odd number and we say that **every code word *e*(*b*) has an odd weight.**

The function $e : B^m \to B^{m+1}$ is called *parity* $(m, m+1)$ *check code.*

It should be noted that if the received word has an odd weight, then we can say that the code word was transmitted correctly.

The following illustration explains this concept more clearly. Consider the encoding function $e : B^3 \to B^4$ defined by

$e(000) = 0000$	If $b = 000$, then
$e(001) = 0011$	$\| b \|$ = Number of 1's in 000 = zero (even)
$e(010) = 0101$	
$e(011) = 0110$	If $b = 010$, then
$e(100) = 1001$	$\| b \|$ = Number of 1's in 010 = one (odd)
$e(101) = 1010$	
$e(110) = 1100$	
$e(111) = 1111$	

Further, if the received word has even weight, then we cannot conclude that the code word was transmitted correctly, since the encoding function $e : B^m \to B^n$ does not detect an even number of errors.

8.42. HAMMING DISTANCE

Consider $x, y \in B^m$. Then, the *Hamming distance* $d(x, y)$ between x and y is the weight $| x \oplus y |$ of $x \oplus y$. *i.e.,* the distance between $x = x_1, x_2 ... x_m$ and $y = y_1, y_2 ..., y_m$ is the number of values of i such that $x_i \neq y_i$, that is, the number of positions in which x and y differ. Also the function $d(x, y)$ is known as *distance function.*

Theorem XXI. Properties of distance function

The distance function d(x, y) satisfies the following properties.

(*i*) $d(x, y) = d(y, x)$ (*ii*) $d(x, y) \geq 0$

(*iii*) $d(x, y) = 0 \iff x = y$ (*iv*) $d(x, y) \leq d(x, z) + d(z, y) \; \forall \; x, y, z \in B^m$.

Proof. (*i*) Let $x, y \in B^m$, then, by definition,

$$d(x, y) = |\, x \oplus y \,| = \text{The weight of } x \oplus y$$
$$= \text{The weight of } y \oplus x = |\, y \oplus x \,| = d(y, x).$$

(*ii*) $d(x, y) = |\, x \oplus y \,|$ = The weight of $x \oplus y$, which is always non-negative.

$\Rightarrow \quad d(x, y) \geq 0 \quad \forall \quad x, y \in B^m$.

(*iii*) $\quad d(x, y) = 0 \iff |\, x \oplus y \,| = 0$

$\iff$ The weight of $x \oplus y = 0 \iff x = y$

(*iv*) $d(x, y) = |\, x \oplus y \,| = |\, x \oplus \bar{0} \oplus y \,| = |\, x \oplus z \oplus z \oplus y \,|$

$= |\, x \oplus z \,| + |\, z \oplus y \,|$

$= d(x, z) + d(z, y)$

If $a \in B^m$, then $a \oplus a = \bar{0}$, the identity element in B^m

8.43. MINIMUM DISTANCE

The minimum distance of an encoding function $e : B^m \to B^n$ is the minimum of the distances between all distinct pairs of code words. That is, min $\{d(e(x), e(y) : x, y \in B^m\}$

Theorem XXII. *An (m, n) encoding function* $e : B^m \to B^n$ *can detect k or fewer errors iff its minimum distance is at least k + 1.*

Proof. Assume that the minimum distance between any two code words is at least $k + 1$. We show that the encoding function e can detect k or fewer errors.

Let $b \in B^m$ and let $x = e(b) \in B^n$ be the code word representing b, then x will be transmitted and received as x_t.

If $x_t \neq x$, then $d(x, x_t) \geq k + 1$. Hence x would be transmitted with $k + 1$ or more errors, that is, if x is transmitted with k or fewer errors, then x_t cannot be a code word. This means that e can detect k or fewer errors.

Converse. Suppose that the minimum distance between code words is r where $r \leq k$. Let $x, y \in B^m$ such that $d(x, y) = r$. If $x_t = y$, then x is transmitted and received as y, then $r \leq k$ errors have been committed and have not been detected. Thus, it is not true that e can detect k or fewer errors.

Cor. A code can correct k or fewer errors iff the minimum distance between any two code words is at least $2k + 1$.

8.44. GROUP CODES

An (m, n) encoding function $e : B^m \to B^n$ is called a group code if

$e(B^m) = \{e(b) : b \in B^m\}$ is a subgroup of B^n.

ILLUSTRATIVE EXAMPLES

Example 1. *Find the weight of the given words*

(*i*) *1011* (*ii*) *0110* (*iii*) *1110* (*iv*) *011101*

(*v*) *11111* (*vi*) *010101*

Sol. (*i*) Let $x = 1011$, then the weight of x is given by $|x|$ = Number of 1's in $x = 1011 = 3$

(*ii*) Let $x = 0110$, then the weight of x is given by $|x|$ = Number of 1's in $x = 0110 = 2$

(*iii*) If $x = 1110$, then $|x| = 3$

(*iv*) If $x = 011101$, then $|x| = 4$

(*v*) If $x = 11111$, then $|x| = 5$

(*vi*) If $x = 010101$, then $|x| = 3$.

Example 2. *Find the Hamming distance between* x and y.

(*i*) $x = 1100010$, $y = 1010001$ (*ii*) $x = 0100110$, $y = 0110010$

(*iii*) $x = 00111001$, $y = 10101001$ (*iv*) $x = 11010010$, $y = 00100111$

Sol. (*i*) The required distance between x and y is given by $d(x, y)$. (Note that $1 \oplus 1 = 0$, $0 \oplus 1 = 0$, $1 \oplus 0 = 0$, $0 \oplus 0 = 0$)

Here $x = 1100010$, $y = 1010001$

$\therefore$ $x \oplus y = 0110011$

$$\begin{array}{r} 1100010 \\ +\ 1010001 \\ \hline 0110011 \end{array}$$

$\Rightarrow$ $|x \oplus y|$ = Number of 1's in $0110011 = 4$

(*ii*) Here $x = 0100110$, $y = 0110010$

$\therefore$ $|x \oplus y| = 0010100$

$$\begin{array}{r} 0100110 \\ +\ 0110010 \\ \hline 0010100 \end{array}$$

$\Rightarrow$ $|x \oplus y|$ = Number of 1's in $0010100 = 2$

(*iii*) Here $x = 00111001$, $y = 10101001$

$\therefore$ $x \oplus y = 10010000$

$$\begin{array}{r} 00111001 \\ +\ 10101001 \\ \hline 10010000 \end{array}$$

$\Rightarrow$ $|x \oplus y|$ = Number of 1's in $10010000 = 2$

(*iv*) Here $x = 11010010$, $y = 00100111$

$\therefore$ $x \oplus y = 11110101$

$$\begin{array}{r} 11010010 \\ +\ 00100111 \\ \hline 11110101 \end{array}$$

$\Rightarrow$ $|x \oplus y|$ = Number of 1's in $11110101 = 6$.

Example 3. *Find the minimum distance of the (2, 4) encoding function* $e : B^2 \to B^4$ *defined as*

$$e(00) = 0000, \quad e(10) = 0110$$
$$e(01) = 1011, \quad e(11) = 1100$$

(P.T.U. B.Tech. Dec. 2012)

Sol. We first find the distances between pairs of code words.

$d(0000, 0110) = |(0000) \oplus (0110)| = 2$

$d(0000, 1011) = |(0000) \oplus (1011)| = 3$

$d(0000, 1100) = |(0000) \oplus (1100)| = 2$

$d(0110, 1011) = |(0110) \oplus (1011)| = 3$

$d(0110, 1100) = |(0110) \oplus (1100)| = 2$

$$\begin{array}{r} 0110 \\ +\ 1011 \\ \hline 1101 \end{array} \qquad \begin{array}{r} 0110 \\ +\ 1100 \\ \hline 1010 \end{array} \qquad \begin{array}{r} 1011 \\ +\ 1100 \\ \hline 0111 \end{array}$$

$\therefore$ The required minimum distance is 2.

Example 4. *Consider (2, 6) encoding function* $e: B^2 \to B^6$ *defined as*

$$e(00) = 000000; \qquad e(10) = 101010$$
$$e(01) = 011110; \qquad e(11) = 111000$$

(a) Find the minimum distance of e

(b) How many errors will e detect ?

Sol. (*a*) We first find the distances between each pairs of code words.

$d(000000, 101010) = |(000000) \oplus (101010)| = 3$

$d(000000, 011110) = |(000000) \oplus (011110)| = 4$

$d(000000, 111000) = |(000000) \oplus (111000)| = 3$

$d(101010, 011110) = |(101010) \oplus (011110)| = 3$

$d(101010, 111000) = |(101010) \oplus (111000)| = 2$

$$\begin{array}{r} 101010 \\ +\ 011110 \\ \hline 110100 \end{array} \qquad \begin{array}{r} 101010 \\ +\ 111000 \\ \hline 010010 \end{array}$$

The required minimum distance is 2.

(*b*) The code will detect k or fewer errors iff its minimum distance is at least $k + 1$. Since the minimum distance is 2, we have

$$2 \geq k + 1 \quad \text{or} \quad k + 1 \leq 2 \quad \text{or} \quad k \leq 1.$$

The code will detect one or fewer error.

Example 5. *Consider the encoding function* $e : B^2 \to B^6$ *defined as*

$$e(00) = 001000\ ; \qquad e(01) = 010100$$
$$e(10) = 100010\ ; \qquad e(11) = 110001$$

How many errors it can detect and correct ?

Sol. Proceeding as in example 4, the minimum distance of $e : B^2 \to B^6$ is 3.

The code can detect k or fewer errors iff the minimum distance is at least $k + 1$. Here the minimum distance is 3.

Therefore, $3 \geq, k + 1$ or $k + 1 \leq 3$ or $k \leq 2$.

Thus, the code will detect 2 or fewer errors.

The code will correct k or fewer errors iff the minimum distance is at least $2k + 1$. Here the minimum distance is 3.

So $\quad 3 \geq 2k + 1 \quad$ or $\quad 2k + 1 \leq 3 \quad$ or $\quad 2k \leq 2 \quad$ or $\quad k \leq 1$.

Thus, the code will correct 1 or fewer errors.

Example 6. Consider $e : B^3 \to B^8$ defined by

$$\left.\begin{aligned} e(000) &= 00000000 \\ e(001) &= 10111000 \\ e(010) &= 00101101 \\ e(011) &= 10010101 \\ e(100) &= 10100100 \\ e(101) &= 10001001 \\ e(110) &= 00011100 \\ e(111) &= 00110001 \end{aligned}\right\} \textit{Code words}$$

(a) How many errors will e detect ?

(b) How many errors will e correct ?

Sol. We first find the distances between pairs of code words. There are 28 distinct pairs of code words. Proceeding as in example 4, the minimum distance is 3.

(*a*) The code will detect k or fewer errors iff the minimum distance is at least $k + 1$. Here the minimum distance is 3

$\therefore \quad 3 \geq k + 1 \quad \text{or} \quad k + 1 \leq 3 \quad \text{or} \quad k \leq 2$

Thus, the code will detect 2 or fewer errors.

(*b*) The code will correct k or fewer errors iff the minimum distance is at least $2k + 1$. Here the minimum distance is 3,

so $\quad 3 \geq 2k + 1 \quad \text{or} \quad 2k + 1 \leq 3 \quad \text{or} \quad 2k \leq 2 \quad \text{or} \quad k \leq 1$

Thus, the code will correct 1 or fewer errors.

Example 7. *Consider* $e : B^2 \to B^5$ *defined by*

$$\left.\begin{aligned} e(00) &= 00000 \\ e(01) &= 01110 \\ e(10) &= 10101 \\ e(11) &= 11011 \end{aligned}\right\} \text{code words}$$

Show that the encoding function $e : B^2 \to B^5$ *is a group code.*

Sol. Let H = {00000, 01110, 10101, 11011} be the set of all code words. Consider the following table.

⊕	00000	01110	10101	11011
00000	00000	01110	10101	11011
01110	01110	00000	11011	10101
10101	10101	11011	00000	01110
11011	11011	10101	01110	00000

From the above table, we observe that every element in the interior of the table belongs to the set H. Hence H is closed under ⊕.

The first row inside the table coincides with the elements listed in H = {00000, 01110, 10101, 11011}

$\therefore$ Identity element of B^5 is 00000.

Also ⊕ is an associative operation. *i.e.* For $x, y, z \in$ H, $x \oplus (y \oplus z) = (x \oplus y) \oplus z$ is true.

Finally, each element of H is its own inverse. Since

$$\begin{aligned} 00000 \oplus 00000 &= 00000 \\ 01110 \oplus 01110 &= 00000 \\ 10101 \oplus 10101 &= 00000 \\ 11011 \oplus 11011 &= 00000 \end{aligned} \qquad \begin{array}{r} 01110 \\ +\,01110 \\ \hline 00000 \end{array}$$

Hence H is a subgroup of B^5 and the given encoding function is a group code.

8.45. MORE APPLICATIONS OF GROUPS (*P.T.U. B. Tech. Dec. 2005*)

By using group code, we can find the solutions of many coding problems by using the concept of Group theory. A coding problem is a problem which is used to represent distinct messages by means of a sequence of letters from a given alphabet. A sequence of letter from an alphabet is called a word. A code is a collection of word that is used to represent distinct messages. A word in a code is called **codeword.** A **blackcode** is a code consisting of words that are of the same length. The concept of **'Group Theory'** can be applied in many situations which arise in coding problem which is clear from the following discussion.

In error correction : Suppose a codeword is transmitted from its origin to its destination. During the course of transmission, some of the information might cause some of the 1's in the codeword to be received such as noises). Let A denote the set of all binary sequences of length n. Let $\oplus$ be a binary operation on A such that for $x, y \in$ A. Let $x \oplus y$ denotes the sequence of length n that has 1's when x and y differ and has 0's when x and y are same. We observe that (A, $\oplus$) is a group with all zero word as its identity and every word is its own inverse. Further for $x, y \in$ A, we define the distance between x and y denoted by $d(x, y)$ to be the weight of $x \oplus y$, say, $w(x \oplus y)$. The weight of x means the number of 1's in x. For example, the weight of 1110000 is 3.

By using the **Minimum Distance Decoding Criterian**, it has been proved that a code of distance $2t + 1$ can correct t or lesser transmission error.

Group codes: A subset G of A is called a group code if (G, $\oplus$) is a subgroup of (A, $\oplus$) where A is the set of binary sequences of length n.

By using **Minimum Distance Decoding Criterion**, we can determine the transmitted word corresponding to a received word. Let (G, $\oplus$) is a group code.

Let y be a received word and $d(x_i, y)$ = The distance between and x_i and $y = w(x_i \oplus y)$.

The weight of the word is the coset G $\oplus$ y are the distances between the code words in G and y. Let e denotes one of the words of smallest weight. Then, according to the **Minimum Distance Decoding Criterian**, $e \oplus y = x_j$ is the transmitted code word. Thus, by using the axioms of group theory, we can find the transmitted code words and group codes.

TEST YOUNOWLEDGE 8.6

1. Find the weight of each of the following words in B^5.

   (a) 01000 (b) 11100 (c) 00000 (d) 11111

2. Find the Hamming distance between x and y :

   (a) $x = 110110$, $y = 000101$ (b) $x = 001100$, $y = 010110$

3. Consider the following (2, 5) encoding function $e : B^2 \to B^5$ *defined as*

$$\left.\begin{aligned} e(00) &= 00000 \\ e(10) &= 00111 \\ e(01) &= 01110 \\ e(11) &= 11111 \end{aligned}\right\} \text{Code words.}$$

   Find the minimum distance of the encoding function e.

4. Consider the (3, 8) encoding function $e : B^3 \to B^8$ defined by

$$\left.\begin{aligned} e(000) &= 00000000 \\ e(001) &= 10111000 \\ e(010) &= 00101101 \\ e(011) &= 10010101 \\ e(100) &= 10100100 \\ e(101) &= 10001001 \\ e(110) &= 00011100 \\ e(111) &= 00110001 \end{aligned}\right\} \text{Code words}$$

   How many errors e will detect ?

5. Consider the (3, 9) encoding function $e : B^3 \to B^9$ *defined as*

$e(000) = 000000000$; $e(100) = 010011010$

$e(001) = 011100101$; $e(101) = 111101011$

$e(010) = 010101000$; $e(110) = 001011000$

$e(011) = 110010001$; $e(111) = 110000111$

(*a*) Find the minimum distance of *e*.

(*b*) How many errors will *e* detect ?

6. Show that the (3, 7) encoding function $e : B^3 \to B^7$ defined by

$e(000) = 0000000$; $e(100) = 1000101$

$e(001) = 0010110$; $e(101) = 1010011$

$e(010) = 0101000$; $e(110) = 1101101$

$e(011) = 0111110$; $e(111) = 1111011$

is a group code.

Answers

1. (*a*) 1 (*b*) 3 (*c*) 0 (*d*) 5
2. (*a*) 4 (*b*) 3 **3.** 2
4. 2 or fewer errors. **5.** (*a*) 3 (*b*) 2 or fewer.

MULTIPLE CHOICE QUESTIONS (MCQs)

1. The matrices $\begin{bmatrix} \cos\theta & -\sin\theta \\ \sin\theta & \cos\theta \end{bmatrix}$ and $\begin{bmatrix} a & 0 \\ 0 & b \end{bmatrix}$ commute under the multiplication

(*a*) if $a = b$ or $\theta = n\pi$, n is an integer (*b*) always

(*c*) never (*d*) if $a. \cos\theta \neq b. \sin\theta$.

2. The set of integers Z with the binary operations '*' defined as $a * b = a + b + 1$ for $a, b \in Z$, is a group. The identity element of this group is

(*a*) 0 (*b*) 1

(*c*) – 1 (*d*) 12.

3. Which of the following is TRUE ?

(*a*) The set of all rational negative numbers forms a group under multiplication

(*b*) The set of all non-singular matrices forms a group under multiplication

(*c*) The set of all matrices forms a group under multiplication

(*d*) Both (*b*) and (*c*) are true.

4. Which of the following statement is FALSE ?

(*a*) The set of rational numbers is an abelian group under addition.

(*b*) The set of rational integers is an abelian group under addition.

(*c*) The set of rational numbers form an abelian group under multiplication

(*d*) None of these.

5. Let A be the set of all non-singular matrices over real numbers and let $*$ be the matrix multiplication operator. Then

(*a*) A is closed under $*$ but <A, *> is not a semi group

(*b*) <A, *> is a semi group but not a monoid

(*c*) <A, *> is a monoid but not a group

(*d*) <A, *> is a group but not an abelian group.

6. Some group (G, 0) is knwon to be abelian. Then, which one of the following is TRUE for G ?

(*a*) $g = g^{-1}$ for every $g \in G$ (*b*) $g = g^2$ for every $g \in G$

(*c*) $(goh)^{-2} = g^2oh^2$ for every $g, h \in G$ (*d*) G is of finite order.

7. For ring $Z_{10} = \{0, 1, 2, ..., 9\}$ of integers modulo 10, units of Z_{10} are

(*a*) 0, 1, 3 and 7 (*b*) 1, 3, 7 and 9

(*c*) 1, 2, 3, and 7 (*d*) 3, 7 and 9.

8. If a and b are positive integers, define $a * b = a$ where $a * b \equiv \alpha$ (modulo 7), with this $*$ operation, the inverse of 3 in group G {1, 2, 3, 4, 5, 6} is

(*a*) 3 (*b*) 1

(*c*) 5 (*d*) 4.

9. If the binary operation $*$ is defined on a set of ordered pairs of real numbers as $(a, b) * (c, d) = (ad + bc, bd)$ and is associative, then $(1, 2) * (3, 5) * (3, 4) =$

(*a*) (74, 40) (*b*) (32, 40)

(*c*) (23, 11) (*d*) (7, 11).

10. G = {*e*, *a*, *b*, *c*} is an abelian group with '*e*' as identity element. The order of the other elements are

(*a*) 2, 2, 2 (*b*) 3, 3, 3

(*c*) 2, 2, 4 (*d*) 2, 3, 4.

Answers and Explanations

1. (*a*) Let $$A = \begin{pmatrix} \cos\theta & -\sin\theta \\ \sin\theta & \cos\theta \end{pmatrix}, B = \begin{pmatrix} a & 0 \\ 0 & b \end{pmatrix}$$

Then $$AB = \begin{pmatrix} \cos\theta & -\sin\theta \\ \sin\theta & \cos\theta \end{pmatrix}\begin{pmatrix} a & 0 \\ 0 & b \end{pmatrix} = \begin{pmatrix} a\cos\theta & -b\sin\theta \\ a\sin\theta & b\cos\theta \end{pmatrix}$$

$$BA = \begin{pmatrix} a & 0 \\ 0 & b \end{pmatrix}\begin{pmatrix} \cos\theta & -\sin\theta \\ \sin\theta & \cos\theta \end{pmatrix} = \begin{pmatrix} a\cos\theta & -a\sin\theta \\ b\sin\theta & b\cos\theta \end{pmatrix}$$

Now, the matrices A and B commute under matrix multiplication if AB = BA

i.e., $$\begin{pmatrix} a\cos\theta & -b\sin\theta \\ a\sin\theta & b\cos\theta \end{pmatrix} = \begin{pmatrix} a\cos\theta & -a\sin\theta \\ b\sin\theta & b\cos\theta \end{pmatrix}$$

$\Rightarrow$ $-b\sin\theta = -a\sin\theta,\ a\sin\theta = b\sin\theta$

$\Rightarrow$ $(a - b)\sin\theta = 0$

$\Rightarrow$ either $a - b = 0$ or $\sin\theta = 0$

$\Rightarrow$ either $a = b$ or $\theta = n\pi,\ n \in Z$.

2. (c) Let e is the identity element such that $a * e = a$

$$\Rightarrow \quad a + e + 1 = a \Rightarrow e + 1 = 0 \Rightarrow e = -1.$$

3. (b) If $a = -\frac{2}{3}, b = -\frac{4}{3}$, then $ab = \frac{8}{9}$, which is not a negative rational number. Hence the set of negative rational numbers cannot form a group under multiplication.Hence the alternative (a) is not true. Also the alternative (c) is not true since singular matrices do not have inverses.

4. (c) If we take $a = \frac{0}{1}$, then a^{-1} does not exist. Hence, the alternative (c) is false.

5. (d) Since A is the set of all non-singular matrices over real numbers, then, if $A_1, A_2 \in A \Rightarrow A_1 . A_2 \in A$. Hence A is closed under matix multiplication. Also matrix multiplication is associative. Therefore (A, .) is a semigroup under matrices multiplication.

Also, $\begin{pmatrix} 1 & 0 \\ 0 & 1 \end{pmatrix}$ is the multiplicative identity. Therefore, A is a monoid. Also, since $|A| \neq 0$

$\therefore$ A^{-1} exists. Hence A is a group. But $A_1 A_2 \neq A_2 A_1$ $\therefore$ A is not an abelian group.

6. (c)

7. (b) The units of Z_{10} are those elements of Z_{10} which are prime to 10. Hence 1, 3, 7 and 9 are units of Z_{10}.

8. (c) Since $3 * 5 = 15 = 1 \quad \therefore \quad 3^{-1} = 5$

9. (a) $(1, 2) * ((3, 5) * (3, 4)) = (1, 2) * (12 + 15, 20)$

$$= (1, 2) * (27, 20)$$

$$= (20 + 54, 40) = (74, 40).$$

10. (a) If $a, b = c, a.c = b$ then $a.a = e \quad \therefore \quad o(a) = 2$

If $b.c. = a$, then $b . b = e \quad \therefore \quad o(b) = 2$

Similarly, $o(c) = 2$.

.	e	a	b	c
e	e	a	b	c
a	a	e	c	b
b	b	c	e	a
c	c	b	a	e

9 RINGS

9.1. RING

(P.T.U. B.Tech. Dec. 2009, May 2008, 2007, Dec. 2006, May 2005)

Let R be a non-empty set with two binary compositions, addition (+) and multiplication (.). Then R is called a ring iff it satisfies the following :

I. R is an additive group under + *i.e.,*

(*i*) For $a, b \in$ R $\Rightarrow$ $a + b \in$ R *i.e.,*
R is closed under addition

(*ii*) For $a, b, c \in$ R, $a + (b + c) = (a + b) + c$ *i.e.,*
Associativity under addition holds in R.

(*iii*) For each $a \in$ R, $\exists\ 0 \in$ R such that $a + 0 = a = 0 + a$ *i.e.,*
R has additive identity.

(*iv*) For each $a \in$ R, $\exists\ -a \in$ R such that $a + (-a) = 0$ *i.e.,*
R has an additive inverse.

(*v*) For each $a, b \in$ R, $a + b = b + a$ *i.e.,*
R is additive.

II. For each $a, b \in$ R, $a.\ b \in$ R *i.e.,*
R is closed under multiplication.

III. For $a, b, c \in$ R, $a.\ (b\ .\ c) = (a.\ b)\ .\ c$ *i.e.,*
Associativity under multiplication holds in R.

IV. For $a, b, c \in$ R,

(*i*) $a\ .\ (b + c) = a.\ b + a.\ c$ (Left distributive law)

(*ii*) $(a + b)\ .\ c = a.\ c + b\ .\ c.$ (Right distributive law)

Remark : The additive identity 0 of R is unique. We call it 'zero' of the ring. The additive inverse is also unique.

9.2. COMMUTATIVE RING

A ring R is called a commutative ring if $a\ .\ b = b\ .\ a \quad \forall\ a, b \in$ R.

9.3. RING WITH UNITY

(P.T.U. B.Tech. Dec. 2005)

A ring R is called ring with unity if for each $x \in$ R, $\exists\ 1 \in$ R such that $1\ .\ x = x = x\ .\ 1$. The element '1' is called multiplicative identity of R.

9.4. FINITE AND INFINITE RING

A ring R with finite number of elements, is known as finite ring, otherwise it is known as infinite ring.

Example. (*i*) $Z_6 = \{0, 1, 2, 3, 4, 5, +6, \times 6\}$ is a finite commutative ring

(*ii*) The quanterian ring given by $J = \{\pm 1, \pm i, \pm j, \pm k\}$ satisfying

$i^2 = j^2 = k^2 = -1; i \cdot j = k; j \cdot i = -k$ etc. is a finite non-commutative ring

(*iii*) (Z, +, .), (Q, +, .), (R, + , .) and (C, + , .) are infinite commutative rings with unity.

9.5. (a) RING WITH ZERO DIVISORS

Let R be a ring and $0 \neq a, b \in R$. Then R is called *ring with zero divisors* if $a . b = 0$. *i.e.,*

If product of two non-zero elements in a ring R is zero, then R is called ring with zero divisors. Also we say that the element a is a zero divisor of b or b is a zero divisor of a.

9.5. (b) RING WITHOUT ZERO DIVISORS

A ring R is called ring without zero divisors if whenever

$$a . b = 0 \quad \Rightarrow \quad a = 0 \quad \text{or} \quad b = 0 \,\forall\, a, b \in R.$$

ILLUSTRATIVE EXAMPLES

Example 1. *Let Z be the set of integers, then (Z, +, .) is a ring. Also Z is a commutative ring with unity.*

Sol. We know that Z is an additive group under +. (See Chapter on 'Groups').

Also for $a, b \in Z \Rightarrow a. b \in Z \quad \forall\, a, b \in Z$ *i.e.,*

Z is closed under multiplication.

For $a, b, c \in Z$, $a . (b . c) = (a. b) . c \qquad \forall\, a, b, c \in Z$ *i.e.,*

Associativity under multiplication holds in Z.

For $a, b, c \in Z$, $\quad a . (b + c) = a. b + a . c$

$$(a + b) . c = a . c + b . c \;\forall\, a, b, c \in Z$$

Hence we can say that Z is a ring.

Further, for $a, b \in Z$, $\quad a . b = b . a \qquad \forall\, a, b \in Z$

$\therefore$ Z is commutative also.

Also for $a \in Z$, $\exists\, 1 \in Z$ such that

$$1 . a = a = a . 1 \qquad \forall\, a \in Z.$$

$\therefore$ Z is a ring with unity (multiplicative identity).

Example 2. *Show that E, the set of even integers is a commutative ring without unity.*

Sol. Consider $\qquad E = \{..., -4, -2, 0, 2, 4, 6, ...\}$

For $a, b \in E$, $a + b \in E \quad \forall\, a, b \in E$. $\quad$ *i.e.;*

E is closed under addition.

F or $a, b, c \in E$, $a + (b + c) = (a + b) + c \;\forall\, a, b, c \in E$. *i.e.,*

E is closed under association.

For $a \in E$, there exists $0 \in E$ such that

$$a + 0 = a = 0 + a \quad \forall\, a \in E. \text{ i.e.,}$$

0 is the additive identity of E.

For $a \in$ E, there exists $-a \in$ E such that $a + (-a) = 0 = (-a) + a \quad \forall a \in$ E. *i.e.*, E has additive inverse.

Also, for $a, b \in E \Rightarrow a + b = b + a \quad \forall a, b \in E$.

Hence E is an additive group.

Further, for $a, b \in E, a \cdot b \in E \quad \forall a, b \in E$. *i.e.*, E is closed under multiplication.

For $a, b, c \in E, \quad a \cdot (b \cdot c) = (a \cdot b) \cdot c \quad \forall a, b, c \in E$. *i.e.*, E is closed under association w.r.t. multiplication.

Also for $a, b, c \in E, \quad a \cdot (b + c) = a \cdot b + a \cdot c$

$$(a + b) \cdot c = a \cdot c + b \cdot c \quad \forall a, b, c \in E.$$

Also for $a, b \in E, \quad a \cdot b = b \cdot a \; \forall a, b \in R$

$\therefore$ E is a commutative ring.

But for $a \in$ E, there exists no $1 \in$ E such that $a \cdot 1 = a = 1 \cdot a$

$\therefore$ E is a commutative ring without unity (multiplicative identity).

Example 3. *Show that the set M of 2 × 2 matrices over integers form a non-commutative ring with unity under matrix addition and multiplication.*

Sol. Let $M = \left\{ \begin{pmatrix} a & b \\ c & d \end{pmatrix} : a, b, c, d \in Z \right\}$. We show M is a ring under matrix addition and multiplication.

Let $A, B \in M \Rightarrow A + B \in M \; \forall A, B \in M$.

Since sum of two matrices of the same order is again a matrix. Therefore M is closed under matrix addition.

For $A, B, C \in M, \quad A + (B + C) = (A + B) + C \quad \forall A, B, C \in M$.

i.e., matrix addition is associative.

For $A \in M$, there exists $O \in M$ such that

$$A + O = A = O + A$$

The element O is called additive identity of M.

For $A \in M$, there exists $-A \in M$ such that

$$A + (-A) = O \; \forall A \in M.$$

The element $-$ A is called additive inverse of A.

For $A, B \in M, A + B = B + A \;\; \forall A, B \in M$ *i.e.*, matrix addition is commutative.

$\therefore$ M is an additive group.

We know that matrix multiplication is associative. *i.e.*,

$$A + (B + C) = (A + B) + C \; \forall A, B, C \in M$$

Also $\quad A \cdot (B + C) = A \cdot B + A \cdot C$

$$(A + B) \cdot C = A \cdot C + B \cdot C \; \forall A, B, C \in M$$

i.e., left distributive law and right distributive law also hold.

Hence M is a ring under matrix addition and multiplication.

But matrix multiplication in general, is not commutative. *i.e.*, we can have $A, B \in M$ for which $AB \neq BA$.

Hence M is non-commutative ring.

Lastly, For $A = \begin{pmatrix} a & b \\ c & d \end{pmatrix} \in M$, there exists $I = \begin{pmatrix} 1 & 0 \\ 0 & 1 \end{pmatrix} \in M$ such that

$A \cdot I = A = I \cdot A$ *i.e.*,

I is the multiplicative identity of M.

Hence M is a non-commutative ring with unity.

9.6. RING OF INTEGERS MODULO *m* ($m \geq 1$)

The set $Z_m = \{0, 1, 2, 3, ..., m - 1, +_m, \times_m\}$ under the operation of addition and multiplication modulo *m* is a ring, known as **ring of integers** modulo *m*.

Example 4. *Consider the set* $X = \{0, 1, 2, 3, 4, 5 ; +_6, \times_6\}$*. then X is a commutative ring with unity under addition and multiplication modulo 6.*

Sol. Consider the addition modulo table shown in Table. I

Table I

$+_6$	*0*	*1*	*2*	*3*	*4*	*5*
0	0	1	2	3	4	5
1	1	2	3	4	5	0
2	2	3	4	5	0	1
3	3	4	5	0	1	2
4	4	5	0	1	2	3
5	5	0	1	2	3	4

From Table I, we observe that each element inside the Table I is also in X. It means X is closed under addition modulo 6. Also addition modulo 6 is associative. *i.e.*,

$$a +_6 (b +_6 c) = (a +_6 b) +_6 c \qquad \forall\ a, b, c \in X.$$

The first row inside the table coincides with the top most row of the Table I. It means 0 is the additive identity of X.

From Table I, we observe that each element of X has an additive inverse. For *e.g.*,

Inverse of 1 is 5 (the element which is at the intersection of 1 and 5 is 0).

Similarly, Inverse of 2 is 4 etc. ($2 +_6 4 = 0$)

Also Table I is symmetrical w.r.t. $+_6$ It means $a +_6 b = b +_6 a \quad \forall\ a, b \in X$.

$\therefore$ X is an additive group under addition modulo 6 ($+_6$).

Consider the multiplication modulo table as shown in Table II.

Table II

$\times_6$	*0*	*1*	*2*	*3*	*4*	*5*
0	0	0	0	0	0	0
1	0	1	2	3	4	5
2	0	2	4	0	2	4
3	0	3	0	3	0	3
4	0	4	2	0	4	2
5	0	5	4	3	2	1

From Table II, we observe that each element inside the table is also in X. It means that X is closed under multiplication modulo 6 (X_6)

i.e., for $a, b \in X, \quad a \times_6 b \in X \quad \forall\ a, b \in X.$

Also multiplication modulo 6 is associative *i.e.*,

for $a, b, c \in X,\ a \times_6 (b \times_6 c) = (a \times_6 b) \times_6 c\ \forall\ a, b, c \in X$

Further, $a \times_6 (b +_6 c) = a \times_6 b + a \times_6 c$

$$(a +_6 b) \times_6 c = a \times_6 c + b \times_6 c \ \forall \ a, b, c \in X$$

i.e., left distributive law and right distributive law also hold.

Hence X is a ring under addition modulo 6 and multiplication modulo 6. Also Table II is symmetrical w.r.t. $\times_6$. It means that X is a commutative ring. The second row inside Table II coincides with the top most row of Table II. It means 1 is the multiplicative identity of X.

Example 5. *Consider the set Z together with binary compositions ⊕ and ⊙ defined by*

$$a \oplus b = a + b - 1$$

$$a \odot b = a + b - ab.$$ (P.T.U. B. Tech. Dec. 2010)

Show that (Z, ⊕, ⊙) *is a ring.*

Sol. We first show that Z is an additive group under ⊕. Let $a, b \in Z$. Since Z is a group under +, we have $a + b - 1 \in Z \quad \forall \ a, b \in Z$

$\Rightarrow \qquad a \oplus b \in Z \ \forall \ a, b \in Z$, *i.e.*,

Z is closed under ⊕.

Let $a, b, c \in Z$ and consider

$$a \oplus (b \oplus c) = a \oplus (b + c - 1) = a + b + c - 1 - 1$$

$$= a + b + c - 2 \qquad ...(1)$$

Also $\qquad (a \oplus b) \oplus c = (a + b - 1) \oplus c = a + b - 1 + c - 1$

$$= a + b + c - 2 \qquad ...(2)$$

Hence from (1) and (2), $a \oplus (b \oplus c) = (a \oplus b) \oplus c \forall \ a, b, c \in Z$

$\therefore$ Z is closed w.r.t. association under ⊕.

For $a \in Z$, consider $a \oplus 1 = a + 1 - 1 = a$

$$1 \oplus a = 1 + a - 1 = a$$

$\Rightarrow \qquad a \oplus 1 = a = 1 \oplus a$

i.e. 1 is the additive identity.

Also for $a, b \in Z$, consider $a \oplus b = 1$

$\Rightarrow \qquad a + b - 1 = 1$

$\Rightarrow \qquad b = 2 - a$

i.e., for $a \in Z$, $2 - a$ is the additive inverse of a. Since,

$$a \oplus 2 - a = a + 2 - a - 1 = 1$$

$$2 - a \oplus a = 2 - a + a - 1 = 1$$

$\Rightarrow \qquad a \oplus 2 - a = 1 = 2 - a \oplus a.$

Further for each $a, b \in Z$, we have

$$a \oplus b = a + b - 1 = b + a - 1 = b \oplus a$$

$\therefore$ Z is an additive group under ⊕.

Let $a, b \in Z \Rightarrow a \,.\, b = a + b - ab \in Z \ \forall \ a, b \in Z$

$\Rightarrow a \,.\, b \in Z$ *i.e.*, Z is closed under ⊙

Also $\qquad a \odot (b \odot c) = a \odot (b + c - bc) = a + b + c - bc - a\,(b + c - bc)$

$$= a + b + c - bc - ab - ac + abc$$

$$(a \odot b) \odot c = (a + b - ab) \odot c = a + b - ab + c - (a + b - ab)\, c$$

$$= a + b + c - ab - bc - ac + abc$$

$$= a \odot (b \odot c)$$

i.e., Z is closed w.r.t. association under ⊙.

Finally, For $a, b, c \in Z$, consider

$$a \odot (b \oplus c) = a \odot (b + c - 1) = a + b + c - 1 - a\,(b + c - 1)$$
$$= a + b + c - 1 - a\,b - a\,c + a = 2a + b + c - ab - ac - 1 \quad ...(3)$$

Also $\quad a \odot b = a + b - ab$

$$a \odot c = a + c - ab$$

$\therefore \quad a \odot b \oplus a \odot c = a + b - ab \oplus a + c - ac = a + b - ab + a + c - ac - 1$

$$= 2a + b + c - a\,b - ac - 1 \quad ...(4)$$

From (3) and (4), we get $a \odot (b \oplus c) = a \odot b \oplus a \odot c$

Similarly, $\quad (a \oplus b) \odot c = a \odot c \oplus b \odot c$

i.e., left distributive law and right distributive law also hold.

$\therefore$ Z is a ring under the binary composition $\oplus$ and $\odot$.

9.7. BOOLEAN RING

A ring R is called a Boolean ring if $x^2 = x \;\forall\; x \in R$

For example: The ring $R = \{0, 1, +_2, \times_2\}$ under addition and multiplication modulo 2 is a Boolean ring.

Example 6. *If R is a ring such that $a^2 = a \;\forall\; a \in R$. Show that*

(i) $a + a = 0 \;\forall\; a \in R$

(ii) $a + b = 0 \;\Rightarrow\; a = b$

(iii) R is commutative. (P.T.U. B.Tech., May 2009)

Sol. *(i)* Given $a^2 = a \;\forall\; a \in R$

$\Rightarrow \quad (a + a)^2 = a + a$

$\Rightarrow \quad (a + a)(a + a) = a + a$

$\Rightarrow \quad a\,.\,(a + a) + a\,.\,(a + a) = a + a$ | Distributive law

$\Rightarrow \quad a\,.\,a + a\,.\,a + a\,.\,a + a\,.\,a = a + a$

$\Rightarrow \quad a + a + a + a = a + a$ | $a^2 = a \;\Rightarrow\; a\,.\,a = a$

$\Rightarrow \quad (a + a) + (a + a) = (a + a) + 0$ | Left cancellation law

$\Rightarrow \quad a + a = 0 \;\forall\; a \in R$

(ii) Let $\quad a + b = 0 = a + a$ | By part *(a)*

$\Rightarrow \quad b = a$ | Left cancellation law

(iii) Given $\quad a^2 = a \;\forall\; a \in R$

$\Rightarrow \quad (a + b)^2 = a + b$

$\Rightarrow \quad (a + b)(a + b) = a + b$

$\Rightarrow \quad (a + b)\,.\,a + (a + b)\,.\,b = a + b$

$\Rightarrow \quad a\,.\,a + b\,.\,a + a\,.\,b + b\,.\,b = a + b$

$\Rightarrow \quad a^2 + b\,.\,a + a\,.\,b + b^2 = a + b$

$\Rightarrow \quad a + b\,.\,a + a.\,b + b = a + b$ | $a^2 = a \;\forall\; a \in R$

$\Rightarrow \quad b\,.\,a + a\,.\,b + b = b$ | Left cancellation law

$\Rightarrow \quad b\,.\,a + a\,.\,b = 0$ | Right cancellation law

$\Rightarrow \quad b\,.\,a = a\,.\,b$ | From part *(b)*, $a + b = 0 \Rightarrow a = b$

Hence R is commutative.

9.8. DIRECT PRODUCT OF RINGS

Let $R_1, R_2, \ldots R_n$ be n rings under the operations $+_1, +_2, \ldots, +_n$ and $\cdot_1, \cdot_2, \ldots \cdot_n$ respectively. The direct products of these n rings is defined by P, where

$$P = \prod_{i=1}^{n} R_i = R_1 \times R_2 \times R_3 \times \ldots \times R_n.$$

Theorem I. *If $R_1, R_2, \ldots R_n$ be n rings under the operation of $+_1, +_2, \ldots +_n$ and $\cdot_1, \cdot_2, \ldots \cdot_n$, respectively. Then the direct product of $R_1, R_2, \ldots R_n$ is also a ring under the operation of componentwise addition and multiplication.*

Proof. Let $P = \prod_{i=1}^{n} R_i$ be the direct product of $R_1, R_2, \ldots R_n$.

Let $a, b \in P$, then $a = (a_1, a_2, \ldots a_n), a_i \in R, 1 \le i \le n$

$b = (b_1, b_2, \ldots b_n), b_i \in R, 1 \le i \le n$

We first show P is an additive group under the operation of componentwise addition defined by

$$a + b = (a_1 +_1 b_1, a_2 +_2 b_2, \ldots a_n +_n b_n)$$

As R_i $(1 \le i \le n)$ is a ring, it must be closed under $+_i$. *i.e.*, if $a_i, b_i \in R_i \Rightarrow a_i +_i b_i \in R_i \; \forall i$

$\Rightarrow \quad (a_1 +_1 b_1, a_2 +_2 b_2, \ldots a_n +_n b_n) \in R_1 \times R_2 \times \ldots \times R_n$

$\Rightarrow \quad a + b \in P$

Hence P is closed under the operation of componentwise addition.

If $a, b, c \in P$, then $a = (a_1, a_2, \ldots a_n), b = (b_1, b_2, \ldots b_n), c = (c_1, c_2, \ldots c_n)$

Consider $a + (b + c) = (a_1 +_1 (b_1 +_1 c_1), a_2 +_2 (b_2 +_2 c_2), \ldots a_n +_n (b_n +_n c_n))$

$= ((a_1 +_1 b_1) +_1 c_1, (a_2 +_2 b_2) +_2 c_2, \ldots (a_n +_n b_n) +_n c_n)$

$= (a + b) + c$

| Each R_i is a ring. Therefore associativity holds in R_i

Hence associativity holds in P.

For $a \in P$, consider

$a + 0 = (a_1, a_2, \ldots a_n) + (0_1, 0_2, \ldots 0_n) = (a_1 +_1 0_1, a_2 +_2 0_2, \ldots a_n +_n 0_n)$

$= (a_1, a_2, \ldots a_n) = a$

Similarly, $0 + a = a$

Hence $a + 0 = a = 0 + a$

i.e., $0 = (0_1, 0_2, \ldots 0_n)$ is the additive identity of P

Further for $a \in P$, consider

$a + (-a) = (a_1, a_2, \ldots a_n) + (-a_1, -a_2, \ldots -a_n) = (a_1 +_1 (-a_1), a_2 +_2 (-a_2), \ldots a_n +_n (-a_n))$

$= (0_1, 0_2, \ldots 0_n) = 0$

Similarly, $(-a) + a = 0$

$\therefore$ P has an additive inverse.

Lastly, For $a, b \in P$, we have

$a + b = (a_1 +_1 b_1, a_2 +_2 b_2, \ldots a_n +_n b_n) = (b_1 +_1 a_1, b_2 +_2 a_2, \ldots, b_n +_n a_n)$

$= b + a$ | Each R_i is additive group

Hence P is an additive group under the operation of componentwise addition.

We next show that associative law under multiplication holds in P.

Define $a \cdot b = (a_1 \cdot_1 b_1, a_2 \cdot_2 b_2, \ldots a_n \cdot_n b_n)$

For $a, b, c \in P$, consider

$$a \cdot (b \cdot c) = (a_1 \cdot_1 (b_1 \cdot_1 c_1), a_2 \cdot_2 (b_2 \cdot_2 c_2), \ldots a_n \cdot_n (b_n \cdot_n c_n))$$
$$= ((a_1 \cdot_1 b_1) \cdot_1 c_1, (a_2 \cdot_2 b_2) \cdot_2 c_2, \ldots (a_n \cdot_n b_n) \cdot_n c_n)$$
$$= (a \cdot b) \cdot c \qquad | \text{ Each } R_i \text{ is associative under } \cdot_i$$

Thus associativity under multiplication also holds in P.

We now show that left distributive law and right distributive law also holds in P.

Consider $$a \cdot (b + c) = (a_1 \cdot_1 (b_1 +_1 c_1), a_2 \cdot_2 (b_2 +_2 c_2), \ldots a_n \cdot_n (b_n +_n c_n))$$
$$= (a_1 \cdot_1 b_1 +_1 a_1 \cdot_1 c_1, a_2 \cdot_2 b_2 +_2 a_2 \cdot_2 c_2, \ldots a_n \cdot_n b_n +_n a_n \cdot_n c_n) \quad \ldots(1)$$

Also $$a \cdot b + a \cdot c = (a_1 \cdot_1 b_1 +_1 a_1 \cdot_1 c_1, a_2 \cdot_2 b_2 +_2 a_2 \cdot_2 c_2, \ldots a_n \cdot_n b_n +_n a_n \cdot_n c_n) \quad \ldots(2)$$

From (1) and (2), we get

$$a \cdot (b + c) = a \cdot b + a \cdot c \quad \forall\, a, b, c \in P$$

Similarly, $$(a + b) \cdot c = a \cdot c + b \cdot c \quad \forall\, a, b, c \in P$$

Hence P is a ring under the operation of addition and multiplication defined by

$$a + b = (a_1 +_1 b_1, a_2 +_2 b_2, \ldots a_n +_n b_n)$$
$$a \cdot b = (a_1 \cdot_1 b_1, a_2 \cdot_2 b_2, \ldots a_n \cdot_n b_n).$$

Cor. If $R_1, R_2, \ldots R_n$ are commutative rings with unity, then $P = R_1 \times R_2 \times R_3 \times \ldots \times R_n$ is also commutative ring with unity (1, 1, 1, ... 1).

Example 7. *If $\{Z_4, +_4, \times_4\}$ and $\{Z_3, +_3, \times_3\}$ are rings, then $Z_4 \times Z_3$ is also a ring. Find the unity, if it exists in $Z_4 \times Z_3$.*

Sol. We know that if R_1, R_2, are rings under the operation of componentwise addition and multiplication then $R_1 \times R_2$ is also a ring.

Here $Z_4 = \{0, 1, 2, 3, \cdot +_4, \times_4\}$

$Z_3 = \{0, 1, 2, +_3, \times_3\}$ are rings under the addition modulo 4 and multiplication modulo 3.

$\therefore$ $Z_4 \times Z_3$ is also a ring under the componentwise addition and multiplication.

| See Theorem I. Direct Product of rings

To find the unity: Let $(x, y) \in Z_4 \times Z_3$ and consider $(x, y)\,(m, n) = (x, y) = (m, n)\,(x, y)$

$$\Rightarrow \quad (x \times_4 m, y \times_3 n) = (x, y) = (m \times_4 x, n \times_3 y)$$
$$\Rightarrow \quad x \times_4 m = x = m \times_4 x \quad \ldots(1)$$
and $$y \times_3 n = y = n \times_3 y \quad \ldots(2)$$

The only elements m in Z_4 and n in Z_3 which satisfies (1) and (2) are $m = 1, n = 1$

Hence the unity of $Z_4 \times Z_3$ is (1, 1).

9.9. MORPHISM OF RINGS

The word 'morphism' is a combination of various terms like ring homomorphism, ring isomorphism etc.

9.9.1. Ring Homomorphism

Let $(R, +, \cdot)$ and $[R', +', \cdot']$ be two rings. Then a mapping $f : R \to R'$ is called a ring homomorphism

(*i*) $f(a + b) = f(a) +' f(b) \quad \forall\, a, b \in R$

(*ii*) $f(a \cdot b) = f(a) \cdot' f(b) \quad \forall\, a, b \in R.$

Also, a ring homomorphism and one-one is called monomorphism. A ring homomorphism and onto is called epimorphism.

Further, if in addition, f is one-one and onto, then f is called on isomorphism and R and R′ are said to be isomorphic and we write $R \cong R'$.

Remarks : To check whether the two *rings are isomorphic,* we should check the following :

(*a*) Both rings should have same cardinality.

(*b*) Both rings should be commutative.

(*c*) Both rings should have same unity.

(*d*) If there exists an equation which is solvable in one ring, but not solvable in another ring, then two rings cannot be isomorphic.

Lemma: *If θ is a ring homomorphism from a ring R to a ring R′. Then*

(*i*) $\theta(0) = 0,\ 0 \in R'$ (*ii*) $\theta(-a) = -\theta(a)\ \forall\ a \in R$

Proof. (*i*) Consider

$$\theta(a) = \theta(a + 0) = \theta(a) + \theta(0) \quad | \ \theta \text{ is a ring homomorphism}$$

$$\Rightarrow \quad \theta(a) + 0 = \theta(a) + \theta(0) \quad | \text{ Adding } 0 \in R \text{ on L.H.S.}$$

$$\Rightarrow \quad 0 = \theta(0) \quad | \text{ Left cancellation law}$$

$$\Rightarrow \quad \theta(0) = 0$$

(*ii*) Consider $\theta(a + (-a)) = \theta(a - a) = \theta(0) = 0$ | Using part (*i*)

$$\Rightarrow \quad \theta(a) + \theta(-a) = 0$$

$$\Rightarrow \quad \theta(-a) = -\theta(a)$$

Example 8. *Consider the rings [Z, + , ·] and [2Z, + , ·] and define*

$f : Z \to 2Z$ by $f(n) = 2n\ \forall\ n \in Z$ (P.T.U. B. Tech. Dec. 2010)

Is f a group homomorphism ? Is f a ring isomorphism ?

Sol. Z and 2Z are groups under addition.

Consider $f : Z \to 2Z$ defined by $f(n) = 2n\ \forall\ n \in Z$

For $m, n \in Z$, consider

$$f(m + n) = 2(m + n)$$

$$= 2m + 2n = f(m) + f(n)\ \forall\ m, n \in Z$$

Hence $f : Z \to 2Z$ is a group homomorphism.

To check whether f is a ring isomorphism.

For $m, n \in Z$, consider $f(mn) = 2\,mn$

and $f(m)\,f(n) = 2m \,.\, 2n = 4\,mn.$

$\therefore \quad f(mn) \neq f(m)\,f(n)\ \forall\ m, n \in Z$

$\therefore \quad f : Z \to 2Z$ cannot be a ring isomorphism.

Alternatively: We know that Z is a ring with unity and 2Z is also a ring but without unity. Hence both rings donot have same unity. Therefore Z and 2Z are not isomorphic, *i.e.*, $Z \not\cong 2Z$.

Example 9. *Examine whether [2Z, + , ·] and [3Z, +, ·] are isomorphic rings ?*

Sol. Consider the equation $x + x = x \,.\, x$ ...(1).

This equation makes sense in both rings.

For $x = 2$, (1) gives $2 + 2 = 2 \,.\, 2 \Rightarrow 4 = 4$, which is true.

Thus equation (1) has a solution $x = 2 \in 2Z$.

For $x = 3$, (1) gives $3 + 3 \neq 3 \,.\, 3$. This means $x = 3$ is not a solution of (1).

Hence we conclude that equation (1) has a solution in 2Z, but does not have a solution in 3Z. Therefore 2Z and 3Z cannot be isomorphic rings. (See Remark (*d*) of Art. 9.9.1)

Example 10. *Show that following rings are not isomorphic.*

(*i*) $[Z, +, \cdot]$ *and* $[M_{2\times 2}(R), +, \cdot]$ (*ii*) $[3Z, +, \cdot]$ *and* $[4Z, +, \cdot]$

(*iii*) $[R, +, \cdot]$ *and* $[Q, +, \cdot]$ (*iv*) $[Z_2 \times Z_2, +, \cdot]$ *and* $[Z_4, +, \cdot]$.

Sol. (*i*) We know that Z is a commutative ring and $M_{2\times2}$ (R) is a non-commutative ring. (since for A, B $\in M_{2\times2}$ (R), AB = BA is not true)

$\therefore$ The rings [Z, + , ·] and [$M_{2\times2}$ (R) , + ,.] cannot be isomorphic rings. (Remark (*a*) of Art. 9.9.1)

(*ii*) Consider the equation $x + x + x = x \cdot x$ ...(1)

This equation maks sense in both rings 3Z and 4Z.

For $x = 3$, (1) gives $3 + 3 + 3 = 3.3 \Rightarrow 9 = 9$, which is true.

Hence equation (1) has a solution $x = 3$ in 3Z.

For $x = 4$, (1) gives $4 + 4 + 4 \neq 4 \cdot 4$

$\Rightarrow \quad 12 \neq 16,$

Hence equation (1) has does not have a solution in 4Z

Therefore, 3Z and 4Z cannot be isomorphic rings. (Remark (*d*) of Art. 9.9.1)

(*iii*) We know that the set of real numbers R is uncountable and the set of rationals Q is countable (see chapter on 'sets').

Hence R and Q cannot have same cardinality and therefore cannot be isomorphic.

(*iv*) By definition, $Z_2 = [0, 1, +_2, \times_2]$

$\therefore \quad Z_2 \times Z_2 = [(0, 0), (0, 1), (1, 0), (1, 1)]$

For $(m, n) \in Z_2 \times Z_2$, consider $(m, n) \cdot (1, 1) = (m, n) = (1, 1) \cdot (m, n)$

i.e., (1, 1) is the unity (multiplicative identity) of $Z_2 \times Z_2$.

Now $Z_4 = [0, 1, 2, 3, +_4, \times_4]$

1 is the unity of Z_4.

Thus $Z_2 \times Z_2$ and Z_4 do not have same unity. Therefore they cannot be isomorphic rings.

Example 11. *Let R and R′ be two rings. Define* $\theta : R \to R'$ *by* $\theta(a) = 0 \ \forall \ a \in R$. *Show that* θ *is a ring homomorphism.*

Sol. Let $a, b \in$ R. Since R is a ring and $a, b \in$ R it implies $a + b \in$ R and $ab \in$ R. Consider

$$\theta(a + b) = 0 = 0 + 0$$
$$= \theta(a) + \theta(b)$$

Also
$$\theta(ab) = 0$$
$$= 0.0$$
$$= \theta(a)\, \theta(b)$$

Hence θ is a ring homomorphism.

Example 12. *Let R be a commutative ring and suppose* $px = 0 \ \forall \ x \in R$, *p is prime. Show that the mapping* $f : R \to R$ *defined by* $f(x) = x^p$, $x \in R$ *is a ring homomorphism.*

Sol. Let $x, y \in$ R and consider

$$f(x + y) = (x + y)^p, \text{ using Binomial theorem,}$$

$$= p_{c_0}\, x^p + p_{c_1}\, x^{p-1} y + p_{c_2}\, x^{p-2} y^2 + \ldots + p_{c_p}\, y^p$$

$$= x^p + p_{c_1}\, x^{p-1} y + p_{c_2}\, x^{p-2} y^2 + \ldots + y^p$$

For $x, y \in \text{R} \Rightarrow x^{p-1} y \in$ R. (Since R is a ring)

Using $px = 0 \ \forall \ x \in$ R, we have $px^{p-1} y = 0$

As we know that $p \mid p_{c_r} \Rightarrow p_{c_r} = kp$ for some k.

Therefore, $p_{c_2}\, x^{p-2} y^2 = kpx^{p-2} y^2 = 0$ for some k, etc.

Hence $f(x + y) = x^p + y^p = f(x) + f(y)$

Further, $f(xy) = (xy)^p = x^p y^p$

$= f(x) f(y)$

$\therefore$ f is a ring homomorphism.

9.9.2. Kernal *f*

If $f : R \to R'$ is a ring homomorphism, then the kernel of f, is the set of all those elements whose image is the zero element of R′. Thus

$Ker f = \{r \in R : f(r) = 0\}$.

Example 13. *Let $\theta : R \to R'$ be a ring homomorphism from a ring R to the ring R′. Show that Ker θ is a subgroup of R under addition.*

Sol. By definition, Ker $\theta = \{r \in R : f(r) = 0\}$

Since $\theta : R \to R'$ is a ring homomorphism,

$\theta(0) = 0,\ 0 \in R'$

$\Rightarrow$ $Ker\ \theta \neq \phi$

Let $x, y \in Ker\ \theta$ $\Rightarrow \theta(x) = 0, \theta(y) = 0$.

Consider $\theta(x - y) = \theta(x) - \theta(y) = 0 - 0 = 0$

$\Rightarrow$ $x - y \in Ker\ \theta\ \forall\ x, y \in R$

$\Rightarrow$ $Ker\ \theta$ is a subgroup of R.

9.10. SUBRING (*P.T.U. Dec. 2005*)

Let [R, + , ·] be a ring and S be a subset of R. Then S is called a subring of R iff S is itself a ring under the operations of R.

Example. *Let E = the ring of even integers, Z = the ring of integers. Then* E is a subring of Z. Also $Z \subseteq Q$, therefore Z is a subring of Q.

Theorem II. *A non-empty subset of a ring R is a subring of R iff*

(*i*) $a, b \in S \Rightarrow a - b \in S\ \forall\ a, b \in S$

(*ii*) $a, b \in S \Rightarrow ab \in S\ \forall\ a, b \in S$.

Proof. Let S be a subring of R. We prove (*i*) and (*ii*).

As S is a subring of R, S is itself a ring under the operations of R.

Hence S is additive group under +. that is, S is closed under addition. *i.e.,*

For $a, b \in S, a + b \in S\ \forall\ a, b \in S$

Also for each $b \in S$, there exists $-b \in S$ such that $-b$ is the additive inverse of b.

Now $a \in S, -b \in S \Rightarrow a + (-b) \in S$

$\Rightarrow$ $a - b \in S$, which proves (*i*)

Further, as S is a subring of R, it must be a ring under the operations of R. Thus, S is closed under multiplication *i.e.*;

For $a, b \in S \Rightarrow a . b \in S\ \forall\ a, b \in S$, which proves (*ii*)

Converse. Let (*i*) and (*ii*) hold. We show S is a subring of R under the operations of R.

For $a, a \in S \Rightarrow a - a \in S \Rightarrow 0 \in S$ | Using (*i*)

i.e., S has additive identity.

Again $0 \in S, a \in S \Rightarrow 0 - a \in S \Rightarrow -a \in S$ | Using (*i*)

i.e., S has additive inverse.

For $a \in S, b \in S \Rightarrow -b \in S$ (Proved above)

From (*i*), $a - (-b) \in S$

$\Rightarrow a + b \in S \ \forall \ a, b \in S$

i.e., S is closed under addition.

Since $S \subseteq R$, elements of S are also in R

$\therefore$ Associativity under addition holds in S

For $a, b \in S \subseteq R \Rightarrow a, b \in R$

$\therefore a + b = b + a$ | R is additive group

Hence we can say that S is an additive group.

From (*ii*), $a, b \in S \Rightarrow a . b \in S \ \forall \ a, b \in S$

i.e., S is closed under multiplication.

Finally, $a, b, c \in S \subseteq R \Rightarrow a, b, c \in R$

$\therefore a. (b + c) = a . b + a . c$

$(a + b) . c = a . c + b . c$ | Distributive laws hold in R

i.e., left distributive law and right distributive law holds in S.

Hence S is a ring under the operations of R.

Example 14. *The set of integers Z is subring of Q.*

Sol. We know that "*A non-empty subset S of a ring R is a subring of R*

iff (*i*) $a, b \in S \Rightarrow a - b \in S \ \forall \ a, b \in S$

(*ii*) $a, b \in S \Rightarrow a . b \in S \ \forall \ a, b \in S$.

Since $Z \subseteq Q$ *i.e.*, Z is a subset of Q.

For $a, b \in Z \Rightarrow a - b \in Z \ \forall \ a, b \in Z$ is true.

Also for $a, b \in Z \Rightarrow a . b \in Z \ \forall \ a, b \in Z$. | Theorem II

Hence Z is a subring of Q.

Example 15. (*a*) *Show that 3Z is a subring of Z.*

(*b*) *Find all subrings of Z_8.*

Sol. (*a*) We know that "*A non-empty subset S of a ring R is a subring of R*

iff (*i*) $a, b \in S \Rightarrow a - b \in S \ \forall a, b \in S$

(*ii*) $a, b \in S \Rightarrow a . b \in S \ \forall a, b \in S$

Let $x, y \in 3Z \Rightarrow x = 3m, m \in Z, \ y = 3n, n \in Z$

Consider $x - y = 3m - 3n = 3(m - n) \in 3Z$

| Since $m, n \in Z \Rightarrow m - n \in Z$ (Z is a ring)
| $\Rightarrow 3(m - n) \in 3Z$

$\Rightarrow x - y \in 3Z$

Also $x.y = 3m . 3n = 3(3mn) = 3k \in 3Z$

where $k = 3\,mn$ | since $3, m, n \in Z$ and Z is a ring $\therefore$ $3mn \in Z$ |

Hence we can say that 3Z is a subring of Z.

(*b*) Consider the following subsets of Z_8.

$Z_2 = [0, 1, +_2, \times_2]$; $Z_3 = [0, 1, 2, +_3, \times_3]$

$Z_4 = [0, 1, 2, 3, +_4, \times_4]$; $Z_5 = [0, 1, 2, 3, 4, +_5, \times_5]$

$Z_6 = [0, 1, 2, 3, 4, 5, +_6, \times_6]$; $Z_7 = [0, 1, 2, 3, 4, 5, 6, +_7, \times_7]$

For $a, b \in Z_2$, we observe that $a - b \in Z_2 \ \forall\ a, b \in Z_2$

Also $a \cdot b \in Z_2 \quad \forall\ a, b \in Z_2$

Hence Z_2 is a subring of Z_8

For $a, b \in Z_3$, we observe that $a - b \in Z_3 \ \forall\ a, b \in Z_3$

Also $a \cdot b \in Z_3 \quad \forall\ a, b \in Z_3$

Hence Z_3 is a subring of Z_8

Similarly, we can prove that Z_4, Z_5, Z_6, Z_7 are all subrings of Z_8.

9.11. UNITS

Let (R, + ,.) be a ring with unity. An element $a \in R$ is said to be a unit (or invertible) if for $0 \neq a \in R$, $\exists\ b \in R$ such that $a \cdot b = 1 = b \cdot a$ or

An element is a unit if a has multiplicative inverse, $a^{-1} \in R$ such that $aa^{-1} = 1 = a^{-1}a$

Consider the rings (R, + ,.) and (Q, + ,.). Every non-zero element in R (set of reals) and Q (set of rationals) has a multiplicative inverse. For example, $\frac{4}{3} \in R$ has multiplicative inverse $\frac{3}{4} \in R$ since $\frac{4}{3} \cdot \frac{3}{4} = 1$.

The only elements in Z that have multiplicative inverses are – 1 and 1.

Theorem III. *An element a in Z_n is a unit iff a and n are relatively prime.*

Proof. By definition, $Z_n = [0, 1, 2, 3, \ldots n - 1, +_n, \times_n]$

Let $a \in Z_n$ be a unit. It means there exists an element $b \in Z_n$ such that $a \times_n b = 1$ *i.e.*, when ab is divided by n, the remainder is 1. *i.e.*, $a \times_n b = 1 \Rightarrow ab = nq + 1$, where q is the quotient.

$\Rightarrow \quad ab - nq = 1$ | For $a, b \in R$, if $\exists\ x, y \in R$ such that $ax - by = 1$, then g.c.d. $(a, b) = 1$

$\Rightarrow \quad (a, n) = 1$

Example 16. *Determine the units (those elements which have multiplicative inverses) for each of the following rings :*

(*a*) $[Z, +, \cdot]$
(*b*) $[Q, +, \cdot]$
(*c*) $[C, +, \cdot]$
(*d*) $[M_{2\times 2}(R), +, \cdot]$
(*e*) $[Z_2, +_2, \times_2]$
(*f*) $[Z_6, +_6, \times_6]$
(*g*) $[Z_8, +_8, \times_8]$
(*h*) $[Z_5, +_5, \times_5]$
(*i*) $[Z \times Z, +, \cdot]$
(*j*) $[Z_2^{\,3}, +, \cdot]$.

Sol. (*a*) The only units of Z are – 1 and 1. Since for each $a \in Z$, $a \cdot 1 = a = 1 \cdot a$

$a \cdot (-1) = -a = (-1) \cdot a$

(*b*) Every non-zero element in Q has multiplicative inverse (unit)

(*c*) Every non-zero complex number in C has multiplicative inverse (unit)

(*d*) Every non-singular or invertible matrix is a unit of $M_{2\times 2}(R)$.

(*e*) Consider $Z_2 = [0, 1, +_2, \times_2]$

For each $x \in Z_2$, we have $x \times_2 1 = x = 1 \times_2 x$

$\therefore$ The only unit of Z_2 is 1.

(*f*) $Z_6 = [0, 1, 2, 3, 4, 5, +_6, \times_6]$

The only elements in Z_6 which are relatively prime to 6 are 1 and 5.

($\because$ $(1, 6) = 1, (5, 6) = 1$

Hence the units of Z_6 are 1 and 5.

Also $1 \times_6 1 = 1,\ 5 \times_6 5 = 1$ | Theorem III

(*g*) $Z_8 = [0, 1, 2, 3, 4, 5, 6, 7, +_8, \times_8]$

The only units of Z_8 are those elements which are relative prime to 8.

The elements relative prime to 8 are 2, 3, 5, 7.

Hence the units of Z_8 are [1, 3, 5, 7].

(*h*) $Z_5 = [0, 1, 2, 3, 4, +_5, \times_5]$

The only units of Z_5 are those elements which are relative prime to 5. The elements relative prime to 5 are 1, 2, 3, 4.

Hence the units of Z_5 are [1, 2, 3, 4].

(*i*) The units of Z are 1 and – 1. Therefore units of Z × Z are

$$(1, 1)\ (1, -1), (-1, 1), (-1, -1).$$

(*j*) $Z_2^3 = Z_2 \times Z_2 \times Z_2$ where $Z_2 = [0, 1, +_2, \times_2]$

The only unit of Z_2 is 1. Therefore the only unit of Z_2^3 is (1, 1, 1).

Theorem IV. *For all a, b ∈ R,*

(*i*) $a.0 = 0.a = 0$ (*ii*) $a.(-b) = (-a).b = -a.b$

(*iii*) $(-a).(-b) = a.b$ (P.T.U. B.Tech. May 2012)

(*iv*) *If R has a unit element 1, then* $(-1).a = -a,\ (-1).(-1) = 1$.

Proof. (*i*) For $a \in$ R, consider

$$a.0 = a.(0+0) \quad \text{| 0 is the identity}$$
$$= a.0 + a.0 \quad \text{| Left distribution law}$$
$$\Rightarrow \quad a.0 + (-a).0 = a.0 + a.0 + (-a).0$$
$$\Rightarrow \quad (a + (-a)).0 = a.0 + (a + (-a)).0 \quad \text{| Right distributive law}$$
$$\Rightarrow \quad 0 = a.0 + 0$$
$$\Rightarrow \quad 0 = a.0$$

Similarly, $0 = 0.a$ (Try yourself)

(*ii*) Consider $a.(b + (-b)) = a.0 = 0$ | by part (*i*)

$$\Rightarrow \quad a.b + a.(-b) = 0 \quad \text{| Left distributive law}$$
$$\Rightarrow \quad a.(-b) = -a.b$$

Similarly, $(-a).b = -a.b$ (Try yourself)

(*iii*)
$$(-a).(-b) = -(a.(-b))$$
$$= -(-(a.b)) \quad \text{| by part (ii)}$$
$$= a.b$$

(*iv*) If R has a unit element 1, then

$$a + (-1).a = 1.a + (-1).a \quad \text{| } 1.a = a$$
$$= (1 + (-1)).a \quad \text{| Left distributive law}$$
$$= 0.a = 0 \quad \text{| by part (i)}$$
$$\Rightarrow \quad (-1).a = -a$$

In particular, if $a = -1$, then

$$(-1).(-1) = -(-1) = 1.$$

9.12. INTEGRAL DOMAIN

Zero divisor

A non-zero element $a \in R$ is called a zero divisor if there exists a non-zero element $b \in R$ such that $ab = 0$

A commutative ring R is called an integral domain if for every

$$0 \neq a,\ b \in R,\ ab = 0 \ \Rightarrow\ a = 0 \ \text{ or } \ b = 0$$

Thus, a commutative ring R is called an integral domain if R has no zero divisor.

Example (*i*) Z, the ring of integers is an integral domain. Also, Q, R, C are integral domains.

(*ii*) Z_5, Z_8, Z_{10} etc. are not integral domains. Z_3, Z_5, Z_7 are integral domains.

Theorem V. *The element a in the ring* $[Z_n, +_n, \times_n]$ *is a zero divisor iff a is not relative prime to n (i.e.; g.c.d.* $(a, n) \neq 1$*).*

Proof. The proof of above theorem is beyond the scope of this book.

Theorem VI. $(Z_p, +_p, \times_p]$ *has no zero divisors iff p is a prime number.*

Proof. Let Z_p has no zero divisors. We show p is prime.

For if, $p = ab$, $\quad 1 < a < p,\ 1 < b < p$

$\Rightarrow \quad a \times_p b = 0$ where a, b are non-zero numbers

$\Rightarrow$ Z_p has a zero divisor, a contradiction, hence p is prime.

Converse. Let p is prime, we show Z_p has no zero divisor. Let $a \times_p b = 0$ for $a, b \in Z_p$

$\Rightarrow \quad ab \equiv 0 \pmod p \ \Rightarrow\ p/ab \ \Rightarrow\ p/a \ \text{ or } \ p/b \qquad | \ p$ is prime

But $a, b \in Z_p \quad \therefore \quad a, b < p.$

Hence $\quad a = 0 \ \text{ or } \ b = 0$

$\therefore$ Z_p has no zero divisor.

Example 17. *Consider* $(Z_8, +_8, \times_8)$*. Find the zero divisors of* Z_8*. Is* Z_8 *an integral domain ?*

Sol. $\quad Z_8 = [0, 1, 2, 3, 4, 5, 6, 7].$

Since $\quad 4 \times_8 2 = 0 = 2 \times_8 4$

$\therefore$ 2 and 4 are zero divisors.

Also $\quad 4 \times_8 6 = 0 = 6 \times_8 4$

$\therefore$ 4 and 6 are zero divisors.

Z_8 cannot be an integral domain since $0 \neq 2, 4 \in Z_8 \ \Rightarrow\ 2 \times_8 4 = 0$ *i.e.*, Z_8 is a ring with zero divisors.

Example 18. *Show that Z, the set of integers is an integral domain.*

Sol. We know that Z is a commutative ring. Also if $a, b \in Z$ then, gives $a \cdot b = 0$

either $a = 0$ or $b = 0$

$\therefore$ Z is a commutative ring without zero divisors *i.e.*, Z is an integral domain.

Example 19. *Consider* $m = \left[\begin{pmatrix} a & b \\ c & a \end{pmatrix} a, b, c, d \in R \right]$ *as a ring under matrix addition and matrix multiplication. Show that* $A = \begin{pmatrix} 0 & 0 \\ 0 & 1 \end{pmatrix}$ *and* $B = \begin{pmatrix} 0 & 1 \\ 0 & 0 \end{pmatrix}$ *are zero divisors.*

Sol. Given $$A = \begin{pmatrix} 0 & 0 \\ 0 & 1 \end{pmatrix} \neq O, \; B = \begin{pmatrix} 0 & 1 \\ 0 & 0 \end{pmatrix} \neq O$$

But $$AB = \begin{pmatrix} 0 & 1 \\ 0 & 0 \end{pmatrix}\begin{pmatrix} 0 & 1 \\ 0 & 0 \end{pmatrix} = \begin{pmatrix} 0 & 0 \\ 0 & 0 \end{pmatrix} = O$$

∴ A and B are zero divisors.

Example 20. *Show that the ring Z_{29} of integers modulo 29 is an integral domain.*

Sol. By using theorem VI, Z_p has no zero divisors iff p is prime. Here $p = 29$, which is a prime. Therefore, Z_{29} has no zero divisors. Consequently, Z_{29} is an integral domain.

Example 21. *Show that the ring Z_{105} of the integers modulo 105 is not an integral domain.*

Sol. By using theorem VI, Z_p has no zero divisors iff p is prime. Here $p = 105$, which is a composite number. Consequently, Z_{105} has zero divisors.

∴ Z_{105} cannot be an integral domain.

Theorem VII. *The cancellation laws hold in a ring (R, +,.) iff R has no zero divisors.*

Or

A commutative ring R is an integral domain iff for a, b, c ∈ R (0 ≠ a)

$$ab = ac \;\Rightarrow\; b = c.$$

Proof. Let R be an integral domain and consider $ab = ac$ $(0 \neq a)$

$\Rightarrow \quad ab - ac = 0$

$\Rightarrow \quad a.(b - c) = 0$ | Left distribution law

$\Rightarrow \quad$ either $a = 0$ or $b - c = 0$ | Since R is an integral domain, so it has no zero divisors

$\Rightarrow \quad b - c = 0 \quad | \; a \neq 0$

$\Rightarrow \quad b = c.$

Converse. Let cancellation laws hold in R. We show R is an integral domain.

Let $a, b \in$ R and $0 \neq a$. Consider

$a \,.\, b = 0$

$\Rightarrow \quad a \,.\, b = a \,.\, 0$ | $a \,.\, 0 = 0$

$\Rightarrow \quad b = 0$ | Cancellation law

Hence R is an integral domain.

Example 22. *Consider $X = [0, 2, 4, 6, 8, +_{10}, \times_{10}]$. Is X an integral domain ? Justify your answer.*

Sol. We first check whether X is a commutative ring under addition modulo 10 and multiplication modulo 10. The addition modulo 10 table is shown in Table I.

Table I

$+_{10}$	0	2	4	6	8
0	0	2	4	6	8
2	2	4	6	8	0
4	4	6	8	0	2
6	6	8	0	2	4
8	8	0	2	4	6

From Table I, we observe that every element inside the table is also in X. It means that X is closed under addition modulo 10. *i.e.*,

$$a, b \in X \Rightarrow a +_{10} b \in X \; \forall a, b \in X$$

Addition modulo 10 is associative *i.e.*,

For $a, b, c \in X, a +_{10} (b +_{10} c) = (a +_{10} b) +_{10} c \; \forall \; a, b, c \in X$

The first row inside the table coincides with the topmost row of the Table I. It means 0 is the additive identity of X.

Also each element of X has an additive inverse. For example

Inverse of 2 is 8 (the intersection of 2 and 8 at zero) | $2 +_{10} 8 = 0$

Inverse of 4 is 6 etc. | $4 +_{10} 6 = 0$

Table I is symmetrical w.r.t. $+_{10}$. It means $a +_{10} b = b +_{10} a \; \forall \; a, b \in X$.

Hence X is an additive group under $+_{10}$.

Now consider the multiplication modulo 10 table as shown in Table II.

Table II

$\times_{10}$	*0*	*2*	*4*	*6*	*8*
0	0	0	0	0	0
2	0	4	8	2	6
4	0	8	6	4	2
6	0	2	4	6	8
8	0	6	2	8	4

From Table II, we observe that each element inside the table is also in X. It means that X is closed under multiplication modulo 10 *i.e.*,

For $a, b \in X, a \times_{10} b \in X \; \forall \; a, b \in X$

Multiplication modulo 10 is associative. Also for $a, b, c \in X$,

$a \times_{10} (b +_{10} c) = a \times_{10} b +_{10} a \times_{10} c$ | Left distributive law

$(a +_{10} b) \times_{10} c = a \times_{10} c +_{10} a \times_{10} c \; \forall \; a, b, c \in X$ | Right distributive law

Hence X is a ring under addition modulo 10 and multiplication modulo 10.

Now to check commutativity of X, from Table II, we observe that the table is symmetrical w.r.t. $\times_{10}$. It means X is a commutative ring.

Finally, X is a ring without zero divisors as it is clear from Tabe II, *i.e.*, there do not exist non-zero elements whose product is zero.

Hence $(X, +_{10}, \times_{10})$ is an integral domain.

Example 23. *Consider* $X = [0, 1, 2, 3, 4, 5, +_6, \times_6]$. *Is X an integral domain ? Justify your answer.*

Sol. Proceeding as in example 4, we can prove that X is a commutative ring.

Also $2, 3 \in X$ and $2 \times_6 3 = 0$ *i.e.*, product of two non-zero elements in X is a zero element. Thus, X is a commutative ring with zero divisors. So X cannot be an integral domain.

Example 24. *(i) Give an example of a finite integral domain*

(ii) Give an example of an infinite integral domain ?

(iii) Give counter example to illustrate the fact that product of two integral domain may not be an integral domain ?

Sol. *(i)* We know that Z_p is an integral domain iff p is prime. Thus $Z_2 = [0, 1, +_2, \times_2]$, $Z_3 = [0, 1, 2, +_3, \times_3]$ are finite integral domain.

(ii) Z, the set of integers is an example of an infinite integral domain.

(iii) Consider $Z_2 = [0, 1, +_2, \times_2]$; $Z_3 = [0, 1, 2, +_3, \times_3]$

Clearly, Z_2 and Z_3 are integral domains as 2 and 3 are primes.

Consider the product $Z_2 \times Z_3$. We know that Z_2 and Z_3 are commutative rings with unity so their product $Z_2 \times Z_3$ is also a commutative ring with unity.

But $(1, 0), (0, 2) \in Z_2 \times Z_3$ are two non-zero elements and $(1, 0) . (0, 2) = (0, 0)$. *i.e.*, $Z_2 \times Z_3$ has zero divisors and hence cannot be an integral domain.

Example 25. *Find all zero divisors of Z_{15}, Z_6, Z_{20}.*

Sol. *(i)* $Z_{15} = [0, 1, 2, ... 14, +_{15}, \times_{15}]$

We know that an element m, in $[Z_n, +_n, \times_n]$ is a zero divisor iff m is not relative prime to n.

Here $n = 15$. The only elements which are not relative prime to 15 are 3, 5, 6, 9, 10, 12.

Hence 3, 5, 6, 9, 10, 12, are zero divisors.

Also $3 \times_{15} 5 = 0$, $9 \times_{15} 10 = 0$, $5 \times_{15} 6 = 0$ $10 \times_{15} 12 = 0$ etc.

(ii) $Z_6 = [0, 1, 2, 3, 4, 5, +_6, \times_6]$

The only elements which are not relative prime to 6 are 2, 3, 4

∴ The zero divisors of Z_6 are 2, 3, 4

Also $2 \times_6 3 = 0$, $3 \times_6 4 = 0$ etc.

(iii) $Z_{20} = [0, 1, 2, 3, ... 19, +_{20}, \times_{20}]$

The only elements which are not relative prime to 20 are 2, 4, 5, 6, 8, 10, 12, 14, 16, 18.

Hence the zero divisors of Z_{20} are 2, 4, 5, 6, 8, 10, 12, 14, 16, 18.

Example 26. *Consider the ring $Z_{10} = \{0, 1, 2, 3, ..., 9\}$ of integers modulo 10.*

(a) Find the unit of Z_{10}

(b) Find -3, -8, 3^{-1}

(c) Let $f(x) = 2x^2 + 4x + 4$. Find the roots of $f(x)$ over Z_{10}. By finding roots of $f(x)$, Conclude that can a polynomial of degree n have more than n roots ?

Sol. *(a)* The units of Z_{10} are those integers which are relatively prime to 10. Clearly, the units of Z_{10} are 1, 3, 7, 9.

(b) By $-a$ in a ring, we mean that element such that $a + (-a) = 0 = (-a) + a$. Therefore,

$-3 = 7$ (Since $3 + 7 = 0 = 7 + 3$)

Also, $-8 = 2$ (Since $8 + 2 = 0 = 2 + 8$)

By a^{-1} in a ring, we mean that element such that $a . a^{-1} = 1 = a^{-1} a$

Therefore, $3^{-1} = 7$ (Since $3.7 = 1 = 7.3$)

(c) The roots of $f(x)$ will be those elements from 0 to 9 which will yield 0.

Put, $x = 0$, $f(0) = 4$,

$x = 1$, $f(1) = 2 + 4 + 4 = 10 = 0$

$x = 2$, $f(2) = 8 + 8 + 4 = 20 = 0$

$x = 3$, $f(3) = 4$,

$x = 4$, $f(4) = 2, f(5) = 4, f(6) = 0, f(7) = 0, f(8) = 4, f(9) = 2$

Thus, $f(1) = 0$, $f(2) = 0, f(6) = 0, f(7) = 0$

Hence, the roots of $f(x)$ are 1, 2, 6, 7.

Conclusion. This example shows that a polynomial of degree n can have more than n roots over an arbitrary ring. But this cannot happen if the ring is a field.

TEST YOUR KNOWLEDGE 9.1

1. Consider the following sets. The operations involved are the usual operations defined on the sets.
   (*a*) [Z, + , ·] (*b*) [Q, + , ·] (*c*) [C, + , ·]
   (*d*) [M_{2+2} (R), + , ·] (*e*) [Z_2, $+_2$, $\times_2$] (*f*) [Z_6, $+_6$, $\times_6$]
   (*g*) [Z_8, $+_8$, $\times_8$] (*h*) [Z_5, $+_5$, $\times_5$] (*i*) [Z × Z, + , ·]
   (*j*) [$Z_2{}^3$, + , ·]
   (*i*) Which of the above sets are rings ?
   (*ii*) Which of the above rings are commutative ? Are they rings with unity? Determine the unity of the above rings.
2. Perform the indicated operations on the [Z_8; $+_8$, $\times_8$]:
   (*i*) $2 \times_8 (-4)$ (*ii*) $(-3) \times_8 5$
   (*iii*) $(-2) \times_8 (-4)$ (*iv*) $(-3) \times_8 5 +_8 (-3) \times_8 (-5)$
3. (*a*) Determine all solutions of the equation $x^2 - 5x + 6 = 0$ in Z_{12}. Find all elements of Z_{12} which satisfy this equation.
   (*b*) Find all solutions of the equation $x^2 - 5x + 6 = 0$ in Z. Can there be any more than two solutions to this equations in Z ?
4. Solve the equation $x^2 - 4x + 4 = 0$
   (*a*) in Z_{12} (*b*) in Z
   (*c*) in $M_{2\times 2}$ (R) (*d*) in Z_3.
5. For any ring [R; + , ·], simplify
   (*i*) $(a + b)(c + d)$ for $a, b, c, d \in$ R
   (*ii*) If R is commutative, show that $(a + b)^2 = a^2 + 2ab + b^2 \ \forall \ a, b \in$ R
   (*iii*) Simplify $(a + b)^5$ in Z_5.
6. Consider the ring Z_{10} = [0, 1, 2, 3, ... 9] of integers modulo 10.
   (*a*) Find the units of Z_{10} *(P.T.U. B.Tech. May 2008)*
   (*b*) Find – 3, – 8 and 3^{-1}
   (*c*) Let $f(x) = 2x^2 + 4x + 4$. Find the roots of $f(x)$ over Z_{10} *(P.T.U. B.Tech. May 2008)*
7. Consider Z_{30} = [0, 1, 2, ... 29, $+_{30}$, $\times_{30}$]
   (*a*) Find – 2, – 7 and – 11 (*b*) Find 7^{-1}, 11^{-1} and 26^{-1}.
8. Suppose $a^2 = a$ for every $a \in$ R (such a ring is called a *Boolean ring.*) Prove that R is commutative given that $x + y = 0 \ \Rightarrow \ x = y$ for all $x, y \in$ R. *(P.T.U. B.Tech. May 2009)*
9. Let G be any additive group. Define a multiplication in G by $a.b = 0$ for every $a, b \in$ G. Show that this makes G into a ring.
10. Let R be a ring with a unity element. Show that R*, the set of units in R is a group under multiplication.
11. Prove that if $x^2 = 1$ in an integral domain D, then $x = 0$ or $x = 1$.
12. If R is a ring with unity, then this unity is unique.
13. Prove that the ring $Z_2 \times Z_3$ is commutative and has unity.

Answers

1. (*i*) All are rings

(*ii*) except (*d*), all rings are commutative. The unity for (*a*), (*b*), (*c*), (*e*), (*f*), (*g*), (*h*) is 1.

The unity for (*d*) is $I_2 = \begin{pmatrix} 1 & 0 \\ 0 & 1 \end{pmatrix}$. The unity for (*i*) is (1, 1). The unity for (*j*) is (1, 1, 1).

2. (*i*) 0 (*ii*) 1 (*iii*) 0 (*iv*) 0.

3. (*a*) $x = 2, 3, x = 2, 3, 6, 11$ (*b*) $x = 2, 3$, No

4. (*a*) $x = 4, 10$ (*b*) $x = -2$ (*c*) Any 2×2 matrix satisfying $(X + 2I)^2 = 0$

(*d*) $x = 1$ **5.** (*iii*) $(a + b)^5 = a^5 + b^5$ **6.** (*a*) 1, 3, 7, 9

(*b*) $-3 = 7, -8 = 2, 3^{-1} = 7$ (*c*) 1, 2, 6, 7.

7. (*a*) $-2 = 28, -7 = 23, -11 = 19$

(*b*) $7^{-1} = 13, 11^{-1} = 11, 26^{-1}$ does not exist since 26 is not a unit.

Hints

2. By $-a$ in a ring Z_8, we mean that element, say, a, such that $a + (-a) = 0 = (-a) + a$

$\therefore \quad -4 = 4, -3 = 5$ etc.

3. (*b*) $Z_{12} = [0, 1, 2, 3, \ldots\ldots 11, +_{12}, \times_{12}]$

If $x = 6$, then $x^2 - 5x + 6 = 36 - 30 + 6 = 12 = 0$

If $x = 11$, then $x^2 - 5x + 6 = 121 - 55 + 6 = 72 = 0$

4. (*a*) $x^2 + 4x + 4 = 0 \Rightarrow (x + 2)(x + 2) \Rightarrow x = -2, -2$. But $-2 = 10$

Also if $x = 4$, then $x^2 + 4x + 4 = 16 + 16 + 4 = 36 = 0$

(*d*) $-2 = 1 \quad \therefore \quad x = 1$ is the only solution.

5. (*i*) $(a + b)(c + d) = a.(c + d) + b.(c + d) = a.c + a.d + b.c + b.d$ | Left distributive law

(*ii*) $(a + b)^2 = (a + b)(a + b) = a.(a + b) + b(a + b)$

$= a.a + a.b + b.a + b.b = a^2 + a.b + b.a + b^2$

$= a^2 + a.b + a.b + b^2 \; a^2 + 2a.b + b^2$ | R is commutative

$= a^2 + 2a.b + b^2$

(*iii*) $(a + b)^5 = 5c_0 a^5 + 5c_1 a^4b + 5c_2 a^3b^2 + 5c_3 a^2b^3 + 5c_4 ab^4 + 5c_5 b^5$

$= a^5 + 5a^4b + 10a^3b^2 + 5ab^4 + b^5$

$= a^5 + b^5$ | In Z_5, $5 = 0, 10 = 0$ etc.

8. Given $a^2 = a$ for all $a \in R$ ...(1)

Let $a, b \in R$. Since R is a ring, it is closed under addition. $\therefore \quad a + b \in R$

Using (1), $(a + b)^2 = a + b$

$\Rightarrow \quad (a + b)(a + b) = a + b$

$\Rightarrow \quad (a + b).a + (a + b).\, b = a + b$ | Right distributive law

$\Rightarrow \quad a.a + b.a + a.b + b.b = a + b$

$\Rightarrow \quad a + b.a + a.b + b = a + b$ | $a.a = a^2 = a$

$\Rightarrow \quad b.\, a + a\, .\, b + b = b$ | Left cancellation law

$\Rightarrow \quad b.\, a + a\, .\, b = 0$ | Right cancellation law

$\Rightarrow \quad a.b = b.\, a$ | $\because$ If $x + y = 0 \Rightarrow x = y$

9. Given (G, +) is an additive group. Also $a.b = 0 \in G \; \forall \; a, b \in G \; \therefore$ G is closed under multiplication.

For $a, b, c \in G$, $a.(b.c) = a.0 = 0$ ...(1)

Also $(a.b) \, . \, c = 0.c = 0$ ...(2)

From (1) and (2), $a.(b.c) = (a.b) \, . \, c \; \forall \; a, b, c \in G$ $\quad | \; b, c \in G \;\Rightarrow\; b.c = 0$

$\therefore$ Associativity holds in G.

Also $a.(b + c) = a.b + a.c = 0 + 0 = 0$

$\Rightarrow$ $a.b + a.c = 0 + 0 = 0$ $\quad | \; a, c \in G \;\Rightarrow\; a.c = 0$

$\therefore$ $a.(b + c) = a.b + a.c$

Similarly, $(a + b) \, . \, c = a.c + b.c$

$\therefore$ (G, +,.) is a ring.

10. Let R* be the set of units in R. We show R* is a group under multiplication. Let $a, b \in R^*$ *i.e.*, a and b are units in R $\Rightarrow$ there exist a^{-1}, $b^{-1} \in$ R such that $aa^{-1} = 1 = a^{-1} a$ and $bb^{-1} = 1 = b^{-1}b$

Consider $(ab) \, (b^{-1} a^{-1}) = a(bb^{-1}) \, a^{-1} = a \, . \, 1 \, . \, a^{-1} = aa^{-1} = 1$

Also $(b^{-1}a^{-1}) \, (ab) = b^{-1}(a^{-1} a) \, b = b^{-1} \, . \, 1 \, . \, b = b^{-1}b = 1$

$\therefore$ $(ab) \, (b^{-1}a^{-1}) = 1 = (b^{-1} a^{-1}) \, (ab)$

Hence ab is also a unit in R. Consequently $ab \in R^*$ *i.e.*, R* is closed under multiplication.

Since R is associative and elements of R* are from R

$\therefore$ R* is associative under multiplication. Finally, if a is a unit in R, then a^{-1} is also a unit in R. Consequently $a^{-1} \in R^*$. Hence R* is a group under multiplication.

9.13. FIELD

(P.T.U. B.Tech. Dec. 2009)

A commutative ring F with unity such that each non-zero element has a multiplicative inversy is called a field. It is denoted by F. Alternatively, F is a field if its non-zero elements form a group under multiplication.

ILLUSTRATIVE EXAMPLES

Example 1. *Show that the following sets are fields.*

(i) [Q; +, .] *(ii) [R; +, .]* *(iii) [C; +, .].*

Sol. (*i*), (*ii*), (*iii*). We know that the sets Q, R and C and commutative ring with unity (see prob. 1. Exercise 9.1). Also each non-zero element in Q, R and C has multiplicatve inverse. Hence they form fields.

Example 2. *Consider the set M of all 2 × 2 matrices of the type* $\begin{pmatrix} a & b \\ -\bar{b} & \bar{a} \end{pmatrix}$ *where* $\bar{a}$, $\bar{b}$ *are the conjugates of a and b. Is M a field ? Justify your answer.*

Sol. Consider A, B $\in$ M where $A = \begin{pmatrix} 2 & 3 \\ 3 & 2 \end{pmatrix}$, $B = \begin{pmatrix} 1 & 1 \\ -1 & 1 \end{pmatrix}$

Then $$AB = \begin{pmatrix} 2 & 3 \\ 3 & 2 \end{pmatrix} \begin{pmatrix} 1 & 1 \\ -1 & 1 \end{pmatrix} = \begin{pmatrix} -1 & 5 \\ 1 & 5 \end{pmatrix}$$

Also $$BA = \begin{pmatrix} 1 & 1 \\ -1 & 1 \end{pmatrix} \begin{pmatrix} 2 & 3 \\ 3 & 2 \end{pmatrix} = \begin{pmatrix} 5 & 5 \\ 1 & -1 \end{pmatrix} \neq AB$$

Hence M is not commutative and therefore cannot be field.

Example 3. *Consider $Z_7 = [0, 1, 2, 3, ... 6, +_7, \times_7]$. Show that Z_7 is a field.*

Sol. Consider the addition modulo 7 table as shown in Table I.

Table I

$+_7$	*0*	*1*	*2*	*3*	*4*	*5*	*6*
0	0	1	2	3	4	5	6
1	1	2	3	4	5	6	0
2	2	3	4	5	6	0	1
3	3	4	5	6	0	1	2
4	4	5	6	0	1	2	3
5	5	6	0	1	2	3	4
6	6	0	1	2	3	4	5

We first show that Z_7 is a ring under addition modulo 7 and multiplication modulo 7.

From Table I, we observe that each element inside the table is also in Z_7. It means that Z is closed under $+_7$.

Addition modulo is always associative

The first row inside the table coincides with the top most row of Table I. It means 0 is the additive identity.

Each element of Z_7 has additive inverse.

For example, Inverse of 1 is 6. Inverse of 2 is 5 etc. $\quad | 1 +_7 6 = 7 = 0$

$| \ 2 +_7 5 = 7 = 0$

Also Table I is symmetrical w.r.t. $+_7$. It means Z_7 is additive w.r.t. $+_7$ *i.e.*,

For $\quad a, b \in Z_7, a +_7 b = b +_7 a \ \forall \ a, b \in Z_7$.

$\therefore \quad Z_7$ is an additive group w.r.t $+_7$.

Now consider the multiplication modulo 7 table as shown in Table II.

Table II

$\times_7$	0	1	2	3	4	5	6
0	0	0	0	0	0	0	0
1	0	1	2	3	4	5	6
2	0	2	4	6	1	3	5
3	0	3	6	2	5	1	4
4	0	4	1	5	2	6	3
5	0	5	3	1	6	4	2
6	0	6	5	4	3	2	1

From Table II, we observe that each element inside the table is also in Z_7. It means Z_7 is closed w.r.t. $\times_7$. *i.e.*, for $a, b \in Z_7 \ \Rightarrow \ a\times_7 b \in Z_7 \ \forall \ a, b \in Z_7$

Finally, For $a, b, c \in Z_7$,

$$a \times_7 (b +_7 c) = a\times_7 b +_7 a \times_7 c$$

$$(a +_7 b) \times_7 c = a \times_7 c +_7 6\times_7 c \text{ is true for all } a, b, c \in Z_7.$$

Hence Z_7 is a ring w.r.t. addition modulo 7 and multiplication modulo 7.

Also the Table II is symmetrical w.r.t. $\times_7$. It means that Z_7 is commutative *i.e.*,

$$a \times_7 b = b \times_7 a \ \forall \ a, b \in Z_7$$

Further, the second row inside the table coincides with the topmost row of Table II. It means 1 is the multiplicative identity of Z_7.

Hence, we have shown that Z_7 is a commutative ring with unity. To show Z_7 is a field, we show each non-zero element of Z_7 has multiplicative inverse.

The units of Z_7 are those elements which are relative primes to 7. (See Topic on 'units')

The elements which are prime to 7 are 1, 2, 3, 4, 5, 6. Hence the units of Z_7 are 1, 2, 3, 4, 5, 6. We can also check the elements which are units as below :

$$1 \times_7 1 = 1; \ 2 \times_7 4 = 1; \ 3 \times_7 5 = 1;$$
$$4 \times_7 2 = 1; \ 5 \times_7 3 = 1; \ 6 \times_7 6 = 1.$$

Hence, each non-zero element of Z_7 has multiplicative inverse. Therefore Z_7 is a field.

9.14. GAUSSIAN INTEGERS

Any number of the form $a + ib$, $a, b \in Z$ is called a Gaussian integer.

Example 4. *Show that the set J[i] of Gaussian integers form a ring under addition and multiplication. Is it an integral domain ? Is it field ?*

Sol. Let X = $[a + ib, \ a, b \in Z]$ be the set of Gaussian integers. Then X is a ring.

We check X for integral domain.

Let $a + ib$, $c + id \in X$ such that a, b, c, d are non-zero integers.

Consider $(a + ib)(c + id) = 0$

$\Rightarrow \quad ac - bd + i(ad + bc) = 0 = 0 + 0i$

$\Rightarrow \quad ac - bd = 0, \quad ad + bc = 0,$

which is possible if either $a = 0 = b$ or $c = 0 = d$ *i.e.*, if either $a + ib = 0$ or $c + id = 0$

Hence X is without zero divisor. Therefore, X is an integral domain.

Further, if $0 \neq a + ib \in X$ be any non-zero element of X where $a, b \in Z$, then the multiplicative inverse of $a + ib$ is

$$\frac{1}{a+ib} = \frac{1}{a+ib} \times \frac{a-ib}{a-ib} = \frac{a-ib}{a^2+b^2} = \frac{a}{a^2+b^2} - i\,\frac{b}{a^2+b^2} \notin J[i]$$

Since $\dfrac{a}{a^2+b^2}$ is not necessary an integer.

$\therefore$ X cannot be a field.

Example 5. *The set of numbers of the form $[a + b\sqrt{2}, \ a, b \in Q]$ is a field.*

Sol. Let $\quad X = [a + b\sqrt{2}; \ a, b \in Q]$. We show X is a ring.

Let $x, y \in X \ \Rightarrow \ x = a_1 + b_1\sqrt{2}, \ a_1, b_1 \in Q$

$y = a_2 + b_2\sqrt{2}, \ a_2, b_2 \in Q$

$\Rightarrow \quad x + y = a_1 + b_1\sqrt{2} + a_2 + b_2\sqrt{2} = a_1 + a_2 + (b_1 + b_2)\sqrt{2} \in X$

$\mid \because \ a_1, a_2 \in Q \ \Rightarrow \ a_1 + a_2 \in Q$

i.e., X is closed under addition.

Addition of rationals is associative.

Further, $0 + 0\sqrt{2} \in X$ is the additive identity since

$$a + b\sqrt{2} + 0 + 0\sqrt{2} = a + 0 + (b + 0)\sqrt{2} = a + b\sqrt{2}$$

Also $\quad 0 + 0\sqrt{2} + a + b\sqrt{2} = 0 + a + (0 + b)\sqrt{2} = a + b\sqrt{2}$

Also for $a + b\sqrt{2} \in X$, $-(a + b\sqrt{2})$ is the additive inverse of $a + b\sqrt{2}$, since

$$a + b\sqrt{2} + (-a - b\sqrt{2}) = a - a + (b - b)\sqrt{2} = 0 + 0\sqrt{2}$$

Also for $a + b\sqrt{2}$, $c + d\sqrt{2} \in X$, we have

$$(a + b\sqrt{2}) + (c + d\sqrt{2}) = a + c + (b + d)\sqrt{2} = c + a + (d + b)\sqrt{2} = (c + d\sqrt{2}) + (a + b\sqrt{2})$$

Hence X is an additive group.

Further, $\quad (a + b\sqrt{2})(c + d\sqrt{2}) = ac + 2bd + (ad + bc)\sqrt{2} \in X$

i.e., X is closed under multiplication.

In Q, the associative laws and distributive laws hold good. Multiplication in Q is commutative and $1 + 0\sqrt{2}$ is the multiplicative identity.

$\therefore$ X is a commutative ring with unity. We lastly show that each non-zero element in X has a multiplicative inverse.

Let $a + b\sqrt{2}$, $a, b \neq 0$ be any element of X. Let $c + d\sqrt{2}$ is the multiplicative inverse of $a + b\sqrt{2}$ such that

$$(a + b\sqrt{2})(c + d\sqrt{2}) = 1 + 0\sqrt{2} = (c + d\sqrt{2})(a + b\sqrt{2})$$

$$\Rightarrow \qquad c + d\sqrt{2} = \frac{1}{a + b\sqrt{2}} = \frac{1}{a + b\sqrt{2}} \times \frac{a - b\sqrt{2}}{a - b\sqrt{2}}$$

$$= \frac{a - b\sqrt{2}}{a^2 - 2b^2} = \frac{a}{a^2 - 2b^2} - \frac{b\sqrt{2}}{a^2 - 2b^2}$$

$$= \frac{a}{a^2 - 2b^2} \times \frac{b}{2b^2 - a^2}\sqrt{2} \in X \qquad \Big| \because \frac{a}{a^2 - 2b^2}, \frac{b}{2b^2 - a^2} \in Q$$

Hence X is a field.

Example 6. *The set of real numbers of the form [a + b$\sqrt{2}$, a, b ∈ Z] is an integral domain. Is it a field ?* (P.T.U. B. Tech. Dec. 2010)

Sol. Proceeding as in example 5, we can show that $X = [a + b\sqrt{2}, a, b \in Z]$ is a commutative ring. We show X is an integral domain. *i.e.*, X has no zero divisor.

Let $a + b\sqrt{2}$, $c + d\sqrt{2} \in X$ where $a, b, c, d \in Z$

Consider $\quad (a + b\sqrt{2})(c + d\sqrt{2}) = 0$

$$\Rightarrow \qquad ac + 2bd + (ad + bc)\sqrt{2} = 0 = 0 + 0\sqrt{2}$$

$$\Rightarrow \qquad ac + 2bd = 0,\ ad + bc = 0,$$

which is possible only if either $a = b = 0$ or $c = d = 0$ *i.e.*, either $a + b\sqrt{2} = 0$ or $c + d\sqrt{2} = 0$

Hence X is an integral domain.

Finally, we check X for multiplicative inverse. Consider $0 \neq 5 + 3\sqrt{2} \in X$ and $c + d\sqrt{2} \in X$ such that

$$(5 + 3\sqrt{2})\,(c + d\sqrt{2}) = 1 + 0\sqrt{2} = (c + d\sqrt{2})\,(5 + 3\sqrt{2})$$

$$\Rightarrow \qquad c + d\sqrt{2} = \frac{1}{5 + 3\sqrt{2}} = \frac{1}{5 + 3\sqrt{2}} \times \frac{5 - 3\sqrt{2}}{5 - 3\sqrt{2}}$$

$$= \frac{5 - 3\sqrt{2}}{25 - 18} = \frac{5}{7} - \frac{3}{7}\sqrt{2} \notin X. \text{ Since } \frac{5}{7}, \frac{3}{7} \notin Z$$

$\therefore$ X cannot be a field.

Example 7. *Let D be the ring of all real 2 × 2 matrices of the form* $\begin{pmatrix} a & -b \\ b & a \end{pmatrix}$. *Show that D is isomorphic to the complex number C where D is a field.*

Sol. Given D is a ring of real 2 × 2 matrices of the form $\begin{pmatrix} a & -b \\ b & a \end{pmatrix}$. Define $f : D \to C$, by

$$f\begin{pmatrix} a & -b \\ b & a \end{pmatrix} = a + ib$$

We show f is homomorphism, one-one and onto. Let $\begin{pmatrix} a & -b \\ b & a \end{pmatrix}, \begin{pmatrix} c & -d \\ d & c \end{pmatrix} \in D$

Consider $$f\left[\begin{pmatrix} a & -b \\ b & a \end{pmatrix} + \begin{pmatrix} c & -d \\ d & c \end{pmatrix}\right] = f\begin{pmatrix} a+c & -(b+d) \\ b+d & a+c \end{pmatrix}$$

$$= a + c + i(b + d) = a + ib + c + id$$

$$= f\begin{pmatrix} a & -b \\ b & a \end{pmatrix} + f\begin{pmatrix} c & -d \\ d & c \end{pmatrix}$$

Thus f is a homomorphism.

Further, Let $f\begin{pmatrix} a & -b \\ b & a \end{pmatrix} = f\begin{pmatrix} c & -d \\ d & c \end{pmatrix}$

$$\Rightarrow \qquad a + ib = c + id$$

Equating real and imaginary part, we get $a = c$, $b = d$

Thus $a + ib = c + id$. $\therefore$ f is one-one.

Hence $f : D \to C$, defined by

$$f\begin{pmatrix} a & -b \\ b & a \end{pmatrix} = a + ib \text{ is an isomorphism.}$$

Theorem VIII. *Every field is an integral domain. But the converse is not true.*

(P.T.U. B.Tech. May 2010, May 2012)

Proof. Let F be a field. We show F is an integral domain. As F is a field, F must be commutative. We show F is without zero divisors.

Let $a, b \in F$ such that $a \,.\, b = 0$ ...(1)

If $a \neq 0$, then as F is a field, each non-zero element of F has multiplicative inverse. *i.e.*, for $a \in F$, there exists $a^{-1} \in F$ such that $a\,a^{-1} = 1 = a^{-1}\,a$

From (1), $a.b = 0$

$\Rightarrow \quad a^{-1}.(a.b) = a^{-1}\,.\,0 = 0 \qquad | \; a.0 = 0 \;\forall\; a \in R$

$\Rightarrow \quad (a^{-1}\,a)\,.\,b = 0$

$\Rightarrow \quad 1.b = 0$

$\Rightarrow \quad b = 0$

Hence if $a \neq 0$, then $b = 0$

Similarly, if $b \neq 0$, then $a = 0$

Hence F is without zero divisors. Consequently F is an integral domain.

The converse is, however, not true.

Example of an integral domain, which is not a field. (P.T.U. B.Tech. Dec. 2013)

Z is an integral domain. But for $2 \in Z$, there is no $a \in Z$ such that $2a = 1 = a.2$ *i.e.*, 2 has no multiplicative inverse.

Therefore, Z cannot be a field.

Theorem IX. *Any finite non-zero integral domain is a field.*

Proof. Let $D = [a_1, a_2, \ldots\ldots a_n]$ be a finite non-zero integral domain where each a_i's are distinct. We show D is a field. For this, we show F is commutative ring with unity and each non-zero element of D has multiplicative inverse.

Since D is an integral domain, D must be commutative.

Let $0 \neq a \in D$ and consider the set $[aa_1, aa_2, \ldots\ldots aa_n]$. We claim

$$D = [aa_1, aa_2, \ldots\ldots aa_n]$$

Now $\quad a_i \in D\ (1 \leq i \leq n)$ and $a \in D \;\Rightarrow\; aa_i \in D \;\forall\; 1 \leq i \leq n.$

$\Rightarrow \quad aa_1, aa_2, \ldots\ldots aa_n \in D$

$\Rightarrow \quad (aa_1, aa_2, \ldots\ldots aa_n) \subseteq D$

$\therefore \quad D = [aa_1, aa_2, \ldots\ldots aa_n]$

We next claim that the elements $aa_1, aa_2, \ldots\ldots aa_n$ are all distinct. For if, $aa_i = aa_j$

$\Rightarrow \quad aa_i - aa_j = 0 \;\Rightarrow\; a(a_i - a_j) = 0$

But $0 \neq a \in D$ and D is without zero divisors

$\therefore \quad a_i - a_j = 0 \;\Rightarrow\; a_i = a_j$, a contradiction as each a_i's are distinct.

Hence $\quad D = (aa_1, aa_2, \ldots\ldots aa_n)$ has n distinct elements.

Now $\quad a \in D \;\Rightarrow\; a = aa_{i_o},\ i_o \in (1, 2, \ldots\ldots n)$

We show a_{i_o} is the multiplicative identity of x, $x \in D$.

For this, we show that $xa_{i_o} = x = a_{i_o}\,x$

As $x \in D = (aa_1, aa_2, \ldots\ldots aa_n) \;\Rightarrow\; x = aa_j,\ 1 \leq j \leq n$

Consider $xa_{i_o} = (aa_j)\,a_{i_o} = a(a_j a_{i_o})$

$= a(a_{i_o}\,a_j) \qquad |$ D is commutative

$= (aa_{i_o})\,a_j \qquad |$ Associativity

$= aa_j \qquad | \; aa_{i_o} = a$

$= x.$

Similarly, $a_{i_o} x = a_{i_o} (aa_j) = (a_{i_o} a)\, a_j$

$= (aa_{i_o})\, a_j$ | D is commutative

$= aa_j = x$

Hence $xa_{i_o} = x = a_{i_o} x$ *i.e.*, a_{i_o} is the multiplicative identity of x.

Lastly, $1 \in D = [aa_1, aa_2, ... aa_n]$

$\Rightarrow$ $1 = aa_j,\ 1 \le j \le n$

$= a_j a$ | D is commutative

$\Rightarrow$ $aa_j = 1 = a_j a$

i.e., a_j is the inverse of a that is, each non-zero element of D has multiplicative inverse.

Hence D is a field.

Example 8. *$[Z_p, +_p, \times_p]$, p is prime, is a field.*

Sol. We know that Z_p , p is prime, is an integral domain. Since Z_p is finite and non-zero, therefore Z_p is a field (use theorem IX above)

TEST YOUR KNOWLEDGE 9.2

1. Write out the addition, multiplication and inverse (additive) table for the following fields:
   (*a*) $[Z_2; +_2, \times_2]$ (*b*) $[Z_3 ; +_3, \times_3]$
   (*c*) $[Z_5; +_5, \times_5]$
2. Find all numbers in $Z_2 = [0, 1, +_2, \times_2]$ that satisfy the following equations:
   (*a*) $x^2 - 1 = 0$ (*b*) $x^2 + 1 = 0$
   (*c*) $x^3 + x^2 + x + 1 = 0$ (*d*) $x^3 + x + 1 = 0$.
3. Determine all values of x from the given fields which satisfy the given equations:
   (*a*) $x + 1 = -1$ over Z_2, Z_3, Z_5 (*b*) $2x + 1 = 2$ over Z_3, Z_5
   (*c*) $3x + 1 = 2$ over Z_5.
4. Write out the operation tables for $[Z_2{}^2, +_2, \times_2]$. Is $Z_2{}^2$ a ring ? an integral domain ? *a* field? Justify your answers.

Answers

1. (*a*)

$+_2$	0	1
0	0	1
1	1	0

Addition Table

$\times_2$	0	1
0	0	0
1	0	1

Multiplication Table

a	a^{-1}
1	1

Inverse Table

(*b*)

$+_3$	0	1	2
0	0	1	2
1	1	2	0
2	2	0	1

Addition Table

$\times_3$	0	1	2
0	0	0	0
1	0	1	2
2	0	2	1

Multiplication Table

a	a^{-1}
1	2
2	1

Inverse Table

(c)

$+_5$	0	1	2	3	4
0	0	1	2	3	4
1	1	2	3	4	0
2	2	3	4	0	1
3	3	4	0	1	2
4	4	0	1	2	3

Addition Table

$\times_5$	0	1	2	3	4
0	0	0	0	0	0
1	0	1	2	3	4
2	0	2	4	1	3
3	0	3	1	4	2
4	0	4	3	2	1

Multiplication Table

a	a^{-1}
1	4
2	3
3	2
4	1

Inverse (additive) Table

2. (a) $x = 1$ (b) $x = 1$ (c) $x = 1$ (d) None.
3. (a) 0 (over Z_2), 1 (over Z_3), 3(over Z_5) (b) 2(over Z_3), 3(over Z_5) (c) 2(over Z_5)
4.

$+_2$	(0, 0)	(0, 1)	(1, 0)	(1, 1)
(0, 0)	(0, 0)	(0, 1)	(1, 0)	(1, 1)
(0, 1)	(0, 1)	(0, 0)	(1, 1)	(1, 0)
(1, 0)	(1, 0)	(1, 1)	(0, 0)	(0, 1)
(1, 1)	(1, 1)	(1, 0)	(0, 1)	(0, 0)

$\times_2$	(0, 0)	(0, 1)	(1, 0)	(1, 1)
(0, 0)	(0, 0)	(0, 0)	(0, 0)	(0, 0)
(0, 1)	(0, 0)	(0, 1)	(0, 0)	(0, 1)
(1, 0)	(0, 0)	(0, 0)	(1, 0)	(1, 0)
(1, 1)	(0, 0)	(0, 1)	(1, 0)	(1, 1)

Z_2^2 is a ring.
Z_2^2 is not an integral domain. Since (1, 0), (0, 1) $\in Z_2^2$ be any two non-zero elements and (1, 0) . (0, 1) = (0, 0) *i.e.,* Z_2^2 is a ring with zero divisors.
Z_2^2 can not be a field. For if, Z_2^2 is a field, it must be an integral domain, which is not so.

9.14. IDEALS

(P.T.U. B.Tech. Dec. 2013)

Left Ideal. A non-empty subset I of a ring R is called a left ideal of R if

(i) For $a, b \in I \Rightarrow a - b \in I \ \forall \ a, b \in I$

(ii) For $a \in I, r \in R \Rightarrow ra \in I$

Right Ideal. A non-empty subset J of a ring R is called a right ideal of R if

(i) For $a, b \in J \Rightarrow a - b \in J \ \forall \ a, b \in J$

(ii) For $a \in J, r \in R \Rightarrow ar \in J$.

Ideal. A non-empty set K of a ring R is called an ideal (or two sided ideal) of R iff K is both left ideal and right ideal of R *i.e.*,

(*i*) For $a, b \in K \Rightarrow a - b \in K \; \forall a, b \in K$

(*ii*) For $a \in K, r \in R \Rightarrow ra \in K, ar \in K$.

Proper and Improper Ideals. (*P.T.U. B.Tech. Dec. 2007*)

Every ideal other than <0> and R are known as proper ideals. The ideals <0> and R are improper ideals of R.

Example. (*i*) Let E = the set of even integers, Z = the ring of integers. Since $E \subseteq Z$, E is an ideal of Z. But $Z \subseteq Q$ and Z is not an ideal of Q.

ILLUSTRATIVE EXAMPLES

Example 1. *Show that {0} is an ideal in any ring R.* (P.T.U. B.Tech May 2010)

Sol. Let $0 \in \{0\}$ and consider $0 - 0 = 0 \in \{0\}$

For $r \in R$, $r \cdot 0 = 0 \in \{0\}$

$0 \cdot r = 0 \in \{0\}$

$\therefore$ {0} is an ideal of R.

Example 2. *Let M be the ring of 2 × 2 matrices over reals. Give an example of a left ideal, which is not a right ideal and an example of a right ideal, which is not a left ideal.*

Sol. Consider $L = \left\{\begin{pmatrix} 0 & a \\ 0 & b \end{pmatrix} : a, b \in R\right\}$

Let A, B ∈ L such that $A = \begin{pmatrix} 0 & a \\ 0 & b \end{pmatrix}, B = \begin{pmatrix} 0 & c \\ 0 & d \end{pmatrix}$, where $a, b, c, d \in R$

Consider $A - B = \begin{pmatrix} 0 & a \\ 0 & b \end{pmatrix} \begin{pmatrix} 0 & c \\ 0 & d \end{pmatrix} = \begin{pmatrix} 0 & a-c \\ 0 & b-d \end{pmatrix} \in L \quad \forall\, a, b, c, d \in R$

Let R ∈ M be any matrix over reals such that

$$R = \begin{pmatrix} \alpha & \beta \\ \gamma & \delta \end{pmatrix}, \alpha, \beta, \gamma, \delta \in R$$

Consider $RA = \begin{pmatrix} \alpha & \beta \\ \gamma & \delta \end{pmatrix} \begin{pmatrix} 0 & a \\ 0 & b \end{pmatrix} = \begin{pmatrix} 0 & \alpha a + \beta b \\ 0 & \gamma a + \delta b \end{pmatrix} \in L$

But $AR = \begin{pmatrix} 0 & a \\ 0 & b \end{pmatrix} \begin{pmatrix} \alpha & \beta \\ \gamma & \delta \end{pmatrix} = \begin{pmatrix} a\gamma & a\delta \\ b\gamma & b\delta \end{pmatrix} \in L$

Hence L is a left ideal, but not a right ideal of M

If we take $K = \left\{\begin{pmatrix} a & b \\ 0 & 0 \end{pmatrix} : a, b \in R\right\}$, then we can show that K is a right ideal, but not a left ideal of M.

Example 3. *The set of even integers is an ideal of Z.*

Sol. Consider E, the set of even integers given by $E = [2m : m \in Z]$

Let $a, b \in E \Rightarrow a = 2m, m \in Z, b = 2n, n \in Z$

$\therefore$ $a - b = 2m - 2n = 2(m - n) \in E$ | $\because$ $m, n \in Z \Rightarrow m - n \in Z$

For $r \in Z$, consider

$ra = r(2m) = 2(rm) \in E$

$ar = (2m)\, r = 2(mr) \in E$

| $m \in Z, r \in Z \Rightarrow rm \in Z$; Similarly, $mr \in Z$

Hence E is an ideal of Z.

Example 4. *Let M be a ring of 2 × 2 matrices over integers. Consider the set* (P.T.U. B. Tech. Dec. 2010)

$L = \left[\begin{pmatrix} a & 0 \\ b & 0 \end{pmatrix} : a, b \in Z\right]$. *Show that L is a left ideal of M. Is L right ideal of M?*

Sol. Since $\begin{pmatrix} 0 & 0 \\ 0 & 0 \end{pmatrix} \in L \Rightarrow L \neq \phi$ *i.e.,* L is non-empty set of M.

Let $A, B \in L \Rightarrow A = \begin{pmatrix} a & 0 \\ b & 0 \end{pmatrix}$; $a, b \in Z$; $B = \begin{pmatrix} c & 0 \\ d & 0 \end{pmatrix}$; $c, d \in Z$

$$A - B = \begin{pmatrix} a & 0 \\ b & 0 \end{pmatrix} - \begin{pmatrix} c & 0 \\ d & 0 \end{pmatrix} = \begin{pmatrix} a-c & 0 \\ b-d & 0 \end{pmatrix} \in L \quad \Big| \; a, c \in Z \Rightarrow a - c \in Z; \; b, d \in Z \Rightarrow b - d \in Z$$

For α, β, γ, δ ∈ Z, Let $R = \begin{pmatrix} \alpha & \beta \\ \gamma & \delta \end{pmatrix} \in M$ and consider

$$RA = \begin{pmatrix} \alpha & \beta \\ \gamma & \delta \end{pmatrix} \begin{pmatrix} a & 0 \\ b & 0 \end{pmatrix} = \begin{pmatrix} \alpha a + \beta b & 0 \\ \gamma a + \delta b & 0 \end{pmatrix} \in L$$

| ∵ $\alpha, \beta, a, b \in Z \Rightarrow \alpha a + \beta b \in Z$
$\gamma, \delta, a, b \in Z \Rightarrow \gamma a + \delta b \in Z$

Hence L is a left ideal of M. Also

$$AR = \begin{pmatrix} a & 0 \\ b & 0 \end{pmatrix} \begin{pmatrix} \alpha & \beta \\ \gamma & \delta \end{pmatrix} = \begin{pmatrix} a\alpha & a\beta \\ b\alpha & b\beta \end{pmatrix} \notin L$$

∴ L is not a right ideal of M.

Example 5. *Every ideal of a ring R is a subring of R. But the converse is not true.*

Sol. Let K be an ideal of R, then we have

For $a, b \in K \Rightarrow a - b \in K \;\forall\; a, b \in K$ ...(1)

For $r \in R$, $ra \in K$, $ar \in K$ ...(2)

We show K is a subring of R. *i.e.,*

We show (*i*) for $a, b \in K \Rightarrow a - b \in K \;\forall\; a, b \in K$ ...(3)

(*ii*) for $a, b \in K \Rightarrow a \,.\, b \in K \;\forall\; a, b \in K$ ...(4)

Now (3) is trivial by using (1). To show (4),

Let $b \in K \subseteq R \Rightarrow b \in R$. Since K is an ideal of R and $a \in K$, we have

$$a\,b \in K \qquad | \text{ Using (2))}$$

Hence K is a subring of R.

Consider Z, the set of integers and Q, the set of rationals. We know that Z is a subring of Q. We show Z is not an ideal of Q.

Take $3 \in Z, \frac{1}{2} \in Q$, then $3 \,.\, \frac{1}{2} = \frac{3}{2} \notin Z$ *i.e.,* Z cannot be an ideal of Q.

Example 6. *If K is an ideal of R and 1 ∈K, then K = R*

Sol. Given K is an ideal of R, so K is a subset of R *i.e.,* $K \subseteq R$. Also $r \in R$ and $1 \in K$

$\Rightarrow \quad r.1 \in K$ | K is an ideal of R

$\Rightarrow \quad r \in K \Rightarrow R \subseteq K$

$\Rightarrow \quad K = R$

Example 7. *If F is a field, then F has no proper ideals.*

Or

If F is a field, then the only ideals of F are <0> and F itself.

Sol. Let, if possible. S is any proper ideal of F and $0 \neq a \in S$. As $S \subseteq F \Rightarrow 0 \neq a \in F$. But F is a field and $0 \neq a \in F$. There exists $a^{-1} \in F$ (Every non-zero element of F has a multiplicative inverse).

Now $a \in S$, $a^{-1} \in F$ and since S is an ideal of F, then $aa^{-1} \in S \Rightarrow 1 \in S$

Hence S = F. | see example 6 above

Hence the only ideals of F are 0 and F itself.

Theorem X. *Intersection of two ideals of a ring R is an ideal of R.*

(P.T.U. B. Tech. May 2013)

Proof. Let A and B are two ideals of R, then, $\phi \neq A \subseteq R$, $\phi \neq \beta \subseteq R \Rightarrow \phi \neq A \cap B \subseteq R$ *i.e.*, $A \cap B$ is a non-empty subset of R. We show $A \cap B$ is an ideal of R.

Let $x, y \in A \cap B \Rightarrow x, y \in A$ and $x, y \in B$. As A and B are ideals of R

$\therefore$ $x - y \in A, x - y \in B \Rightarrow x - y \in A \cap B$

For $r \in R, x \in A \Rightarrow rx \in A$ and $xr \in A$ | A is an ideal of R.

Also for $r \in R$, $x \in B \Rightarrow rx \in B$ and $xr \in B$ | B is an ideal of R

Hence $rx \in A$, $rx \in B \Rightarrow rx \in A \cap B$

Also $xr \in A$, $xr \in B \Rightarrow xr \in A \cap B$

Hence the theorem.

Example 8. *Let $T : R \to S$ be a ring homomorphism. Show that the kernel of T is a two sided ideal of R.* (P.T.U. B.Tech. May 2013)

Sol. By definition:

$$\text{Ker } T = \{r \in R : f(r) = 0, \text{ where } 0 \in S\}$$

Let $x, y \in \text{Ker } T \Rightarrow T(x) = 0, T(y) = 0$

Consider $T(x - y) = T(x) - T(y) = 0 - 0 = 0$

$\Rightarrow$ $x - y \in \text{Ker }(T)$

Further, if $r \in R$ and $x \in \text{Ker } T$

Then
$$T(rx) = T(r)\,T(x) = T(r) \cdot 0 = 0$$

Hence $rx \in \text{Ker } T$

Similarly,
$$T(xr) = T(x)\,T(r) = 0 \cdot T(r) = 0$$

$\Rightarrow$ $xr \in \text{Ker } T$

Hence, kernel of T is a two sided ideal of R.

9.15. SUM OF IDEALS

Let A and B be two ideals of a ring R, then the sum of the ideals A and B, denoted by A + B, is defined by $A + B = [a + b : a \in A, b \in B]$

Theorem XI. *If A and B are two ideals of R, then A + B is an ideal of R.*

Proof. $0 = 0 + 0 \in A + B \Rightarrow A + B \neq \phi$ *i.e.*, A + B is non-empty subset of R.

Let $x, y \in A + B \Rightarrow x = a_1 + b_1, a_1 \in A, b_1 \in B$

$\Rightarrow$ $y = a_2 + b_2, a_2 \in A, b_2 \in B$

$\therefore$ $x - y = a_1 + b_1 - (a_2 + b_2) = a_1 - a_2 + b_1 - b_2 \in A + B$

$\because$ $a_1, a_2 \in A$ and A is an ideal of R

$\therefore$ $a_1 - a_2 \in A$. Similarly $b_1 - b_2 \in B$

Further, for $r \in R$, $x \in A + B$, consider $rx = r(a_1 + b_1) = ra_1 + rb_1 \in A + B$

| $\because$ $r \in R$, $a_1 \in A$ and A is an ideal of R. $\therefore$ $ra_1 \in A$ Similarly, $rb_1 \in B$

Also $xr = (a_1 + b_1)\, r = a_1 r + b_1 r \in A + B$

| $\because$ $r \in R$, $a_1 \in A$ and A is an ideal of R $\therefore$ $a_1 r \in A$. Similarly $b_1 r \in B$

Hence A + B is an ideal of R.

9.16. QUOTIENT RING

(P.T.U. B.Tech. May 2010, Dec. 2007, 2006, May 2005)

Let R be a ring and I be an ideal of R. Define R/I, by

$R/I = [x + I : x \in R]$

Then R/I is also a ring under the addition and multiplication defined by

$$(r + I) + (s + I) = r + s + I \;\forall\; r, s \in R$$

$$(r + I)(s + I) = rs + I \quad \forall\; r, s \in R.$$

The ring defined by R/I is known as quotient ring.

9.17. FUNDAMENTAL THEOREM OF RING HOMOMORPHISM

Statement If f : *$R \to R'$ be a ring homomorphism. Then $R \,|\, Ker\, f \to R'$*

Proof. Define $\theta : R \,|\, \text{Ker}\, f \to R'$ such that

$$\theta(x + K) = f(x) \;\forall\; x \in R \text{ where } K = \text{Ker}\, f.$$

We show θ is well-defined, homomorphism, one-one and onto. Since Ker θ is a two sided ideal of R. Therefore, R | Ker θ is defined

θ is well-defined: Consider

$x + K = y + K \Rightarrow x - y \in K = \text{Ker}\, f$

$\Rightarrow$ $f(x - y) = 0 \Rightarrow f(x) - f(y) = 0$

$\Rightarrow$ $f(x) = f(y) \Rightarrow \theta(x + K) = \theta(y + K)$

$\therefore$ θ is well-defined

θ is homomorphism: Consider

$\theta(x + K + y + K) = \theta(x + y + K) = f(x + y) = f(x) + f(y)$ | f is a ring homomorphism

$= \theta(x + K) + \theta(y + K)$

Also $\theta((x + K)(y + K)) = \theta(xy + K) = f(xy) = f(x)\, f(y)$ | f is a ring homomorphism

$= \theta(x + K)\, \theta(y + K)$

$\therefore$ θ is homomorphism

θ is one-one

Let $\theta(x + K) = \theta(y + K)$

$\Rightarrow$ $f(x) = f(y)$

$\Rightarrow$ $f(x) - f(y) = 0$

$\Rightarrow \quad f(x-y) = 0$

$\Rightarrow \quad x - y \in \text{Ker} f = K$

$\Rightarrow \quad x + K = y + K$

$\Rightarrow$ θ is one-one

θ is onto: Let $r' \in R'$. Then there exists $r \in R$ such that $f(r) = r'$

$\Rightarrow \quad \theta(r + K) = r'$

$\therefore$ θ is onto.

Hence $R \mid \text{Ker} f \cong R'$

Example 9. *(i) If R is commutative, then R/I is also commutative.*

(ii) If R is a ring with identity, then R/I is also a ring with I as identity, where I is an ideal of R.

Sol. *(i)* By definition, if I is an ideal of R, then $R/I = \{a + I : a \in R\}$

Let $a + I,\ b + I \in R/I$ and consider

$$(a + I)(b + I) = ab + I$$
$$= ba + I \qquad \mid a, b \in R \text{ and R is commutative}$$
$$= (b + I)(a + I)$$

$\Rightarrow$ R/I is commutative.

(ii) Let $1 \in R$ be the identity and if $a + I \in R/I$, then consider

$$(a + I)(1 + I) = a \,.\, 1 + I = a + I$$
$$= 1 \,.\, a + I = (1 + I)(a + I)$$

Thus 1 + I is the identity of R/I.

Example 10. *Let $H_4 = \{4n : n \in Z\}$ then show that H_4 is an ideal of Z and hence $Z \mid H_4$ is a quotient ring given by $Z \mid H_4 = \{H_4, H_4 + 1, H_4 + 2, H_4 + 3\}$*

Sol. Let $x, y \in H_4 \Rightarrow x = 4m, y = 4n$ for some: $m, n \in Z$.

Consider $x - y = 4n - 4m = 4(n - m) \in H_4$

Also for $r \in Z$, $x \in H_4$, we have

$$rx = r(4m)$$
$$= 4rm \in H_4$$
$$xr = (4m)\, r$$
$$= 4mr \in H_4$$

Hence H_4 is an ideal of Z.

Further $\quad H_4 = \{... -8, -4, 0, 4, 8, ...\}$

$H_4 + 1 = \{... -7, -3, 1, 5, 9 ...\}$

$H_4 + 2 = \{... -6, -2, 2, 6, 10, ...\}$

$H_4 + 3 = \{... -5, -1, 3, 7, 11, ...\}$

$H_4 + 4 = \{... -4, 0, 4, 8, 12, ...\} = H_4$

$H_4 + 5 = \{... -3, 1, 5, 9, ...\} = H_4 + 1$

$H_4 + 6 = H_4 + 2$

$H_4 + 7 = H_4 + 3$

$H_4 + 8 = H_4$ and so on

$\therefore \quad Z \mid H_4 = \{H_4, H_4 + 1, H_4 + 2, H_4 + 3\}$.

TEST YOUR KNOWLEDGE 9.3

1. Define proper and improper ideals.
2. Define quotient ring with example. *(P.T.U. B.Tech. Dec. 2006)*
3. Let I be an ideal of a ring R and $R/I = \{x + I = x \in R\}$ under the addition and multiplication defined by

$$(x + I) + (y + I) = x + y + I$$
$$(x + I)(y + I) = xy + I \quad \forall\, x, y \in R.$$

Then R/I is a ring, called the quotient ring.

4. If J be an ideal of a ring R and $u \in J$ be a unit in J. Then J = R.
5. If J and K are ideals in a ring R, then J + K and $J \cap K$ are also ideals in R.

Hint

2. Consider $H_4 = \{4n\,;\, n \in Z\}$

We show H_4 is an ideal of Z, the ring of integers.

Let $a, b \in H_4 \Rightarrow a = 4n, n \in Z, \quad b = 4m, m \in Z$

$\therefore \quad a - b = 4n - 4m = 4(n - m) \in H_4\, n, \qquad | \; m \in Z \Rightarrow n - m \in Z$

Also, If $r \in Z$, $a \in H_4$, then

$$ar = (4n)\, r = 4(nr) \in H_4 \qquad | \; n, r \in Z \Rightarrow nr \in Z$$
$$ra = r(4n) = 4(rn) \in H_4 \qquad | \; r, n \in Z \Rightarrow rn \in Z$$

Hence H_4 is an ideal of Z

Further,

$$H_4 = \{ \ldots\ldots 8, -4, 0, 4, 8, \ldots\ldots\}$$
$$H_4 + 1 = \{\ldots\ldots -7, -3, 1, 5, 9, \ldots\ldots\}$$
$$H_4 + 2 = \{\ldots\ldots -6, -2, 2, 6, 10, \ldots\ldots\}$$
$$H_4 + 3 = \{\ldots\ldots -5, -1, 3, 7, 11, \ldots\ldots\}$$
$$H_4 + 4 = \{\ldots\ldots -4, 0, 4, 8, \ldots\ldots\} = H_4$$

Thus $Z/H_4 = \{H_4, H_4 + 1, H_4 + 2, H_4 + 3\}$ is a quotient ring.

9.18. PRINCIPAL IDEAL

Let R be a commutative ring with an identity element 1. Let $a \in R$, then the set $= \{ra : r \in R\}$ is an ideal, called the *principal ideal generated by a.*

9.19. PRINCIPAL IDEAL DOMAIN (P.I.D.)

A ring R is called a principal ideal domain if

(*i*) R is an integral domain

(*ii*) Every ideal in R is principal.

Example. *Show that Z, the ring of integers, is a P.I.D.*

Sol. We know that Z, the set of integers is an integral domain. Let J is an ideal in Z. We show J is a principal ideal.

Case I. If J = {0}, then it is principal ideal and hence the result.

Case II. If $J \neq \{0\}$, Let $0 \neq x \in J$, then

$$-x = (-1)\, x \in J \text{ for some positive } x.$$

Hence J contains at least one positive integer. Let a be the smallest positive integer in J.

We claim $J = \{ra : r \in Z\}$

For $x \in J$, using division algorithm,

$$x = qa + r,\ 0 \le r \le a,\ q \in Z$$

But J is an ideal and $a \in J$, $q \in Z$

$\therefore\quad q\,a \in J$ and $x - q\,a \in J \quad \Rightarrow \quad r \in J.$

But a is the smallest positive integer in J satisfying $0 \le r \le a$. Hence we must have $r = 0$.

$$\therefore \qquad x = qa \quad i.e.,$$

$$J = \{qa : q \in Z\}$$

$\therefore$ Z is a P.I.D.

9.20. EUCLIDEAN DOMAIN

(P.T.U. B.Tech. May 2007, May 2006)

Let R be an integral domain and suppose that for every $0 \neq a \in R$, there exists a non-negative number $d(a)$ satisfying the following:

(*i*) For $a, b \in R$, $a \neq 0$, $b \neq 0$, $d(ab) \geq d(a)$

(*ii*) For $0 \neq a \in R$, $b \in R$, there exists $q, r \in R$ such that $b = aq + r$, where either $r = 0$ or $d(r) < d(a)$.

Then R is called Euclidean domain.

Remark. The condition (1) can be replaced by $d(ab) \geq d(b)$.

ILLUSTRATIVE EXAMPLES

Example 1. *The ring of integers is a Euclidean domain.*

Sol. For $0 \neq a \in Z$, define $d(a) = |\, a \,|$ (non-negative integer)

Let $\quad 0 \neq a,\ 0 \neq b \in Z$. consider

$$d(ab) = |\, ab \,| = |\, a \,|\ |\, b \,| \geq |\, a \,| = d(a)$$

$$\Rightarrow \qquad d(ab) \geq d(a)$$

Also for $0 \neq a$, $b \in Z$, we can find $q, r \in Z$ such that $b = qb + r$, where $r = 0$ or $d(r) < d(a)$.

Hence Z is a Euclidean domain.

Example 2. *Every field is a Euclidean domain.*

Sol. For $0 \neq 0 \in F$, define $d(a) = 1$

| Since if $0 \neq a \in F$ and F is a field, we can find $a^{-1} \in F$ such that $aa^{-1} = 1 = d(a) \in F$

Let $\quad 0 \neq a,\ 0 \neq b \in F \ \Rightarrow\ ab \in F$

Now $\quad d(ab) = 1 \geq d(a) \ \Rightarrow\ d(ab) \geq d(a)$

Also for $0 \neq b \in F$, we can write

$$a = (ab^{-1})\, b + 0 = qb + r \quad \text{where} \quad q = ab^{-1},\ r = 0$$

Hence every field is an integral domain.

9.21. ASSOCIATE

Let R be a commutative ring with unity. An element $a \in$ R is said to be an associate of $b \in$ R if $a = bu$, where u is a unit in R. If a is an associate of b, we write $a \sim b$.

Example 3. (*a*) *Find the associates of 4 in Z_{10}, where Z_{10} denotes the integers modulo 10.*

(*b*) *Find the associates of 5 in Z_{10}.*

(*c*) *Find the associates of n in Z.*

Sol. (*a*) The units of Z_{10} are those numbers which are relative prime to 10.

$\therefore$ The units of Z_{10} are 1, 3, 7, 9.

Multiplying each of the units by 4, we get

$$1.4 = 4,\ 3.4 = 12 = 2,\ 7.4 = 28 = 8,\ 9.4 = 36 = 6$$

Thus, 2, 4, 6, 8 are the associates of 4 in Z_{10}.

(*b*) Similarly, the associates of 5 can be obtained by multiplying each of the units by 5, we get

$$1.5 = 5,\ 3.5 = 15 = 5,\ 7.5 = 35 = 5,\ 9.5 = 45 = 5.$$

Thus, the only associate of 5 in Z_{10} is 5 itself.

(*c*) The units of Z are – 1 and 1. Multiplying each of the units by n, we get $-n$ and n, as the associates of n in Z.

Example 4. *Consider the ring*

Z_{12} = {0, 1, 2, 3, ..., 11} of integers modulo 12.

(*a*) *Find the units of Z_{12}*

(*b*) *Find the roots of $f(x) = x^2 + 4x + 4$ over Z_{12}*

(*c*) *Find the associates of 2.*

Sol. (*a*) The units of Z_{12} are those elements which are relative prime to 12.

Hence, the units are 1, 5, 7, 11.

(*b*) Put $x = 0, 1, 2, \ldots, 11$ in $f(x) = x^2 + 4x + 4$, we get

when $x = 0$, $f(0) = 4$

$x = 1$, $f(1) = 1 + 4 + 4 = 9$

$x = 2$, $f(2) = 4 + 8 + 4 = 16 = 4$

$x = 3$, $f(3) = 9 + 12 + 4 = 25 = 1$

$x = 4$, $f(4) = 16 + 16 + 4 = 36 = 0$ etc.

Also, when $x = 10$, $f(10) = 100 + 40 + 4 = 144 = 0$

$\therefore$ The roots of $f(x)$ are $x = 4, 10$

(*c*) The units of Z_{12} are 1, 5, 7, 11

The associates of 2 are obtained by multiplying each unit by 2, we get

$$1.2 = 2,\ 5.2 = 10,\ 7.2 = 14 = 2,\ 11.2 = 22 = 10$$

$\therefore$ The associates of 2 are {2, 7}.

Theorem XII. *Show that relation of being associates is an equivalence relation in a ring R.*

Proof. Reflexive. For $x \in$ R, we can write $x = x\ .\ 1$ where 1 is a unit in R. Hence $x \sim x$. *i.e.*, $\sim$ is reflexive.

Symmetric. Let $a \sim b \Rightarrow$ there exist some unit $u \in$ R such that $b = au$.

Now $au^{-1} = (bu)u^{-1} = b(uu^{-1}) = b$

$\Rightarrow$ $b = au^{-1}$, u^{-1} is also a unit in R

$\Rightarrow$ $b \sim a$

Hence ~ is symmetric.

Transitive. If $a \sim b$ $\Rightarrow$ there exists some unit $u \in$ R such that $a = bu$

If $b \sim c$ $\Rightarrow$ there exists some unit $u_1 \in$ R such that $b = u_1c$. We show $a \sim c$.

Now $a = bu = (u_1c)u = c(u_1u)$

$\Rightarrow$ $a \sim c$.

| As u, u_1 are units in R and units form a subgroup. $\therefore$ $u_1u \in$ R

Hence ~ is transitive.

$\therefore$ Relation of being associates is an equivalence relation.

MULTIPLE CHOICE QUESTIONS (MCQs)

1. Consider the ring $Z_{10} = \{0, 1, 2, ... 9\}$ of integers modulo 10. The 3^{-1} is
   (*a*) 3 (*b*) 7
   (*c*) 6 (*d*) none.
2. Consider the ring $Z_8 = \{0, 1, 2, 3, ... 7\}$ of integers modulo 8 and $f(t) = t^2 + 6t$ be a polynomial in Z_8. Then the number of roots of $f(t)$ is
   (*a*) 2 (*b*) 1
   (*c*) 4 (*d*) none.
3. Consider the ring $Z_{12} = \{0, 1, 2, ... 11\}$ of integers modulo 12. The associates of 2 are
   (*a*) 2, 10 (*b*) 2, 2
   (*c*) 2, 4 (*d*) none.
4. Consider the ring $Z_{10} = \{0, 1, 2, ... 9\}$ of integers modulo 10. The units of Z_{10} are
   (*a*) 1, 3, 7, 9 (*b*) 3, 7, 9
   (*c*) 3. 1 (*d*) 1.
5. Consider the ring $Z_7 = \{0,1, 2, ... 6\}$ of integers modulo 7, which of the following is true.
   (*a*) Z_7 is an integer domain (*b*) Z_7 is a field
   (*c*) Z_7 has no zero divisors (*d*) all of the above.
6. If F is a field, then the number of proper ideals of F is
   (*a*) 2 (*b*) 1
   (*c*) 0 (*d*) none.
7. If I and J are two ideals of a ring R. Which is false ?
   (*a*) I $\cap$ J is an ideal of R (*b*) I $\cup$ J is an ideal of R
   (*c*) I $\cup$ J may or may not be an ideal of R (*d*) none.
8. Which of the following is true ? The ring of integers $(Z, +, .)$.
   (*a*) is an integral domain (*b*) is a field
   (*c*) is a division ring (*d*) none.

9. Which of the following statement is not true ?
   (*a*) Every field is an integral domain
   (*b*) Any finite non-zero integral domain is a field
   (*c*) Every ideal of a ring R is a subring of R
   (*d*) All are true.

10. Which of the following is a false statement ?
   (*a*) The ring of integers is an integral domain
   (*b*) The ring of integers is a Euclidean domain
   (*c*) The ring of integers is a P.I.D.
   (*d*) None.

Answers and Explanations

1. (*b*) By a^{-1} in a ring R, we mean that element such that $a \cdot a^{-1} = 1 = a^{-1}a$

   Here $3.7 = 1 = 7.3 \quad \therefore \quad 3^{-1} = 7$

2. (*c*)

$$f(t) = t^2 + 6t$$
$$f(0) = 0, f(2) = 4 + 12 = 16 = 0$$
$$f(4) = 16 + 24 = 40 = 0$$
$$f(6) = 36 + 36 = 72 = 0$$

   Hence the roots of $f(t)$ are 0, 2, 4, 6

3. (*d*) The units of Z_{12} are those numbers which are relative prime to 12.

   $\therefore$ The units of Z_{12} are

$$1, 5, 7, 11$$

   Multiplying each of the units by 2, we get

$$1 \cdot 2 = 2, \quad 5 \cdot 2 = 10 = 2, \quad 7 \cdot 2 = 14 = 2$$
$$11 \cdot 2 = 22 = 10 = 2$$

   $\therefore$ The associates of 2 are 2, 7

4. (*a*) The units of Z_{10} are those elements which are relative prime to 10. Therefore, the units of Z_{10} are 1, 3, 7, 9

5. (*d*) Z_p is a field iff p is a prime number. Also every field is an integral domain.

6. (*d*) A field has no proper ideals.

7. (*b*) **8.** (*a*) **9.** (*d*) **10.** (*d*).

10 BOOLEAN ALGEBRA

10.1. PARTIALLY ORDERED RELATION

Consider a relation R on a set S satisfying the following properties :

1. R is reflexive *i.e.*, $x\mathrm{R}x$ for every $x \in \mathrm{S}$.
2. R is antisymmetric *i.e.*, if $x\mathrm{R}y$ and $y\mathrm{R}x$, then $x = y$.
3. R is transitive *i.e.*, if $x\mathrm{R}y$ and $y\mathrm{R}z$, then $x\mathrm{R}z$.

Then R is called a partially order relation and the set S together with partial order relation R is called a *partially order set* or simply, an *ordered set* or POSET and is denoted by (S, ≤).

For example :

1. The set N of natural numbers form a poset under the relation '≤' because firstly $x \le x$, secondly, if $x \le y$ and $y \le x$, then we have $x = y$ and lastly if $x \le y$ and $y \le z$, it implies $x \le z$ for all $x, y, z \in \mathrm{N}$.

2. The set N of natural numbers under divisibility *i.e.*, 'x divides y' forms a poset because x/x for every $x \in \mathrm{N}$. Also if x/y and y/x, we have $x = y$. Again if x/y, y/z we have x/z, for every $x,y,z \in \mathrm{N}$.

3. Consider a set S = {1, 2} and power set of S is P(S). The relation of set inclusion ⊆ is a partial order relation. Since, for any sets A, B, C in P(S), firstly we have A ⊆ A, secondly, if A ⊆ B and B ⊆ A, then we have A = B. Lastly, if A ⊆ B and B ⊆ C, then A ⊆ C. Hence, (P(S), ⊆) is a poset.

ILLUSTRATIVE EXAMPLES

Example 1. *Consider the set Z of integers. Define aRb by $b = a^r$, for some positive integer r. Show that R is a partial order relation on Z*

Sol. (*i*) Since $a = a^1$ $\therefore$ $a\mathrm{R}a$ *i.e.*, R is reflexive

(*ii*) Let $a\mathrm{R}b$ and $b\mathrm{R}a$. Then there exist positive integers r and s such that

$$b = a^r \text{ and } a = b^s$$

Now $$a = b^s = (a^r)^s = a^{rs}$$

Consider the following possibilities

(*a*) If $rs = 1$, then $r = 1$, $s = 1$ and hence $a = b$. Therefore, R is antisymmetric

(*b*) If $a = 1$, then $b = 1^r = 1 = a$. Therefore, R is antisymmetric (*c*) If $b = 1$, then $a = 1^s = 1 = a$

∴ R is antisymmetric.

(*d*) If $a = -1$, then $b = (-1)^r = -1$ (since $b \neq 1$) and hence $a = b$. Therefore, R is antisymmetric.

Hence, in all the cases, R is antisymmetric.

(*iii*) Let $a\text{R}b$, $b\text{R}c$, then there exist positive integers r and s such that

$$b = a^r,\ c = b^s$$

∴ $c = b^s = (a^r)^s = a^{rs}$, for some positive integer rs. Hence $a\text{R}c$

∴ R is transitive also.

Therefore, R is a partial order relation.

10.2. COMPARABLE ELEMENTS

Consider an ordered set A. Two elements a and b of set A are called comparable if $a \leq b$ or $b \leq a$ where $\leq$ is a partial order relation.

10.3. NON-COMPARABLE ELEMENTS

Consider an ordered set A. Two elements a and b of set A are called non-comparable if neither $a \leq b$ nor $b \leq a$.

10.4. LINEARLY ORDERED SET OR TOTALLY ORDERED SET

Consider an ordered set A. The set A is called linearly ordered set or **totally ordered set,** if every pair of elements in A are comparable.

For example : The set of positive integers I_+ with the usual order $\leq$ is a linearly ordered set.

Example 2. *Let N = {1, 2, 3, 4, ...} is ordered by divisibility. State whether each of the following subsets of N are linearly (or totally ordered).*

(*a*) *{24, 2, 6}* (*b*) *{3, 15, 5}* (*c*) *{1, 2, 3, ...}*

(*d*) *{5}* (*e*) *{2, 8, 32, 16}.*

Sol. (*a*) Consider the set S = {24, 2, 6}

Since 2/6, 6/24 ∴ S is linearly ordered.

(*b*) 3 does not divide 5. Therefore, the set {3, 15, 5} is not linearly ordered.

(*c*) Let S = {1, 2, 3, ...}

As 2 does not divide 3 and 4 does not divide 7 ∴ This set is not linearly ordered.

(*d*) Any set consisting of one element is linearly ordered.

(*e*) As 2 divides 8, 8 divides 16, 16 divides 32,

∴ This set is linearly ordered set.

Example 3. *Consider N = {1, 2, 3, 4 ...} with the relation "less than or equal to". Find all linearly ordered subsets of N.*

Sol. We know that the set N is a linearly ordered set under '≤'. Therefore, every subset of N is linearly ordered.

Example 4. *Let A = {2, 3, 6, 8, 9, 18} is ordered by divisibility.*

(a) Find the non-comparable pairs of elements of A.

(b) Find the linearly ordered subsets of A with three or more elements.

Sol. *(a)* The non-comparable pairs are

{2, 3}, {2, 9}, {3, 8}, {6, 8}, {6, 9}, {8, 9}, {8, 18}

(b) We know that a set is called linearly ordered if every pair of its elements are comparable. The linearly ordered set of A with three or more elements are {2, 6, 18}, {3, 9, 18}. Note that the set {2, 6, 8} is not linearly ordered since the elements 6 and 8 are not comparable.

Example 5. *Consider the set Z of integers under the partial order relation defined by $aRb \iff b = a^r$, for some +ve integers. Examine which of the following subsets of Z are linearly ordered*

(i) {2, 4, 64} *(ii) {3, 9, 18}* *(iii) {8}*

(iv) {3, 27, 729}

Sol. *(i)* As $4 = 2^2$, $64 = 2^6$ $\therefore$ The set {2, 4, 64} is linearly ordered.

(ii) $9 = 3^2$, $18 \neq 3^r$, for some positive value r.

$\therefore$ The set {3, 9, 18} is not linearly ordered set.

(iii) Any set with one element is always linearly ordered. Hence the set {8} is linearly ordered.

(iv) $27 = 3^3$, $729 = 3^6$

$\therefore$ {3, 27, 729} is also linearly ordered set.

10.5. HASSE DIAGRAMS

It is a useful tool, which completely describes the associated partial order. Therefore, it is also called an **ordering diagram**. It is very easy to convert a **directed graph** of a relation on a set A to an equivalent **Hasse diagram**. Therefore, while drawing a **Hasse diagram,** the following points must be remembered.

1. The vertices in **Hasse diagram** are denoted by points rather than by circles.

2. Since a partial order relation is reflexive, hence each vertex of A must be related to itself, so the edges from a vertex to itself are deleted in **Hasse diagram**.

3. Since a partial order relation is transitive, hence whenever aRb, bRc, we have aRc. Eliminate all edges that are implied by the transitive property in Hasse diagram *i.e.,* Delete edge from a to c but retain the other two edges.

4. If a vertex 'a' is connected to vertex 'b' by an edge *i.e., a*R*b*, then vertex 'b' appears above vertex 'a'. Therefore, the arrows may be ommitted from the edges in **Hasse diagram**.

The **Hasse diagram** is much simpler than the **directed graph** of the partial order relation.

Example 6. *Describe the diagram of a partially ordered set.*

Sol. The diagram of a partially ordered set S is the directed graph whose vertices are the elements of S and there is an edge from a to b whenever aRb is S. (Instead of drawing an arrow from a to b, we can place b higher than a and draw a line between them.)

Example 7. *Let A = {1, 2, 3, 4, 6, 8, 9, 12, 18, 24} be ordered by divisibility. Draw the Hasse diagram of A.*

Sol. The relation "x divides y" is given as:

R = {(1, 1), (1, 2), (1, 3), (1, 4), (1, 6), (1, 8), (1, 9), (1, 12), (1, 18), (1, 24), (2, 2), (2, 4), (2, 6), (2, 8), (2, 12), (2, 18), (2, 24), (3, 3), (3, 6), (3, 9), (3,12), (3, 18), (3, 24), (4, 4), (4, 8), (4, 12), (4, 18), (4, 24), (6, 6), (6, 12), (6, 18), (6, 24), (8, 8), (8, 24), (9, 9), (9, 18), (12, 12), (12, 24), (18, 18), (18, 24), (24, 24)}

To draw the Hasse diagram, write down the elements 1, 2, 3, 4, 6, 8, 9, 12, 18, 24. Draw an arrow from the integer x to the integer y if x divides y.

The Hasse diagram of A is shown below.

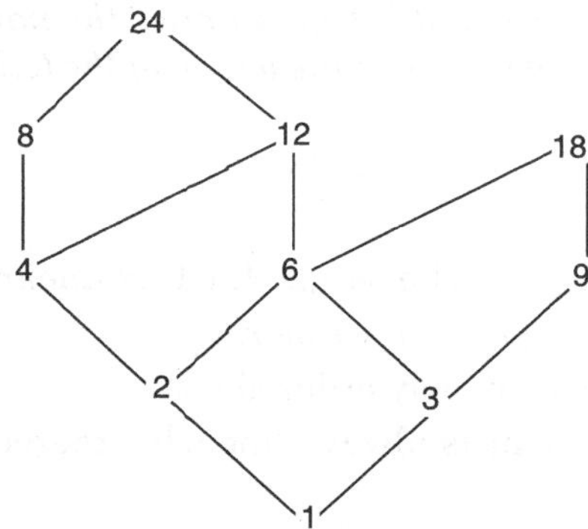

Example 8. *Consider a set S = {a, b, c}. Is the relation of set inclusion '⊆' is a partial order relation on P(S) where P(S) is a power set of S ?*

Sol. The power set of S is

$$P(S) = \{\{a\}, \{b\}, \{c\}, \{a, b\}, \{a, c\}, \{b, c\}, \{a, b, c\}, \phi\}$$

Now consider any sets A, B and C in P(S).

1. Since every A ⊆ A, hence it is reflexive.
2. If A ⊆ B and B ⊆ A, we have A = B. Hence it is antisymmetric.
3. If A ⊆ B, B ⊆ C, we have A ⊆ C. Hence it is transitive.

∴ (P(S), ⊆) is a poset.

Example 9. *Consider a set A = {4, 9, 16, 36}. Is the relation 'divides' a partial order relation.*

Sol. The relation 'divides' is a partial order relation if it satisfies the property of reflexivity, antisymmetry and transitivity.

1. Since for every $a \in$ A, we have a/a. Hence, 'divides' is reflexive.
2. If a/b and b/a, we have $a = b$ for any a, $b \in$ A. Hence, 'divides' is antisymmetric.
3. If a/b and b/c, we have a/c for any a, b, $c \in$ A. Hence, the relation 'divides' is a partial order relation and (A, /) is a poset.

Example 10. *Consider A = {1, 2, 3, 5, 6, 10, 15, 30} is ordered by divisibility. Determine all the comparable and non-comparable pairs of elements of A.*

Sol. The comparable pairs of elements of A are :

{1, 2}, {1, 3}, {1, 5}, {1, 6}, {1, 10}, {1, 15}, {1, 30}
{2, 6}, {2, 10}, {2, 30}
{3, 6}, {3, 15}, {3, 30}
{5, 10}, {5, 15}, {5, 30}
{6, 30}, {10, 30}, {15, 30}.

The non-comparable pair of elements of A are :

{2, 3}, {2, 5}, {2, 15}

{3, 5}, {3, 10}, {5, 6}, {6, 10}, {6, 15}, {10, 15}.

Example 11. *Consider the set I = {1, 2, 3,......} is ordered by divisibility. Determine whether each of the following subsets of I are linearly ordered or not.*

(i) {2, 4, 8} *(ii) {3, 6, 9, 11}*

(iii) {1} *(iv) {2, 4, 6, 8, 10, ...}.*

Sol. *(i)* The subset is linearly ordered, since every pair of elements is comparable *i.e.,* 2 | 4 | 8.

(ii) The subset is not linearly ordered, since the pair (3, 11) is not comparable.

(iii) The subset is linearly ordered, since the set containing one element is always linearly ordered.

(iv) The subset is not linearly ordered since every pair of elements is not comparable *i.e.,* neither 4/6 nor 6/4.

Example 12. *Consider the set A = {4, 5, 6, 7}. Let R be the relation ≤ on A. Draw the directed graph and the Hasse diagram of R.*

Sol. The relation ≤ on the set A is given by

R = {{4, 5}, {4, 6}, {4, 7}, {5, 6}, {5, 7}, {6, 7}, {4, 4}, {5, 5}, {6, 6}, {7, 7}}

The directed graph of the relation R is as shown in Fig. 10.1.

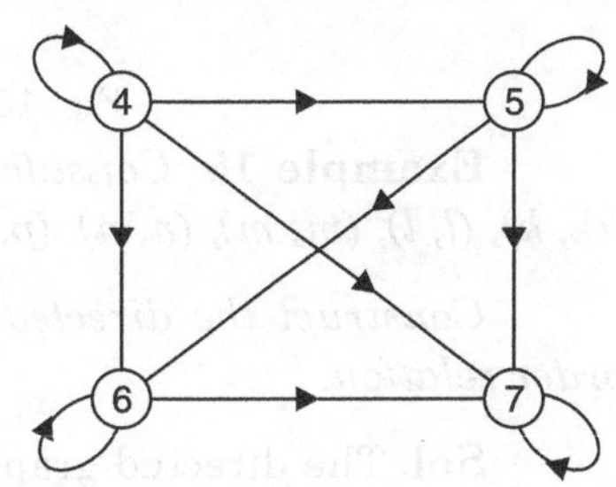

Fig. 10.1

To draw the **Hasse diagram** of partial order, apply the following points :

1. Delete all edges implied by reflexive property *i.e.,* (4, 4), (5, 5), (6, 6), (7, 7).
2. Delete all edges implied by transitive property *i.e.,* (4, 7), (5, 7) and (4, 6).
3. Replace the circles representing the vertices by dots.
4. Omit the arrows.

The **Hasse diagram** is as shown in Fig. 10.2.

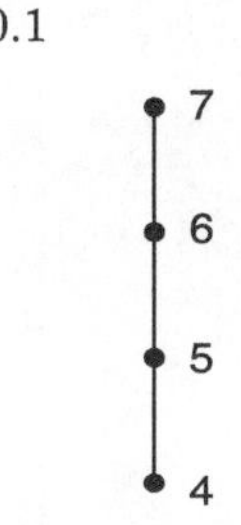

Fig. 10.2

Example 13. *Draw the directed graph of relation determined by the* **Hasse diagram** *on the set A = {1, 4, 6, 8} as shown in Fig. 10.3.*

Sol. The directed graph is shown in Fig. 10.4.

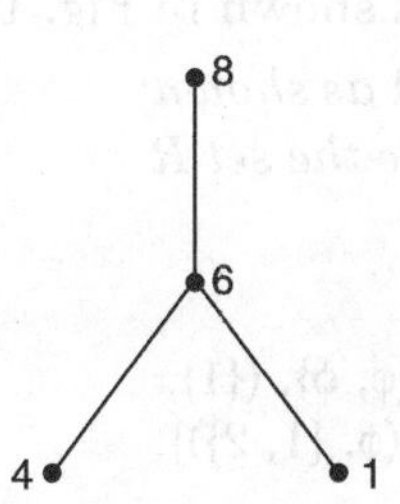

Fig. 10.3

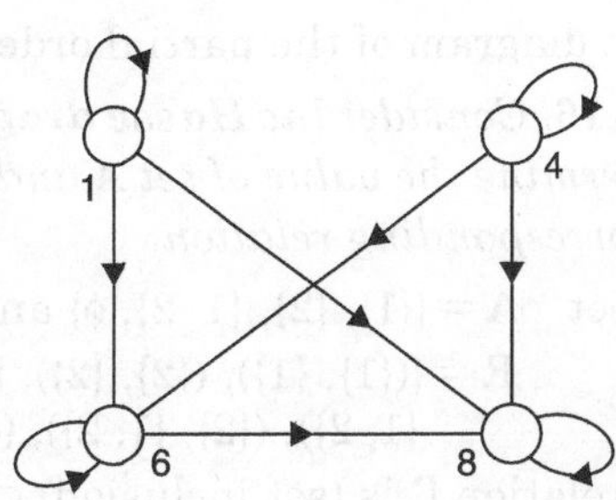

Fig. 10.4

Example 14. *Determine the **Hasse diagram** of the partial order relation having the directed graph as shown in Fig. 10.5.*

Sol. The Hasse diagram of the given partial order relation determined by the directed graph is as shown in Fig. 10.6.

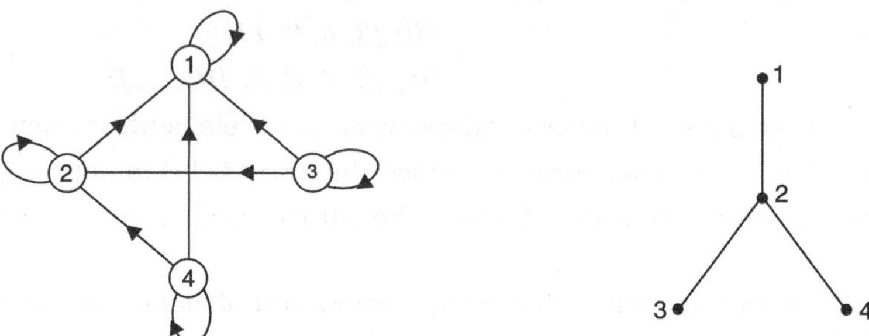

Fig. 10.5 Fig. 10.6

Example 15. *Consider the set A = {k, l, m, n, p} and the corresponding relation R = {(k, k), (l, l), (m, m), (n, n), (p, p), (k, m), (k, l), (k, n), (k, p), (m, n), (m, p), (n, p), (l, p)}.*

*Construct the directed graph and the corresponding **Hasse diagram** of this partial order relation.*

Sol. The directed graph of the partial order relation is as shown in Fig. 10.7.

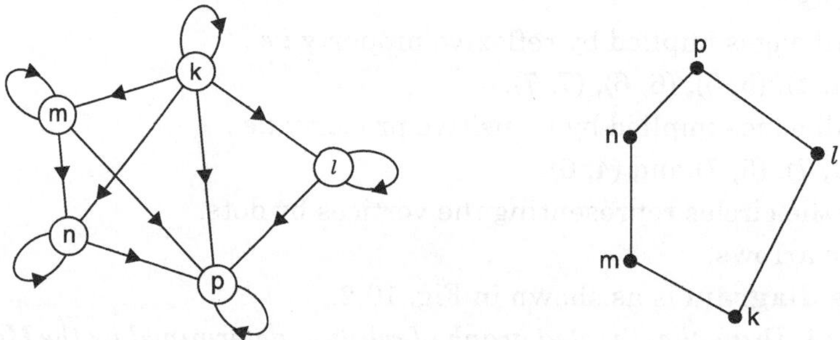

Fig. 10.7 Fig. 10.8

The Hasse diagram of the partial order relation is as shown in Fig. 10.8.

Example 16. *Consider the **Hasse diagram** on a set A as shown in Fig. 10.9. Determine the value of set A and also determine the set R where R is the corresponding relation.*

Sol. The set A = {{1}, {2}, {1, 2}, ϕ} and

R = {({1}, {1}), ({2}, {2}), ({1, 2}, {1, 2}), {ϕ, ϕ}, ({1}, {1, 2}), ({2}, {1, 2}), (ϕ, {1}), (ϕ, {2}), (ϕ, {1, 2})}.

Here the relation R is (set inclusion) $\subseteq$.

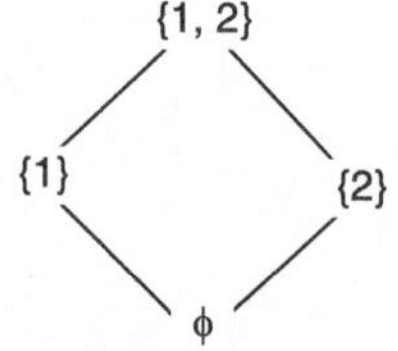

Fig. 10.9

Example 17. *Let A = {1, 2, 3, 4}, and let r be the relation ≤ on A. Draw the digraph and the Hasse diagram of r.*

Sol. The elements of the relation ≤ on A which means first element is less than or equal to the second element is as follows.

$$r = \{(1, 1), (2, 2), (3, 3), (4, 4), (1, 2), (1, 3), (1, 4), (2, 3), (2, 4), (3, 4)\}$$

Now, draw the directed graph of the relation *r* as shown in Fig. 10.10.

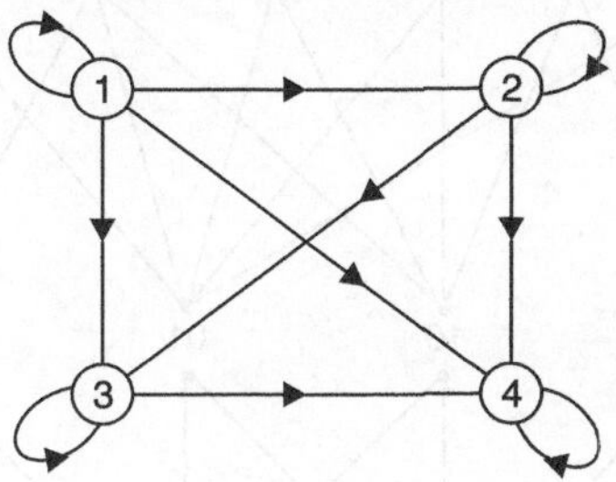

Fig. 10.10

After, drawing the directed graph, convert it into the Hasse diagram. The Hasse diagram is shown in Fig. 10.11.

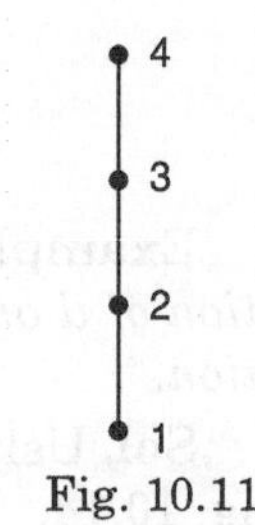

Fig. 10.11

Example 18. *Let B = {2, 3, 4, 6, 12, 36, 48}, and let s be the relation, "divides" on B. Draw the Hasse diagram of S.*

Sol. The Hasse diagram of the relation S on set B is shown in Fig. 10.12.

The easy way to draw the Hasse diagram is to find the elements of relation *r*, then draw directed graph of it and after that convert it into Hasse diagram.

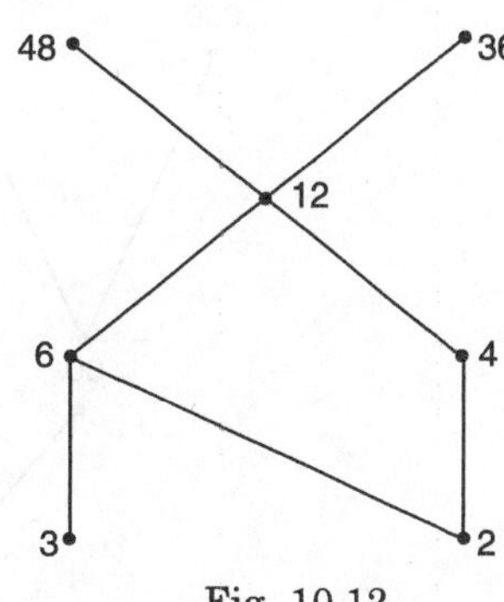

Fig. 10.12

Example 19. *Let A be the set of strings of 0's and 1's of length 3 or less. Define the relation of d on A by xdy if x is contained within y. e.g. (01)d(101). Draw the Hasse diagram for this relation.*

Sol. Firstly find all the elements of the set A as given below.

A = {0, 1, 00, 01, 10, 11, 000, 001, 010, 011, 100, 101, 110, 111}.

Now to draw the Hasse diagram, write all the elements having length 3 in one row at the top as these elements are not contained in any other element. Then in the second row, write all the elements having length 2 and connect them with elements having length 3 and are contained in them. Repeat the above process for elements having length 1.

The Hasse diagram of the relation d is shown in Fig. 10.13.

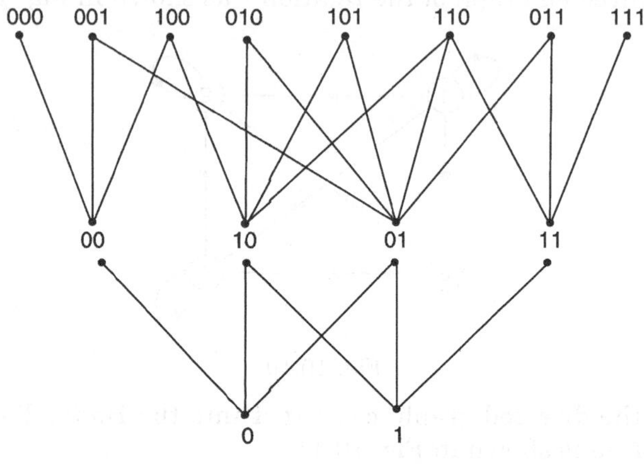

Fig. 10.13

Example 20. *Let A be the set of strings of 0's and 1's of length 3 or less. Define the relation of d on A by xdy if x is a prefix of y e.g. (10)d(101). Draw the Hasse diagram for this relation.*

Sol. Using the procedure followed in the previous example, the Hasse diagram is shown in Fig. 10.14.

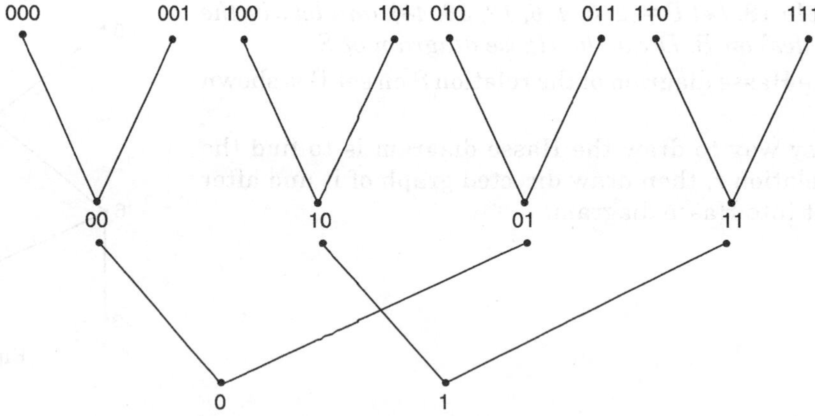

Fig. 10.14

TEST YOUR KNOWLEDGE 10.1

1. Let L = {1, 3, 5, 7, 15, 21, 35, 105} and let ≤ be a relation/(divides) on L. Then (L, ≤) is a poset.
2. Let M = P(A), the power set of A where A = {*a*, *b*, *c*} and ⊆ be a relation on M. Show that (M, ⊆) is a poset. Also draw its Hasse diagram.
3. Let D_{30} = {1, 2, 3, 5, 6, 10, 15, 30} and let the relation/(divides) be a partial order relation on D_{30}. Draw the Hasse Diagram of D_{30}.
4. List the elements of the sets D_8, D_{50}, D_{1001}
5. If a poset L has a least element, then this element is unique.
6. Draw the Hasse Diagrams of

   (*a*) D_{12} (*b*) D_{15} (*c*) D_{16} (*d*) D_{17}

Answers

2. See Fig. 10.15

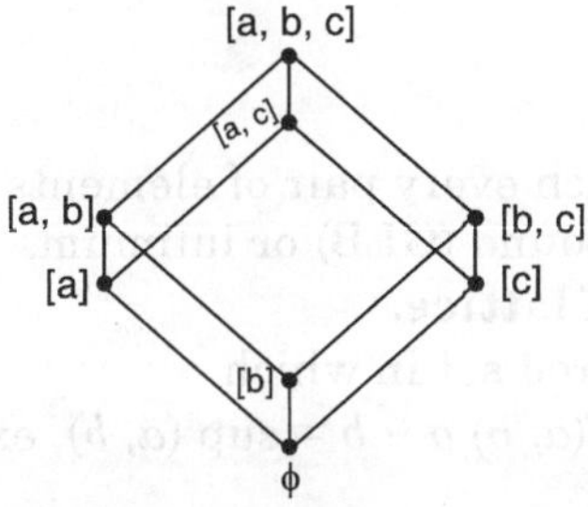

Fig. 10.15

3. See Fig. 10.16

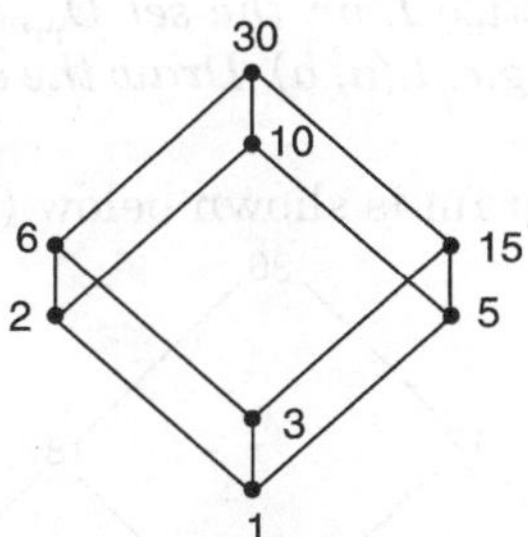

Fig. 10.16

4. D_8 = {1, 2, 4, 8};
   D_{50} = {1, 2, 5, 10, 25, 50};
   D_{1001} = {1, 7, 11, 13, 77, 91, 143, 1001}

6.

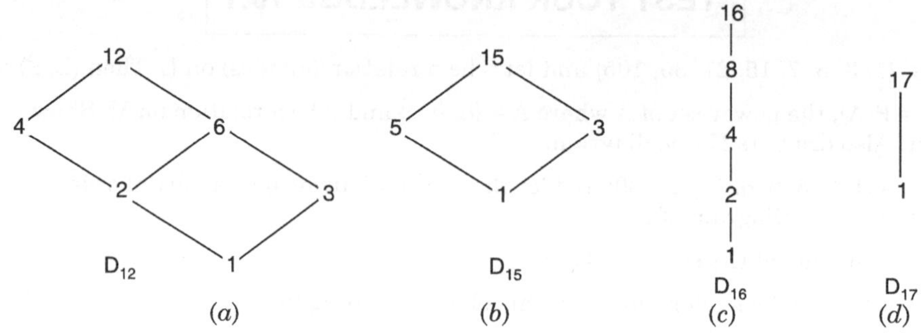

Hint

5. If a and b are distinct least elements, then

$a \leq b$ since a is least element

Also $b \leq a$ since b is least element

$\Rightarrow$ $a = b$ due to antisymmetry, a contradiction

Hence **a = b.**

10.6. LATTICE

A lattice L is a poset in which every pair of elements has a least upper bound (LUB) or supremum and a greatest lower bound (GLB) or infimum.

Alternative definition of lattice.

A lattice is a partially ordered set in which

$$a \wedge b = \text{Inf}(a, b) \; a \vee b = \sup (a, b), \text{ exists}$$

for any pairs of elements a and b.

ILLUSTRATIVE EXAMPLES

Example 1. *Consider the lattice L on the set D_m, the divisors of a positive integer m, where $a \vee b = l.c.m.(a, b)$ and $a \wedge b = g.c.d.(a, b)$. Draw the diagram of the partial order induced by L for m = 36.*

Sol. The required Hasse diagram is shown below (Fig. 10.17).

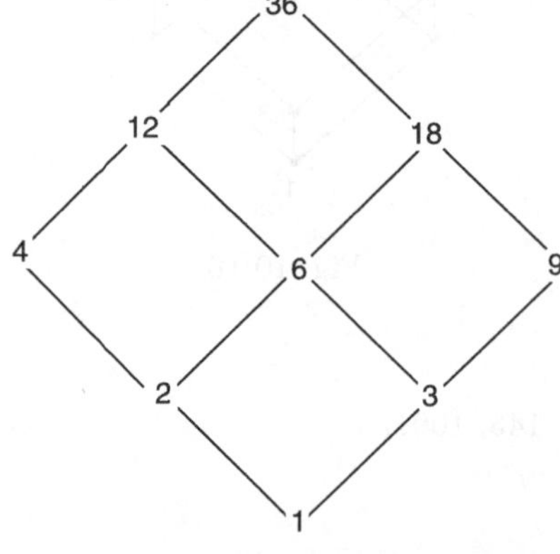

Fig. 10.17

Example 2. *Let P(S) be the power set of the set S = {1, 2, 3}. Construct the Hasse diagram of the partial order relation induced on P(S) by the lattice (P(S), ∧, ∨).*

Sol. The Hasse diagram obtained by a lattice is same as obtained under the partial ordering of set inclusion. In the lattice, $a \leq b$ whenever $a \wedge b = a$. Thus, in the above case $a \leq b$, whenever $a \wedge b = a$.

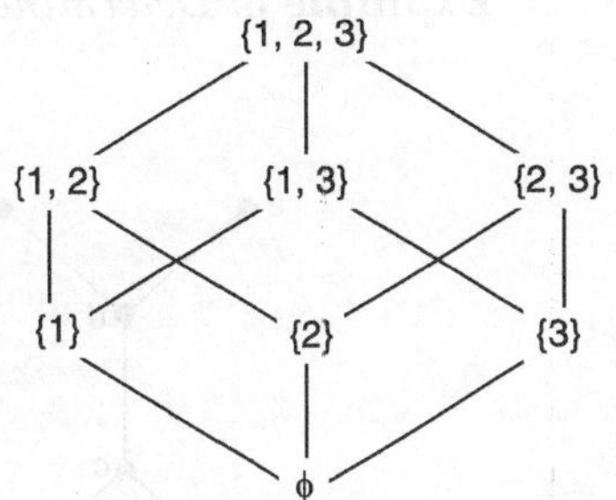

Fig. 10.18

The required Hasse diagram is shown in Fig. 10.36.

Example 3. *Determine which of the posets shown in Fig. 10.19 are lattices.*

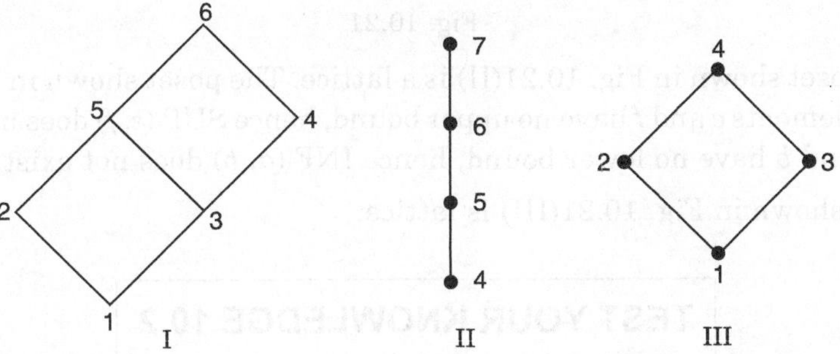

Fig. 10.19

Sol. All the posets shown in Fig. 10.19 are lattices.

Example 4. *Determine whether the posets shown in Fig. 10.20 are lattices or not.*

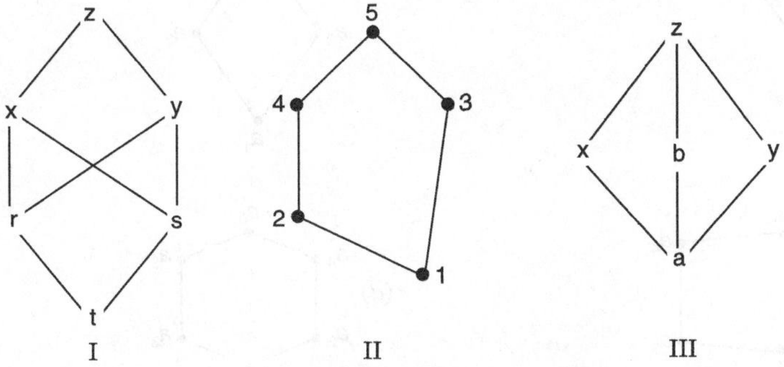

Fig. 10.20

Sol. The posets shown in Figs. 10.20(II) and 10.20(III) are lattices. The poset shown in Fig. 10.20(I) is not lattice as (x, y) has three lower bounds *i.e.*, r, s and t but the Inf (x, y) does not exist. Also, Sup (r, s) does not exist.

Example 5. *Determine whether the posets shown in Fig. 10.21 are lattices or not.*

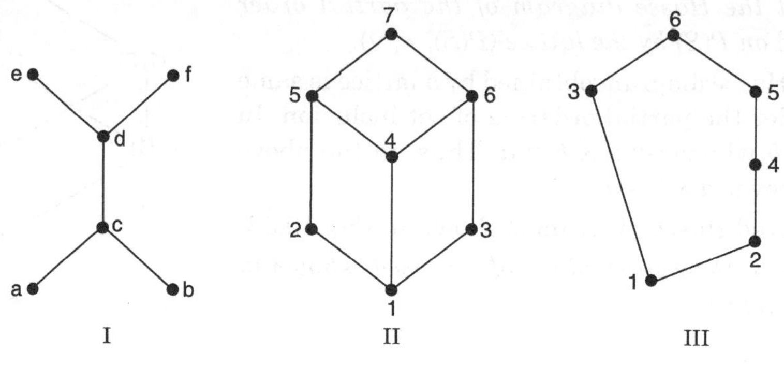

Fig. 10.21

Sol. The poset shown in Fig. 10.21(II) is a lattice. The poset shown in Fig. 10.21(I) is not lattice since the elements *e* and *f* have no upper bound, hence SUP (*e*, *f*) does not exist. Similarly, the elements *a* and *b* have no lower bound, hence INF (*a*, *b*) does not exist.

The poset shown in Fig. 10.21(III) is lattice.

TEST YOUR KNOWLEDGE 10.2

1. Consider the Hasse Diagrams of posets in the given figures:

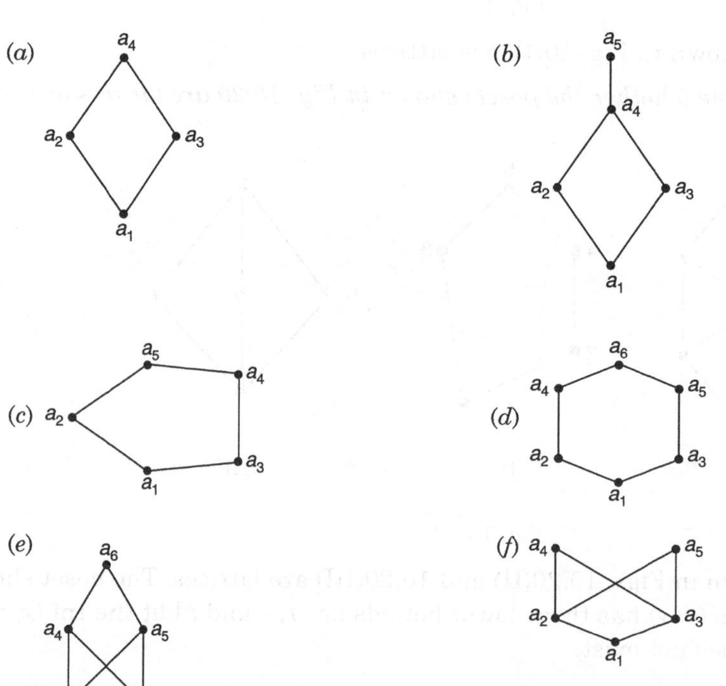

(g)

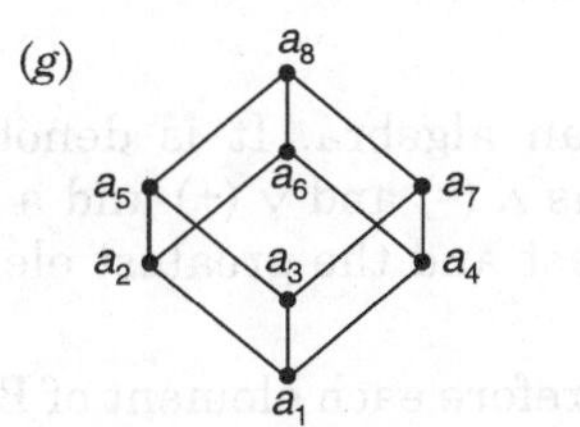

(h)

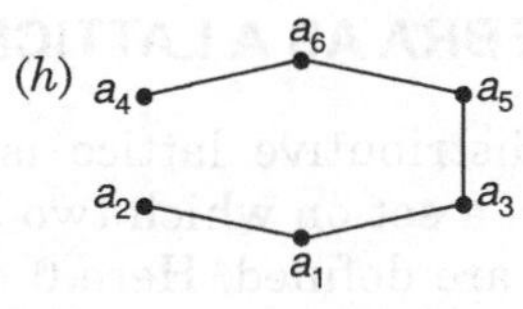

(i) Which of the posets are lattices ?

Answer

1. (i) (a), (b), (c), (d) and (h) are lattices,

10.7. BOOLEAN ALGEBRA

George Boolean an English mathematician, developed an algebra, which was useful in designing logic circuits in the computer systems. This algebra is now known as *Boolean algebra.* Boolean algebra provides us a straight forward approach in designing and analysing of switching circuits.

10.8. UNARY OPERATION

Let S be a non-empty set. Then a function from S into S is called a unary operation on S. A unary operation may be denoted by the symbols f, g, ϕ, ψ etc.

Thus, if f is a unary operation on a set S, then $f(a)$ is unique element of S for all $a \in$ S.

For example. Let N be the set of natural numbers, and f be a function from N to N defined by $f(n) = n^2$ for all $n \in$ N. Then f is a unary operation on N. Also, for this unary operation, $f(3) = 9$, $f(8) = 64$ etc.

10.9. BINARY OPERATION

Let S be a non-empty set. A function from S × S into S is called a **binary operation.** if $*$ is a binary operation on S, it means $*$ is a function from S × S into S.

If $(a, b) \in$ S × S, then the image $*\ (a, b)$ of (a, b) under the binary operation '$*$' is written as $a * b$. Thus, in order to check, whether '$*$' is a binary operation on S, we should check.

(i) $a * b \in$ S for all $a, b \in$ S

(ii) $a * b$ is unique.

For example, ordinary addition '+' is a binary operation on Z, since, '+' is a function from Z × Z into Z. Thus, if 4, 7, ∈ Z, then + (4, 7) = 4 + 7 = 11 ∈ Z.

On the other hand, ordinary subtraction '–' is not a binary operation on N, since difference of two natural numbers may not be a natural number. For example, if 4, 7 ∈ N, then

$$(4, 7) = 4 - 7 = -3 \notin \text{N}.$$

Ordinary multiplication is a binary operation on N, Z, Q, R, C.

10.10. BOOLEAN ALGEBRA AS A LATTICE

A complemented distributive lattice is called a Boolean algebra. It is denoted by (B, $\wedge$, $\vee$,′, 0, 1), where B is a set on which two binary operations $\wedge$ (∗) and $\vee$ (+) and a unary operation ′ (complement) are defined. Here 0 and 1 are the least and the greatest elements of B respectively.

Since (B, $\wedge$, $\vee$) is a complemented distributive lattice, therefore each element of B has a unique complement.

10.11. ALTERNATE DEFINITION OF BOOLEAN ALGEBRA *(P.T.U. B.Tech Dec. 2013)*

Consider a set B on which two binary operations ∗ and + and a uniary operation′ (complement) are defined. Also let 0 and 1 are two distinct elements of B. Then it is called a Boolean algebra, if the following laws are satisfied for any elements a, b and c of the set B by it.

1. Commutative Laws

(*i*) $a + b = b + a$

(*ii*) $a * b = b * a$

2. Distributive Laws

(*i*) $a + (b * c) = (a + b) * (a + c)$

(*ii*) $a * (b + c) = (a * b) + (a * c)$

3. Identity Laws

(*i*) $a + 0 = a$

(*ii*) $a * 1 = a$

4. Complement Laws

(*i*) $a + a' = 1$

(*ii*) $a * a' = 0$

The Boolean algebra is denoted by (B, +, ∗, ′, 0, 1). The elements 0 and 1 are called *zero element* and *unit element* of B.

Note 1. If a finite lattice L does not contain 2^n element for some positive integer n, then L cannot be a Boolean algebra.

2. If $|L| = 2^n$, then L may or may not be a Boolean algebra.

Remark. The binary operations '+' and '.' used above have got nothing to do with ordinary addition and multiplication of numbers. In fact, +, *, . are just symbols to represent Binary operations. Similarly '0' and '1' are two specific elements of B satisfying certain properties. These two numbers have no bearing with the numbers 0 and 1 of the set of integers.

ILLUSTRATIVE EXAMPLES

Example 1. *Let S be a non-empty set and B denotes the set of all subsets of S. Show that {B, ∪, ∩, ′} is a Boolean algebra.* (P.T.U., B.Tech. May, 2013)

Sol. Let X, Y, ∈ B, Then X ∪ Y, X ∩ Y ∈ B since X ∪ Y, X ∩ Y are also subsets of S.

∴ ∪ and ∩ are binary operations on B.

If X′ denotes the complement of X, then we write X′ = S – X ∈ B.

We verify the following laws :

Commutative laws. Let X, Y, ∈ B, then

X ∪ Y = Y ∪ X and X ∩ Y = Y ∩ X are always true. Hence commutative laws hold.

Distributive laws. Let X, Y, Z ∈ B, then

$$X \cup (Y \cap Z) = (X \cup Y) \cap (X \cup Z)$$

$$X \cap (Y \cup Z) = (X \cap Y) \cup (X \cap Z),$$

are always true. Hence, distributive laws hold.

Identity laws. Let Q, S, X ∈ B, then $X \cup \phi = X$ and $X \cap S = X$,

are always true. Hence, identity laws hold. Here ϕ is the zero element and S is the unit element of B.

Complement laws. Let X ∈ B, then

$$X \cup X' = X \cup (S - X) = S$$

$$X \cap X' = X \cap (S - X) = \phi,$$

are always true. Hence, complement laws also hold. Hence, {B, ∩, ∪, '} is a Boolean algebra.

Example 2. *Let B = {0, 1}. Define the binary operations '+' and '.' on B as below :*

+	0	1
0	0	1
1	1	1

.	0	1
0	0	0
1	0	1

Let the unary operation ''' is defined by 0′ = 1 and 1′ = 0. Show that {B, +, ., '} is a Boolean algebra.

Sol. From the Table, it is clear that $a + b$ and $a \,.\, b$ are unique elements of B. Therefore, '+' and '.' are binary operations on B. We verify the following :

Commutative laws. Consider the following Tables. I and II.

Table I

a	b	$a + b$	$b + a$
0	0	0	0
1	0	1	1
0	1	1	1
1	1	1	1

Table II

a	b	$a.b$	$b.a$
0	0	0	0
1	0	0	0
0	1	0	0
1	1	1	1

From the above Tables, $a + b = b + a \ \forall\ a, b \in B$

Also $a \,.\, b = b \,.\, a \ \forall\ a, b \in B$

Hence commutative laws hold.

Distributive laws. Consider the following tables. III and IV.

Table III

a	b	c	$b.c$	$a + (b.c)$	$a + b$	$a + c$	$(a + b).(a + c)$
0	0	0	0	0	0	0	0
1	0	0	0	1	1	1	1
0	1	0	0	0	1	0	0
0	0	1	0	0	0	1	0
1	1	0	1	1	1	1	1
1	0	1	0	1	1	1	1
0	1	1	1	1	1	1	1
1	1	1	1	1	1	1	1

From Table III, $a + (b \, . \, c) = (a + b) \, . \, (a + c)$

From Table IV, $a \, . \, (b + c) = a \, . \, b + a \, . \, c \;\forall\; a, b, c \in B$

Table IV

a	b	c	$b + c$	$a.(b + c)$	$a.b$	$a.c$	$(a.b) + (a.c)$
0	0	0	0	0	0	0	0
1	0	0	0	0	0	0	0
0	1	0	1	0	0	0	0
0	0	1	1	0	0	0	0
1	1	0	1	1	1	0	1
1	0	1	1	1	0	1	1
0	1	1	1	0	0	0	0
1	1	1	1	1	1	1	1

∴ Distributive laws hold.

Identity laws. Consider the following Tables V and VI.

Table V

a	0	$a + 0$
0	0	0
1	0	1

Table VI

a	1	$a.1$
0	1	0
1	1	1

From Table V, $a + 0 = a \;\forall\; a \in B$

From Table VI, $a.1 = a \;\forall\; a \in B$

∴ Identity laws hold. Here '0' is the zero element and '1' is the unit element.

Complement laws. Consider the following Tables VII and VIII.

Table VII

a	a'	$a + a'$
0	1	1
1	0	1

Table VIII

a	a'	$a.a'$
0	1	0
1	0	0

From Table VII, $a + a' = 1 \;\forall\; a \in B$

From Table VIII, $a.a' = 0 \;\forall\; a \in B$

∴ {B, +, ., '} is a Boolean algebra.

Example 3. *Let $B = \{1, 2, 3, 6\}$ be the set of positive factors of 6. Let the binary operations '+' and '.' on B are defined as follows.*

$a + b = l.c.m\,(a, b)$ and $a.b = g.c.d\,(a, b) \;\forall\; a, b \in B$. Let the unary operation "'" on B is defined by $a' = \dfrac{6}{a} \;\forall\; a \in B$. Show that $\{B, +, ., '\}$ is a Boolean algebra.

Sol. Given B = {1, 2, 3, 6} and the binary operations are '+' and '.'. The composition tables for the binary operations are given as :

Table I

+	*1*	*2*	*3*	*6*
1	1	2	3	6
2	2	2	6	6
3	3	6	3	6
6	6	6	6	6

Table II

.	*1*	*2*	*3*	*6*
1	1	1	1	1
2	1	2	1	2
3	1	1	3	3
6	1	2	3	6

Each element inside the Table I and II are also element of B = {1, 2, 3, 6}. Therefore, '+' and '.' are binary operations on B. We verify the following :

Commutative laws. From Table I, it is clear that table I is symmetrical. Also Table II is symmetrical. Therefore,

$$a + b = b + a \quad \text{and} \quad a.b = b.a \ \forall \ a, b \in \text{B}$$

Identity laws. For $a \in$ B, consider the following tables. III and IV

Table III

a	*1*	$a + 1$
1	1	1
2	1	2
3	1	3
6	1	6

Table IV

a	*6*	$a.6$
1	6	1
2	6	2
3	6	3
6	6	6

From table III, $a + 1 = a \ \forall \ a \in$ B

From table IV, $a.6 = a \ \forall \ a \in$ B

$\therefore$ 1 is the zero element and 6 is the unit element. Hence Identity laws hold.

Table V

a	a'	$a + a'$
1	6	6
2	3	6
3	2	6
6	1	6

Table VI

a	a'	$a.a'$
1	6	1
2	3	1
3	2	1
6	1	1

Complement laws. From table V, $a + a' = 6$ and from table VI $a.a' = 1$

$\therefore$ Complement laws hold

Distributive laws. For $a, b, c \in$ B,

$$a + (b.c) = (a + b) \, . \, (a + c)$$

and $$a \, . \, (b + c) = a.b + a.c \text{ also hold}$$

For example, $a = 3, b = 6, c = 2$.

Then $a + (b.c) = 3 + (6 \,.\, 2) = 3 + 2 = 6$ | $6 \,.\, 2 = 2 = \text{g.c.d}\ (2, 6)$; $3 + 2 = 6 = \text{l.c.m}\ (2, 3)$

Also $(a + b) \,.\, (a + c) = (3 + 6) \,.\, (3 + 2) = 6 \,.\, 6 = 6$

$\therefore \quad a + (b \,.\, c) = (a + b) \,.\, (a + c)$

Further, $a \,.\, (b + c) = 3 \,.\, (6 + 2) = 3 \,.\, 6 = 3$

$a.b + a.c = 3.6 + 3.2 = 3 + 1 = 3$

$\therefore \quad a \,.\, (b + c) = a.b + a.c$

Hence, distributive laws hold

$\therefore$ B is a Boolean algebra

Example 4. *Let D_{70} = {1, 2, 5, 7, 10, 14, 35, 70}, the divisors of 70. Show that D_{70} is a boolean algebra under the binary operations '+', '.' and ''' defined by*

$$\left.\begin{aligned} a + b &= l.c.m \quad (a, b) \\ a \cdot b &= g.c.d \quad (a, b) \end{aligned}\right\} \qquad ...(1)$$

$$a' = \frac{70}{a}, \text{ where 1 is a zero element and 70 is a unit element.}$$

Sol. We verify the following :

I. Commutative laws.

Given $a + b = l.c.m\ (a, b) = l.c.m\ (b, a) = b + a$

Also $a \cdot b = g.c.d.\ (a, b) = g.c.d.\ (b, a) = b \cdot a$

$\therefore$ Commutative laws hold.

II. Identity laws. Consider the following composition tables of the binary operations '+' and '.' defined in (1)

+	1	2	5	7	10	14	35	70
1	1	2	5	7	10	14	35	70
2	2	2	10	14	10	14	70	70
5	5	10	5	35	10	70	35	70
7	7	14	35	7	70	14	35	70
10	10	10	10	70	10	70	70	70
14	14	14	70	14	70	14	70	70
35	35	70	35	35	70	70	35	70
70	70	70	70	70	70	70	70	70

.	1	2	5	7	10	14	35	70
1	1	1	1	1	1	1	1	1
2	1	2	1	1	2	2	1	2
5	1	1	5	1	5	1	5	5
7	1	1	1	7	1	7	7	7
10	1	2	5	1	10	2	5	10
14	1	2	1	7	2	14	7	14
35	1	1	5	7	5	7	35	35
70	1	2	5	7	10	14	35	70

From the above tables, we observe that

$$a + 1 = a \quad \forall \quad a \in D_{70}$$

and $$a.70 = a \;\forall\; a \in D_{70}$$

$\therefore$ Identity laws hold

Also, 1 is the zero element and 70 is the unit element.

III. Distributive laws. For $a, b, c \in D_{70}$,

$$a + (b \,.\, c) = (a + b) \,.\, (a + c)$$

and $$a \,.\, (b + c) = a.b + a.c$$ are true.

For example, Take $a = 2$, $b = 5$, $c = 7$

Consider $a + (b \,.\, c) = 2 + (5 \,.\, 7) = 2 + 1 = 2$

Also $(a + b) \,.\, (a + c) = (2 + 5) \,.\, (2 + 7)$

$= 10 \,.\, 14 = 2$

| $5.7 = 1 = \text{g.c.d.}\ (5, 7)$ |
| $2 + 1 = 2 = \text{l.c.m.}\ (2, 1)$ |

$\therefore$ $a + (b \,.\, c) = (a + b) \,.\, (a + c)$

Also $a \,.\, (b + c) = 2 \,.\, (5 + 7) = 2.35 = 1$

and $a.b + a.c = 2.5 + 2.7 = 1 + 1 = 1$

$\therefore$ $a \,.\, (b + c) = a.b + a.c$

IV. Complement laws. Consider the following table :

a	$a' = \dfrac{70}{a}$	$a + a' = l.c.m.\ (a, a')$	$a.a' = g.c.d.\ (a, a')$
1	70	70	1
2	35	70	1
5	14	70	1
7	10	70	1
10	7	70	1
14	5	70	1
35	2	70	1
70	1	70	1

From the table, we observe that

$$a + a' = 70 \quad \text{and} \quad a.a' = 1 \;\forall\; a \in D_{70}$$

Hence, D_{70} is a Boolean algebra with '1' as zero element and '70' as unit element.

Example 5. *Consider the Boolean algebra D_{70} in above example. Find the value of*

(i) $x = 35 \,.\, (2 + 7')$ *(ii)* $y = (35 \,.\, 10) + 14'$

(iii) $z = (2 + 7) \,.\, (14 \,.\, 10)'$

Sol. We know that D_{70} is a Boolean algebra under the binary operations '+', '.' and ''' defined by (see example 4 above)

$$a + b = l.c.m.\ (a, b)$$

$$a \,.\, b = g.c.d.\ (a, b)$$

$$a' = \frac{70}{a}$$

(*i*) $x = 35 \,.\, (2 + 7') = 35 \,.\, (2 + 10)$ $\quad | \; 7' = \frac{70}{7} = 10$

$= 35.10$ $\quad 2 + 10 = l.c.m. \,(2, 10)$

$= 5$ $\quad 35.10 = g.c.d. \,(35, 10) = 5$

(*ii*) $y = (35 \,.\, 10) + 14'$

$= 5 + 14' = 5 + 5 = 5$ $\quad | \; 14' = \frac{70}{14} = 5$

(*iii*) $z = (2 + 7) \,.\, (14.10)' = 14.2' = 14.35 = 7$ $\quad | \; 14.10 = g.c. \, 1 \,(14, 10) = 2$

Remark. The set B of all positive factors (or divisors) of a positive integer m w.r.t. the operations defined as :

$$a + b = l.c.m. \,(a, b),\ a.b = g.c.d. \,(a, b) \ \forall\ a, b \in \mathrm{B} \quad \text{and} \quad a' = \frac{m}{a}$$

is a Boolean algebra if m can be expressed as a product of distinct primes,

For example, if B = {1, 2, 3, 4, 6, 12}, the set of divisors of 12. The binary operations '+', '.' are defined as usual and ''' is defined by $a' = \frac{12}{a}$. Since $12 = 3.4 \neq$ Product of distinct primes, hence B cannot a Boolean algebra. We can also verify that complement laws do not hold on B.

Let $2 \in \mathrm{B}$, then $2' = \frac{12}{2} = 6$ and $2 + 2' = 2 + 6 = 6 \neq 12$, the unit element.

TEST YOUR KNOWLEDGE 10.3

1. Let B = {1, 5, 7, 35} be the set of all positive factors of 35. Two binary operations '+' and '.' are defined as follows :

$$a + b = l.c.m. \,(a, b),\ a.b = g.c.d. \,(a, b) \ \forall\ b \in \mathrm{B}.$$

A unary operation ''' on B is defined as $a' = \frac{35}{a} \ \forall\ a \in \mathrm{B}$. Show that (B, + , . , ') is a Boolean algebra.

2. Let B be the set of all positive factors of 105. Two binary operations '+' and '.' are defined as follows :

$$a + b = l.c.m. \,(a, b),\ a.b = g.c.d. \,(a, b) \ \forall\ a, b \in \mathrm{B}.$$

A unary operation ''' on B is defined as $a' = \frac{105}{a} \ \forall\ a \in \mathrm{B}$. Show that (B, +, ., ') is a Boolean algebra.

3. Let B = (1, 2, 4) be the set of all positive factors of 4. Two binary operations '+' and '.' are defined as follows :

$a + b = l.c.m.\ (a, b)$, $a.b = g.c.d.\ (a, b)\ \forall\ a, b \in B$. A unary operation ''' on B is defined as

$$a' = \frac{4}{a}\ \forall a \in B.$$ Show that (B, +, ., ') is not a Boolean algebra.

4. Let B = {1, 2, 4, 8} be the set of positive factors of 8. Two binary operations '+' and '.' are defined as follows :

$$a + b = l.c.m.\ (a, b),\ a.b = g.c.d.\ (a, b)\ \forall\ a, b \in B.$$

A unary operation ''' on B is defined as $a' = \frac{8}{a}\ \forall\ a \in B$. Show that (B, +, ., ') is not a Boolean algebra.

5. Let B = {φ, {1}, {2}, {3}, {1, 2}, {1, 3}, {2, 3}, {1, 2, 3}}. Show that (B, ∪, ∩, ') is a Boolean algebra.

6. In the Boolean algebra (B, +, ., '), show that

(i) $a + b + c.a' = a + b + c\ \forall\ a, b, c \in B$

(ii) $a.b + c.\ (a' + b') = a.b + c\ \forall\ a, b, c \in B$.

7. In the Boolean algebra (B, +, ., '), simplify : $x.(x + y) + ((y' + x).y)'$.

8. If a, b, c are elements of the Boolean algebra (B, +, ., '), then find the complements of the following expressions :

(i) $a + b'$ (ii) $a''.b'$

(iii) $a.b.c'$ (iv) $b + c.a'$.

9. State and prove the following laws in a Boolean algebra :

(i) Idempotent laws (ii) Boundedness laws

(iii) Absorption laws (iv) Involution law.

10. State and prove the following laws in a Boolean algebra :

(i) Associative law of addition (ii) Associative law of multiplication.

11. In the Boolean algebra (B, +, ., '), show that

$(a + b)\ .\ (c + d) = a.c + a.d + b.c + b.d\ \forall\ a, b, c, d \in B$.

12. In the Boolean algebra (B, +, ., ') show that

$(a + b)\ .\ (a' + c) = a.c + a'.b\ \forall\ a, b, c \in B$.

13. In the Boolean algebra (B, +, ., '), show that for $a, b \in B$

(i) $a + b = 0$ if and only if $a = 0, b = 0$

(ii) $a.b' = 0$ if and only if $a' + b = 1$.

14. In the Boolean algebra (B, +, ., '), show that $(a + b)\ .\ [a.b' + b]' = 0\ \forall\ a, b \in B$.

15. In the Boolean algebra (B, +, ., '), show that

(i) $(a + b + c)' = a'.b'.c'\ \forall\ a, b, c \in B$ (ii) $(a.b.c)' = a' + b' + c'\ \forall\ a, b, c \in B$

(iii) $a + a\ .\ (b + 1) = a\ \forall\ a, b \in B$ (iv) $(a + b)\ .\ (a + 1) = a + b\ \forall\ a, b \in B$.

16. Let (B, +, ., ') be a Boolean algebra. Prove that for $a, b \in B$:

$a.b' = 0$ if and only if $a.b = a$.

Answers

8. (*i*) $a'.b$ (*ii*) $a' + b$ (*iii*) $a' + b' + c$ (*iv*) $b'.(c' + a)$.

Hints

6. (*i*) L.H.S $= (a + c.a') + b = (a + c) . (a + a') + b$

(*ii*) L.H.S $= a.b + c.(a.b)' = (a.b + c) . (a.b)'$

7. $x.(x + y) + ((y' + x).y)' = x + (y'.y + x.y)'$

$= x + (0 + x.y)' = x + (x.y)' = x + (x' + y') = (x + x') + y'$

$= 1 + y' = 1.$

12. L.H.S. $= (a + b).a' + (a + b).c = a'.a + a'.b + c.(a + b) . (a + a')$

$= a'.b + c.(a + b.a') = a'.b + c.a + c.b.a'$

$= a'.b.1 + a'.b.c. + a.c = a'.b.(1 + c) + a.c$

13. (*i*) Let $a + b = 0$. Then

$$a = a + 0 = a + b.b' = (a + b) . (a + b') = 0 \cdot (a + b') = 0.$$

10.12. BOOLEAN SUB-ALGEBRA

(P.T.U. B.Tech. Dec. 2013)

Let C be a non-empty subset of a Boolean algebra B. Then, C is said to be a Boolean sub-algebra of B if C is itself a Boolean algebra w.r.t. the same operations as of B. or C is a sub-algebra of B iff C is closed under the three operations, namely, '+', '.' and '''.

ILLUSTRATIVE EXAMPLES

Example 1. *Determine whether or not each of the following subsets of D_{70}, is a sub-algebra ?*

(*i*) $A = \{1, 5, 10, 70\}$ (*ii*) $C = \{1, 2, 35, 70\}$

Sol. (*i*) Consider the following composition tables :

Table I

+	1	5	10	70
1	1	5	10	70
5	5	5	10	70
10	10	10	10	70
70	70	70	70	70

Table II

.	1	5	10	70
1	1	1	1	1
5	1	5	5	5
10	1	5	10	10
70	1	5	10	70

The binary operations are

$$a + b = l.c.m.\ (a, b)$$
$$a \cdot b = g.c.d.\ (a, b)$$
$$a' = \frac{70}{a}$$

From Table I, each element inside the table is from the set A = {1, 5, 10, 70}. Therefore, A is closed under '.'.

From Table II, each element inside the table is from A = {1, 5,10, 70}. Therefore, A is closed under '.'.

For complement, consider the following table :

a	$a' = \frac{70}{a}$
1	70
5	14
10	7
70	1

Here complement of $5 = 5' = 14 \notin A$

∴ A is not closed under "′".

Hence, A is not a sub algebra of 70.

(*ii*) Consider the composition tables given below :

Table III

+	1	2	35	70
1	1	2	35	70
2	2	2	70	70
35	35	70	35	70
70	70	70	70	70

Table IV

.	1	2	35	70
1	1	1	1	1
2	1	2	1	2
35	1	1	35	2
70	1	2	35	70

From table III, each element inside the table III is from C. Therefore, C is closed under '+'. From table IV, each element inside the table IV is from C. Therefore, C is closed under '.'.

Also, consider the following table V.

Table V

a	$a' = \frac{70}{a}$
1	70
2	35
35	2
70	1

From table V, 1′ = 70, 2′ = 35, 35′ = 2, 70′ = 1. Hence, complement of each element of C exists. Hence C is a Boolean algebra.

10.13. ATOMS OF A BOOLEAN ALGEBRA

A non-zero element in a Boolean Algebra {B, –, ∨, ∧} is called an atom if for every $x \in B$, there exists $a \in B$ such that

$$x \wedge a = a \quad \text{or} \quad x \wedge a = 0 \ \forall \ a$$

For example, consider the Boolean algebra $\{D_{30}, -, \vee, \wedge\}$. We find the set of atoms of D_{30}

Since D_{30} = {1, 2, 3, 5, 6, 10, 15, 30}

Here $2 \wedge 1$ = g.c.d of 2 and 1 = 1

Also $2 \wedge 2 = 2$, $2 \wedge 3 = 1$, $2 \wedge 5 = 1$, $2 \wedge 6 = 2$, $2 \wedge 10 = 2$, $2 \wedge 15 = 1$, $2 \wedge 30 = 2$

Thus, for all $x \in D_{30}$, we have $x \wedge 2 = 1$ or $x \wedge 2 = 2$

∴ 2 is an atom of D_{30}

Similarly, $x \wedge 3 = 1$ or $x \wedge 3 = 3$

∴ 3 is also an atom of D_{30}. Also 5 is an atom of D_{30}

∴ The set of atoms of $D_{30} = \{2, 3, 5\}$

Alternative. The atoms of D_m are the distinct prime divisors of m.

Here $D_{30} = \{1, 2, 3, 5, 6, 10, 15, 30\}$

The prime divisors of 30 are 2, 3, 5

∴ 2, 3, 5 are atoms of D_{30}.

Example 2. *Consider the Boolean algebra* D_{210}.

(*a*) *List its elements and draw its Hasse diagram.*

(*b*) *Find the set A of atoms.*

(*c*) *Find two Boolean subalgebra with eight elements.*

(*d*) *Is* $X = \{1, 2, 6, 210\}$ *a sublattice of* D_{210} *? A subalgebra ?*

(*e*) *Is* $Y = \{1, 2, 3, 6\}$ *a sublattice of* D_{210} *? A Boolean subalgebra ?*

Sol. (*a*) The divisors of 210 are 1, 2, 3, 5, 6, 7, 10, 14, 15, 21, 30, 35, 42, 70, 105, 210

∴ $D_{210} = \{1, 2, 3, 5, 6, 7, 10, 14, 15, 21, 30, 35, 42, 70, 105, 210\}$

The Hasse diagram of D_{210} is shown in Fig. 10.22.

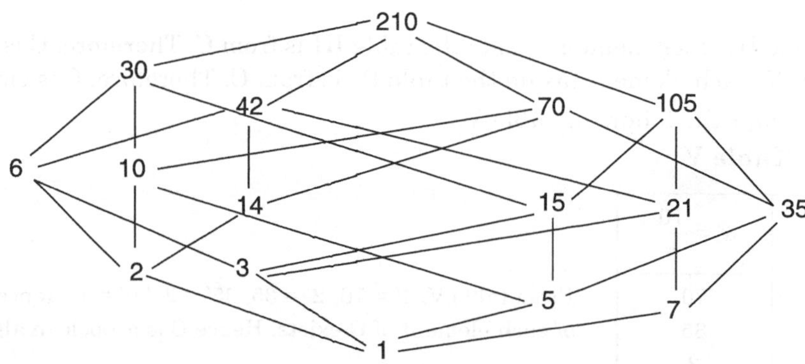

Fig. 10.22

(*b*) The prime divisors of 210 are 2, 3, 5, 7. Hence the set A of atoms is A = {2, 3, 5, 7}

(*c*) (*i*) By definition (complement), a set B is called subalgebra if it contains the elements 0 and 1 and is closed under ∧, ∨ and ' '.

Take $B = \{1, 2, 3, 6, 35, 70, 105, 210\}$

The least element is 1 and the largest element is 210. Also B is closed under ∧ and ∨.

Since $2 \wedge 3 =$ g.c.d. of 2 and $3 = 1 \in B$, $2 \wedge 70 = 2 \in B$ etc.

Also $2 \vee 3 =$ l.c.m. of 2 and $3 = 6 \in B$, $2 \vee 70 = 70 \in B$

Further $2 \wedge 105 =$ g.c.d. of 2 and $105 = 1$

$2 \vee 105 =$ l.c.m. of 2 and $105 = 210$

$\therefore \quad \overline{2} = 105,\ \overline{105} = 2$. Thus, $x \vee 2,\ x \wedge 2 \in B$ for all $x \in B$

Also $\quad 3 \wedge 70 = 1,\ 3 \vee 70 = 210$

$\therefore \quad \overline{3} = 70,\ \overline{70} = 3$

Also $\quad 6 \wedge 35 = 1,\ 6 \vee 35 = 210$

$\therefore \quad \overline{6} = 35,\ \overline{35} = 6.$

Finally, $\quad 1 \wedge 210 = 1,\ 1 \vee 210 = 210$

$\therefore \quad \overline{1} = 210,\ \overline{210} = 1$

Thus, each element of B has its complement

$\therefore$ The set B = {1, 2, 3, 6, 35, 70, 105, 210} is a subalgebra.

(*ii*) Take C = {1, 5, 6, 7, 30, 35, 42, 210}

Proceeding as in part (*i*), we can show that C is also a subalgebra.

(*d*) Here X = {1, 2, 6, 210}. We know that a set X is called sublattice if

$a \vee b \in X,\ a \wedge b \in X$ for all $a, b \in X$

Since, $\quad 1 \vee 2 =$ l.c.m of 1 and $2 = 2 \in X$
$\quad 1 \wedge 2 =$ g.c.d of 1 and $2 = 1 \in X$

Also, $\quad 1 \vee 6 = 6,\ 1 \wedge 6 = 1,\ 1 \vee 210 = 210,\ 1 \wedge 210 = 1$

Further, $2 \vee 1 = 2,\ 2 \wedge 1 = 1,\ 2 \vee 6 = 6,\ 2 \wedge 6 = 2,\ 2 \vee 210 = 210,\ 2 \wedge 210 = 2$ etc.

$\therefore$ X is a sublattice of D_{210}

But complement of $2 = 105 \notin X$ $\therefore$ X is not a subalgebra

(*e*) Here Y = {1, 2, 3, 6}

Let $a, b \in Y$, then $a \vee b,\ a \wedge b \in Y$ for all $a, b \in Y$

$\therefore$ Y is a sublattice of D_{210}

Also $\quad \overline{2} = 105 \notin Y \quad \therefore$ Y is not a subalgebra.

Example 3. *Find the number of subalgebras of* D_{210}.

Sol. A subalgebra of D_{210} must contain two, four, eight or sixteen elements.

(*i*) There can be only one *two-element* subalgebra which consists of greatest element 210 and least element 1, *i.e.*, {1, 210}

(*ii*) Any *four-element* subalgebra is of the form $\{1, x, \overline{x}, 210\}$, *i.e.*, consists of the greatest element, least element, and a nonbound element and its complement.

There are fourteen nonbound elements in D_{210} which are

2, 3, 5, 6, 7, 10, 14, 15, 21, 30, 35, 42, 70, 105. Hence there are 14/2 = 7 pairs $(x, \overline{x})$.

Thus D_{210} has seven *four-element* subalgebras.

(*iii*) An *eight-element* of subalgebra S will contain three elements s_1, s_2, s_3 where s_1, s_2 are any two of the four atoms of D_{210} and s_3 is the product of the other two atoms.

Here the atoms are 2, 3, 5, 7.

If $s_1 = 2$, $s_2 = 3$, then $s_3 = 5.7 = 35$.

There are ${}^4C_2 = 6$ ways to choose s_1 and s_2 from the four atoms of D_{210}.

Hence D_{210} has six *eight-element* subalgebra.

(*iv*) Since D_{210} contains 16 elements. ∴ The only *sixteen-element* subalgebra is D_{210} itself.

∴ Total number of subalgebras of D_{210}

$$= 1 + 7 + 6 + 1 = \mathbf{15}.$$

Example 4. *Consider the lattice D_{1001}.*

(*a*) *Draw the Hasse diagram of D_{1001}.*

(*b*) *Find the complement of each element of D_{1001}.*

(*c*) *Find the set A of atoms.*

(*d*) *Find the number of subalgebras of D_{1001}.*

Sol. (*a*) The divisors of 1001 are 1, 7, 11, 13, 77, 91, 143, 1001

∴ $$D_{1001} = \{1, 7, 11, 13, 77, 91, 143, 1001\}$$

The Hasse diagram of D_{1001} is shown in Fig. 10.23.

1001
77
91
143
11
7
13
1

Fig. 10.23

(*b*) $7 \vee 143$ = l.c.m of 7 and 143 = 1001 (greatest element)

$7 \wedge 143$ = g.c.d of 7 and 143 = 1 (least element)

∴ $$\overline{7} = 143,\ \overline{143} = 7$$

Also $$11 \vee 91 = 1001,\ 11 \wedge 91 = 1$$

∴ $$\overline{11} = 91,\ \overline{91} = 11,$$

Also $$13 \vee 77 = 1001,\ 13 \wedge 77 = 1$$

∴ $$\overline{13} = 77,\ \overline{77} = 13$$

Also $$1 \vee 1001 = 1001,\ 1 \wedge 1001 = 1$$

∴ $$\overline{1} = 1001,\ \overline{1001} = 1$$

Thus, $\overline{1} = 1001$, $\overline{7} = 143$, $\overline{11} = 91$, $\overline{13} = 77$, $\overline{77} = 13$, $\overline{91} = 11$, $\overline{143} = 7$, $\overline{1001} = 1$

(*c*) By definition, the atoms of D_{1001} are prime divisors of 1001.

∴ Here the prime divisors of 1001 are 7, 11, 13 which are the required atoms.

(*d*) A subalgebra of D_{1001} must contain two, four or eight elements.

(*i*) There can be only one, *two-element* subalgebra which consists of least element and greatest element *i.e.*, {1, 1001}

(*ii*) Any *four-element* subalgebra is of the form $\{1, x, \overline{x}, 1001\}$ *i.e.*, consists of least element, greatest element and a nonbound element and its complement.

There are six nonbound elements which are 7, 11, 13, 77, 91, 143. Hence there are

$\frac{6}{2} = 3$ pairs $(x, \overline{x})$. Thus, D_{1001} has three *four-element* subalgebra.

(*iii*) Since D_{1001} contains 8 elements. The only *eight-element* subalgebra is D_{1001} itself.

$\therefore$ Total number of subalgebras

$$= 1 + 3 + 1 = \mathbf{5}.$$

10.14. ISOMORPHIC BOOLEAN ALGEBRAS

Two Boolean algebras B and B_1 are called isomorphic if there is a one-to-one correspondence $f : B \to B_1$ which preserves the three operations +, * and ′ for any elements a, b in B *i.e.*,

$$f(a + b) = f(a) + f(b)$$

$$f(a * b) = f(a) * f(b) \quad \text{and} \quad f(a') = f(a)'.$$

For example, the following are two distinct Boolean algebras with two elements, which are isomorphic.

1. The first one is a Boolean algebra that is derived from a power set P(S) under $\subseteq$ (set inclusion), *i.e.*, let S = {a}, then B = {P(S), ∪, ∩, ′} is a boolean algebra with two elements P(S) = {ϕ, {a}}.

2. The second one is a Boolean algebra {B, ∨, ∧, ′} with two elements 1 and p {here p is a prime number} under operation divides, *i.e.*, let B = {1, p}. So, we have $1 \wedge p = 1$ and $1 \vee p = p$ also $1' = p$ and $p' = 1$.

10.15. REPRESENTATION THEOREM

Let B be a finite algebra : Let A be the set of atoms of B and let P(A) be the Boolean algebra of all subsets of the set A of atoms. Let $0 \neq x \in$ B such that $x = a_1 + a_2 + a_3 + \ldots\ldots + a_r$ then the function

$$f = B \to P(A) \text{ defined by } f(x) = \{a_1, a_2, a_3, \ldots\ldots, a_r\} \text{ is an isomorphism.}$$

Proof. If $a_1, a_2, \ldots\ldots, a_r$ are atoms, then by definition, $a_i\, a_j = \begin{cases} 0, & i \neq j \\ a_i^2, & i = j \end{cases}$

Let $x, y \in$ B such that $x = a_1 + a_2 + \ldots\ldots + a_r + b_1 + b_2 + \ldots\ldots + b_s$

$$y = b_1 + b_2 + \ldots\ldots + b_s + c_1 + c_2 + \ldots\ldots + c_t$$

where, $\quad A = \{a_1, a_2, \ldots\ldots, a_r, b_1, b_2, \ldots\ldots, b_s, c_1, c_2, \ldots\ldots, c_t, d_1, d_2, \ldots\ldots, d_k\}$

is the set of atoms of B.

Consider $\quad x + y = (a_1 + a_2 + \ldots\ldots + a_r + b_1 + b_2 + \ldots\ldots + b_s)$

$$+ (b_1 + b_2 + \ldots\ldots + b_s + c_1 + c_2 + \ldots\ldots + c_t)$$

$$xy = (a_1 + a_2 + \ldots\ldots + a_r + b_1 + b_2 + \ldots\ldots + b_s)(b_1 + b_2 + \ldots\ldots + b_s + c_1 + c_2 + \ldots\ldots + c_t)$$

$$= b_1 + b_2 + \ldots\ldots + b_s \qquad \left|\; \begin{array}{l} a_i^2 = a_i \\ a_i a_j = 0, \quad i \neq j \end{array}\right.$$

Hence, $\quad f(x + y) = \{a_1, a_2, \ldots\ldots, a_r, b_1, b_2, \ldots\ldots, b_s, c_1, c_2, \ldots\ldots, c_t\}$

$$= \{a_1, a_2, \ldots, a_r, b_1, b_2, \ldots\ldots, b_s\} \cup \{b_1, b_2, \ldots\ldots, b_s, c_1, c_2, \ldots\ldots, c_t\}$$

$$= f(x) \cup f(y)$$

Also $f(xy) = \{b_1, b_2, ..., b_s\}$

$= \{a_1, a_2, - , a_r, b_1, b_2, - b_s\} \cap \{b_1, b_2, - b_s, c_1, c_2, - c_t\}$

$= f(x) \cap f(y)$

Hence, f is a homomorphism.

Further, since the representation is unique, therefore f is one-one and onto. Hence, $f: B \to P(A)$ is an isomorphism.

10.16. LAWS OF BOOLEAN ALGEBRA

I. Uniqueness of zero element unit element and complement element

Theorem. *If {B, +, ., ') is a Boolean algebra, then*

(i) The zero element is unique

(ii) The unit element is unique

(iii) The complement element is unique. (P.T.U. B.Tech. Dec. 2010, May 2010)

Proof. (*i*) If 0_1 and 0_2 are two zero elements of B, we show $0_1 = 0_2$

Now 0_1 is zero element of B and $0_2 \in B$, then

$$0_2 + 0_1 = 0_2 = 0_1 + 0_2 \quad ...(1)$$

Similarly, 0_2 is zero element of B and $0_1 \in B$, then

$$0_1 + 0_2 = 0_1 = 0_2 + 0_1 \quad ...(2)$$

From (1) and (2), $0_1 = 0_2$

Hence, the zero element is unique.

(*ii*) Let I_1 and I_2 be two unit elements of B, we show $I_1 = I_2$

Now I_1 is unit element of B and $I_2 \in B$, then

$$I_2 + I_1 = I_2 = I_1 + I_2 \quad ...(3)$$

Similarly, I_2 is unit element of B and $I_1 + B$, then

$$I_1 + I_2 = I_1 = I_2 + I_1 \quad ...(4)$$

From (3) and (4), $I_1 = I_2$

Hence, unit element is unique.

(*iii*) Let $x \in B$ be such that

$$a + x = 1 \quad ...(1)$$

$$ax = 0 \quad ...(2)$$

i.e., $x \in B$ be the complement of a.

If a' be the another complement of a, then by definition of Boolean algebra,

$$a + a' = 1 \quad ...(3)$$

$$a \cdot a' = 0 \quad ...(4)$$

Consider $x = x \cdot 1$ | Identity law

$= x \cdot (a + a')$ | Using (3)

$= x \cdot a + x \cdot a'$ | Distributive law

$= a \cdot x + x \cdot a'$ | Using (2)

$= 0 + x \cdot a'$ | Using (4)

$= a . a' + x \cdot a'$ | Distributive law

$= (a + x) \cdot a'$ | Using (1)

$= 1 \cdot a' = a'$ which proves the uniqueness of the complement | Identity law

II. Theorem. *Let $\{B, +, ., '\}$ is a Boolean algebra. For $a \in B$, if $x \in B$ be such that $a + x = 1$ and $a \,.\, x = 0$, then $x = a'$. Also $0' = 1$ and $1' = 0$.*

Proof. Given $a + x = 1$ ...(1)

$a.x = 0$ ...(2)

Let a' be the complement of a, then by definition of Boolean algebra,

$a + a' = 1$ ...(3)

$a \,.\, a' = 0$ ...(4)

Consider

$$
\begin{aligned}
x &= x.1 && \text{| Indentity law}\\
&= x \,.\, (a + a') && \text{| Using (3)}\\
&= x \,.\, a + x \,.\, a' && \text{| Distributive law}\\
&= a \,.\, x + x \,.\, a' && \\
&= 0 + x.a' && \text{| Using (2)}\\
&= a.a' + x.a' && \text{| Using (4)}\\
&= (a + x) \,.\, a' && \text{| Distributive law}\\
&= 1.a' && \text{| Using (1)}\\
&= a' && \text{| Identity law}
\end{aligned}
$$

Further, 1 is unit element of B, then

$1.1' = 1' = 1'.1$ ...(5)

Also, by complement laws, $1.1' = 0$ ...(6)

From (5) and (6) $1' = 0$

Again, by complement law, $0 + 0' = 1$ ...(7)

and if 0 is zero element of B, then

$0 + 0' = 0' = 0' + 0$ ...(8)

From (7) and (8), $0' = 1$

III. Idempotent laws *(P.T.U. B.Tech. May 2010)*

Theorem. *If $\{B, +, ., '\}$ is a Boolean algebra, then*

(i) $a + a = a \;\forall\; a \in B$ *(ii)* $a.a = a \;\forall\; a \in B$.

Proof. (*i*) If 0 is zero element, then

$$
\begin{aligned}
a &= a + 0 && \text{| Identity law}\\
&= a + a.a' && \text{| Complement law}\\
&= (a + a) \,.\, (a + a') && \text{| Distributive law}\\
&= (a + a) \,.\, 1 && \text{| Complement law}\\
&= a + a \;\forall\; a \in B && \text{| Identity law}
\end{aligned}
$$

(*ii*) If 1 is unit element of B, then

$$
\begin{aligned}
a &= a.1 && \text{| Identity law}\\
&= a \,.\, (a + a') && \text{| Complement law}\\
&= a.a + a.a' && \text{| Distributive law}\\
&= a.a + 0 && \text{| Complement law}\\
&= a.a && \text{| Identity law}
\end{aligned}
$$

IV. Boundedness law *(P.T.U. B.Tech. May 2010)*

Theorem. *If $\{B, +, ., '\}$ is a Boolean algebra, then*

(i) $a + 1 = 1 \;\forall\; a \in B$ *(ii)* $a.0 = 0 \;\forall\; a \in B$.

Proof. (*i*) 1 is a unit element of B, therefore,

$a + 1 = (a + 1) . 1$

$= (a + 1) . (a + a')$ | Complement law

$= a + (1 . a')$ | Distributive law

$= a + a'$ | Identity law

$= 1$ | Complement law

$\therefore \quad a + 1 = 1 \ \forall \ a \in B$

(*ii*) 0 is zero element of B, therefore,

$a.0 = a.0 + 0$

$= a.0 + a.a'$ | Complement law

$= a.(0 + a')$ | Distributive law

$= a.(a' + 0)$

$= a.a'$ | Identity law

$= 0$ | Complement law

$\therefore \quad a.0 = 0 \ \forall \ a \in B$

V. Absorption laws

Theorem. *If {B, +, ., '} is a Boolean algebra, then*

(*i*) $a + a.b = a \ \forall \ a, b \in B$

(*ii*) $a.(a + b) = a \ \forall \ a, b \in B$.

Proof. (*i*) $a + a.b = a.1 + a.b$ | Identity law

$= a.(1 + b)$ | Distributive law

$= a.(b + 1)$

$= a.1$ | Boundedness law

$= a$ | Identity law

(*ii*) $a.(a + b) = (a + 0).(a + b)$ | Identity law

$= a + 0.b$ | Distributive law

$= a + b.0$

$= a + 0$ | Boundedness law

$= a$ | Identity law

VI. Involution laws

Theorem. *If {B, +, ., '} is a Boolean algebra, then*

$a'' = a \ \forall \ a \in B$ *where* $a'' = (a')'$.

Proof. Consider $a'' = a'' + 0$ | Identity law

$= a'' + a.a'$ | Complement law

$= (a'' + a).(a'' + a')$ | Distributive law

$= (a'' + a).(a' + a'')$

$= (a'' + a).1$ | Complement law

$= (a + a'').(a + a')$ | Complement law

$= a + a''.a'$ | Distributive law

$= a + a'.a''$

$= a + 0$ | Complement law

$= a$ | Identity law

$\therefore \quad \mathbf{a'' = a \ \forall \ a \in B}$

VII. Associative laws. *If* $\{B, +, ., '\}$ *is a Boolean algebra, then*

(*i*) $a + (b + c) = (a + b) + c \ \forall\ a, b, c \in B$

(*ii*) $a.(b.c) = (a.b).c \ \forall\ a, b, c \in B$.

Proof.

(*i*) $a + (b + c) = (a + (b + c)) . 1 = (a + (b + c)) . (c + c')$ | Complement law

$= [(a + (b + c)).c] + [(a + (b + c)).c']$ | Distributive law

$= [c.(a + (b + c))] + [c'.(a + (b + c))]$

$= [(c.a) + (c.(b + c))] + [(c'.a) + (c'.(b + c))]$ | Distributive law

$= [(c.a) + c] + [(c'.a) + ((c'.b) + (c'.c))]$ | Absorption law

$= [(c.a) + (c.1)] + [(c'.a) + (c'.b) + 0]$ | Complement law

$= [c.(a + 1) + c'.(a + b)$ | Distributive law

$= (c.1) + c'.(a + b)$ | Boundedness law

$= c + c'.(a + b)$

$= (c + c').(c + (a + b))$ | Distributive

$= 1.((a + b) + c)$ | Complement law

$= (a + b) + c$

$\therefore$ $\mathbf{a + (b + c) = (a + b) + c \ \forall\ a, b, c \in B.}$

(*ii*) $a.(b.c) = a.(b.c) + 0 = (a.(b.c)) + (c.c')$ | Complement law

$= [(a.(b.c)) + c].[(a.(b.c)) + c']$ | Distributive law

$= [c + (a.(b.c))].[c' + (a.(b.c))]$

$= [(c + a).(c + (b.c))].[(c' + a).(c' + (b.c))]$ | Distributive law

$= [(c + a).c].[(c' + a).((c' + b).(c' + c))]$ | Absorption law

$= [(c + a).(c + 0)].[(c' + a).((c' + b).1)]$ | Complement law

$= [c + (a.0)].[c' + (a.b)]$ | Distributive law

$= (c + 0).[c' + (a.b)]$

$= c.(c' + (a.b))$

$= (c.c') + (c.(a.b))$

$= 0 + ((a.b).c)$

$= (a.b).c$

$\therefore$ $\mathbf{a.(b.c) = (a.b).c \ \forall\ a, b, c \in B.}$

VIII. De Morgan's laws

Theorem. *If* $\{B, +, ., '\}$ *is a Boolean algebra, then*

(*i*) $(a + b)' = a'.b'$ (*ii*) $(a.b)' = a' + b' \ \forall\ a, b \in B$.

Proof. (*i*) By theorem II, For $a \in$ B, if $x \in$ B is such that $a + x = 1$ and $a.x = 0$, then $x = a'$

Consider $(a + b) + a'.b' = ((a + b) + a').((a + b) + b')$ | Distributive

$= (a + (b + a')).(a + (b + b'))$ | Associative law

$= (a + (a' + b)).(a + 1)$ | Complement law and commutative law

$= ((a + a') + b).1$ | Boundedness law

$= (1 + b)$ | Complement law and identity law

$= 1$ ...(1) | Boundedness law

Also $(a + b).(a'.b') = (a'.b').(a + b)$ | Commutative law

$= ((a'.b').a) + ((b'.b').b)$ | Distributive law

$= ((b'.a').a) + (a'.(b'.b))$ | Associative law

$= (b'.(a'.a)) + (a'.(b'.b))$
$= (b'.0) + (a'.0)$ | Complement law
$= 0 + 0 = 0$ ...(2) | Boundedness law

From (1) and (2),

$(a + b)' = a'.b'$

(*ii*) From Part (*i*), $(a + b)' = a'.b'$. Applying Principle of duality, $(a.b)' = a' + b'$.

IX. Cancellation laws. *If {B, +, ., '} is a Boolean algebra, then*

(*i*) *If* $a + b = a + c$ *and* $a' + b = a' + c$, *then* $b = c$

(*ii*) *If* $a.b = a.c$ *and* $a'.b = a'.c$, *then* $b = c$.

Proof. (*i*) Consider $b = b + 0$ | Identity law
$= b + (a.a')$ | Complement law
$= (b + a).(b + a')$ | Distributive law
$= (a + b).(a' + b)$ | Commutative law
$= (a + c).(a' + c)$ | Given
$= (c + a).(c + a')$ | Commutative law
$= c + (a.a')$ | Distributive law
$= c + 0$ | Complement law
$= c$ | Identity law

(*ii*) Consider $b = b.1$ | Identity law
$= b.(a + a')$ | Complement law
$= (b + a) . (b + a')$ | Distributive law
$= (a + b).(a' + b)$ | Commutative law
$= (a + c).(a' + c)$ | Given
$= (c + a).(c + a')$ | Commutative law
$= c + (a.a')$ | Distributive law
$= c + 0$ | Complement law
$= c.$ | Identity law

Now we prove one more important theorem of Boolean algebra.

Theorem X. *If {B, +, ., '} is a Boolean algebra. Then the following are equivalent*

(*i*) $a + b = b$ (*ii*) $a.b = a$

(*iii*) $a' + b = 1$ (*iv*) $a.b' = 0$.

Proof. (*i*) ⇒ (*ii*) Let $a + b = b$ and we show $a.b = a$

Consider $a = a.(a + b)$ | Absorption law
$= a.b$ | Given

Let $a.b = a$ and we show $a + b = b$

Consider $b = b + (b.a)$ | Absorption law
$= b + (a.b)$ | Commutative law
$= b + a$ | Given
$= a + b$ | Commutative law

Hence (*i*) and (*ii*) are equivalent

(*i*) ⇒ (*iii*) Let $a + b = b$ and we show $a' + b = 1$

Consider $a' + b = a' + (a + b)$ | Given
$= (a' + a) + b$ | Associative law

$= 1 + b$ | Complement law

$= 1$ | Boundedness law

If $a' + b = 1$, we show $a + b = b$

Consider $a + b = 1.(a + b)$ | Identity law

$= (a' + b).(a + b)$ | Given

$= (a'.a) + b$ | Distributive law

$= 0 + b$ | Complement law

$= b$ | Identity law

Hence (*i*) and (*iii*) are equivalent

(*iii*) $\Rightarrow$ (*iv*) Let $a' + b = 1$ and we show $a.b' = 0$

Consider $0 = 1' = (a' + b)' = a''.b'$ | De Morgan's law

$= a.b'$ | Involution law

$\therefore$ $a.b' = 0$

If $a.b' = 0$, we show $a' + b = 1$

Consider $1 = 0' = (a.b')' = a' + b''$ | De Morgan's law

$= a' + b$ | Involution law

Hence (*iii*) and (*iv*) are equivalent

Accordingly, (*i*), (*ii*), (*iii*) and (*iv*) are equivalent.

Example 5. *Using Boolean Postulates and theorem, simplify the following expression :*

$$a + ab + abc + abcd + \bar{a} + \bar{a}b + \bar{a}bc + \bar{a}bcd.$$ (P.T.U. B.Tech. May 2009)

Sol. Given expression is

$f(a, b, c, d) = a + ab + abc + abcd + \bar{a} + \bar{a}b + \bar{a}bc + \bar{a}bcd$

$= a(1 + b) + abc(1 + d) + \bar{a}\,(1 + b) + \bar{a}\,bc(1 + d)$

$= a\,.\,1 + abc\,.\,1 + \bar{a}\,.\,1 + \bar{a}\,bc\,.\,1$ For any Boolean algebra $1 + x = x \quad \forall\, x \in B$

$= a + abc + \bar{a} + \bar{a}\,bc$ | Idendity law

$= a + a\,.\,(bc) + \bar{a} + \bar{a}\,.\,(bc)$

$= a + \bar{a}$ Absorption law : $a + a\,.\,b = a \quad \forall a, b \in B$

$= 1.$ | Complement law

10.17. PRINCIPLE OF DUALITY

The dual of any expression E is obtained by interchanging the operations + and * and also interchanging the corresponding identity elements 0 and 1, in original expression E.

Example 6. *Write the dual of the following Boolean expressions :*

(*i*) $(x_1 * x_2) + (x_1 * \bar{x}_3)$ (*ii*) $(1 + x_2) * (x_1 + 1)$

(*iii*) $(a \wedge (b \wedge c))$.

Sol. (*i*) $(x_1 + x_2) * (x_1 + \bar{x}_3)$ (*ii*) $(0 * x_2) + (x_1 * 0)$

(*iii*) $(a \vee (b \wedge c))$.

Note. The dual of any theorem in a Boolean algebra is also a theorem.

TEST YOUR KNOWLEDGE 10.4

1. Show that {M; ∨, ∧, –} is a Boolean algebra where M = P(A), the power set of A and A = {a, b, c}.
2. Express each Boolean expression E (x, y, z) as a sum of products and then its complete sum of products form

   (a) $E = x(xy' + x'y + y'z)$ (b) $E = z(x' + y) + y'$
3. Express $E(x, y, z) = (x' + y)' + x'y$ in its complete sum of products form.
4. State the dual of

   (a) $a \vee (b \wedge a) = a$ (b) $a \vee \overline{\left(\left(\overline{b} \vee a\right) \wedge b\right)} = 1$

   (c) $\overline{\left(a \wedge \overline{b}\right)} \wedge b = a \vee b$
5. Write the dual of each Boolean expression

   (a) $a(a' + b) = ab$ (b) $(a + 1)(a + 0) = a$

   (c) $(a + b)(b + c) = ac + b$
6. Let [B; +, ., ′] be a Boolean algebra and let $f(x, y)$ be a Boolean function of the variables x and y. By using the following table, find the Boolean expression.

x	y	$f(x, y)$
1	1	0
1	0	1
0	1	1
0	0	0

7. Let [B; +, ., ′] be a Boolean algebra and let $f(x, y, z)$ be a Boolean function of the variables x, y and z. By using the following table, find the Boolean expression $f(x, y, z)$.

x	y	z	$f(x, y, z)$
1	1	1	1
1	1	0	0
1	0	1	0
0	1	1	1
1	0	0	0
0	1	0	1
0	0	1	0
0	0	0	0

8. Suppose B is a Boolean algebra with at least 100 elements. How many elements can B have ?
9. Using Boolean algebra, show that $abc + ab\overline{c} + a\overline{b}c + \overline{a}bc = ab + bc + ca$.

   (P.T.U. B. Tech. May 2008)
10. Consider the lattices $\mathbf{D}_m$ of divisors of m (where $m > 1$).

    (a) Show that $\mathbf{D}_m$ is a Boolean algebra if and only if m is *square-free*, that is, if m is a product of distinct primes. *(P.T.U. B. Tech. Dec. 2013)*

    (b) If $\mathbf{D}_m$ is a Boolean algebra, show that the atoms are the distinct prime divisors of m.

11. Consider the following lattices: (*a*) $\mathbf{D}_{20}$; (*b*) $\mathbf{D}_{55}$; (*c*) $\mathbf{D}_{99}$; (*d*) $\mathbf{D}_{130}$. Which of them are Boolean algebras, and what are their atoms ?

12. Consider the Boolean algebra $\mathbf{D}_{110}$. (*a*) List its elements and draw its diagram. (*b*) Find all its subalgebras. (*c*) Find the number of sublattices with four elements. (*d*) Find the set A of atoms of $\mathbf{D}_{110}$. (*e*) Give the isomorphic mapping $f : \mathbf{D}_{110} \to P(A)$ as defined in representation theorem.

13. Let B be a Boolean algebra. Show that :

(*a*) For any x in B, $0 \le x \le 1$.

(*b*) $a < b$ if and only if $b' < a'$.

14. An element x in a Boolean algebra is called a *maxterm* if the identify 1 is its only successor. Find the maxterms in the Boolean algebra $\mathbf{D}_{20}$.

15. Let B be a Boolean algebra. (*a*) Show that complements of the atoms of B are the maxterms. (*b*) Show that any element x in B can be expressed uniquely as a product of maxterms.

16. Let B be a 16-element Boolean algebra and let S be an 8-element subalgebra of B. Show that two of the atoms of S must be atoms of B.

17. Let B = (B, +, ∗, ′, 0, 1) be a Boolean algebra. Define an operation Δ on B (called the symmetric difference) by

$$x \,\Delta\, y = (x * y') + (x' * y).$$

Prove that R = (B, D, ∗) is a commutative Boolean ring.

18. Let R = (R, ⊕, .) be a Boolean ring with identity $1 \neq 0$. Define

$$x' = 1 \oplus x,\ x + y = x \oplus y \oplus x \,.\, y, \quad x * y = x \,.\, y$$

Prove that B = (R, + , ∗, ′, 0, 1) is a Boolean algebra.

Answers

2. (*a*) (*i*) $xy' + xy'z$ (*ii*) $xy'z + xy'z'$

(*b*) (*i*) $x'z + yz + y'$ (*ii*) $xyz + xy'z + xy'z' + x'yz + x'y'z + x'y'z'$

3. $xy'z + xy'z' + x'yz + x'yz'$

4. (*a*) $a \wedge (b \vee a) = a$ (*b*) $a \wedge \overline{(\bar{b} \wedge a)} \vee b = 0$

(*c*) $\overline{(a \vee \bar{b})} \vee b = a \wedge b$

5. (*a*) $a + a'b = a + b$ (*b*) $a.0 + a.1 = a$

(*c*) $ab + ac = (a + c)b$ **6.** $f(x, y) = x \,.\, y' + x' \,.\, y$

7. $f(x, y, z) = (x' + z) \,.\, y.$ **8.** 2, 4, 6, 8, 16, 32 or 64.

11. (*b*) $\mathbf{D}_{55}$; atoms 5 and 11. (*d*) $\mathbf{D}_{130}$; atoms 2, 5 and 13.

12. (*a*) There are eight elements 1, 2, 5, 10, 11, 22, 55, 110. See Fig. (*a*).

(*b*) There are five subalgebras : {1, 110}, {1, 2, 55, 110}, {1, 5, 22, 110}, {1, 10, 11, 110}, $\mathbf{D}_{110}$.

(*c*) There are 15 sublattices which include the above three subalgebras.

(*d*) A = {2, 5, 11}.

(*e*) See Fig. (*b*).

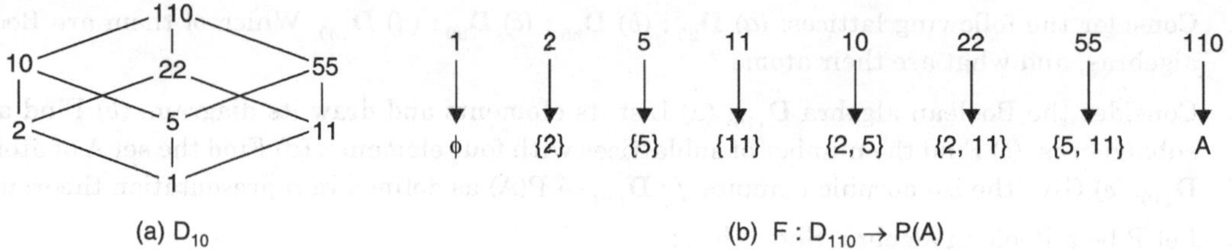

(a) D_{10} (b) $F : D_{110} \to P(A)$

Hints

1. We know that a lattice which contains a least element, a greatest element and which is both complemented and distributive is called Boolean Algebra.

Here the least element is ϕ, the greatest element is A. The complements of each element of B is given below :

B	*Complement of B*
ϕ	A
{*a*}	{*b*, *c*}
{*b*}	{*c*, *a*}
{*c*}	{*a*, *b*}
{*a*, *b*}	{*c*}
{*b*, *c*}	{*a*}
{*c*, *a*}	{*b*}
A	ϕ

Also {M, ∨, ∧, –} is a distributive lattice.

2. (*a*) Given $E = x(xy' + x'y + y'z)$

Express E as a sum of products forms, we get

$$E = xxy' + xx'y + xy'z = xy' + xy'z \text{ is the required sum of products forms.}$$

Further $E = xy'(z + z') + xy'z$ $\quad | \; xx' = 0$

$= xy'z + xy'z' + xy'z$

$= xy'z + xy'z',$

is the required complete sum of products form. $\quad | \; a + a = a$

(*b*) Given $E = z(x' + y) + y'$

$= x'z + zy + y'$, is the required sum of products form

Further, $E = x'z\,(y + y') + (x + x')zy + (x + x')\,y'\,(z + z')$

$= x'zy + x'zy' + xzy + x'zy + y'\,(xz + xz' + x'z + x'z')$

$= x'zy + x'zy' + xzy + x'zy + xy'z + xy'z' + x'y'z + x'y'z'$

$= x'yz + x'y'z + xyz + xy'z + xy'z' + x'y'z'.$

3. $E = (x' + y)' + x'y = xy' + x'y$, is the sum of products form $\quad | \;$ De Morgan's law

Further,

$E = xy'(z + z') + x'y\,(z + z')$

$= xy'z + xy'z' + x'yz + x'yz'$, is the required complete sum-of-products form of E.

10.18. BOOLEAN EXPRESSION OR BOOLEAN FUNCTION

Let {B, +, ., '} is a Boolean algebra. An expression involving the variables $x_1, x_2, \ldots x_n$ and the binary operations '+', '.' and ''' is called Boolean expression. The Boolean expression or Boolean function is denoted by $f(x_1, x_2, \ldots, x_n)$.

For example,

$$f(x, y, z) = (x + y'z)' + (xyz' + x'y)'$$

$$f(x, y, z) = (xy'z' + y)' + (x'z)',$$

are Boolean expression.

10.19. LITERAL

A *literal* is a variable or complemented variable like x, x', y, y' etc in the Boolean expression.

10.20. FUNDAMENTAL PRODUCT

By *Fundamental Product,* we mean a literal or a product of two or more literals in which no two literals involve the same variable. For example,

$xy'z, x, y', yz', x'yz$ are fundamental products. But $xyx'z, xyzy$ are not fundamental products as the first expression contains x and x' whereas the second expression contains y at two places.

10.21. SUM-OF-PRODUCTS FORM OR SOP FORM (*P.T.U. B.Tech. Dec. 2013*)

Let $f(x_1, x_2, \ldots, x_n)$ be a Boolean expression. Then, f is said to be in *sum-of-products* form or *minterm* form if f is a fundamental product or sum of two or more fundamental products, none of which is contained in another.

If P_1 and P_2 are fundamental products. Then, P_1 is said to be contained in P_2 if the literals of P_1 are also literals of P_2. For example, $x'z$ is contained in $x'yz$ since x' and z are literals in $x'yz$. However, $x'z$ is not contained in $xy'z$, since x' is not a literal in $xy'z$.

10.22. COMPLETE SUM-OF-PRODUCTS FORM

A non-zero Boolean expression $f(x_1, x_2, \ldots, x_n)$ is said to be in *Complete-sum-of-products* form if f is in sum-of-products form and each product involves all the variables.

Remarks 1. Every non-zero Boolean expression $f(x_1, x_2, \ldots, x_n)$ can be put into *Complete-sum-of-Products* form and this form is unique.

2. A Boolean expression $f(x_1, x_2, \ldots, x_n)$, in n variables can have a maximum of $2n$ such products.

ILLUSTRATIVE EXAMPLES

Example 1. *Reduce the following Boolean products to either 0 or a fundamental product*

(i) $xyx'z$ and $xyzy$ *(ii) $xyz'xyx$ and $xyz'xy'z'$.*

Sol. (*i*) $xyx'z = xx'yz$ | Commutative law

$= 0$ | $xx' = 0$ (Complement laws)

Also, $xyzy = xzyy$ | Commutative law

$= xzy$ | Idempotent law

(ii) $xyz'xyx = xxxyyz'$ | Commutative law

$= xyz'$ | Idempotent law

Also, $xyz'xy'z' = xxyy'z'z'$ | Commutative law

$= xyy'z'$ | Idempotent law

$= 0.$ | $yy' = 0$ (Complement law)

Example 2. *Let {B, +, ., '} is a Boolean algebra. Show that*

(i) $a.b + a.b' + a'.b + a'.b' = 1$ *(ii)* $(a + b).a'.b' = 0$

(iii) $(a + b)' + (a + b')' = a'$ *(iv)* $a.b + ((a + b').b)' = 1.$

Sol. *(i)* $a.b + a.b' + a'.b + a' . b'$

$= a.(b + b') + a'.(b + b')$ | Distributive law

$= a.1 + a'.1$ | Complement law

$= a + a'$ | Identity law

$= 1$ | Complement law

(ii) $(a + b).a'.b' = (a + b).(a + b)'$ | De Morgan's law

$= 0$ | Complement law

(iii) $(a + b)' + (a + b')'$

$= a'.b' + a'.b''$ | De Morgan's law

$= a'.(b' + b'')$ | Distributive law

$= a'.(b' + b)$ | Involution law

$= a'.1.$ | Complement law

$= a'.$ | Identity law

(iv) Please Try Yourself.

Example 3. *Prove that a Boolean algebra cannot have three elements.*

Sol. Let, if possible, {B, +, ., '} has three elements, say, 0, 1 and a (other than 0 and 1)

We claim $a' \neq 0$ and $a' \neq 1$

For if, $a' = 0 \Rightarrow a'' = 0' = 1 \Rightarrow a = 1,$

which is a contradiction. Hence $a' \neq 0$ | Involution law

Similarly, if $a' = 1 \Rightarrow a'' = 1' \Rightarrow a = 0,$

which is also a contradiction. Hence $a' \neq 1$. Therefore, we must have $a' = a$

Now $a.a' = 0$ | Complement law

$\Rightarrow$ $a.a = 0$ | $a' = a$

$\Rightarrow$ $a = 0$ | Idempotent law

a contradiction as $a \neq 0$.

Hence, there cannot be a Boolean algebra containing three elements.

Example 4. *Consider the Boolean expressions*

$$f_1 = xz' + y'z + xyz'$$

$$f_2 = xz' + x'yz' + xy'z$$

Reduce these expressions into sum-of-products form

Sol. $f_1 = xz' + y'z + xyz'$

is not in a *sum-of-products* form since xz' is contained in xyz'

$\therefore$ $f_1 = xz' + y'z + xyz'$

$= xz' + xyz' + y'z$

$= xz' + y'z$ | Absorption law

which is a *sum-of-products* form

Also $f_2 = xz' + x'yz' + xy'z$ is already in sum-of-products form.

Example 5. *Consider a Boolean expression*

$$f_1(a, b, c) = ((ab)'\, c)'\, ((a' + c)\,(b' + c'))' \text{ into a sum-of-products form}$$

(P.T.U. B.Tech. Dec. 2013)

Sol. $f_1(a, b, c) = ((ab)'\, c)'\, ((a' + c)\,(b' + c'))'$

$= ((ab)'' + c')\,((a' + c)' + (b' + c')')$ | De Morgan's law

$= (ab + c')\,((a' + c)' + (b' + c')')$ | Involution law

$= (ab + c')\,(a''c' + b''c'')$ | De Morgan's law

$= (ab + c')\,(ac' + bc)$ | Involution law

$= abac' + abbc + c'ac' + c'bc$

$= aabc' + abbc + ac'c' + bc'c$ | Commutative law

$= abc' + abc + ac' + 0$ | By idempotent law $aa = a$

$= abc' + abc + ac'$ | By complement law, $c'c = 0$

$= abc' + ac' + abc$

$= ac' + abc$ | Absorption law

Example 6. *Consider a Boolean expression $f(x, y, z) = x(y'z)'$. Reduce it to complete sum-of-products form.*

Sol. $f(x, y, z) = x(y'z)'$

$= x(y'' + z')$ | De Morgan's law

$= x(y + z')$ | Involution law

$= xy + xz'$, which is in sum-of-products form

$= xy(z + z') + x(y + y')z'$ | By complement law, $a + a' = 1$

$= xyz + xyz' + xyz' + xy'z'$

$= xyz + xyz' + xy'z'$ | $a + a = a$ (Idempotent law)

which is in complete-sum-of-products form.

Example 7. *Consider the Boolean expression*

(i) $f_1(x, y, z) = z(x' + y) + y'$ (ii) $f_2(x, y, z) = (x' + y)' + x'y$

(iii) $f_3(x, y, z) = x(xy' + x'y + y'z)$ (iv) $f_4(x, y, z) = (x + y)'\,(xy')'$.

Reduce to sum-of-products form and hence to complete sum-of-product forms.

Sol. (i) $f_1(x, y, z) = z(x' + y) + y'$

$= (x' + y)\,z + y'$ | Commutative law

$= x'z + yz + y'$, which is sum-of-products form

$= x'z(y + y)' + (x + x')\,yz + (x + x')y'(z + z')$

| By complement law, $a + a' = 1$

$= x'zy + x'zy' + xyz + x'yz + (xy' + x'y')\,(z + z')$

$= x'yz + x'y'z + xyz + x'yz + xy'z + zy'z' + x'y'z + x'y'z'$

| Commutative law

$= x'yz + x'y'z + xyz + xy'z + xy'z' + x'y'z'$,

which is in complete sum-of products form.

(*ii*) $f_2(x, y, z) = (x' + y)' + x'y$

$= x''y' + x'y$ | De Morgan's law

$= xy' + x'y,$ | Involution law

which is a sum-of-products form.

$= xy'(z + z') + x'y(z + z')$ | By complement law, $a + a' = 1$

$= xy'z + xy'z' + x'yz + x'yz'$

which is in complete-*sum-of-products* form.

(*iii*) $f_3(x, y, z) = x(xy' + x'y + y'z)$

$= xxy' + xx'y + xy'z$

$= xy' + 0 + xy'z$ | Complement law and Idempotent law

$= xy' + xy'z$, which is in sum-of products form.

$= xy'(z + z') + xy'z$

$= xy'z + xy'z' + xy'z$

$= xy'z + xy'z',$ | $a + a = a$ (Idempotent law)

which is in complete sum-of-products form.

(*iv*) Please try yourself **Ans.** $x'y'z + x'y'z'$.

Example 8. *Let* $E = xy' + xyz' + x'yz'$, *prove the following :*

(*i*) $xz' + E = E$

(*ii*) $x + E \neq E$ (*iii*) $z' + E \neq E$.

Sol. We know that $A + E = E, A \neq 0$, *iff* the summands in the *Complete sum-of-products* form for A are among the summands in the *complete sum-of-products form* for E.

Now $E = xy' + xyz' + x'yz'$, reduce it to Complete-sum-of-products form, we have

$E = xy'(z + z') + xyz' + x'yz'$

$= xy'z + xy'z' + xyz' + x'yz'$

(*i*) Express xz' in complete-sum-of-products form.

$xz' = x(y + y')z' = xyz' + xy'z'$

Since the summands of xz' are among those of E, therefore $xz' + E = E$

(*ii*) Express x in the *complete-sum-of-products* form, we have

$x = x(y + y')(z + z') = x(yz + yz' + y'z + y'z')$ | $y + y' = 1 = z + z'$ (Complement law)

$= xyz + xyz' + xy'z + xy'z'$, which is in *complete-sum-of-products form*

Now the summand xyz of x is not a summand of E. Therefore, $x + E \neq E$

(*iii*) Express z' in *Complete sum-of-products* form, we have

$z' = (x + x')(y + y')z'$ | $x + x' = 1 = y + y'$ (Complement law)

$= (xy + xy' + x'y + x'y')z'$

$= xyz' + xy'z' + x'yz' + x'y'z',$

which is in *Complete, sum-of-products* form.

But the summand $x'y'z'$ of z' is not a summand of E. Therefore, $z' + E \neq E$.

10.23. MINTERM

Let $\{B, +, \cdot, '\}$ is a Boolean algebra. Let $x_1, x_2, ..., x_n$ are n variables. A product of the form $y_1 \cdot y_2 ...., y_n$, where $y_i = x_i$ or x_i', $i = 1, 2, ..., n$, is called a **minterm** in n variables $x_1, x_2, ..., x_n$. Total number of minterms in a Boolean function of n variables $= 2^n$. For example, (*i*) Minterms in variables x, y are $x.y$, $x.y'$, $x'.y$, $x'.y'$ (4 terms)

(*ii*) Minterms in variables x, y, z are $x.y.z$, $x.y.z'$, $x.y'.z$, $x'.y.z$, $x.y'.z'$, $x'.y'.z'$, $x'.y'.z$, $x'.y'.z'$ (8 terms)

The following theorem (*without proof*) gives a method of expressing a Boolean function in terms of the minterms in the corresponding variables.

10.24. BOOLEAN EXPANSION THEOREM

If {B, +, .,′, 0, 1} is a Boolean algebra and $f(x_1, x_2)$ is a Boolean function in 2 variables, x_1, x_2, then $f(x_1, x_2) = f(1, 1).x_1.x_2 + f(0, 1).x_1'.x_2 + f(1, 0).\ x_1.x_2' + f(0, 0).x_1'.x_2'$

Similarly, for 3 variables x_1, x_2, x_3, we have

$$\begin{aligned} f(x_1, x_2, x_3) = f(1, 1, 1).x_1.x_2.x_3 + f(0, 1, 1).x_1'.x_2.x_3 + f(1, 0, 1).x_1.x_2'.x_3 \\ + f(1, 1, 0)x_1.x_2.x_3' + f(0, 0, 1).x_1'.x_2'.x_3 + f(0, 1, 0).x_1'.x_2.x_3' \\ + f(1, 0, 0).x_1.x_2'.x_3' + f(0, 0, 0).x_1'.x_2'.x_3' \end{aligned}$$

Example 9. *Let {B, +, ., ′, 0, 1} be a Boolean algebra and let f(x, y) be a Boolean function of the variables x and y. Find the Boolean expression from the truth table given below :*

x	*y*	*f(x, y)*
1	*1*	*0*
1	*0*	*1*
0	*1*	*0*
0	*0*	*1*

Sol. From the truth Table,

$$f(1, 1) = 0,\ f(1, 0) = 1,\ f(0, 1) = 0,\ f(0, 0) = 1$$

∴ By Boolean expansion theorem,

$$\begin{aligned} f(x, y) &= f(1, 1).x.y + f(1, 0).x.y' + f(0, 1).x'.y + f(0, 0).x'.y' \\ &= 0 + x.y' + 0 + x'.y' = x.y' + x'.y'. \\ &= y'.x + y'.x' && \text{| Commutative law} \\ &= y'.(x + x') && \text{| Distributive law} \\ &= y'.1 = y' && \text{| Complement law} \end{aligned}$$

Example 10. *Let {B, +, ., ′, 0, 1} be a Boolean algebra and f(x, y, z) be a Boolean function of the variables x, y, z. Find the Boolean expression from the truth table given below :*

x	*y*	*z*	*f(x, y, z)*
1	*1*	*1*	*1*
1	*1*	*0*	*0*
1	*0*	*1*	*1*
0	*1*	*1*	*0*
1	*0*	*0*	*0*
0	*1*	*0*	*0*
0	*0*	*1*	*1*
0	*0*	*0*	*1*

Sol. From the truth table,

$$f(1, 1, 1) = 1, f(1, 1, 0) = 0, f(1, 0, 1) = 1, f(0, 1, 1) = 0$$

$$f(1, 0, 0) = 0, f(0, 1, 0) = 0, f(0, 0, 1) = 1, f(0, 0, 0) = 1.$$

By **Boole's expansion theorem,** we have

$$\begin{aligned} f(x, y, z) &= f(1, 1, 1).x.y.z + f(1, 1, 0).x.y.z' + f(1, 0, 1).x.y'.z + f(0, 1, 1).x'y.z \\ &\quad + f(1, 0, 0).x.y'.z' + f(0, 1, 0).x'.y.z' + f(0, 0, 1).x'.y'.z \\ &\quad + f(0, 0, 0).x'y'z' \\ &= 1.x.y.z + 0.x.y.z' + 1.x.y'.z + 0.x'.y.z + 0.x.y'.z' + 0.x'.y.z' \\ &\quad + 1.x'.y'.z + 1.x'.y'.z' \\ &= x.y.z + 0 + x.y'.z + 0 + 0 + 0 + x'.y'.z + x'.y'.z' \end{aligned}$$

$$\therefore \quad \mathbf{f(x, y, z) = x.y.z + x.y'.z + x'.y'.z + x'.y'.z'}$$

$$\begin{aligned} &= (x.z.y + x.z.y') + (x'.y'.z + x'.y'.z') \\ &= x.z.(y + y') + x'.y'.(z + z') \\ &= x.z.1 + x'.y'.1 = x.z + x'.y'. \end{aligned}$$

| Complement law

10.25. DISJUNCTIVE NORMAL FORM OR SUM-OF-PRODUCTS (OR SOP) FORM

A Boolean expression over ({0, 1}, $\vee$, $\wedge$, ′) is said to be in disjunctive normal form if it is a join of minterms *e.g.*,

$$(x_1' \wedge x_2' \wedge x_3') \vee (x_1' \wedge x_2 \wedge x_3') \vee (x_1 \wedge x_2 \wedge x_3)$$

is a Boolean expression in disjunctive normal form.

Since there are three minterms $x_1' \wedge x_2' \wedge x_3'$, $x_1' \wedge x_2 \wedge x_3$ and $x_1 \wedge x_2 \wedge x_3$.

Maxterm. A Boolean expression of n variables $x_1, x_2, \ldots, x_n$ is said to be a maxterm if it is of the form $\tilde{x}_1 \vee \tilde{x}_2 \vee \ldots\ldots \vee \tilde{x}_n$, where $\tilde{x}_i$ is used to denote x_i or x_i'.

10.26. CONJUNCTIVE NORMAL FORM OR PRODUCTS OF SUMS (OR POS) FORM

A Boolean expression over ({0, 1}, $\vee$, $\wedge$, ′) is said to be in conjunctive normal form if it is a meet of maxterms *e.g.*,

$$(x_1 \vee x_2 \vee x_3) \wedge (x_1 \vee x_2 \vee x_3) \wedge (x_1 \vee x_2 \vee x_3) \wedge (x_1' \vee x_2 \vee x_3') \wedge (x_1' \wedge x_2' \wedge x_3)$$

is a Boolean expression in conjunctive normal form consisting of five maxterms.

10.27. (a) OBTAINING A DISJUNCTIVE NORMAL FORM

Consider a function from $\{0, 1\}^n$ to $\{0, 1\}$. A Boolean expression can be obtained in disjunctive normal form corresponding to this function by having a minterm corresponding to each ordered n-tuple of 0's and 1's for which the value of function is 1.

10.27. (b) OBTAINING A CONJUNCTIVE NORMAL FORM

Consider a function from $\{0, 1\}^n$ to $\{0, 1\}$. A Boolean expression can be obtained in conjunctive normal form corresponding to this function by having a maxterm corresponding to each ordered n-tuple of 0's and 1's at which the value of function is 0.

Example 11. *Express the following function in*

(i) Disjunctive normal form *(ii) Conjunctive normal form*

	f		*f*
(0, 0, 0)	*1*	*(1, 0, 0)*	*0*
(0, 0, 1)	*0*	*(1, 0, 1)*	*1*
(0, 1, 0)	*1*	*(1, 1, 0)*	*0*
(0, 1, 1)	*0*	*(1, 1, 1)*	*1*

Sol. (*i*) $(x_1' \wedge x_2' \wedge x_3') \vee (x_1' \wedge x_2 \wedge x_3') \vee (x_1 \wedge x_2' \wedge x_3) \vee (x_1 \wedge x_2 \wedge x_3)$

(*ii*) $(x_1' \vee x_2' \vee x_3') \wedge (x_1' \vee x_2 \vee x_3) \wedge (x_1 \vee x_2' \vee x_3') \wedge (x_1 \vee x_2 \vee x_3')$.

Example 12. *If* $f(x, y, z) = (x \vee y) \wedge (x \vee y') \wedge (x' \vee z)$ *be a given Boolean function. Determine its DN form.*

Sol. First determine all values of $f(x, y, z)$ when x, y, z take values 0, 1 and then from this table, we will find the DN form

x	y	z	$f(x, y, z)$	$(x \vee y) \wedge (x \vee y') \wedge (x' \vee z)$
0	0	0	0	$= (0 \vee 0) \wedge (0 \vee 1) \wedge (1 \vee 0)$
0	0	1	0	$= (0 \vee 0) \wedge (0 \vee 1) \wedge (1 \vee 1)$
0	1	0	0	$= (0 \vee 1) \wedge (0 \vee 0) \wedge (1 \vee 0)$
0	1	1	0	$= (0 \vee 1) \wedge (0 \vee 0) \wedge (1 \vee 1)$
1	0	0	0	$= (1 \vee 0) \wedge (1 \vee 1) \wedge (0 \vee 0)$
1	0	1	1	$= (1 \vee 0) \wedge (1 \vee 1) \wedge (0 \vee 1)$
1	1	0	0	$= (1 \vee 1) \wedge (1 \vee 0) \wedge (0 \vee 0)$
1	1	1	1	$= (1 \vee 1) \wedge (1 \vee 0) \wedge (0 \vee 1)$

The disjunctive normal form of the function is

$$f(x, y, z) = (x \wedge y' \wedge z) \vee (x \wedge y \wedge z).$$

Example 13. *Determine the disjunctive normal form of the following Boolean expression :*

$$x \wedge (y \vee z).$$

Sol. First determine all values of $f(x, y, z)$ when x, y, z take values 0, 1 and then from this table we will write disjunctive form. Thus,

x	y	z	$f(x, y, z)$	$x \wedge (y \vee z)$
0	0	0	0	$= 0 \wedge (0 \vee 0)$
0	0	1	0	$= 0 \wedge (0 \vee 1)$
0	1	0	0	$= 0 \wedge (1 \vee 0)$
0	1	1	0	$= 0 \wedge (1 \vee 1)$
1	0	0	0	$= 1 \wedge (0 \vee 0)$
1	0	1	1	$= 1 \wedge (0 \vee 1)$
1	1	0	1	$= 1 \wedge (1 \vee 0)$
1	1	1	1	$= 1 \wedge (1 \vee 1)$

Thus, the DNF of the function is

$$f(x, y, z) = (x \wedge y \wedge z) \vee (x \wedge y' \wedge z) \vee (x \wedge y \wedge z').$$

10.28. PRIME IMPLICANTS

We know that a *fundamental product* is a literal or a product of two or more literals in which no two literals involve the same variable. A fundamental product, say, P, is called a prime implicant of a Boolean expression E if P + E = E, but no other fundamental product contained in P has this property.

We now discuss a method known as **Karnaugh Map** method for finding prime implicants and minimal forms for Boolean expression involving atmost six variables. But in this chapter, we will only discuss the cases of two, three and four variables.

10.29. KARNAUGH MAP

It is a graphical tool used to simplify a logic equation or to convert a truth table to its corresponding logic circuit. A **Karnaugh Map**, hence forth, will be abbreviated as **K-map.**

10.30. ADJACENT FUNDAMENTAL PRODUCTS

Let P_1 and P_2 are two fundamental products. Then P_1 and P_2 are said to be adjacent if P_1 and P_2 have the same variables and if they differ in exactly one literal, it must be a complemented variable in one product and uncomplemented in another.

ILLUSTRATIVE EXAMPLES

Example 1. *Is $P = xz'$, a prime implicant of $E = xy' + xyz' + x'yz'$?*

Sol. We have proved in example 8 above,

$$xz' + E = E, \text{ but } x + E \neq E,\ z' + E \neq E$$

$\therefore$ $P = xz'$ is the prime implicant of E.

Example 2. *Find the sum of adjacents products P_1 and P_2 where*

(i) $P_1 = xyz'$, $P_2 = xy'z'$ *(ii)* $P_1 = x'yzt$, $P_2 = xyz't$

(iii) $P_1 = x'yzt$, $P_2 = xyz't$ *(iv)* $P_1 = xyz'$, $P_2 = xyzt$.

Sol. (*i*) $P_1 + P_2 = xyz' + xy'z' = xz'y + xz'y'$ | Commutative law

$= xz'(y + y')$

$= xz'.1$ | Complement law

$= xz'$

Note. That sum of two adjacent products (squares) P_1 and P_2 is a fundamental product with one less literal :

(*ii*) $P_1 + P_2 = x'yzt + x'yz't = x'ytz + x'ytz'$

$= x'yt\,(z + z')$

$= x'yt.1$ | Complement law

$= x'yt$

(*iii*) P_1 and P_2 are not adjacent since they have two variables x and z. Here x and z are complemented in P_1 and uncomplimented in P_2.

(*iv*) P_1 and P_2 are not adjacent since they have different variables. Thus, in particular, they will not appear as squares in the same Karnaugh Map.

10.31. KARNAUGH MAP FOR TWO VARIABLES

The Karnaugh map corresponding to Boolean expression $f(x, y)$ of two variables is shown in Table given below :

	y	y'
x	xy	xy'
x'	$x'y$	$x'y'$

Also, any complete sum-of-products Boolean expression $f(x, y)$ of two variables, can be represented in Karnaugh map by placing 1's in the corresponding squares. A prime implicant of $f(x, y)$ will be either a pair of adjacent squares or an isolated square (a square which is not adjacent to any square of $f(x, y)$).

Also, the four possible fundamental products with two literals are :

$$xy,\ xy',\ x'y,\ x'y'.$$

10.32. KARNAUGH MAP FOR THREE VARIABLES

The Karnaugh map corresponding to Boolean expression $f(x, y, z)$ of three variables is shown in the Table given below :

	yz	yz'	$y'z'$	$y'z$
x				
x'				

Also, the eight possible fundamental products with three literals are :

$$xyz,\ xyz',\ xy'z',\ xy'z,\ x'yz,\ x'yz',\ x'y'z',\ x'y'z.$$

10.33. KARNAUGH MAP FOR FOUR VARIABLES

The Karnaugh map corresponding to Boolean expression $f(x, y, z, t)$ of four variables is shown in the Table given below :

	zt	zt'	$z't'$	$z't$
xy				
xy'				
$x'y'$				
$x'y$				

Also, the sixteen possible fundamental with four literals are :

$$xyzt,\ xyzt',\ xyz't',\ xyz't\ ;$$
$$xy'zt,\ xy'zt',\ xy'z't',\ xy'z't\ ;$$
$$x'y'zt,\ x'y'zt',\ x'y'z't',\ x'y'z't\ ;$$
$$x'yzt,\ x'yzt',\ x'yz't',\ x'yz't.$$

10.34. LOOPING

The expression for the output Boolean expression can be simplified by properly combining those squares in the K-map which contain 1's. This process for combining these 1's called *looping*.

10.35. LOOPING GROUPS OF TWO

Consider a K-map for there variables as shown below (Fig. 10.24). This map contains a pair of 1s that are vertically adjacent to each other.

	$\bar{z}$	z
$\bar{x}\bar{y}$	1	1
$\bar{x}y$	1	0
xy	1	0
$x\bar{y}$	0	0

Fig. 10.24

The first represents $\bar{x}y\bar{z}$ and the second represents $xy\bar{z}$. We note that in these two terms, only x appears both in complemented and uncomplimented form while y and $\bar{z}$ remain unchanged.

These two terms can be looped (combined) to give a resultant that eliminates the variable x as it appears both in complemented and uncomplimented form. Thus, the corresponding Boolean expression is

$$f(x, y) = \bar{x}y\bar{z} + xy\bar{z} = (\bar{x} + x)\, y\bar{z} = 1.\, y\bar{z} = y\bar{z} \quad | \text{ Complement law}$$

Consider another example of a K-map of three variables as shown below (Fig. 10.25) :

	$\overline{\text{C}}$	C
$\overline{\text{A}}\overline{\text{B}}$	0	0
$\overline{\text{A}}\text{B}$	1	1
AB	0	0
$\text{A}\overline{\text{B}}$	0	0

Fig. 10.25

This map contains a pair of 1s that are horizontally adjacent. The first represents $\overline{\text{A}}\text{B}\overline{\text{C}}$ and the second represents $\overline{\text{A}}\text{BC}$.

These two terms can be looped (combined) to give a resultant that eliminates the variable C as it appears both in complemented and uncomplimented form. Thus, the corresponding Boolean expression is

$$f(\text{A, B}) = \overline{\text{A}}\text{B}\overline{\text{C}} + \overline{\text{A}}\text{BC} = \overline{\text{A}}\text{B}\,(\overline{\text{C}} + \text{C}) = \overline{\text{A}}\text{B}.1 = \overline{\text{A}}\text{B}$$

Consider another example of a K-map for three variables as shown below (Fig. 10.26).

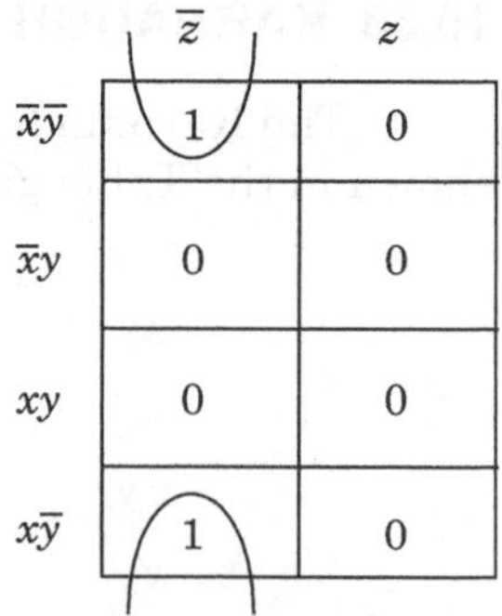

Fig. 10.26

This map contains 1s in the top row and bottom row respectively. These two squares are also considered to be adjacent. These two 1s can be looped (combined) to give a resultant that eliminates the variables x as it appears both in complemented and uncomplimented form. Thus, the corresponding Boolean expression is

$$f(x, y, z) = \bar{x}\,\bar{y}\,\bar{z} + x\bar{y}\,\bar{z} = (\bar{x} + x)\,\bar{y}\,\bar{z} = 1.\,\bar{y}\,\bar{z} = \bar{y}\,\bar{z}$$

Hence, from above discussion, we conclude that

"Looping a pair of adjacent 1s in a K-map eliminates the variable that appears in complemented and uncomplimented form".

10.36. LOOPING GROUPS OF FOUR

A K-map may contain a group of four 1's that are adjacent to each other. This group is called a **quad**. Consider the following Fig. 10.27.

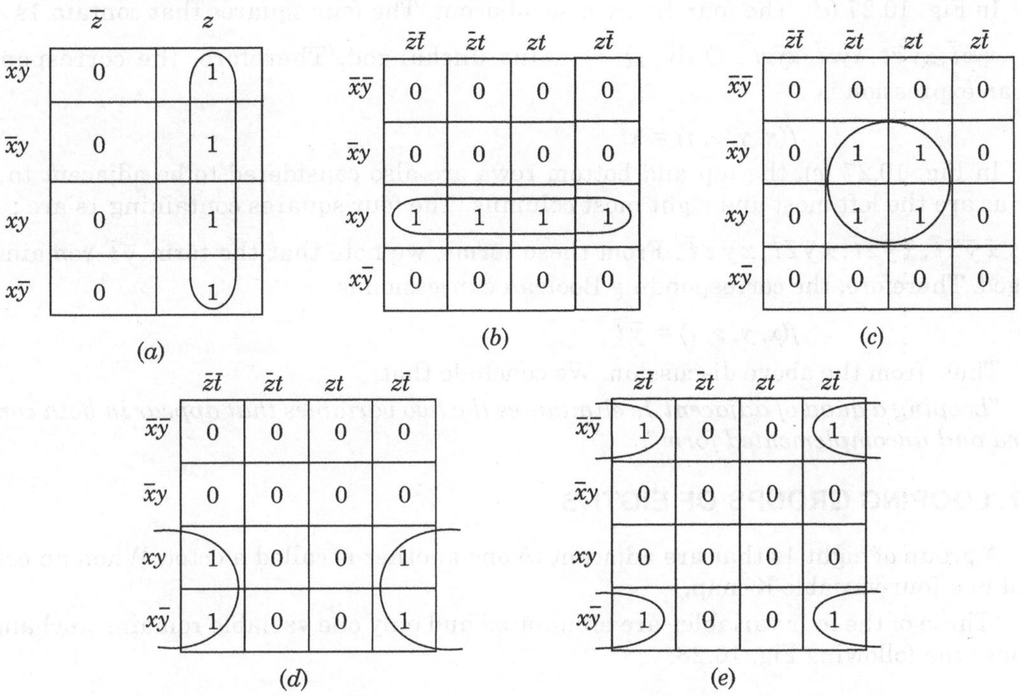

Fig. 10.27

In Fig. 10.27(*a*), the four 1s are vertically adjacent. The four squares that contain *a* '1' are $\bar{x}\,\bar{y}z, \bar{x}yz, xyz, x\bar{y}z$. From these terms, we rate that the only variable that remains unchanged is *z* (as both *x* and *y* appear in complemented and uncomplimented form). Thus, the corresponding Boolean expression is $f(x, y, z) = z$. We can also prove it mathematically as below :

Here $f(x, y, z) = \bar{x}\,\bar{y}z + \bar{x}yz + xyz + x\bar{y}z$

$= \bar{x}z\bar{y} + \bar{x}yz + xyz + xz\bar{y}$ | Commutative law

$= \bar{x}z\,(\bar{y} + y) + xz\,(y + \bar{y})$

$= \bar{x}z.1 + xz.1 = \bar{x}z + xz$ | Complement law

$= (\bar{x} + x).z = 1.z = z$

In Fig. 10.27 (*b*), the four 1s are horizontally adjacent. The four squares that contain 1 are

$xy\bar{z}\,\bar{t}, xy\bar{z}t, xyzt, xyz\bar{t}$. From these terms, we note that the only variable that remains unchanged is *xy* (as *z* and *t* appear in both complemented and uncomplimented form).

Thus, the corresponding Boolean expression is $f(x, y, z) = xy$. We can also prove it mathematically as below.

Here $f(x, y, z) = xy\bar{z}\bar{t} + xy\bar{z}t + xyzt + xyz\bar{t}$

$= xy\bar{z}\,(\bar{t} + t) + xyz\,(t + \bar{t})$

$= xy\bar{z}.1 + xyz.1 = xy\bar{z} + xyz = xy\,(\bar{z} + z) = xy.1 = xy$

In Fig. 10.27 (*c*). The four 1s are in a square and there are considered to be adjacent to each other. The corresponding Boolean expression is $f(x, y, z, t) = yt$ (as *x* and *z* appear in both complemented and uncomplimented form).

In Fig. 10.27 (*d*). The four 1s are also adjacent. The four squares that contain 1s are : $xy\bar{z}\bar{t}, x\bar{y}\bar{z}\bar{t}, xyz\bar{t}, x\bar{y}z\bar{t}$. Only $x\bar{t}$ remains unchanged. Therefore, the corresponding Boolean expression is

$$f(x, y, z, t) = x\bar{t}$$

In Fig. 10.27 (*e*), the top and bottom rows are also considered to be adjacent to each other as are the left most and right most columns. The four squares containing 1s are : $\bar{x}\bar{y}\bar{z}\bar{t}, \bar{x}\bar{y}z\bar{t}, x\bar{y}\bar{z}\bar{t}, x\bar{y}z\bar{t}$. From these terms, we note that the term $\bar{y}\bar{t}$ remains unchanged. Therefore, the corresponding Boolean expression is

$$f(x, y, z, t) = \bar{y}\bar{t}$$

Thus, from the above discussion. We conclude that

"Looping a quad of adjacent 1s eliminates the two variables that appear in both complemented and uncomplemented form".

10.37. LOOPING GROUPS OF EIGTHS

A group of eight 1s that are adjacent to one another is called a octet. When an octet is looped in a four-variable K-map,

"Three of the four variables are eliminated and only one variable remains unchanged". Consider the following Fig. 10.28.

	$\bar{z}\bar{t}$	$\bar{z}t$	zt	$z\bar{t}$
$\bar{x}\bar{y}$	1	0	0	0
$\bar{x}y$	1	1	1	1
xy	1	1	1	1
$x\bar{y}$	0	0	0	0

(*a*)

	$\bar{z}\bar{t}$	$\bar{z}t$	zt	$z\bar{t}$
$\bar{x}\bar{y}$	1	1	0	0
$\bar{x}y$	1	1	0	0
xy	1	1	0	0
$x\bar{y}$	1	1	0	0

(*b*)

	$\bar{z}\bar{t}$	$\bar{z}t$	zt	$z\bar{t}$
$\bar{x}\bar{y}$	1	0	0	1
$\bar{x}y$	1	0	0	1
xy	1	0	0	1
$x\bar{y}$	1	0	0	1

(*c*)

	$\bar{z}\bar{t}$	$\bar{z}t$	zt	$z\bar{t}$
$\bar{x}\bar{y}$	1	1	1	1
$\bar{x}y$	0	0	0	0
xy	0	0	0	0
$x\bar{y}$	1	1	1	1

(*d*)

Fig. 10.28

In Fig. 10.28 (*a*), only y remains unchanged

$\therefore \quad f(x, y, z, t) = y$

In Fig. 10.28 (*b*), $\bar{z}$ remains unchanged

$\therefore \quad f(x, y, z, t) = \bar{z}$

In Fig. 10.28 (*c*), $\bar{t}$ remains unchanged

$f(x, y, z, t) = \bar{t}$

In Fig. 10.28 (*d*), $f(x, y, z, t) = \bar{y}$.

10.38. KARNAUGH MAP METHOD FOR FINDING PRIME IMPLICANTS AND MINIMAL FORM FOR A BOOLEAN EXPRESSION

First express the given Boolean expression in complete sum of products form.

The following algorithm can be used by which minimized Boolean expression can be obtained.

1. Identify the ones which cannot be combined with any other ones and encircle them.
2. Identify the ones that can be combined in groups of two in only one way and encircle them as groups.
3. Identify the ones that can be combined with three other ones, to make a group of four adjacent ones, in only one way and encircle them as groups.
4. Identify the ones that can be combined with seven other ones, to make a group of eight adjacent ones, in only way and encircle them as groups.
5. After identifying the essential groups of 2, 4 and 8 ones, if there still remains some ones which have not been encircled, then these are to be combined with each other or with other already encircled ones *i.e.*, we should combine the left over ones in largest possible groups and in as few groupings as possible.

Example 3. *Use Karnaugh maps to find the prime implicants and minimal form of the following Boolean expression*

(i) $f_1(x, y) = xy + xy'$ *(ii)* $f_2(x, y) = xy + x'y + x'y'$

(iii) $f_3(x, y) = xy + x'y'$.

Sol. *(i)* We first express $f_1(x, y)$ into complete sum-of-product form.

Here $f_1(x, y) = xy + xy'$, which is already in complete sum-of-products form.

The K-map for the two variables is shown below (Fig. 10.29)

Put 1s in the cells corresponding to the terms xy and xy' and 0 elsewhere.

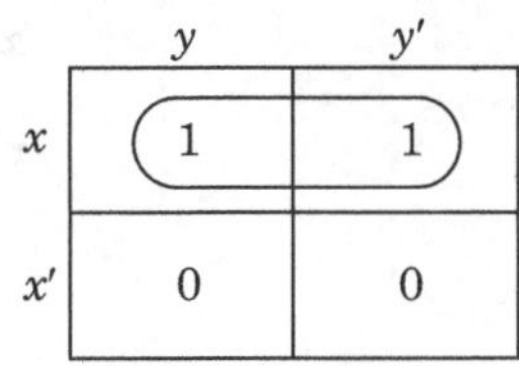

Fig. 10.29

The two squares containing 1s are xy and xy'. From these terms, we note that the variable x remains unchanged (as y appears in complemented and uncomplimented form). Hence, the prime implicant of $f_1(x, y)$ is x. Hence

$f_1(x, y) = x$, is its minimal sum

(ii) $f_2(x, y) = xy + x'y + x'y'$, which is already in complete sum-of-products form.

The K-map for the two variables is shown below (Fig. 10.30).

Put 1s in the cells corresponding to the terms xy, $x'y$ and as else where. This map contains two pairs of adjacent squares. (designated by the two loops). The vertices pair represents y and the horizontal pair represents x'. Hence y and x' are the prime implicants.

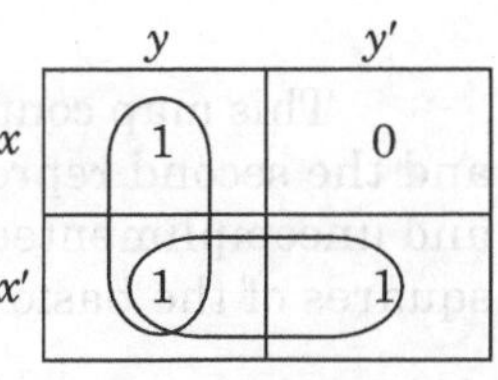

Fig. 10.30

Therefore,

$$f_2(x, y) = y + x', \text{ is its minimal sum.}$$

(*iii*) $f_3(x, y) = xy + x'y'$

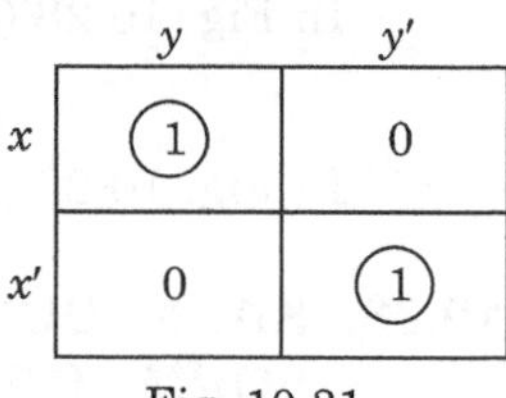

Fig. 10.31

Put 1s in the cells corresponding to the terms containing xy and $x'y'$ and 0 elsewhere. This map consists of two isolated squares which represent xy and $x'y'$ (Fig. 10.31). Hence xy and $x'y'$ are prime implicants. Therefore

$$f_3(x, y) = xy + x'y' \text{ is its minimal sum.}$$

10.39. BASIC RECTANGLE FOR THREE VARIABLE K-MAP

A *basic rectangle* in a Karnaugh map of three variables is a square or two adjacent squares or four squares which form a *one-by-four* or *two-by-two rectangle.* These basic rectangles corresponds to fundamental products of three, two and one literal respectively. Hence, the fundamental product represented by a basic rectangle is the product of just those literals that appear in every square-of the rectangle.

Example 4. *Find the fundamental product represented by each basic rectangle in the following (Fig. 10.32) Karnaugh maps*

	yz	yz'	$y'z'$	$y'z$
x	0	0	0	0
x'	0	1	1	0

(*a*)

	yz	yz'	$y'z'$	$y'z$
x	1	0	0	1
x'	0	0	0	0

(*b*)

	yz	yz'	$y'z'$	$y'z$
x	1	0	0	1
x'	1	0	0	1

(*c*)

Fig. 10.32

Sol. (*a*) The two 1s in Fig. 10. 32 (*a*) can be looped (combined) as shown in Fig. 10.33.

	yz	yz'	$y'z'$	$y'z$
x	0	0	0	0
x'	0	1	1	0

Fig. 10.33

This map contains a pairs of 1s that are horizontally adjacent. The first represents $x'yz'$ and the second represents $x'y'z'$. In these two forms, the variable y appears in complemented and uncomplimented form, where $x'z'$ remains unchanged. That is to say, $x'z'$ appear in all the squares of the basic rectangle. Therefore, by definition, the fundamental product is $P = x'z'$.

(b) The two 1s in the given Fig. 10.32 (b) can be combined or looped as shown in Fig. 10.34.

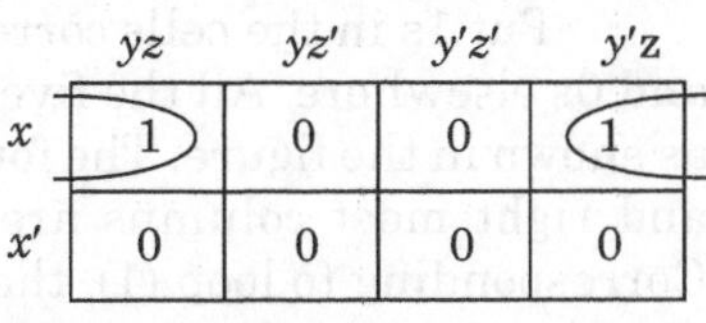

Fig. 10.34

This map contains a pair of 1s in the left row and right row. These two squares are also considered to be adjacent.

The first square represents xyz and the second represents $xy'z$. In these two terms, y appears in complemented and uncomplimented form, while xz remains unchanged. That is to say, xz appears in all the squares of the basic rectangle. Therefore, by definition, the fundamental product P is given by

$$P = xz$$

(c) The four 1s in given Fig. 10.32 (c) can be looped or combined as shown in Fig. 10.35.

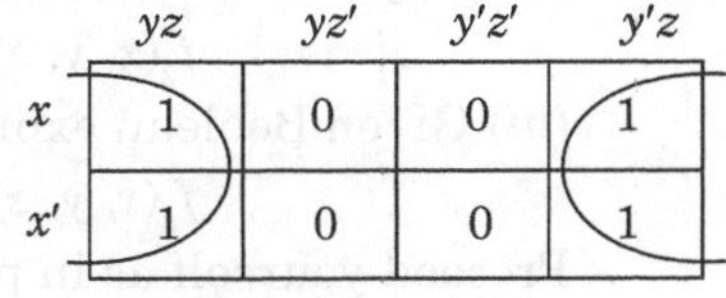

Fig. 10.35

The four 1s in this map are considered to be adjacent since the leftmost and rightmost columns are considered to be adjacent. The fours squares that contain 1s are xyz, $x'yz$, $xy'z$, $x'y'z$. In these terms, we note that x and y appears in complemented and uncomplement form whereas z remains unchanged. That is to say, z appears in all the squares. Therefore, by definition, the fundamental product P is given by

$$P = z$$

Example 5. *Use Karnaugh map to find the prime implicants and minimal form for each of the following complete-sum-of-products form*

(i) $f_1(x, y, z) = xyz + xyz' + x'yz' + x'y'z$

(ii) $f_2(x, y, z) = xyz + xyz' + xy'z + x'yz + x'y'z$

(iii) $f_3(x, y, z) = xyz + xyz' + x'yz + x'y'z' + x'y'z.$

Sol. (i) Given Boolean expression, in complete-sum-of-products form is

$$f_1(x, y, z) = xyz + xyz' + x'yz' + x'y'z \quad ...(1)$$

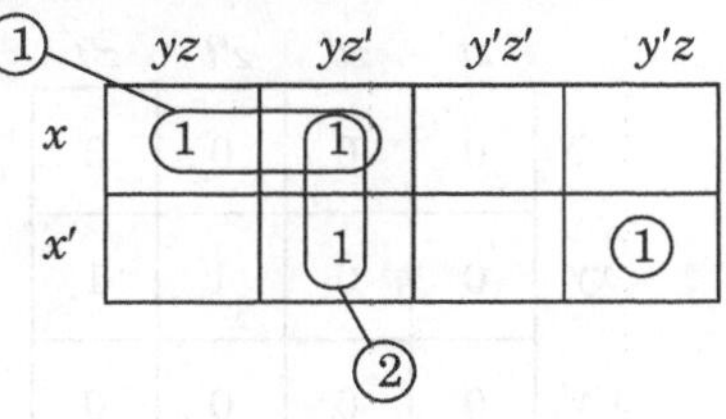

Fig. 10.36

The K-map for three variables is shown in Fig. 10.36 given below :

Put 1s in the cells corresponding to the terms xyz, xyz', $x'yz'$, $x'y'z$ and 0s elsewhere.

All the four 1s can be looped (combined) as shown in the figure.

Corresponding to loop (1), the variable xy appears in both the squares since z appears in complemented and uncomplimented form. Therefore, xy is prime implicant of $f_1(x, y, z)$. Corresponding to loop (2), the term yz' appears in both the squares since x appears in complemented and uncomplemented form. Therefore, yz' is prime implicant of $f_1'(x, y, z)$.

Corresponding to loop (3), the prime implicant is $x'y'z$. Hence, the required minimal form is

$$f_1(x, y, z) = xy + yz' + x'y'z$$

(ii) Given Boolean expression in complete-sum-of-products form is

$$f_2(x, y, z) = xyz + xyz' + xy'z + x'yz + x'y'z \quad ...(1)$$

The K-map for the three variables is shown in Fig. 10.37.

Put 1s in the cells corresponding to the terms in (1) and 0s elsewhere. All the five 1s can be looped (combined) as shown in the figure. The four 1s in the leftmost columns and right most columns are considered to be adjacent. Corresponding to loop (1), the variables x and y appear in complemented and uncomplimented form while z remains uncharged. That is to say, z appears in all the square. Hence, the prime implicant of $f_2(x, y, z)$ is z.

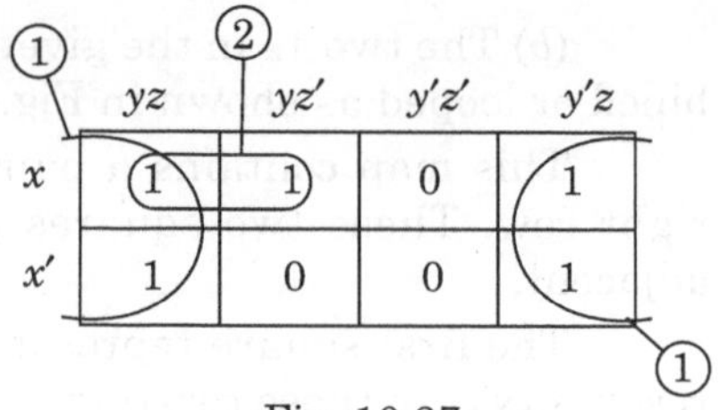

Fig. 10.37

Corresponding to loop (2), the variable z appears in complemented and uncomplemented form and xy appears in both the squares. Therefore, the prime implicant of $f_2(x, y, z)$ is xy. Therefore, required Boolean expression in minimal form is

$$f_2(x, y, z) = z + xy$$

(*iii*) Given Boolean expression in complete-sum-of-products form is

$$f_3(x, y, z) = xyz + xyz' + x'yz' + x'y'z' + x'y'z$$

Proceed yourself as in part (*i*) and (*ii*), the minimal form is

$$f_3(x, y, z) = xy + yz' + x'y' \quad \text{(Corresponding to K-map of Fig. 10.38)}$$

$$= xy + x'z' + x'y' \quad \text{(Corresponding to K-map of Fig. 10.39)}$$

	yz	yz'	$y'z'$	$y'z$
x	1	1	0	0
x'	0	1	1	1

Fig. 10.38

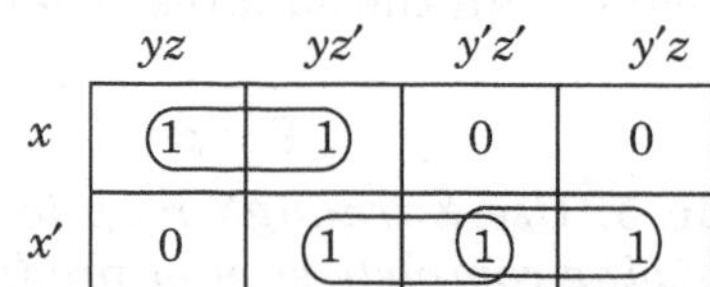

Fig. 10.39

Example 6. *Find the fundamental product represented by each basic rectangle in the K-maps shown below (Fig. 10.40)*

	zt	zt'	$z't'$	$z't$
xy	0	0	0	0
xy'	0	0	1	1
$x'y'$	0	0	0	0
$x'y$	0	0	0	0

(*a*)

	zt	zt'	$z't'$	$z't$
xy	1	1	0	0
xy'	0	0	0	0
$x'y'$	0	0	0	0
$x'y$	1	1	0	0

(*b*)

	zt	zt'	$z't'$	$z't$
xy	1	0	0	1
xy'	1	0	0	1
$x'y'$	1	0	0	1
$x'y$	1	0	0	1

(*c*)

Fig. 10.40

Sol. (*a*) In Fig. 10.40 (*a*), the two 1s can be looped (or combined) as shown in Fig. 10.41 given in next page.

The variable t appears in complemented and uncomplemented form, while x, y', z' appears in both the squares. Hence,

The fundamental product P = $xy'z'$

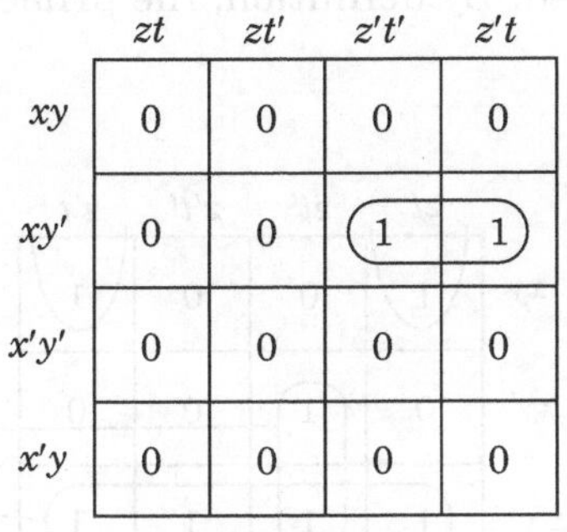

Fig. 10.41

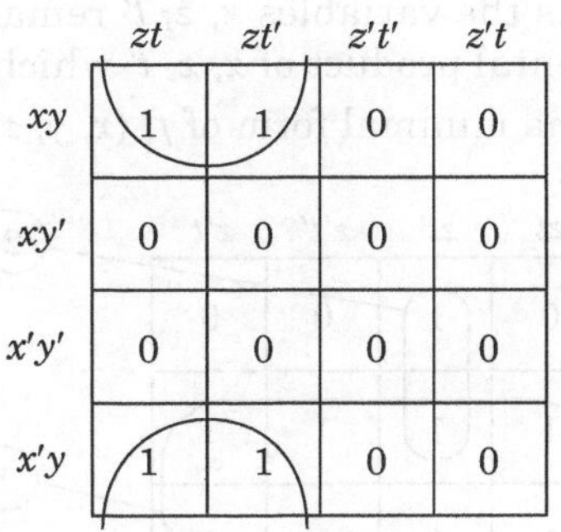

Fig. 10.42

(*b*) In Fig. 10.39 (*b*), all the four 1s can be looped (combined) as shown in Fig. 10.42.

The four squares contain the terms *xyzt, xyzt′, x′yzt, x′yzt′*. From these terms, we note that, the variables *x* and *t* appear in complemented and uncomplemented form whereas *y* and *z* appear in all the four squares. Hence, the fundamental product P is given by

$$P = yz$$

(*c*) In Fig. 10.39 (*c*), all the 1s can be looped (combined) as shown in Fig. 10.43 given below :

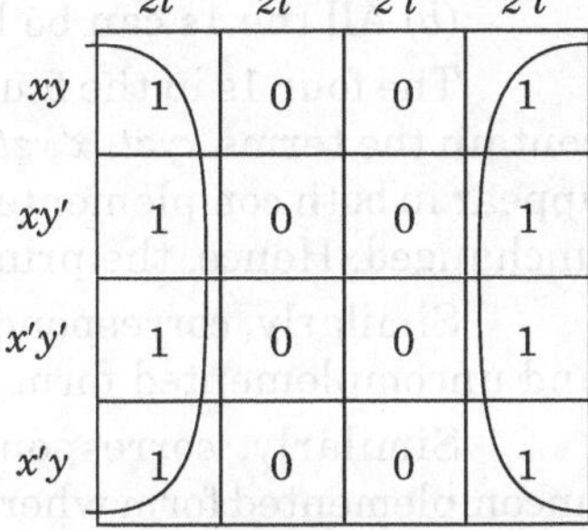

Fig. 10.43

The left most column and right most column are considered to be adjacent. In these squares, the variables *x, y, z* appear in complemented and uncomplemented form whereas the variable *t* appears in all the eight squares. Hence, the fundamental product P is given by P = *t*.

Example 7. *Find the minimal sum for the Boolean expressions f_1 and f_2 whose K-maps are given below Fig. 10.44.*

	zt	*zt′*	*z′t′*	*z′t*
xy	0	1	0	0
xy′	1	1	1	1
x′y′	1	1	1	1
x′y	0	0	0	0

(*a*) f_1

	zt	*zt′*	*z′t′*	*z′t*
xy	1	0	0	1
xy′	0	1	0	0
x′y′	1	1	1	1
x′y	1	0	0	1

(*b*) f_2

Fig. 10.44

Sol. (*a*) All the 1s can be looped (combined) as shown in Fig. 10.44 given in next page. Corresponding to loop (1), the variables *x, z, t* appear in complemented and uncomplement form whereas the variable *y′* appears in all the eight squares. Therefore, *y′* 1s is the prime implicant of $f_1(x, y, z, t)$.

Corresponding to loop (2), the variable y appears in complemented and uncomplemented form whereas the variables x, z, t' remains unchanged. By definition, the prime implicant is the fundamental product of x, z, t' which is $P = xzt'$

$\therefore$ The minimal form of $f_1(x, y, z) = y' + xzt'$

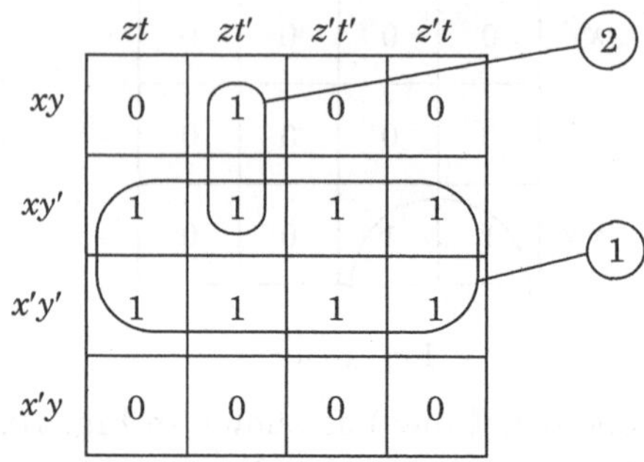

Fig. 10.45

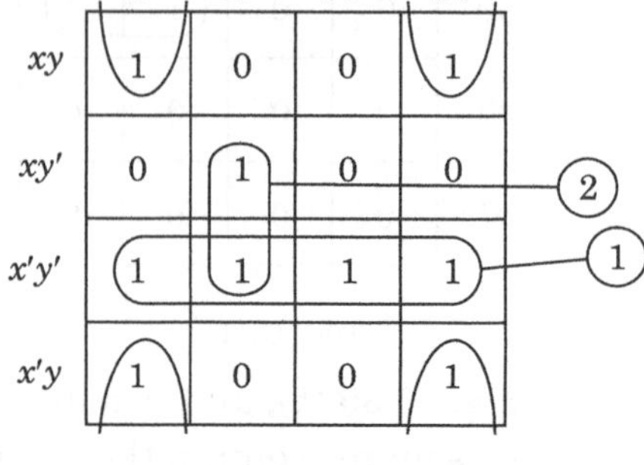

Fig. 10.46

(*b*) All the 1s can be looped (combined) as shown in Fig. 10.46.

The four 1s in the four corner squares are considered to be adjacent. These four squares contain the terms $xyzt$, $x'yzt$, $xyz't$, $x'yz't$. From these terms, we note that the variables x and z appear in both complemented and uncomplemented form whereas the variables y and t remains unchanged. Hence, the prime implicant is the fundamental product of y and t, *i.e.*, yt.

Similarly, corresponding to loop (1), the variables z and t appear in both complemented and uncomplemented form whereas x', y' remain unchanged. The prime implicant is $x'y'$.

Similarly, corresponding to loop (2), the variable x appears in complemented and uncomplemented form where *as* y', z, t' remains unchanged. The prime implicant is $y'zt'$. Hence, the required minimal form is

$$f_2(x, y, z, t) = yt + x'y' + y'zt'.$$

Example 8. *Use Karnaugh map to find the minimal sum for*

$$f(x, y, z, t) = xy' + xyz + x'y'z' + x'yzt'.$$

Sol. The Karnaugh map for four variables is shown as follows (Fig. 10.47).

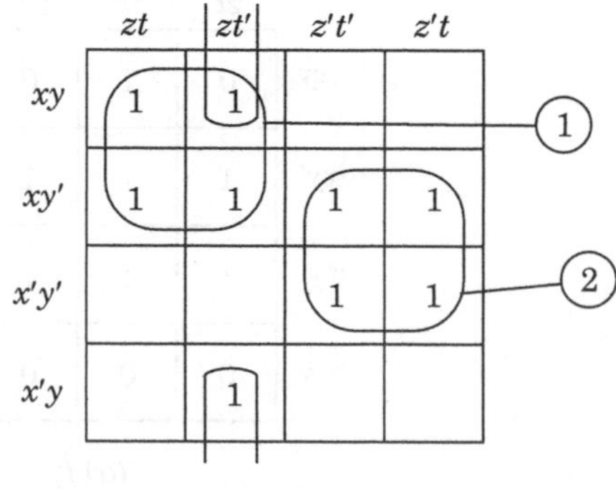

Fig. 10.47

Given $f(x, y, z, t) = xy' + xyz + x'y'z' + x'yzt'$ ...(1)

We express (1) in complete-sum-of-products form. Consider

$$\begin{aligned} xy' &= xy'(t + t')(z + z') \\ &= xy'(tz + tz' + t'z + t'z') \\ &= xy'tz + xy'tz' + xy't'z + xy't'z' \\ &= xy'zt + xy'z't + xy'zt' + xy'z't' \end{aligned}$$

Also $xyz = xyz(t + t') = xyzt + xyzt'$ ...(2)

$x'y'z' = x'y'z'(t + t') = x'y'z't + x'y'z't'$ ...(3)

From (1), (2), and (3), we have

$$f(x, y, z, t) = xy'zt + xy'z't + xy'zt' + xy'z't' + xyzt + xyzt' + x'y'z't + x'y'z't' + x'yzt' \quad ...(4)$$

Enter 1s in the cells corresponding to the terms in (4), and 0s elsewhere. All 1s in the K-map can be looped (combined) as shown in Fig. 10.47.

Corresponding to loop (1), the variables y and t appear in complemented and uncomplemented form whereas the variables x and z remains unchanged, that is to say, x and z appear in all the four squares. Therefore, the prime implicant is the fundamental product of these variable which appear in all the squares of the basic rectangle. Hence, the prime implicant corresponding to loop (1) is xz. Corresponding to loop (2), the variables x and t appear in complemented and uncomplemented form whereas the variables y' *and* z' remains unchanged. That is to say, y' and z' appear in all the four squares of the basic rectangle.

Therefore, the prime implicant corresponding to loop (2) is $y'z'$. Also, the two 1s in the top row and bottom row are also considered to be adjacent. The variable x appears in complemented and uncomplemented form whereas the variables y, z and t' remains unchanged. The prime implicant is the fundamental product of these variables which is yzt'.

Hence, the minimal sum is given by

$$f(x, y, z, t) = xz + y'z' + yzt'.$$

Example 9. *Construct the K-map for the Boolean function whose truth table is given below. Express the Boolean function, so obtained in its minimal sum*

x	y	$f(x, y)$
0	0	1
0	1	0
1	0	0
1	1	1

Sol. By **Boolean Expansion Theorem**, the Boolean function corresponding to the variables x, y is given as

$$f(x, y) = f(1, 1).x.y + f(1, 0).x.y' + f(0, 1).x'.y + f(0, 0).x'.y'$$
$$= 1.x.y + 0 + 0 + 1.x'.y'$$
$$= xy + x'y' \quad ...(1)$$

The K-map for the Boolean expression (1) is given below (Fig. 10.48) :

	y	y'
x	(1)	0
x'	0	(1)

Fig. 10.48

Enter 1s in the cells corresponding to the terms xy and $x'y'$ and 0s elsewhere. The prime implicants are xy and $x'y'$. The required minimal sum is $f(x, y) = xy + x'y'$.

Example 10. *Construct a K-map for the Boolean function whose truth table is given below. Express the Boolean function, so obtained in its minimal sum.*

x	y	z	$f(x, y, z)$
0	0	0	0
0	0	1	1
0	1	0	0
0	1	1	1
1	0	0	0
1	0	1	1
1	1	0	0
1	1	1	1

Sol. By **Boolean Expansion Theorem**, the Boolean function corresponding to the variables x, y, z is given as

$$f(x, y, z) = f(1, 1, 1)\, x.y.z + f(0, 1, 1).x'.y.z + f(1, 0, 1).x.y'.z + f(1, 1, 0).x.y.z' + f(0, 0, 1).x'.y'.z + f(0, 1, 0).x.y'.z + f(1, 0, 0).x.y'.z' + f(0, 0, 0).x'.y'.z'$$

$$= 1.x.y.z + 1.x'yz + 0 + 0 + 1.x'.y'.z + 1.x.y'.z + 0 + 0$$

$$= xyz + x'yz + x'y'z + xy'z \qquad ...(1)$$

The K-map for the Boolean function $f(x, y, z)$ is shown in Fig. 10.49.

Enter 1s corresponding to the terms in (1) and 0s elsewhere. All four 1s can be looped (combined) as shown in Fig. 10.49. All four squares in left most column and right most column are considered to be adjacent.

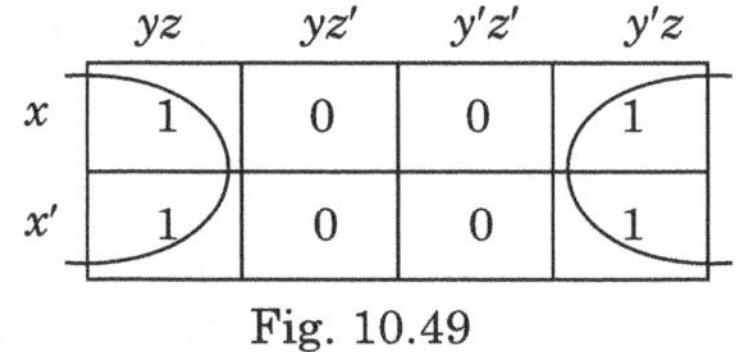

Fig. 10.49

These squares contain the terms xyz, $x'yz$, $xy'z$ and $x'y'z$. The variables x and y appear in complemented and uncomplemented form whereas z remains unchanged. Therefore, the prime implicant is z. Hence, the required minimal sum is

$$f(x, y, z) = z.$$

EXERCISE 10.5

1. Let R be a basic rectangle in a Karnaugh map for four variables x, y, z, t. Write the number of literals in the fundamental product P corresponding to R interms of the number of squares in R.
2. Find the fundamental product P represented by each basic rectangle R in the Karnaugh map in the figure given below :

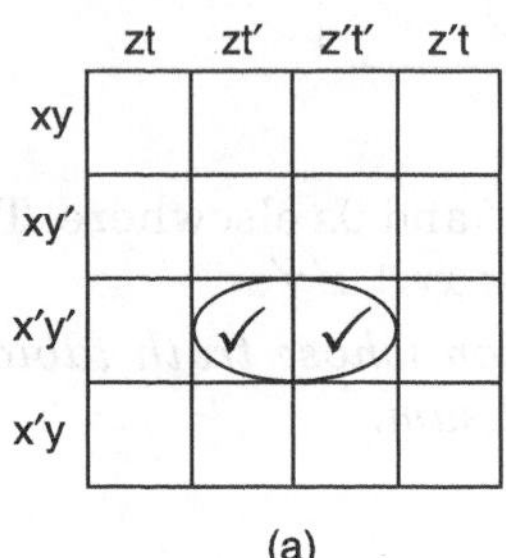

(a)

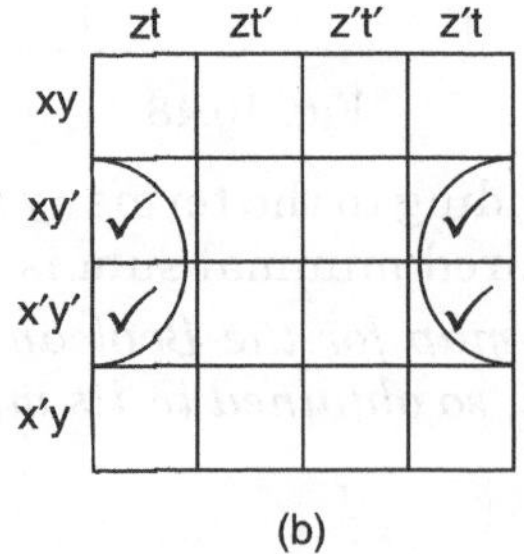

(b)

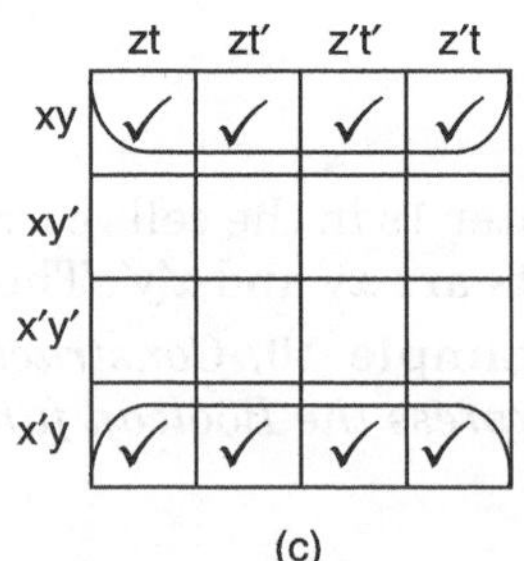

(c)

3. Let E denotes the Boolean expression given in the Karnaugh map as shown below :
(*a*) Write E as a complete sum of products form.
(*b*) Find the minimal form of E.

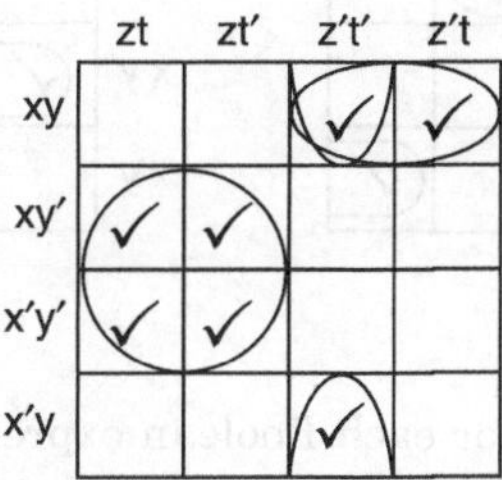

4. Consider the Boolean expressions E_1 and E_2 in variables *x*, *y*, *z* and *t* which are given by Karnaugh maps in the following figure. Find the minimal forms of E_1 and E_2.

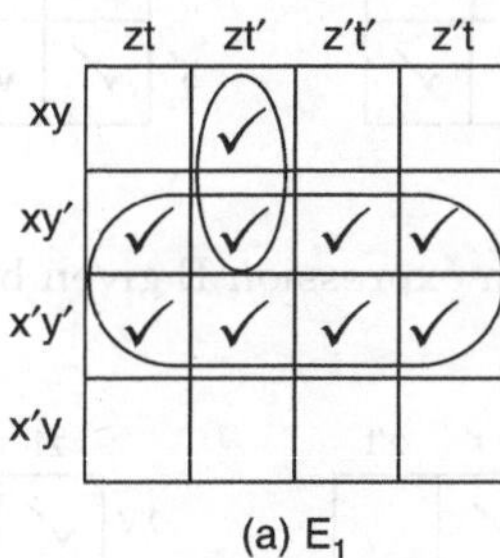

(a) E_1

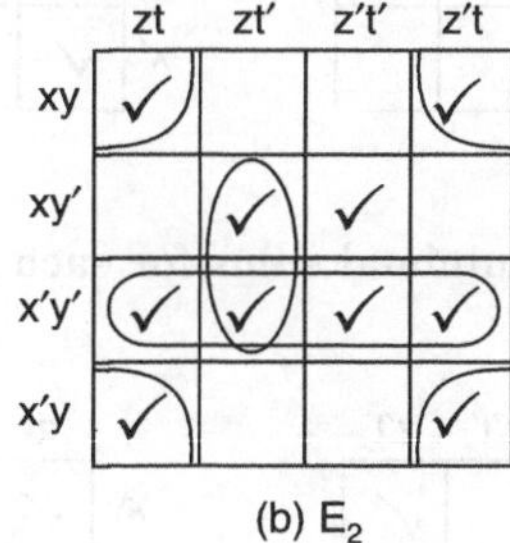

(b) E_2

5. Consider the Boolean expressions E_1 and E_2 in variables *x*, *y*, *z*, *t* which are given by the Karnaugh maps in the Figure. Find a minimal sum for (*a*) E_1 ; (*b*) E_2.

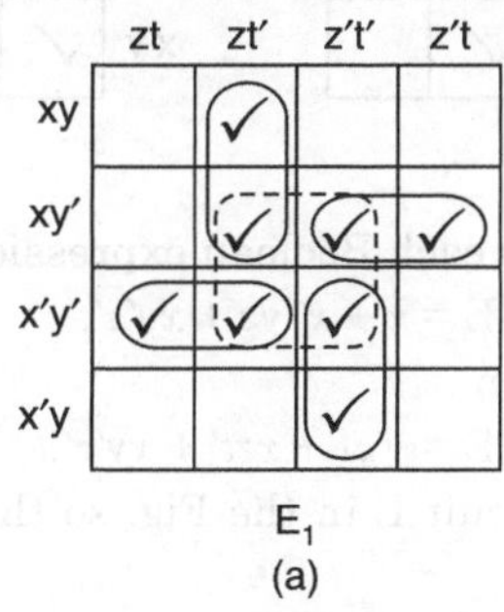

E_1
(a)

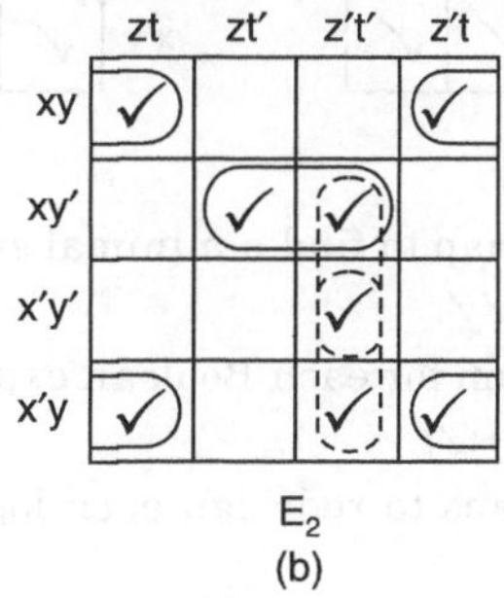

E_2
(b)

6. Use a Karnaugh map to find a minimal sum for :
(*a*) $E_1 = x'yz + x'yz't + y'zt' + xyzt'$. (*P.T.U. B.Tech. Dec. 2013*)
(*b*) $E_2 = y't' + y'z't + x'y'\,zt + yzt'$.

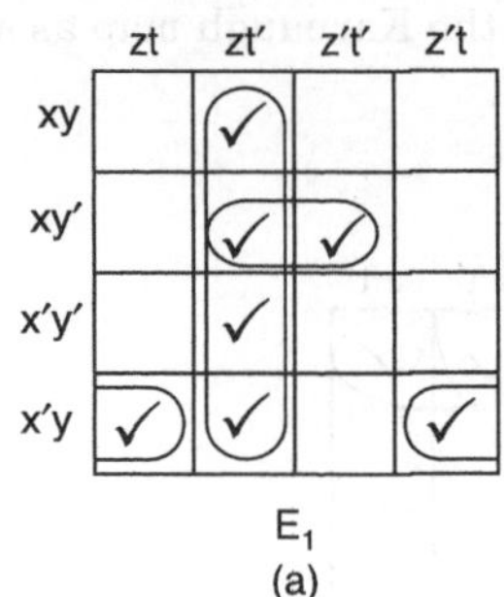
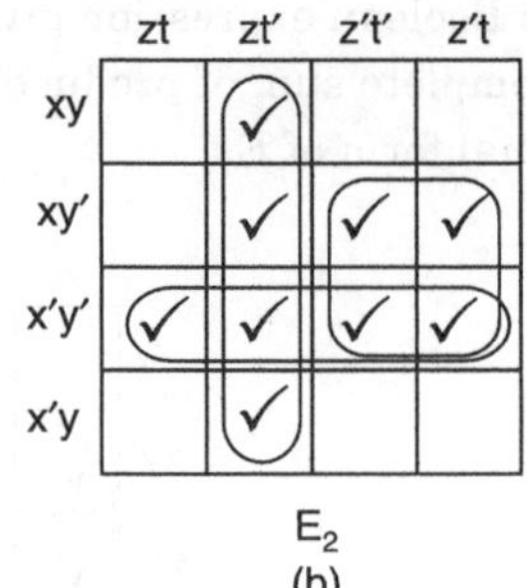

E_1
(a)

E_2
(b)

7. Find all possible minimal sums for each Boolean expression E given by the Karnaugh maps in the Fig.

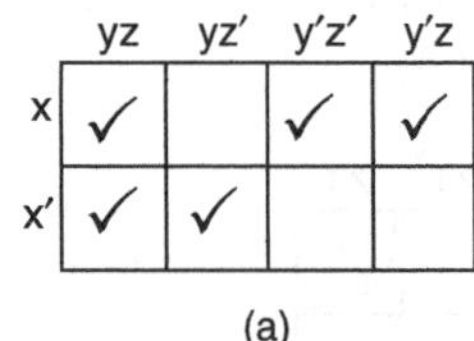
(a)

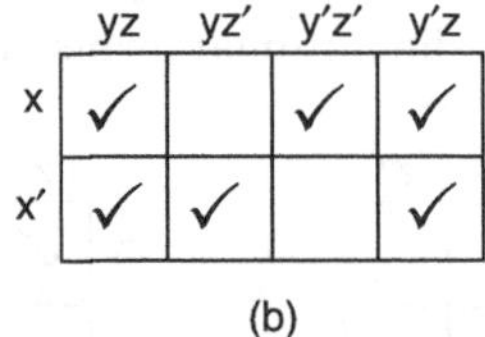
(b)

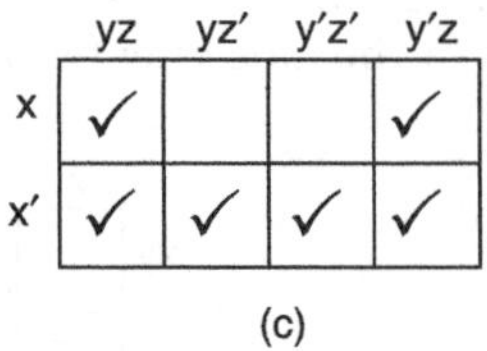
(c)

8. Find all possible minimal sums for each Boolean expression E given by the Karnaugh maps in the Fig.

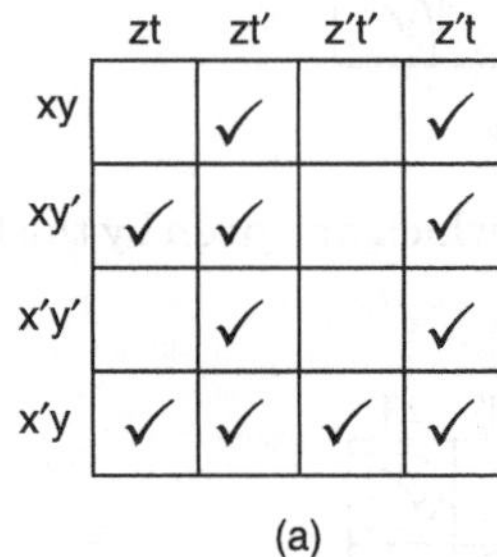
(a)

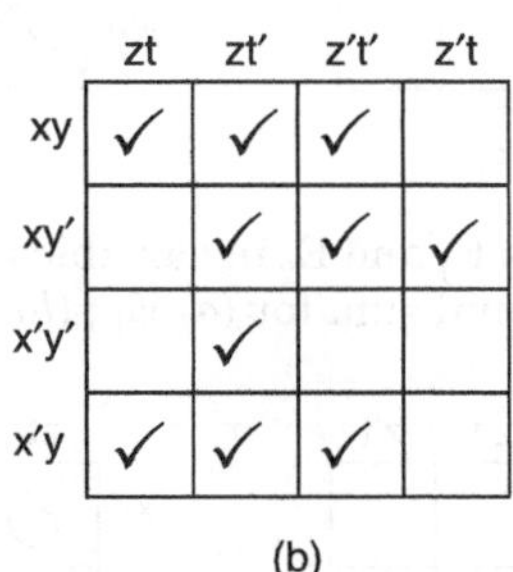
(b)

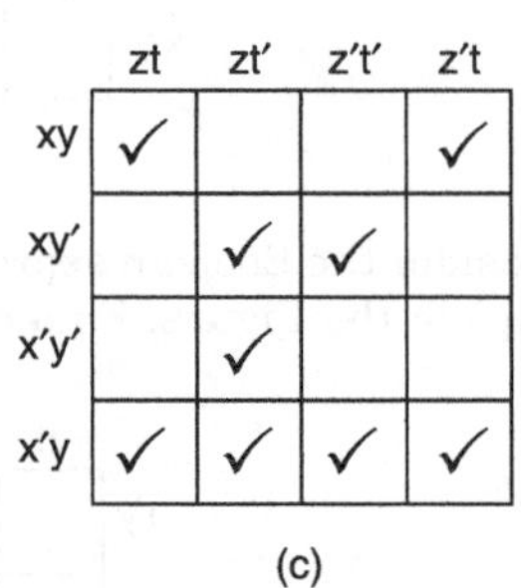
(c)

9. Use a Karnaugh map to find a minimal sum for each Boolean expression :
   (*a*) $E = xy + x'y + x'y'$. (*b*) $E = x + x'\,yz + xy'z'$.
10. Find a minimal sum for each Boolean expression :
   (*a*) $E = y'z + y'z't' + z't$. (*b*) $E = y'zt + xzt' + xy'z'$.
11. Use Karnaugh maps to redesign each logic circuit L in the Fig. so that it becomes a minimal AND-OR circuit.

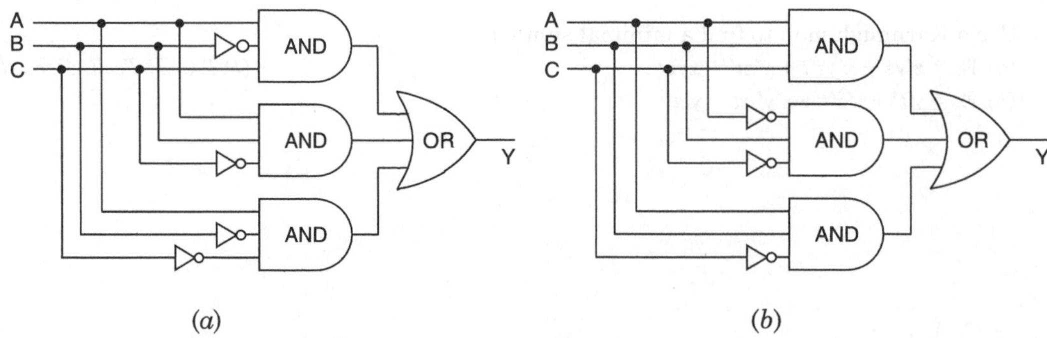

Fig. 10.50

Answers

1.

R	P
8 (squares)	1 (literal)
4 (squares)	2 (literals)
2 (squares)	3 (literals)
1 (square)	4 (literals)

2. (*a*) $P = x'yt'$ (*b*) $P = y't$ (*c*) $P = y$.
3. (*a*) $E = x'y'zt + x'\ y'zt' + x'yz't'$ (complete sum of product from)
   (*b*) $E = y'z + xyz' + yz't$ (minimal form)
4. (*a*) $E_1 = y' + xzt'$ (*b*) $E_2 = yt + x'y' + y'zt'$
5. (*a*) $E_1 = xzt' + xy'z' + x'y'z + x'z't'$
   (*b*) $E_2 = x'y + yt + xy't' + xy't' + y'z't' = x'y + yt + xy't' + x'z't.$
6. (*a*) $E_1 = zt' + xy't' + x'yt$ (*b*) $E = zt' + xy't' + x'y't.$
7. (*a*) $E = xy' + x'y + yz = xy' + x'y + xz'.$ (*b*) $E = xy' + x'y + z.$ (*c*) $E = x' + z.$
8. (*a*) $E = x'y + zt' + xz't + xy'z = x'y + zt' + xz't + xy't.$
   (*b*) $E = yz + yt' + zt' + xy'z'.$
   (*c*) $E = x'y + yt + xy't' + x'zt = x'y + yt + xy't' + y'zt.$
9. (*a*) $E = x' + y;$ (*b*) $E = xz' + yz.$
10. (*a*) $E = y' + z't;$ (*b*) $E = xy' + zt' + y'zt.$
11.

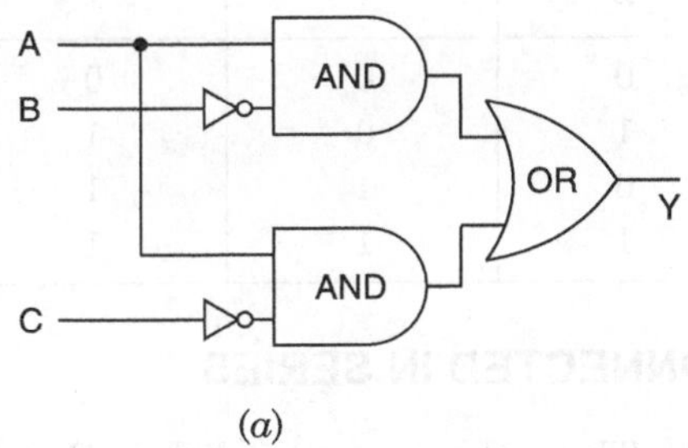

(*a*)

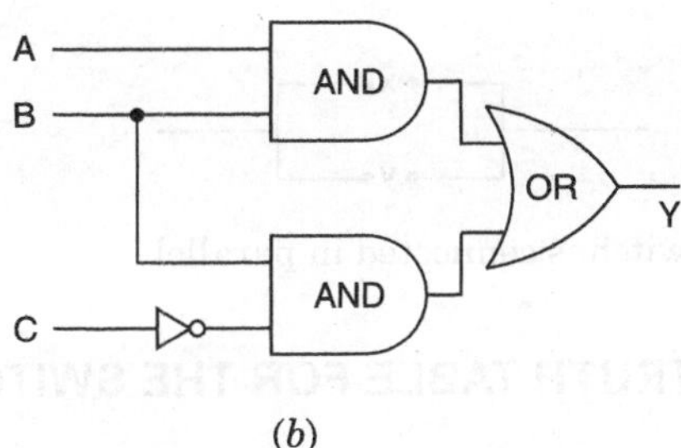

(*b*)

Fig. 10.51

10.40. APPLICATIONS OF BOOLEAN ALGEBRA TO SWITCHING CIRCUITS

Switching circuit. It is an arrangement of wires and switches connected together to the terminal of the battery. A **Switch** *is a two state* device used for allowing current to pass through it or not to pass through it.

If current is allowed to pass through a switch, then it is said to be **'closed'** or **'ON'**. If the current is not allowed to pass through a switch, them it is said to be 'Open' or 'OFF'. The switches are denoted by the letters *x, y, z* or *a, b, c*.

Methods of connecting Two switches. There are basically two methods of connecting two switches :

(*i*) Connecting switches in parallel

(*ii*) Connecting switches in series.

If x and y are two switches connected in parallel, then we say, the bulb is 'ON' iff atleast one of the switches x and y are closed. (Fig. 10.52)

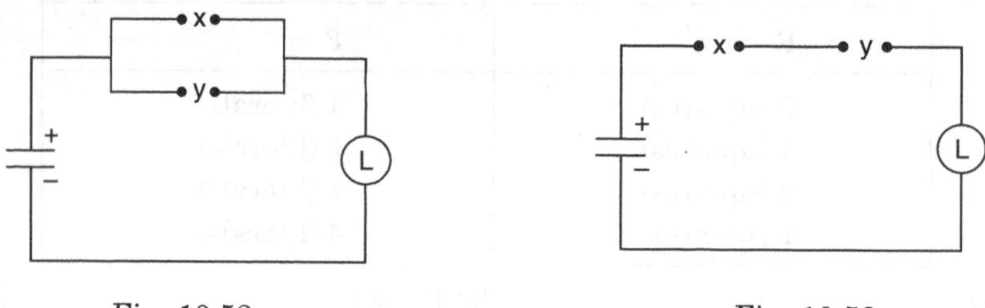

Fig. 10.52 Fig. 10.53

If x and y are two switches connected in series, then we say, the bulb is 'ON' iff x and y are both closed. (Fig. 10.53)

Further, if a switch is 'ON', then its value is denoted by '1' and if a switch is 'OFF'. Then its value is denoted by '0'.

10.41. TRUTH TABLE FOR THE SWITCHES CONNECTED IN PARALLEL

Let the switches x and y are connected in parallel. Then, the current will flow through the circuit of x and y only when at least one switch is 'ON'. In other words, the value of the circuit of x and y is 1 when at least one switch has value 1. The circuit having switches x and y in parallel is denoted by $x + y$. The truth table for the switches connected in parallel is shown as :

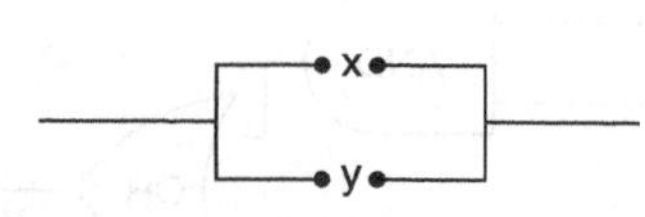

Switches connected in parallel

Truth table

x	y	$x + y$
0	0	0
1	0	1
0	1	1
1	1	1

10.42. TRUTH TABLE FOR THE SWITCHES CONNECTED IN SERIES

Let the switches x and y are connected in series. Then, the current will flow through the circuit of x and y only when both switches are 'ON'. In other words, the value of the circuit of x and y is 1 when both switches have value 1. The circuit having switches x and y in series is denoted by $x.y$.

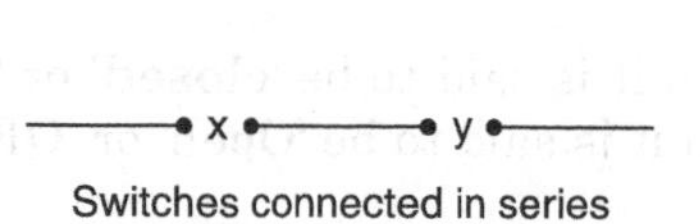

Switches connected in series

Truth table

x	y	$x.y$
0	0	0
1	0	0
0	1	1
1	1	1

If x and y are two switches such that y is closed when x is open and y is open when x is closed, then y is written as x'.

ILLUSTRATIVE EXAMPLES

Example 1. *Consider the circuit as shown below. Express the circuit as a Boolean function of the Boolean algebra of switching circuit (Fig. 10.54).*

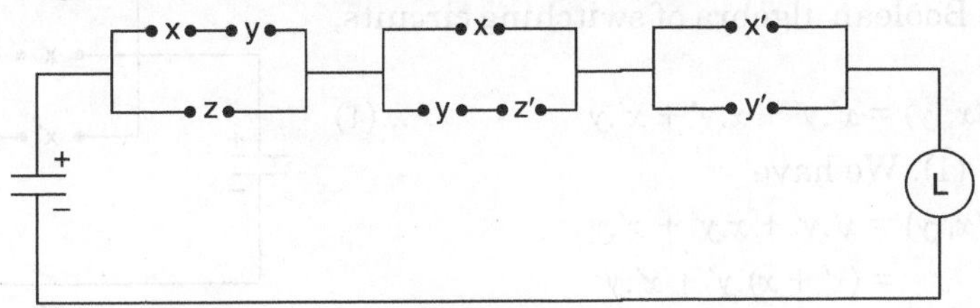

Fig. 10.54

Sol. From the given circuit, the first portion represents $x.y + z$

The second portion represents $x + y.z'$

The third portion represents $x' + y'$.

All these three portion are connected in series.

∴ The Boolean function of the given circuit is

$$f(x, y, z) = (xy + z)\ (x + yz')\ (x' + y').$$

Example 2. *Consider the Boolean function $f(x, y, z) = (x.y.z') + x'.(y + z')$. Construct a circuit representing $f(x, y, z)$ as a Boolean algebra of switching circuits.*

Sol. Given Boolean function is

$$f(x, y, z) = x.y.z' + x'.(y + z') \quad ...(1)$$

From (1), the circuits corresponding to $x.y.z'$ and $x'.(y + z')$ are connected in parallel.

The term $x.y.z'$ implies that the switches x, y and z' are connected in series.

The terms $x'.(y + z')$ implies that the switch x' and the circuit corresponding to $y + z'$ are connected in series. The required circuit is shown in Fig. 10.55.

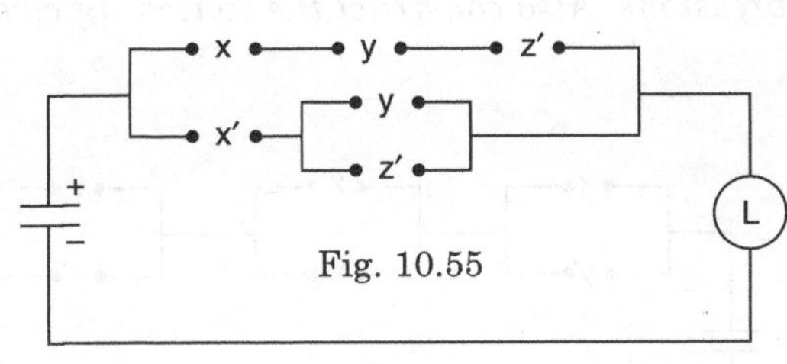

Fig. 10.55

Example 3. *Consider the Boolean function*

$$f(x, y, z) = (x.y. + z).(x' + y.z').(x' + z)$$ (P.T.U. May 2012)

Construct the circuit corresponding to the Boolean function of the Boolean algebra of switching circuits.

Sol. Given Boolean function is

$$f(x, y, z) = (x.y + z).(x' + y.z').(x' + z) \quad ...(1)$$

From (1), we observe that the circuits corresponding to $x.y + z$; $x' + y.z'$ and $x' + z$ are connected in series. The term, $x.y + z$ implies that the circuit corresponding to $x.y$ and the switch z are connected in parallel. The term, $x' + y.z'$ implies that the switch x' and the circuit corresponding to $y.z'$ are connected in parallel.

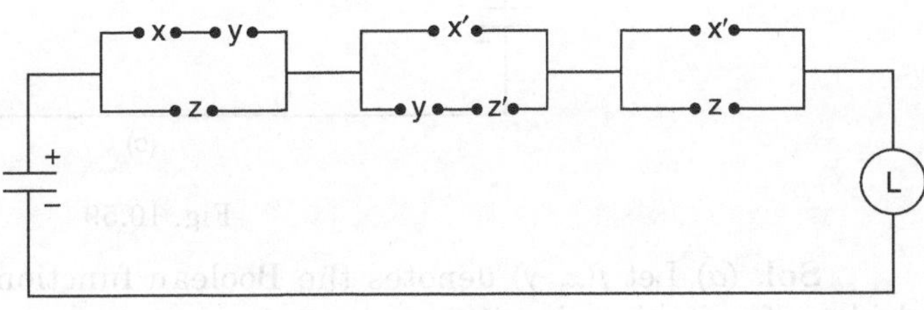

Fig. 10.56

The term, $x' + z$ implies that the switches x' and z are connected in parallel. Hence, the circuit of the given Boollean function is shown in Fig. 10.56.

Example 4. *Simplify the circuit as shown below. (Fig. 10.57)*

Sol. Let $f(x, y)$ denotes the Boolean function of the given circuit in the Boolean algebra of switching circuits, then

$$f(x, y) = x'.y' + x.y' + x'.y \quad \ldots(1)$$

Fig. 10.57

We simplify (1). We have

$$\begin{aligned} f(x, y) &= x'.y' + x.y' + x'.y \\ &= (x' + x).y' + x'.y \\ &= 1.y' + x'.y \quad \text{| Complement law} \\ &= y' + x'.y \\ &= (y' + x').(y' + y) \quad \text{| Disributive law} \\ &= (y' + x').1 \quad \text{| Complement law} \\ &= y' + x' \\ &= x' + y' \quad \ldots(2) \\ & \quad \text{| Commutative law} \end{aligned}$$

From (2), we can say that the given circuit is equivalent to the circuit in which the switches x' and y' are connected in parallel. (Fig. 10.58)

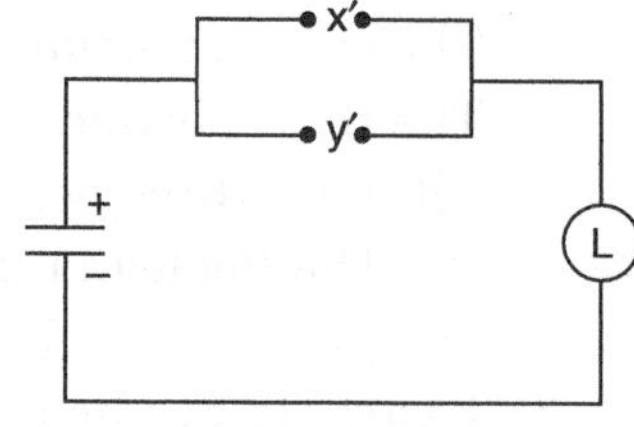

Fig. 10.58

Example 5. *Simplify the circuits given below Fig. 10.59 by obtaining the Boolean expressions. Also construct the switching circuit for the simplified Boolean expression.*

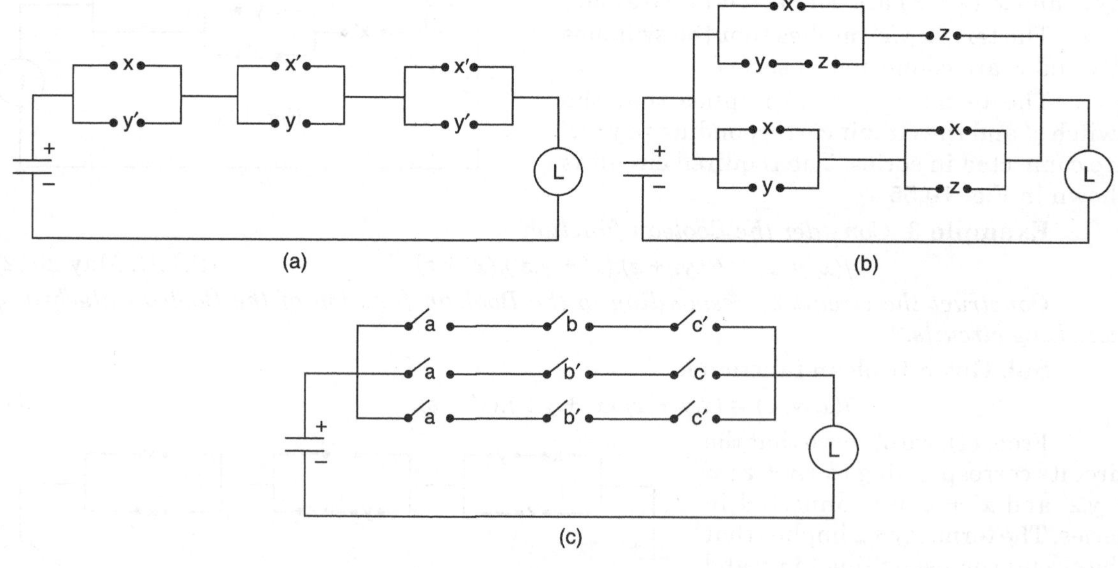

Fig. 10.59

Sol. (*a*) Let $f(x, y)$ denotes the Boolean function of the given circuit in the Boolean algebra of switching circuits.

First portion of the circuit represents $x + y'$

Second portion of the circuit represents $x' + y$

Third portion of the circuit represents $x' + y'$

All these three portions are connected in series. Therefore, the Boolean function is

$$f(x, y) = (x + y').(x' + y).(x' + y') \quad ...(1)$$

Now to simplify (1), we have

$$\begin{aligned} f(x, y) &= (x + y').(x' + y).(x' + y') \\ &= (x + y').(x' + y.y') && | \text{ Distributive law} \\ &= (x + y').(x' + 0) && | \text{ Complement law} \\ &= (x + y').x' \\ &= x.x' + y'.x' \\ &= 0 + y'.x' && | \text{ Complement law} \\ &= x'.y' \quad ...(2) && | \text{ Commutative law} \end{aligned}$$

Hence from (2), we conclude that the given circuit is equivalent to the circuit in which the switches x' and y' are connected in series. The circuit diagram for (2) is given in Fig. 10.60.

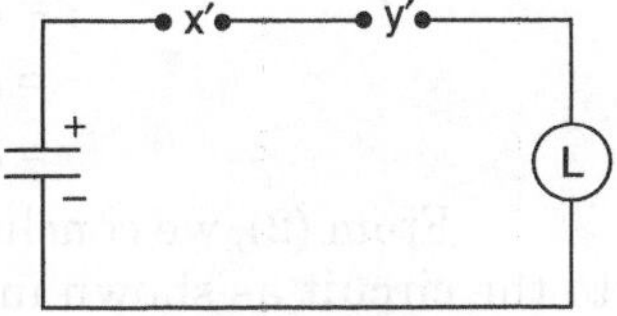

Fig. 10.60

(*b*) Let $f(x, y, z)$ denotes the Boolean expression of the given circuit in the Boolean algebra of the switching circuits.

In the top row, the first portion represents that the switch x and the circuit $y.z$ are connected in parallel and hence the top row represents the circuit

$$(x + y.z).z \quad ...(2)$$

In the bottom row of the circuit, First portion represents that the switches x and y are connected in parallel and the second portion represents that the switches x and z are connected in parallel. Both the circuits $(x + y)$ {corresponding to the first portion of the bottom row} and $(x + z)$ {corresponding to the second portion of the bottom row) are connected in series and hence the corresponding circuit is

$$(x + y).(x + z) \quad ...(3)$$

The circuits represented by (2) and (3) are connected in parallel. Therefore, the Boolean function is

$$f(x, y, z) = (x + y.z).z + (x + y).(x + z) \quad ...(4)$$

Now to simplify (4), we have

$$\begin{aligned} f(x, y, z) &= (x + y.z).z + (x + y).(x + z) \\ &= x.z + y.z.z + y.z && | \text{ Distributive law} \\ &= x.z + y.z + x + y.z && | \text{ Idempotent law} \\ &= x + x.z + y.z && | \text{ Idempotent law} \\ &= x + y.z \quad ...(5) && | \text{ Absorption law} \end{aligned}$$

From (5), we conclude that the given circuit is equivalent to the circuit in which the switch x and the circuit $y.z$ are connected in parallel. The simplified circuit is shown in Fig. 10.61.

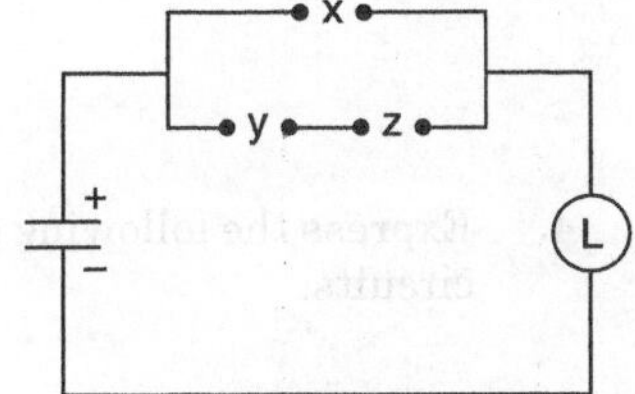

Fig. 10.61

(c) Let $f(a, b, c)$ denoted the Boolean function of the given circuit in the Boolean algebra of the switching circuits. Proceeding yourself as in part (*i*) and (*ii*), we have

$$f(a, b, c) = a.b.c' + a.b'.c + a.b'.c' \quad ...(1)$$

To simplify (1), we have

$$\begin{aligned} f(a, b, c) &= a.b.c' + a.b'.c + a.b'.c' \\ &= a.b.c' + a.b'.(c + c') && \text{| Distributive law} \\ &= a.b.c' + a.b'.1 && \text{| Complement law} \\ &= a.b.c' + a.b' \\ &= a.(b.c' + b') && \text{| Distributive law} \\ &= a.(b + b').(c' + b') \\ &= a.1.(c' + b') && \text{| Complement law} \\ &= a.(c' + b') \\ &= a.(b' + c') && ...(2) \end{aligned}$$

From (2), we conclude that the given circuit is equivalent to the circuit as shown in Fig. 10.62.

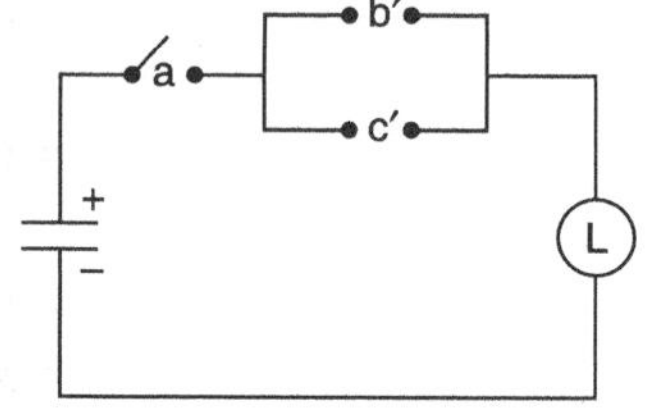

Fig. 10.62

TEST YOUR KNOWLEDGE 10.6

1. Express the following circuit (Fig. 10.112) as a Boolean function of the Boolean algebra of switching circuits.

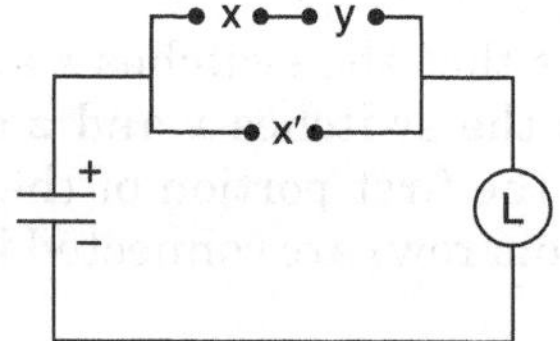

Fig. 10.63

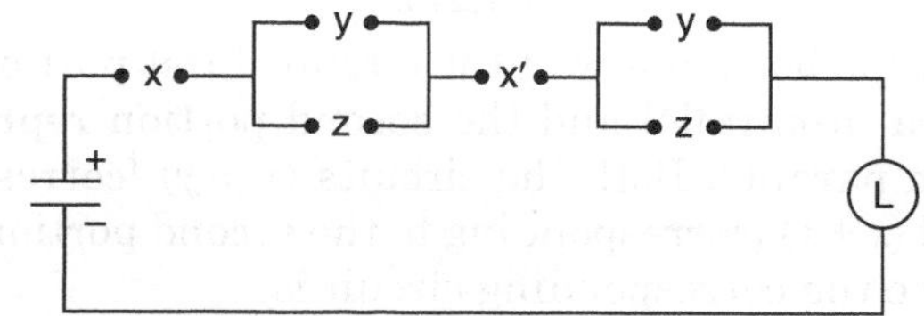

Fig. 10.64

2. Express the above circuit (Fig. 10.64) as a Boolean function of the Boolean algebra of switching circuits.
3. Express the following circuit (Fig. 10.65) as a Boolean function of the Boolean algebra of switching circuits.

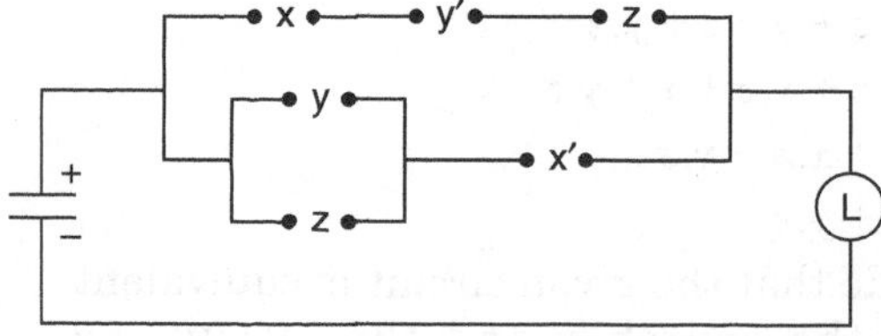

Fig. 10.65

4. Express the following circuit (Fig. 10.66) as a Boolean function of the Boolean algebra of switching circuits.

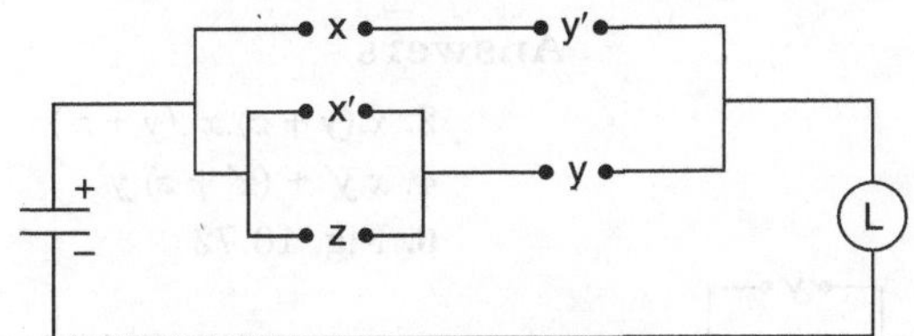

Fig. 10.66

5. Construct the circuit corresponding to the Boolean function $x.(y' + z) + y.z'$ of the Boolean algebra of switching circuits.
6. Construct the circuit corresponding to the Boolean function $x.y + x'.(x + y + y')$ of the Boolean algebra of switching circuits.
7. Construct the circuit corresponding to the Boolean function $x.y + x.y' + x'.y$ of the Boolean algebra of switching circuits.
8. Construct the circuit corresponding to the Boolean function $x.y'.z + (y + z).x'$ of the Boolean algebra of switching circuits.
9. Simplify the circuit (Fig. 10.67) given below :

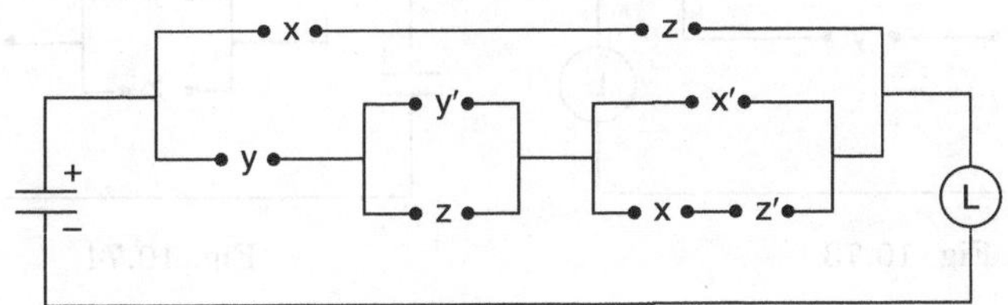

Fig. 10.67

10. Simplify the circuit (Fig. 10.68) given below :

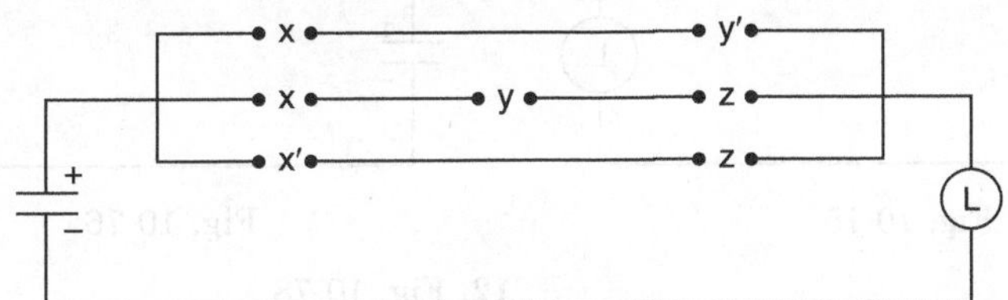

Fig. 10.68

11. Simplify the circuit (Fig. 10.69) :

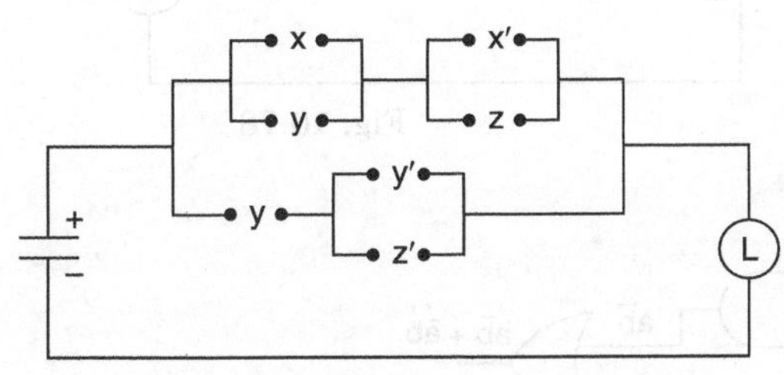

Fig. 10.69

12. Simplify the circuit (Fig. 10.70) given below :

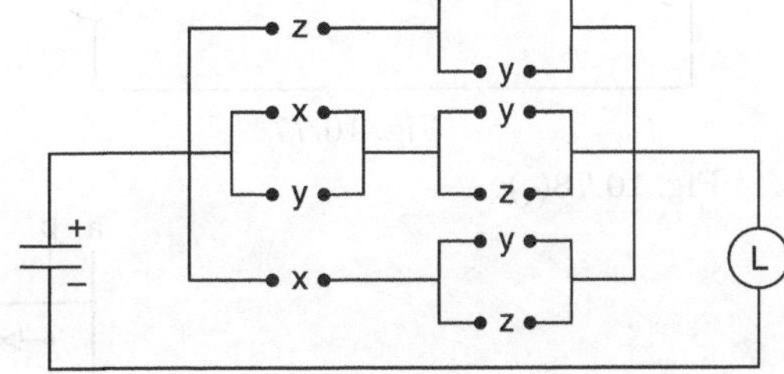

Fig. 10.70

13. Draw the logic circuit for $ab' + a'b$. (*P.T.U., B. Tech. Dec. 2012*)

Answers

1. $x.y + x'$
2. $x.(y + z).x'.(y + z')$
3. $x.y'. z + (y + z) . x'$
4. $x.y' + (x' + z).y$
5. Fig. 10.71
6. Fig. 10.72

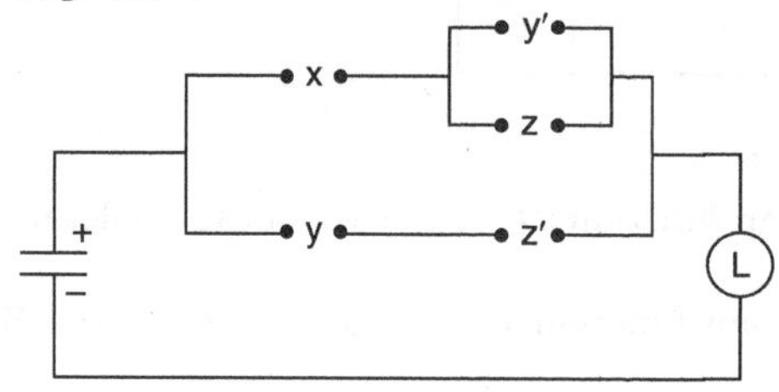

Fig. 10.71

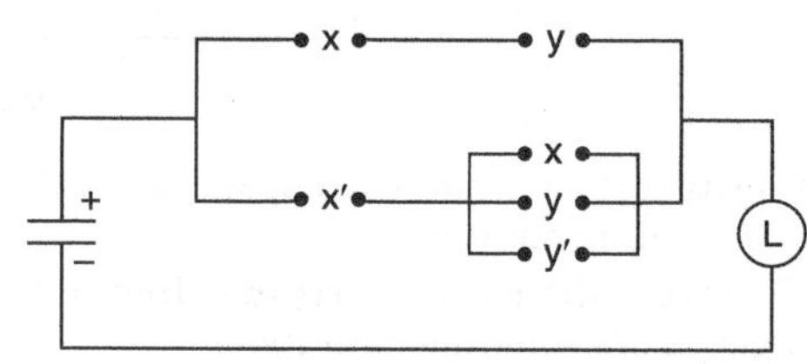

Fig. 10.72

7. Fig. 10.73
8. Fig. 10.74

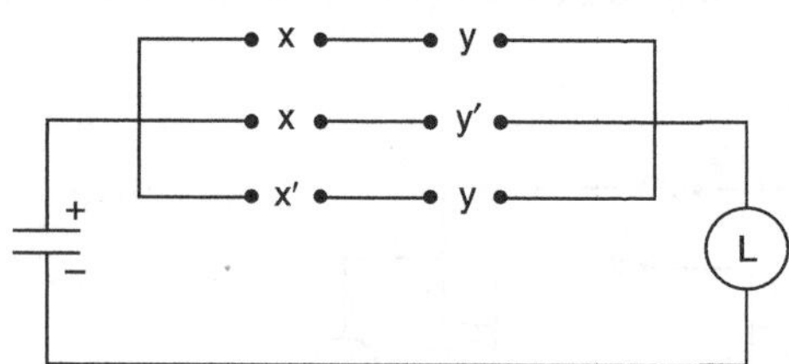

Fig. 10.73

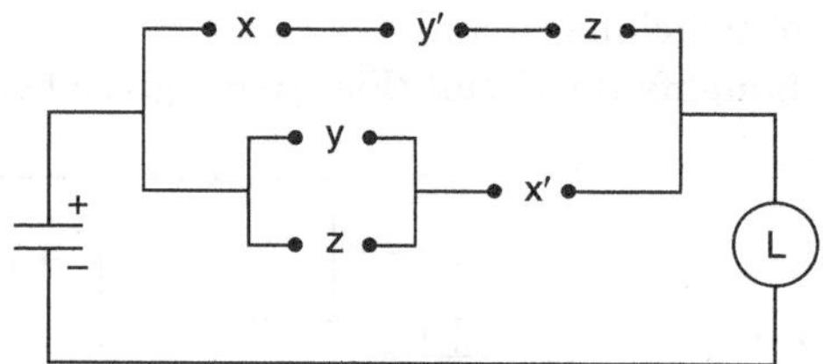

Fig. 10.74

9. Fig. 10.75
10. Fig. 10.76

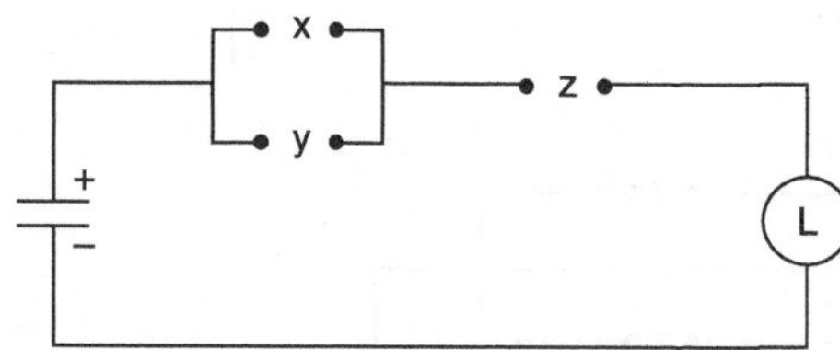

Fig. 10.75

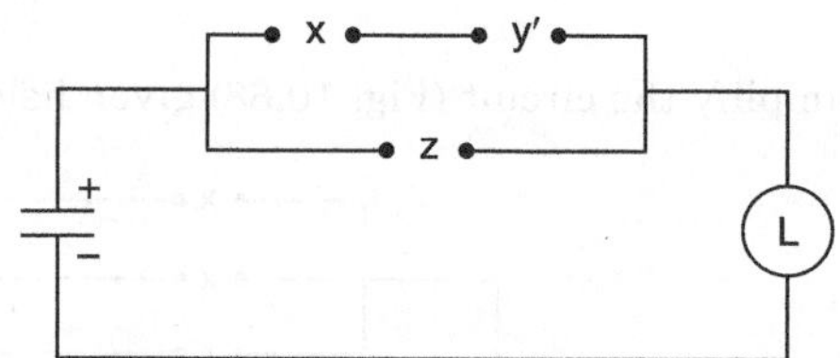

Fig. 10.76

11. Fig. 10.77
12. Fig. 10.78

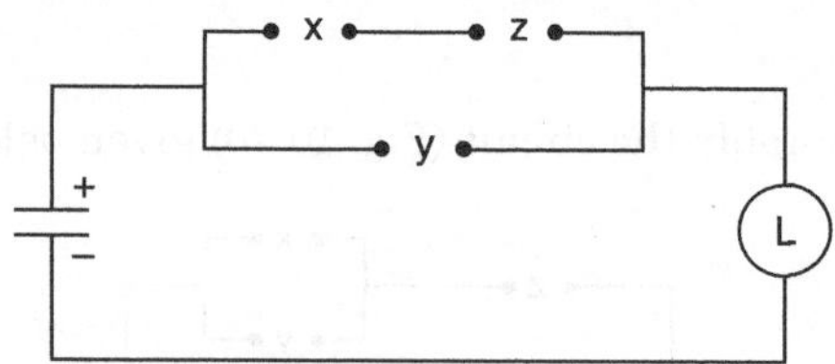

Fig. 10.77

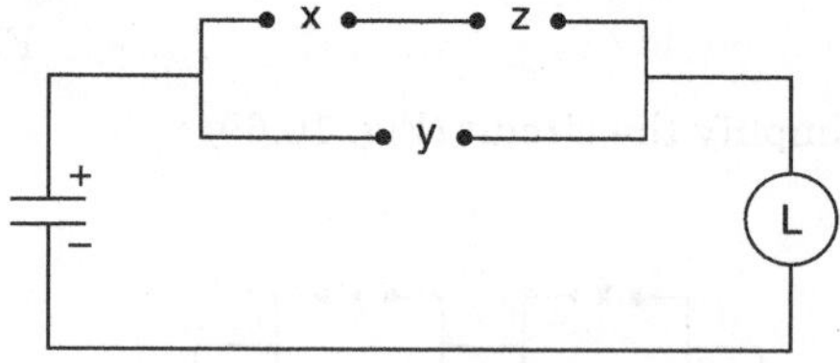

Fig. 10.78

13. Fig. 10.78(*a*)

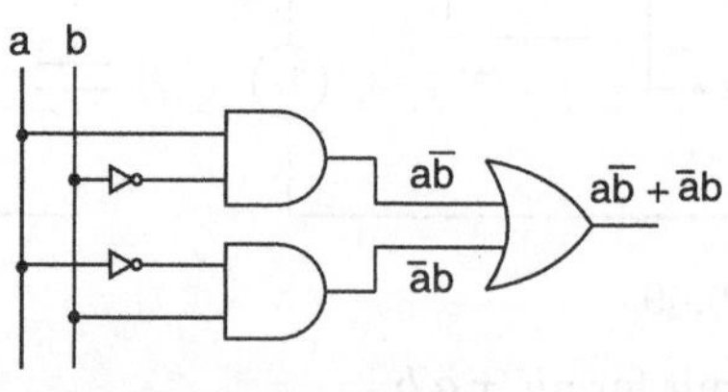

Fig. 10.78(*a*)

10.43. APPLICATION OF BOOLEAN ALGEBRA TO LOGIC CIRCUITS

Logic gates. A logic gate is simply an electronic circuit which operate on one or more input signals to produce standard output signals. These logic gates are the building blacks of all the circuits in a computer.

There are, basically, three logic gates, namely,

(*i*) OR gate (*ii*) AND gate (*iii*) NOT gate.

We shall use the fact that the lines entering the gate symbol from the left are input lines and the single line on the right is the output line.

OR gate. An OR gate is an electronic circuit that generates the output signal of 1 if any of the input signals is 1. Two or more switches connected in parallel behave as an OR gate.

(i) (ii)

If the switches x and y are connected in parallel then the input current will reach the output point when any one of the two switches are in the ON (1) state. There will be no output when both the switches x and y are in the off (0) state. The OR gate for two input signals x and y is shown in Fig. (*ii*).

Also the truth table for OR gate is shown below. An output of signal is 1 when any of the input signals is 1 and output is zero when both the inputs are zero.

Truth table for an OR gate

Input		*Output*
x	y	$x + y$
0	0	0
1	0	1
0	1	1
1	1	1

An OR gate may have more than two inputs. The figure given below shows an OR gate having three inputs. The truth table for OR gate with three inputs x, y, z is shown below :

Truth table for an OR gate

Inputs			*Output*
x	y	z	$x + y + z$
0	0	0	0
0	0	1	1
0	1	0	1
1	0	0	1
0	1	1	1
1	0	1	1
1	1	0	1
1	1	1	1

AND gate. An AND gate is an electronic circuit that generates the output signal of 1 if all the input signals are 1. Two or more switches connected in series behave as an AND gate.

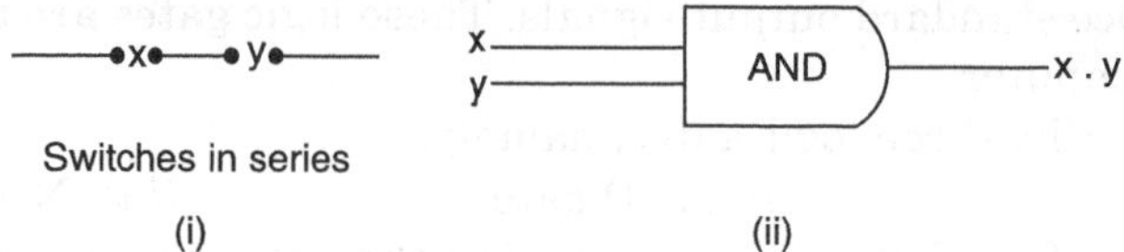

Switches in series

(i) (ii)

If the switches x and y are connected in series, then the input current will reach the output point only when both the switches are in ON (1) state. There will be no output when at least one of the switches x and y is in the off (0) state the AND gate for two input signals x and y is shown in Fig. (*ii*). An output of signal is 1 when both input signals are 1. Output is 0 when at least one of the inputs is 0.

Truth table for an AND gate

	Input	*Output*
x	y	$x.y$
0	0	0
1	0	0
0	1	0
1	1	1

An AND gate may have more than two inputs. The truth table for an AND gate with three inputs is given below. The figure given below shows an AND gate having three inputs. The AND gate with inputs x, y, z has output 1 only if x, y, z have value 1.

Truth table for an AND gate

Input			*Output*
x	y	z	$x.y.z$
0	0	0	0
0	0	1	0
0	1	0	0
1	0	0	0
0	1	1	0
1	0	1	0
1	1	0	0
1	1	1	1

x y z

x
y
z
AND
x . y . z

11.27. DISCONNECTED DIGRAPH

A directed graph is called disconnected if its undirected graph is disconnected.

Example 28. *Consider the graphs shown in Figs. 11.81, 11.82 and 11.83 which of the graphs are*

(i) Unilaterally connected digraph *(ii) Weakly connected digraph*

(iii) Strongly connected digraph

(iv) Disconnected digraph (also find its connected components).

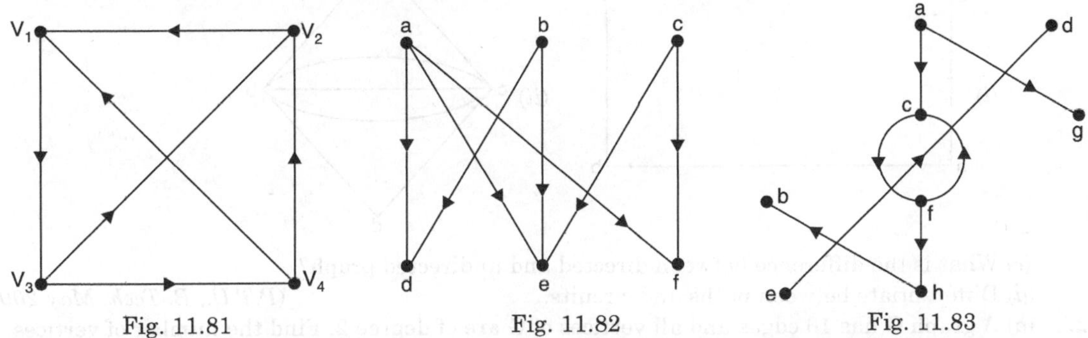

Fig. 11.81 Fig. 11.82 Fig. 11.83

Sol. The graph shown in Fig. 11.81 is strongly connected because there is a path from every vertex u to v and also there is a path from v to u. It is also weakly and unilaterally connected because a strongly connected digraph is both weakly and unilaterally connected.

The graph shown in Fig. 11.82 is weakly connected but not unilaterally connected because there is no directed path from vertex a to b or a to c etc. but its undirected graph is connected.

The graph shown in Fig. 11.83 is a disconnected graph. The components of this graph are $\{g, a, c, f, h, b\}$ and $\{e, d\}$.

11.28. DIRECTED COMPLETE GRAPH

A directed complete graph G = (V, E) on n vertices is a graph in which each vertex is connected to every other vertex by an arrow. It is denoted by K_n.

Example 29. *Draw directed complete graphs K_3 and K_5.*

Sol. Place the number of vertices at appropriate place and then draw an arrow from each vertex to every other vertex as shown in Figs. 11.84 and 11.85.

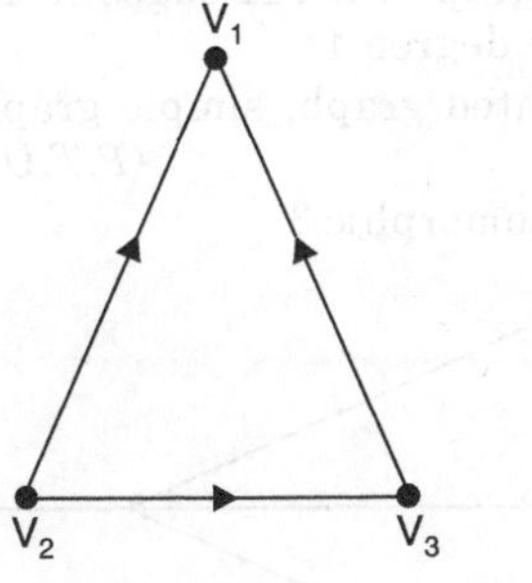

Fig. 11.84. K_3.

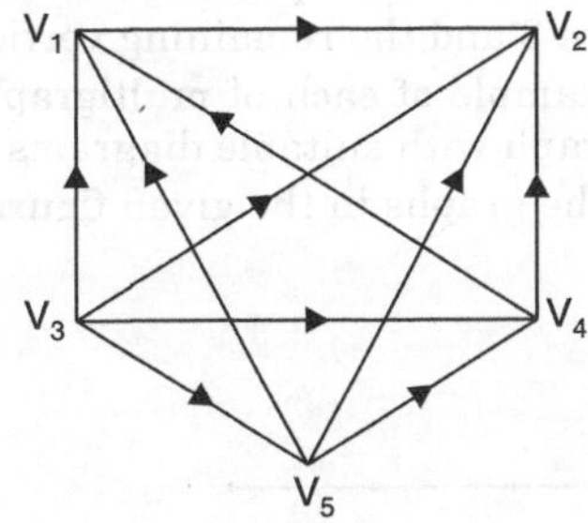

Fig. 11.85. K_5.

TEST YOUR KNOWLEDGE 11.1

1. (*a*) If V = {1, 2, 3, 4, 5} and E = {(1, 2), (2, 3), (3, 3), (3, 4), (4, 5)}. Find the number of edges and size of graph G = (V, E)

   (*b*) Find the order and size of the graph G shown in the figure below :

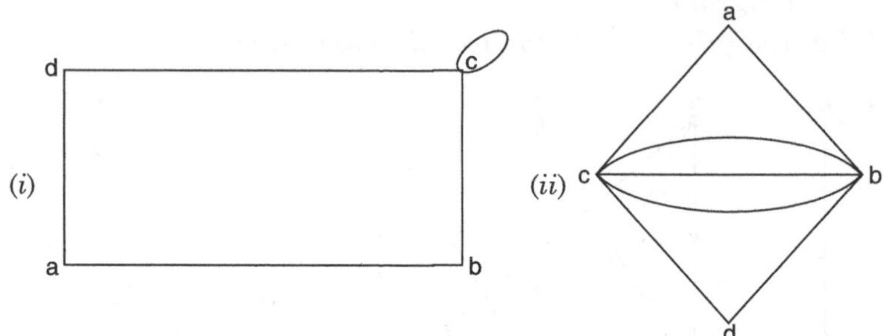

   (*c*) What is the difference between directed and undirected graph?

   (*d*) Differentiate between paths and circuits. (*P.T.U., B. Tech. May 2007*)

2. (*a*) A graph G has 16 edges and all vertices of G are of degree 2. Find the number of vertices.

   (*b*) A graph G has 21 edges, 3 vertices of degree 4 and other vertices are of degree 3. Find the number of vertices in G.

   (*c*) A graph G has 5 vertices, 2 of degree 3 and 3 of degree 2. Find the number of edges.

3. (*a*) How many nodes (vertices) are required to construct a graph with exactly 6 edges in which each node is of degree 2?

   (*b*) Show that there does not exist a graph with 5 vertices with degrees 1, 3, 4, 2, 3 respectively.

   (*c*) Can there be a graph with 8 vertices and 29 edges?

   (*d*) How many vertices are there is a graph with 10 edges if each vertex has degree 2?

   (*e*) Does there exist a graph with two vertices each of degree 4? If so, draw it.

4. (*a*) Draw a simple graph with 3 vertices

   (*b*) Draw a simple graph with 4 vertices

   (*c*) Give an example for

   (*i*) simple graph (*ii*) non-simple graph (*iii*) Multigraph, with suitable diagrams

5. Show that the maximum number of edges in a graph with n vertices and no multiple edges are $\frac{n(n-1)}{2}$.

6. Prove Handshaking theorem which states that the sum of degree of the vertices of a graph is equal to twice the number of edges.

7. (*a*) Determine whether it is possible to construct a graph with 12 edges such that 2 of the vertices have degree 3 and the remaining vertices have degree 4.

   (*b*) Give an example of each of multigraph, weighted graph, simple graph, non-simple graph, directed graph with suitable diagrams. (*P.T.U., B.Tech. May 2005*)

8. (*a*) Which of the graphs in the given figures are isomorphic ?

(*i*) 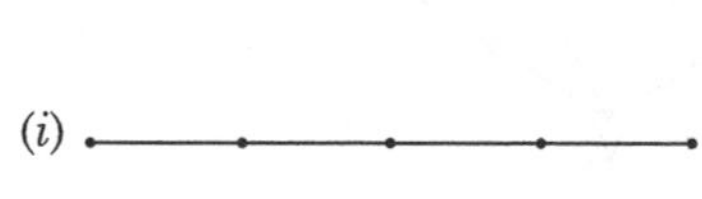

(*ii*) 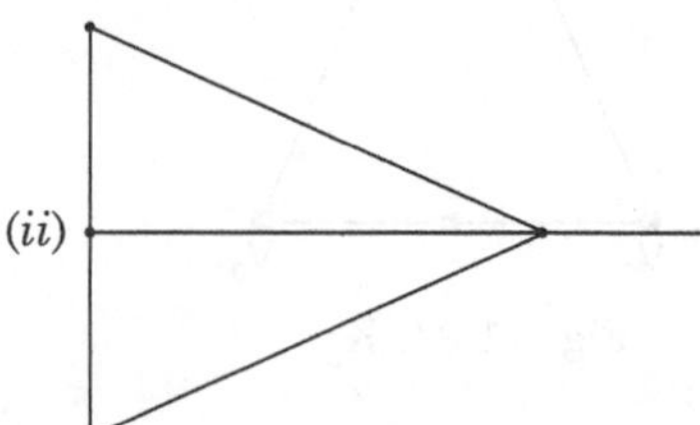

(iii) (iv)

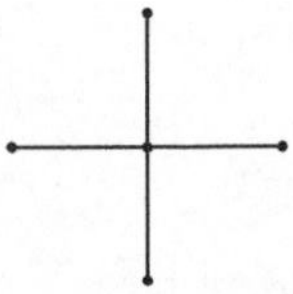

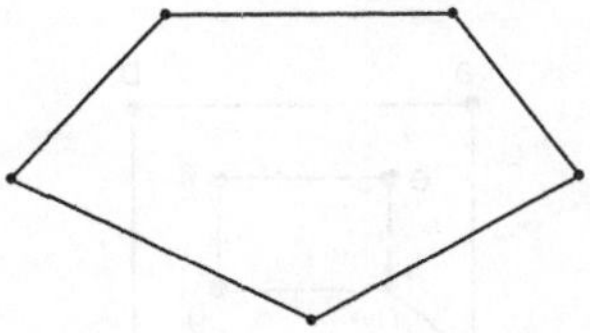

(v) 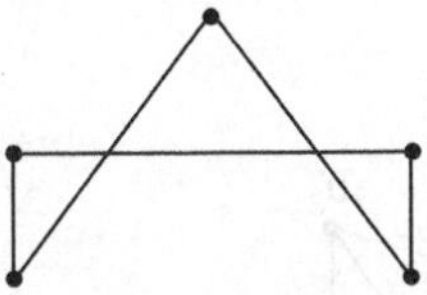 (vi)

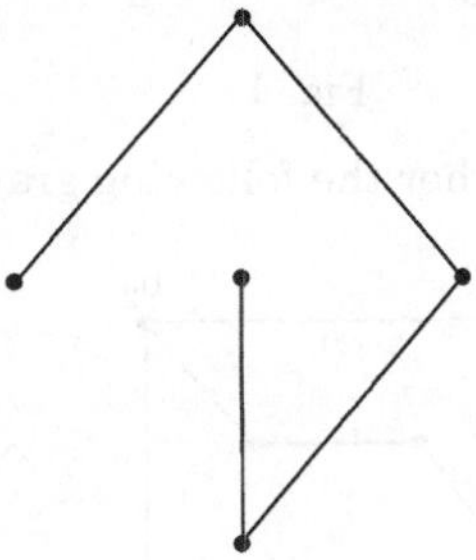

(vii) 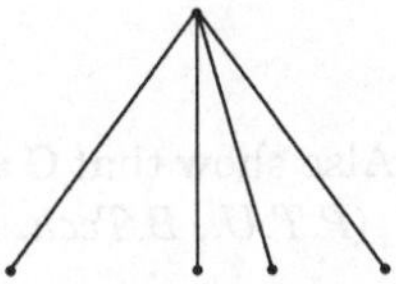 (viii)

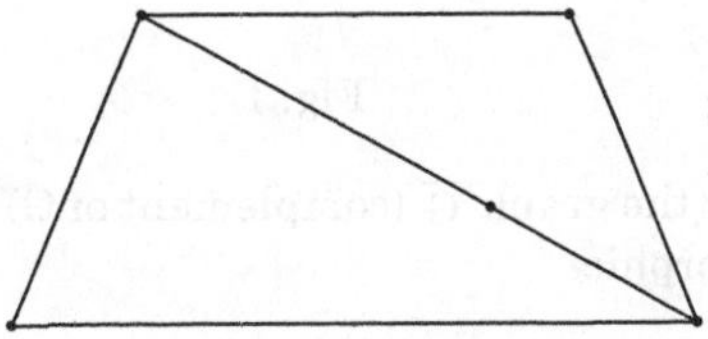

(ix) 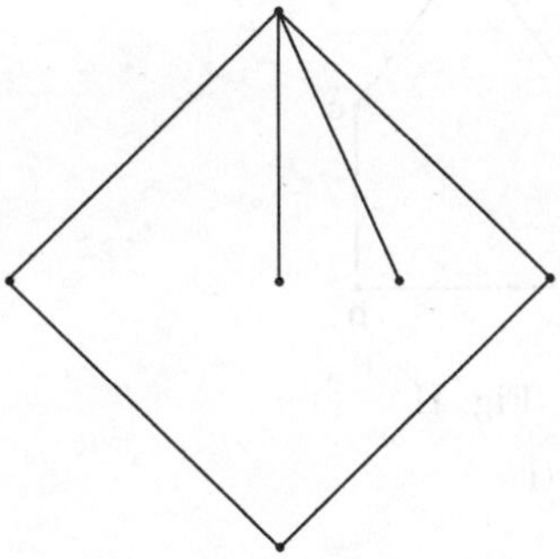 (x)

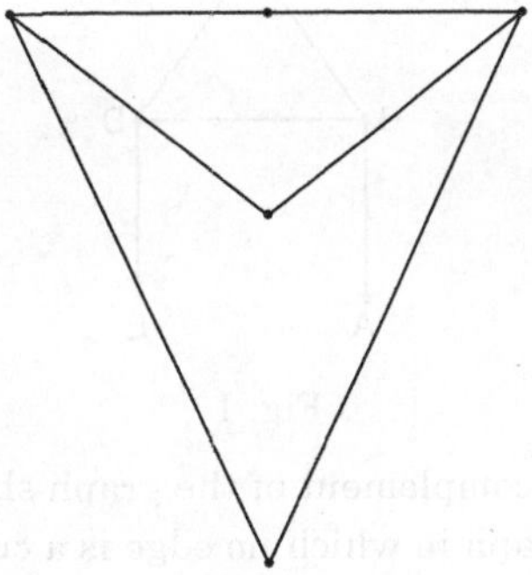

(b) Show that the following graphs are not isomorphic.

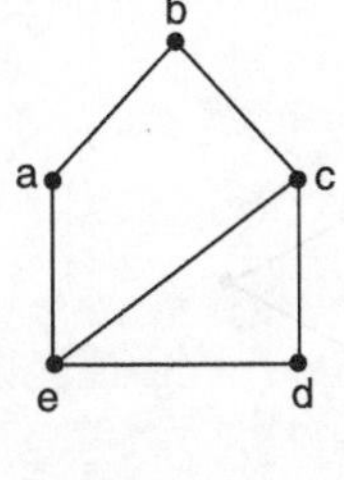

Fig. I

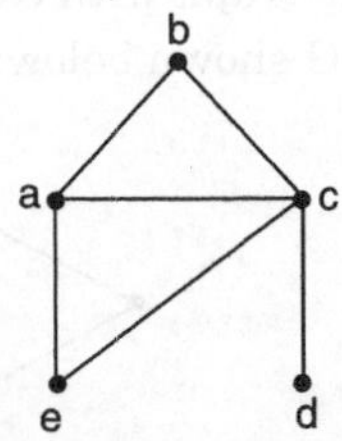

Fig. II

(c) Determine whether the following graphs are isomorphic or not.

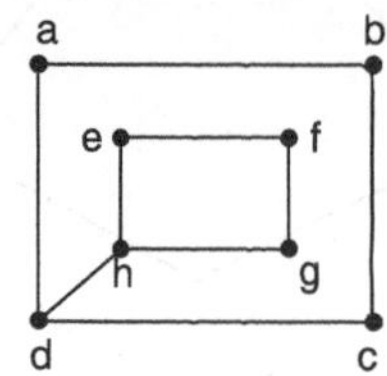

Fig. I

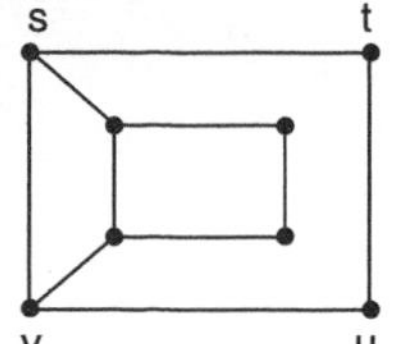

Fig. II

(d) Determine whether the following graphs are isomorphic or not

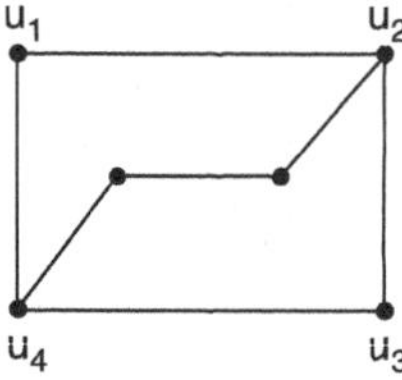

Fig. I

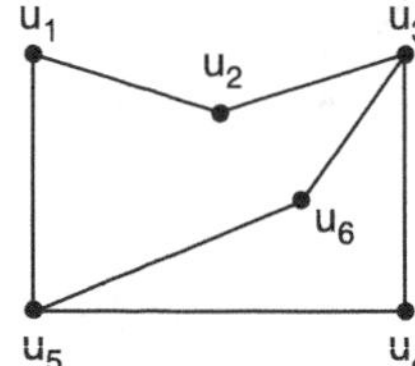

Fig. II

9. (a) Draw the graph $\overline{G}$ (complement of G) of the graph shown below. Also show that G and $\overline{G}$ are isomorphic. *(P.T.U., B.Tech. Dec. 2002)*

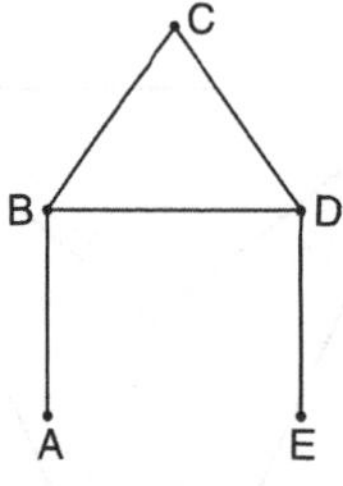

Fig. I

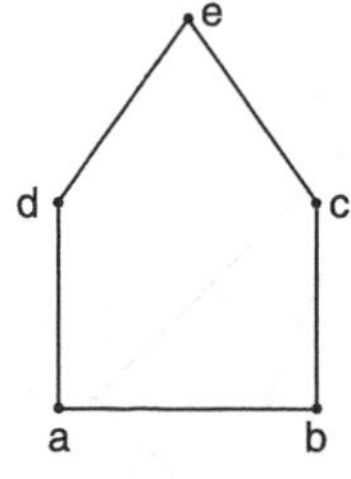

Fig. II

(b) Draw the complement of the graph shown in the fig. II

10. Draw (a) a graph in which no edge is a cut edge.

(b) a graph in which every edge is a cut edge.

(c) only one cut vertex.

11. Find k, if a k-regular graph with 8 vertices has 12 edges. Also draw k-regular graph.

12. Consider the graph G shown below:

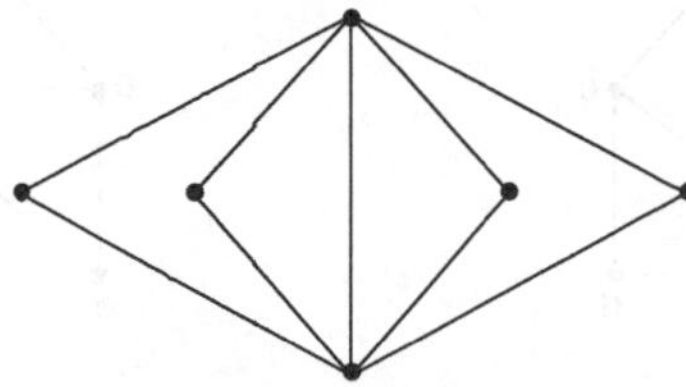

(a) Is G simple?

(b) What is order and size of incidence matrix for G?

(c) Find minimum and maximum degree for G.

13. Prove that in a simple graph with n vertices, each vertex has maximum degree $(n - 1)$.

14. Prove that maximum degree of edges in a graph G with n vertices and no multiple edges are $\frac{n(n-1)}{2}$.

15. Suppose a directed graph has m vertices. Show that if there is a path from vertex u to v, then there is a path ρ' of length $m - 1$ or less from u to v.

16. Construct a graph that has six vertices and five edges but is not a tree

(P.T.U., B.Tech. Dec. 2012)

Answers

1. (*a*) Number of edges = 4, Size of graph = 5 (*b*) (*i*) Order = 4, Size = 6, (*ii*) Order = 4, Size = 7

2. (*a*) 6 (*b*) 13

3. (*a*) 6 (*c*) No

(*d*) 10. (*e*)

4. (*a*)

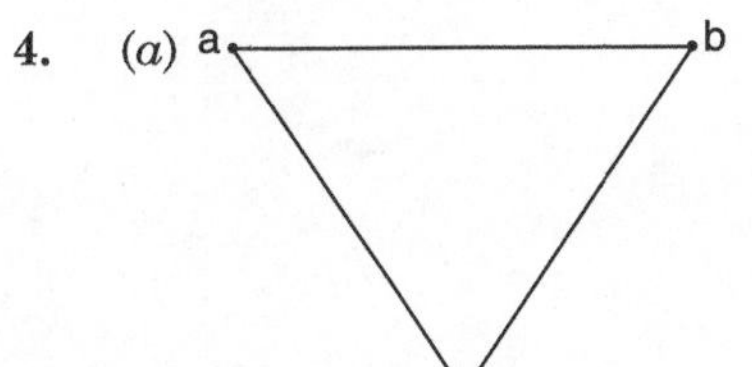

or

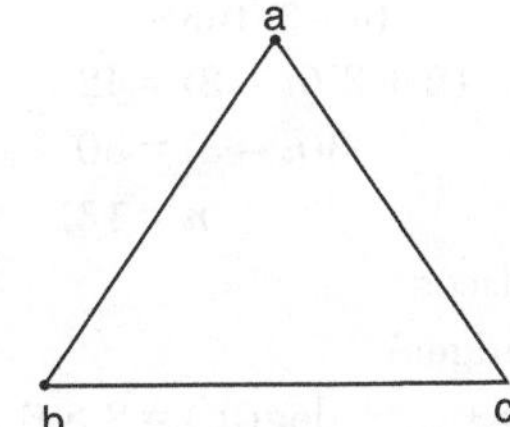

(*b*) d c a b

7. (*a*) Not possible

8. (*a*) (*i*) and (*vi*) are isomorphic (*ii*) and (*ix*) are isomorphic

(*iii*) and (*vii*) are isomorphic (*iv*) and (*v*) are isomorphic

(*viii*) and (*ix*) are isomorphic

(*c*) Not isomorphic (*d*) Isomorphic

9. (*b*) , Since $\overline{G} = K_5 - G$ = –

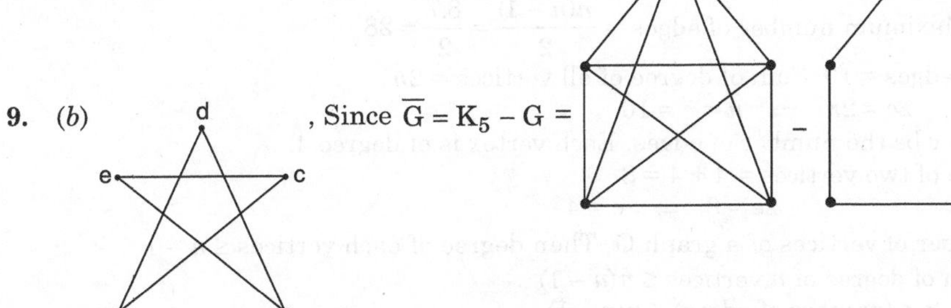

10. (*a*)

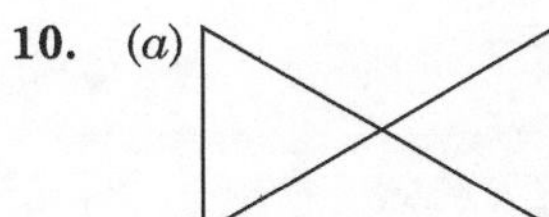

(*b*)

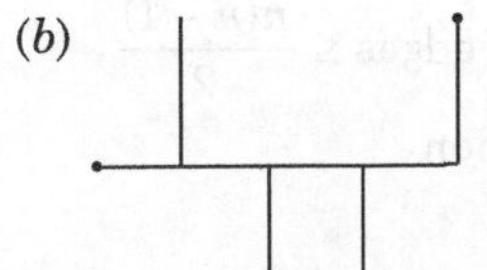

(*c*)

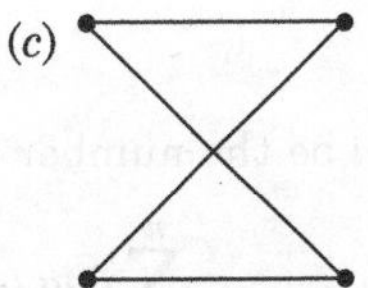

11. 3, k_3 =

12. (*a*) Yes (*b*) 6, 9 (*c*) 5, 2

Hints

2. (*a*) Let $v_1, v_2 \ldots v_n$ be n vertices such that $\deg(v_i) = 2$, $1 \le i \le n$.

Since sum of degree of all vertices is equal to twice the number of edges, *i.e.*,

$$\therefore \quad \Sigma\deg(v_i) = 2 \text{ (Number of edges)}$$
$$\Rightarrow \quad \deg(v_1) + \deg(v_2) + \ldots + \deg(v_n) = 2 \times 16$$
$$\Rightarrow \quad 2 + 2 + \ldots + 2 \; (n \text{ times}) = 32$$
$$\Rightarrow \quad 2n = 32$$
$$\Rightarrow \quad n = 16$$

(*b*) Let n be the number of vertices in G.

Since, sum of degree of all vertices is equal to twice the number of edges *i.e.*,

$$\sum_{i=1}^{n} \deg(v_i) = 2 \times \text{number of edges}$$
$$\Rightarrow \quad \deg(v_1) + \deg(v_2) + \ldots + \deg(v_n) = 2 \times 21$$
$$\Rightarrow \quad \underbrace{(4 + 4 + 4)}_{3 \text{ times}} + \underbrace{(3 + 3 + \ldots + 3)}_{(n-3) \text{ times}} = 42$$
$$\Rightarrow \quad 12 + 3\,(n - 3) = 42$$
$$\Rightarrow \quad 3(n - 3) = 30$$
$$\Rightarrow \quad n = \mathbf{13}.$$

3. (*a*) Let n be the number of vertices

Since, $\Sigma\deg(v_i) = 2$ (number of edges)

$$\Rightarrow \quad \deg(v_1) + \deg(v_2) + \ldots + \deg(v_n) = 2 \times 6$$
$$\Rightarrow \quad 2 + 2 + \ldots + 2 \; (n \text{ times}) = 12$$
$$\Rightarrow \quad 2n = 12$$
$$\Rightarrow \quad n = 6$$

(*b*) Using, $\Sigma\deg(v_i) = 2 \times$ number of edges

$$\Rightarrow \quad 1 + 3 + 4 + 2 + 3 = 2 \text{ (number of edges)}$$
$$\Rightarrow \quad \text{number of edges} = \frac{13}{2}, \text{ which is not possible.}$$

(*c*) $n = 8$, $e = 29$. Maximum number of edges $= \dfrac{n(n-1)}{2} = \dfrac{8.7}{2} = 28$

(*d*) e = number of edges = 10. Sum of degree of all vertices = $2n$

Also $\quad 2e = 2n \;\Rightarrow\; n = e = 10$

(*e*) Here $n = 2$. Let e be the number of edges. Each vertex is of degree 4.

$\therefore$ Sum of degree of two vertices $= 4 + 4 = 8$

$$\Rightarrow \quad 2e = 8 \;\Rightarrow\; e = 4$$

5. Let n be the number of vertices of a graph G. Then degree of each vertices $\le n - 1$

$$\Rightarrow \quad \text{sum of degree of } n \text{ vertices} \le n(n-1)$$
$$\Rightarrow \quad 2 \times (\text{number of edges}) \le n(n-1)$$
$$\Rightarrow \quad \text{number of edges} \le \frac{n(n-1)}{2}.$$

7. Let n be the number of vertices, then

$$\sum_{i=1}^{n} \deg(v_i) = 2 \times \text{number of edges}$$
$$\Rightarrow \quad (v_1 + v_2) + (v_3 + v_4 + \ldots (n-2)) = 2 \times 12$$
$$\Rightarrow \quad (3 + 3) + (4 + 4 + \ldots (n-2)) = 24$$

$\Rightarrow \quad 6 + 4(n-2) = 24$

$\Rightarrow \quad 4(n-2) = 18$

$\Rightarrow \quad n - 2 = \frac{18}{4} = \frac{9}{2}$

$\Rightarrow \quad n = \frac{9}{2} + 2 = \frac{13}{2}$, not possible.

8. (*b*) Both graphs have five vertices and six edges. However the graph (Fig. II) has a vertex of degree 1 namely, *e*, but the graph (Fig. I) has no vertex of degree 1. Hence the given graphs are not isomorphic.

(*c*) Both graphs (Fig. I and Fig. II) have 8 vertices and 10 edges. Also both graphs have four vertices of degree 2 and four vertices of degree 3. But deg(a) = 2, it must correspond to either *x*, *y*, *z*. *t* or 4 in Fig. II. All these vertices in Fig. II is adjacent to another vertex of degree 2. But this is not true for the vertex *a* in Fig. I. Hence they cannot be isomorphic.

11. Given, V = 8, E = 12

$\therefore$ Sum of degree of all vertices = $2 \times 12 = 24$

$\Rightarrow$ degree of each vertex = $\frac{24}{8} = 3$

13. A simple graph is a graph without parallel edges or self loops.

If G = [V, E] is a simple graph with only one vertex, then the number of edges in G is zero

$\therefore$ maximum degree of a vertex in G = 1 – 1 = (0)

If G = [V, E] be a simple graph with 2 vertices, then the number of edges in G is 1 = 2 – 1 and degree of each vertex is (2 – 1).

....

If G = [V, E] be a simple graph with n vertices, then maximum degree of each vertex = $n - 1$.

14. Let G be a simple graph with n vertices, then degree of each vertex in G is $\leq n - 1$

$\therefore$ sum of degrees of n vertices in G $\leq n(n-1)$

$\Rightarrow \quad 2e \leq n(n-1) \quad \Rightarrow \quad e \leq \frac{n(n-1)}{2}$ where e is the number of edges in G.

15. Suppose there is a path from the vertex u to the vertex v. Let $[u_1, u_2, ..., u_i, u_{i+1}, ... v]$ be the sequence of vertices that the path meets when it is traced from u to v. If there are k edges in the path. There are $k + 1$ vertices in the sequence. Choose a number $k > m - 1$ such that the vertex, say, u_l, should appear more than once in the sequence, that is, $(u, ... u_i, ... u_e, ... u_l, ... v)$.

Deleting the edges in the path that leads u_e back to u_l, we have a path from u to v that has fewer edges than the original one. Repeating this process until we have a path that has $(m - 1)$ or fewer edges. Hence we have proved that in a directed graph with m vertices, if there is a path from vertex u to v, then there is also a path from u to v of length $m - 1$ or less edges.

16. The required graph is

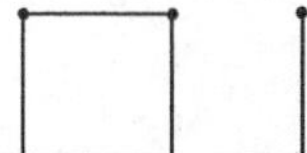

11.29. LABELLED GRAPHS

A graph G = (V, E) is called a labelled graph if its edges are labelled with some name or data. So, we can write these labells in place of an ordered pair in the edge set. For *e.g.*,

The graphs shown in Figs. 11.86 and 11.87 are labelled graphs.

G = {{1, 2, 3, 4, 5}, $\{e_1, e_2, e_3, e_4, e_5\}$}. G = {{$a$, b, c, d}, $\{e_1, e_2, e_3, e_4\}$}.

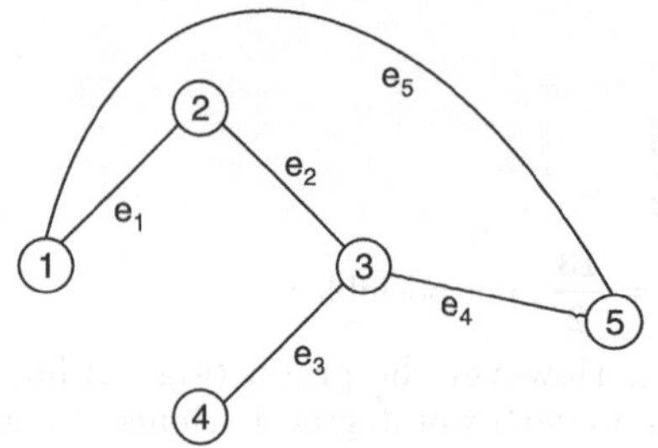

Fig. 11.86. Undirected Labelled Graph.

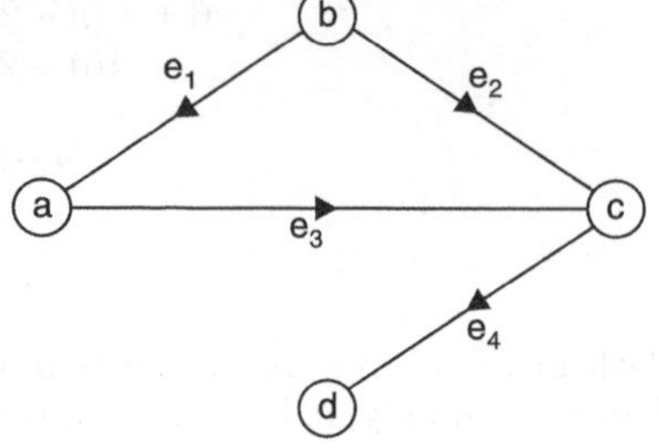

Fig. 11.87. Directed Labelled Graph.

11.30. WEIGHTED GRAPHS (*P.T.U., B.Tech. May 2007, Dec. 2006, May 2005*)

A graph G = (V, E) is called a weighted graph if each edge of graph G is assigned a positive number w called the weight of the edge e. For example,

The graph shown in Figs. 11.88 and 11.89 is a weighted graph.

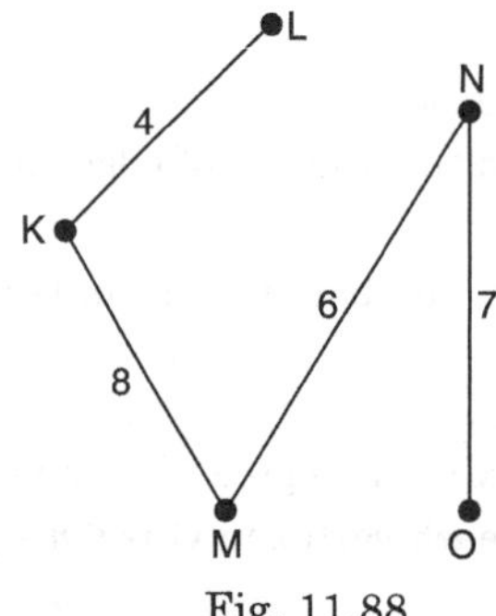

Fig. 11.88

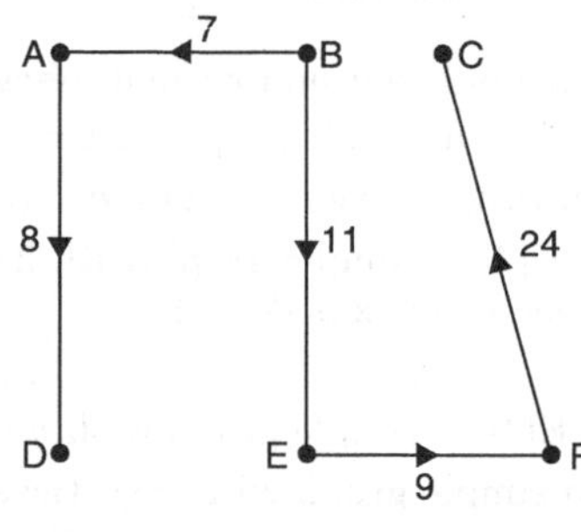

Fig. 11.89

11.31. MULTIPLE EDGES

Two edges e_1 and e_1' which are distinct are said to be multiple edges if they connect the end points *i.e.*, if $e_1 = (u, v)$ and $e_1' = (u, v)$ then e_1 and e_1' are multiple edges.

11.32. MULTIGRAPH (*P.T.U., B.Tech. May 2007, Dec. 2006, May 2005*)

A multigraph G = (V, E) consists of a set of vertices V and a set of edges E such that edge set E may contain multiple edges and self loops. For example,

Consider the following graph shown in Fig. 11.90.

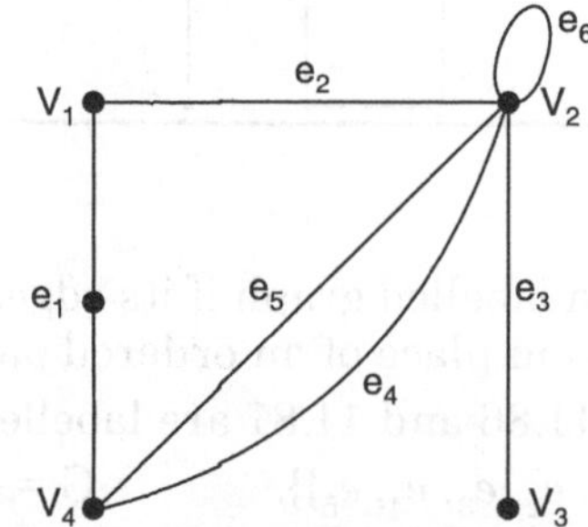

Fig. 11.90. Undirected Multigraph.

In the above Fig. 11.91, e_4 and e_5 are multiple edges, e_6 is a self-loop.

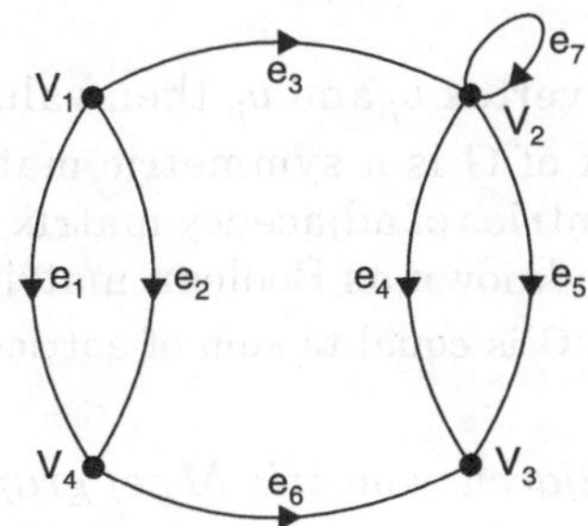

Fig. 11.91. Directed Multigraph.

In the graph shown in Fig. 11.91, the edges e_1, e_2 and e_4, e_5 are multiple edges e_7 is a loop.

11.33. TRAVERSABLE MULTIGRAPHS

Consider a multigraph G = (V, E). If the multigraph G consists of a path which includes all vertices and whose edge list contains each edge of graph exactly once. Then, the multigraph G is called a traversable multigraph.

The sufficient and necessary condition for a multigraph to be traversable is that it should be connected and have either zero or two vertices of odd degree.

Consider the multigraph shown in Fig. 11.92.

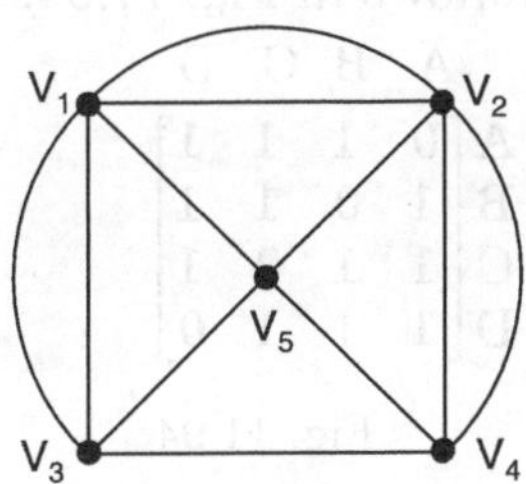

Fig. 11.92

The multigraph has three even degree vertices *i.e.*, V_3, V_4 and V_5 and two odd degree vertices *i.e.*, V_1 and V_2. Hence it is a traversable multigraph.

11.34. REPRESENTATION OF GRAPHS

There are two important ways to represent a graph G with the matrices *i.e.*,

I. Adjacency matrix representation.

II. Incidence matrix representation.

(*a*) Representation of Undirected Graph

(*i*) **Adjacency matrix representation.** If an undirected graph G consists of n vertices, then the adjacency matrix of graph is an $n \times n$ matrix A = $[a_{ij}]$ and defined by

$$a_{ij} = \begin{cases} 1, \text{ if } \{v_i, v_j\} \text{ is an edge } i.e., v_i \text{ is adjacent to } v_j \\ 0, \text{ if there is no edge between } v_i \text{ and } v_j \end{cases}$$

If there exists an edge between vertex v_i and v_j, where i is a row and j is a column then value of $a_{ij} = 1$.

If there is no edge between vertex v_i and v_j, then value of $a_{ij} = 0$.

Note that adjacency matrix of G is a symmetric matrix. Since simple graph does not contain any self loop, so diagonal entries of adjacency matrix are all zero. Further, as adjacency matrix contains 0 or 1, so it is also known as Boolean matrix.

Note. Degree of a vertex v_i in G is equal to sum of entries in the ith row or ith column of the adjacency matrix.

For example, *we find the adjacency matrix M_A of graph G shown in Fig. 11.93.*

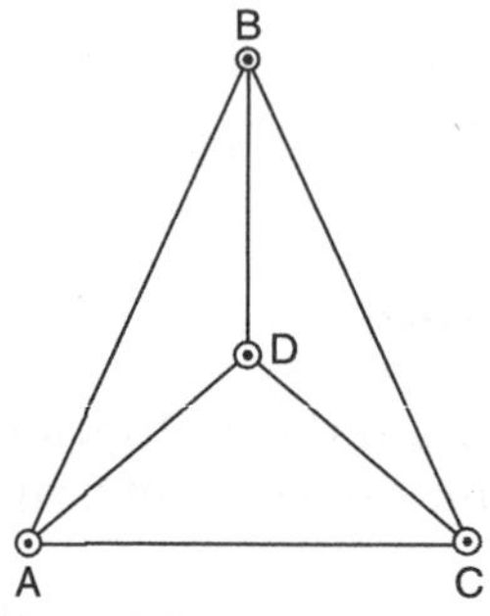

Fig. 11.93

Since the graph G consists of four vertices. Therefore, the adjacency matrix will be a 4 × 4 matrix. The adjacency matrix is as follows in Fig. 11.94.

$$M_A = \begin{array}{c} \\ A \\ B \\ C \\ D \end{array} \begin{array}{c} \begin{array}{cccc} A & B & C & D \end{array} \\ \begin{bmatrix} 0 & 1 & 1 & 1 \\ 1 & 0 & 1 & 1 \\ 1 & 1 & 0 & 1 \\ 1 & 1 & 1 & 0 \end{bmatrix} \end{array}$$

degree of vertex 'c' is 3 which is equal to sum of entries in third row 6/4 mm of adjacency matrix.

Fig. 11.94

Adjacency List. In a adjacency list of a graph, we list each vertex followed by the vertices adjacent to it. First write vertices of graph in a vertical column, then after each vertex, write the vertices adjacent to it.

Consider the graph shown in Fig. 11.95 the adjacency list is given below :

v_1; v_2, v_3

v_2; v_1, v_3

v_3; v_1, v_2, v_4

v_4; v_3

V_1 V_2 V_3 V_4

Fig. 11.95

(*ii*) **Incidence matrix or Binary matrix representation.** If an undirected graph consists of n vertices and m edges, then the incidence matrix is an $n \times m$ matrix $C = [c_{ij}]$ defined by

$$c_{ij} = \begin{cases} 1, \text{ if the vertex } v_i \text{ incident by edge } e_j \\ 0, \text{ otherwise} \end{cases}$$

There is a row for every vertex and a column for every edge in the incidence matrix.

Note that incidence matrix of a graph need not be a square matrix. Entries in a row are added to give degree of corresponding vertex.

For example ;

Consider the graph $G = [V, E.]$

where $V = [v_1, v_2, v_3, v_4]$, $E = [e_1, e_2, e_3]$ as shown in Fig. 11.96.

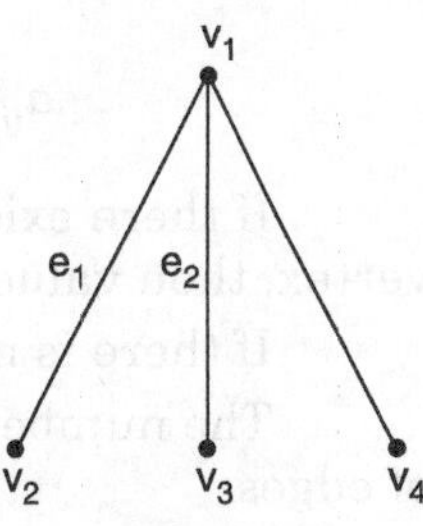

Fig. 11.96

The incidence matrix M_1 for G is shown below :

$$M_1 = \begin{array}{c} \\ v_1 \\ v_2 \\ v_3 \\ v_4 \end{array} \begin{array}{c} \begin{array}{ccc} e_1 & e_2 & e_3 \end{array} \\ \begin{bmatrix} 1 & 1 & 1 \\ 1 & 0 & 0 \\ 0 & 1 & 0 \\ 0 & 0 & 1 \end{bmatrix} \end{array}$$

Since each edge in the graph is incident on v_1,

$\therefore$ first row for v_1 has all entries 1.

$\therefore$ degree $v_1 = 1 + 1 + 1 = 3$

Also v_2, v_3, v_4 are pendant vertices.

In incidence matrix of a graph, sum of entries in column is not degree of vertex. As an edge is incident on two vertices in a graph, therefore, each column of incidence matrix will have two 1's.

The number of one's in an incidence matrix of undirected graph (without loops) is equal to the sum of degrees of all the vertices of the graph.

For example : *Consider the undirected graph G as shown in Fig. 11.97. We find its incidence matrix M_I.*

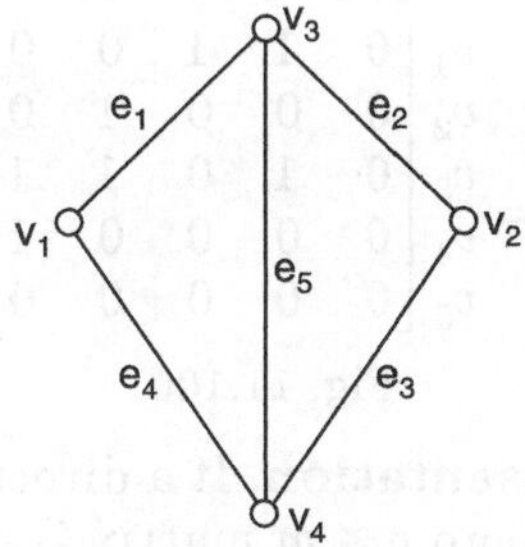

Fig. 11.97

Sol. The undirected graph consists of four vertices and five edges. Therefore, the incidence matrix is a 4 × 5 matrix, which is shown in Fig. 11.98

$$M_I = \begin{array}{c} \\ v_1 \\ v_2 \\ v_3 \\ v_4 \end{array} \begin{array}{c} \begin{array}{ccccc} e_1 & e_2 & e_3 & e_4 & e_5 \end{array} \\ \begin{bmatrix} 1 & 0 & 0 & 1 & 0 \\ 0 & 1 & 1 & 0 & 0 \\ 1 & 1 & 0 & 0 & 1 \\ 0 & 0 & 1 & 1 & 1 \end{bmatrix} \end{array}$$

Fig. 11.98

(*b*) Representation of Directed Graph

(*i*) **Adjacency matrix representation.** If a directed graph G consists of n vertices, then the adjacency matrix of graph is an $n \times n$ matrix $A = [a_{ij}]$ defined by

$$a_{ij} = \begin{cases} 1, \text{ if } v_i, v_j \text{ is an edge } i.e., \text{ if } v_i \text{ is initial vertex and } v_j \text{ is final vertex} \\ 0, \text{ if there is no edge between } v_i \text{ and } v_j \end{cases}$$

If there exists an edge between vertex v_i and v_j with v_i as initial vertex and v_j as final vertex, then value of $a_{ij} = 1$.

If there is no edge between vertex v_i and v_j then value of $a_{ij} = 0$.

The number of one's in the adjacency matrix of a directed graph is equal to the number of edges.

For example : *Consider the directed graph shown in Fig. 11.99. We determine its adjacency matrix M_A.*

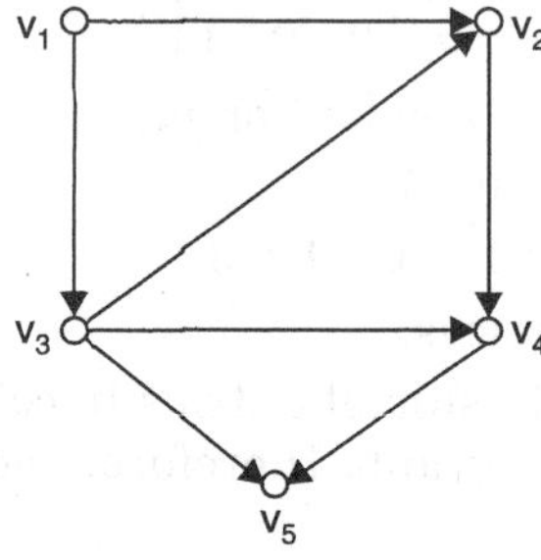

Fig. 11.99

Sol. Since the directed graph G consists of five vertices. Therefore, the adjacency matrix will be a 5 × 5 matrix. The adjacency matrix of the directed graph is as follows in Fig. 11.100.

$$M_A = \begin{array}{c} \\ v_1 \\ v_2 \\ v_3 \\ v_4 \\ v_5 \end{array} \begin{array}{c} \begin{matrix} v_1 & v_2 & v_3 & v_4 & v_5 \end{matrix} \\ \begin{bmatrix} 0 & 1 & 1 & 0 & 0 \\ 0 & 0 & 0 & 1 & 0 \\ 0 & 1 & 0 & 1 & 1 \\ 0 & 0 & 0 & 0 & 1 \\ 0 & 0 & 0 & 0 & 0 \end{bmatrix} \end{array}$$

Fig. 11.100

(ii) **Incidence matrix representation.** If a directed graph consists of n vertices and m edges then the incidence matrix is an $n \times m$ matrix $C = [c_{ij}]$, defined by

$$c_{ij} = \begin{cases} 1, \text{ if } v_i \text{ is initial vertex of edge } e_j \\ -1, \text{ if } v_i \text{ is final vertex of edge } e_j \\ 0, \text{ if } v_i \text{ is not incident on edge } e_j \end{cases}$$

The number of one's in the incidence matrix is equal to the number of edges in the graph.

For example, Consider the directed graph G shown in Fig. 11.101. Find its incidence matrix M_I.

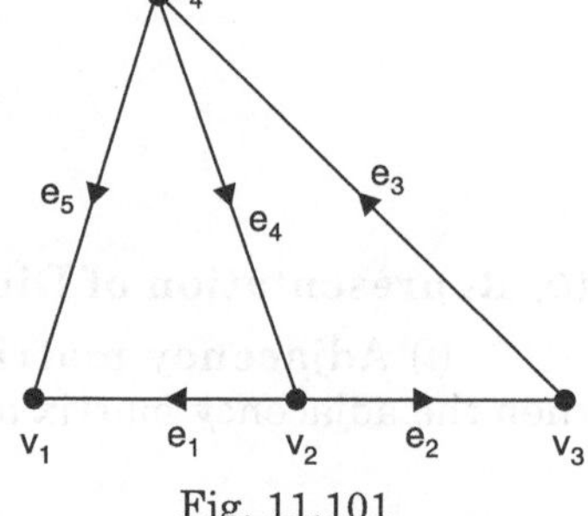

Fig. 11.101

Sol. The directed graph consists of four vertices and five edges. Therefore, the incidence matrix is a 4 × 5 matrix which is shown in Fig. 11.102.

$$M_I = \begin{array}{c} \\ v_1 \\ v_2 \\ v_3 \\ v_4 \end{array} \begin{array}{c} \begin{array}{ccccc} e_1 & e_2 & e_3 & e_4 & e_5 \end{array} \\ \begin{bmatrix} -1 & 0 & 0 & 0 & -1 \\ 1 & 1 & 0 & -1 & 0 \\ 0 & -1 & 1 & 0 & 0 \\ 0 & 0 & -1 & 1 & 1 \end{bmatrix} \end{array}$$

Fig. 11.102.

(c) Representation of Multigraph

Represented only by adjacency matrix representation.

(*i*) **Adjacency matrix representation of multigraph.** If a multigraph G consists of n vertices, then the adjacency matrix of graph is an $n \times n$ matrix $A = [a_{ij}]$ and is defined by

$$a_{ij} = \begin{cases} \text{N, If there are one or more than one edges between vertex } v_i \text{ and } v_j, \text{ where N is the number of edges.} \\ 0, \text{ otherwise.} \end{cases}$$

If there exists one or more than one edges between vertex v_i and v_j then a_{ij} = N, where N is the number of edges between v_i and v_j.

If there is no edge between vertex v_i and v_j then value of $a_{ij} = 0$. For *e.g.*,

For example : *Consider the multigraph shown in Fig. 11.103. We determine its adjacency matrix.*

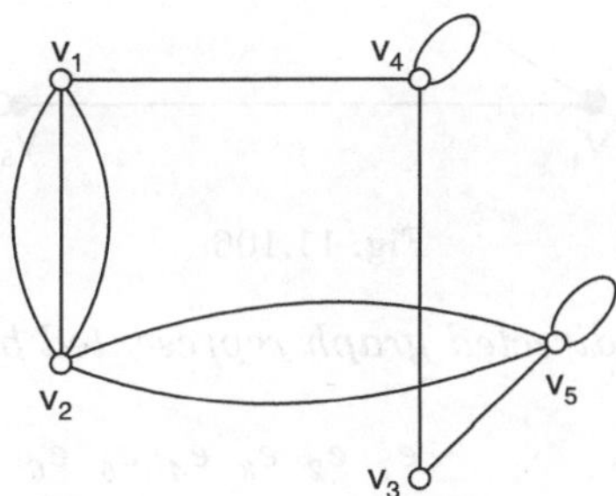

Fig. 11.103

Sol. Since the multigraph consists of five vertices. Therefore, the adjacency matrix will be an 5 × 5 matrix. The adjacency matrix of the multigraph is as follows in Fig. 11.104.

$$M_A = \begin{array}{c} \\ v_1 \\ v_2 \\ v_3 \\ v_4 \\ v_5 \end{array} \begin{array}{c} \begin{array}{ccccc} v_1 & v_2 & v_3 & v_4 & v_5 \end{array} \\ \begin{bmatrix} 0 & 3 & 0 & 0 & 1 \\ 3 & 0 & 0 & 0 & 2 \\ 0 & 0 & 0 & 1 & 1 \\ 1 & 0 & 1 & 1 & 0 \\ 0 & 2 & 1 & 0 & 1 \end{bmatrix} \end{array}$$

Fig. 11.104.

ILLUSTRATIVE EXAMPLES

Example 1. *Draw the undirected graph represented by adjacency matrix M_A shown in Fig. 11.105.*

$$M_A = \begin{array}{c} \\ v_1 \\ v_2 \\ v_3 \\ v_4 \\ v_5 \end{array}\begin{array}{c} \begin{array}{ccccc} v_1 & v_2 & v_3 & v_4 & v_5 \end{array} \\ \begin{bmatrix} 0 & 1 & 1 & 0 & 0 \\ 1 & 0 & 1 & 0 & 0 \\ 1 & 1 & 0 & 1 & 0 \\ 0 & 0 & 1 & 0 & 1 \\ 0 & 0 & 0 & 1 & 1 \end{bmatrix} \end{array}$$

Fig. 11.105.

Sol. The graph represented by adjacency matrix M_A is shown in Fig. 106.

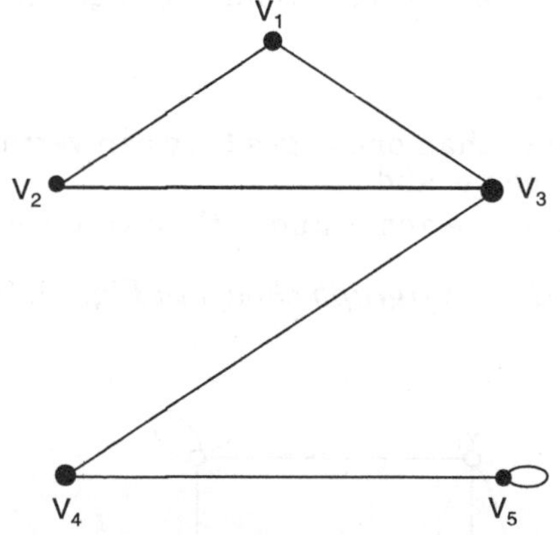

Fig. 11.106.

Example 2. *Draw the undirected graph represented by incidence matrix M_I shown in Fig. 11.107.*

$$M_I = \begin{array}{c} \\ a \\ b \\ c \\ d \\ e \end{array}\begin{array}{c} \begin{array}{cccccc} e_1 & e_2 & e_3 & e_4 & e_5 & e_6 \end{array} \\ \begin{bmatrix} 0 & 1 & 0 & 0 & 1 & 1 \\ 0 & 1 & 1 & 0 & 1 & 0 \\ 0 & 0 & 0 & 1 & 0 & 1 \\ 1 & 0 & 0 & 0 & 0 & 0 \\ 1 & 0 & 0 & 1 & 0 & 0 \end{bmatrix} \end{array}$$

Fig. 11.107.

Sol. The graph represented by incidence matrix M_I is shown in Fig. 11.108.

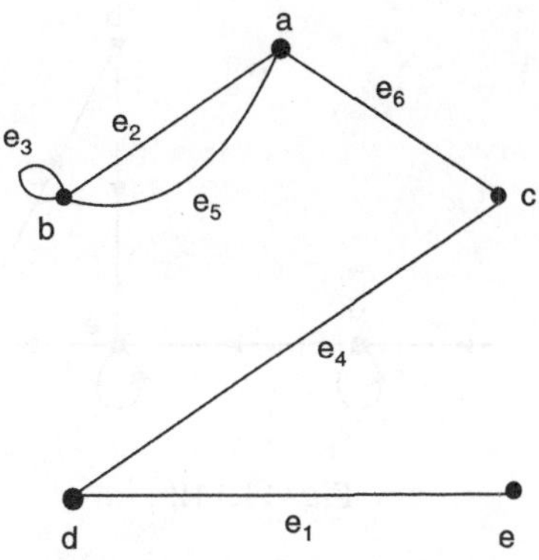

Fig. 11.108.

Example 3. *Draw the multigraph G whose adjacency matrix M_A is shown in Fig. 11.109.*

$$M_A = \begin{bmatrix} 1 & 3 & 0 & 0 \\ 3 & 0 & 1 & 1 \\ 0 & 1 & 2 & 2 \\ 0 & 1 & 2 & 0 \end{bmatrix}$$

Fig. 11.109.

Sol. The multigraph corresponding to the adjacency matrix M_A is shown in Fig. 11.110.

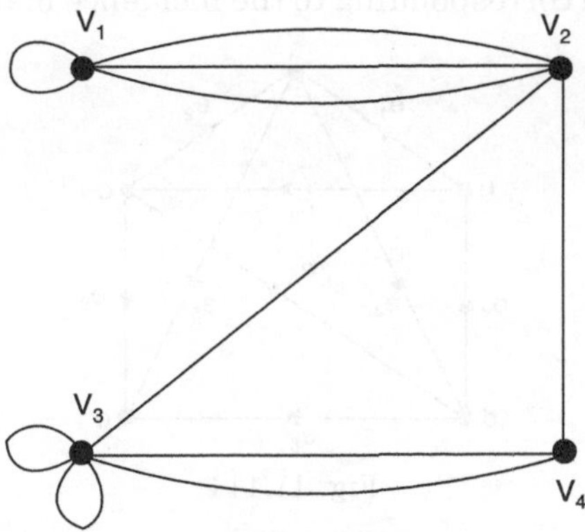

Fig. 11.110

Example 4. *Draw the directed graph G whose adjacency matrix M_A is shown in Fig. 11.111.*

$$M_A = \begin{array}{c} \\ a \\ b \\ c \\ d \\ e \\ f \end{array} \begin{array}{c} \begin{array}{cccccc} a & b & c & d & e & f \end{array} \\ \begin{bmatrix} 0 & 0 & 1 & 0 & 0 & 0 \\ 0 & 0 & 0 & 0 & 1 & 0 \\ 0 & 0 & 0 & 1 & 0 & 0 \\ 0 & 0 & 0 & 1 & 1 & 0 \\ 0 & 0 & 0 & 0 & 1 & 1 \\ 0 & 1 & 0 & 0 & 0 & 1 \end{bmatrix} \end{array}$$

Fig. 11.111

Sol. The directed graph corresponding to the adjacency matrix M_A is shown in Fig. 11.112.

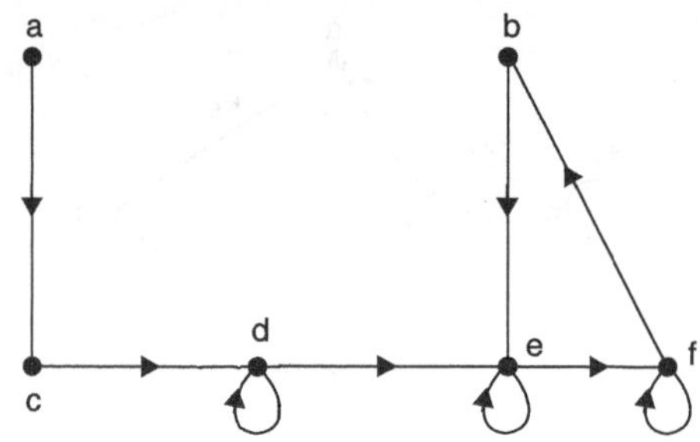

Fig. 11.112

Example 5. *Draw the directed graph G whose incidence matrix M_I is shown in Fig. 11.113.*

$$M_I = \begin{array}{c} \\ a \\ b \\ c \\ d \\ e \end{array}\begin{array}{c} \begin{array}{ccccccccc} e_1 & e_2 & e_3 & e_4 & e_5 & e_6 & e_7 & e_8 & e_9 \end{array} \\ \begin{bmatrix} -1 & -1 & 0 & 1 & 1 & 0 & 0 & 0 & 0 \\ 1 & 0 & 1 & 0 & 0 & 0 & -1 & 0 & 0 \\ 0 & +1 & -1 & 0 & 0 & -1 & 0 & 0 & 1 \\ 0 & 0 & 0 & -1 & 0 & 1 & 1 & 1 & 0 \\ 0 & 0 & 0 & 0 & -1 & 0 & 0 & -1 & -1 \end{bmatrix} \end{array}$$

Fig. 11.113

Sol. The directed graph corresponding to the incidence matrix M_I is shown in Fig. 11.114

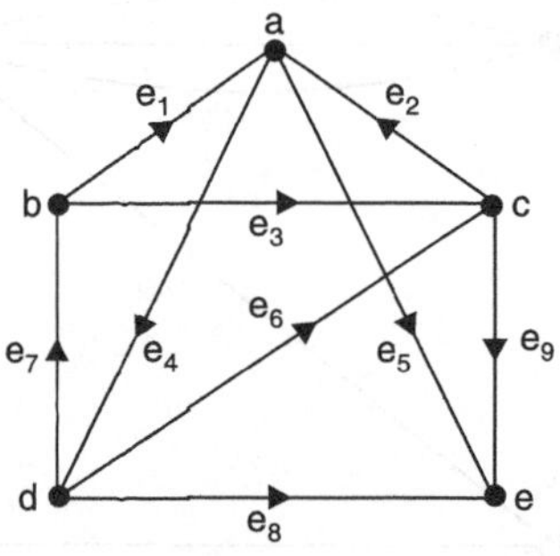

Fig. 11.114

11.35. OTHER IMPORTANT GRAPHS

(*a*) **Bipartite Graph** *(P.T.U., M.C.A. May 2008, 2007)*

A graph G = (V, E) is called a bipartite graph if its vertices V can be partitioned into two subsets V_1 and V_2 such that each edge of G connects a vertex of V_1 to a vertex of V_2. In other words, no edge joining two vertices in V_1 or two vertices in V_2. It is denoted by $K_{m, n}$, where m and n are number of vertices in V_1 and V_2 respectively.

For *e.g.*; consider the graph shown in Fig. 11.115. This graph is a bipartite graph.

Here $\quad V = [a, b, c, d]$

Let $\quad V_1 = [a, b], V_2 = [c, d]$

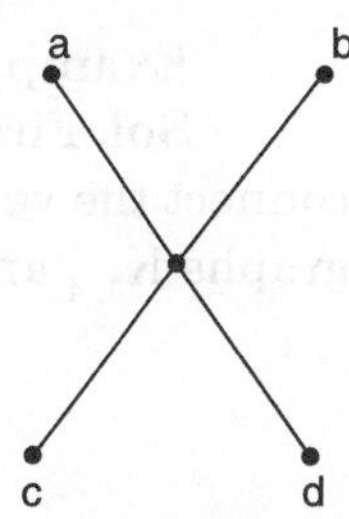

Fig. 11.115

$\therefore \quad V_1 \cup V_2 = [a, b, c, d] = V$ and $V_1 \cap V_2 = \phi$

$\therefore \quad [V_1, V_2]$ is a partition of V.

Consider another graph shown in Fig. 11.116

This graph is also a bipartite graph.

Here $\quad V = [a, b, x, y, z]$

Let $\quad V_1 = [a, b], V_2 = [x, y, z]$

$\therefore \quad V_1 \cup V_2 = V$ and $V_1 \cap V_2 = \phi$

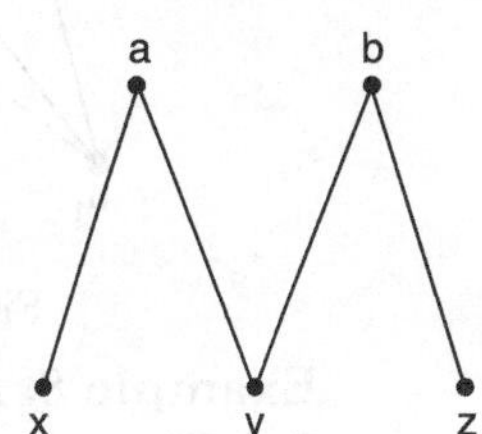

Fig. 11.116

Bipartite graphs are also called 2-colourable graphs as one can think vertices in V_1 of one colour and in V_2 of another colour and vertex is joined by an edge to a vertex of same colour.

(*b*) **Complete Bipartite Graph.** A graph G = (V, E) is called a complete bipartite graph if its vertices V can be partitioned into two subsets V_1 and V_2, such that each vertex of V_1 is connected to each vertex of V_2. The number of edges in a complete bipartite graph is $m. n$ as each of the m vertices is connected to each of the n vertices. It is denoted by $K_{m, n}$ and $m \le n$.

For example, the graphs shown in Fig. 11.117 are complete bipartite graphs.

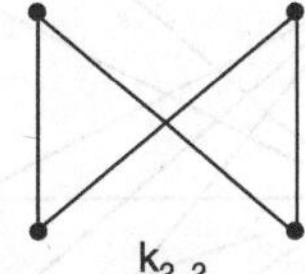

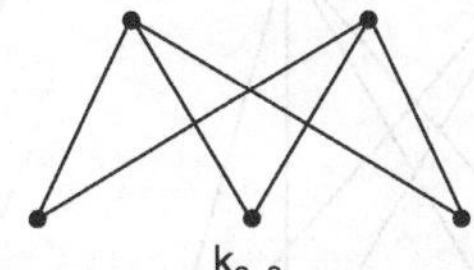

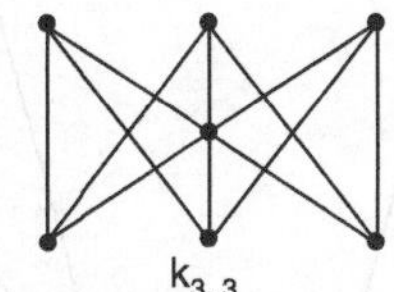

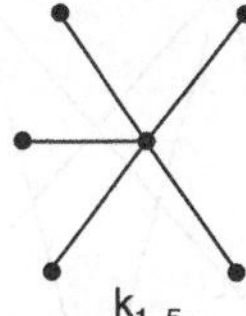

Fig. 11.117

A complete Bipartite graph $K_{1, n}$ is called a **star graph**.

Example 6. *Draw the bipartite graphs $K_{2,4}$ and $K_{3, 4}$. Assuming any number of edges.*

Sol. First draw the appropriate number of vertices on two parallel columns or rows and connect the vertices in one column or row with the vertices in other column or row. The bipartite graphs $K_{2, 4}$ and $K_{3, 4}$ are shown in Fig. 11.118 and Fig. 11.119 respectively.

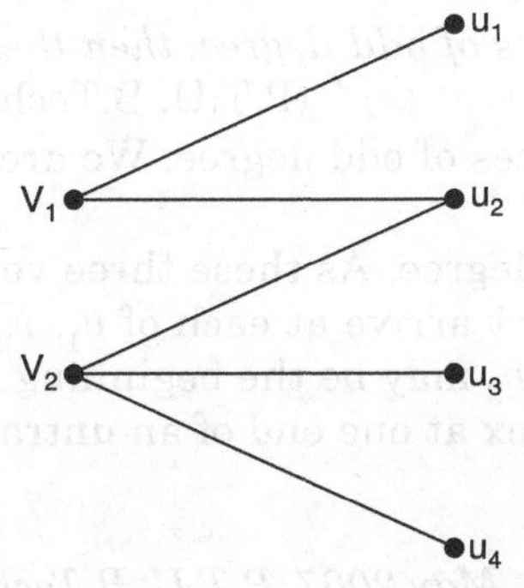

Fig. 11.118. Bipartite Graph $K_{2, 4}$

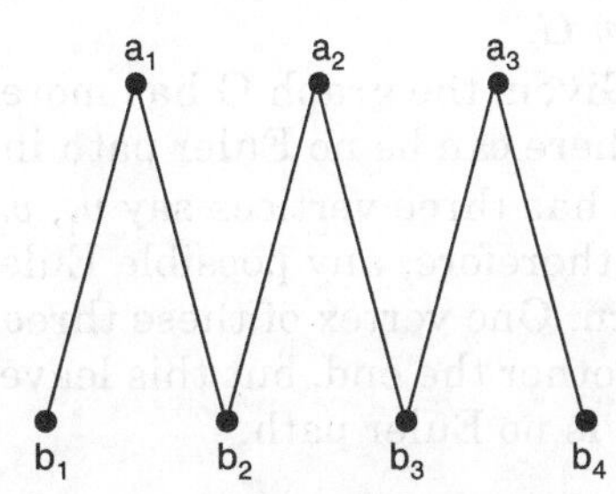

Fig. 11.119 Bipartite Graph $K_{3, 4}$

Example 7. *Draw the complete bipartite graphs $K_{3,4}$ and $K_{1,5}$.*

Sol. First draw the appropriate number of vertices in two parallel columns or rows and connect the vertices in first column or row with all the vertices in second column or row. The graphs $K_{3,4}$ and $K_{1,5}$ are shown in Fig. 11.120 and Fig. 11.121 respectively.

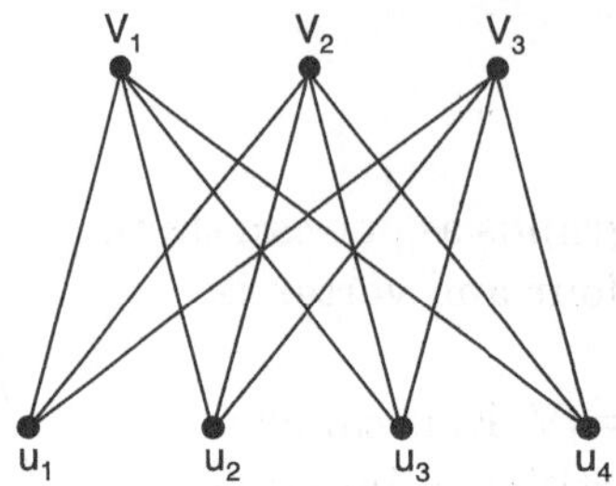

Fig. 11.120 $K_{3,4}$.

Fig. 11.121 $K_{1,5}$.

Example 8. *Draw the complete bipartite graphs $K_{2,3}$, $K_{2,4}$ and $K_{2,5}$.*

Sol. The complete bipartite graphs $K_{2,3}$, $K_{2,4}$ and $K_{2,5}$ are shown in Figs. 11.122, 11.123 and 11.124 respectively.

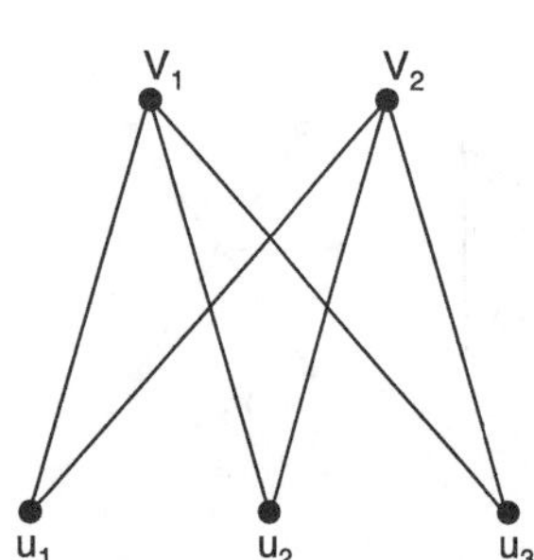

Fig. 11.122 $K_{2,3}$.

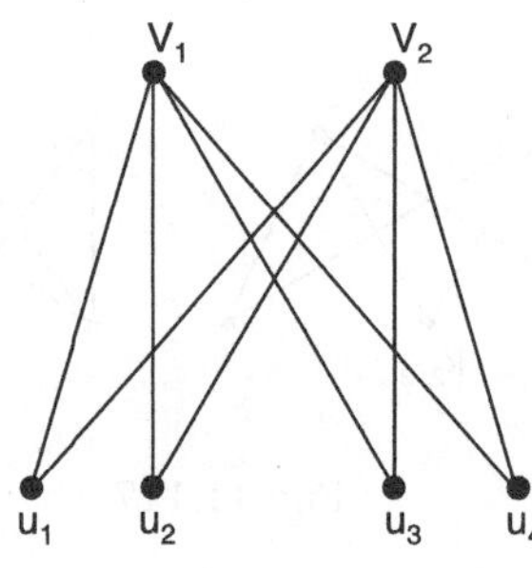

Fig. 11.123 $K_{2,4}$.

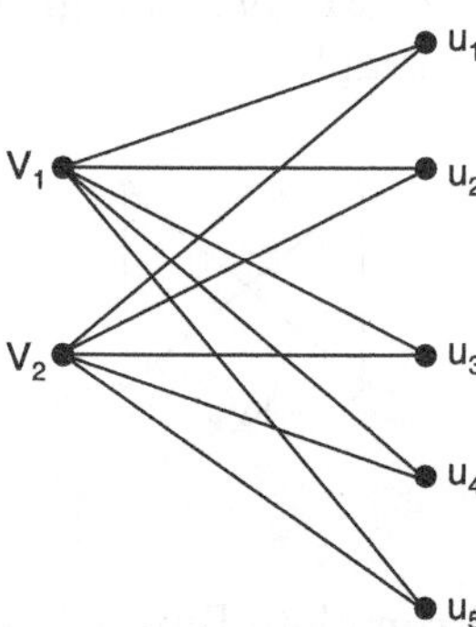

Fig. 11.124. $K_{2,5}$

11.36. EULER PATH (OR CHAIN) *(P.T.U., B.Tech. Dec. 2007, Dec. 2006)*

An Euler path (or chain) through a graph is a path whose edge list contains each edge of the graph exactly once.

Example. *If a graph G has more than two vertices of odd degree, then there can be no Euler path in G.* (P.T.U. B.Tech. May 2008)

Sol. Given, the graph G has more than two vertices of odd degree. We are required to prove that there can be no Euler path in G.

Let G has three vertices say v_1, v_2 and v_3 of odd degree. As these three vertices are of odd degree, therefore, any possible Euler path in G must arrive at each of v_1, v_2, v_3 with no way to return. One vertex of these three vertices v_1, v_2, v_3 may be the beginning of the Euler path and another the end, but this leaves the third vertex at one end of an untravelled edge. Hence there is no Euler path.

11.37. EULER CIRCUIT (OR CYCLE) *(P.T.U., M.C.A. May 2007, P.T.U. B.Tech. Dec. 2006)*

An Euler circuit (or cycle) is a path through a graph, in which the initial vertex appears second time as the terminal vertex.

11.38. EULER GRAPH

An Euler graph is a graph that possesses an Euler circuit. An Euler circuit uses every edge exactly once but vertices may be repeated.

Example 9. *The graph shown in Fig. 11.125 is an Euler graph. Determine Euler circuit for this graph.*

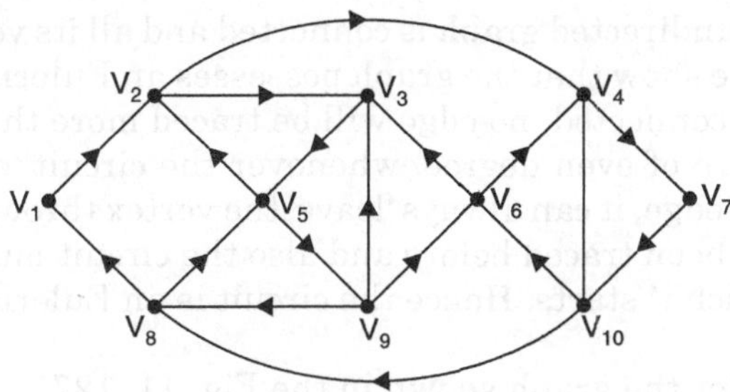

Fig. 11.125 Euler Graph.

Sol. The Euler circuit for this graph is

$V_1, V_2, V_3, V_5, V_2, V_4, V_7, V_{10}, V_6, V_3, V_9, V_6, V_4, V_{10}, V_8, V_5, V_9, V_8, V_1$.

We can produce an Euler circuit for a connected graph with no vertices of odd degree. The following algorithm called Fleury's algorithm is helpful to construct an Euler's Path.

Theorem I. *An undirected graph possesses an Eulerian path iff it is connected and has either zero or two vertices of odd degree. Give suitable example.*

Proof. Let the undirected graph possesses an Eulerian path. Then, by definition, the graph must be connected. Now, since the graph has Eulerian path, it means that every time the path meets a vertex, it goes through two edges which are incident with the vertex and have not been traced before.

Thus, except for the two vertices at the two ends of the path, the degree of all other vertices in the graph must be even.

If the two vertices at the two ends of the Eulerian path are distinct, then there are only two vertices with odd degree.

Converse. Let the undirected graph is connected and two of its vertices are of odd degree. We show the graph possesses an Eulerian path. Since the graph is connected, no edge will be traced more than once. For a vertex of even degree, whenever the path 'enters' the vertex through an edge, it can always 'leave' the vertex through another edge that has not been traced before. Therefore, when the construction is completed, we must have reached the other vertex of odd degree. Tracing all the edges in this way, we will get an Eulerian path. If not all of the edges in the graph were traced, we shall remove those edges that have been traced and obtain a subgraph formed by the remaining edges. The degrees of the vertices of this subgraph will be even. Starting from one of these vertices, we can again construct a path that passes through the edges. Because the degrees of the vertices are all even, this path must return to the vertex at which it starts. Combining this path with the path we have constructed to obtain one which starts and ends at the two vertices of odd degree, the path so obtained is Eulerian path.

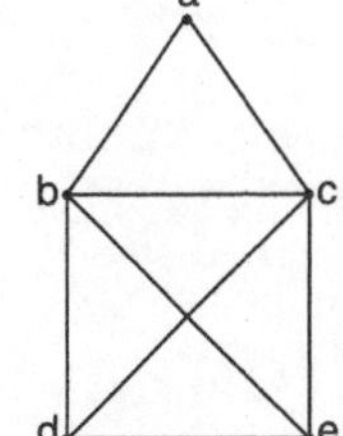

Fig. 11.126

For example, consider the graph (Fig. 11.126) as shown below.

This graph has only two vertices of odd degree, by above theorem, it has an Eulerian path.

Theorem II. *An undirected graph possesses an Eulerian circuit iff it is connected and its vertices are all of even degree. Prove with the help of suitable example.*

Proof. Let the undirected graph possess an Eulerian circuit. Then, by definition, the graph must be connected. Now since the graph has Eulerian circuit, it means that every time the graph meets a vertex, it goes through edges which are incident with the vertex and have not been traced before. Thus, the degree of all vertices in the graph must be even.

Converse. Let the undirected graph is connected and all its vertices are of even degree. We show that the graph possesses an Eulerian circuit. Since the graph is connected, no edge will be traced more than once. As all the vertices are of even degree, whenever the circuit 'enters' the vertex through an edge, it can always 'leave' the vertex through another edge that has not been traced before and also the circuit must return to the vertex at which it starts. Hence the circuit is an Eulerian circuit.

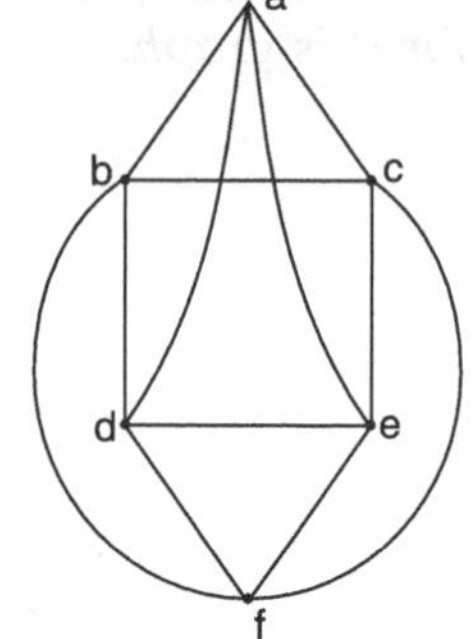

Fig. 11.127

For example, consider the graph shown in the Fig. 11. 127.

The graph is connected and all vertices are of even degree. Therefore, it has an Eulerian circuit.

Theorem III. *Let G be a connected graph such that each vertex is of degree 2. Prove that G is a cycle.*

Proof. Because G is a connected graph such that every vertex is of even degree, it follows that G has an Euler circuit. This circuit contains all the vertices and all the edges of G. Because the degree of each vertex is 2, it follows that in the above circuit, a vertex, except the starting vertex, cannot appear more than once. Hence, the above circuit is a cycle. This shows that graph G is a cycle.

11.39. FLEURY'S ALGORITHM

Input

Let G = (V, E) be a connected graph with every vertex of even degree.

Step 1. Select a vertex v_0 of V as the starting vertex to construct the circuit and designate P : v_0 as the starting of the path to be constructed.

Step 2. Consider that P : $v_0, v_1, v_2, ..., v_k$ as been constructed so far. If at v_k there is only one edge say $\{v_k, v_{k+1}\}$, then extend P to P : $v_0, v_1, v_2 ..., v_{k+1}$. Remove $\{v_k, v_{k+1}\}$ from edge set E and v_{k+1} from V. If at v_k there are several edges, choose one that is not a bridge to the remaining graph, say $\{v_k, v_{k+1}\}$. Extend P to P : $v_0, v_1, v_2, ..., v_k, v_{k+1}$ and remove $\{v_k, v_{k+1}\}$ from edge set E.

Step 3. Repeat step 2 until no edges left in E.

Example 10. *Use Fleury's algorithm to construct an Euler circuit for the graph shown in Fig. 11.128.*

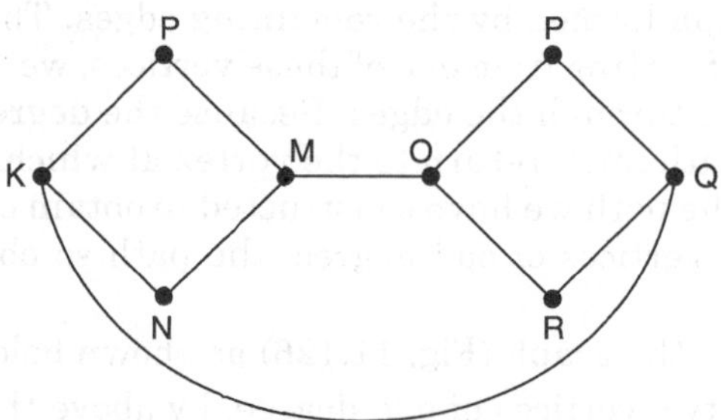

Fig. 11.128

Sol. Start from any vertex, as per step 1. So, choose vertex k as the starting vertex. The summary of the results applying step 2 repeatedly is shown in the table—

Current path	*Next edge to traverse*	*Reason*
P : k	$\{k, l\}$	No edge from k is a bridge. Choose any one.
P : k, l	$\{l, m\}$	Only one edge from l.
P : k, l, m	$\{m, n\}$	No edge from m is a bridge. Choose any one.
P : k, l, m, n	$\{n, k\}$	Only one edge from n.
P : k, l, m, n, k	$\{k, m\}$	No edge from k is a bridge. Choose any one.
P : k, l, m, n, k, m	$\{m, o\}$	Only one edge from m remains.
P : k, l, m, n, k, m, o	$\{o, p\}$	No edge from k is a bridge. Choose any one.
P : k, l, m, n, k, m, o, p	$\{p, q\}$	Only one edge from k remains.
P : $k, l, m, n, k, m, o, p, q$	$\{q, r\}$	No edge from k is a bridge. Choose any one.
P : $k, l, m, n, k, m, o, p, q, r$	$\{r, o\}$	Only one edge from r remains.
P : $k, l, m, n, k, m, o, p, q, r, o$	$\{o, q\}$	Only one edge from o remains.
P : $k, l, m, n, k, m, o, p, q, r, o, q$	$\{q, k\}$	Only one edge from q remains.
P : $k, l, m, n, k, m, o, p, q, r, o, q, k$		

11.40. HAMILTONIAN PATH (OR CHAIN) *(P.T.U., B.Tech. Dec. 2007)*

A Hamiltonian path (or chain) through a graph is a path whose vertex list contain each vertex of the graph exactly once, except if path is a circuit.

11.41. HAMILTONIAN CIRCUIT (OR CYCLE)

(P.T.U., M.C.A. May 2007 ; B.Tech. Dec. 2007)

A Hamiltonian circuit (or cycle) is a path in which the initial vertex appears a second time as the terminal vertex.

11.42. HAMILTONIAN GRAPH *(P.T.U., B.Tech. May 2007)*

A Hamiltonian graph is a graph that possesses a Hamiltonian path. A Hamiltonian path uses each vertex exactly once but edges may not be included.

Theorem IV. *A graph G has a Hamilton circuit if* $e \geq \frac{n^2 - 3n + 6}{2}$ *where n is the number of vertices and e the number of edges in G.* (P.T.U., M.C.A. May 2007)

Proof. Let, if possible, the graph G is non-Hamilton and we show $e \leq \frac{n^2 - 3n + 6}{2}$.

By **Dirac's theorem** which states that the connected graph G with $|\ v\ | \geq n$ vertices has a Hamiltonian circuit if $\deg(v_i) \geq n/2$ for each vertex n_i) there, exists a pair of non-adjacent vertices u and v such that

$$\deg(u) + \deg(v) \leq n - 1 \qquad ...(1)$$

Let H be the subgraph of G obtained by deleting vertices u and v from G. So graph H has $n - 2$ vertices and $e - [\deg(u) + \deg(v)]$ edges. Thus maximum number of edges in H will be ${}^{n-2}C_2$ and hence

$$e - [\deg(u) + \deg(v)] \leq {}^{n-2}C_2$$

$$= \frac{(n-2)!}{2!(n-2-2)!} = \frac{(n-2)!}{2(n-4)!} = \frac{(n-2)(n-3)(n-4)!}{2(n-4)!} = \frac{1}{2}(n^2 - 5n + 6)$$

$\therefore$ $$e \leq \frac{1}{2}(n^2 - 5n + 6) + [\deg(u) + \deg(v)]$$

$\Rightarrow$ $$e \leq \frac{1}{2}(n^2 - 5n + 6) + (n - 1)$$

$\Rightarrow$ $$e \leq \frac{n^2 - 5n + 6 + 2n - 2}{2} = \frac{n^2 - 3n + 4}{2}$$

$$\leq \frac{n^2 - 3n + 6}{2}$$

Thus $e \leq \frac{1}{2}(n^2 - 3n + 6)$. Hence the theorem.

Remark. The converse of above theorem, however, is not true.

11.43. RULES FOR CONSTRUCTING HAMILTON PATHS (OR CHAINS) AND HAMILTON CIRCUITS (OR CYCLES) IN A GRAPH

Rule I. If a graph G has n vertices, then a Hamilton path in G must contain exactly $(n - 1)$ edges and a Hamilton circuit in G must contain exactly n edges.

Rule II. In a Hamilton circuit, there cannot be more than three or more edges incident with one vertex. *i.e.*, every vertex V in a Hamilton circuit will contain exactly 2 edges incident on V.

Also, If V is a vertex in G, then a Hamilton path must contain atleast one edge incident on V and atmost 2 edges incident on V.

Remarks

(*i*) Multigraph cannot have Hamilton circuit.

(*ii*) Hamilton path, if it exists, is the longest simple path in a graph.

(*iii*) Each graph which has Hamilton cycle will have Hamilton path, the converse, however, is not true.

(*iv*) Every complete graph K_n is Hamilton for $n \geq 3$.

(*v*) A Hamilton graph with n vertices must have atleast n edges.

Example 11. *The graph shown in Fig. 11.129 is a Hamiltonian graph. Determine Hamiltonian circuit for this graph.*

Sol. The Hamiltonian circuit is shown in Fig. 11.130.

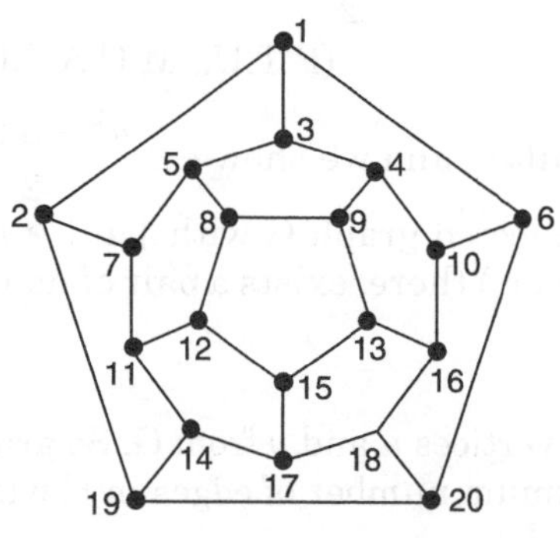

Fig. 11.129 Hamiltonian Graph.

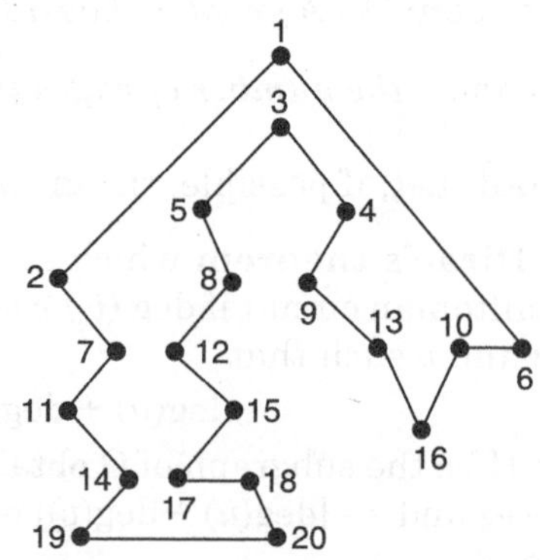

Fig. 11.130 Hamiltonian Circuit.

Example 12. *Give an example of a graph that has an Euler circuit which is also a Hamiltonian circuit.*

Sol. The graph having an Euler circuit which is also a Hamiltonian circuit is shown in Fig. 11.131.

Fig. 11.131

In this graph V_1, V_2, V_3, V_4, V_1 is both an Euler circuit as well as Hamiltonian circuit. Since using this path, we can traverse both vertices and edges exactly once.

Example 13. *Give an example of a graph that has an Euler circuit and a Hamiltonian circuit, which are distinct.* (P.T.U., M.C.A. Dec. 2006)

Sol. The graph having an Euler circuit and a Hamiltonian circuit which are distinct is shown in Fig. 11.132.

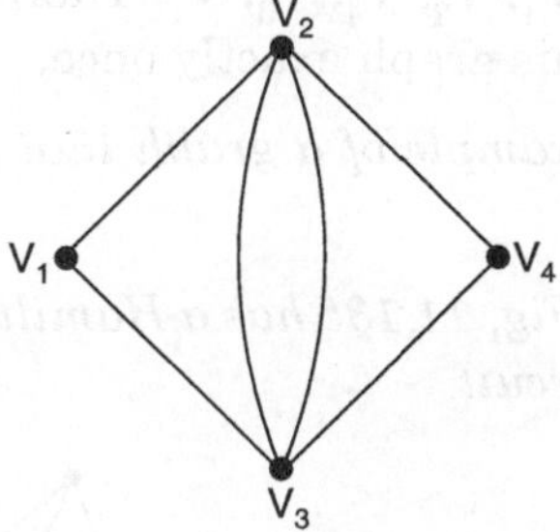

Fig. 11.132

The Euler circuit is V_1, V_3, V_2, V_3, V_4, V_2, V_1, which visits each edge exactly once.

The Hamiltonian circuit is V_1, V_2, V_4, V_3, V_1, which visits each vertex exactly once.

Example 14. *Give an example of a graph which has an Euler circuit but not a Hamiltonian circuit.*

Sol. The graph having an Euler circuit but not a Hamiltonian circuit is shown in Fig. 11.133.

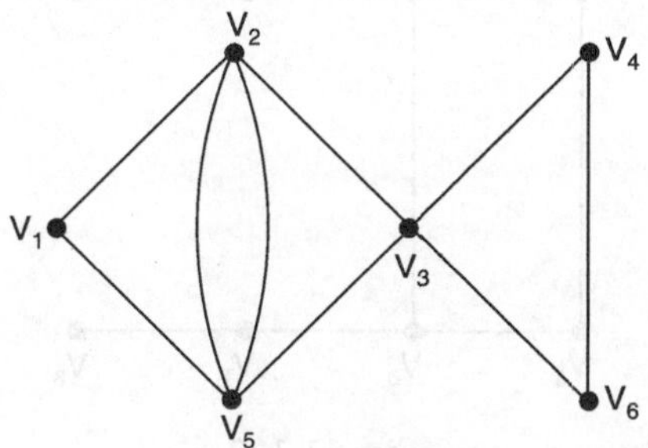

Fig. 11.133

The Euler circuit is V_1, V_5, V_2, V_5, V_3, V_4, V_6, V_3, V_2, V_1.

There is no Hamiltonian circuit. Since it is not possible to traverse each vertex of this graph exactly once.

Example 15. *Give an example of a graph which has a Hamiltonian circuit but not an Euler circuit.* *(P.T.U., M.C.A. May 2008)*

Sol. The graph having a Hamiltonian circuit but not an Euler circuit is shown in Fig. 11.134.

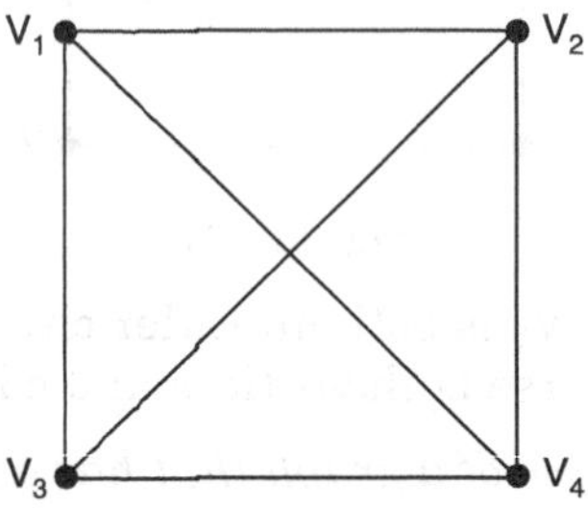

Fig. 11.134

The Hamiltonian circuit is V_1, V_2, V_4, V_3, V_1. There is no Euler circuit. Since it is not possible to traverse each edge of this graph exactly once.

Example 16. (*a*) *Give an example of a graph that has neither an Euler circuit nor a Hamiltonian circuit.*

(*b*) *Show that the graphs in Fig. 11.135 has a Hamiltonian circuit where as the graph in Fig. 11.136 has no Hamiltonian circuit.*

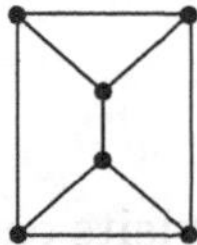

Fig. 11.135

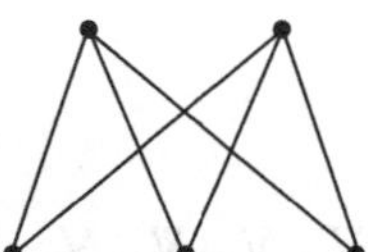

Fig. 11.136

Sol. (*a*) The graph having neither an Euler circuit nor a Hamiltonian circuit is shown in Fig. 11.137. It does not contain Euler circuit since each vertex is not of even degree.

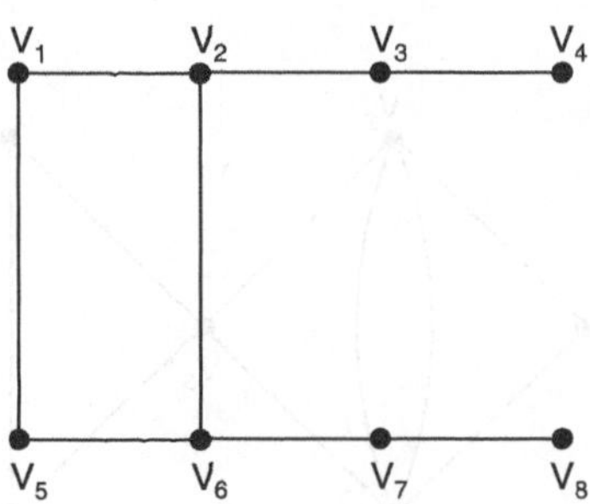

Fig. 11.137

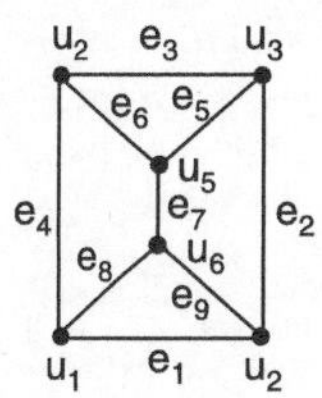

Fig. 11.138

(*b*) We know that if there is a path in a graph G that uses each vertex of the graph exactly once, except initial vertex that appears twice as the terminal vertex, then such a path is called a **Hamiltonian circuit.**

The graph in Fig. 11.138 has Hamiltonical circuit given as

$$u_1, e_1, u_2, e_2, u_3, e_3, u_2, e_6,, u_5, e_7, u_6, e_8, u_1$$

Also, we know that a connected graph G with $| v | \geq n$ vertices has a Hamiltonian circuit if $\deg(v_i) \geq n/2$ for each vertex n_i.

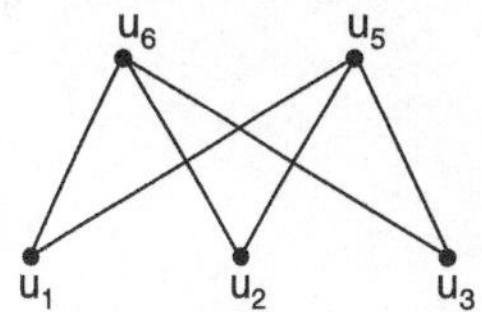

Fig. 11.139

The graph in Fig. 11.139 has 5 vertices. Here $\deg(u_1) = 2 \not\geq 5/2$

The graph in Fig. 11.139 has no Hamiltonian circuit.

Example 17. *Consider the graph G shown in the following Fig. 11.140.*

(*a*) *Find the Euler's path if it exists.*

(*b*) *Is the graph G Eulerian?*

(*c*) *Is the graph bipartite?*

(*d*) *Is the graph hamiltonian?*

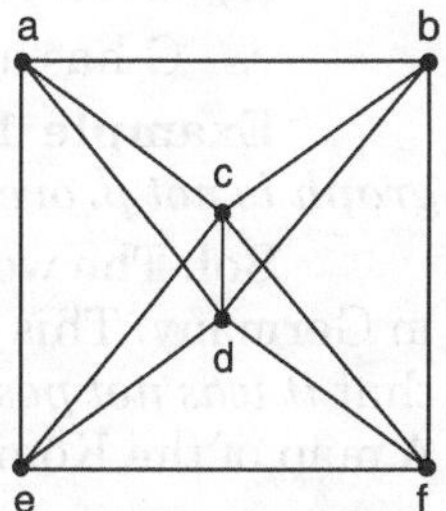

Fig. 11.140

Sol. (*a*) Consider the vertices '*c*' and '*d*'. The degree of these vertices is 5(odd)

∴ G has Euler's path between *c* and *d*.

The Euler's path is *c, a, b, d, f, b, d, f, e, a, d, e, c, d*

(*b*) A graph G can have Euler's circuit if each vertices of G are of even degree. Since the vertices '*c*' and '*d*' are of odd degree, therefore, G cannot have Euler's circuit.

(*c*) The graph G has odd cycle *a, c, d, a*

∴ it cannot be a bipartite graph

(*d*) Yes, the Hamiltonian circuit is *a, b, f, e, d, c, a*.

Example 18. *Consider the graph G shown below in Fig. 11.141.*

(*a*) *Is it a complete graph?*

(*b*) *Is G connected and regular?*

(*c*) *Is it a planar graph? If so, find the number of regions.*

(*d*) *Is G Eulerian?*

Sol. (*a*) Since the edge between *a* and *d* is not present in the given graph. It cannot be complete graph. (Every pair of vertices must be joined by an edge.)

(*b*) Given graph is a 4-regular graph. Also it is connected because there is a path from every vertex to other.

(*c*) The given graph is planar graph as it can be re-drawn as shown in Fig. 11.142 in which no two edges cross

Here $V = 6, E = 12$

By Euler's theorem, $V - E + R = 2$

$\Rightarrow \quad 6 - 12 + R = 2$

$\Rightarrow \quad R = 8$

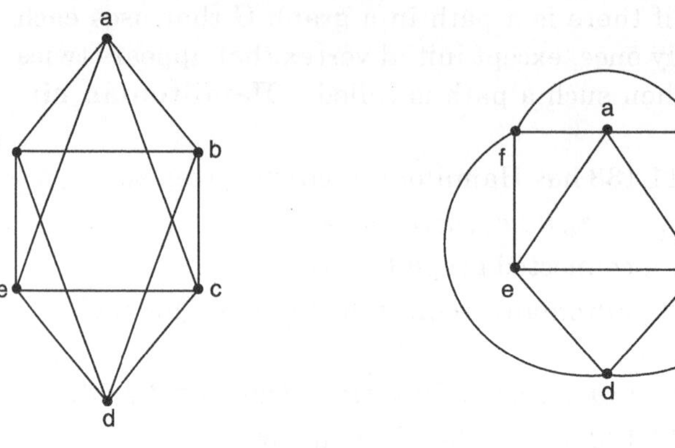

Fig. 11.141 Fig. 11.142

(*d*) Given graph is 4-regular. $\therefore$ every vertex is of degree 4 (even)

$\therefore$ G has an Euler's circuit and hence G is an Eulerian graph.

Example 19. *State and prove Eulerian theorem on graph to show that Königsberg's graph is not proved to a solution.*

Sol. The word Königsberg is the name of a town, situated on the bank of a river, Pregel in Germany. This city has seven bridges. In 1736, L. Euler, the father of graph theory, proved that *it was not possible to cross each of the seven bridges once and only once in a walking tour.* A map of the Königsberg is shown in the following Fig. 11.143.

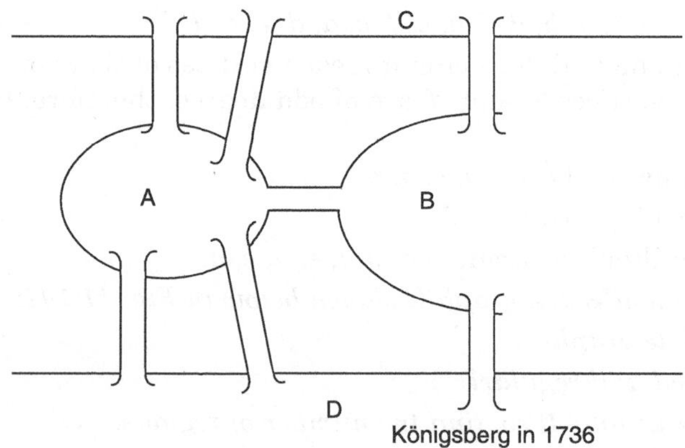

Fig. 11.143

Euler replaced the islands and the two sides of the river by points and the bridges by curves as shown in Fig. 11.144.

Figure 11.144 is a multigraph. A multigraph is a said to be **traversable** if it can be drawn without any breaks in the curve and without repeating any edges. *i.e.,* if there is a path which includes all vertices and uses each edge exactly once and such a path is called Travesable Trial.

According to Euler, the walk in Königsberg is possible iff the multigraph in Fig. 11.144 is traversable. But Euler proved that the multigraph in Fig. 11.144 is not traversable and hence the walk in Königsberg is impossible. We prove it.

We know that a vertex is even or odd according as its degree is even or odd. Suppose a multigraph is travesable and that a traversable trail does not begin or end at a vertex, say, P. We claim that P is an even vertex. For whenever the traversable trail enters P by an edge, there must always be an edge not previously used by which the trail can leave P. Thus, the edges in the trail incident with P must appear in pairs and so P is an even vertex. Further, if a vertex, Q is odd, the traversable trail must begin or end at Q. Hence, a multigraph with more than two odd vertices cannot be traversable.

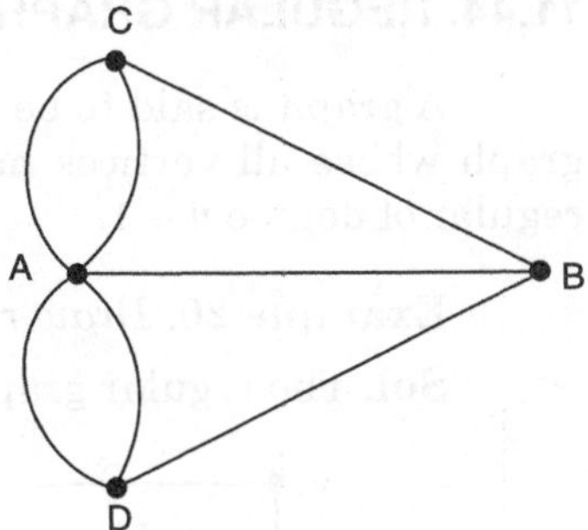

Fig. 11.144

Now the multigraph corresponding to the Königsberg bridge problem has four odd vertices. Thus, one cannot walk through Königsberg so that each bridge is crossed exactly once.

Euler actually proved the converse of the above statement, which is contained in the following theorem, called Euler theorem.

Theorem V. *A finite connected graph is Eulerian iff each vertex has even degree.*

(P.T.U., M.C.A. May 2008, B.Tech. Dec. 2012)

Proof. We know that a graph G is called an Eulerian graph if there exists a closed traversable Trial, called an Eulerian Trial. Suppose G is Eulerian and T is a closed Eulerian trial. Let v be any vertex of G. We show the vertex v is of even degree. Since the trail T enters and leaves the vertex v the same number of times without repeating any edge. $\therefore$ v is of even degree.

Conversly, Let each vertex of G has even degree. We construct an Eulerian Trial. Start with a trial T_1 at any edge e. Extend T_1 by adding one edge after the other. If T_1 is not closed *i.e.*, If T_1 begins at u and ends at $v \neq u$, then only an odd number of edges incident on v appear in T_1. Hence we can extend T_1 by another edge incident on v. Thus, we can continue to extend T_1 until T_1 returns to its initial vertex u. *i.e.*, until T_1 is closed.

If T_1 includes all the edges in G, then T_1 is the required Eulerian Trial.

If T_1 does not include all edges of G, consider the graph H obtained by deleting all edges of T_1 from G. Now H has each vertex of even degree (since T_1 contains an even number of the edges incident on any vertex). Since G is connected, there is an edge e' of H which has an end point u' in T_1. We construct a trail T_2 in H begining at u' and using e'. Since all the vertices in H have even degree, we can continue to extent T_2 in H until T_2 returns to u' as shown in the following Fig. 11.145. We can clearly put T_1 and T_2 together to form a larger closed trial in G. Proceeding the above process until all the edges of G are used, we finally obtain an Eulerian trial and hence G is Eulerian.

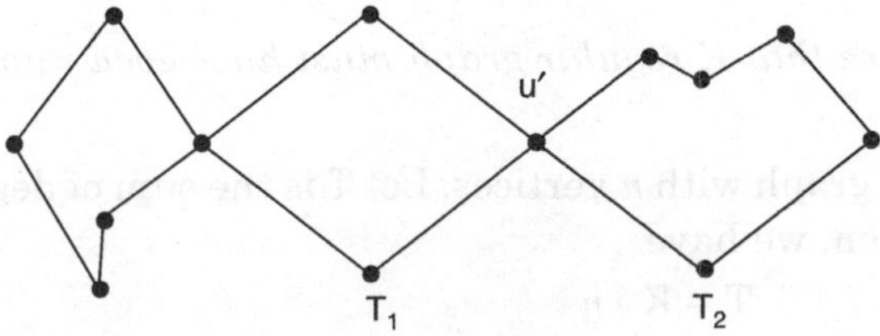

Fig. 11.145

11.44. REGULAR GRAPH

A graph is said to be regular or K-regular if all its vertices have the same degree K. A graph whose all vertices have degree 2 is called 2-regular graph. A complete graph K_n is regular of degree $n - 1$.

Example 20. *Draw regular graphs of degree 2 and 3.*

Sol. The regular graphs of degree 2 and 3 are shown in Figs. 11.146 and 11.147.

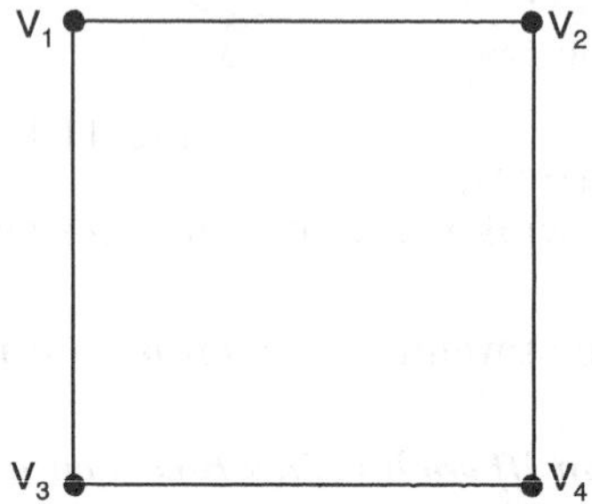

Fig. 11.146. 2-regular Graph.

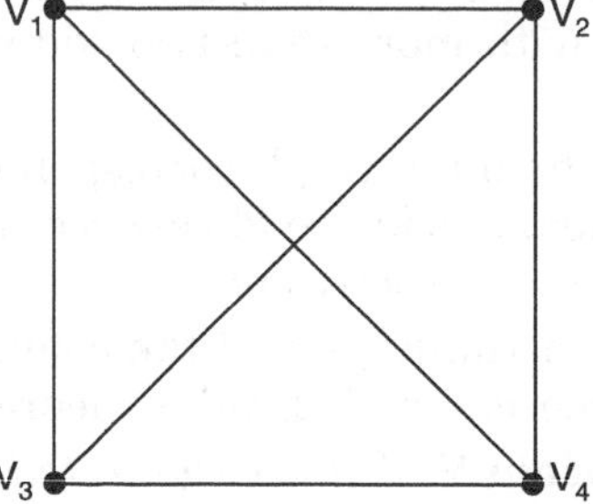

Fig. 11.147. Regular Graph.

Example 21. *Draw a 2-regular graph of five vertices.*

Sol. The 2-regular graph of five vertices is shown in Fig. 11.148

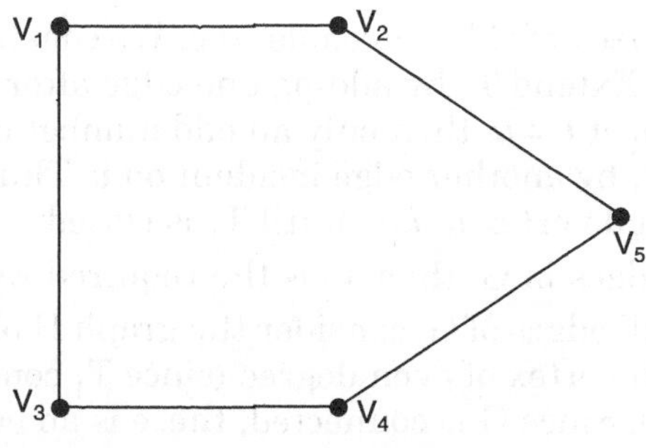

Fig. 11.148

Example 22. *Draw a 3-regular graph of five vertices.*

Sol. It is not possible to draw 3-regular graph of five vertices. The 3-regular graph must have an even number of vertices.

Theorem VI. *Prove that K-regular graph must have even number of vertices when the value of K is odd.*

Proof. Consider a graph with n vertices. Let T is the sum of degrees of all the n vertices of a K-regular graph. Then, we have

$$T = K \cdot n$$

The sum T must be even (from the theorem V).

Now, suppose that K is odd, so the value of n must be even.

11.45. PLANAR GRAPH

(P.T.U. B.Tech. Dec. 2013)

A graph is said to be planar if it can be drawn in a plane so that no edges cross.

For *e.g.*, the graph is a planar graph. Also K_4 = is a planar graph because it can be re-drawn as in which edges do not cross each other.

For example: The graphs shown in Fig. 11.149 and Fig. 11.150 are planar graphs.

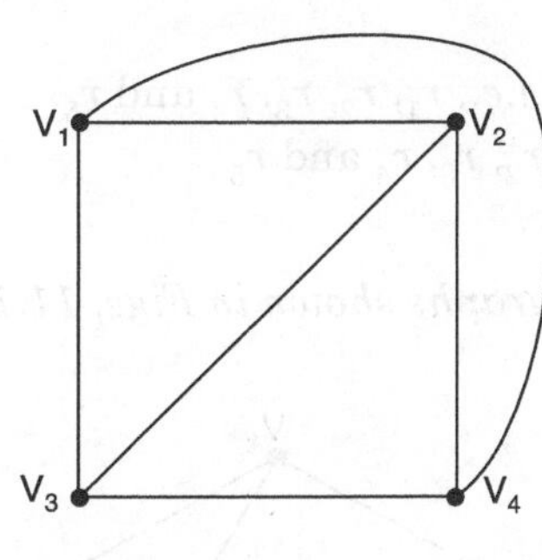

Fig. 11.149

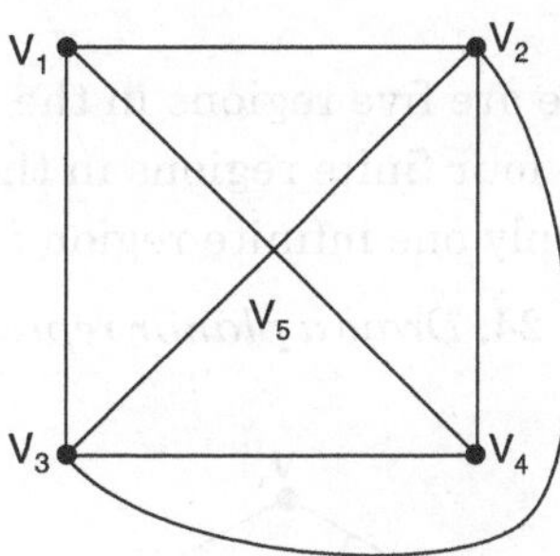

Fig. 11.150

Theorem VII. *A planar and connected graph has a vertex of degree less than or equal to 5.*

Proof. Let G be connected and planar and suppose, if possible, degree of each vertex $x \in G$ is greater than 5.

i.e., $\deg x > 5 \Rightarrow \deg x \geq 6$ *i.e.*, sum of degree of all vertices $\geq 6v$

$\Rightarrow$ $2e \geq 6v$, where e and v are the number of edges and vertices respectively.

$\Rightarrow$ $e \geq 3v$, which contradicts $e \leq 3u - 6 < 3v$.

Hence $\deg x \leq 5$.

11.46. REGION OF A GRAPH

Consider a planar graph G = (V, E). A region is defined to be an area of the plane that is bounded by edges and cannot be further subdivided. A planar graph divides the plane into one or more regions. One of these regions will be infinite.

(*a*) **Finite Region.** If the area of the region is finite, then that region is called finite region.

(*b*) **Infinite Region.** If the area of the region is infinite, that region is called infinite region. A planar graph has only one infinite region.

Example 23. *Consider the graph shown in Fig. 11.151. Determine the number of regions, finite regions and an infinite region.*

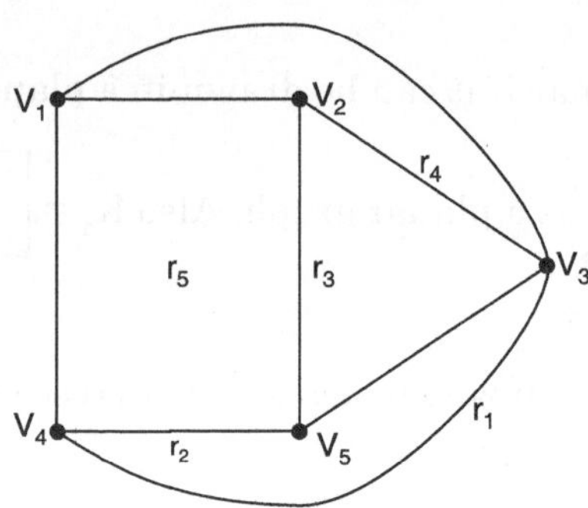

Fig. 11.151

Sol. There are five regions in the above graph *i.e.,* r_1, r_2, r_3, r_4 and r_5.
There are four finite regions in the graph *i.e.,* r_2, r_3, r_4 and r_5.
There is only one infinite region *i.e.,* r_1.

Example 24. *Draw a planar representation of graphs shown in Figs. 11.152 and 11.153.*

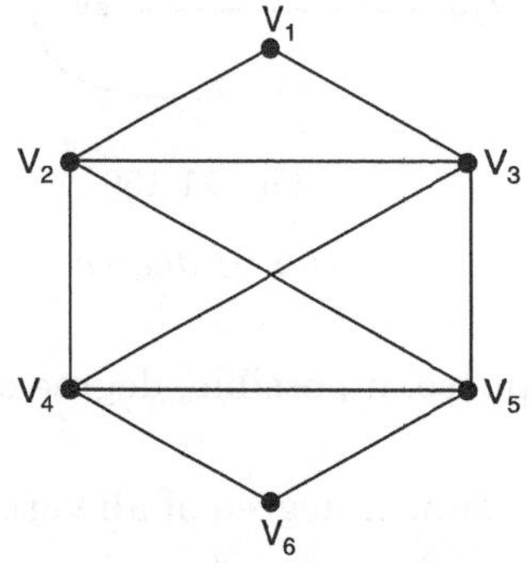

Fig. 11.152

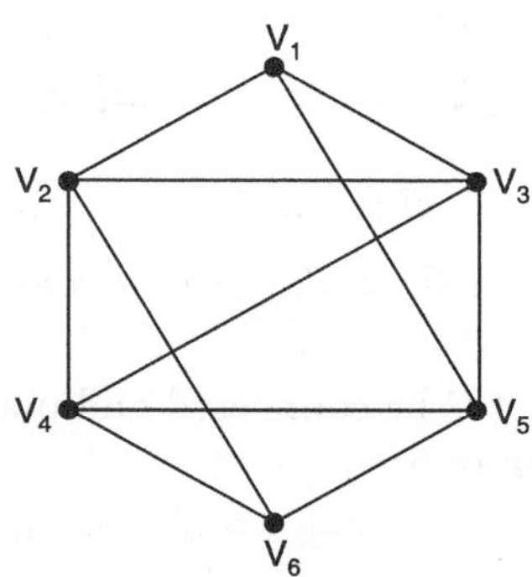

Fig. 11.153

Sol. The planar representation of graph shown in Fig. 11.154 is shown in Fig. 11.155.

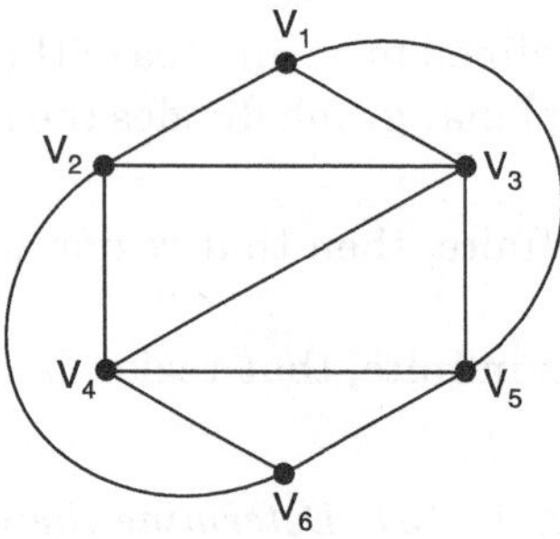

Fig. 11.154

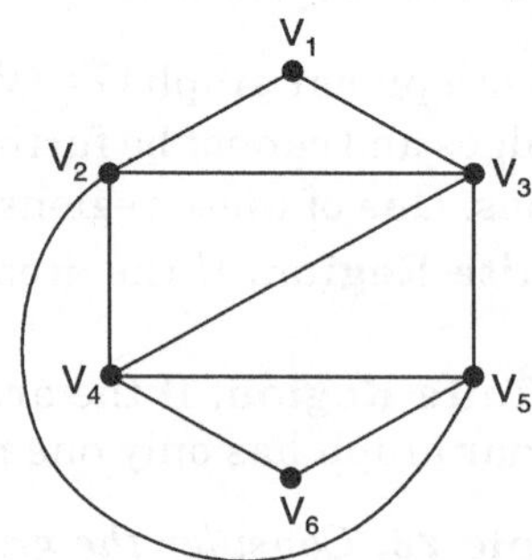

Fig. 11.155

11.47. PROPERTIES OF PLANAR GRAPHS

(P.T.U. B.Tech. Dec. 2013)

Theorem I. *If a connected planar graph G has e edges and r regions, then* $r \leq \frac{2}{3}e$.

Theorem II. *If a connected planar graph G has e edges and v vertices, then* $3v - e \geq 6$.

Theorem III. *A complete graph* K_n *is planar if and only if* $n < 5$.

Theorem IV. *A complete bipartite graph* $K_{m, n}$ *is planar if and only if* $m < 3$ *or* $n > 3$.

I. Proof. In a connected planar graph, each region is bounded by at least 3 regions

$\therefore$ r regions are bounded by minimum $3r$ edges

$\Rightarrow$ Number of edges in graph $\geq 3r$

But number of edge in the graph = $2e$ (as each edge belongs to two regions)

$$\therefore \quad 2e \geq 3r$$

$$\Rightarrow \quad r \leq \frac{2e}{3}$$

II. Let r be the no. of regions in a planar representation of G. $\therefore$ By Euler formula

$$v + r - e = 2 \quad \text{...(1)}$$

Now sum of degrees of the regions = $2e$. But each region has degree 3 or more.

$$\therefore \quad 2e \geq 3r \quad \Rightarrow \quad r \leq \frac{2e}{3}.$$

From (1) we get $\quad 2 = v + r - e \leq v + \frac{2e}{3} - e = v - \frac{e}{3}$

$$6 \leq 3v - e$$

$\Rightarrow \quad e \leq 3v - 6$ Hence proved.

III. If G has one or two vertices, then the result is true. If G has at least 3 vertices then

$$e \leq 3v - 6 \quad \text{or} \quad 2e \leq 6v - 12 \quad \text{...(1)}$$

If degree of every vertex were at least 6, then using $2e = \sum_{v \in V} \deg v$, we would have $2e \geq 6v$, which contradicts the inequality (1), Hence there must be a vertex with degree not greater than 5.

IV. Proof is beyond the scope of the book.

Example 25. *Prove that complete graph* K_4 *is planar.*

Sol. The complete graph K_4 contains 4 vertices and 6 edges.

We know that for a connected planar graph $3v - e \geq 6$. Hence for K_4, we have

$3 \times 4 - 6 = 6$ which satisfies the property (3).

Thus K_4 is a planar graph. Hence proved.

11.48. STATE AND PROVE EULER'S THEOREM ON GRAPHS

(P.T.U., B.Tech. Dec. 2007, 2006, May 2006)

Statement. *Consider any connected planar graph G = (V, E) having R regions, V vertices and E edges. Then*

$V + R - E = 2.$ (P.T.U., B.Tech. May 2012, May 2005; M.C.A. Dec. 2005)

Proof. Use induction on the number of edges to prove this theorem.

Assume that the edges $e = 1$. Then we have two cases, graphs of which are shown in Figs. 11.156 and 11.157.

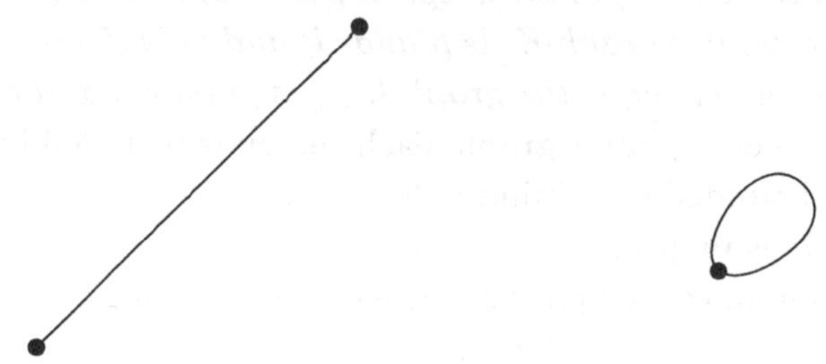

Fig. 11.156 Fig. 11.157

In Fig. 11.156 we have V = 2 and R = 1. Thus 2 + 1 – 1 = 2

In Fig. 11.157 we have V = 1 and R = 2. Thus 1 + 2 – 1 = 2. Hence, the result holds for $e = 1$.

Let us assume that the formula holds for connected planar graphs with K edges.

Let G be a graph with K + 1 edges.

Firstly, we suppose that G contains no circuits. Now, take a vertex v and find a path starting at v. Since G is circuit free, whenever we find an edge, we have a new vertex. At last we will reach a vertex v with degree 1. So we cannot move further as shown in Fig. 11.158.

Now remove vertex v and the corresponding edge incident on v. So, we are left with a graph G* having K edges as shown in Fig. 11.159.

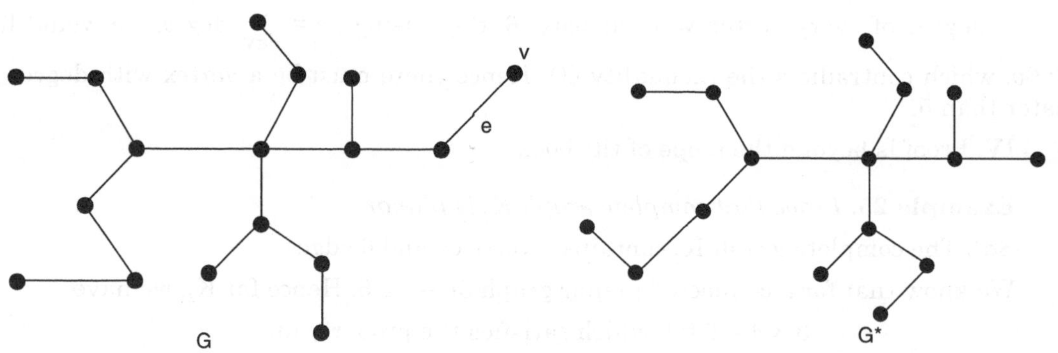

Fig. 11.158. G. Fig. 11.159. G*.

Hence, by inductive assumption, Euler's formula holds for G*.

Now, since G has one more edge than G*, one more vertex than G* with same number of regions as in G*. Hence, the formula also holds for G.

Secondly, we assume that G contains a circuit and e is an edge in the circuit shown in Fig. 11.160.

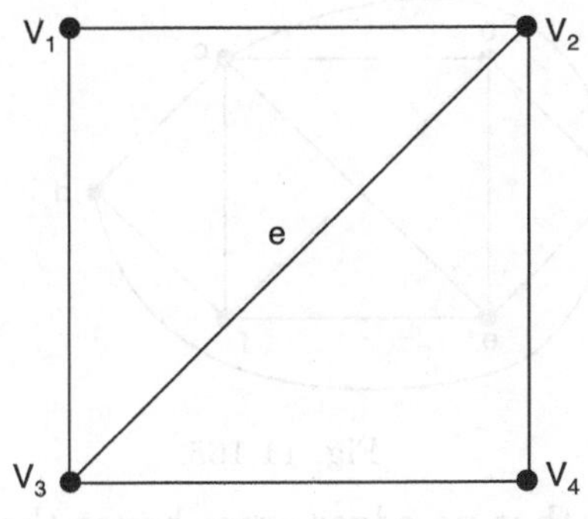

Fig. 11.160

Fig. 11.161

Now, as e is the part of a boundary for two regions. So, we only remove the edge and we are left with graph G* having K edges (Fig. 11.161).

Hence, by inductive assumption, Euler's formula holds for G*.

Now, since G has one more edge than G*, one more region than G* with same number of vertices as G*. Hence the formula also holds for G which, verifies the inductive step and hence proves the theorem.

Example 26. *Show that $V - E + R = 2$ for the connected planar graphs shown in Figs. 11.162 and 11.163.*

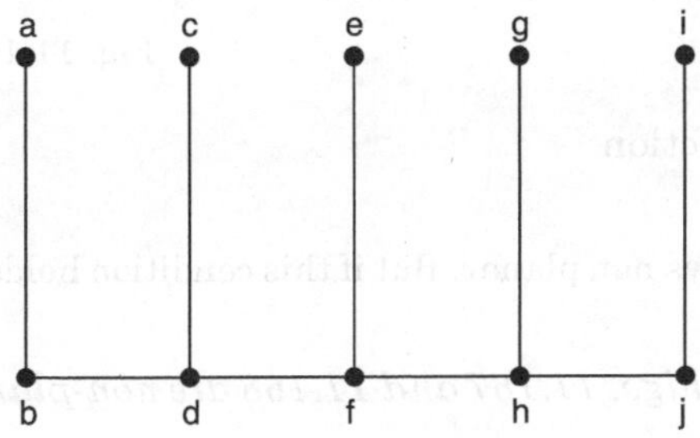

Fig. 11.162

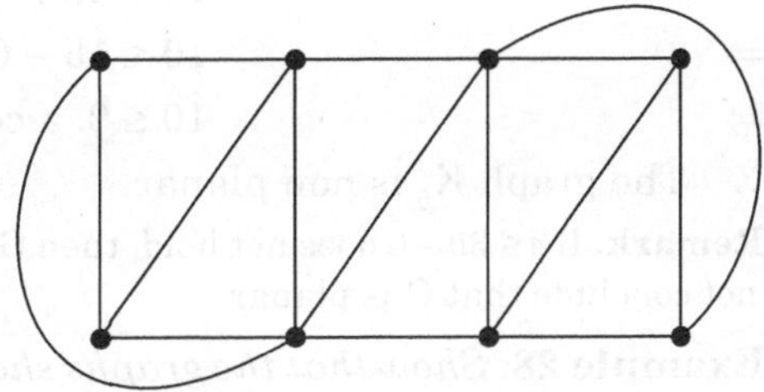

Fig. 11.163

Sol. (*i*) The graph shown in Fig. 11.162 contains vertices V = 10, edges E = 9 and regions R = 1. Putting the values, we have 10 – 9 + 1 = 2. Hence proved.

(*ii*) The graph shown in Fig. 11.163 contains vertices V = 8, edges E = 15 and regions R = 9. Putting the values, we have 8 – 15 + 9 = 2. Hence proved.

11.49. NON PLANAR GRAPHS

A graph is said to be non planar if it cannot be drawn in a plane so that no edges cross.

For example : The graphs shown in Figs. 11.164 and 11.165 are non planar graphs.

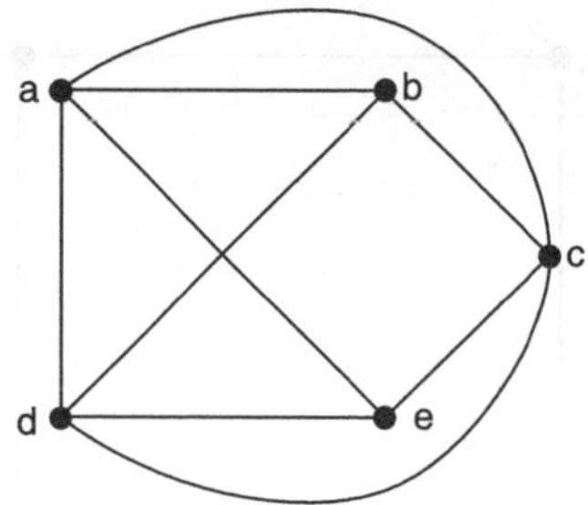

Fig. 11.164

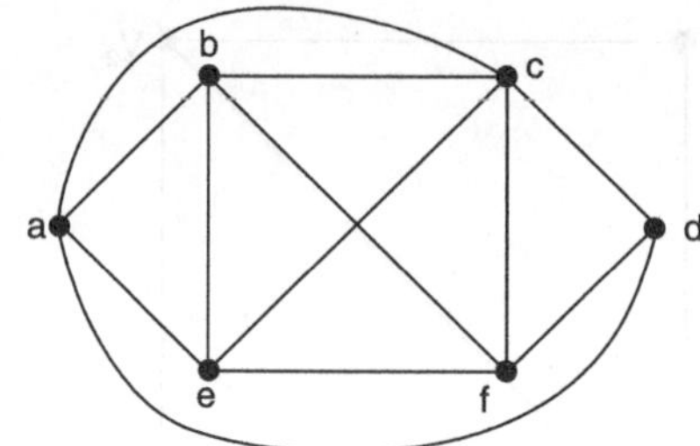

Fig. 11.165

These graphs cannot be drawn in a plane so that no edges cross hence they are non planar graphs.

11.50. PROPERTIES OF NON PLANAR GRAPHS

A graph is non-planar if and only if it contains a subgraph homeomorphic to K_5 or $K_{3,\,3}$ [KURATOWSKI'S THEOREM].

Example 27. *Show that K_5 is non-planar. Fig. 11.166.*

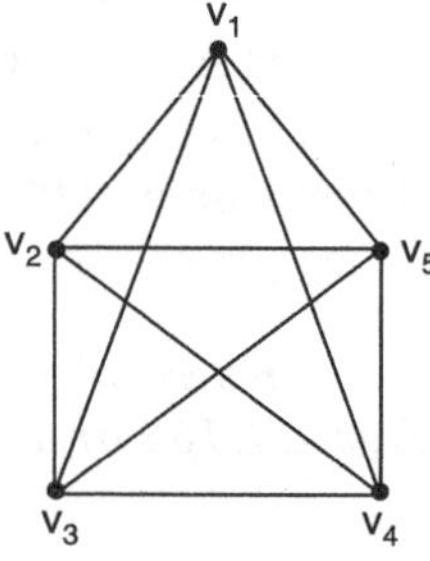

Fig. 11.166

Sol. Clearly K_5 is a connected. Also we show K_5 is non planar. For,

$$v = 5,\ e = 10$$

If, K_5 is planar then, $\quad e \le 3v - 6$

$\Rightarrow \quad 10 \le 3(5) - 6$

$\Rightarrow \quad 10 \le 15 - 6$

$\Rightarrow \quad 10 \le 9$, a contradiction

$\therefore$ The graph K_5 is non planar.

Remark. If $e \le 3v - 6$ does not hold, then G is always non planar. But if this condition holds, then we can not conclude that G is planar.

Example 28. *Show that the graphs shown in Figs. 11.167 and 11.168 are non-planar by finding a subgraph homeomorphic to K_5 or $K_{3,\,3}$.*

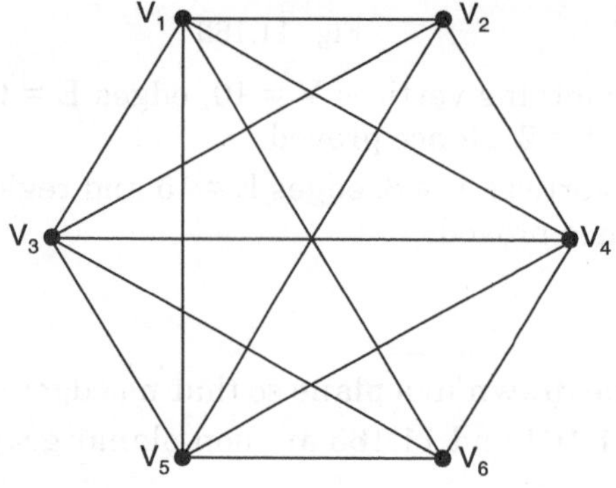

Fig. 11.167. G_1.

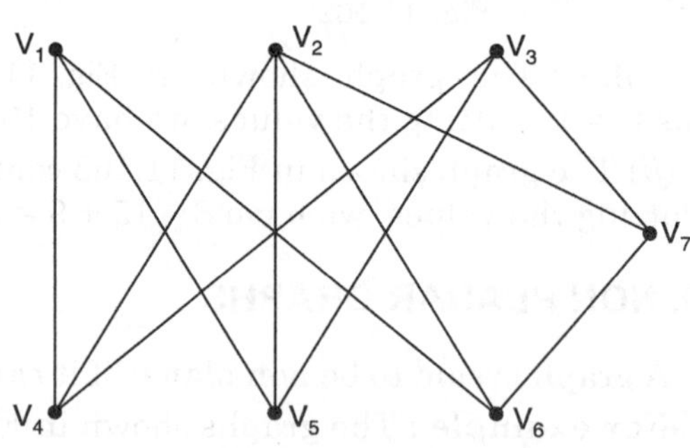

Fig. 11.168. G_2.

Sol. If we remove the edges (V_1, V_4), (V_3, V_4) and (V_5, V_4), the graph G_1 becomes homeomorphic to K_5. Hence it is non planar.

If we remove the edge (V_2, V_7), the graph G_2 becomes homeomorphic to $K_{3,3}$. Hence it is non-planar.

Theorem. *Prove that every planar graph has at least one vertex of degree 5 or less than 5.*

Proof. Consider a graph G, whose all vertices are of degree 6 or more, then the sum of the degrees of all the vertices would be greater than or equal to $6v$. We know that the sum of the degrees of the vertices is twice the number of edges. Therefore, we have

$$6v \le 2e$$

or $$v \le \frac{e}{3} \qquad \text{...(1)}$$

But, any planar graph have the property,

$$r \le \frac{2e}{3} \qquad \text{...(2)}$$

Also, from Euler's formula, we have

$$2 = v - e + r \qquad \text{...(3)}$$

Now, putting the value of v and r from (1) and (2) in (3), we have

$$2 \le \frac{e}{3} - e + \frac{2e}{3} = 0$$

Since, the statement $2 \le 0$ is not true, hence we conclude that there must exist some vertex in G with degree 5 or less than 5.

Example 29. *Let G = [V, E] be a graph having at least 11 vertices. Prove that G or its complement $\overline{G}$ is non planar.*

Sol. Let if possible, G and $\overline{G}$ are planar. We know that G and $\overline{G}$ have same number of vertices say, n. $\therefore$ we have $n \ge 11$

Now $$\overline{G} = K_n - G$$

$$\Rightarrow \quad \overline{G} + G = K_n$$

$$\Rightarrow \quad |\overline{E}| + |E| = \text{set of edges in } K_n = \frac{n(n-1)}{2} \qquad \text{...(1)}$$

Since G and $\overline{G}$ are planar

$$\therefore \quad |E| \le 3n - 6, \quad |\overline{E}| \le 3n - 6$$

$$\Rightarrow \quad |E| + |\overline{E}| \le 3n - 6 + 3n - 6$$

$$\Rightarrow \quad \frac{n(n-1)}{2} \le 6n - 12 \qquad | \text{ Using (1)}$$

$$\Rightarrow \quad n(n-1) \le 12n - 24, \text{ which is not true for } n \ge 11$$

Hence G or $\overline{G}$ is non-planar.

TEST YOUR KNOWLEDGE 11.2

1. Consider the graph shown in the given Fig. I.
   (*a*) Find all simple paths from A to F.
   (*b*) All Trials (distinct edges) from A to F.
   (*c*) *d*(A, F), the distance from A to F.
   (*d*) Diam (G), the diameter of G.
   (*e*) All cycles which include vertex A.
   (*f*) All cycles in G.

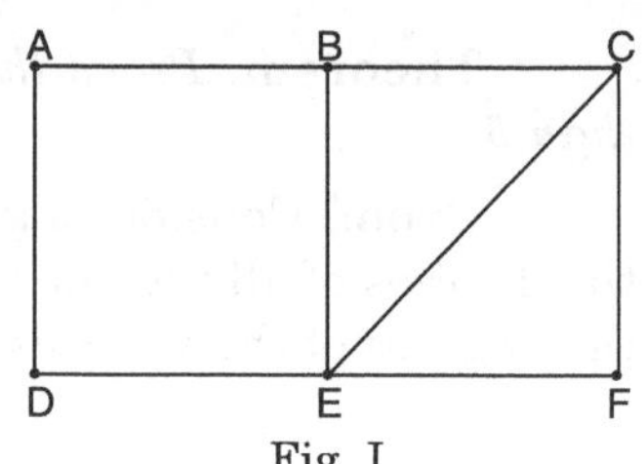

Fig. I

2. Consider the graph shown in the given Fig. II Find.
   (*a*) All sample paths from A to G.
   (*b*) All trials (distinct edges) from B to C.
   (*c*) *d*(A, C), the distance from A to C.
   (*d*) Diam G, the diameter of G.

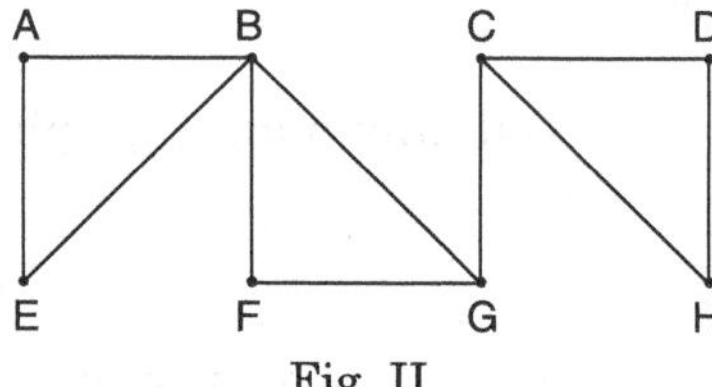

Fig. II

3. Find the adjacency matrix A = $\{a_{ij}\}$ of the graphs shown below :

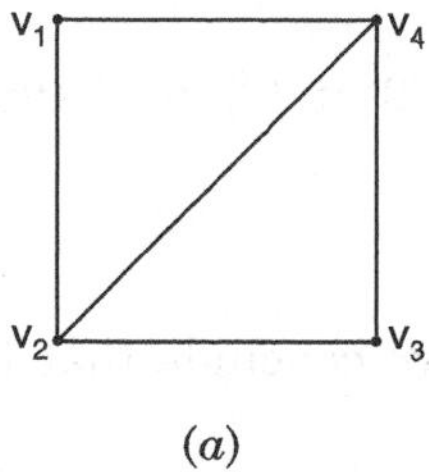

(*a*)

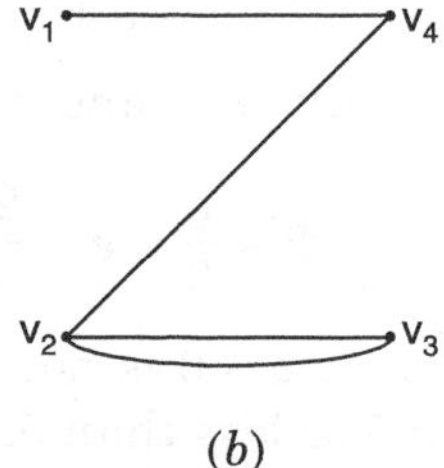

(*b*)

4. Draw the graph G corresponding to each adjacency matrix.

$$(a)\ A = \begin{bmatrix} 0 & 1 & 0 & 1 & 0 \\ 1 & 0 & 0 & 1 & 1 \\ 0 & 0 & 0 & 1 & 1 \\ 1 & 1 & 1 & 0 & 1 \\ 0 & 1 & 1 & 1 & 0 \end{bmatrix} \qquad (b)\ A = \begin{bmatrix} 1 & 3 & 0 & 0 \\ 3 & 0 & 1 & 1 \\ 0 & 1 & 2 & 2 \\ 0 & 1 & 2 & 0 \end{bmatrix}$$

5. (*a*) Consider the graph (Fig. I) G show in the given figure. Verify Euler Theorem *i.e.,* V + R – E = 2.

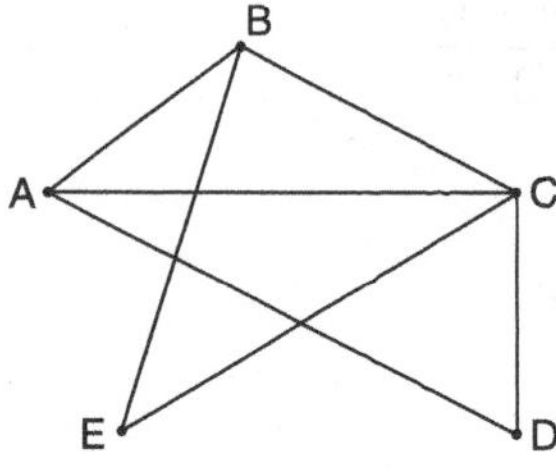

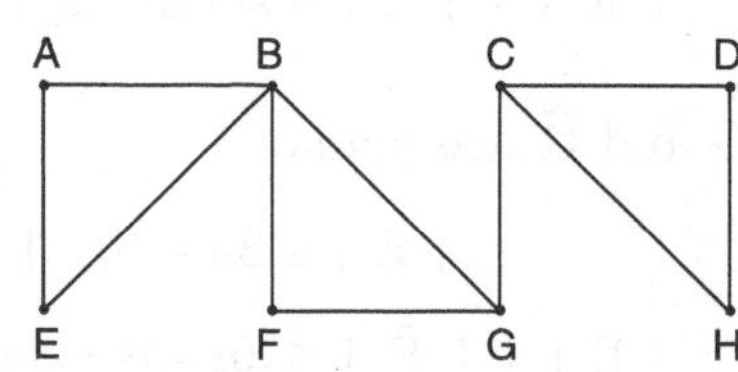

Fig. I.   Fig. II.

(*b*) Verify Euler Theorem *i.e.,* V + R – E = 2 for the graph Fig. II.

6. Consider each graph as shown below :

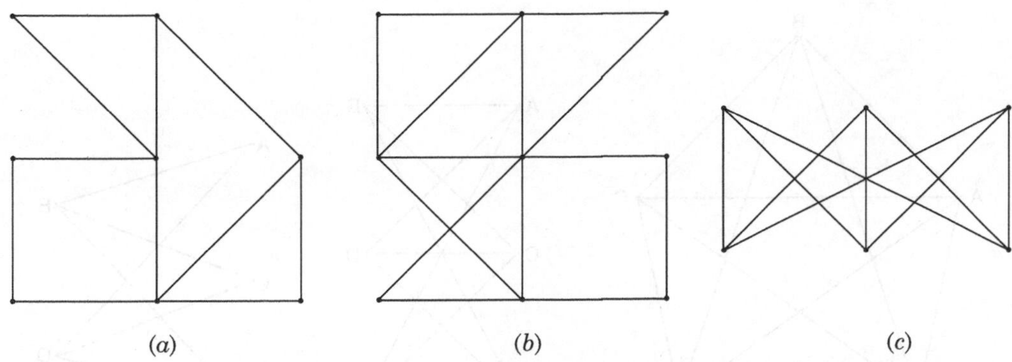

(*i*) Which of the graphs (*a*), (*b*), (*c*) has Euler path ? Which have Euler circuit. If not, explain why ?

(*ii*) Which of the graphs (*a*), (*b*), (*c*) have a Hamiltonian circuit ? If not, explain why ?

7. (*a*) Verify Euler's formula for the following graphs :

(*b*) Show that if G is a bipartite simple graph with u vertical and e edges, then $e = \frac{u}{4}$.

(P.T.U., M.C.A. Dec 2006)

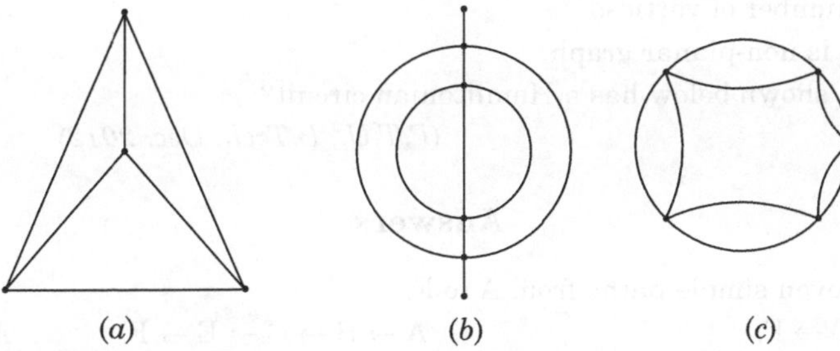

8. Let G be a finite connected planner graph with at least three vertices. Show that G has at least one vertex of degree 5 or less.

9. (*a*) Suppose a graph G contains two distinct paths from a vertex a to a vertex b. Show that G has a cycle.

(*b*) If a graph G has more than two vertices of odd degree, then prove that there can be no Euler Path. *(P.T.U., B.Tech. May 2008)*

10. Show that a connected graph G with n vertices must have atleast $(n - 1)$ edges.

11. Consider each graph G in the given figures:

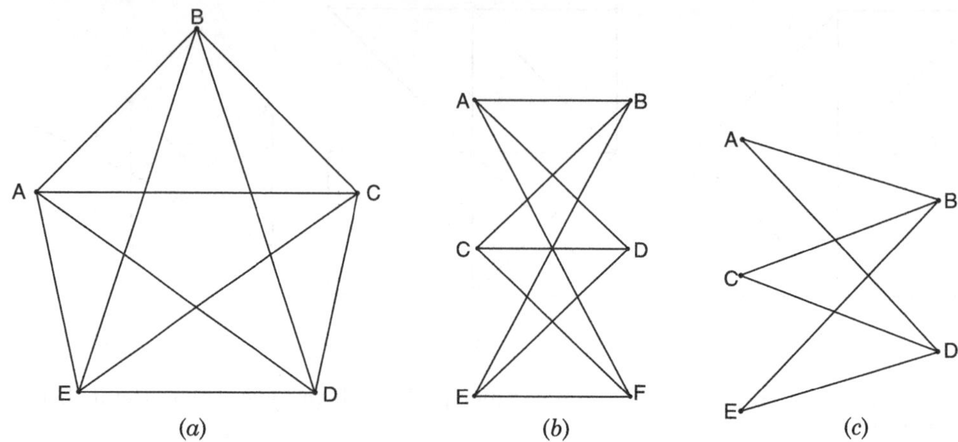

(*i*) Find an Euler path or Euler circuit, if it exists. If not, explain why ?

(*ii*) Find a Hamilton path or a Hamilton circuit, if it exists. If not, explain why ?

12. Draw the following graphs

(*a*) $K_{2,\ 5}$ (*b*) K_4 (*c*) $K_{2,\ 3}$

13. If G is a simple, connected and planner graph with more than one edge, then

(*i*) $2 \mid E \mid \geq 3 \mid R \mid$

(*ii*) $\mid E \mid \leq 3 \mid V \mid - 6$, where $\mid E \mid$ denotes the number of edges, $\mid R \mid$, the number of regions and $\mid V \mid$, the number of vertices.

14. Show that $K_{3,\ 3}$ is non-planar graph.

15. Does the graph shown below has a Hamiltonian circuit?

(*P.T.U., B.Tech. Dec. 2012*)

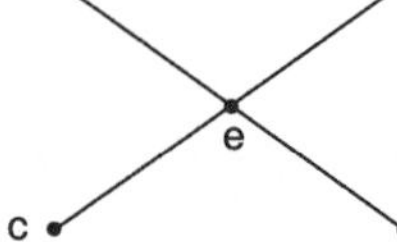

Fig. Q.15

Answers

1. (*a*) There are seven simple paths from A to F.

$A \to B \to C \to F$, $A \to B \to C \to E \to F$, $A \to B \to E \to F$,

$A \to B \to E \to C \to F$, $A \to D \to E \to F$,

$A \to D \to E \to B \to C \to F$, $A \to D \to E \to C \to F$

(*b*) There are nine trials. The seven simple paths of part (*a*) and

$A \to D \to E \to B \to C \to E \to F$; $A \to D \to E \to C \to B \to E \to F$

(*c*) $d(A, F) = 3$ (*d*) $d(G) = 3$

(*e*) There are three cycles including the vertex A;

$A \to B \to E \to D$; $A \to B \to C \to E \to D \to A$, $A \to B \to C \to F \to E \to D \to A$

(*f*) There are six cycles in G. The three cycles are of part (*e*) and

$B \to C \to E \to B$; $C \to F \to E \to C$; $B \to C \to F \to E \to B$

2. (*a*) ABG, ABFG, AEBG, AEBFG (*b*) BGC, BFGC, BAEBGC, BAEBFGC

(*c*) $d(A, C) = 3$ (*d*) $\dim(G) = 4$

3. (*a*) $A = \begin{bmatrix} 0 & 1 & 0 & 1 \\ 1 & 0 & 1 & 1 \\ 0 & 1 & 0 & 1 \\ 1 & 1 & 1 & 0 \end{bmatrix}$ (*b*) $A = \begin{bmatrix} 1 & 0 & 0 & 1 \\ 0 & 0 & 2 & 1 \\ 0 & 2 & 0 & 0 \\ 1 & 1 & 0 & 1 \end{bmatrix}$

4. (a)

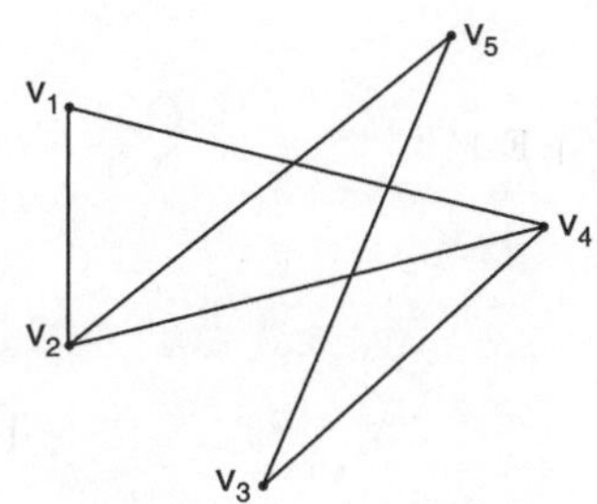

(b)

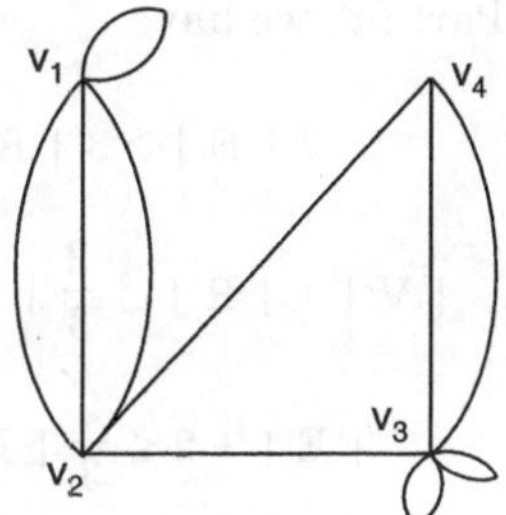

6. (a) Has Euler path (b) Has Euler circuit
   (c) Has not Euler path

11. (i) (a) Is Eulerian since all vertices are even. The Eulerian path is ABCDEACEBDA
    (b) Not an Eulerian Path, not an Eulerian circuit
    (c) Has Eulerian path, BADCBED
    (ii) (a) ABCDEA (b) ABCDEFA
    (c) has not Hamilton path and Hamilton circuit

12. (a)

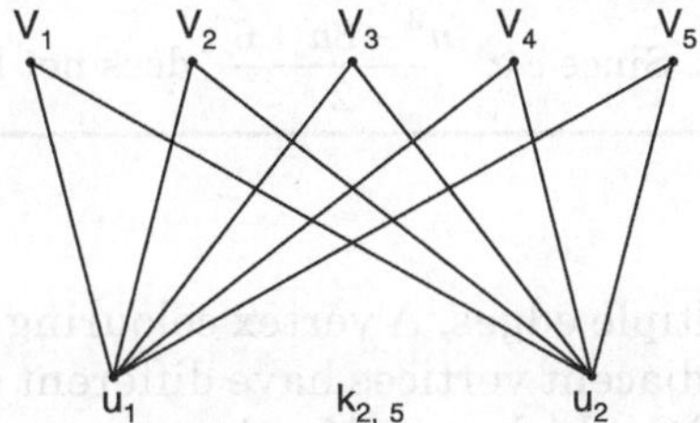

(b)

u_3 u_4

u_1 u_2

K_4

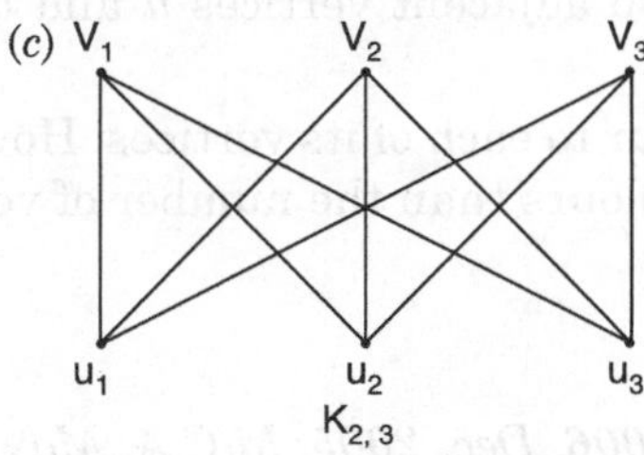

Hints

5. Here deg(A) = 3, deg(B) = 3, deg(C) = 4, deg(D) = 2, deg(E) = 2

   $\therefore$ Sum of degrees = 3 + 3 + 4 + 2 + 2 = 14

   Also number of edges = 7

6. A graph G has an Euler path iff 0 or 2 vertices have odd degree. A graph G has an Euler circuit if all the vertices are of even degree.

13. (a) Assume $|E| > 1$. If G has only one region (unbounded), then $|R| = 1$.

    Since $|E| > 1 \Rightarrow |E| \geq 2$

    $\therefore$ $2|E| \geq 3|R|$ is true

If $|R| > 1$, then each region is bounded by at least 3 edges. But in a planar graph, each edge touches at most 2 region. Thus $2|E| \geq 3|R|$

(*b*) From Part (*a*), we have

$$2|E| \geq 3|R| \Rightarrow |R| \leq \frac{2}{3}|E|$$

$$\Rightarrow \quad |V| + |R| \leq \frac{2}{3}|E| + |V|$$

$$\Rightarrow \quad |E| + 2 \leq \frac{2}{3}|E| + |V| \qquad | \text{ Euler's formula}$$

$$\Rightarrow \quad 3|E| + 6 \leq 2|E| + 3|V|$$

$$\Rightarrow \quad |E| \leq 3|V| - 6.$$

14. If $K_{3,3}$ is a planar graph, then we must have $2|E| \geq 3|R|$. Where each region is bounded by at least three edges. But for $K_{3,3}$, each region is bounded by at least 4 edges $\therefore$ We have

$2|E| \geq 4|R|$

$$\Rightarrow \quad 2|E| \geq 4\{|E| - |V| + 2\} \qquad | \text{ Euler's formula}$$

$$\Rightarrow \quad 2 \times 9 \geq 4(9 - 6 + 2) \qquad \text{For } K_{3,3}, |E| = 9, |V| = 6$$

$$\Rightarrow \quad 18 \geq 20, \text{ a contradiction}$$

Hence $K_{3,3}$ is non-planar.

15. No, the graph does not contain Hamiltonian circuit. Since $e \geq \frac{n^2 - 3n + 6}{2}$ does not hold.

11.51. GRAPH COLOURING

Suppose that $G = (V, E)$ is a graph with no multiple edges. A vertex colouring of G is an assignment of colours to the vertices of G such that adjacent vertices have different colours. A graph G is M-colourable if there exists a colouring of G which uses M-colours.

Proper Colouring. A colouring is proper if any two adjacent vertices u and v have different colours otherwise it is called improper colouring.

A graph can be coloured by assigning a different colour to each of its vertices. However, for most graphs a colouring can be found that uses fewer colours than the number of vertices in the graph.

11.52. CHROMATIC NUMBER OF G

(P.T.U., B.Tech. Dec. 2013, May 2006, Dec. 2005; M.C.A. May 2007)

The minimum number of colours needed to produce a proper colouring of a graph G is called the chromatic number of G and is denoted by $\chi(G)$.

The graph shown in Fig.11.169 is minimum 3-colourable, hence $\chi(G) = 3$.

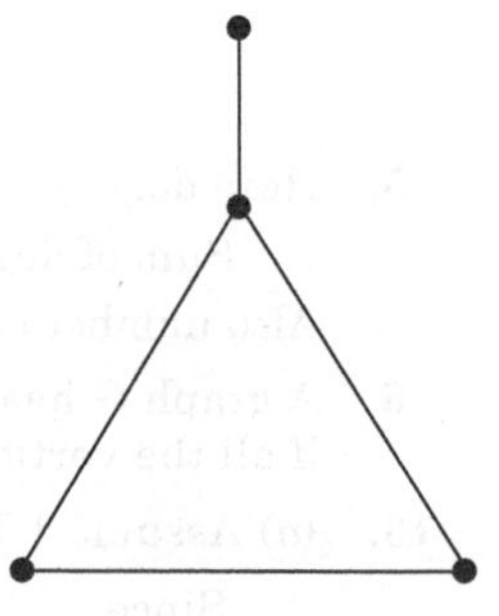

Fig. 11.169

Similarly, for the complete graph K_6 we need six colours to colour K_6 since every vertex is adjacent to every other vertex and we need a different colour for each vertex. $\therefore$ The chromatic number for K_6 is $\chi(K_6) = 6$. Similarly, the chromatic number of K_{10} is $\chi(K_{10}) = 10$.

ILLUSTRATIVE EXAMPLES

Example 1. *The chromatic number of K_n is n.* (P.T.U., B.Tech. Dec. 2012)

Sol. A colouring of K_n can be constructed using n colours by assigning a different colour to each vertex. No two vertices can be assigned the same colour, since every two vertices of this graph are adjacent. Hence the chromatic number of $K_n = n$.

Example 2. *The chromatic number of complete bipartite graph* $K_{m, n}$, *where m and n are positive integers is two.* (P.T.U., B.Tech. Dec. 2013)

Sol. The number of colours needed does not depend upon m and n. However, only two colours are needed to colour the set of m vertices with one colour and the set of n vertices with a second colour. Since, edges connect only a vertex from the set of m vertices and a vertex from the set of n vertices, no two adjacent vertices have the same colour.

Note 1. Every connected bipartite simple graph has a chromatic number of 2 or 1.

2. Conversely, every graph with a chromatic number of 2 is bipartite.

Example 3. *The chromatic number of graph* c_n, *where* c_n *is the cycle with n vertices is either 2 or 3.* (P.T.U., B.Tech. Dec. 2013)

Sol. Two colours are needed to colour c_n, when n is even. To construct such a colouring, simply pick a vertex and colour it black. Then move around the graph in clockwise direction colouring the second vertex white, the third vertex black, and so on. The nth vertex can be coloured white since the two vertices adjacent to it, namely the $(n - 1)$th and the first are both coloured black as shown in Fig. 11.170.

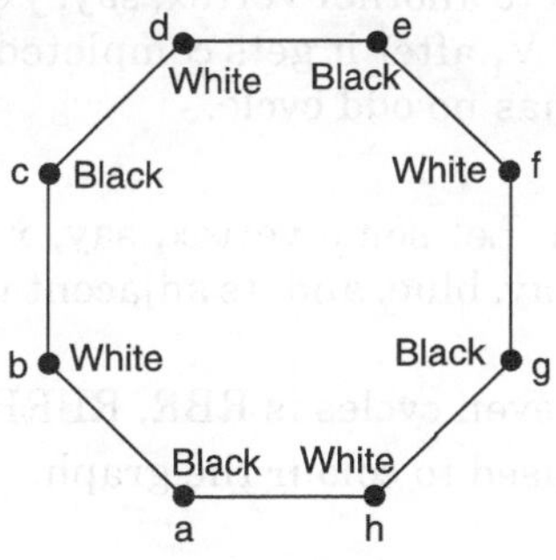

Fig. 11.170

When n is odd and $n > 1$, the chromatic number of c_n is 3. To construct such a colouring, pick an initial vertex. First use only two colours and alternate colours as the graph is traversed in a clockwise direction. However, the nth vertex reached is adjacent to two vertices of different colours, the first and $(n - 1)$th. Hence, a third colour is needed. (Fig. 11.171)

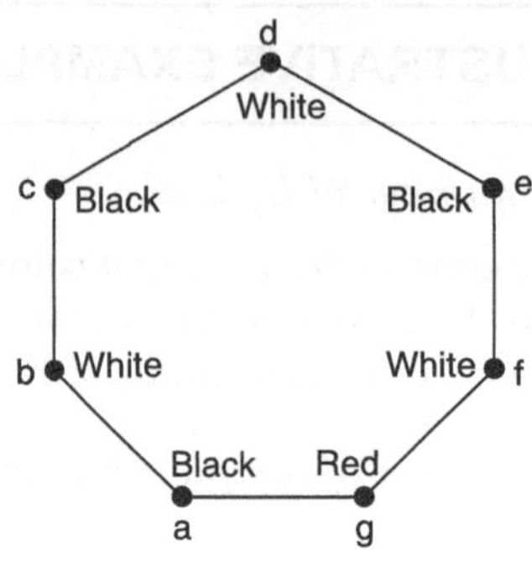

Fig. 11.171

Theorem I. *The following are equivalent for a graph G :*

(i) G is 2-colourable

(ii) G is bipartite

(iii) Every cycle of G has even length.

Proof. *(i)* ⇒ *(ii)*

If G is 2-colourable, then G has two sets of vertices V_1 and V_2 with different colours, say, red and blue respectively.

Since no vertices of V_1 or V_2 are adjacent (being of same colour)

∴ $\{V_1, V_2\}$ is a partition of G ⇒ G is bipartite.

(ii) ⇒ *(iii)*

Let G be bipartite and $[V_1, V_2]$ be partition of vertices of G. Let $x \in V_1$ be any vertex and a cycle begins at x. Join this vertex to another vertex, say, $y \in V_2$ and then to a vertex in V_1 and so on. This cycle will return to $x \in V_1$ after it gets completed and will be of even length. (Since G is a bipartite graph). Hence G has no odd cycle.

(iii) ⇒ *(i)*

Let each cycle in G is even. Let some vertex, say, x is coloured red, then its adjacent vertex will have different colour, say, blue, and its adjacent vertex will have red colour because every cycle has even length.

∴ sequence of vertices of even cycles is RBR, RBRBR and so on.

Thus only two colours are used to colour the graph.

∴ G is 2-colourable.

Example 4. *Determine the chromatic number of the graphs shown in Fig. 11.172.*

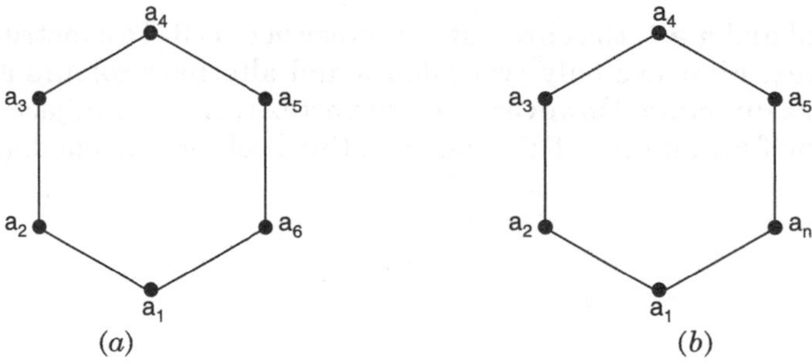

Fig. 11.172

Sol. The graphs shown in Fig. 11.172(*a*), has the chromatic number $\chi(G) = 2$.

The graph shown in Fig. 11.172(*b*) has the chromatic number $\chi(G) = 2$, when n is an even number and $\chi(G) = 3$, where n is odd.

Theorem II. *If an undirected graph has a subgraph K_3, then its chromatic number is at least three.*

Proof. Let G be an undirected graph. As G contains a complete graph K_3, which is 3-colourable. $\therefore$ G cannot be coloured with one or two colours

$\therefore \quad \psi(G) \geq 3.$

Four Colour Theorem. Every planar graph is four colourable.

Five Colour Theorem. Every planar graph has chromatic number ≤ 5.

Theorem III. *The vertices of every planar graph can be properly coloured with five colours.*

Proof. We will prove this theorem by induction. All the graphs with 1, 2, 3, 4 or 5 vertices can be properly coloured with five colours. Now let us assume that every planar graph with $n - 1$ vertices can be properly coloured with five colours. Next, if we prove that any planar graph G with n vertices will require no more than five colours, we have done.

Consider the planar graph G with n vertices.

Since G is planar, it must have at least one vertex with degree five or less as shown in theorem V. Assume this vertex to be 'u'.

Let G_1 be a graph of $n - 1$ vertices obtained from G by deleting vertex 'u'. The G_1 graph requires no more than five colours (Induction hypothesis). Consider that the vertices in G_1 have been properly coloured and now add to it 'u' and all the edges incident on u. If the degree of u is 1, 2, 3, or 4, a proper colour to u can be easily assigned.

Now, we have one case left, in which the degree of u is 5, and all the 5 colours have been used in colouring the vertices adjacent to u, as shown in Fig. 11.173.

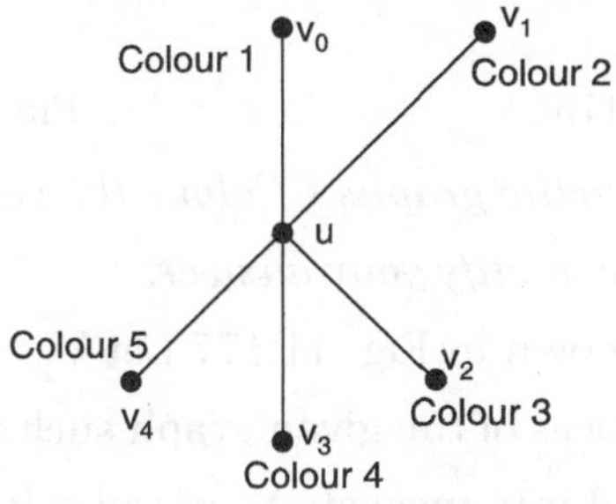

Fig. 11.173

Suppose that there is a path in G_1 between vertices v_0 and v_3 coloured alternately with colours 1 and 4 as shown in Fig. 11.174.

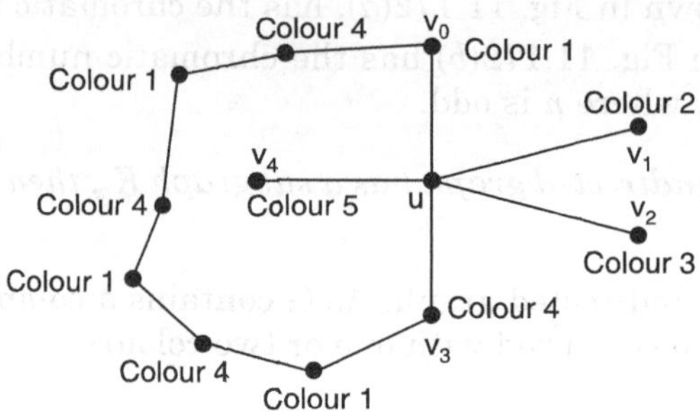

Fig. 11.174

Then a similar path between v_4 and v_2, coloured alternately with colours 5 and 3, can not exist ; otherwise, these two paths will intersect and cause G to be non-planar.

Thus, if there is no path between v_4 and v_2 coloured alternately with colour 5 and 3 of all vertices connected to v_2 through vertices of alternating colours 5 and 3. This interchange will colour vertex v_2 with colour 5 and yet keep G, properly coloured. As vertex v_4 is still with colour 5, the colour 3 is left over with which to colour vertex u which proves the theorem.

Example 5. *Consider the following graphs*

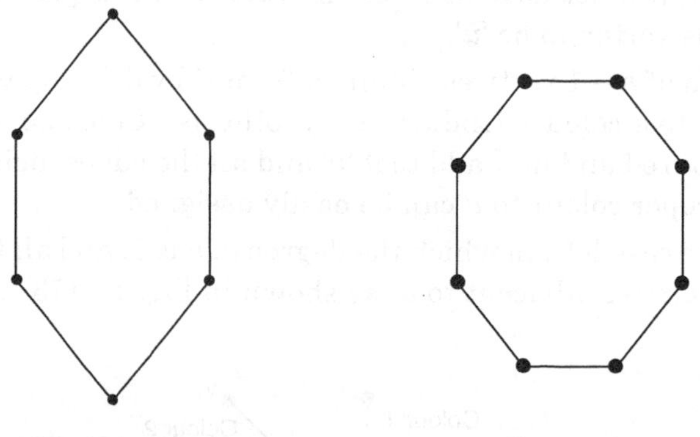

Fig. 11.175. Fig. 11.176

(a) which graph(s) are bipartite graphs ? Colour the vertices of the bipartite graph.

(b) If graph is not bipartite, justify your answer.

Sol. Consider the graph shown in Fig. 11.177 Let V_1 = [R, R, R], V_2 = [B, B, B]

∴ Let V be the set of vertices of the given graph such that $V = V_1 \cup V_2$ and $V_1 \cap V_2 = \phi$

Hence V can be partitioned into two sets V_1 of red colours and V_2 of blue colours.

∴ This graph is a bipartite graph.

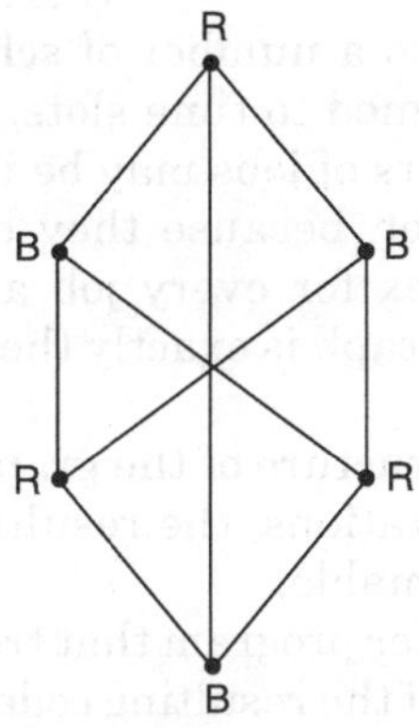

Fig. 11.177

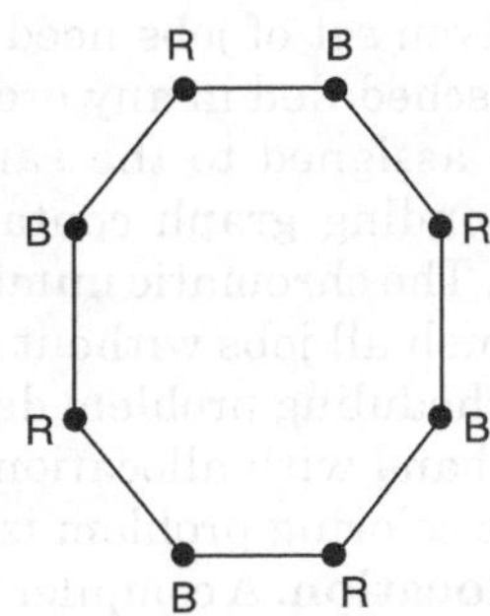

Fig. 11.178

(*b*) However, the graph shown in Fig. 11.178 is not a bipartite graph as it is not 2-colourable. If we try to label the vertices using two colours red (R) and Blue (B), we get the graph (Fig 11.178) with adjacent vertices of same colour.

Example 6. *Consider the graph (Fig. 11.179) G shown below.*

(*a*) *Find the shortest and longest simple path between A and F*

(*b*) *What is the diameter of the graph ?*

(*c*) *Is there Euler's path in G?*

(*d*) *Is G planar and connected?*

(*e*) *Find chromatic number of G.*

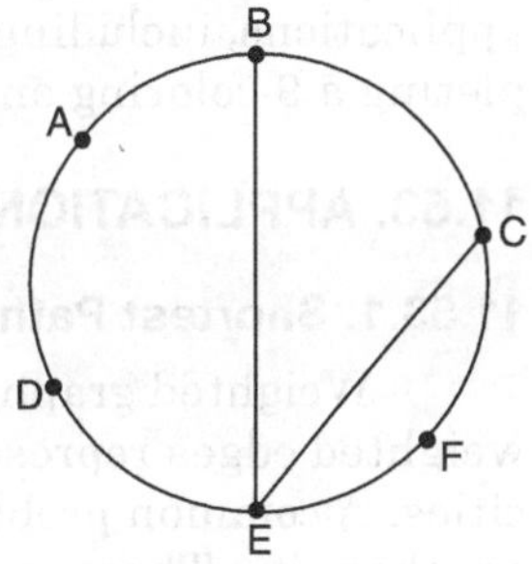

Fig. 11.179

Sol. (*a*) The shortest path between A and F is A, D, E, F. It is of length 3.

The longest path between A and F is A, D, E, B, C, F. It is of length 5.

(*b*) We know that the diameter of a graph is the maximum distance between any two vertices.

∴ diameter (G) = 3

(*c*) Since the vertices B and C are of odd degree.

∴ G has an Euler path between B and C.

(*d*) Since no two edges in the graph intersect and there is a path between each pair of vertices.

∴ G is connected and planar

(*e*) Consider the subgraph EBC(K_3)

∴ The chromatic number of the graph is ≥ 3

But three colours are sufficient to paint the vertices property (see Fig. 11.180). A(Red), B(Blue), C(Red), D(Green), E(Green), F(Blue)

∴ required chromatic number = 3

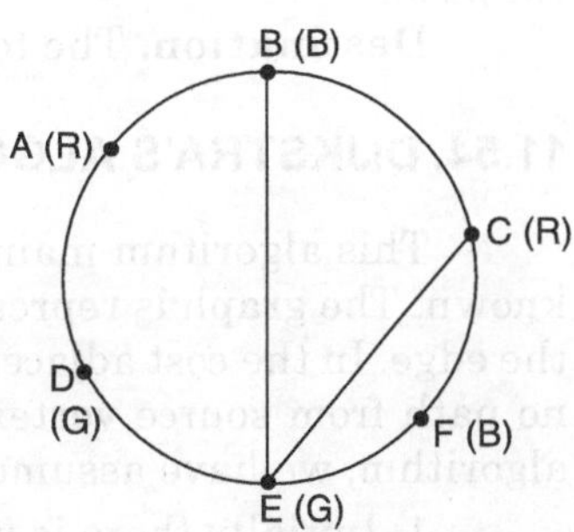

Fig. 11.180

Example 7. *Write any three applications of colouring of graph.*

(P.T.U. B.Tech. Dec. 2009)

Sol. (*i*) Scheduling. Vertex coloring models to a number of scheduling problems. In the cleanest form, a given set of jobs need to be assigned to time slots, each job requires one such slot. Jobs can be scheduled in any order, but pairs of jobs may be in conflict in the sense that they may not be assigned to the same time slot, because they both rely on a shared resource. The corresponding graph contains a vertex for every job and an edge for every conflicting pair of jobs. The chromatic number of the graph is exactly the minimum makespan, the optimal time to finish all jobs without conflicts.

Details of the scheduling problem define the structure of the graph. For example, when assigning aircrafts in band with allocation to radio stations, the resulting conflict graph is a unit disk graph, so the coloring problem is 3-approximable.

(*ii*) Register allocation. A compiler is a computer program that translates one computer language into another. To improve the execution time of the resulting code, one of the techniques of compiler optimization is register allocation, where the most frequently used values of the compiled program are kept in the fast processor register. Ideally, values are assigned to registers so that they can all reside in the registers when they are used.

The textbook approach to this problem is to model it as a graph coloring problem. The compiler constructs an interference graph, were vertices are symbolic registers and an edge connects two nodes if they are needed at the same time. If the graph can be colored with *k* color, then the variables can be stored in *k* registers.

(*iii*) Other applications. The problem of coloring a graph has found a number of applications, including pattern matching. The recreational puzzle Sudoku can be seen as completing a 9-coloring on given specific graph with 81 vertices.

11.53. APPLICATIONS OF GRAPH THEORY

11.53.1. Shortest Path in Weighted Graphs *(P.T.U., B.Tech. May 2007)*

Weighted graphs can be used to represent highways connecting the different cities. The weighted edges represent the distance between different cities and the vertices represent the cities. A common problem with this type of graph is to find the shortest path from one city to another city. There are many ways to tackle this problem one of which is as follows :

Shortest Paths from Single Source. We will find shortest paths from a single vertex to all other vertices of the graph. The first algorithm was proposed by **E. Dijkstra** in 1959. Some common terms related with this algorithm are as follows :

Path Length. The length of a path is the sum of the weights of the edges on that path.

Source. The starting vertex of the graph from which we have to start to find the shortest path.

Destination. The terminal or last vertex upto which we have to find the path.

11.54. DIJKSTRA'S ALGORITHM FOR SHORTEST PATH *(P.T.U., M.C.A. Dec. 2005)*

This algorithm maintains a set of vertices whose shortest path from source is already known. The graph is represented by its cost adjacency matrix, where cost being the weight of the edge. In the cost adjacency matrix of the graph, all the diagonal values are zero. If there is no path from source vertex V_s to any other vertex V_i, then it is represented by $+\infty$. In this algorithm, we have assumed all weights are positive.

1. Initially there is no vertex in sets.

2. Include the source vertex V_s in S. Determine all the paths from V_s to all other vertices without going through any other vertex.

3. Now, include that vertex in S which is nearest to V_s and find shortest paths to all the vertices through this vertex and update the values.

4. Repeat the step 3 until $n - 1$ vertices are not included in S if there are n vertices in the graph.

After completion of the process, we get the shortest paths to all the vertices from the source vertex.

Example 8. *Find the shortest path between K and L in the graph shown in Fig. 11.181 by using Dijkstra's Algorithm.*

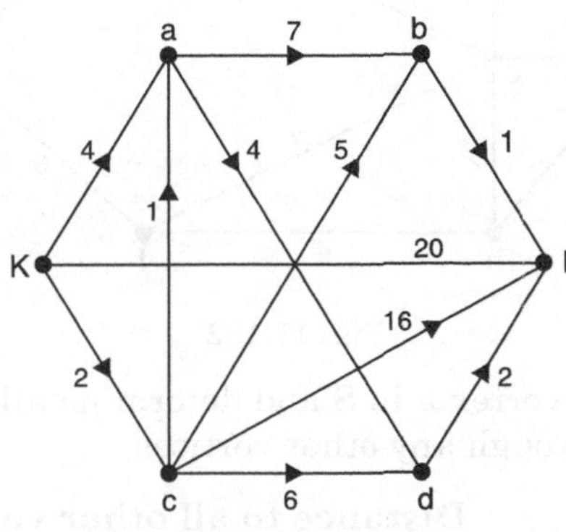

Fig. 11.181

Sol. Step I. Include the vertex K in S and determine all the direct paths from K to all other vertices without going through any other vertex.

S	K	*a*	*b*	*c*	*d*	L
	Distance to all other vertices					
K	0	4(K)	∞	2(K)	∞	20(K)

Step II. Include the vertex in S which is nearest to K and determine shortest paths to all vertices through this vertex and update the values. The nearest vertex is *c*.

S	K	*a*	*b*	*c*	*d*	L
	Distance to all other vertices					
K, *c*	0	3(K, *c*)	7(K, *c*)	2(K)	8(K, *c*)	18(K, *c*)

Step III. The vertex which is 2nd nearest to K is *a*, included in S.

S	K	*a*	*b*	*c*	*d*	L
	Distance to all other vertices					
K, *c*, *a*	0	3(K, *c*)	7(K, *c*)	2(K)	7(K, *c*, *a*)	18(K, *c*)

Step IV. The vertex which is 3rd nearest to K is *b*, is included in S.

S	K	*a*	*b*	*c*	*d*	L
	Distance to all other vertices					
K, *c*, *a*, *b*	0	3(K, *c*)	7(K, *c*)	2(K)	7(K, *c*, *a*)	8(K, *c*, *b*)

Step V. The vertex which is next nearest to K is *d*, is included in S.

S	K	*a*	*b*	*c*	*d*	L
	Distance to all other vertices					
K, *c*, *a*, *b*, *d*	0	3(K, *c*)	7(K, *c*)	2(K)	7(K, *c*, *a*)	8(K, *c*, *b*).

Since, $n-1$ vertices included in S. Hence we have found the shortest distance from K to all other vertices.

Thus, the shortest distance between K and L is 8 and the shortest path is K, *c*, *b*, L.

Example 9. *Find the shortest path between a and z in the graph shown in Fig. 11.182.*

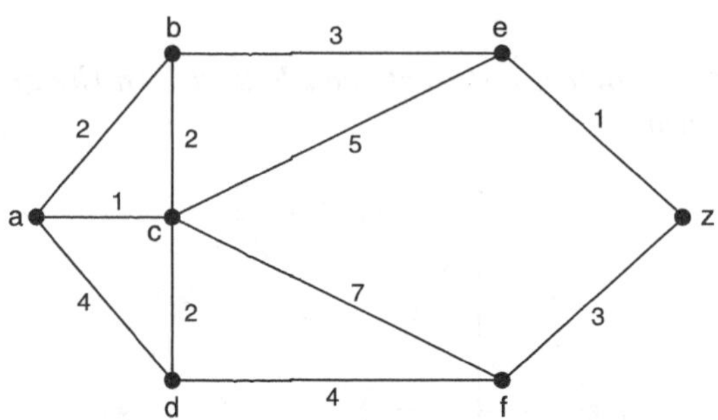

Fig. 11.182

Sol. Step I. Include the vertex a in S and determine all the direct paths from *a* to all other vertices without going through any other vertices.

S	Distance to all other vertices						
a	*a*	*b*	*c*	*d*	*e*	*f*	*z*
	0	2(*a*)	1(*a*)	4(*a*)	∞	∞	∞

Step II. Include the vertex in S which is nearest to *a* and determine shortest path to all the vertices through this vertex. The nearest vertex is *c*.

S	Distance to all other vertices						
a, *c*	*a*	*b*	*c*	*d*	*e*	*f*	*z*
	0	2(*a*)	1(*a*)	3(*a*, *c*)	6(*a*, *c*)	8(*a*, *c*)	∞

Step III. Include the vertex in S which is 2nd nearest to S and determine shortest path to all the vertices through this vertex. The 2nd nearest vertex is *b*.

S	Distance to all other vertices						
a, *c*, *b*	*a*	*b*	*c*	*d*	*e*	*f*	*z*
	0	2(*a*)	1(*a*)	3(*a*, *c*)	5(*a*, *b*)	8(*a*, *c*)	∞

Step IV. Next vertex included in S is *d*.

S	Distance to all other vertices						
a, *c*, *b*, *d*	*a*	*b*	*c*	*d*	*e*	*f*	*z*
	0	2(*a*)	1(*a*)	3(*a*, *c*)	5(*a*, *b*)	7(*a*, *c*)	∞

Step V. Next vertex included in S is *e*.

S	Distance to all other vertices						
a, *c*, *b*, *d*, *e*	*a*	*b*	*c*	*d*	*e*	*f*	*z*
	0	2(*a*)	1(*a*)	3(*a*, *c*)	5(*a*, *b*)	7(*a*, *c*)	6(*a*, *b*, *e*)

Step VI. Next vertex included in S is *f*.

S	Distance to all over vertices						
a, c, b, d, e, f	a	b	c	d	e	f	z
	0	2(a)	1(a)	3(a, c)	5(a, b)	7(a, c)	6(a, b, e).

This ends our procedure as $n - 1$ vertices are included in S. Thus, the shortest distance between a and z is 6 and the shortest path is *a, b, e, z.*

Example 10. *Using either breadth first search algorithm or Dijkstra's algorithm, find the shortest path from s to t in the following weighted graph. (Fig. 11.183)*

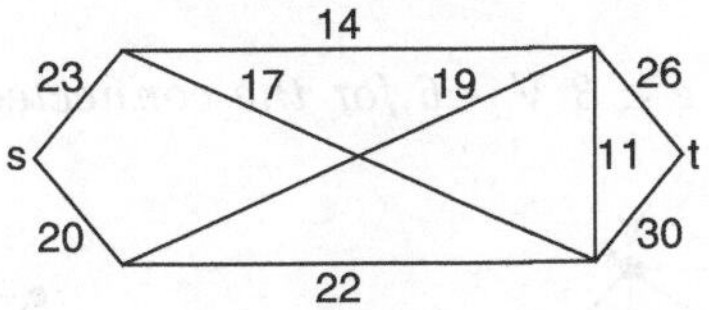

Fig. 11.183

Sol. Let S be the set of vertices of the given weighted graph. *i.e.*, S = {*s, a, b, c, d, t*} as shown in the Fig. 11.184.

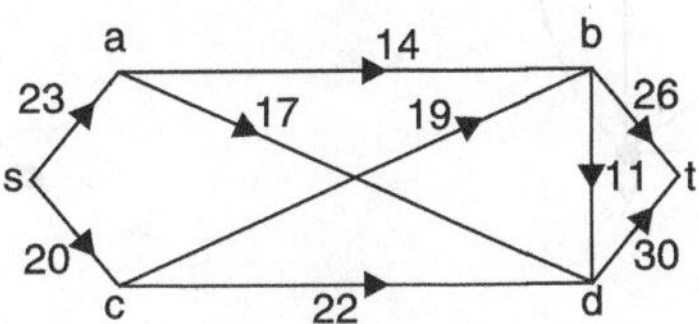

Fig. 11.184

Step I. Include the vertex s in S and determine all the direct paths from s to all other vertices without going through any other vertex.

S	Distance to all other vertices					
	s	a	b	c	d	t
s	0	23(s)	∞	20(s)	∞	∞

Step II. Include the vertex in S which is nearest to s and determine shortest paths to all vertices through this vertex and update the values. The nearest vertex is C.

S	Distance to all other vertices					
	s	a	b	c	d	t
s, c	0	23(s)	39 (s,c)	20(s)	42(s, c)	∞

Step III. The vertex which is 2nd nearest to s is a. Include this vertex in S

S	Distance to all other vertices					
	s	a	b	c	d	t
s, c, a	0	23(s)	37(s, a)	20(s)	42(s, c)	∞

Step IV. The vertex which is 3rd nearest to s is b. Include this vertex in S.

S	Distance to all other vertices					
	s	a	b	c	d	t
s, c, a, b	0	23(s)	37(s, a)	20(s)	40(s, a)	63(s, a, b)

Step V. The vertex which is next nearest to s is (d). Include this vertex in S.

S	Distance to all other vertices					
	s	a	b	c	d	t
s, c, a, b, d	0	23(s)	37(s, a)	20(s)	40(s, a)	63(s, a, b)

Since $n - 1$ = S vertices are included in S. Hence we have found the shortest distance from s to all other vertices. Thus, the shortest distance between s and t is 63 and the shortest path is s, a, b, t.

Example 11. *Show that $e \geq 3\,V - 6$ for the connected planar graphs shown in Figs. 11.185 and 11.186.*

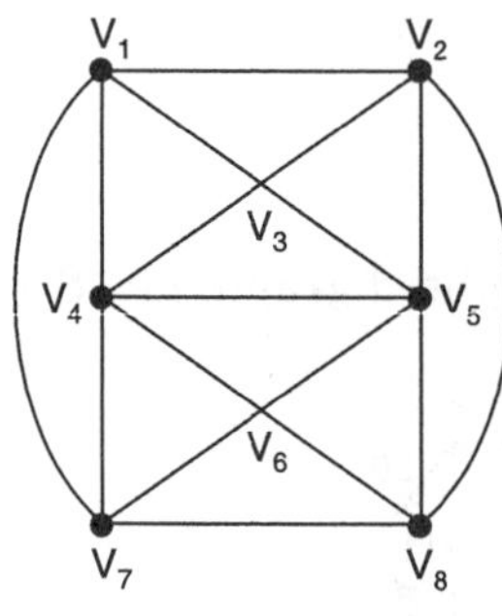

Fig. 11.185

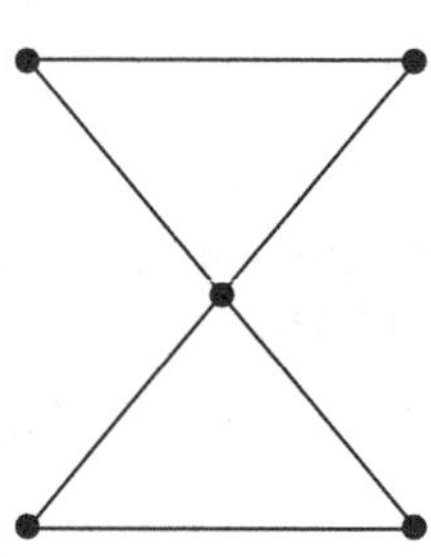

Fig. 11.186

Sol. (*i*) The graph shown in Fig. 11.185 contains vertices V = 8 and edges e = 17. Putting the values we have $e = 3 \times 8 - 6 = 18 \geq 17$. Hence proved.

(*ii*) The graph shown in Fig. 11.186 contains vertices V = 5 and edges e = 6. Putting the values, we have $3 \times 5 - 6 = 11 > 6$. Hence proved.

TEST YOUR KNOWLEDGE 11.3

1. Find the chromatic numbers of the following graphs.

(*a*) 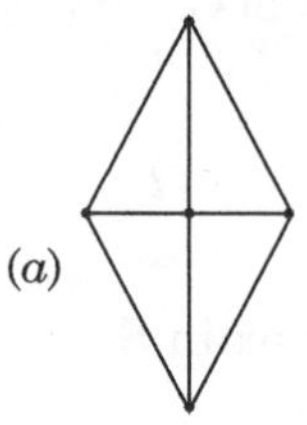

(*b*) 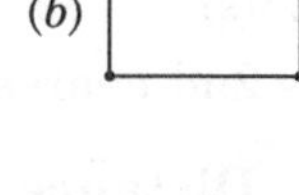

(*c*) 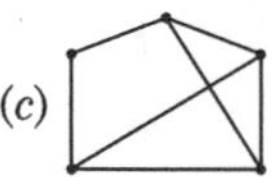

(*d*) 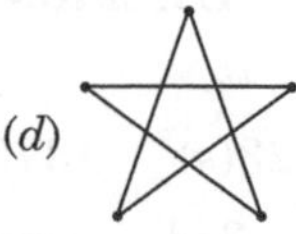

(*e*) 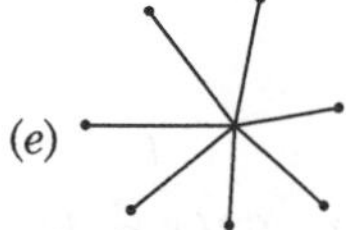

(*f*) 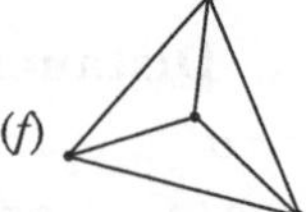

(*g*) Complete bipartite graph $K_{3,4}$ (*P.T.U., B.Tech. May 2013*)

2. Find the shortest path, by using either Breadth first search or Dijkstra's algorithm, from P to Q in the following weighted graph.

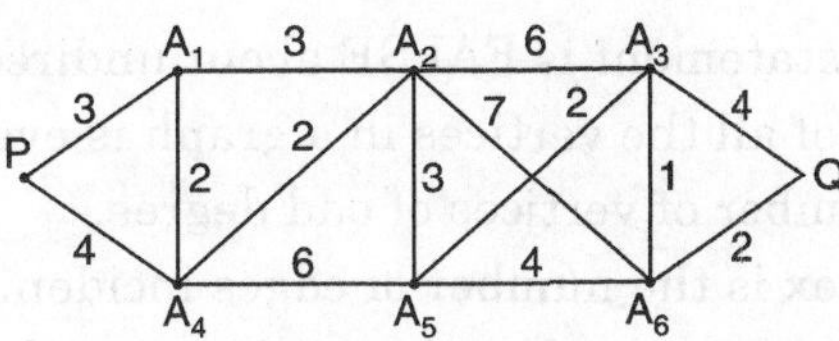

3. Find the chromatic number of the following graphs.

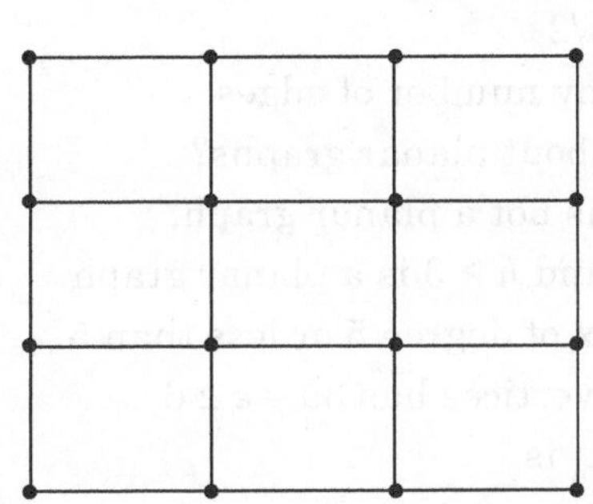

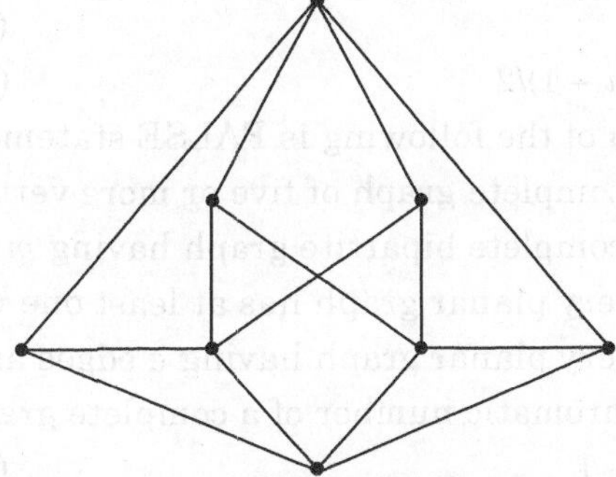

4. Find the shortest path and its length from *s* to *t* by using Dijkstra's algorithm in the following graph.

(*a*)

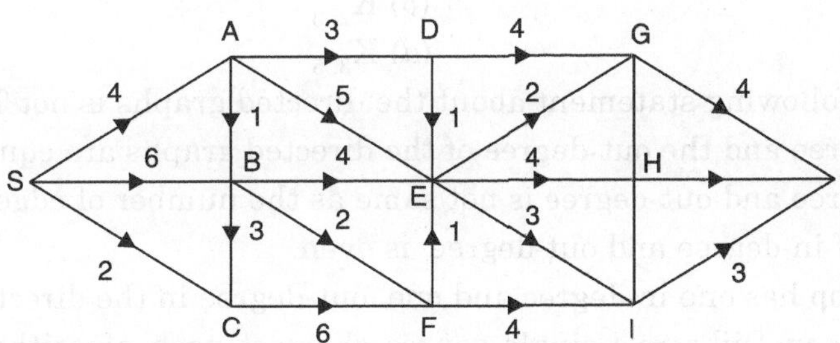

(*b*)

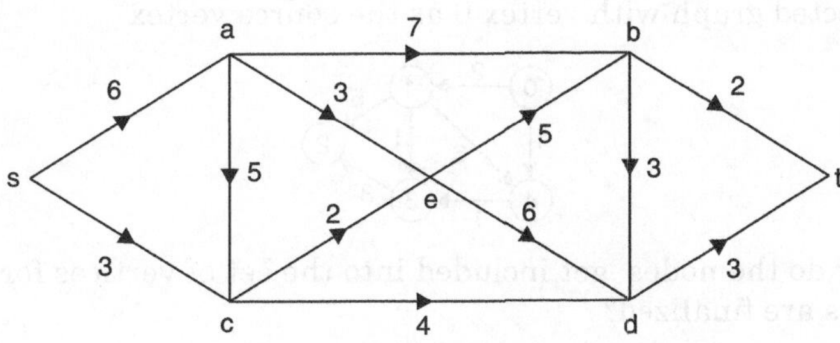

Answers

1. (*a*) 3 (*b*) 3 (*c*) 3 (*d*) 3 (*e*) 2 (*f*) 4 (*g*) 2

2. P A_1 A_2 A_5 A_3 A_6 Q.

3. 2

4. (*a*) *s* – A – D – E – H – *t*, length = 13

(*b*) *r* – C – *d* – *t*.

MULTIPLE CHOICE QUESTIONS (MCQs)

1. Which of the following statement is FALSE about undirected graphs?
   (*a*) The sum of degrees of all the vertices in a graph is even.
   (*b*) There is an even number of vertices of odd degree.
   (*c*) The degree of a vertex is the number of edges incident on it.
   (*d*) The self loop is counted once, when degree is counted.
2. How many edges do a complete graph contains having *n* vertices?
   (*a*) $2n$ (*b*) $2n/2$
   (*c*) $n(n-1)/2$ (*d*) Any number of edges.
3. Which of the following is FALSE statement about planar graphs?
   (*a*) A complete graph of five or more vertices is not a planar graph.
   (*b*) A complete bipartite graph having $m \geq 3$ and $n \geq 3$ is a planar graph.
   (*c*) Every planar graph has at least one vertex of degree 5 or less than 5.
   (*d*) Every planar graph having ***e*** edges and v vertices has $3v - e \geq 6$.
4. The chromatic number of a complete graph K_n is
   (*a*) $n - 1$ (*b*) n
   (*c*) 2 (*d*) Any number.
5. Which of the following bipartite graph has Hamiltonian circuit?
   (*a*) $K_{1, 2}$ (*b*) $K_{2, 3}$
   (*c*) $K_{3, 3}$ (*d*) $K_{3, 5}$
6. Which of the following statement about the directed graphs is not TRUE?
   (*a*) The in-degree and the out-degree of the directed graphs are equal.
   (*b*) The in-degree and out-degree is not same as the number of edges.
   (*c*) The sum of in-degree and out-degree is even.
   (*d*) The self loop has one in-degree and one out-degree in the directed graph.
7. Suppose you run Dijkstra's single source shortest path algorithm on the following weighted directed graph with vertex 0 as the source vertex

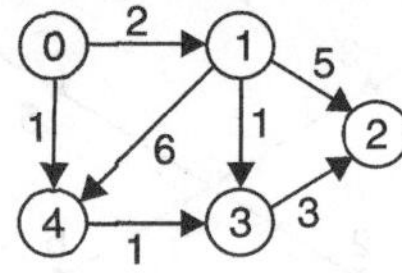

   In what order do the nodes get included into the set of vertices for which the shortest path distances are finalized?
   (*a*) 0, 1, 2, 3, 4 (*b*) 0, 4, 1, 3, 2
   (*c*) 0, 4, 3, 2, 1 (*d*) 0, 1, 4, 3, 2
8. How many edges are there in a graph with 20 vertices and the sum of the degrees (in-degree and out-degree) is 100?
   (*a*) 50 (*b*) 100
   (*c*) 20 (*d*) 40

9. If G is a directed graph with 10 vertices, how many Boolean values will be needed to represent G using an adjacency matrix?

(*a*) 50 (*b*) 100

(*c*) 200 (*d*) 1000

10. Which of the following is Not TRUE about the directed graphs?

(*a*) If a digraph is reflexive, then the diagonal elements of the adjacency matrix are 1.

(*b*) If G is a simple digraph whose adjacency matrix is A then the adjacency matrix of G^c, is the transpose of A.

(*c*) The diagonal elements of $A \cdot A^T$ show the out degree of the vertices.

(*d*) The adjacency matrix of a directed graph is a symmetric matrix.

Answers and Explanation

1. (*d*) The self loop is counted twice, when degree is counted.

2. (c) The number of edges in a complete graph is $n(n - 1)/2$

3. (*b*) A complete bipartite graph having $m \geq 3$ and $n \geq 3$ is not a planar graph.

4. (*b*) The chromatic number of a complete graph of n vertices is n.

5. (c) $K_{3,\,3}$ has Hamiltonian circuit.

6. (*b*) The in-degree and out-degree is same as the number of edges.

7. (*b*) Apply the algorithm and it will be 0, 4, 1, 3, 2

8. (*a*) One edge contributes two degree.

9. (*b*) The adjacency matrix is of N × N dimension, where n is the number of vertices.

10. (*d*) The adjacency matrix of a directed graph is not a symmetric matrix.

12 TREES

12.1. INTRODUCTION

In this chapter, we will discuss a special class of graphs, called trees. The concept of trees is frequently used in both mathematics and sciences. To understand the concept of trees, it is essential to know the various common types of trees. Their basic properties and applications.

12.2. TREE

A graph which has no cycle is called an acyclic graph. A tree is an acyclic graph or graph having no cycles.

A tree or general tree is defined as a non-empty finite set of elements called vertices or nodes having the property that each node can have minimum degree 1 and maximum degree n. It can be partitioned into $n + 1$ disjoint subsets such that the first subset contains the root of the tree and the remaining n subsets contains the elements of the n subtree. (Fig. 12.1)

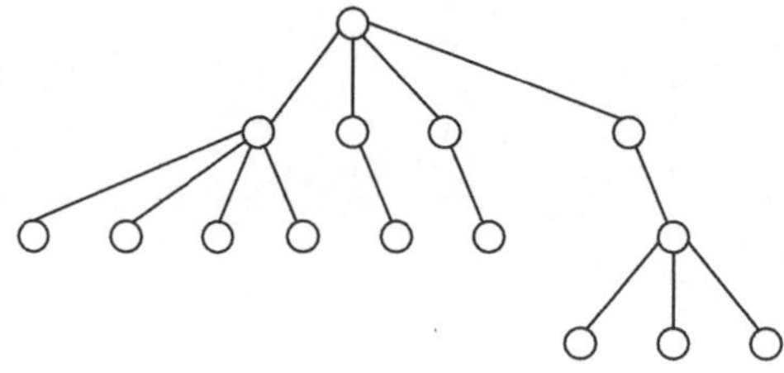

Fig. 12.1. General Tree.

12.3. DIRECTED TREES

A directed tree is an acyclic directed graph. It has one node with indegree 1, while all other nodes have indegree 1 as shown in Figs. 12.2 and 12.3.

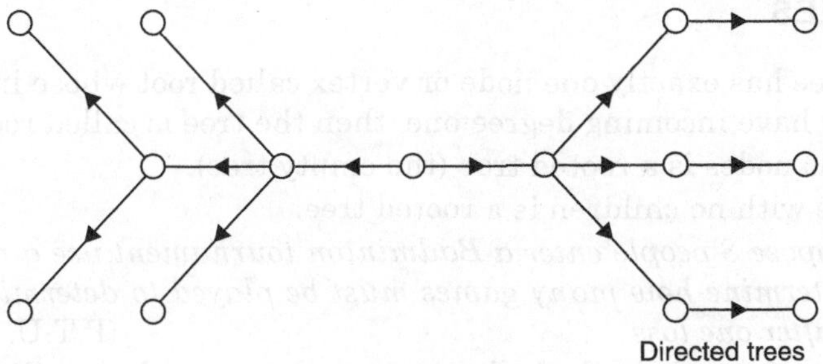

Fig. 12.2

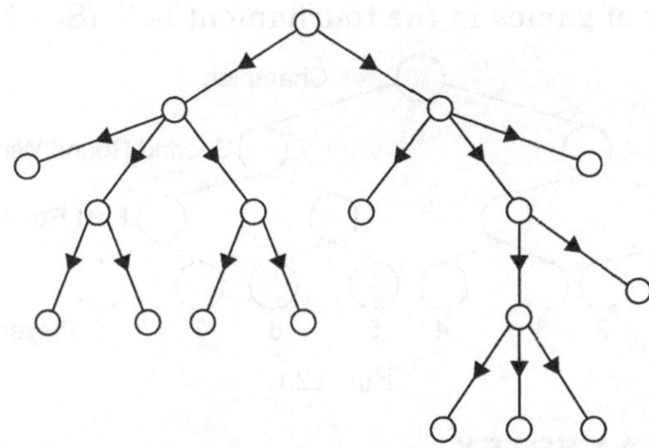

Fig. 12.3

The node which has outdegree 0 is called an external node or a terminal node or a leaf. The nodes which has outdegree greater than or equal to one are called internal nodes or branch nodes.

12.4. ORDERED TREES

If in a tree at each level, an ordering is defined, then such a tree is called an ordered tree.

e.g., the trees shown in Figs. 12.4 and 12.5 represent the same tree but have different orders.

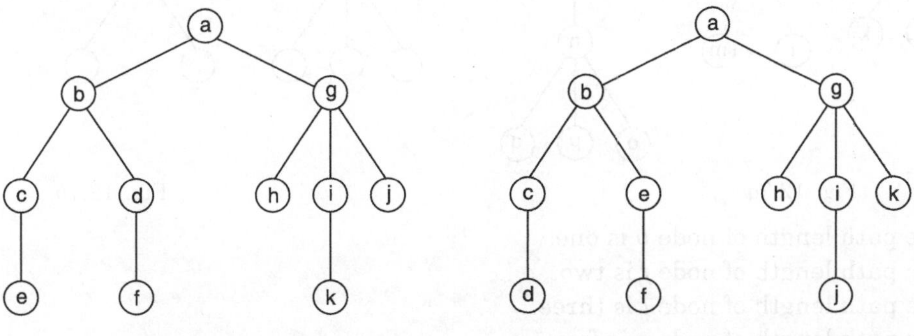

Fig. 12.4 Fig. 12.5

12.5. ROOTED TREES

If a directed tree has exactly one node or vertex called root whose incoming degree is 0 and all other vertices have incoming degree one, then the tree is called rooted tree.

* A tree with no nodes is a rooted tree (the empty tree).

* A single node with no children is a rooted tree.

Example. *Suppose 8 people enter a Badminton tournament use a rooted tree model of the tournament to determine how many games must be played to determine a champion if a player is eliminated after one loss.* (P.T.U. B.Tech. May 2009)

Sol. As there are 8 people in the badminton tournament, there will be four games to be played in the first round two games to be played in the second round, one game to be played in the final round

Hence, total number of games in the tournament is 7. (See Fig. 12.6).

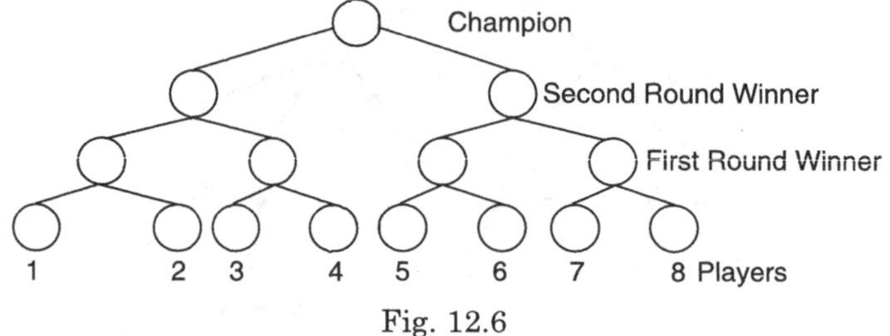

Fig. 12.6

12.6. PATH LENGTH OF A VERTEX

The path length of a vertex in a rooted tree is defined to be the number of edges in the path from the root to the vertex.

For example, we find the path lengths of the nodes b, f, l, q in Fig. 12.7a.

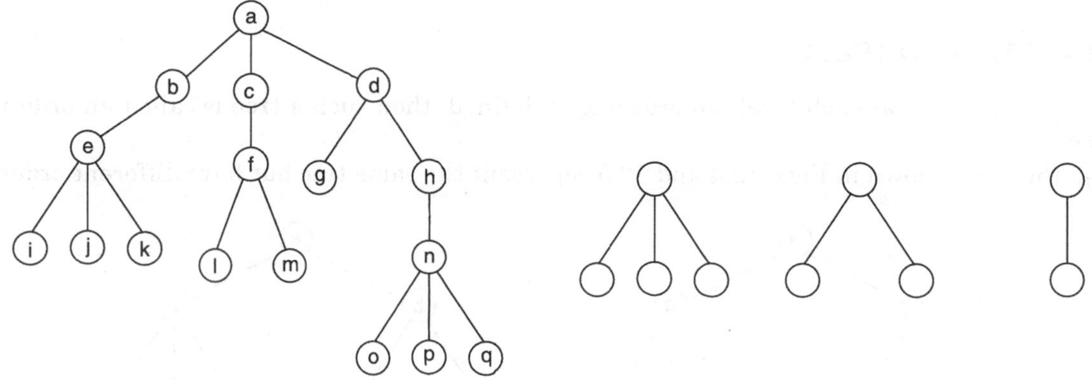

Fig. 12.7*a* Fig. 12.7*b*

The path length of node *b* is one.
The path length of node *f* is two.
The path length of node *l* is three.
The path length of node *q* is four.

Theorem I. *Prove that there is one and only one path between every pair of vertices in a tree T.*

Proof. We know that T is a connected graph, in which there must exist at least one path between every pair of vertices. Now assume that there exists two different paths from some node a to some node b of T. The union of these two paths will contain a cycle and therefore T cannot be a tree. Hence, there is only one path between every pair of vertices in a tree.

12.7. FOREST

If the root and the corresponding edges connecting the nodes are deleted from a tree, we obtain a set of disjoint trees. This set of disjoint trees is called a forest. (Fig. 12.7*b*)

12.8. BINARY TREE

If the outdegree of every node is less than or equal to 2, in a directed tree then the tree is called a binary tree. A tree consisting of no nodes (empty tree) is also a binary tree.

12.9. BASIC TERMINOLOGY

(*a*) **Root.** A binary tree has a unique node called the root of the tree.

(*b*) **Left Child.** The node to the left of the root is called its left child.

(*c*) **Right Child.** The node to the right of the root is called its right child.

(*d*) **Parent.** A node having left child or right child or both is called parent of the nodes.

(*e*) **Siblings.** Two nodes having the same parent are called siblings.

(*f*) **Leaf.** A node with no children is called a leaf. The number of leaves in a binary tree can vary from one (minimum) to half the number of vertices (maximum) in a tree.

(*g*) **Ancestor.** If a node is the parent of another node, then it is called ancestor of that node. The root is an ancestor of every other node in the tree.

(*h*) **Descendent.** A node is called descendent of another node if it is the child of the node or child of some other descendent of that node. All the nodes in the tree are descendents of the root.

(*i*) **Left Subtree.** The subtree whose root is the left child of some node is called the left subtree of that node.

(*j*) **Right Subtree.** The subtree whose root is the right child of some node is called the right subtree of that node.

(*k*) **Level of a Node.** The level of a node is its distance from the root. The level of root is defined as zero. The level of all other nodes is one more than its parent node. The maximum number of nodes at any level N is 2^N.

(*l*) **Depth or Height of a Tree.** The depth or height of a tree is defined as the maximum number of nodes in a branch of tree. This is one more than the maximum level of the tree *i.e.*, the depth of root is one. The maximum number of nodes in a binary tree of depth d is $2^d - 1$, where $d \geq 1$.

(*m*) **External Nodes.** The nodes which has no children are called external nodes or terminal nodes.

(*n*) **Internal Nodes.** The nodes which has one or more than one children are called internal nodes or non-terminal nodes.

Theorem II. *Let G be a graph with more than one vertex. Then the following are equivalent :*

(*i*) *G is a tree.*

(*ii*) *Each pair of vertices is connected by exactly one simple path.*

(*iii*) *G is connected, but if any edge is deleted then the resulting graph is not connected.*

(iv) G is cycle tree, but if any edge is added to the graph then the resulting graph has exactly one cycle.

Proof. To prove this theorem, we prove that *(i)* ⇒ *(ii)*, *(ii)* ⇒ *(iii)*, *(iii)* ⇒ *(iv)* and finally *(iv)* ⇒ *(i)*. The complete proof is as follows :

(i) ⇒ *(ii)* Let us assume two vertices u and v in G. Since G is a tree, so G is connected and there is at least one path between u and v. More over, there can be only one path between u and v, otherwise G will contain a cycle.

(ii) ⇒ *(iii)* Let us delete an edge $e = (u, v)$ from G. It means e is a path from u to v. Suppose the graph result from $G - e$ has a path p from u to v. Then P and e are two distinct paths from u to v, which is a contradiction of our assumption. Thus, there does not exist a path between u and v in $G - e$, so $G - e$ is disconnected.

(iii) ⇒ *(iv)* Let us suppose that G contains a cycle c which contains an edge $e = \{u, v\}$. By hypothesis, G is connected but $G' = G - e$ is disconnected with u and v belonging to different components of G′. This contradicts the fact that u and v are connected by the path $P = C - e$, which lies in G′. Hence G is cycle free.

Now, Let us take two vertices x and y of G and let H be the graph obtained by adjoining the edge $e = \{x, y\}$ to G. Since G is connected, there is a path P from x to y in G ; hence C = Pe forms a cycle in H. Now suppose H contains another cycle C_1. Since G is cycle free, C_1 must contain the edge e, say $C_1 = P_1e$.

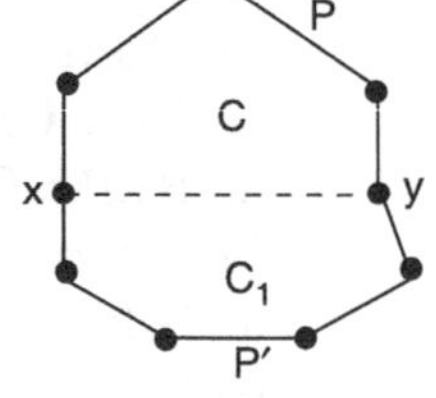

Fig. I

Then P and P_1 are two paths in G from x to y as shown in Fig. I. Thus, G contains a cycle, which contradicts the fact that G is cycle free. Hence H contains only one cycle.

(iv) ⇒ *(i)* By adding any edge $C = (x, y)$ to G produces a cycle, the vertices x and y must be connected already in G. Thus, G is connected and is cycle is free *i.e.,* G is a tree.

Theorem III. *Let G be a finite graph with n > 1 vertices. Then the following are equivalent :*

(i) G is a tree.

(ii) G is cycle free and has n – 1 edges.

(iii) G is connected and has n – 1 edges.

To prove this.

Proof. We use induction on the number of vertices n of G.

Let us assume $n = 1$ *i.e.,* G has only one vertex. Then G has O edges and so G is connected and cycle free. Thus the theorem holds for $n = 1$.

Now, assume that $n > 1$ *i.e.,* G has more than one vertex. Assume that *(i)*, *(ii)* and *(iii)* are equivalent for all graphs with less than n vertices.

We have to show that they are equivalent for G.

(i) ⇒ *(ii)* Suppose G is a tree. Then G is cycle free, so we have to show only that G has $n - 1$ edges. We know that G has a vertex of degree 1. Deleting this vertex and its edge, we obtain a tree T which has $n - 1$ vertices. Thus the theorem holds for T, so T has $n - 2$ edges. Hence G has $n - 1$ edges.

(ii) ⇒ *(iii)* Suppose G is cycle free and has $n - 1$ edges. We have to show only that G is connected. Suppose G is disconnected and has k components, T_1, T_2, ..., T_k, which are trees since each is connected and cycle free, *i.e.*, Ti has ni vertices and $ni < n$. Hence the theorem holds for Ti, so Ti has $ni - 1$ edges. Thus,

$$n = n_1 + n_2 + ... + n_k$$

and
$$n - 1 = (n_1 - 1) + (n_2 - 1) + ... + (n_k - 1)$$
$$= n_1 + n_2 + ... + n_k - k = n - k$$

Hence $k = 1$. But it contradicts our assumption that G is disconnected and has $k > 1$ components. Hence G is connected.

(iii) ⇒ *(i)* Suppose G is connected and has $n - 1$ edges. We have to show only that G is cycle free. Suppose G has a cycle containing an edge e. Deleting e, we obtain the graph H = G – e, which is also connected.

But H has n vertices and $n - 2$ edges and therefore must be disconnected. Thus G is cycle free and hence it is a tree.

ILLUSTRATIVE EXAMPLES

Example 1. *For the tree as shown in Fig. 12.8.*

(i) Which node is the root ?

(ii) Which nodes are leaves ?

(iii) Name the parent node of each node.

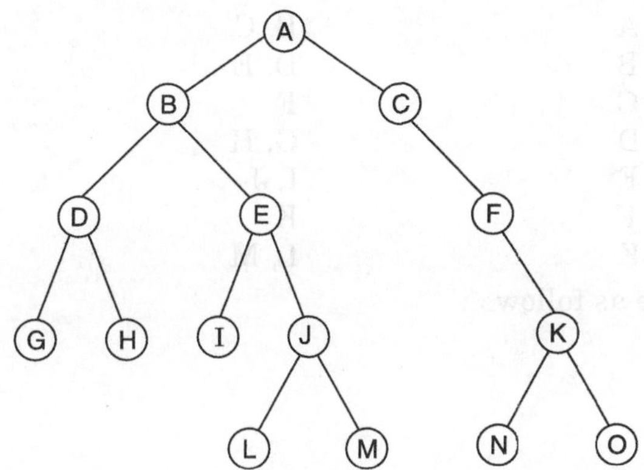

Fig. 12.8

Sol. *(i)* The node A is the root node.

(ii) The nodes G, H, I, L, M, N, O are leaves.

(iii)

Nodes	Parent
B, C	A
D, E	B
F	C
G, H	D
I, J	E
K	F
L, M	J
N, O	K

Example 2. *For the tree as shown in Fig. 12.9.*

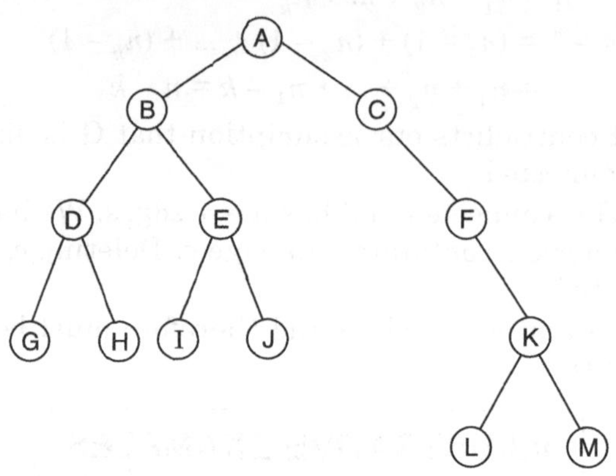

Fig. 12.9

(*i*) *List the children of each node.* (*ii*) *List the siblings.*
(*iii*) *Find the depth of each node.* (*iv*) *Find the level of each node.*

Sol. (*i*) The children of each node is as follows :

Node	Children
A	B, C
B	D, E
C	F
D	G, H
E	I, J
F	K
K	L, M

(*ii*) The siblings are as follows :

Siblings
B and C
D and E
G and H
I and J
L and M are all siblings.

(*iii*)

Node	Depth or Height
A	1
B, C	2
D, E, F	3
G, H, I, J, K	4
L, M	5

(*iv*)

Node	Level
A	0
B, C	1
D, E, F	2
G, H, I, J, K	3
L, M	4

Example 3. *Show that if in a graph G there exists one and only one path between every pair of vertices, then G is a tree.*

Sol. The graph G is connected since there is a path between every pair of vertices. A cycle in a graph exists if there is at least one pair of vertices (v_1, v_2) such that there exist two distinct paths from v_1 to v_2. But the graph G has one and only one path between every pair of vertices. Thus, G contains no cycle. Hence, G is a tree.

Example 4. *Draw two different binary trees with five nodes having only one leaf.*

Sol. The two trees out of many possible trees with five nodes having only one leaf is shown in Fig. 12.10.

Fig. 12.10

Example 5. (*a*) *Draw two different binary trees with five nodes having maximum number of leaves.*

(*b*) *Let T be a tree with n vertices. Determine the number of leaf nodes in a tree.*

(P.T.U. B.Tech. Dec. 2008)

Sol. (*a*) There are many possible trees, out of which two different binary trees are shown in Fig. 12.11.

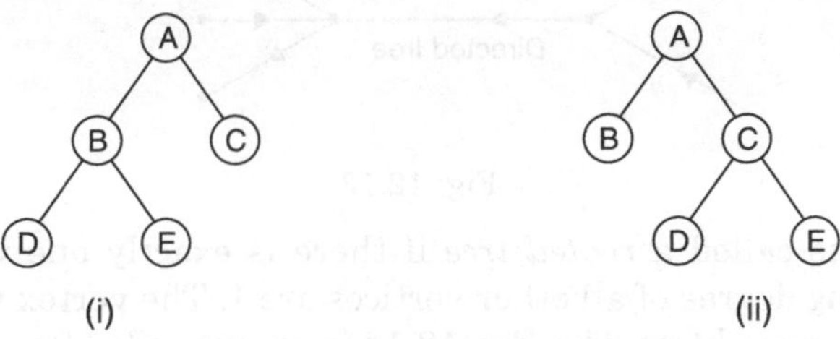

Fig. 12.11

(*b*) Given T is a binary tree with *n* vertices. Therefore, the tree with *n* vertices has $(n - 1)$ edges. Also, if L denotes the number of leaves and I be the number of internal nodes, then

$$n = \text{L} + 1$$

But $$\text{I} = \frac{n-1}{2}$$

Using (2) in (1), we have

$$n = \text{L} + \frac{n-1}{2} \Rightarrow \text{L} = n - \frac{n-1}{2} = \frac{2n - n + 1}{2} = \frac{n+1}{2}$$

Example 6. (*a*) *How will you differentiate between a general tree and a binary tree ?*

(*b*) *Define a rooted tree with an example and show how it may be viewed as directed graph.*

Sol. (*a*)

General Tree	*Binary Tree*
1. There is no such tree having zero nodes or an empty general tree.	1. There may be an empty binary tree.
2. If some node has a child, then there is no such distinction.	2. If some nodes has a child, then it is distinguished as a left child or a right child.
3. The trees shown in Fig. 12.12 are same, when we consider them as general trees.	3. The trees shown in Fig. 12.12 are distinct, when we consider them as binary trees, because in (*i*), 4 is right child of 2 while in (*ii*), 4 is left child of 2.

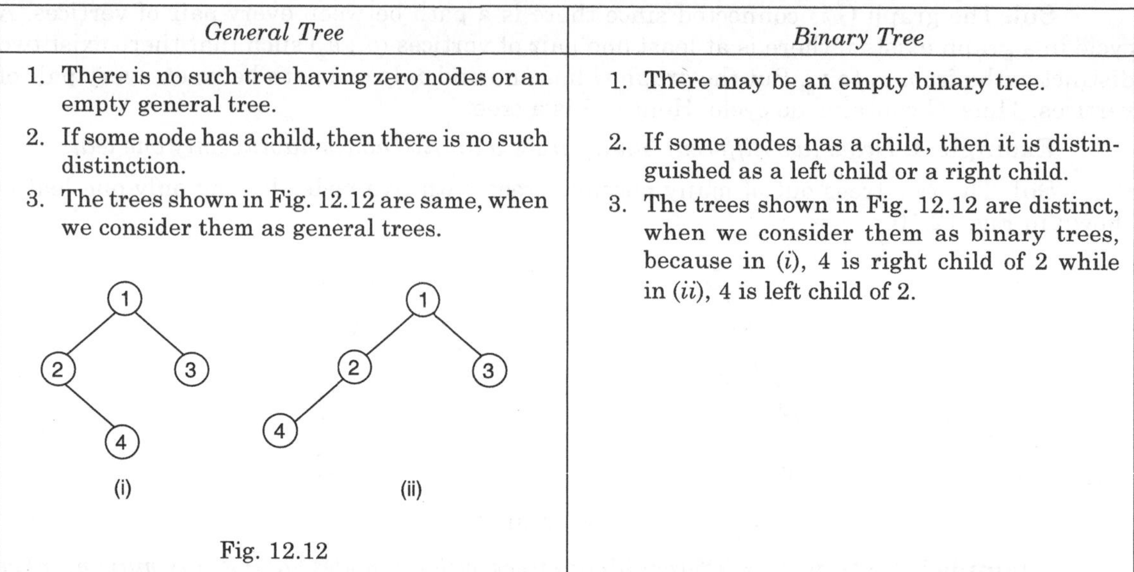

Fig. 12.12

(*b*) **Rooted tree:** We first define the term '*directed tree*'. A directed graph is said to be a directed tree if it becomes a tree when the directions of the edges are ignored. For example, the Fig. 12.13 is a *directed tree.*

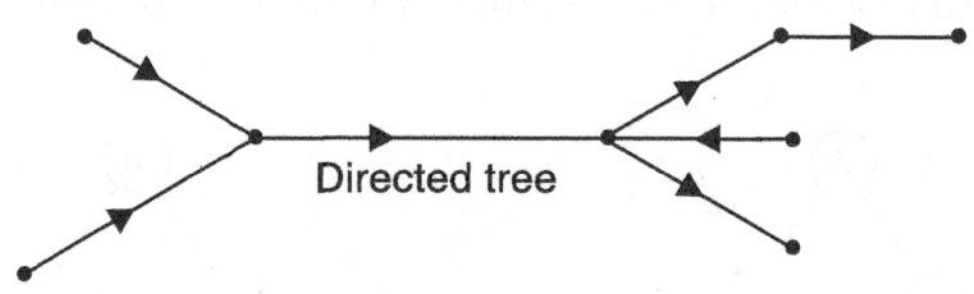

Fig. 12.13

A directed tree is called a *rooted tree* if there is exactly one vertex whose incoming degree is 0 and incoming degree of all other vertices are 1. The vertex with incoming degree 0 is called the *root* of the *rooted tree.* The Fig. 12.14 is an example of a rooted tree.

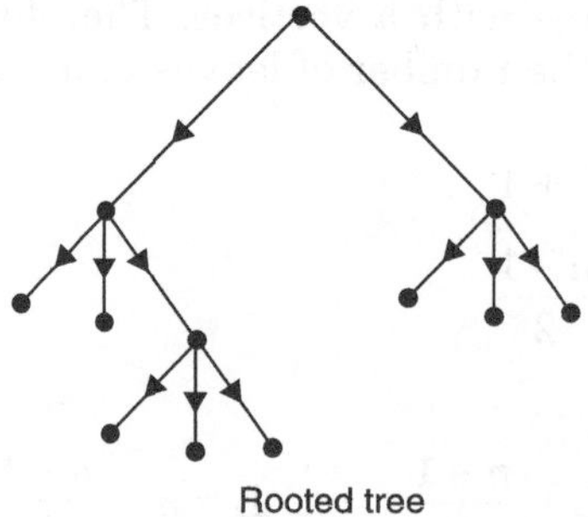

Fig. 12.14

In a rooted tree, a vertex whose outgoing degree is 0 is called a *leaf* or a *terminal code* and a vertex whose *outgoing degree is non zero, is called a branch node or an internal node.*

Rooted tree may be viewed as directed graph. We know that a tree is a graph which is connected and without any cycles. A rooted tree T is a tree with a designated vertex r, called the root of the tree. Since there is a unique simple path from the root r to any other vertex v in T, this determines a direction to the edges of T. Thus T may be viewed as a directed graph.

12.10. BINARY EXPRESSION TREES

Algebraic expression can be conveniently expressed by its expression tree. An expression having binary operators can be decomposed into

< left operand or expression > (operator) < right operand or expression >

depending upon precedence of evaluation.

The expression tree is a binary tree whose root contains the operator and whose left subtree contains the left expression and right subtree contains the right expression.

Example 7. *Construct the binary expression tree for the expression (a + b) * (d/c).*

Sol. The binary expression tree for the expression $(a + b) * (d/c)$ is shown in Fig. 12.15.

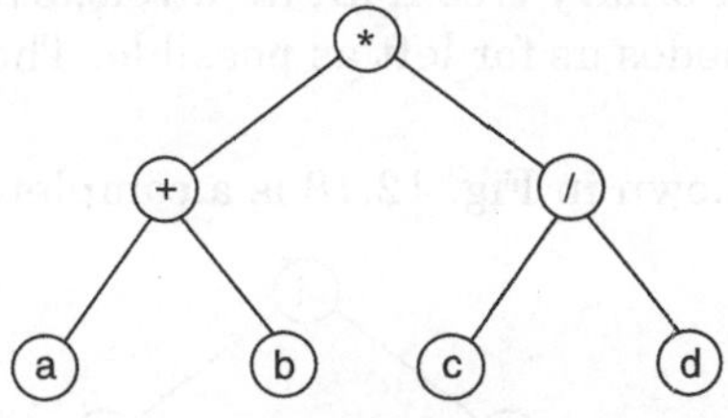

Fig. 12.15

Example 8. *Determine the value of expression tree shown in Fig. 12.16.*

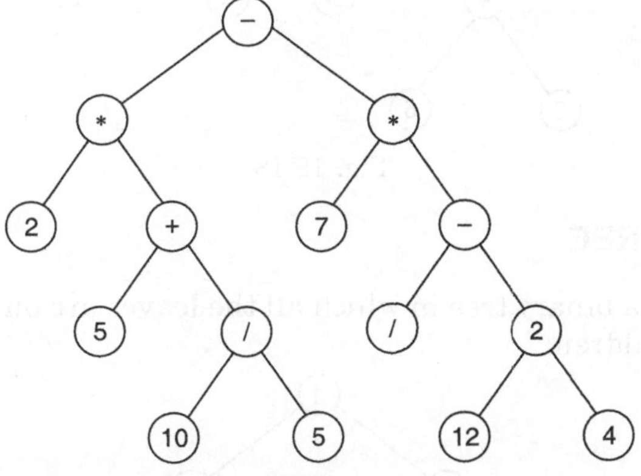

Fig. 12.16

Sol. The value of expression tree is 2.

Example 9. *Determine the value of expression tree shown in Fig. 12.17.*

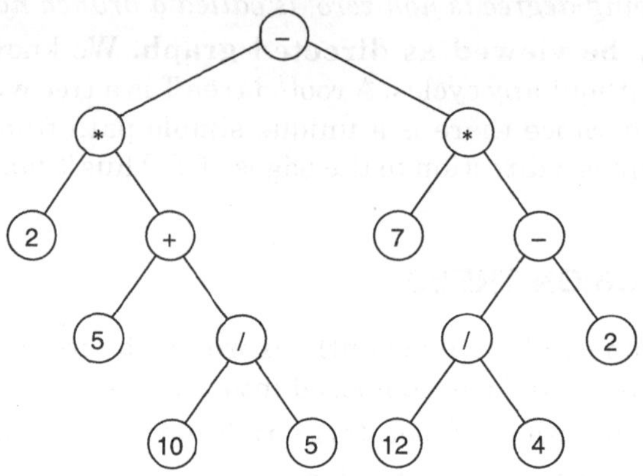

Fig. 12.17

Sol. The value of expression tree is 7.

12.11. COMPLETE BINARY TREE

Complete binary tree is a binary tree if all its levels, except possibly the last, have the maximum number of possible nodes as for left as possible. The depth of complete binary tree having n nodes is $\log_2 n + 1$.

For example : The tree shown in Fig. 12.18 is a complete binary tree.

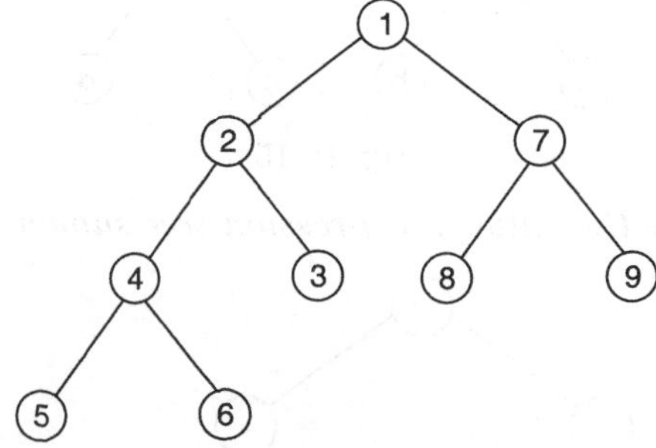

Fig. 12.18

12.12. FULL BINARY TREE

Full binary tree is a binary tree in which all the leaves are on the same level and every non-leaf node has two children.

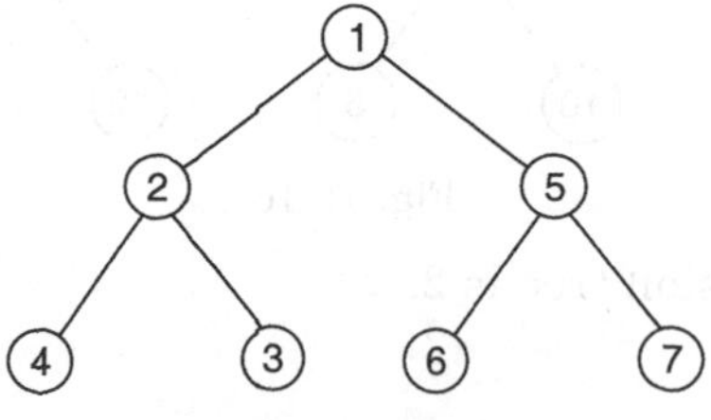

Fig. 12.19

For example : The tree shown in Fig. 12.19 is a full binary tree.

Theorem I. *Prove that the maximum number of nodes on level n of a binary tree is 2^n, where $n \geq 0$.*

Proof. This can be proved by induction.

Basis of Induction. The only node at level $n = 0$ is the root node. Thus, the maximum number of nodes on level $n = 0$ is $2^0 = 1$.

Induction Hypothesis. Now assume that it is true for level j, where $n \geq j \geq 0$. Therefore, the maximum no. of nodes on level j is 2^j.

Induction Step. By induction hypothesis, the maximum number of nodes on level $j - 1$ is 2^{j-1}. Since, we know that each node in binary tree has maximum degree 2. Therefore, the maximum number of nodes on level j is twice the maximum number of level $j - 1$.

Hence, at level j, the maximum number of nodes is

$$= 2 \cdot 2^{j-1}$$

$$= 2^j. \text{ Hence proved.}$$

Theorem II. *Prove that the maximum number of nodes in a binary tree of depth d is $2^d - 1$, where $d \geq 1$.*

Proof. This can be proved by induction.

Basis of Induction. The only node at depth $d = 1$ is the root node. Thus, the maximum number of nodes on depth $d = 1$ is $2^1 - 1 = 1$.

Induction Hypothesis. Now assume that it is true for depth K, $d > k \geq 1$. Therefore, the maximum number of nodes on depth K is $2^k - 1$.

Induction Step. By induction hypothesis, the maximum number of nodes on depth K – 1 is $2^{k-1} - 1$. Since, we know that each node in a binary tree has maximum degree 2, therefore, the maximum number of nodes on depth d = K is twice the maximum number of nodes on depth K – 1.

So, at depth d = K, the maximum number of nodes is $= (2 \cdot 2^{K-1}) - 1$

$$= 2^{K-1+1} - 1 = 2^K - 1. \text{ Hence proved.}$$

Theorem III. *Prove that in a binary tree, if n_E is the number of external nodes or leaves and n_I is the no. of internal nodes, then $n_E = n_I + 1$.*

Proof. Let n be the total number of nodes in the tree. Then, we may have three types of nodes in the tree.

n_E = the number of nodes having zero degree.

n_I = the number of nodes having two degree.

n_0 = the number of nodes having one degree.

So, we have $\quad n = n_E + n_I + n_0 \quad$ ...(1)

Let us assume that the number of edges of the tree is E. So, with these E edges we can connect E + 1 nodes.

Hence $\quad n = E + 1 \quad$ ...(2)

Since, all edges are either from a node of degree one or from a node of degree two, therefore, $\quad E = n_0 + 2n_I \quad$ ...(3)

Put this value in the eq. (2), we have

$$n = n_0 + 2n_I + 1 \qquad ...(4)$$

Subtract eq. (4) from eq. (1), we get

$$n_E = n_I + 1$$

Hence proved.

12.13. TRAVERSING BINARY TREES

Traversing means to visit all the nodes of the tree. There are three standard methods to traverse the binary trees. These are as follows :

1. Preorder traversal
2. Postorder traversal
3. Inorder traversal.

1. Preorder traversal. The preorder traversal of a binary tree is a recursive process. The preorder traversal of a tree is

(*i*) Visit the root of the tree.

(*ii*) Traverse the left subtree in preorder.

(*iii*) Traverse the right subtree in preorder.

2. Postorder traversal. The postorder traversal of a binary tree is a recursive process. The postorder traversal of a tree is

(*i*) Traverse the left subtree in postorder.

(*ii*) Traverse the right subtree in postorder.

(*iii*) Visit the root of the tree.

3. Inorder traversal. The inorder traversal of a binary tree is a recursive process. The inorder traversal of a tree is

(*i*) Traverse in inorder the left subtree.

(*ii*) Visit the root of the tree.

(*iii*) Traverse in inorder the right subtree.

Example 10. *Determine the preorder, postorder, and inorder traversal of the binary tree as shown in Fig. 12.20.*

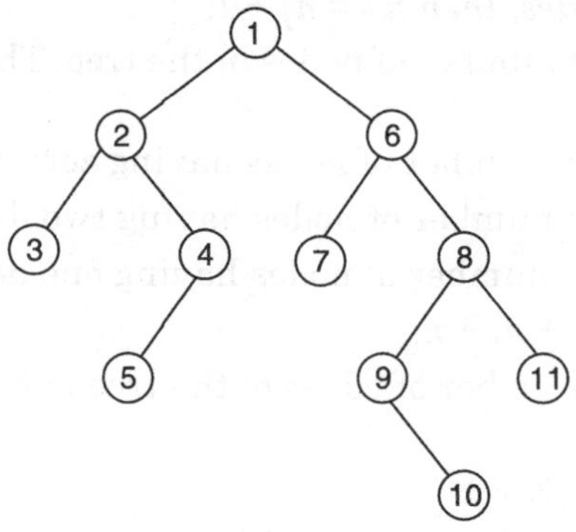

Fig. 12.20

Sol. The preorder, postorder and inorder traversal of the tree is as follows :

Preorder	:	1	2	3	4	5	6	7	8	9	10	11
Postorder	:	3	5	4	2	7	10	9	11	8	6	1
Inorder	:	3	2	5	4	1	7	6	9	10	8	11.

Example 11. *Give the preorder, inorder and postorder traversals of the tree shown in Fig. 12.21.*

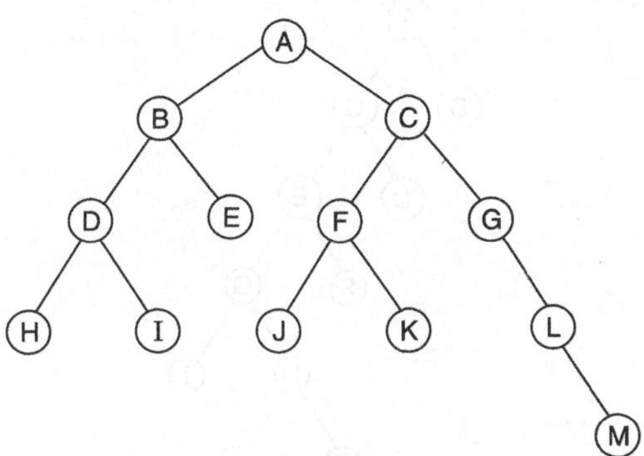

Fig. 12.21

Sol.

Preorder	:	A,	B,	D,	H,	I,	E,	C,	F,	J,	K,	G,	L,	M
Inorder	:	H,	D,	I,	B,	E,	A,	J,	F,	K,	C,	G,	L,	M
Postorder	:	H,	I,	D,	E,	B,	J,	K,	F,	M,	L,	G,	C,	A.

12.14. ALGORITHMS

(*a*) Algorithm to Draw a Unique Binary Tree When Inorder and Preorder Traversal of the Tree is Given

1. We know that the root of the binary tree is the first node in its preorder. Draw the root of the tree.

2. To find the left child of the root node, first use the inorder traversal to find the nodes in the left subtree of the binary tree. (All the nodes that are left to the root node in the inorder traversal are the nodes of the left subtree). After that the left child of the root is obtained by selecting the first node in the preorder traversal of the left subtree. Draw the left child.

3. In the same way, use the inorder traversal to find the nodes in the right subtree of the binary tree. Then the right child is obtained by selecting the first node in the preorder traversal of the right subtree. Draw the right child.

4. Repeat the steps 2 and 3 with each new node until every node is not visited in preorder. Finally, we obtain the unique tree.

Example 12. *Draw the unique binary tree when the inorder and preorder traversal is given as follows :*

Inorder	:	*B*	*A*	*D*	*C*	*F*	*E*	*J*	*H*	*K*	*G*	*I*
Preorder	:	*A*	*B*	*C*	*D*	*E*	*F*	*G*	*H*	*J*	*K*	*I.*

Sol. We know that the root of the binary tree is the first node in preorder traversal. Now check A, in the inorder traversal, all the nodes that are left of A, are nodes of left subtree and all the nodes that are right of A, are nodes of right subtree. Read the next node in preorder and check its position against the root node, if it is left of root node, then draw it as left child, otherwise draw it as right child. Repeat the above process for each new node until all the nodes of preorder traversal are read and finally we obtain the binary tree as shown in Fig. 12.22.

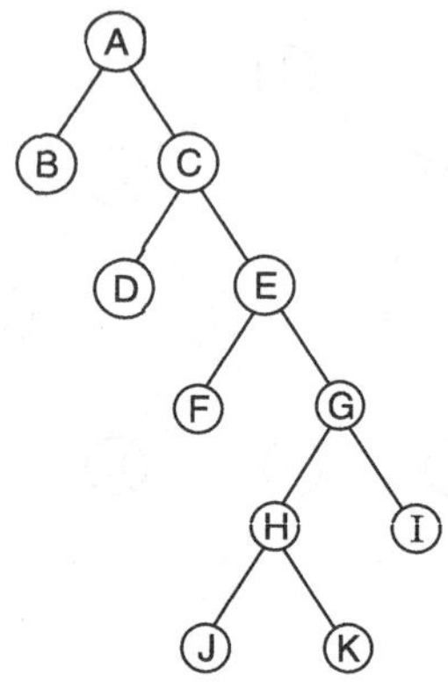

Fig. 12.22

Example 13. *Draw the unique binary tree when the following is given :*

Inorder	:	*d*	*b*	*h*	*e*	*a*	*f*	*c*	*i*	*j*	*g*
Preorder	:	*a*	*b*	*d*	*e*	*h*	*c*	*f*	*g*	*i*	*j*

Sol. The first node in preorder is *a* and hence *a* is the root node. Unique binary tree is shown in Fig. 12.23.

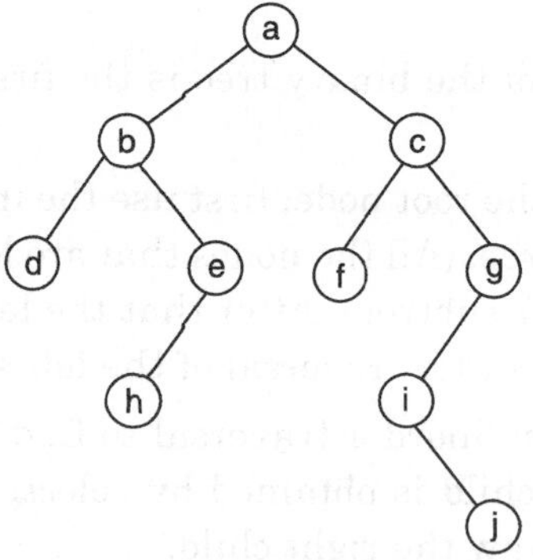

Fig. 12.23

Example 14. *Draw the unique binary tree when inorder and preorder traversal of tree is given as follows :*

Preorder	:	*10*	*9*	*8*	*7*	*6*	*5*	*4*	*3*	*2*	*1*
Inorder	:	*10*	*8*	*9*	*7*	*6*	*4*	*5*	*2*	*3*	*1*

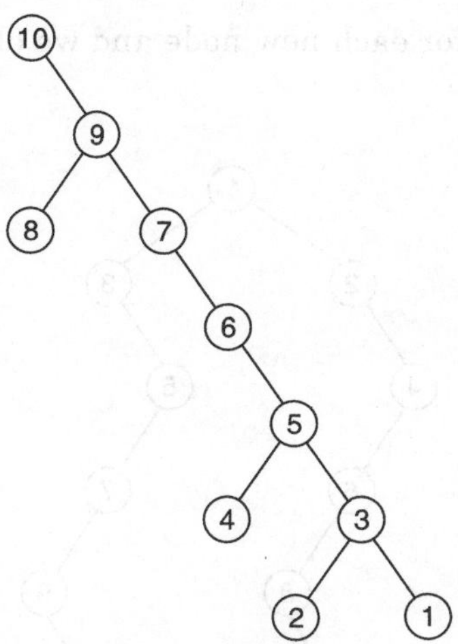

Fig. 12.24

Sol. Unique binary tree is shown in Fig. 12.24.

(*b*) **Algorithm to Draw a Unique Binary Tree When Inorder and Postorder Traversal of the Tree is Given**

1. We know that the root of the binary tree is the last node in its postorder. Draw the root of the tree.

2. To find the right child of the root node, first use the inorder traversal to find the nodes in the right subtree of the binary tree. (All the nodes that are right to the root node in the inorder traversal are the nodes of the right subtree). After that the right child of the root is obtained by selecting the last node in the postorder traversal of the right subtree. Draw the right child.

3. In the same way, use the inorder traversal to find the nodes in the left subtree of the binary tree. Then the left child is obtained by selecting the last node in the postorder traversal of the left subtree. Draw the left child.

4. Repeat the steps 2 and 3 with each new node until every node is not visited in postorder. After visiting the last node, we obtain the unique tree.

Example 15. *Draw the unique binary tree for the given Inorder and Postorder traversal*

Inorder	:	*4*	*6*	*10*	*12*	*8*	*2*	*1*	*5*	*7*	*11*	*13*	*9*	*3*
Postorder	:	*12*	*10*	*8*	*6*	*4*	*2*	*13*	*11*	*9*	*7*	*5*	*3*	*1*

Sol. We know that the root node is the last node in postorder traversal. Hence one is the root node.

Now check the inorder traversal, we know that root is at the centre, hence all the nodes that are left to the root node in inorder traversal are the nodes of left subtree and all that are right to the root node are the nodes of the right subtree.

Now, visit the next node from back in postorder traversal and check its position in inorder traversal, if it is on left of root then draw it as left child and if it is on right, then draw it as right child.

Repeat the above process for each new node and we obtain the binary tree as shown in Fig. 12.25.

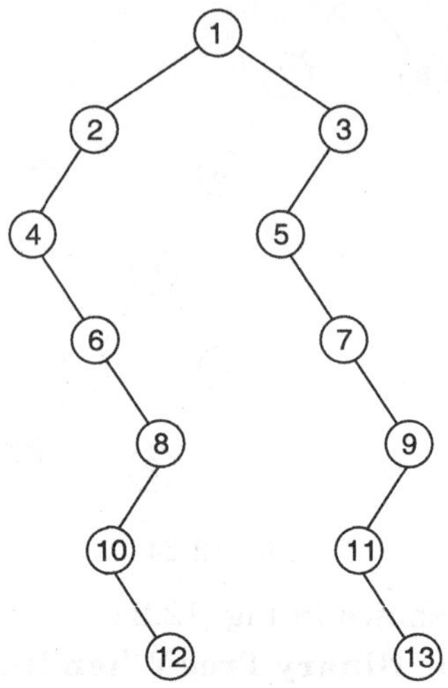

Fig. 12.25

Example 16. *Draw the binary tree when Inorder and Postorder traversal is given :*

Inorder :	*m*	*k*	*n*	*j*	*o*	*l*	*u*	*s*	*v*	*q*	*t*	*p*	*r*
Postorder :	*m*	*n*	*k*	*o*	*u*	*v*	*s*	*t*	*q*	*r*	*p*	*l*	*j*.

Sol. We know that the last node in Postorder is the root node, hence *j* is the root. Now applying the algorithm as above, we obtain the tree shown in Fig. 12.26.

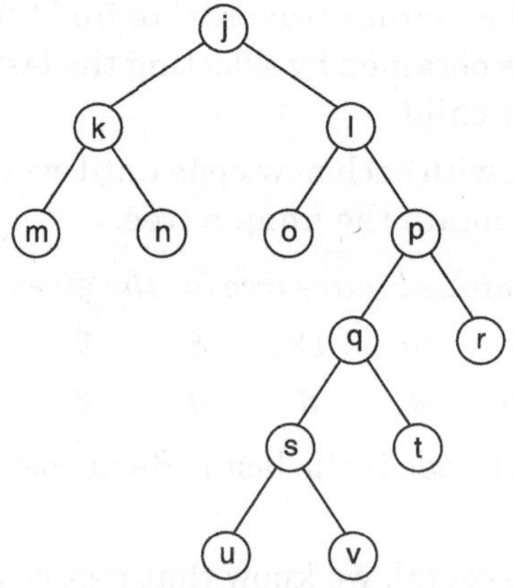

Fig. 12.26

Example 17. *Draw the unique binary tree when inorder and preorder traversal of tree is given as follows :*

Inorder :	3	*	*a*	+	*b*	*	6	*	*c*	+	*d*
Preorder :	*	+	*	3	*a*	*b*	+	*	6	*c*	*d*

Sol. Unique binary tree is shown in Fig. 12.27.

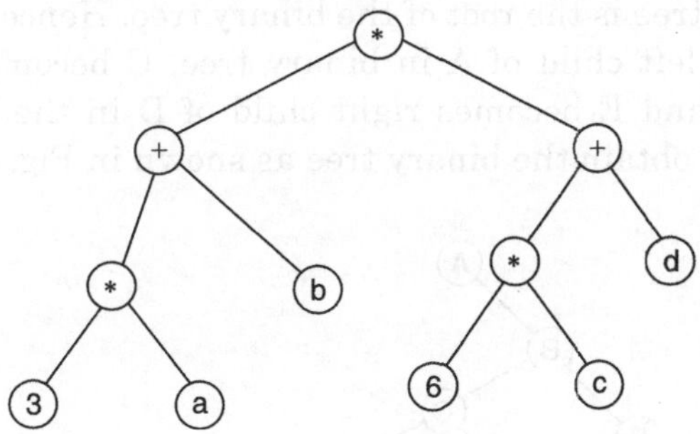

Fig. 12.27

Example 18. *Draw the binary expression tree, when inorder and postorder traversal of the tree is given as follows :*

Postorder :	*k*	*l*	– *m*	+	*n*	*p*	*q*	*r*	*t* – /	+	– *
Inorder :	*k*	–	*l* + *m*	*	*n*	–	*p*	+	*q*/*r*	–	*

Sol. Binary expression tree is shown in Fig. 12.28.

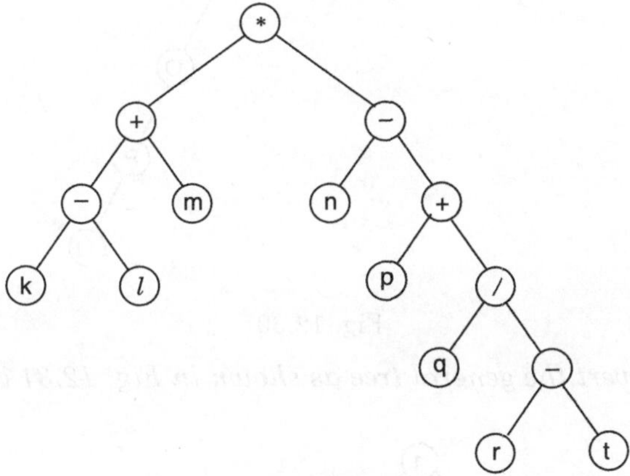

Fig. 12.28

(*c*) Algorithm to Convert General Tree into the Binary Tree

1. Starting from the root node, the root of the tree is also the root of the binary tree.

2. The first child C_1 (from left) of the root node in the tree is the left child C_l of the root node in binary tree and the sibling of the C_1 is the right child of C_l and so on.

3. Repeat the step 2 for each new node.

Example 19. *Convert the following tree as shown in Fig. 12.29 into binary tree.*

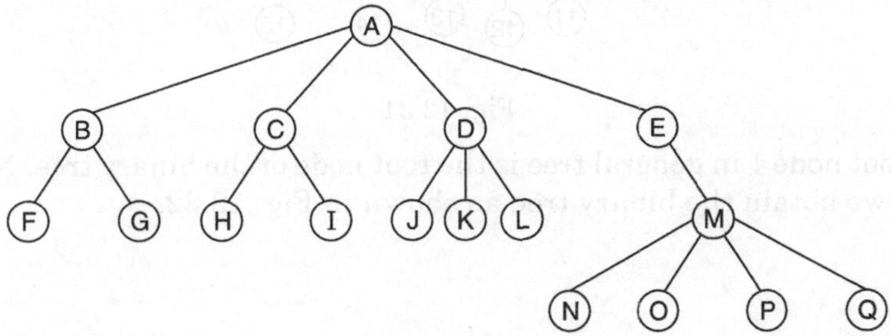

Fig. 12.29

Sol. The root of the tree is the root of the binary tree. Hence A is the root of the binary tree. Now B becomes the left child of A in binary tree, C becomes the right child of B, D becomes right child of C and E becomes right child of D in the binary tree and similarly applying the algorithm we obtain the binary tree as shown in Fig. 12.30.

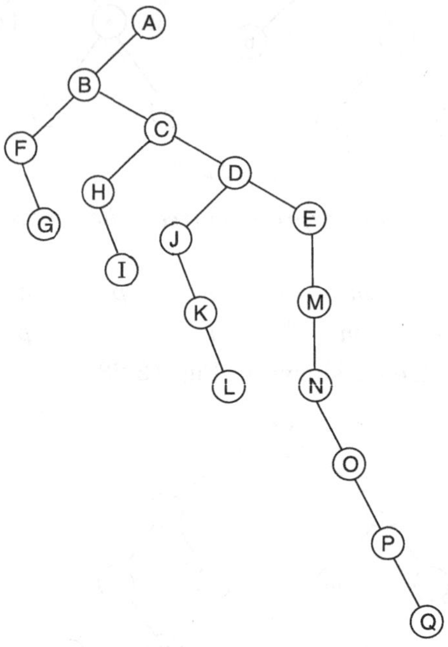

Fig. 12.30

Example 20. *Convert the general tree as shown in Fig. 12.31 into binary tree.*

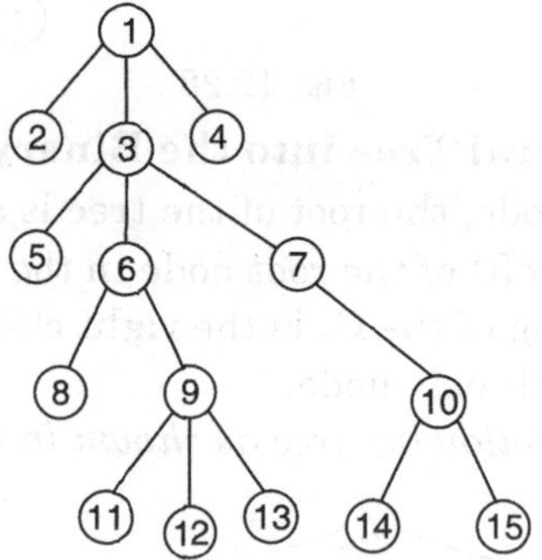

Fig. 12.31

Sol. The root node 1 in general tree is the root node of the binary tree. Now applying the above algorithm we obtain the binary tree as shown in Fig. 12.32.

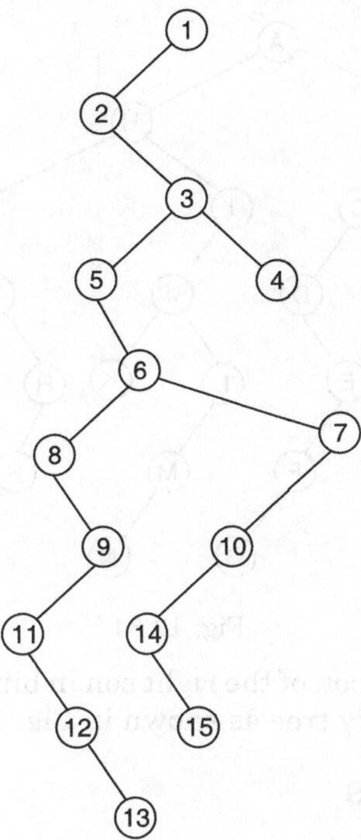

Fig. 12.32

Example 21. *Convert the forest shown in Fig. 12.33 into binary tree.*

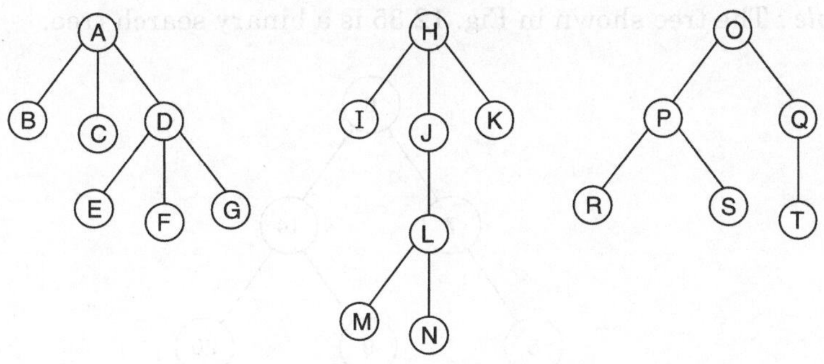

Fig. 12.33

Sol. The root of the binary tree is the root node of the first tree (from left) and the root node of the second tree becomes the right son of the root node in binary tree and the root node

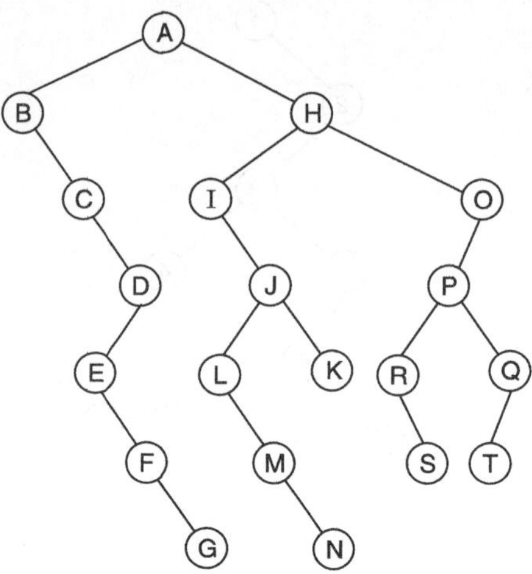

Fig. 12.34

of the third tree becomes the right son of the right son in binary tree. Repeat this procedure for each level and we obtain the binary tree as shown in Fig. 12.34.

12.15. BINARY SEARCH TREES

Binary search trees has the property that the node to the left contains a smaller value than the node pointing to it and the node to the right contains a larger value than the node pointing to it.

It is not necessary that a node in a 'Binary Search Tree' point to the nodes whose values immediately precede and follow it.

For example : The tree shown in Fig. 12.35 is a binary search tree.

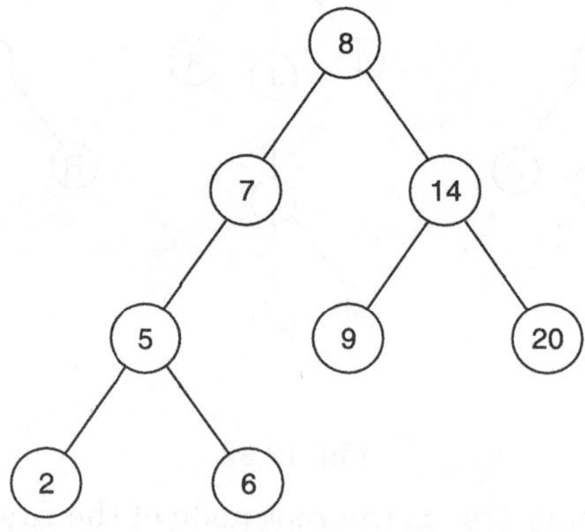

Fig. 12.35

12.16. INSERTING INTO A BINARY SEARCH TREE

Consider a binary search tree T. Suppose we have given an ITEM of information to insert in T. The ITEM is inserted as a leaf in the tree. The following steps explains a procedure to insert an ITEM in the binary search tree T.

1. Compare the ITEM with the root node.

2. If ITEM > ROOT NODE, proceed to the right child and it becomes root node for the right subtree.

3. If ITEM < ROOT NODE, proceed to the left child.

4. Repeat the above steps until we meet a node which has no left and right subtree.

5. Now if the ITEM is greater than node, then the ITEM is inserted as the right child and if the ITEM is less than node, then the ITEM is inserted as the left child.

Deleting in a Binary Search Tree. Consider a binary search tree T. Suppose we want to delete a given ITEM from binary search tree. To delete an ITEM from a binary search tree, we have three cases, depending upon the number of children of the deleted node.

1. Deleted Node has no Children. Deleting a node which has no children is very simple, as just replace the node with null.

2. Deleted Node has Only One Child. Replace the value of deleted node with the only child.

3. Deleted Node has Two Children. In this case, replace the deleted node with the node that is closest in the value to the deleted node. To find the closest value, we move once to the left and then to the right as far as possible. This node is called immediate predecessor. Now replace the value of deleted node with immediate predecessor and then delete the replaced node by using case 1 or 2.

Example 22. *Show the binary search tree after inserting 3, 1, 4, 6, 9, 2, 5, 7 into an initially empty binary search tree.*

Sol. The insertion of the above nodes in the empty binary search tree is shown in Fig. 12.36.

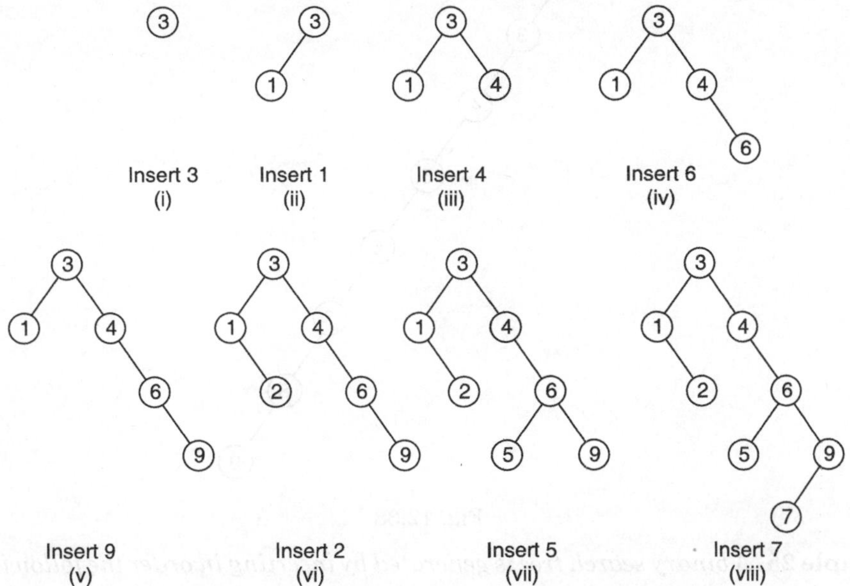

Fig. 12.36

Example 23. *Show the binary tree shown in Fig. 12.36 (viii) after deleting the root node.*

Sol. To delete the root node, first replace the root node with the closest element of the root. For this, first move one step left and then to the right as far as possible to the node. Then delete the replaced node. The tree after deletion is shown in Fig. 12.37.

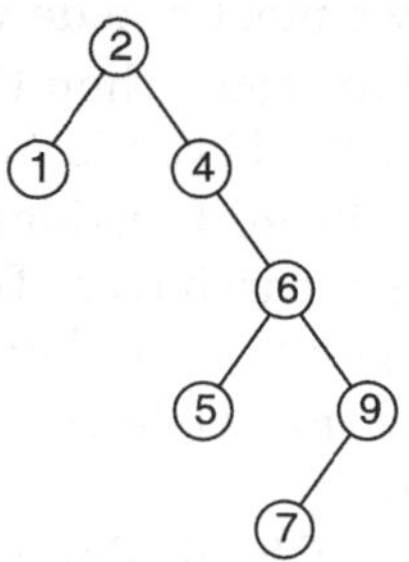

Fig. 12.37

Example 24. *A binary search tree contains the values 1, 2, 3, 4, 5, 6, 7, 8, 9. The tree is traversed in pre-order and the values are printed out. Determine the sequence of the print out values.*

Sol. First of all draw the binary search tree shown in Fig. 12.38.

Now traverse the tree in pre-order and get the output as below 1, 2, 3, 4, 5, 6, 7, 8, 9.

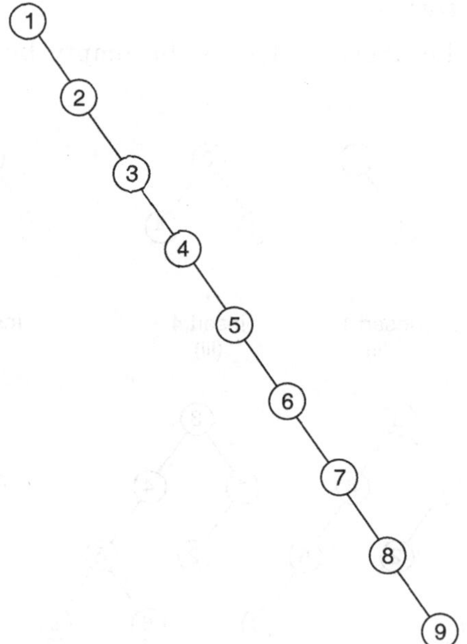

Fig. 12.38

Example 25. *A binary search tree is generated by inserting in order the following integers :*

50, 15, 62, 5, 20, 58, 91, 3, 8, 37, 60, 24.

Determine the number of nodes in the left subtree and right subtree of the root.

Sol. First of all draw the binary search tree as shown in Fig. 12.39.

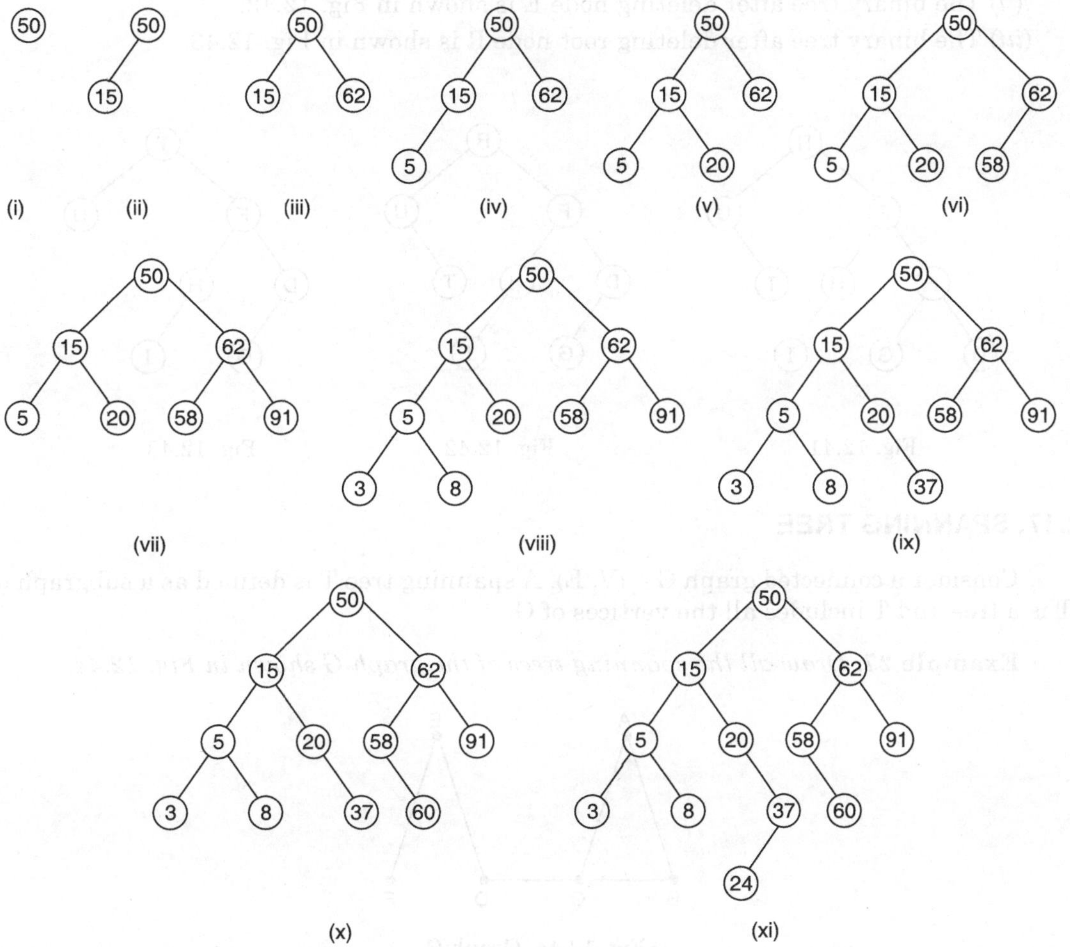

50
15
62
5
20
58
91
3
8
37
60
24
(i)
(ii)
(iii)
(iv)
(v)
(vi)
(vii)
(viii)
(ix)
(x)
(xi)

Fig. 12.39

Thus, the number of nodes in left subtree of the root is 7 and right subtree of the root is 4.

Example 26. *Consider the binary tree as shown in Fig. 12.40. Draw the binary tree for each of the following operations, if applied to the binary tree.*

(i) Delete the node V *(ii) Delete the node E*

(iii) Delete the root node R.

R
F
U
E
H
T
V
D
G
I

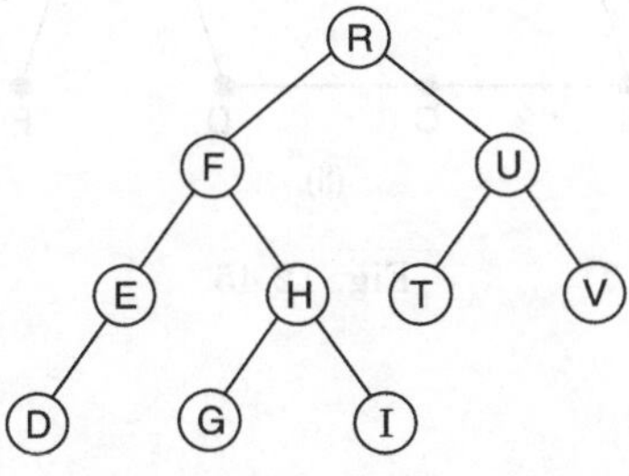

Fig. 12.40

Sol. (*i*) The binary tree after deleting node V is shown in Fig. 12.41.

(*ii*) The binary tree after deleting node E is shown in Fig. 12.42.

(*iii*) The binary tree after deleting root node R is shown in Fig. 12.43.

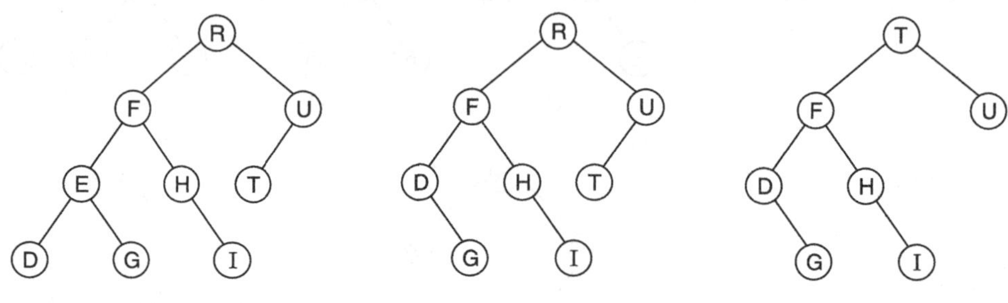

Fig. 12.41 Fig. 12.42 Fig. 12.43

12.17. SPANNING TREE

Consider a connected graph G = (V, E). A spanning tree T is defined as a subgraph of G if T is a tree and T includes all the vertices of G.

Example 27. *Draw all the spanning trees of the graph G shown in Fig. 12.44.*

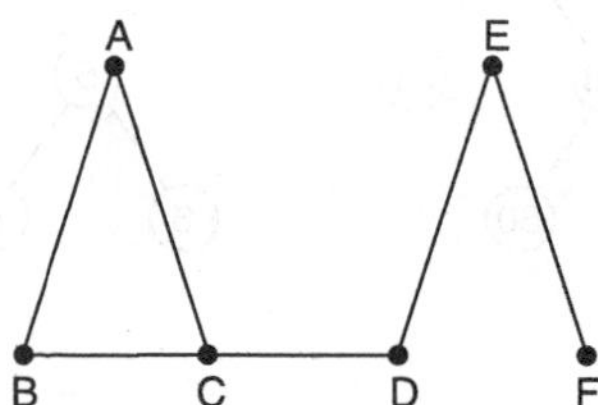

Fig. 12.44. Graph G.

Sol. All the spanning trees of graph G is as shown in Fig. 12.45.

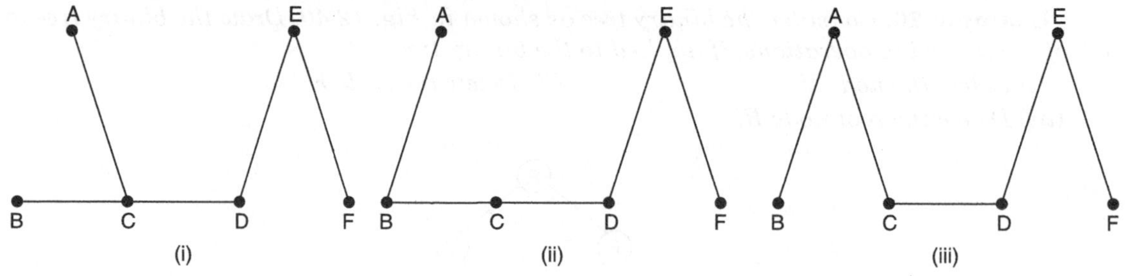

Fig. 12.45

12.18. APPLICATIONS OF TREES

12.18.1. Minimum Spanning Tree

Consider a connected weighted graph G = (V, E). A minimal spanning tree T of the graph G is a tree whose total weight is smallest among all the spanning trees of the graph G. The total weight of the spanning tree is the sum of the weights of the edges of the spanning trees.

The minimum weight of the spanning tree is unique but the spanning tree may not be unique because more than one spanning tree are possible when more than one edges exist having the same weight.

Theorem IV. *Prove that a simple graph is connected iff it has a spanning tree.*

(P.T.U. B.Tech. Dec. 2008)

Proof. First of all, suppose that a simple graph G has a spanning tree T. The tree T contains every vertex of G. Further, there is a path in T between any two of its vertices. Since T is subgraph of G, there is a path in G between any two of its vertices. Hence, G is connected.

Now, suppose that G is connected. If G is not a tree, then it must contain a simple circuit. Remove an edge from one of these simple circuits, the resulting subgraph has one fewer edge but still contains all the vertices of G and is connected. If this subgraph is not a tree, it has a simple circuit, so again remove an edge that is in simple circuit. Repeat this process untill no simple circuits remain. This is possible because there are only a finite number of edges in the graph. The process terminates when no simple circuits remain. A tree is produced since the graph is still connected as edges are removed. This is a spanning tree since it contains every vertex of G. Hence the theorem.

12.19. KRUSKAL'S ALGORITHM TO FIND MINIMUM SPANNING TREE

This algorithm finds the minimum spanning tree T of the given connected weighted graph G.

1. Input the given connected weighted graph G with n vertices whose minimum spanning tree T, we want to find.

2. Order all the edges of the graph G according to increasing weights.

3. Initialise T with all vertices but do not include any edge.

4. Add each of the graph G in T which does not form a cycle until $n - 1$ edges are added.

Example 28. *Determine the minimum spanning tree of the weighted graph shown in Fig. 12.46.*

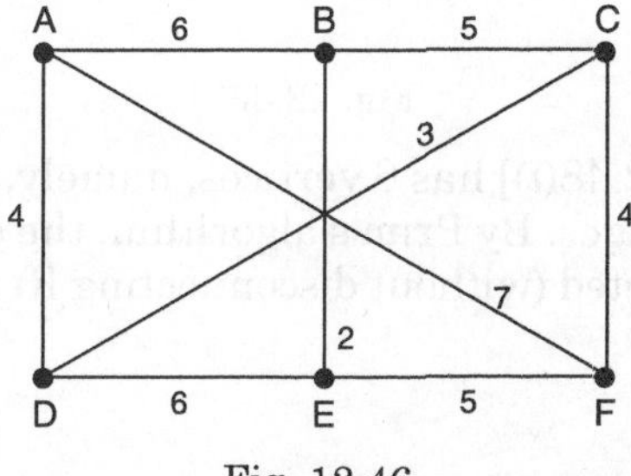

Fig. 12.46

Sol. Using Kruskal's algorithm, arrange all the edges of the weighted graph in increasing order and initialise spanning tree T with all the six vertices of G. Now start adding the edges of G in T which do not form a cycle and having minimum weights until five edges are not added as there are six vertices. (Fig. 12.47).

Edges	Weights	Added or Not
(B, E)	2	Added
(C, D)	3	Added
(A, D)	4	Added
(C, F)	4	Added
(B, C)	5	Added
(E, F)	5	Not added
(A, B)	6	Not added
(D, E)	6	Not added
(A, F)	7	Not added.

Minimum Spanning Tree

Fig. 12.47

Example 29. *Write a shortnote on* ***Prim's*** *and* ***Kruskal's*** *algorithms and execute them by giving a suitable example.* (P.T.U. B.Tech. May 2008)

Sol. Prim's Algorithm. Let R be a symmetric and connected relation with n vertices. The **Prime's algorithm** involves the following steps.

Step I. Choose a vertex v_1 of R. Let V = $\{v_1\}$ and E = { }

Step II. Choose a nearest neighbour v_i of V which is adjacent to v_j where $v_i, v_j \in$ V and for which the edge (v_i, v_j) does not form a cycle with members of E. Add v_i to V and (v_i, v_j) to E.

Step III. Repeat the step II until we get E = $n - 1$

Then V contains all n vertices of R and E contains the edge of a minimum spanning tree for R.

Example 30. *We find the minimal spanning tree for the graph R shown below (Fig. 12.48).*

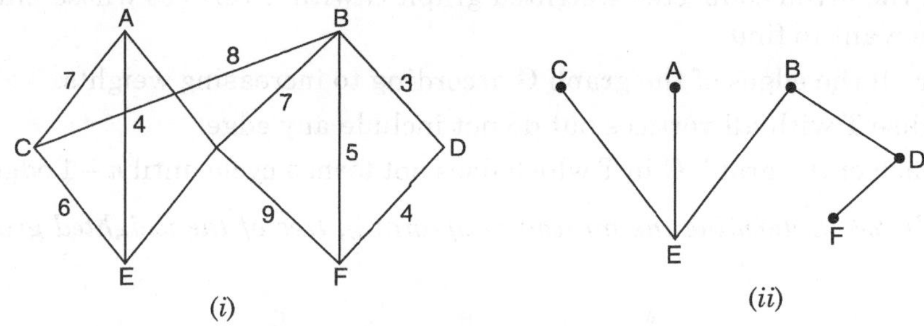

Fig. 12.48

Sol. This graph R [Fig. 12.48(*i*)] has 6 vertices, namely, A, B, C, D, E, F. Therefore, any spanning tree of R will have 5 edges. By Prim's algorithm, the edges are ordered by decreasing lengths and are successively deleted (without disconnecting R) until we have five edges remain. This gives the following data.

Edges	AF	BC	AC	BE	CE	BF	AE	DF	BD
Length	9	8	7	7	6	5	4	4	3
Deleted edges	✓	✓	✓	×	×	✓	×	×	×

Hence the minimal spanning tree of R will contains the edges {BE, CE, AE, DF, BD}. This spaning tree has length 24 as shown in Fig. 12.48(*ii*).

Kruskal's algorithm. Let R be a symmetric and connected relation with n vertices and let S = $[e_1, e_2, ..., e_k]$ be the set of weighted edges of R.

The Kruskal "algorithm involves the following steps".

Step I. Choose an edge e_1 in S of least weight. Let E = $\{e_1\}$. Replace S with S – $\{e\}$

Step II. Select an edge in S of least weight that will not make a cycle with members of $\{e_i\}$ and S with S – $\{e_i\}$.

Step III. Repeat step II until we get E = $n - 1$

Example 31. *Consider the graph R as shown below (Fig. 12.49). We find the minimal spanning tree of R.*

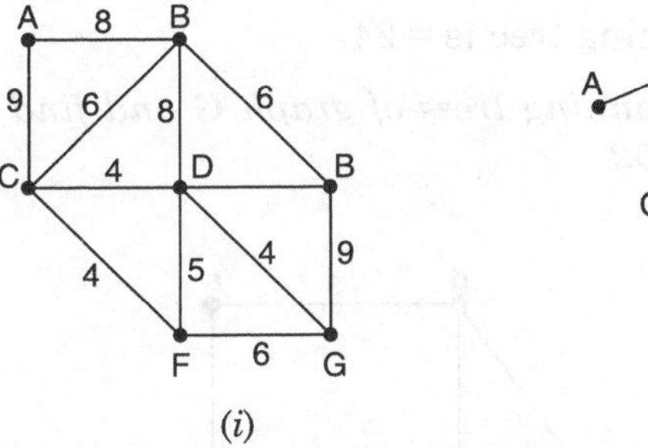

Fig. 12.49

Sol. The graph R [Fig. 12.49(*i*)] has 7 vertices namely A, B, C, D, E, F and G. Therefore, any spanning tree of R will have 6 edges. By Kruskal is algorithm, the edges are odered by increasing length and are successively **added** (without forming any cycle) until 6 edges are included. This gives the following data.

Edges	CD	CF	DG	DF	BC	BE	FG	DE	AB	BD	AC	EG
Length	4	4	4	5	6	6	6	7	8	8	9	9
Added edges	✓	✓	✓	×	✓	✓	×	×	✓	×	×	×

Therefore, the minimal spanning tree of R will contain the edges {CD, CF, DG, BC, BE, AB}. The minimum spanning tree of the graph [Fig. 12.49(*i*)] is shown in Fig. 12.49(*ii*).

Example 32. *Find a minimum spanning tree of the labelled connected graph shown in Fig. 12.50.*

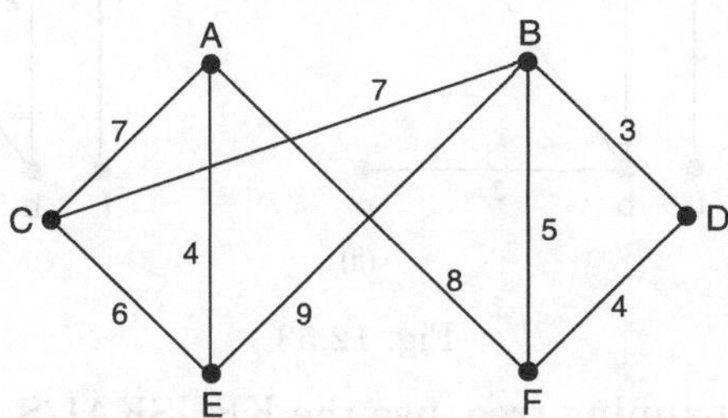

Fig. 12.50

Sol. Using KRUSKAL'S ALGORITHM, arrange all the edges of the graph in increasing order and initialize spanning tree with all the vertices of G. Now, add the edges of G in T which do not form a cycle and have minimum weight until $n - 1$ edges are not added, where n is the number of vertices. The spanning tree is shown in Fig. 12.51.

Edges	Weights	Added or Not
(B, D)	3	Added
(A, E)	4	Added
(D, F)	4	Added
(B, F)	5	Not added
(C, E)	6	Added
(A, C)	7	Not added
(B, C)	7	Added
(A, F)	8	Not added
(E, B)	9	Not added

Minimum Spanning Tree

Fig. 12.51

The minimum weight of spanning tree is = 24.

Example 33. *Find all the spanning trees of graph G and find which is the minimal spanning tree of G shown in Fig. 12.52.*

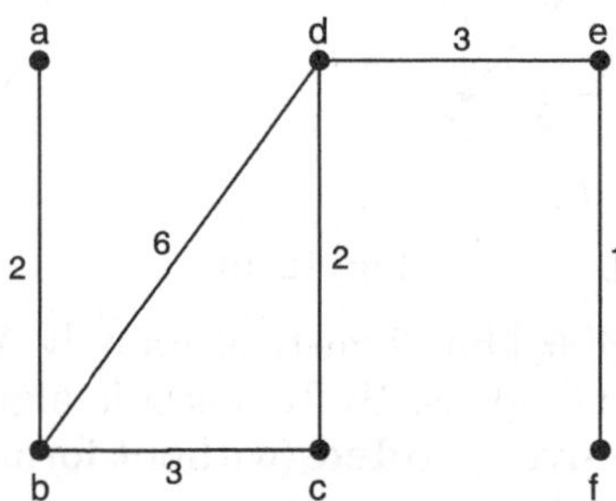

Fig. 12.52

Sol. There are total three spanning trees of the graph G which are as shown in Fig. 12.53.

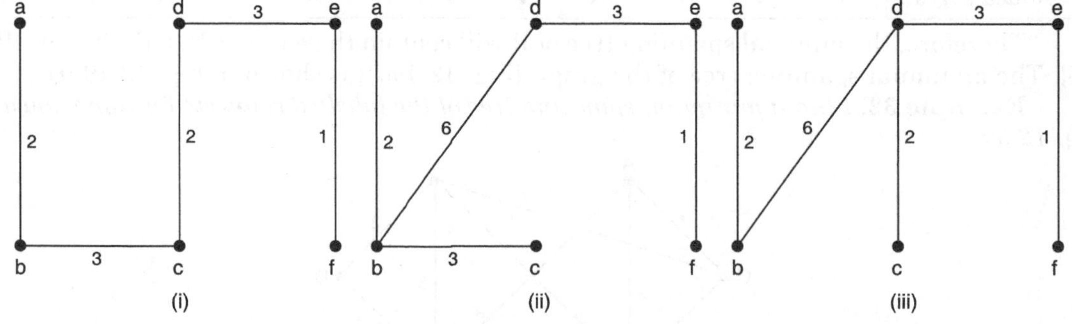

Fig. 12.53

To find the minimal spanning tree, use the KRUSKAL'S ALGORITHM. The minimal spanning tree is shown in Fig. 12.54.

Edges	Weights	Added or Not
(E, F)	1	Added
(A, B)	2	Added
(C, D)	2	Added
(B, C)	3	Added
(D, E)	3	Added
(B, D)	6	Not added.

Minimal Spanning Tree

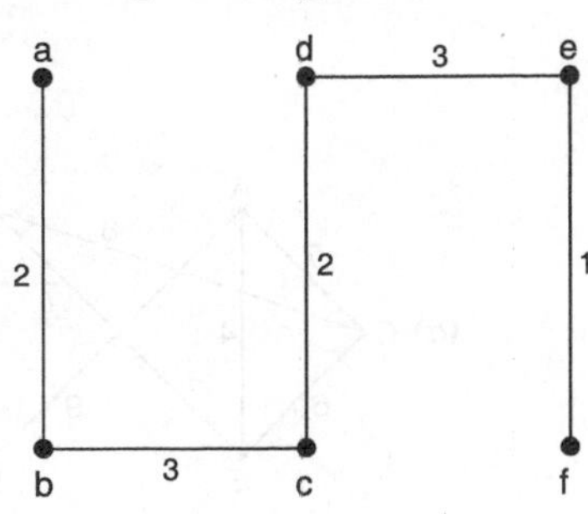

Fig. 12.54

The first one is the minimal spanning having the minimum weight = 11.

Example 34. *What are the properties of minimum spanning tree.*

Sol. Properties of Minimum spanning tree

A minimum spanning tree T of a graph G is a tree whose total weight is the smallest among all the spanning trees of the graph G. It has the following properties.

(*i*) The total weight of the spanning tree is the sum of the weights of the edges of the spanning trees.

(*ii*) The minimum weight of the spanning tree is unique.

TEST YOUR KNOWLEDGE

1. Draw all trees with exactly six vertices. (*P.T.U. B.Tech. Dec. 2013*)
2. Draw all trees with five or fewer vertices.
3. Find the number of trees with seven vertices.
4. Find a minimum spanning tree of the weighted graph shown below :

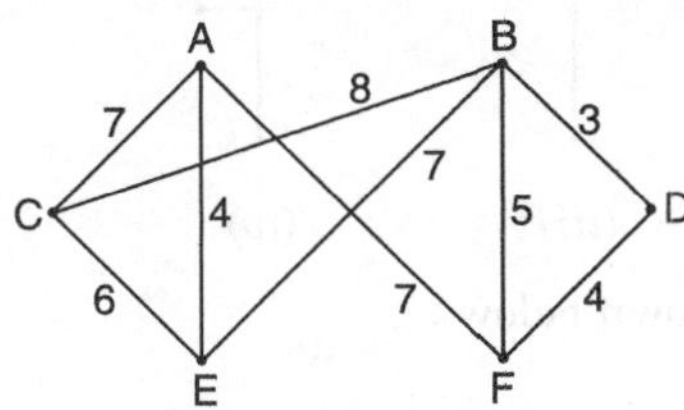

5. Find all spanning trees of the graph shown below :

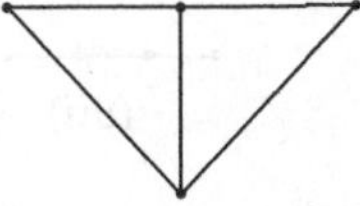

6. Find all spanning trees of the graph shown in the following figure.

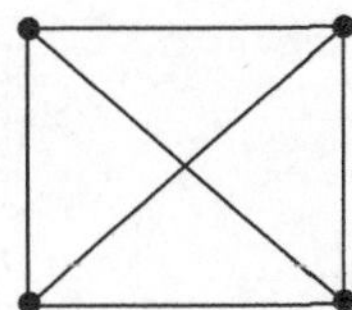

7. Find the minimal spanning tree of the following graph

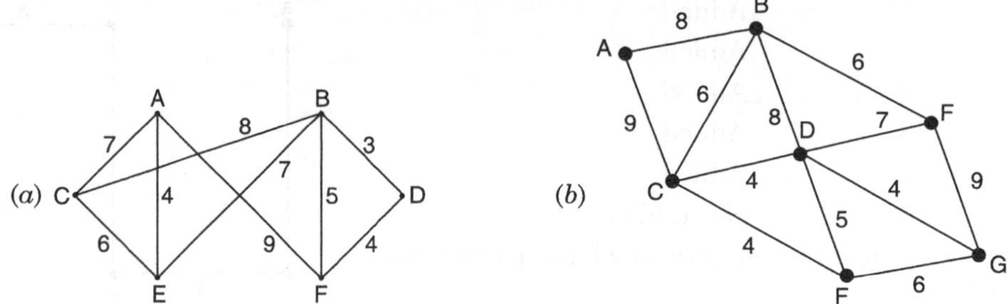

(c) Find the minimal spanning tree T for the weighted graph shown below :

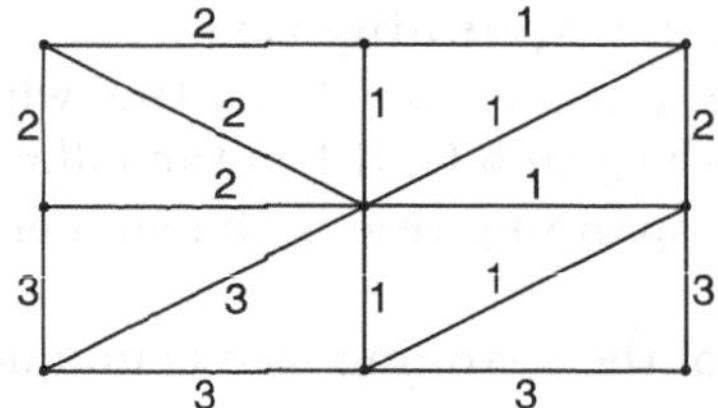

8. Show that the sum of the degrees of the vertices of a tree with n vertices is $2n - 2$.

Answers

1. There are six such trees shown below :

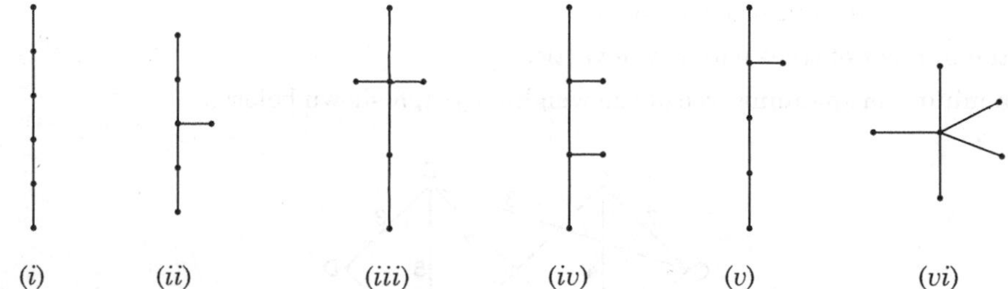

2. There are eight such trees shown below :

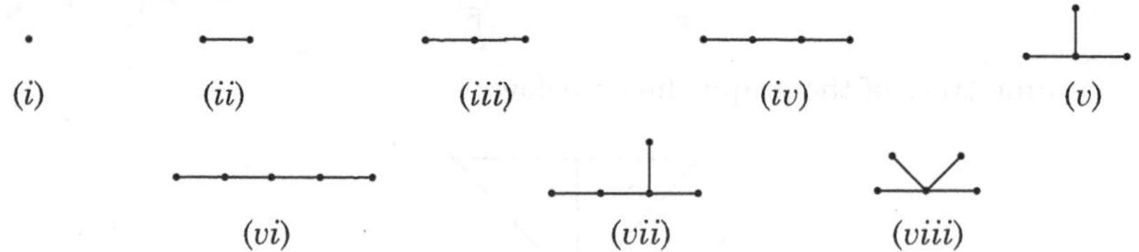

3. 15

4.

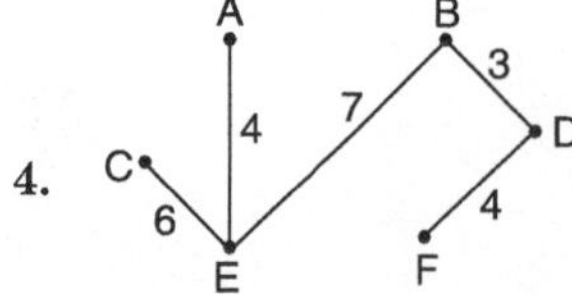

5. There are eight such spanning trees shown below :

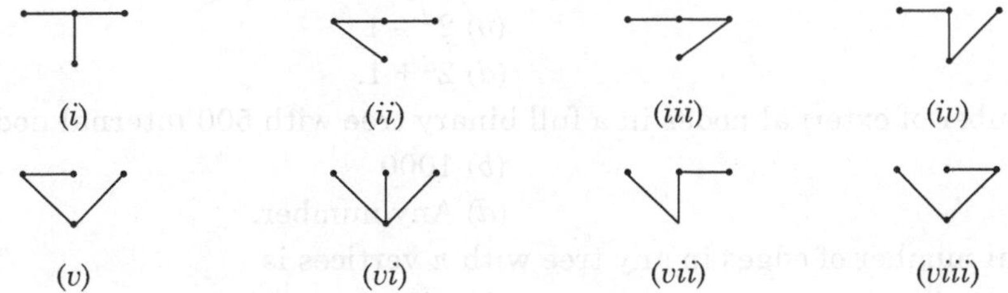

6. There are twelve such spanning trees shown below.

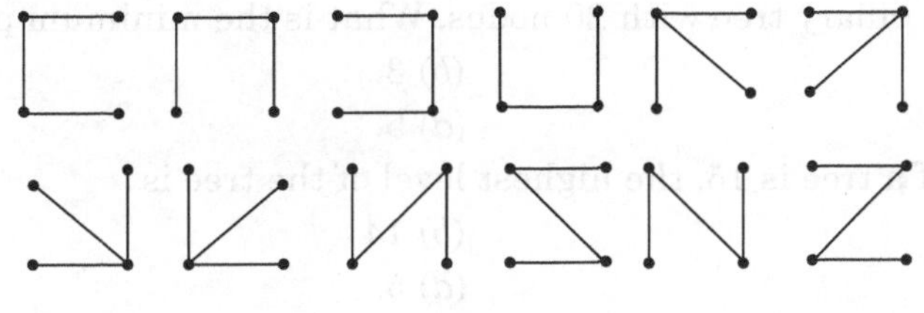

7.

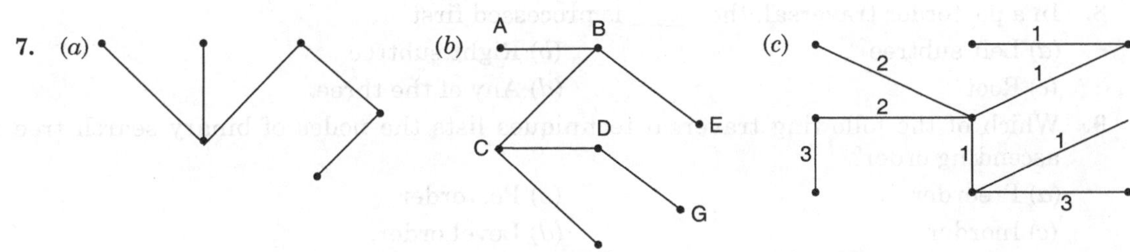

MULTIPLE CHOICE QUESTIONS (MCQs)

1. For the given binary tree, the preorder traversal is

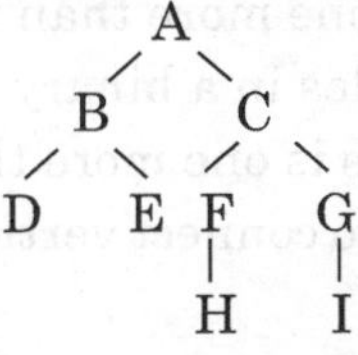

(a) A B D E C F H G I (b) A B C D E F G H I
(c) D E B A C F G H I (d) D B E A H F C I G.

2. Let T be a full binary tree with I internal nodes. Then which of the following statements is TRUE?

(a) T has $2i + 1$ total nodes and $i + 1$ terminal nodes.

(b) T has $2i$ total nodes and $i + 1$ terminal nodes.

(c) T has $2i + 2$ total nodes and $i + 2$ terminal nodes.

(d) T has $2i^2$ total nodes and i^2 terminal nodes.

3. The maximum number of nodes in a binary tree of depth d is
   (*a*) $2d - 1$ (*b*) $2^d - 1$
   (*c*) $2d + 1$ (*d*) $2^d + 1$.
4. The number of external nodes in a full binary tree with 500 internal nodes are
   (*a*) 501 (*b*) 1000
   (*c*) 500 (*d*) Any number.
5. The total number of edges in any tree with n vertices is
   (*a*) $n(n - 1)/2$ (*b*) $n/2$
   (*c*) n (*d*) $n - 1$.
6. Suppose T is a binary tree with 20 nodes. What is the minimum possible depth of T?
   (*a*) 1 (*b*) 3
   (*c*) 4 (*d*) 5.
7. If the height of a tree is 15, the highest level of the tree is
   (*a*) 15 (*b*) 14
   (*c*) 3 (*d*) 5.
8. In a postorder traversal, the _____ is processed first
   (*a*) Left subtree (*b*) Right subtree
   (*c*) Root (*d*) Any of the three.
9. Which of the following traversal techniques lists the nodes of binary search tree in ascending order?
   (*a*) Preorder (*b*) Postorder
   (*c*) Inorder (*d*) Level order.
10. Which traversal of the BST would print result in original order of input?
   (*a*) Preorder (*b*) Postorder
   (*c*) Inorder (*d*) Level order.

Answers and Explanations

1. (*a*) Apply the algorithm and check.
2. (*a*) External nodes in a tree are one more than internal nodes.
3. (*b*) The maximum number of nodes in a binary tree at depth d is one less than 2^d.
4. (*a*) The number of external nodes is one more than number of internal nodes.
5. (*d*) The number of edges needed to connect vertices of a tree is one less than the number of vertices.
6. (*d*) Draw the tree and see.
7. (*b*) The level of the tree is one less than its height.
8. (*a*) First the left subtree, then right subtree and then root.
9. (*c*) Inorder traversal of BST gives nodes in ascending order.
10. (*c*) Preorder traversal of BST gives nodes in original order.

13* PROPOSITIONAL CALCULUS

13.1. BASIC LOGIC OPERATIONS

The following are the main logic operations in this chapter :

(*i*) $p \wedge q$, called as " *conjunction of p and q*" and read as "*p* and *q*".

(*ii*) $p \vee q$, called as "disjunction of *p* and *q*" and read as "*p* or *q*".

(*iii*) $\sim p$, called as "*negation of p*" and read as "not *p*".

(*iv*) T, read as "True"

(*v*) F, read as "False".

13.2. STATEMENT

A statement is any collection of symbols or sounds which is either true or false, but not both. The truth or falsity of a statement is called its **truth value.**

For example, consider the following :

(*a*) Delhi is in France

(*b*) Where are you going ?

(*c*) 2 + 3 = 5

The expression (*a*) is a statement and it is a false statement.

The expression (*b*) is not a statement since it is neither true nor false.

The expression (*c*) is a true statement.

13.3. PROPOSITION

(P.T.U., M.C.A. May 2007)

A proposition is a statement which is either true or false. It is a declarative sentence.

For example : The following statements are all propositions :

(*i*) Jawahar Lal Nehru is the first prime minister of India.

(*ii*) It rained yesterday.

(*iii*) If x is an integer, then x^2 is a + ve integer.

For example : The following statements are not propositions :

(*i*) Please report at 11 a.m. sharp

(*ii*) What is your name ?

(*iii*) $x^2 = 13$.

---

* Not meant for P.T.U. B.Tech., "Discrete Structure" (BTCS–402) Course.

13.4. PROPOSITIONAL VARIABLES

The lower case letters starting from p onwards are used to represent propositions *e.g.*,

p : India is in Asia

$q : 2 + 2 = 4$.

Example. *Classify the following statements as propositions or non-propositions.*

(*i*) *The population of India goes upto 120 million in year 2012.*

(*ii*) $x + y = 30$

(*iii*) *Come here*

(*iv*) *The Intel Pentium-III is a 64-bit computer.*

Sol. (*i*) Proposition (*ii*) Not a proposition

(*iii*) Not a proposition (*iv*) Proposition.

13.5. TRUTH TABLE

The truth value of a proposition depends upon the truth values of its variables. Once the truth values of the variables are known, the truth value of the proposition is also known. The table to show this relationship is known as truth table. For *e.g.*, consider the proposition $\sim (p \wedge \sim q)$. The truth table for this proposition is given in Fig. 13.1.

p	q	$\sim q$	$p \wedge \sim q$	$\sim(p \wedge \sim q)$
T	T	F	F	T
T	F	T	T	F
F	T	F	F	T
F	F	T	F	T

Fig. 13.1

ILLUSTRATIVE EXAMPLES

Example 1. *Construct the truth tables of :*

(*i*) $p \wedge (\sim q)$ (*ii*) $(p \wedge q) \vee (q \wedge r) \vee (r \wedge p)$

(*iii*) $(\sim p) \vee (\sim q)$ (*iv*) $p \vee q \vee r$.

Sol. The truth table for these propositions are as follows: (Fig. 13.2 to Fig. 13.5)

(*i*)

p	q	$\sim q$	$p \wedge (\sim q)$
T	T	F	F
T	F	T	T
F	T	F	F
F	F	T	F

Fig. 13.2

(ii)

p	q	r	$p \wedge q$	$q \wedge r$	$r \wedge p$	$(p \wedge q) \vee (q \wedge r) \vee (r \wedge p)$
T	T	T	T	T	T	T
T	T	F	T	F	F	T
T	F	F	F	F	F	F
T	F	T	F	F	T	T
F	T	T	F	T	F	T
F	F	T	F	F	F	F
F	T	F	F	F	F	F
F	F	F	F	F	F	F

Fig. 13.3

(iii)

p	q	$\sim p$	$\sim q$	$(\sim p) \vee (\sim q)$
T	T	F	F	F
T	F	F	T	T
F	T	T	F	T
F	F	T	T	T

Fig. 13.4

(iv)

p	q	r	$p \vee q \vee r$
T	T	T	T
T	T	F	T
T	F	F	T
T	F	T	T
F	T	T	T
F	F	T	T
F	T	F	T
F	F	F	F

Fig. 13.5

13.6. COMBINATION OF PROPOSITIONS

We can combine the propositions to produce new propositions. There are three fundamental and three derived connectors to combine the propositions. These are explained as follows one by one.

(*a*) **Fundamental Connectors**

1. **Conjunction.** It means ANDing of two statements. Assume p and q be two propositions. Conjunction of p and q to be a proposition which is true when both p and q are true, otherwise false. It is denoted by $p \wedge q$. (Fig. 13.6)

Truth tables are used to determine the truth or falsity of the combined proposition.

p	q	$p \wedge q$
T	T	T
T	F	F
F	T	F
F	F	F

Fig. 13.6. Truth Table of $p \wedge q$.

2. **Disjunction.** It means ORing of two statements. Assume p and q be two propositions. Disjunction of p and q to be a proposition which is true when either one or both p and q are true and is false when both p and q are false. It is denoted by $p \vee q$. (Fig. 13.7)

p	q	$p \vee q$
T	T	T
T	F	T
F	T	T
F	F	F

Fig. 13.7. Truth Table of $p \vee q$.

3. **Negation.** It means opposite of original statement. Assume p be a proposition. Negation of p to be a proposition which is true when p is false, and is false when p is true. It is denoted by $\sim p$. (Fig. 13.8)

p	$\sim p$
T	F
F	T

Fig. 13.8. Truth Table of $\sim p$.

Example 2. *Consider the following :*

p : He is rich

q : He is Generous.

Write the proposition which combines the proposition p and q using conjunction ($\wedge$), disjunction ($\vee$), and negation ($\sim$).

Sol. Conjunction. He is rich and generous *i.e.,* $p \wedge q$.

Disjunction. He is rich or generous *i.e.,* $p \vee q$.

Negation. He is not rich *i.e.,* $\sim p$

He is not generous *i.e.,* $\sim q$.

It is false that he is rich or generous *i.e.,* $\sim (p \vee q)$.

He is neither rich nor generous *i.e.,* $\sim p \wedge \sim q$.

It is false that he is not rich *i.e.,* $\sim (\sim p)$.

Example 3. *Let p be "It is hot day" and q be "The temperature is 45°C". Write in simple sentences the meaning of following :*

(*i*) $\sim p$ (*ii*) $\sim (p \vee q)$ (*iii*) $\sim (p \wedge q)$
(*iv*) $\sim (\sim p)$ (*v*) $p \vee q$ (*vi*) $p \wedge q$
(*vii*) $\sim p \wedge \sim q$ (*viii*) $\sim (\sim p \vee \sim q)$.

Sol. (*i*) It is not a hot day.
(*ii*) It is false that it is hot day or temperature is 45°C.
(*iii*) It is not true that it is hot day and temperature is 45°C.
(*iv*) It is false that it is not a hot day.
(*v*) It is hot day or temperature is 45°C.
(*vi*) It is hot day and temperature is 45°C.
(*vii*) It is neither a hot day nor temperature is 45°C.
(*viii*) It is false that it is not a hot day or temperature is not 45°C.

Example 4. *Consider the following statements :*

p : He is coward.
q : He is lazy.
r : He is rich.

Write the following compound statements in the symbolic form.
(*i*) *He is either coward or poor.*
(*ii*) *He is neither coward nor lazy.*
(*iii*) *It is false that he is coward but not lazy.*
(*iv*) *He is coward or lazy but not rich.*
(*v*) *It is false that he is coward or lazy but not rich.*
(*vi*) *It is not true that he is not rich.*
(*vii*) *He is rich or else he is both coward and lazy.*

Sol. (*i*) $p \wedge \sim r$ (*ii*) $\sim p \wedge \sim q$ (*iii*) $\sim (p \wedge \sim q)$
(*iv*) $(p \vee q) \wedge \sim r$ (*v*) $\sim((p \vee q) \wedge \sim r)$ (*vi*) $\sim (\sim r)$
(*vii*) $r \vee (p \wedge q)$.

(*b*) **Derived Connectors**

1. **NAND.** It means negation after ANDing of two statements. Assume *p* and *q* be two propositions. Nanding of *p* and *q* to be a proposition which is false when both *p* and *q* are true, otherwise true. It is denoted by $p \uparrow q$. (Fig. 13.9)

p	q	$p \uparrow q$
T	T	F
T	F	T
F	T	T
F	F	T

Fig. 13.9

2. **NOR or Joint Denial.** It means negation after ORing of two statements. Assume *p* and *q* be two propositions. NOring of *p* and *q* to be a proposition which is true when both *p* and *q* are false, otherwise false. It is denoted by $p \downarrow q$. (Fig. 13.5)

p	q	$p \downarrow q$
T	T	F
T	F	F
F	T	F
F	F	T

Fig. 13.10

3. **XOR.** Assume p and q be two propositions. XORing of p and q is true if p is true or if q is true but not both and vice versa. It is denoted by $p \oplus q$. (Fig. 13.11)

p	q	$p \oplus q$
T	T	F
T	F	T
F	T	T
F	F	F

Fig. 13.11.

Example 5. *Generate the truth table for following :*
(*i*) $A \oplus B \oplus C$ (*ii*) $A \uparrow B \uparrow C$.

Sol. The -1 truth table for above formulas are as shown in Figs. 13.12 and 13.13.

(*i*)

A	B	C	$A \oplus B$	$A \oplus B \oplus C$
T	T	T	F	T
T	T	F	F	F
T	F	T	T	F
T	F	F	T	T
F	T	T	T	F
F	T	F	T	T
F	F	T	F	T
F	F	F	F	F

Fig. 13.12

(*ii*) Truth table for (*ii*) is

A	B	C	$A \uparrow B$	$A \uparrow B \uparrow C$
T	T	T	F	T
T	T	F	F	T
T	F	T	T	F
T	F	F	T	T
F	T	T	T	F
F	T	F	T	T
F	F	T	T	F
F	F	F	T	T

Fig. 13.13

Example 6. *Prove that* $X \oplus Y \equiv (X \wedge \sim Y) \vee (\sim X \wedge Y)$.

Sol. Construct the truth table for both the propositions. (Fig. 13.14)

X	Y	$X \oplus Y$	$\sim Y$	$\sim X$	$X \wedge \sim Y$	$\sim X \wedge Y$	$(X \wedge \sim Y) \vee (\sim X \wedge Y)$
T	T	F	F	F	F	F	F
T	F	T	T	F	T	F	T
F	T	T	F	T	F	T	T
F	F	F	T	T	F	F	F

Fig. 13.14

As the truth table for both the proposition are same.

$$X \oplus Y \cong (X \wedge \sim Y) \vee (\sim X \wedge Y).$$ Hence proved.

Example 7. *Show that* $(p \oplus q) \vee (p \downarrow q)$ *is equivalent to* $p \uparrow q$.

Sol. Construct the truth table for both the propositions.

p	q	$(p \oplus q)$	$(p \downarrow q)$	$(p \oplus q) \vee (p \downarrow q)$	$p \uparrow q$
T	T	F	F	F	F
T	F	T	F	T	T
F	T	T	F	T	T
F	F	F	T	T	T

Fig. 13.15

Since, the values of $(p \oplus q) \vee (p \downarrow q)$ is same as $p \uparrow q$ as in Fig. 13.15. Hence, they are equivalent.

Example 8. *Show that* $(p \uparrow q) \oplus (p \uparrow q)$ *is equivalent to* $(p \vee q) \wedge (p \downarrow q)$.

Sol. Construct the truth table for both the propositions

p	q	$p \uparrow q$	$(p \uparrow q) \oplus (p \uparrow q)$	$p \vee q$	$p \downarrow q$	$(p \vee q) \wedge (p \downarrow q)$
T	T	F	F	T	F	F
T	F	T	F	T	F	F
F	T	T	F	T	F	F
F	F	T	F	F	T	F

Fig. 13.16

Since, the values of $(p \uparrow q) \oplus (p \uparrow q)$ and $(p \vee q) \wedge (p \downarrow q)$ are same as in Fig. 13.16. Hence, they are equivalent.

(c) Some Other Connectors

1. **Conditional.** Statements of the form "If p then q" are called conditional statements. It is denoted as $p \rightarrow q$ and read as "p implies q" or "q is necessary for p" or "p is sufficient for q".

Conditional statement is true if both p and q are true or if p is false. It is false if p is true and q is false. The proposition p is called hypothesis and the proposition q is called conclusion. The truth table of conditional statement is (Fig. 13.17)

p	q	$p \rightarrow q$
T	T	T
T	F	F
F	T	T
F	F	T

Fig. 13.17. Truth Table of $p \rightarrow q$.

For example : The followings are conditional statements :

1. If $a = b$ and $b = c$, then $a = c$.
2. If I will get money, then I will purchase computer.

13.7. (a) LAWS OF THE ALGEGBRA OF PROPOSITIONS

I. Idempotent Laws

(i) $p \vee p \equiv p$ (ii) $p \wedge p \equiv p$

II. Associative Laws

(i) $(p \vee q) \vee r \equiv p \vee (q \vee r)$ (ii) $(p \wedge q) \wedge r \equiv p \wedge (q \wedge r)$

III. Commutative Laws

(i) $p \vee q \equiv q \vee p$ (ii) $p \wedge q \equiv q \wedge p$

IV. Distributive Laws

(i) $p \vee (q \wedge r) \equiv (p \vee q) \wedge (p \vee r)$ (ii) $p \wedge (q \vee r) \equiv (p \wedge q) \vee (p \wedge r)$

V. Identity Laws

(i) $p \vee f \equiv p$ (ii) $p \wedge t \equiv p$

(iii) $p \vee t \equiv t$ (iv) $p \wedge f \equiv f$

where t denotes the true value and f denotes the false value.

VI. Complement Laws

(i) $p \vee \sim p \equiv t$ (ii) $p \wedge \sim p \equiv f$

(iii) $\sim \sim p \equiv p$ (iv) $\sim t \equiv f, \sim f \equiv t$

VII. De Morgan's Laws

(i) $\sim p(\vee q) \equiv \sim p \wedge \sim q$ (ii) $\sim (p \wedge q) \equiv \sim p \vee \sim q$

Example 9. *Show by using laws of algebra of propositions, the logical equivalence of*

$$(p \vee q) \wedge \sim p \equiv \sim p \wedge q.$$

Sol. $(p \vee q) \wedge \sim p \equiv \sim p \wedge (p \vee q)$ | Commutative law

$\equiv (\sim p \wedge p) \vee (\sim p \wedge q)$ | Distributive law

$\equiv f \vee (\sim p \wedge q)$ | Complement law

$\equiv \sim p \wedge q$ | Identity law

Example 10. *Show that* $p \vee (p \wedge q) \equiv p$ *by using laws of algebra of propositions t*

Sol. $p \vee (p \wedge q) \equiv (p \wedge t) \vee (p \wedge q)$ | Identity law

$\equiv p \wedge (t \vee q)$ | Distributive law

$\equiv p \wedge t$ | Identity law

$\equiv p$ | Identity law

Example 11. *Use the laws of algebra of logic propositions, show that*

$$\sim (p \vee q) \vee (\sim p \wedge q) \equiv \sim p.$$

Sol. $\sim (p \vee q) \vee (\sim p \wedge q) \equiv (\sim p \wedge \sim q) \vee (\sim p \wedge q)$ | De Morgan's law

$\equiv \sim p \wedge (\sim q \vee q)$ | Distributive law

$\equiv \sim p \wedge t$ | Complement law

$\equiv \sim p$ | Identity law.

13.7. (b) VARIATIONS IN CONDITIONAL STATEMENT

Contrapositive. The proposition $\sim q \to \sim p$ is called contrapositive of $p \to q$.

Converse. The proposition $q \to p$ is called the converse of $p \to q$.

Inverse. The proposition $\sim p \to \sim q$ is called the inverse of $p \to q$.

Example 12. *Show that $p \to q$ and its contrapositive $\sim q \to \sim p$ are logically equivalent.*

Sol. Construct truth table for both the propositions. (as in Fig. 13.18)

p	q	$\sim p$	$\sim q$	$p \to q$	$\sim q \to \sim p$
T	T	F	F	T	T
T	F	F	T	F	F
F	T	T	F	T	T
F	F	T	T	T	T

Fig. 13.18

As, the values in both cases are same, hence both propositions are equivalent.

Example 13. *Show that proposition $q \to p$ and $\sim p \to \sim q$ is not equivalent to $p \to q$.*

Sol. Construct truth table for all the above propositions :

p	q	$\sim p$	$\sim q$	$p \to q$	$q \to p$	$\sim p \to \sim q$
T	T	F	F	T	T	T
T	F	F	T	F	T	T
F	T	T	F	T	F	F
F	F	T	T	T	T	T

Fig. 13.19

As the values of $p \to q$ in table is not equal to $q \to p$ and $\sim p \to \sim q$ as in Fig. 13.19. So both of them are not equal to $p \to q$ but they are themselves logically equivalent.

Example 14. *Prove that the following propositions are equivalent to $p \to q$.*

(*i*) $\sim (p \wedge \sim q)$ (*ii*) $\sim p \vee q$ (*iii*) $\sim q \to \sim p$.

Sol. Construct the truth table for all the above propositions :

p	q	$\sim p$	$\sim q$	$\sim p \vee q$	$\sim q \to \sim p$	$(p \wedge \sim q)$	$\sim (p \wedge \sim q)$	$p \to q$
T	T	F	F	T	T	F	T	T
T	F	F	T	F	F	T	F	F
F	T	T	F	T	T	F	T	T
F	F	T	T	T	T	F	T	T

Fig. 13.20

In the above table (Fig. 13.20) the values of $p \to q$ is equivalent to (*i*), (*ii*) and (*iii*), hence they are equivalent to $p \to q$. Hence proved.

2. **Biconditional.** Statements of the form "if and only if" are called biconditional statements.

It is denoted as $p \leftrightarrow q$ and read as "p if and only if q". The proposition $p \leftrightarrow q$ is true if p and q have the same truth values and is false if p and q do not have the same truth values. Fig. 13.21. The name of biconditional comes from the fact that $p \leftrightarrow q$ is equivalent to $(p \to q) \wedge (q \to p)$.

The truth table of $p \leftrightarrow q$ is

p	q	$p \leftrightarrow q$
T	T	T
T	F	F
F	T	F
F	F	T

Fig. 13.21. Truth Table of $p \leftrightarrow q$.

For example : (*i*) Two lines are parallel if and only if they have same slope.

(*ii*) You will pass the exam if and only if you will work hard.

Example 15. *Prove that $p \leftrightarrow q$ is equivalent to $(p \to q) \wedge (q \to p)$.*

Sol. Construct the truth tables of both propositions :

p	q	$p \leftrightarrow q$
T	T	T
T	F	F
F	T	F
F	F	T

Fig. 13.22

p	q	$p \to q$	$q \to p$	$(p \to q) \wedge (q \to p)$
T	T	T	T	T
T	F	F	T	F
F	T	T	F	F
F	F	T	T	T

Fig. 13.23

Since, the truth tables are same, hence they are logically equivalent. (Fig. 13.22 and Fig. 13.23). Hence proved.

13.8. PRINCIPLE OF DUALITY

Two formulas A_1 and A_2 are said to be duals of each other if either one can be obtained from the other by replacing $\wedge$ (AND) by $\vee$ (OR) and $\vee$ (OR) by $\wedge$ (AND). Also if the formula contains T (True) or F (False), then we replace T by F and F by T to obtain the dual.

Note 1. The two connectives $\wedge$ and $\vee$ are called dual of each other.

2. Like AND and OR, $\uparrow$ (NAND) and $\downarrow$ (NOR) are dual of each other.

3. If any formula of proposition is valid, then its dual is also a valid formula.

Example 16. *Determine the dual of each of the following :*

(*a*) $p \wedge (q \wedge r)$ (*b*) $\sim p \vee \sim q$ (*c*) $(p \wedge \sim q) \vee (\sim p \wedge q)$

(*d*) $(p \uparrow q) \uparrow (p \uparrow q)$ (*e*) $((\sim p \vee q) \wedge (q \wedge \sim s)) \vee (p \vee F)$, *here F means false.*

Sol. To obtain the dual of all the above formulas, replace $\wedge$ by $\vee$ and $\vee$ by $\wedge$, and also replace T by F and F by T.

Also replace $\uparrow$ by $\downarrow$ and vice versa.

(*a*) $p \wedge (q \wedge r) = p \vee (q \vee r)$ (*b*) $\sim p \vee \sim q = \sim p \wedge \sim q$

(*c*) $(p \wedge \sim q) \vee (\sim p \wedge q) = (p \vee \sim q) \wedge (\sim p \vee q)$

(*d*) $(p \downarrow q) \downarrow (p \downarrow q)$

(*e*) $((\sim p \vee q) \wedge (q \wedge \sim s)) \vee (p \vee \mathrm{F}) = ((\sim p \wedge q) \vee (q \vee \sim s)) \wedge (p \wedge \mathrm{T})$

13.9. LOGICAL IMPLICATION

A proposition P(*p*, *q*, ...) is said to *logically imply* a proposition Q (*p*, *q*, ...), written as P(*p*, *q*, ...) ⇒ Q (*p*, *q*, ...) if Q(*p*, *q*, ...) is true whenever P(*p*, *q*, ...) is true.

Example 17. *Show that* $p \leftrightarrow q$ *logically implies* $p \leftrightarrow q$.

Sol. Consider the truth tables of $p \leftrightarrow q$ and $p \rightarrow q$ as in the following table. Fig. 13.24 $p \rightarrow q$ is true whenever $p \leftrightarrow q$ is true. Hence $p \leftrightarrow q$ logically implies $p \rightarrow q$

p	q	$p \leftrightarrow q$	$p \rightarrow q$
T	T	T	T
T	F	F	F
F	T	F	T
F	F	T	T

Fig. 13.24

Example 18. *Show that p logically implies* $p \vee q$.

Sol. Consider the truth tables for p and $p \vee q$ as in Fig. 13.25 Now $p \vee q$ is true whenever p is true. Hence p logically implies $p \vee q$.

p	q	$p \vee q$
T	T	T
T	F	T
F	T	T
F	F	F

Fig. 13.25

Example 19. *Show that* $p \wedge q$ *logically implies* $p \leftrightarrow q$.

Sol. Consider the truth tables of $p \wedge q$ and $p \leftrightarrow q$ as in Fig. 13.26. Now $p \leftrightarrow q$ is true whenever $p \wedge q$ is true. Thus $p \wedge q$ logically implies $p \leftrightarrow q$.

p	q	$p \wedge q$	$p \leftrightarrow q$
T	T	T	T
T	F	F	F
F	T	F	F
F	F	F	T

Fig. 13.26

Exmaple 20. *Show that* $p \leftrightarrow \sim q$ *does not logically imply* $p \rightarrow q$.

Sol. Construct the truth tables of $p \leftrightarrow \sim q$ and $p \rightarrow q$ as in Fig. 13.27. Recall that $p \leftrightarrow \sim q$ logically implies $p \rightarrow q$ if $p \rightarrow q$ is true whenever $p \leftrightarrow \sim q$ is true. But $p \leftrightarrow \sim q$ is true when $p \rightarrow q$ is false. Hence $p \leftrightarrow \sim q$ does not logically imply $p \rightarrow q$.

p	q	$\sim q$	$p \leftrightarrow \sim q$	$p \rightarrow q$
T	T	F	F	T
T	F	T	T	F
F	T	F	T	T
F	F	T	F	T

Fig. 13.27

13.10. LOGICALLY EQUIVALENCE OF PROPOSITIONS

Two propositions are said to be logically equivalent if they have exactly the same truth values under all circumstances. The table 1 contains the fundamental logical equivalent expressions :

Table 1.

1. **De Morgan's Laws**
   $\sim (p \wedge q) \equiv \sim p \vee \sim q$
   $\sim (p \vee q) \equiv \sim p \wedge \sim q$
2. **Commutative Properties**
   $p \vee q \equiv q \vee p\ ;\ \ p \wedge q \equiv q \wedge p.$
3. **Associative Properties**
   $(p \vee q) \vee r \equiv p \vee (q \vee r)$
   $(p \wedge q) \wedge r \equiv p \wedge (q \wedge r)$
4. **Distributive Properties**
   $p \wedge (q \vee r) \equiv (p \wedge q) \vee (p \wedge r)$
   $p \vee (q \wedge r) \equiv (p \vee q) \wedge (p \vee r)$
5. **Idempotent Laws**
   $p \vee p \equiv p$ and $p \wedge p \equiv p$
6. **Complement Properties**
   $p \equiv \sim \sim p$
7. **Transposition**
   $(p \to q) \equiv (\sim q \to \sim p)$
8. **Material Implication**
   $(p \to q) \equiv (\sim p \vee q)$
9. **Material Equivalence**
   $(p \leftrightarrow q) \equiv [(p \to q) \wedge (q \to p)]$
   $(p \leftrightarrow q) \equiv [(p \to q) \vee (\sim p \wedge \sim q)]$
10. **Exportation**
    $[(p \wedge q) \to r] \equiv [p \to (q \to r)]$

Example 21. *Consider the following propositions*

$$\sim p \vee \sim q \text{ and } \sim (p \wedge q).$$

Are they equivalent ?

Sol. Construct the truth table for both (as in Fig. 13.28).

p	q	$\sim p$	$\sim q$	$\sim p \vee \sim q$	$p \wedge q$	$\sim (p \wedge q)$
T	T	F	F	F	T	F
T	F	F	T	T	F	T
F	T	T	F	T	F	T
F	F	T	T	T	F	T

Fig. 13.28

Since, the final values of both the propositions are same, hence the two propositions are equivalent.

Example 22. *Show that the propositions* $\sim(p \wedge q)$ *and* $\sim p \vee \sim q$ *are logically equivalent.*

Sol. Construct the truth tables of $\sim (p \wedge q)$ and $\sim p \vee \sim q$ as in Fig. 13.29. Since the truth tables are the same, *i.e.*, both propositions are false in the first case and true in the other three cases, the propositions $\sim (p \wedge q)$ and $\sim p \vee \sim q$ are logically equivalent and we can write

$$\sim (p \wedge q) \equiv \sim p \vee \sim q$$

p	q	$p \wedge q$	$\sim (p \wedge q)$
T	T	T	F
T	F	F	T
F	T	F	T
F	F	F	T

(*a*)

p	q	$\sim p$	$\sim q$	$\sim p \vee \sim q$
T	T	F	F	F
T	F	F	T	T
F	T	T	F	T
F	F	T	T	T

(*b*)

Fig. 13.29

Example 23. *Prove the associative law:* $(p \wedge q) \wedge r \equiv p \wedge (q \wedge r)$.

Sol. Construct the required truth tables as in Fig 13.30. Since the truth tables are identical, the propositions are equivalent.

p	q	r	$p \wedge q$	$(p \wedge q) \wedge r$	$q \wedge r$	$p \wedge (q \wedge r)$
T	T	T	T	T	T	T
T	T	F	T	F	F	F
T	F	T	F	F	F	F
T	F	F	F	F	F	F
F	T	T	F	F	T	F
F	T	F	F	F	F	F
F	F	T	F	F	F	F
F	F	F	F	F	F	F

Fig. 13.30

Example 24. *Prove that disjunction distributes over conjunction; that is, prove the distributive law* $p \vee (q \wedge r) \equiv (p \vee q) \wedge (p \vee r)$.

Sol. Construct the required truth tables as in Fig. 13.31. Since the truth tables are identical, the propositions are equivalent.

p	q	r	$q \wedge r$	$p \vee (q \wedge r)$	$p \vee q$	$p \vee r$	$(p \vee q) \wedge (p \vee r)$
T	T	T	T	T	T	T	T
T	T	F	F	T	T	T	T
T	F	T	F	T	T	T	T
T	F	F	F	T	T	T	T
F	T	T	T	T	T	T	T
F	T	F	F	F	T	F	F
F	F	T	F	F	F	T	F
F	F	F	F	F	F	F	F

Fig. 13.31

13.11. TAUTOLOGIES

(P.T.U., M.C.A. May 2007)

A proposition P is a tautology if it is true under all circumstances. It means it contains only T in the final column of its truth table.

Example 25. *Prove that the statement* $(p \to q) \leftrightarrow (\sim q \to \sim p)$ *is a tautology.*

Sol. Make the truth table of above statement :

p	q	$p \to q$	$\sim q$	$\sim p$	$\sim q \to \sim p$	$(p \to q) \leftrightarrow (\sim q \to \sim p)$
T	T	T	F	F	T	T
T	F	F	T	F	F	T
F	T	T	F	T	T	T
F	F	T	T	T	T	T

Fig. 13.32

As the final column contains all T's, so it is a tautology. (Fig. 13.32)

Example 26. *Prove that $p \vee p \leftrightarrow p$ is a tautology.*

Sol. Construct the truth table of the given statement :

p	$p \vee p$	$p \vee p \leftrightarrow p$
T	T	T
F	F	T

Fig. 13.33

As the last column contains all T's, so it is a tautology. (Fig. 13.33)

13.12. CONTRADICTION

(P.T.U., M.C.A. May 2007)

A statement that is always false is called a contradiction.

Example 27. *Show that the statement $p \wedge \sim p$ is a contradiction.*

p	$\sim p$	$p \wedge \sim p$
T	F	F
F	T	F

Fig. 13.34

Sol. Construct the truth table of above statement.

Since, the last column contains all F's, so it is a contradiction. (Fig. 13.34)

13.13. CONTINGENCY

A statement that can be either true or false depending on the truth values of its variables, is called a contingency.

p	q	$p \to q$	$p \wedge q$	$(p \to q) \to (p \wedge q)$
T	T	T	T	T
T	F	F	F	T
F	T	T	F	F
F	F	T	F	F

Fig. 13.35

Example 28. *Prove that the statement $(p \to q) \to (p \wedge q)$ is a contingency.*

Sol. Construct the truth table of above statement.

As, the value of final column depends on the truth value of the variables, so it is a contingency. (Fig. 13.35)

Example 29. *From the following formulae, find out tautology, contingency and contradiction.*

(i) $A \equiv A \wedge (A \vee B)$ *(ii)* $(p \wedge \sim q) \vee (\sim p \wedge q)$ *(iii)* $\sim (p \vee q) \vee (\sim p \vee \sim q)$.

Sol. (*i*) Construct the truth table for A → A ∧ (A ∨ B).

A	B	$A \vee B$	$A \wedge (A \vee B)$	$A \rightarrow A \wedge (A \vee B)$
T	T	T	T	T
T	F	T	T	T
F	T	T	F	T
F	F	F	F	T

Fig. 13.36

Since, the last column of the table contains all T's, hence it is a tautology. (Fig. 13.36)

(*ii*) Construct the truth table for $(p \wedge \sim q) \vee (\sim p \wedge q)$ as in Fig. 13.37.

p	q	$\sim p$	$\sim q$	$p \wedge \sim q$	$\sim p \wedge q$	$(p \wedge \sim q) \vee (\sim p \wedge q)$
T	T	F	F	F	F	F
T	F	F	T	T	F	T
F	T	T	F	F	T	T
F	F	T	T	F	F	F

Fig. 13.37

Since, the value of the final column depends on the value of the different variables, hence it is a contingency.

(*iii*) Construct the truth table of the proposition $\sim (p \vee q) \vee (\sim p \vee \sim q)$ as in Fig. 13.38.

p	q	$\sim p$	$\sim q$	$p \wedge q$	$\sim (p \wedge q)$	$\sim p \vee \sim q$	$\sim (p \vee q) \vee (\sim p \vee \sim q)$
T	T	F	F	T	F	F	F
T	F	F	T	F	T	T	T
F	T	T	F	F	T	T	T
F	F	T	T	F	T	T	T

Fig. 13.38

Since, the value of final column depends upon the value of different variables, hence it is a contingency.

Example 30. *Verify that proposition $p \vee \sim (p \wedge q)$ is tautology.*

Sol. Construct the truth table for above proposition. (Fig. 13.39)

p	q	$p \wedge q$	$\sim(p \wedge q)$	$p \vee \sim (p \wedge q)$
T	T	T	F	T
T	F	F	T	T
F	T	F	T	T
F	F	F	T	T

Fig. 13.39

Since, the last column contains all T' s, hence it is a tautology.

Example 31. *Determine whether the following is a tautology, contingency and a contradiction :*

(*i*) $p \to (p \to q)$ (*ii*) $p \to (q \to p)$ (*iii*) $p \wedge \sim p$.

Sol. (*i*) Construct truth table for $p \to (p \to q)$ as in Fig. 13.40.

p	q	$p \to q$	$p \to (p \to q)$
T	T	T	T
T	F	F	F
F	T	T	T
F	F	T	T

Fig. 13.40

Since, the value of last column depends on the value of different variables, hence it is a contingency.

(*ii*) Construct truth table for $p \to (q \to p)$ as in Fig. 13.41.

p	q	$q \to p$	$p \to (q \to p)$
T	T	T	T
T	F	T	T
F	T	F	T
F	F	T	T

Fig. 13.41

Since, the last column contains all T's, hence it is a tautology.

(*iii*) Construct truth table for $p \wedge \sim p$ as in Fig. 13.42.

Since, the last column contain all F's, hence it is a contradiction.

p	$\sim p$	$p \wedge \sim p$
T	F	F
F	T	F

Fig. 13.42

13.14. FUNCTIONALLY COMPLETE SETS OF CONNECTIVES

We have three basic and two conditional connectives *i.e.*, $\wedge$, $\vee$, $\sim \Rightarrow$ and $\Leftrightarrow$. If we have given any formula containing all these connectives, we can write an equivalent formula with certain proper subsets of these connectives.

A set of connectives is called functionally complete if every formula can be expresses on terms of an equivalent formula containing the connectives from this set.

Example 32. *Write an equivalent formula for* $P \wedge (R \Leftrightarrow S) \vee (S \Leftrightarrow P)$ *which does not involve biconditional.*

Sol. We know that $P \Leftrightarrow Q \equiv (P \Rightarrow Q) \wedge (Q \Rightarrow P)$ ...(*i*)

So, apply eqn. (*i*) to formula to obtain the formula equivalent to

$$[P \wedge (R \Leftrightarrow S) \vee (S \Leftrightarrow P)],$$

which does not involve biconditional.

$$= [P \wedge ((R \Rightarrow S) \wedge (S \Rightarrow R)) \vee ((S \Rightarrow P) \wedge (P \Rightarrow S))].$$

Example 33. *Write an equivalent formula for $R \vee (S \Leftrightarrow T)$, which does not involve biconditional as well as conditional.*

Sol. We know that

$$(P \Leftrightarrow Q) = (P \Rightarrow Q) \wedge (Q \Rightarrow P) \quad ...(i)$$

$$(P \Rightarrow Q) = \sim P \wedge Q \quad ...(ii)$$

So, applying the eqn. (*i*) and (*ii*) on the above formula, we can obtain an equivalent formula, which does not involve biconditional as well as conditional.

$$[R \vee (S \Leftrightarrow T)] = [R \vee ((S \Rightarrow T) \vee (T \Rightarrow S))]$$
$$= [R \vee ((\sim S \vee T) \vee (\sim T \vee S))].$$

Example 34. *Show that $\{\sim, \wedge\}$ is functionally complete.*

Sol. Take any formula which involve all the five connectives $\wedge, \vee, \sim, \Rightarrow$ and $\Leftrightarrow$. We can obtain an equivalence formula by first replacing biconditional and then replacing conditional and finally we can replace $\vee$.

As $\quad P \Leftrightarrow Q = (P \Rightarrow Q) \wedge (Q \Rightarrow P)$

$$= (\sim P \vee Q) \wedge (\sim Q \vee P) = (\sim (\sim\sim P \wedge \sim Q)) \wedge (\sim (\sim \sim Q \wedge \sim P))$$

Hence, $\{\sim, \wedge\}$ is functionally complete.

Similarly, we can show that $(\sim, \vee\}$ is functionally complete.

Example 35. *Show that $\{\sim, \rightarrow\}$ is functionally complete.*

Sol. We know that $\quad P \Rightarrow Q \cong \sim P \vee Q.$

So, we have $\quad P \vee Q \cong \sim P \Rightarrow Q.$

Since, $\{\sim, \vee\}$ is functionally complete. Hence from above, $\{\sim, \Rightarrow\}$ is also functionally complete.

i.e., given any formula which involve all the five connectives, we can obtain an equivalence formula using $\{\sim, \rightarrow\}$ by first replacing biconditional and then replacing ($\wedge$) ANDing and finally replacing $\vee$.

Example 36. *Express $P \Leftrightarrow Q$ in terms of $\{\sim, \wedge\}$ only.*

Sol. $(P \Leftrightarrow Q) = (P \Rightarrow Q) \wedge (Q \Rightarrow P) \qquad \because (P \Leftrightarrow Q) = (P \Rightarrow Q) \wedge (Q \Rightarrow P)$

$= (\sim P \vee Q) \wedge (\sim Q \vee P) \qquad \because P \Rightarrow Q = \sim P \vee Q$

$= (\sim (\sim \sim P \wedge \sim Q)) \wedge (\sim (\sim \sim Q \wedge \sim P)) \qquad \because P \vee Q = \sim (\sim P \wedge \sim Q).$

Example 37. *Express $(P \wedge \sim Q) \vee (\sim P \wedge \sim Q)$ in terms of $(\sim, \vee)$ only.*

Sol. $(P \wedge \sim Q) \vee (\sim P \wedge \sim Q) = (\sim (\sim P \vee \sim \sim Q)) \vee (\sim (\sim \sim P \vee \sim \sim Q)) \because P \wedge Q = \sim (\sim P \vee \sim Q)$

13.15. ARGUMENT

An argument is an assertion ; that a group of propositions called premises, yields another proposition, called the conclusion. Let $P_1, P_2, P_3, ..., P_n$ is the group of propositions that yields the conclusion Q. Then, it is denoted as $P_1, P_2, P_n, ..., P_n \mid - Q$.

Conclusion. The conclusion of an argument is the proposition that is asserted on the basis of other proposition of the argument.

Premises. The propositions, which are assumed for accepting the conclusion, are called the premises of that argument.

(*a*) Valid Argument

An argument is called valid argument if the conclusion is true whenever all the premises are true.

The argument is also valid if and only if the ANDing of the group of propositions implies conclusion is a tautology *i.e.*, $P(p_1, p_2, p_3,..., p_n) \rightarrow Q$ is a tautology. Where $P(p_1, p_2, p_3,..., p_n)$ is the group of propositions and Q is the conclusion.

Some common valid argument forms are given in Table 2.

(*b*) Falacy Argument

An argument is called falacy or an invalid argument if it is not a valid argument.

Elementary Valid Argument Forms

Table 2.

1. Modus ponens

$p \rightarrow q$

p

$\therefore \quad q$

2. Modus tollens

$p \rightarrow q$

$\sim q$

$\therefore \quad \sim p$

3. Hypothetical syllogism

$p \rightarrow q$

$q \rightarrow r$

$\therefore \quad p \rightarrow r$

4. Disjunctive syllogism

$p \vee q$

$\sim p$

$\therefore \quad q$

5. Constructive dilemma

$(p \rightarrow q) \wedge (r \rightarrow s)$

$p \vee r$

$\therefore \quad q \vee s$

6. Absorption

$p \rightarrow q$

$\therefore \quad p \rightarrow (p \wedge q)$

7. Simplification

$p \wedge q$

$\therefore \quad p$

8. Conjunction

p

q

$\therefore \quad p \wedge q$

9. Addition

p

$\therefore \quad p \vee q.$

Example 38. *Show that the following rule is valid :*

$$p \mid\!- p \vee q \quad or \quad p \quad \therefore \quad p \vee q.$$

Sol. We can prove this rule from the truth table (Fig. 13.43)

p	q	$p \vee q$
T	T	T
T	F	T
F	T	T
F	F	F

Fig. 13.43

p is true in line 1 and 2 and $p \vee q$ is also true in line 1 and 2. Hence, argument is valid.

Example 39. *Show that the rule modus ponens is valid.*

$p \to q$

p

$\therefore \quad q$

Sol. The truth table of this rule is as follows : (Fig. 13.44)

p	q	$p \to q$
T	T	T
T	F	F
F	T	T
F	F	T

Fig. 13.44

The p is the true in line 1 and 2 and $p \to q$ and p both are true in line 1 and $p, p \to q$ and q all are true in line 1. Hence, argument is valid.

Example 40. *Show that the rule of hypothetical syllogism is valid*

$p \to q$

$q \to r$

$\therefore \quad p \to r.$

Sol. The truth table of above rule is as in Fig. 13.45.

p	q	r	$p \to q$	$q \to r$	$p \to r$
T	T	T	T	T	T
T	T	F	T	F	F
T	F	F	F	T	F
T	F	T	F	T	T
F	T	T	T	T	T
F	T	F	T	F	T
F	F	T	T	T	T
F	F	F	T	T	T

Fig. 13.45

$p \to q$ is true in lines 1, 2, 5, 6, 7, 8.

$q \to r$ is true in lines 1, 3, 4, 5, 7, 8.

Both $p \to q$ and $q \to r$ is true in lines 1, 5, 7, 8 $p \to r$ is also true in lines 1, 5, 7, 8. Hence, argument is valid.

Example 41. *Show that the rule of Modus Tollens is valid*

$p \to q$

$\sim q$

$\sim p.$

Sol. The truth table of above propositions are (Fig. 13.46)

p	q	$\sim p$	$\sim q$	$p \rightarrow q$
T	T	F	F	T
T	F	F	T	F
F	T	T	F	T
F	F	T	T	T

Fig. 13.46

$p \rightarrow q$ is true in line 1, 3 and 4. $\sim q$ is true in line 2 and 4. Both $p \rightarrow q$ and $\sim q$ are true in line 4. $\sim p$ is also true in line 4. Hence, the argument is valid.

Example 42. *Show that the rule of disjunctive syllogism is valid*

$$p \vee q$$
$$\sim p$$
$$\therefore \quad q.$$

Sol. The truth table of above rule is as follows : (Fig. 13.47)

p	q	$\sim p$	$p \vee q$
T	T	F	T
T	F	F	T
F	T	T	T
F	F	T	F

Fig. 13.47.

$p \vee q$ is true in line 1, 2 and 3. $\sim p$ is true in line 3. Both $p \vee q$ and $\sim p$ is true in line 3. As q is also true in line 3. Hence, argument is valid.

Example 43. *Show that the rule of simplification is valid*

$$p \wedge q$$
$$\therefore \quad p.$$

Sol. The truth table of above argument is as follows : (Fig. 13.48)

p	q	$p \wedge q$
T	T	T
T	F	F
F	T	F
F	F	F

Fig. 13.48

$p \wedge q$ is true in line 1 and p is also true in line 1. Hence, the argument is valid.

Example 44. *Show that the rule of conjunction is valid.*

$$p$$
$$q$$
$$\therefore \quad p \wedge q.$$

Sol. The argument is valid if $p \wedge q \rightarrow p \wedge q$ is a tautology. The truth table for the above proposition is as follows : (Fig. 13.49)

p	q	$p \wedge q$	$p \wedge q \rightarrow p \wedge q$
T	T	T	T
T	F	F	T
F	T	F	T
F	F	F	T

Fig. 13.49

As the proposition is a tautology. Hence, the argument is valid.

Example 45. *Show that the rule of absorption is valid*

$$p \rightarrow q$$
$$\therefore \quad p \rightarrow (p \wedge q).$$

Sol. We have to show that $(p \rightarrow q) \rightarrow [p \rightarrow (p \wedge q)]$ is tautology. The truth table of the above argument is as follows : (Fig. 13.50)

p	q	$p \wedge q$	$p \rightarrow q$	$p \rightarrow (p \wedge q)$	$(p \rightarrow q) \rightarrow [p \rightarrow (p \wedge q)]$
T	T	T	T	T	T
T	F	F	F	F	T
F	T	F	T	T	T
F	F	F	T	T	T

Fig. 13.50

Since, the argument is a tautology. Hence, it is a valid argument.

13.16. PROOF OF VALIDITY

We can test the validity of any argument by constructing the truth tables. But as the no. of variable statements increases, the truth tables grow unwieldly. So, a more efficient method to test the validity of the argument is to deduce its conclusion from its premises by a sequence of elementary arguments each of which is known to be valid.

Example 46. *Prove that the argument* $p \rightarrow \sim q,\ r \rightarrow q,\ r \vdash \sim p$ *is valid without using truth tables.*

Sol. (*i*) $p \rightarrow \sim q$ (Given)

(*ii*) $r \rightarrow q$ (Given)

(*iii*) $\sim q \rightarrow \sim r$ Contrapositive of (*ii*)

(*iv*) $p \rightarrow \sim r$ Hypothetical syllogism using (*i*) and (*iii*)

(*v*) $r \rightarrow \sim p$ Contrapositive of (*iv*)

(*vi*)	r is true	(Given)
(*vii*)	$\sim p$ is true	Modus ponens using (*v*) and (*vi*).

Example 47. *Prove that the argument* $p \to q, q \to r, r \to s, \sim s, p \vee t \vdash t$ *is valid without using truth tables.*

Sol. (*i*)	$p \to q$	(Given)
(*ii*)	$q \to r$	(Given)
(*iii*)	$r \to s$	(Given)
(*iv*)	$\sim s$	(Given)
(*v*)	$p \vee t$	(Given)
(*vi*)	$p \to r$	Hypothetical syllogism using (*i*) and (*ii*)
(*vii*)	$p \to s$	Hypothetical syllogism using (*vi*) and (*iii*)
(*viii*)	$\sim p$	Modus tollens using (*vii*) and (*iv*)
(*ix*)	t	Disjunctive syllogism using (*v*) and (*viii*).

Example 48. *Prove that the argument* $p, q \vdash (p \vee r) \wedge q$ *is valid without using truth tables.*

Sol. (*i*)	p	(Given)
(*ii*)	$p \vee r$	Rule of addition using (*i*)
(*iii*)	q	(Given)
(*iv*)	$(p \vee r) \wedge q$	Rule of conjunction using (*ii*) and (*iii*).

Example 49. *Prove that the argument* $p \to q, p \wedge r \vdash q$ *is valid without using truth table.*

Sol. (*i*)	$p \to q$	(Given)
(*ii*)	$p \wedge r$	(Given)
(*iii*)	p	Rule of simplification using (*i*)
(*iv*)	q	Modus ponens using (*i*) and (*iii*).

Example 50. *Prove that the argument* $(p \to q) \wedge (r \to s), (p \vee r) \wedge (q \vee r) \vdash q \vee s$ is valid

Sol. (*i*)	$(p \to q) \wedge (r \to s)$	(Given)
(*ii*)	$(p \vee r) \wedge (q \vee r)$	(Given)
(*iii*)	$(p \vee r)$	Simplification using (*iii*)
(*iv*)	$q \vee s$	Constructive dilemma using (*i*) and (*iii*).

Example 51. *Prove that the argument* $(p \wedge q) \vee (r \to s), t \to r, \sim (p \wedge q) \vdash t \to s$ *is valid without using truth tables.*

Sol. (*i*)	$(p \wedge q) \vee (r \to s)$	(Given)
(*ii*)	$t \to r$	(Given)
(*iii*)	$\sim (p \wedge q)$	(Given)
(*iv*)	$r \to s$	Disjunctive syllogism using (*i*) and (*iii*)
(*v*)	$t \to s$	Hypothetical syllogism using (*ii*) and (*iii*).

Example 52. *Prove that the argument* $(p \to q) \wedge (r \to s), q \to s, (q \to s) \to (p \vee r) \vdash q \vee s$ *is valid using deduction method.*

Sol. (*i*)	$(p \to q) \wedge (r \to s)$	(Given)
(*ii*)	$q \to s$	(Given)
(*iii*)	$(q \to s) \to (p \vee r)$	(Given)
(*iv*)	$p \vee r$	Modus ponens using (*iii*) and (*ii*)
(*v*)	$q \vee s$	Constructive dilemma using (*i*) and (*iv*).

Example 53. *Prove that the argument* $p \to (q \vee r), (s \wedge t) \to q, (q \vee r) \to (s \wedge t) \vdash p \to q$ *is valid without using truth table.*

Sol. (*i*)	$p \to (q \vee r)$	(Given)
(*ii*)	$(s \wedge t) \to q$	(Given)
(*iii*)	$(q \vee r) \to (s \wedge t)$	(Given)
(*iv*)	$p \to (s \wedge t)$	Hypothetical syllogism using (*i*) and (*iii*)
(*v*)	$p \to q$	Hypothetical syllogism using (*ii*) and (*iv*).

Example 54. *Prove that the argument* $p \vee (q \to p), \sim p \wedge r \vdash \sim q$ *is valid without using truth tables.*

Sol. (*i*)	$p \vee (q \to p)$	(Given)
(*ii*)	$\sim p \wedge r$	(Given)
(*iii*)	$\sim p$	Rule of simplification using (*ii*)
(*iv*)	$q \to p$	Disjunctive syllogism using (*i*) and (*iii*)
(*v*)	$\sim q$	Modus tollens using (*iv*) and (*iii*).

Example 55. *Test the validity of following argument. If I will select in IAS examination, then I will not be able to go to London. Since, I am going to London, I will not select in IAS examination.*

Sol. Let p be "I will select in IAS examination" and q be "I am going to London". Then the above argument can be written in symbolic form as follows :

$$p \to \sim q$$
$$q$$
$$\therefore \quad \sim p$$

Construct the truth table for above argument (Fig. 13.51)

p	q	$\sim p$	$\sim q$	$p \to \sim q$
T	T	F	F	F
T	F	F	T	T
F	T	T	F	T
F	F	T	T	T

Fig. 13.51

$p \to \sim q$ is true in line 2, 3 and 4. q is true in line 1 and 4 and $\sim p$ is true in line 3 and 4. Hence, all three are true in line 4. So it is a valid statement.

Example 56. *Consider the following argument and determine whether it is valid.*

Either I will get good marks or I will not graduate. If I did not graduate I will go to Canada. I get good marks. Thus, I would not go to Canada.

Sol. Let p be "I will get good marks" and q be "I will graduate" and r be "I will go to Canada". Thus, the above argument can be written in symbolic form as follows :

$$p \vee \sim q$$
$$\sim q \rightarrow r$$
$$p$$
$$\therefore \quad \sim r$$

The truth table of above proposition is (Fig. 13.52)

p	q	r	$\sim q$	$\sim r$	$p \vee \sim q$	$\sim q \rightarrow r$
T	T	T	F	F	T	T
T	T	F	F	T	T	T
T	F	F	T	T	T	F
T	F	T	T	F	T	T
F	T	T	F	F	F	T
F	T	F	F	T	F	T
F	F	T	T	F	T	T
F	F	F	T	T	T	F

Fig. 13.52

$p \vee \sim q$ is true in line 1, 2, 3, 4, 7, 8 and $\sim q \rightarrow r$ is true in line 1, 2, 4, 5, 6, 7 and p is true in line 1, 2, 3 and 4. $\sim r$ is true in 2, 3, 6 and 8. All the above are true in line 2. Hence, the argument is valid.

Example 57. *Determine the validity of the following argument without using truth tables.*

Either I will pass the examination, or, I will not graduate. If I do not graduate, I will go to Canada. I failed : Thus, I will go to Canada.

Sol. Let p be "I will pass the examination" and q be "I will graduate" and t be "I will go to Canada". Thus the above argument, in symbolic form can be written as

$$p \vee \sim q$$
$$\sim q \rightarrow t$$
$$\sim p$$
$$\therefore \quad t$$

Thus to prove the validity of the argument, use the standard results as follows :

(*i*)	$p \vee \sim q$	(Given)
(*ii*)	$\sim q \rightarrow t$	(Given)
(*iii*)	$\sim p$	(Given)
(*iv*)	$\sim q$	Disjunctive syllogism using (*i*) and (*iii*)
(*v*)	t	Modus ponens using (*ii*) and (*iv*)

Hence proved.

Example 58. *Determine the validity of the following argument using deduction method.*

If I study, then I will pass examination. If I do not go to picnic, then I will study. But I failed examination. Therefore, I went to picnic.

Sol. Let p be "I study" and q be "I will pass examination" and t be "I go to picnic". Then the above argument is written in symbolic form as follows :

$$p \to q$$
$$\sim t \to p$$
$$\sim p$$
$$\therefore \quad t$$

Thus to prove the validity of the argument use the rules of inference.

(*i*) $p \to q$ (Given)

(*ii*) $\sim t \to p$ (Given)

(*iii*) $\sim p$ (Given)

(*iv*) $\sim\sim t$ Modus tollens using (*ii*) and (*iii*)

(*v*) t Complement property using (*iv*)

Hence proved.

Example 59. *Prove the validity of the following argument using truth tables as well as deduction method.*

"If the market is free then there is no inflation. If there is no inflation then there are price controls. Since there are price controls, therefore, the market is free".

Sol. Let p be "The market is free" and q be "There is inflation" and r be "There are price controls". Then the above argument can be written in symbolic form as follows :

$$p \to \sim q$$
$$\sim q \to r$$
$$r$$
$$\therefore \quad p$$

Ist Method. By using truth tables

Construct the truth table of above argument (Fig. 13.53)

p	q	r	$\sim q$	$p \to \sim q$	$\sim q \to r$
T	T	T	F	F	T
T	T	F	F	F	T
T	F	F	T	T	F
T	F	T	T	T	T
F	T	T	F	T	T
F	T	F	F	T	T
F	F	T	T	T	T
F	F	F	T	T	F

Fig. 13.53

$p \to \sim q$ is true in line 3, 4, 5, 6, 7 and 8 $\sim q \to r$ is true in line 1, 2, 4, 5, 6, 7 r is true in line 1, 4, 5, 7. All the above three are true in line 4 and 5. Also p is true in line 4. Hence the argument is valid.

IInd Method. Using deduction method

(*i*)	$p \to \sim q$	(Given)
(*ii*)	$\sim q \to r$	(Given)
(*iii*)	$p \to r$	Hypothetical syllogism using (*i*) and (*ii*)
(*iv*)	$\sim p \to \sim r$	Transposition using (*iii*)
(*v*)	r	(Given)
(*vi*)	$\sim \sim p$	Modus tollens using (*iv*) and (*v*)
(*vii*)	p	Complement of (*vi*).

13.17. QUANTIFIERS

The following notation and theorem involving the universal quantifier ∀ and the existential quantifier ∃ will be used throughout this section:

$(\forall\ x \in A)\ p\ (x)$ or $\forall\ x, p(x)$, for every $x \in A$, $p(x)$ is true.

$(\exists\ x \in A)\ p\ (x)$ or $\exists\ x, p(x)$, there exists $x \in A$ such that $p(x)$ is true.

Here $p(x)$ is a propositional function (or open-sentence or condition) on A, that is, $p\ (a)$ is true or false for every a in A.

Theroem (De Morgan): Let $p(x)$ be a propositional function on A. Then

(*i*) $\sim(\forall\ x \in A)\ p\ (x) \equiv (\exists\ x \in A) \sim p(x)$

(*ii*) $\sim(\exists\ x \in A)\ p\ (x) \equiv (\forall\ x \in A) \sim p(x)$

Remark: The above says, in words, that the following two statements are equivalent:

(*i*) "It is not true that, for every $a \in A$, $p\ (a)$ is true".

(*ii*) "There exists an $a \in A$ such that $p(a)$ is false".

and similarly, the following two statements are equivalent:

(*i*) "It is not true that there exists an $a \in A$ such that $p(a)$ is true".

(*ii*) "For all $a \in A$, $p(a)$ is false".

13.18. EXISTENTIAL QUANTIFIER

If $p(x)$ is a proposition over the universe U. Then it is denoted as $\exists x\ p(x)$ and read as "There exists at least one value in the universe of variable x such that $p(x)$ is true. The quantifier ∃ is called the existential quantifier.

There are several ways to write a proposition, with an existential quantifier *i.e.*,

$(\exists x \in A)\ p(x)$ or $\exists x \in A$ such that $p(x)$ or $(\exists x)\ p(x)$ or $p(x)$ is true for some $x \in A$.

13.19. UNIVERSAL QUANTIFIER

If $p(x)$ is a proposition over the universe U. Then it is denoted as $\forall\ x, p(x)$ and read as "For every $x \in U$, $p(x)$ is true". The quantifier ∀ is called the universal quantifier.

There are several ways to write a proposition, with a universal quantifier.

$\forall x \in A, p(x)$ or $p(x), \forall x \in A$

or $\forall x, p(x)$ or $p(x)$ is true for all $x \in A$.

Example 60. *Let A(x) : x has a white colour, B(x) : x is a polar bear, C(x) : x is found in cold regions, over the universe of animals. Translate the following into simple sentences :*

(*i*) $\exists x\ (B(x) \wedge \sim A(x))$ (*ii*) $(\exists x)\ (\sim C(x))$

(*iii*) $(\forall x)\ (B(x) \wedge C(x) \to A(x))$.

Sol. (*i*) There exists a polar bear whose colour is not white.

(*ii*) There exists an animal that is not found in cold regions.

(*iii*) Every polar bear that is found in cold regions has a white colour.

Example 61. *Let K(x) : x is a two-wheeler, L(x) : x is a scooter. M(x) : x is manufactured by Bajaj. Express the following using quantifiers.*

(*i*) *Every two wheeler is a scooter.*

(*ii*) *There is a two wheeler that is not manufactured by Bajaj.*

(*iii*) *There is a two wheeler manufactured by Bajaj that is not a scooter.*

(*iv*) *Every two wheeler that is a scooter is manufactured by Bajaj.*

Sol. (*i*) $(\forall x)\ (K(x) \to L(x))$

(*ii*) $(\exists x)\ (K(x) \wedge M(x))$

(*iii*) $(\exists x)\ (K(x) \wedge M(x) \to \sim L(x))$

(*iv*) $(\forall x)\ (K(x) \wedge L(x) \to M(x))$

Example 62. *Determine the truth value of each of the following statements. (Here **R** is the universal set.)*

(*i*) $\forall x, \mid x \mid = x$ (*ii*) $\exists x, x^2 = x$

(*iii*) $\forall x, + 1 > x$ (*iv*) $\exists x, x + 2 = x.$

Sol. (*i*) False. Note that if $x_0 = -3$ then $\mid x_0 \mid \neq x_0$.

(*ii*) True. For if $x_0 = 1$ then $x_0{}^2 = x_0$.

(*iii*) True. For every real number is a solution to $x + 1 > x$.

(*iv*) False. There is no solution to $x + 2 = x$.

Example 63. Negate each of the statements in equation

Sol. (*i*) $\sim \forall x, \mid x \mid = x \equiv \exists x \sim (\mid x \mid = x) \equiv \exists x, \mid x \mid \neq x$

(*ii*) $\sim\exists x, x^2 = x \equiv \forall x \sim(x^2 = x) \equiv \forall x, x^2 \neq x$

(*iii*) $\sim \forall x, x + 1 > x = \exists x \sim (x + 1 > x) = \exists x, x + 1 \leq x$

(*iv*) $\sim \exists x, x + 2 = x \equiv \forall x \sim(x + 2 = x) \equiv \forall x, x + 2 \neq x$

Example 64. (*a*) *Let A = {1, 2, 3, 4 , 5}. Determine the truth value of each of the following statements:*

(*i*) $(\exists x \in A)\ (x + 3 = 10)$ (*ii*) $(\forall x \in A)\ (x + 3 < 10)$

(*iii*) $(\exists x \in A)\ (x + 3 < 5)$ (*iv*) $(\forall x \in A)\ (x + 3 \leq 7)$

(*b*) *Negate each of the statements.*

Sol. (*a*) (*i*) False. For no number in A is a solution to $x + 3 = 10$.

(*ii*)True. For every number in A satisfies $x + 3 < 10$.

(*iii*)True. For if $x_0 = 1$, then $x_0 + 3 < 5$, *i.e.*, 1 is a solution.

(*iv*)False. For if $x_0 = 5$, then $x_0 + 3 \nleq 7$. In other words, 5 is not a solution to the given condition.

(*b*) (*i*) $\sim (\exists x \in A)\ (x + 3 = 10) \equiv (\forall x \in A) \sim (x + 3 = 10) \equiv (\forall x \in A)\ (x + 3 \neq 10)$

(*ii*) $\sim (\forall x \in A)\ (x + 3 < 10) \equiv (\exists x \in A) \sim (x + 3 < 10) \equiv (\exists x \in A)\ (x + 3 \geq 10)$

(*iii*) $\sim (\exists x \in A)\ (x + 3 < 5) \equiv (\forall x \in A) \sim (x + 3 < 5) \equiv (\forall x \in A)\ (x + 3 \geq 5)$

(*iv*) $\sim(\forall x \in A)\ (x + 3 \leq 7) \equiv (\exists x \in A) \sim (x + 3 \leq 7) \equiv (\exists x \in A)\ (x + 3 > 7)$

Example 65. *Determine the truth value of each of the following statements (where* **R** *is the universal set):*

(*i*) $\forall x, x^2 = x$; (*ii*) $\exists x, 2x = x$; (*iii*) $\forall x, x - 3 < x$;

(*iv*) $\exists x, x^2 - 2x + 5 = 0$

Sol. (*i*) False, for $x = 2$ does not satisfy $x^2 = x$.

(*ii*) True, for $x = 0$ satisfies $2x = x$.

(*iii*) True, since every real number satisfies $x - 3 < x$.

(*iv*) False, since $x^2 - 2x + 5 = 0$ has no real roots.

Example 66. *Let A = {1, 2, 3, 4} be the universal set. Determine the truth value of each statement:*

(*i*) $\forall x, x + 3 < 6$; (*ii*) $\exists x, x + 3 < 6$; (*iii*) $\exists x, 2x^2 + x = 15$

Sol. (*i*) False, since $x = 4$ does not satisfy $x + 3 < 6$.

(*ii*) True, since $x = 1$ satisfies $x + 3 < 6$.

(*iii*) False, since no number in A satisfies $2x^2 + x = 15$.

13.20. NEGATION OF QUANTIFIED PROPOSITIONS

When we negate a quantified proposition *i.e.*, when a universally quantified proposition is negated, we obtain an existentially quantified proposition and when an existentially quantified proposition is negated, we obtain a universally quantified proposition.

The two rules for negation of quantified proposition are as follows. These are also called De Morgan's law.

(*i*) $\sim \exists\, x\, p(x) \cong \forall\, x \sim p(x)$ (*ii*) $\sim \forall\, x\, p(x) \cong \exists\, x \sim p(x)$.

Example 67. *Negate each of the following propositions :*

(*i*) *All boys can run faster than all girls.*

(*ii*) *Some girls are more intelligent than all boys.*

(*iii*) *Some students do not live in hostel.*

(*iv*) *All students pass the semester exams.*

(*v*) *Some of the students are absent and the classroom is empty.*

Sol. (*i*) Some boys can run faster than some girls.

(*ii*) All girls are more intelligent than some boys.

(*iii*) All students live in hostel.

(*iv*) Some students do not pass the semester exams.

(*v*) All students are present and the class-room is full.

Example 68. *Negate each of the following propositions :*

(*i*) $\forall x\, P(x) \wedge \exists\, y\, q(y)$ (*ii*) $\forall\, x\, p(x) \wedge \forall\, y\, q(y)$

(*iii*) $\exists\, x\, p(x) \vee \forall\, y\, q(y)$ (*iv*) $\exists\, x\, p(x) \vee \exists\, y\, q(y)$

(*v*) $(\exists\, x \in U)\, (x + 6 = 25)$ (*vi*) $(\forall x \in U)\, (x < 25)$.

Sol. (*i*)

$$\sim (\forall\, x\, p(x) \wedge \exists\, y\, q(y))$$

$$\cong \sim \forall\, x\, p(x) \vee \sim \exists\, y\, q(y) \qquad (\because \sim (p \wedge q) = \sim p \vee \sim q)$$

$$\cong \exists\, x \sim p(x) \vee \forall\, y \sim q(y)$$

(ii) $\sim (\forall\, x\, p(x) \wedge \forall\, y\, q(y))$

$\cong \sim \forall\, x\, p(x) \vee \sim \forall\, y\, q(y)$ $(\because \;\; \sim (p \wedge q) = \sim p \sim q)$

$\cong \exists\, x \sim p(x) \vee \exists\, y \sim q(y)$

(iii) $\sim (\exists\, x\, p(x) \vee \forall\, y\, q(y))$

$\cong \sim \exists\, x\, p(x) \wedge \sim \forall\, y\, q(y)$ $(\because \;\; \sim (p \vee q) = \sim p \wedge \sim q)$

$\cong \forall\, x \sim p(x) \wedge \exists\, y \sim q(y)$

(iv) $\sim (\exists\, x\, p(x) \vee \exists\, y\, q(y))$

$\cong \sim \exists\, x\, p(x) \wedge \sim \exists\, y\, q(y)$ $(\because \;\; \sim (p \vee q) = \sim p \wedge \sim q)$

$\cong \forall\, x \sim p(x) \wedge \forall\, y \sim q(y)$

(v) $\sim (\exists\, x \in \text{U})\, (x + 6) = 25$

$\cong \forall x \in \text{U} \sim (x + 6) = 25$

$\cong (\forall x \in \text{U})\, (x + 6) \neq 25$

(vi) $\sim (\forall\, x \in \text{U})\, (x < 25)$

$\cong \exists\, x \in \text{U} \sim (x < 25)$

$\cong (\exists\, x \in \text{U})\, (x \geq 25)$

13.21. PROPOSITIONS WITH MULTIPLE QUANTIFIERS

The proposition having more than one variable can be quantified with multiple quantifiers. The multiple universal quantifiers can be arranged in any order without altering the meaning of the resulting proposition. Also the multiple existential quantifiers can be arranged in any order without altering the meaning of proposition.

The proposition which contains both universal and existential quantifiers, the order of these quantifiers can't be exchanged without altering the meaning of proposition *e.g.*, the proposition $\exists\, x\, \forall\, y\, p(x, y)$ means " There exists some x such that $p(x, y)$ is true for every y".

Example 69. *Write the negation for each of the following. Determine whether the resulting statement is true or false. Assume U = R.*

(i) $\forall\, x\, \exists\, m(x^2 < m)$ *(ii)* $\exists\, m\, \forall\, x(x^2 < m)$.

Sol. *(i)* Negation of $\forall\, x\, \exists\, m(x^2 < m)$ is $\exists\, x\, \forall\, m\, (x^2 \geq m)$. The meaning of $\exists\, x\, \forall\, m(x^2 \geq m)$ is that there exists some x such that $x^2 \geq m$, for every m. The statement is true as there is some greatest x such that $x^2 \geq m$, for every m.

(ii) Negation of $\exists\, m\, \forall\, x\, (x^2 < m)$ is $\forall\, m\, \exists\, x\, (x^2 \geq m)$. The meaning of $\forall\, m\, \exists\, x\, (x^2 \geq m)$ is that for every m, there exists some x such that $(x^2 \geq m)$. The statement is true as for every m, there exists some greatest x such that $x^2 \geq m$.

Example 70. *Check the validity of following formula under given interpretation.*

(a) $\forall\, x\, \exists\, y\, P(x, y)$ *under interpretation domain = {1, 2} and* $P(1, 1) = T, P(1, 2) = F, P(2, 1) = T, P(2, 2) = T$ *where T and F refer to true and false respectively.*

(b) $\forall\, x\, P(x) \rightarrow Q(a, f(x))$ *under interpretation domain = {1, 2} and* $a = 1, f(1) = 2, f(2) = 1, P(1) = F, P(2) = T, Q(1, 1) = T, Q(1, 2) = T, Q(2, 1) = F$ *and* $Q(2, 2) = T$. *Here T and F refer to true and false respectively.*

Sol. (*a*) $\forall x \exists y P(x, y)$ means for every x, there exists some y such that $P(x, y)$ is true. Since $P(1, 1) = T$ and $P(2, 1) = T$. Hence, the formula is valid under the domain = {1, 2}.

(*b*) $\forall x P(x) \to Q(a, f(x))$ means for every x, whenever $P(x)$ is true implies $Q(a, f(x))$ is also true. So, under the domain {1, 2}, the different values of $\forall x P(x) \to Q(a, f(x))$ are as follows :

x	P(x)	Q(a, $f(x)$)
1	F	T
2	T	T

So, when $P(x)$ is true, $Q(a, f(x))$ is also true. Hence, the formula is valid.

ILLUSTRATIVE EXAMPLES

Example 1. *Translate the following statements in proposition logic :*

(*i*) *If you study, you will get good marks. If you do not study, you will enjoy. Therefore, either you will get good marks or you will enjoy.*

(*ii*) *If the catalogue is correct, then if the seeds are planted on April, flowers will bloom in July.*

(*iii*) *If John is elected class representative, then either Mary is elected treasurer or Alice is elected vice-treasurer.*

(*iv*) *Either taxes are increased or if expenditures rise, then the debt ceiling is raised.*

Sol. (*i*) Let p be "You study" and q be "You will get good marks" and r be "You will enjoy". Then the proposition logic is

$$p \to q$$
$$\sim p \to r$$
$$\therefore \quad q \vee r$$

(*ii*) Let p be "The catalogue is correct" and q be "Seeds are planted in April" and r be "The flowers bloom in July". Then, proposition logic is

$$p \to (q \to r)$$

(*iii*) Let p be "John is elected class representative" and q be "Mary is elected treasurer" and r be "Alice is elected vice-treasurer". Then, proposition logic is

$$p \to q \vee r$$

(*iv*) Let p be "Taxes are increased" and q be "Expenditures rise" and r be "The debt ceiling is raised". Then proposition logic is

$$p \vee (q \to r).$$

Example 2. *The meaning of proposition $p \to q$ is "If p then q" or "q is a necessary condition for p" or "p only if q" or "p is a sufficient condition for q".*

Write the following statements in terms of above.

(*a*) p : *New Delhi is capital of India* q : *India is in Asia.*

(*b*) $p : x^2 = 4$ $q : x = 2$

(*c*) $p : x^2 = y^2$ $q : x = y.$

(*d*) p = *he works hard* q : *he is a Gold Medalist.*

Sol. (*a*) (*i*) If New Delhi is capital of India, then India is in Asia.

(*ii*) India is in Asia is a necessary condition for New Delhi to be capital of India.

(*iii*) New Delhi is capital of India only if India is in Asia.

(*iv*) The fact that New Delhi is capital of India is a sufficient condition that India is in Asia.

(*b*) (*i*) If $x^2 = 4$, then $x = 2$

(*ii*) $x = 2$ is necessary for $x^2 = 4$

(*iii*) $x^2 = 4$ only if $x = 2$

(*iv*) $x^2 = 4$ is sufficient for $x = 2$.

(*c*) (*i*) If $x^2 = y^2$ then $x = y$

(*ii*) $x = y$ is necessary for $x^2 = y^2$

(*iii*) $x^2 = y^2$ only if $x = y$

(*iv*) $x^2 = y^2$ is sufficient for $x = y$.

(*d*) (*i*) If he works hard then he is a Gold Medalist.

(*ii*) Gold medal is necessary for hard work.

(*iii*) He works hard only if he is a Gold Medalist.

(*iv*) Hard work is sufficient condition for Gold Medalist.

Example 3. *Construct the truth table for the following statements*

(*i*) $(p \to p) \to (p \to \sim p)$ (*ii*) $(\sim q \to \sim p) \to (p \to q)$.

Sol. (*i*) The truth table for $(p \to p) \to (p \to \sim p)$ is as follows. (Fig. 13.54)

p	$\sim p$	$p \to p$	$p \to \sim p$	$(p \to p) \to (p \to \sim p)$
T	F	T	F	F
F	T	T	T	T

Fig. 13.54

(*ii*) The truth table of statement $(\sim q \to \sim p) \to (p \to q)$ is as follows. (Fig. 13.55)

p	q	$\sim p$	$\sim q$	$\sim q \to \sim p$	$p \to q$	$(\sim q \to \sim p) \to (p \to q)$
T	T	F	F	T	T	T
T	F	F	T	F	F	T
F	T	T	F	T	T	T
F	F	T	T	T	T	T

Fig. 13.55

Example 4. *Construct the truth table for following statements*

(*i*) $(p \to (q \to r)) \to ((p \to q) \to (p \to r))$

(*ii*) $p \leftrightarrow (\sim p \vee \sim q)$

(*iii*) $(p \to p) \vee (p \to \sim p)$.

Sol. (*i*) Truth table for $(p \to (q \to r)) \to ((p \to q) \to (p \to r))$ is as follows. (Fig. 13.56)

p	q	r	$q \to r$	$p \to q$	$p \to r$	$\sim(p \to (q \to r))$	$((p \to q) \to (p \to r))$	K → L
T	T	T	T	T	T	T	T	T
T	T	F	F	T	F	F	F	T
T	F	F	T	F	F	T	T	T
T	F	T	T	F	T	T	T	T
F	T	T	T	T	T	T	T	T
F	T	F	F	T	T	T	T	T
F	F	T	T	T	T	T	T	T
F	F	F	T	T	T	T	T	T

Fig. 13.56

(*ii*) Truth table for $p \leftrightarrow (\sim p \vee \sim q)$ is as following. (Fig. 13.57)

p	q	$\sim p$	$\sim q$	$\sim p \vee \sim q$	$p \leftrightarrow (\sim p \vee \sim q)$
T	T	F	F	F	F
T	F	F	T	T	T
F	T	T	F	T	F
F	F	T	T	T	F

Fig. 13.57

(*iii*) The truth table for $(p \to p) \vee (p \to \sim p)$ is as follows. (Fig. 13.58)

p	$\sim p$	$p \to p$	$p \to \sim p$	$(p \to p) \vee (p \to \sim p)$
T	F	T	F	T
F	T	T	T	T

Fig. 13.58

Example 5. *Assume the value of $p \to q$ is false. Determine the value of $(\sim p \vee \sim q) \to q$.*

Sol. Construct the truth table for both the statements and determine the value of $(\sim p \vee \sim q) \to q$ against the false values of $p \to q$. (Fig. 13.59)

p	q	$\sim p$	$\sim q$	$p \to q$	$\sim p \vee \sim q$	$(\sim p \vee \sim q) \to q$
T	T	F	F	T	F	T
T	F	F	T	F	T	F
F	T	T	F	T	T	T
F	F	T	T	T	T	F

Fig. 13.59

So, when $p \to q$ is false, the value of $(\sim p \vee \sim q) \to q$ is also false.

Example 6. *Given the value of $p \to q$ is true. Determine the value of $\sim p \vee (p \leftrightarrow q)$.*

Sol. Construct truth table for both statements. (Fig. 13.60)

p	q	$\sim p$	$p \to q$	$p \leftrightarrow q$	$\sim p \vee (p \leftrightarrow q)$
T	T	F	T	T	T
T	F	F	F	F	F
F	T	T	T	F	T
F	F	T	T	T	T

Fig. 13.60

So, when the value of $p \to q$ is true *i.e.*, in line 1, 3 and 4. The value of $\sim p \vee (p \leftrightarrow q)$ is also true.

Example 7. *Prove that the negation of conditional statement $\sim (p \to q)$ is equivalent of $p \wedge \sim q$.*

Sol. The truth table of the above propositions are as follows. (Fig. 13.61)

p	q	$\sim q$	$p \to q$	$\sim (p \to q)$	$p \wedge \sim q$
T	T	F	T	F	F
T	F	T	F	T	T
F	T	F	T	F	F
F	F	T	T	F	F

Fig. 13.61

As the values of both the propositions are same, hence $\sim (p \to q) \equiv p \wedge \sim q$.

Example 8. *Prove that the negation of biconditional statement $\sim (p \leftrightarrow q)$ is equivalent to $p \leftrightarrow \sim q$ or $\sim p \leftrightarrow q$.*

Sol. The truth table of all the above propositions is as follows. (Fig. 13.62)

p	q	$\sim p$	$\sim q$	$p \leftrightarrow q$	$\sim (p \leftrightarrow q)$	$p \leftrightarrow \sim q$	$\sim p \leftrightarrow q$
T	T	F	F	T	F	F	F
T	F	F	T	F	T	T	T
F	T	T	F	F	T	T	T
F	F	T	T	T	F	F	F

Fig. 13.62

As the values of $\sim (p \leftrightarrow q)$ and $p \leftrightarrow \sim q$ and $\sim p \leftrightarrow q$ are same, hence they all are equivalent.

Example 9. *From the following formulae find out tautology, contingency and contradiction.*

(*i*) $\sim (A \rightarrow B) \vee (\sim A \vee (A \wedge B))$ (*ii*) $(H \rightarrow (I \wedge J)) \rightarrow \sim (H \rightarrow I)$

(*iii*) $(p \leftrightarrow Q) \equiv (p \wedge Q) \vee (\sim p \wedge Q)$.

Sol. (*i*) Construct the truth table for $\sim (A \rightarrow B) \vee (\sim A \vee (A \wedge B))$ as in Fig. 13.63.

A	B	~ A	~ A ∨ (A ∧ B)	A → B	~ (A → B)	~ (A → B) ∨ (~ A ∨ (A ∧ B))
T	T	F	T	T	F	T
T	F	F	F	F	T	T
F	T	T	T	T	F	T
F	F	T	T	T	F	T

Fig. 13.63

As the last column of the table contains all T's, hence it is a tautology.

(*ii*) Construct truth table for $(H \rightarrow I \wedge J) \rightarrow \sim (H \rightarrow I)$ as in (Fig. 13.64)

H	I	J	(I ∧ J)	H → (I ∧ J)	H → I	~ (H → I)	(H → I ∧ J) → ~ (H → I)
T	T	T	T	T	T	F	F
T	T	F	F	F	T	F	T
T	F	F	F	F	F	T	T
T	F	T	F	F	F	T	T
F	T	T	T	T	T	F	F
F	F	T	F	T	T	F	F
F	T	F	F	T	T	F	F
F	F	F	F	T	T	F	F

Fig. 13.64

As the value of last column depends upon the value of the variables, hence it is a contingency.

(*iii*) Construct the truth table for $(p \leftrightarrow q) \rightarrow (p \wedge q) \vee (\sim p \wedge q)$ as in (Fig. 13.65)

p	q	$\sim p$	$p \leftrightarrow q$	$p \wedge q$	$\sim p \wedge q$	$(p \wedge q) \vee (\sim p \wedge q)$	$(p \leftrightarrow q) \rightarrow (p \wedge q) \vee (\sim p \wedge q)$
T	T	F	T	T	F	T	T
T	F	F	F	F	T	T	T
F	T	T	F	F	T	T	T
F	F	T	T	F	F	F	T

Fig. 13.65

As the last column contains all T's, hence the above formulae is a tautology.

Example 10. *Prove that following is a tautology*

$$A \vee (\overline{B \wedge C}) \equiv (A \vee \overline{B}) \vee \overline{C}.$$

Sol. Construct the truth table for $A \vee \overline{(B \wedge C)} \rightarrow (A \vee \overline{B}) \vee \overline{C}$ as in (Fig. 13.66)

A	B	C	$\overline{B}$	$\overline{C}$	$\overline{B \wedge C}$	$A \vee \overline{(B \wedge C)}$	$(A \vee \overline{B})$	$(A \vee \overline{B}) \vee \overline{C}$	$A \vee \overline{(B \wedge C)} \rightarrow (A \vee \overline{B}) \vee \overline{C}$
T	T	T	F	F	F	T	T	T	T
T	T	F	F	T	T	T	T	T	T
T	F	F	T	T	T	T	T	T	T
T	F	T	T	F	T	T	T	T	T
F	T	T	F	F	F	F	F	F	T
F	T	F	F	T	T	T	F	T	T
F	F	T	T	F	T	T	T	T	T
F	F	F	T	T	T	T	T	T	T

Fig. 13.66

As the last column of the table contains all T's, hence it is a tautology.

Example 11. *Determine whether the following are equivalent, using biconditional statement.*

(*i*) $p \leftrightarrow q \equiv (p \wedge q) \vee (\sim p \wedge \sim q)$

(*ii*) $(p \rightarrow q) \rightarrow t \equiv (p \wedge \sim q) \rightarrow t$.

Sol. To prove that the above pairs are equivalent. Prove that $(p \leftrightarrow q) \leftrightarrow ((p \wedge q) \vee (\sim p \wedge \sim q))$ and $((p \rightarrow q) \rightarrow t) \leftrightarrow ((p \wedge \sim q) \rightarrow t)$ are tautologies.

(*i*) Construct truth table for $(p \leftrightarrow q) \leftrightarrow ((p \wedge q) \vee (\sim p \wedge \sim q))$ as in (Fig. 13.67)

p	q	$\sim p$	$\sim q$	$p \leftrightarrow q$	$p \wedge q$	$\sim p \wedge \sim q$	$(p \wedge q) \vee (\sim p \wedge \sim q)$	$(p \leftrightarrow q) \leftrightarrow (p \wedge q) \vee (\sim p \wedge \sim q)$
T	T	F	F	T	T	F	T	T
T	F	F	T	F	F	F	F	T
F	T	T	F	F	F	F	F	T
F	F	T	T	T	F	T	T	T

Fig. 13.67

As the proposition $(p \leftrightarrow q) \leftrightarrow ((p \wedge q) \vee (\sim p \wedge \sim q))$ is a tautology, hence they are equivalent.

(*ii*) Construct the truth table for $((p \rightarrow q) \rightarrow t) \leftrightarrow ((p \wedge \sim q) \rightarrow t)$ as in Fig. 13.68.

p	q	t	$\sim q$	$p \rightarrow q$	$(p \rightarrow q) \rightarrow t$	$p \wedge \sim q$	$p \wedge \sim q \rightarrow t$	$((p \rightarrow q) \rightarrow t) \leftrightarrow ((p \wedge \sim q) \rightarrow t)$
T	T	T	F	T	T	F	T	T
T	T	F	F	T	F	F	T	F
T	F	F	T	F	T	T	F	F
T	F	T	T	F	T	T	T	T
F	T	T	T	T	T	F	T	T
F	T	F	F	T	F	F	T	F
F	F	T	T	T	T	F	T	T
F	F	F	F	T	F	F	T	F

Fig. 13.68

Since, the proposition is not a tautology, hence they are not equivalent.

Example 12. *Determine the dual of each of the following :*

(a) $(P \wedge Q) \vee (\sim P \vee (\sim P \vee Q))$

(b) $\sim P \wedge (\sim Q \wedge R) \vee (Q \wedge R) \vee (P \wedge R)$

(c) $(P \downarrow Q) \uparrow R$.

Sol. The dual of above propositions are as follows :

(a) $(P \wedge Q) \vee (\sim \phi \vee (\sim P \vee Q)) = (P \vee Q) \wedge (\sim P \wedge (\sim P \wedge Q))$

(b) $\sim P \wedge (\sim Q \wedge R) \vee (Q \wedge R) \vee (P \wedge R) = \sim P \vee (\sim Q \vee R) \wedge (Q \vee R) \wedge (P \vee R)$

(c) $(P \downarrow Q) \uparrow R = (P \uparrow Q) \downarrow R$.

Example 13. *Express* $P \Rightarrow Q$ *using Nand* $\uparrow$ *only.*

Sol. $P \Rightarrow Q \cong \sim P \vee Q \cong \sim (\sim \sim P \wedge \sim Q)$ $\quad \because \; P \vee Q \cong \sim (\sim P \wedge \sim Q)$

$\cong \sim (P \wedge \sim Q)$ $\quad \because \; \sim \sim P \equiv P$

$\cong P \uparrow \sim Q$ $\quad \because \; P \uparrow Q \cong \sim (P \wedge Q)$

$\cong P \uparrow Q \uparrow Q$ $\quad \because \; \sim P = P \uparrow P$.

Example 14. *Express* $P \downarrow Q$ *using* $\uparrow$ *only.*

Sol. $P \downarrow Q \cong \sim (P \vee Q) \cong (P \vee Q) \uparrow (P \vee Q)$

$\cong [(P \uparrow P) \uparrow (Q \uparrow Q)] \uparrow [(P \uparrow P) \uparrow (Q \uparrow Q)]$ $\quad \because \; P \vee Q \cong (P \uparrow Q) \uparrow (P \uparrow Q)$

Example 15. *Express the following formula using only* $\sim$ *and* $\wedge$.

$$(P \downarrow Q) \uparrow R$$

where $\downarrow$ *denotes NOR and* $\uparrow$ *denotes NAND.*

Sol. $(P \downarrow Q) \uparrow R \cong \sim (P \vee Q) \uparrow R$ $\quad \because \; P \downarrow Q \cong \sim (P \vee Q)$

$\cong \sim ((\sim P \vee Q) \wedge R)$ $\quad \because \; P \uparrow Q \cong \sim (P \wedge Q)$

$\cong \sim ((\sim \sim \sim (P \wedge \sim Q)) \wedge R)$ $\quad \because \; P \vee Q \cong \sim (\sim P \wedge \sim Q)$

$\cong \sim ((\sim (P \wedge \sim Q)) \wedge R)$ $\quad \because \; \sim \sim P = P$

Note. $\uparrow$ (NAND) and $\downarrow$ (NOR) are dual of each other.

Example 16. *Show that the connective* $\uparrow$ *(Nand) is functionally complete.*

Sol. To show that the connective $\uparrow$ is functionally complete. We have to show that the set of connectives $(\wedge, \sim)$ and $(\vee, \sim)$ can be expressed in terms of $\uparrow$ alone which is shown as follows :

$$\sim P \cong \sim P \vee \sim P \cong \sim (P \wedge P) \cong P \uparrow P$$

$$P \vee Q \cong \sim (\sim P \wedge \sim Q) \cong \sim P \uparrow \sim Q \cong (P \uparrow P) \uparrow (Q \uparrow Q)$$

and $$P \wedge Q \cong \sim (P \uparrow Q) \cong (P \uparrow Q) \uparrow (P \uparrow Q)$$

Since the above expresses $\wedge$, $\vee$ and $\sim$ in terms of $\uparrow$ alone. Hence $\uparrow$ is functionally complete.

Example 17. *Prove the validity of following arguments without using truth tables.*

1. $p \vee q, \sim p \mid\!- q$ $\qquad$ *2.* $p, p \to q, q \to r \mid\!- r$

3. $p, q, (p \wedge q) \to r \mid\!- r$ $\qquad$ *4.* $p, (p \wedge \sim q) \to \sim p \mid\!- p \to q$.

Sol. 1. *(i)* $p \vee q$ (Given)

(ii) $\sim p$ (Given)

(iii) q Disjunctive syllogism

2. (*i*)	p	(Given)
(*ii*)	$p \to q$	(Given)
(*iii*)	$q \to r$	(Given)
(*iv*)	$p \to r$	Hypothetical syllogism using (*ii*) and (*iii*)
(*v*)	r	Modus ponens using (*iv*) and (*i*)
3. (*i*)	p	(Given)
(*ii*)	q	(Given)
(*iii*)	$(p \wedge q) \to r$	(Given)
(*iv*)	$p \wedge q$	Rule of conjunction using (*i*) and (*ii*)
(*v*)	r	Modus ponens using (*iii*) and (*iv*)
4. (*i*)	p	(Given)
(*ii*)	$(p \wedge \sim q) \to \sim p$	(Given)
(*iii*)	$\sim (p \wedge \sim q)$	Modus tollens using (*ii*) and (*i*)
(*iv*)	$\sim \sim (p \to q)$	As $\sim (p \to q) \equiv (p \wedge \sim q)$
(*v*)	$p \to q$	Complement property using (*iv*).

Example 18. *Prove that the argument $(p \wedge q) \to r, p \to q \mid\!- p \to ((p \wedge q) \wedge r)$ is valid without using truth table.*

Sol. (*i*)	$(p \wedge q) \to r$	(Given)
(*ii*)	$p \to q$	(Given)
(*iii*)	$p \to (p \wedge q)$	Rule of absorption using (*ii*)
(*iv*)	$(p \wedge q) \to ((p \wedge q) \wedge r)$	Rule of absorption using (*i*)
(*v*)	$p \to ((p \wedge q) \wedge r)$	Hypothetical syllogism using (*iii*) and (*iv*).

Example 19. *Prove that the following arguments are valid without using truth tables.*

1. $(p \vee q) \to \sim r, r \vee t, p \mid\!- t$

2. $(p \wedge q) \to r, (r \to q), (r \to q) \to (q \wedge r) \mid\!- (p \wedge q) \to (q \wedge r)$

Sol. 1. (*i*)	$(p \vee q) \to \sim r$	(Given)
(*ii*)	$r \vee t$	(Given)
(*iii*)	p	(Given)
(*iv*)	$p \vee q$	Rule of addition using (*iii*)
(*v*)	$\sim r$	Modus ponens using (*i*) and (*iv*)
(*vi*)	t	Disjunctive syllogism using (*ii*) and (*v*)
2. (*i*)	$p \wedge q \to r$	(Given)
(*ii*)	$r \to q$	(Given)
(*iii*)	$(r \wedge q) \to (q \wedge r)$	(Given)
(*iv*)	$r \to (r \wedge q)$	Rule of absorption using (*ii*)
(*v*)	$(p \wedge q) \to (r \wedge q)$	Hypothetical syllogism using (*i*) and (*iv*)
(*vi*)	$(p \wedge q) \to (q \wedge r)$	Hypothetical syllogism using (*v*) and (*iii*)

Example 20. *Prove the validity of following argument using deduction system.*

(1) $A \to B$ (2) $B \to \sim C$

(3) $\sim C \wedge D$ (4) $A \to D$

$\therefore B.$

Sol. (*i*)	$A \to B$	(Given)
(*ii*)	$B \to \sim C$	(Given)
(*iii*)	$\sim C \wedge D$	(Given)
(*iv*)	$A \to D$	(Given)
(*v*)	$A \to \sim C$	Hypothetical syllogism using (*i*) and (*ii*)
(*vi*)	$\sim C$	Simplification of (*iii*)
(*vii*)	$\sim A \to C$	Transposition of (*iv*)
(*viii*)	A	Modus tollens using (*vii*) and (*vi*)
(*ix*)	B	Modus ponens using (*i*) and (*viii*)

Example 21. *Prove the validity of following by deduction method.*

(A) (*i*) $P \to Q$
(*ii*) $\sim Q \vee R$
(*iii*) $\sim (R \wedge \sim S)$
(*iv*) P
$\therefore S$

(B) (*i*) $P \vee Q$
(*ii*) $Q \to R$
(*iii*) $R \wedge S$
(*iv*) $P \to S$
(*v*) P
$\therefore S$

(C) (*i*) $(Q \to R) \wedge (S \to T)$
(*ii*) $(U \to V) \wedge (W \to X)$
(*iii*) $Q \vee U$
$\therefore R \vee V$

Sol. (A) (*i*)	$P \to Q$	(Given)
(*ii*)	$\sim Q \vee R$	(Given)
(*iii*)	$\sim (R \wedge \sim S)$	(Given)
(*iv*)	P	(Given)
(*v*)	Q	Modus ponens using (*i*) and (*iv*)
(*vi*)	$R \vee \sim Q$	Commutative property using (*ii*)
(*vii*)	$\sim R$	Rule of Modus tollens using (*vi*) and (*v*)
(*viii*)	$\sim R \vee S$	De Morgan's Law using (*iii*)
(*ix*)	S	Modus ponens using (*viii*) and (*vii*)
(B) (*i*)	$P \vee Q$	(Given)
(*ii*)	$Q \to R$	(Given)
(*iii*)	$R \wedge S$	(Given)
(*iv*)	$P \to S$	(Given)
(*v*)	P	(Given)
(*vi*)	S	Modus ponens using (*iv*) and (*v*)
(C) (*i*)	$(Q \to R) \wedge (S \to T)$	(Given)
(*ii*)	$(U \to V) \wedge (W \to X)$	(Given)
(*iii*)	$Q \vee U$	(Given)
(*iv*)	$(Q \to R)$	Simplification using (*i*)
(*v*)	$(U \to V)$	Simplification using (*ii*)
(*vi*)	$(Q \to R) \wedge (U \to V)$	Conjunction using (*iv*) and (*v*)
(*vii*)	$R \vee V$	Constructive dilemma using (*vi*) and (*iii*)

Example 22. *Prove the validity of following argument using truth tables.*

"If it rains then it will be cold. If it is cold then I shall stay at home. Since it rains therefore, I shall stay at home".

Sol. Let p be "It rains" and q be "It will be cold" and r be "I shall stay at home". Then the above argument in symbolic form is

$$p \to q$$
$$q \to r$$
$$p$$
$$\therefore \quad r$$

Construct the truth table of $(p \to q) \wedge (q \to r) \wedge p \to r$ as in (Fig. 13.69).

p	q	r	$p \to q$	$q \to r$	$(p \to q) \wedge (q \to r)$	$(p \to q) \wedge (q \to r) \wedge p$	$(p \to q) \wedge (q \to r) \wedge p \to r$
T	T	T	T	T	T	T	T
T	T	F	T	F	F	F	T
T	F	F	F	T	F	F	T
T	F	T	F	T	F	F	T
F	T	T	T	T	T	F	T
F	T	F	T	F	F	F	T
F	F	T	T	T	T	F	T
F	F	F	T	T	T	F	T

Fig. 13.69

As the last column contain all T's, hence the argument is valid.

Example 23. *Translate the following into symbolic form and test the validity of the argument.*

If 6 is even then 2 does not divide 7. Either 5 is not prime or 2 divides 7. But 5 is prime, therefore, 6 is odd.

Sol. Let p be "6 is even" and q be "2 divide 7" and r be "5 is primed". Thus, the above argument in symbolic form can be written as

$$p \to \sim q$$
$$\sim r \vee q$$
$$r$$
$$\therefore \quad \sim p$$

Construct the truth table to test the validity. (Fig. 13.70)

p	q	r	$\sim p$	$\sim q$	$\sim r$	$p \to \sim q$	$\sim r \vee q$
T	T	T	F	F	F	F	T
T	T	F	F	F	T	F	T
T	F	F	F	T	T	T	F
T	F	T	F	T	F	T	T
F	T	T	T	F	F	T	T
F	T	F	T	F	T	T	T
F	F	T	T	T	F	T	F
F	F	F	T	T	T	T	T

Fig. 13.70

$p \to \sim q$ is true in line 3, 4, 5, 6, 7 and 8, $\sim r \vee q$ is true in line 1, 2, 3, 4, 5 and r is true in line 1, 4, 5 and 7. All the three are true in line 4 and 5 and $\sim p$ is true in line 5. Hence, argument is valid.

Example 24. *Determine the negation of the following statements*

(*i*) $\forall x \forall y \forall z, p(x, y, z)$ (*ii*) $\forall x \exists_y, p(x, y)$

(*iii*) $\exists_y \forall_x \forall_z, p(x, y, z)$.

Sol. (*i*) $\sim (\forall_x \forall_y \forall_z, p(x, y, z)) \equiv \exists_x \exists_y \exists_z \sim p(x, y, z)$

(*ii*) $\sim (\forall_x \exists_y, p(x, y)) \equiv \exists_x \forall_y, \sim p(x, y)$

(*iii*) $\sim (\exists_y \forall_x \forall_z, p(x, y, z)) \equiv \forall_y \exists_x \exists_z, \sim p(x, y, z)$.

Example 25. *Determine the negation of the following statements*

(*i*) $\exists_x \forall_y (p(x) \vee q(y))$ (*ii*) $\forall_x \exists_y (p(x, y) \rightarrow q(x, y))$

(*iii*) $\forall_x \forall_y (p(x) \wedge q(y))$.

Sol. (*i*) $\sim (\exists x \forall_y, (p(x) \vee q(y)) \equiv \forall_x \exists_y \sim (p(x) \vee q(y))$

$\equiv \forall_x \exists_y (\sim p(x) \wedge \sim q(y))$

(*ii*) $\sim \forall_x \exists_y (p(x, y) \rightarrow q(x, y)) \equiv \exists_x \forall_y \sim (p(x, y) \rightarrow q(x, y))$

(*iii*) $\equiv \forall_x \forall_y (p(x) \wedge p(y)) \equiv \exists_x \exists_y \sim (p(x) \wedge p(y)) \equiv \exists_x \exists_y (\sim p(x) \vee \sim p(y))$.

Example 26. *Let $U = Q$. Use quantifiers to express the following statements*

(*i*) $\sqrt{5}$ *is not rational.*

(*ii*) *Subtraction of any two rational numbers is rational.*

Sol. (*i*) $\sim(\exists x)(x^2 = 5)$

(*ii*) $(\forall x)(\forall y)$ ($x - y$ is a rational).

TEST YOUR KNOWLEDGE

Statements, Basic Operations

1. Let p be "It is cold" and q be "It is raining". Give a verbal sentence which describes each of the following statements :

(*a*) $\sim p$ (*b*) $p \wedge q$

(*c*) $p \vee q$ (*d*) $q \vee - p$

(*e*) $\sim p \wedge \sim q$ (*f*) $\sim (\sim q)$

2. Let p be "He is tall" and q be "He is handsome". Write the following statements in symbolical form using p and q.

(*a*) He is tall and handsome. (*b*) He is tall but not handsome.

(*c*) It is false that he is short or handsome. (*d*) He is neither tall nor handsome.

(*e*) He is tall or he is short and handsome. (*f*) It is not true that he is short or not handsome.

3. Let p be "Andrey speaks French" and q be" Andrey speaks Danish". Give a simple verbal sentence which describes each of the following

(*a*) $p \vee q$ (*b*) $p \wedge q$

(*c*) $p \wedge \sim q$ (*d*) $\sim p \vee \sim q$

(*e*) $\sim (\sim p)$ (*f*) $\sim (\sim p \wedge \sim q)$ (*P.T.U. M.C.A. Dec. 2002*)

Truth Value of Compound Statement

4. Determine the truth value of each of the following statements:

(*a*) $4 + 2 = 5$ and $6 + 3 = 9$ (*b*) $3 + 2 = 5$ and $6 + 1 = 7$

(*c*) $4 + 5 = 9$ and $1 + 2 = 4$ (*d*) $3 + 2 = 5$ and $4 + 7 = 11$.

5. Determine the truth value of each of the following statements

(*a*) $1 + 1 = 5$ or $2 + 2 = 4$ (*b*) $2 + 5 = 9$ or $3 + 7 = 8$

(*c*) $1 + 1 = 5$ or $3 + 3 = 4$ (*d*) $2 + 5 = 9$ or $1 + 7 = 8$

Truth Tables

6. Find the Truth Table of

   (*a*) $\sim p \wedge q$ (*b*) $\sim (p \vee q)$

   (*c*) $\sim (p \vee \sim q)$ (*d*) $\sim (p \Leftrightarrow q) \equiv p \Leftrightarrow \sim q \equiv \sim p \Leftrightarrow q$.

   (*P.T.U., M.C.A. May 2008*)

7. Find the Truth Table for

   (*a*) $p \vee \sim q$ (*b*) $\sim p \wedge \sim q$

Tautologies and Contradictions

8. Verify that the proposition $p \vee \sim (p \wedge q)$ is a tautology.
9. (*a*) Verify that the proposition $(p \wedge q) \wedge \sim (p \vee q)$ is a contradiction.

   (*b*) Prove that the sentence "It is raining or it is not raining" is a tautology.

   (*P.T.U., M.C.A. Dec. 2006*)

   (*c*) Let *c* denotes contradiction. Then for any statement *p*,

   (*i*) $p \vee c \equiv p$ (*ii*) $p \wedge c \equiv c$. (*P.T.U., M.C.A. Dec. 2006*)

Logical Equivalence

10. Show that the propositions $\sim (p \wedge q)$ and $\sim p \vee \sim q$ are logically equivalent.
11. Show that $p \vee q \equiv \sim (\sim p \wedge \sim q)$ by using Truth Tables.

Exclusive Disjunction, Joint Denial

12. Show that $p \wedge q \equiv (p \downarrow p) \downarrow (q \downarrow q)$ by using Truth Tables.
13. (*a*) Show that $\sim p \equiv p \downarrow p$

    (*b*) $p \vee q \equiv (p \downarrow q) \downarrow (p \downarrow q)$

Negation and De Morgan's Law

14. Using De Morgan's Law. Show

    (*a*) $\sim p \vee \sim q \equiv - (p \wedge q)$ (*b*) $\sim p \wedge \sim q \equiv - (p \vee q)$

    (*c*) $\sim p \vee \sim q \equiv \sim p \wedge q$ (*d*) $\sim (\sim p \wedge q) \equiv p \vee \sim q$

15. Simplify each of the following statements.

    (*a*) It is not true that his mother is English or his father is French.

    (*b*) It is not true that he studies physics but not mathematics.

16. Write the negation of the following statements.

    (*a*) He is tall but handsome. (*b*) He has blond hair or blue eyes.

    (*c*) He is neither rich nor happy.

    (*d*) He has lost his job or he did not go to work today.

    (*e*) Paris is in France and London is in England. (*P.T.U., M.C.A. Dec. 2006*)

Conditional Statements and Variations

17. Determine the contrapositive of each of the following statements.

    (*a*) If John is a poet, then he is poor (*b*) Only if Marc studies he will pass the test.

18. Determine the contrapositive of each statement.

    (*a*) If he has courage, he will win. (*b*) It is necessary to be strong in order to be a sailor.

    (*c*) Only if he does not tire will he win.

    (*d*) It is sufficient for it to be a square in order to be a rectangle.

Biconditional

19. Determine the truth value of the following statements.

    (*a*) $2 + 2 = 4$ iff $3 + 6 = 9$ (*b*) $2 + 2 = 7$ iff $5 + 1 = 2$

    (*c*) $1 + 1 = 2$ iff $3 + 2 = 8$ (*d*) $1 + 2 = 5$ iff $3 + 1 = 4$

20. Find the truth table for $(p \rightarrow q) \vee \sim (p \leftrightarrow \sim q)$

Argument and Statements

21. Show that the following statements are valid statements

(*a*) $p \leftrightarrow q, q \vdash p$ (*b*) $p \rightarrow q, \sim q \vdash \sim p$

22. Determine the validity of the argument $\sim p \rightarrow q, p \vdash \sim q$

Logically Implication

23. Show that $p \leftrightarrow q$ logically implies $p \rightarrow q$

24. (*a*) Show that $p \wedge q$ logically implies $p \leftrightarrow q$

(*b*) Show that $p \Leftrightarrow q = (p \vee q) \rightarrow (p \wedge q)$ using (*a*) truth tables (*b*) algebra of propositions.

(*P.T.U., M.C.A. May 2008*)

Quantifiers

25. Determine the truth value of the following statements.

(*a*) $\forall x, \mid x \mid = x$ (*b*) $\exists x, x^2 = x$

(*c*) $\forall x, x + 1 > x$ (*d*) $\exists x, x + 2 = x$

(*e*) $\forall x, 2x = x$ (*f*) $\forall x, x - 3 < x$

(*g*) $\exists x, x^2 - 2x + 5 = 0$

Answers

1. (*a*) It is not cold. (*b*) It is cold and raining.
(*c*) It is cold or it is raining. (*d*) It is raining or it is not cold.
(*e*) It is not cold and it is not raining.
(*f*) It is not true that it is not raining or it is false that it is not raining.

2. (*a*) $p \wedge q$ (*b*) $p \wedge \sim q$
(*c*) $\sim (\sim p \vee q)$ (*d*) $\sim p \wedge \sim q$
(*e*) $p \vee (\sim p \wedge q)$ (*f*) $\sim (\sim p \vee \sim q)$.

3. (*a*) Andrey speaks French or Danish.
(*b*) Andrey speaks French and Danish.
(*c*) Andrey speaks French but not Danish.
(*d*) Andrey does not speak French or Andrey does not speak Danish.
(*e*) It is false that Andrey does not speak French.
(*f*) It is false that Andrey speaks neither French nor Danish.

4. The Statement "*p* and *q*" is true only when both statements are true.
(*a*) false (*b*) True (*c*) False (*d*) True.

5. The statement "*p* or *q*" is false when both statements are false.
(*a*) True (*b*) False (*c*) False (*d*) True.

6. (*a*)

p	q	$\sim p$	$\sim p \wedge q$
T	T	F	F
T	F	F	F
F	T	T	T
F	F	T	F

(*b*)

p	q	$p \vee q$	$\sim (p \vee q)$
T	T	T	F
T	F	T	F
F	T	T	F
F	F	F	T

(*c*)

p	q	$\sim q$	$p \vee \sim q$	$\sim (p \vee \sim q)$
T	T	F	T	F
T	F	T	T	F
F	T	F	F	T
F	F	T	T	F

(*d*)

p	q	$\sim(p \Leftrightarrow q)$	$p \Leftrightarrow \sim q$	$\sim p \Leftrightarrow q$
T	T	F	F	F
T	F	T	T	T
F	T	T	T	T
F	F	F	F	F

7. (*a*)

p	q	$\sim q$	$p \vee \sim q$
T	T	F	T
T	F	T	T
F	T	F	F
F	F	T	T

(*b*)

p	q	$\sim p$	$\sim q$	$\sim p \wedge \sim q$
T	T	F	F	F
T	F	F	T	F
F	T	T	F	F
F	F	T	T	T

11.

p	q	$p \vee q$	$\sim p$	$\sim q$	$\sim p \wedge \sim q$	$\sim(\sim p \wedge \sim q)$
T	T	T	F	F	F	T
T	F	T	F	T	F	T
F	T	T	T	F	F	T
F	F	F	T	T	T	F

Since the truth tables are identical, the propositions are equivalent

15. (*a*) His mother is not English and his father is not French.
(*b*) He does not study physics or he studies mathematics.

16. (*a*) He is not tall or not handsome.
(*b*) He has neither bland hair nor blue eyes.
(*c*) He is rich or happy.
(*d*) He has not lost his Job and he did go to work Today.
(*e*) Paris is not in France or London is not in England.

17. (*a*) If John is not poor, then he is not a poet.
(*b*) If Marc does not study, then he will not pass the test.

18. (*a*) If he does not win, then he does not have courage.
(*b*) If he is not strong, then he is not a sailor.
(*c*) If he tires, then he will not win.
(*d*) If it is not a rectangle, then it is not a square.

19. $p \leftrightarrow q$ is true whenever p and q have the same truth values.
(*a*) True (*b*) True (*c*) False (*d*) False.

20.

p	q	p	$\rightarrow$	q	$\vee$	$\sim$	p	$\leftrightarrow$	$\sim$	q
T	T	T	T	T	T	T	T	F	F	T
T	F	T	F	F	F	F	T	T	T	F
F	T	F	T	T	T	F	F	T	F	T
F	F	F	T	F	T	T	F	F	T	F

22. False

25. (*a*) False (*b*) True (*c*) True (*d*) False (*e*) True (*f*) True (*g*) False.

MULTIPLE CHOICE QUESTIONS (MCQs)

1. Which of the following is not a proposition?
   (*a*) India is in Asia (*b*) $2 + 2 = 4$
   (*c*) The Intel Pentium-III is a 64-bit processor
   (*d*) $X + Y = 30$.
2. Let p: He is coward and q: He is rich. Which of the following is the symbolic form of the statement? "He is neither coward nor rich".
   (*a*) $\sim (p \wedge q)$ (*b*) $\sim p \wedge \sim q$
   (*c*) $\sim p \sim q$ (*d*) $\sim (p \wedge \sim q)$.
3. Let $p \to q$ is a proposition, which of the following is false?
   (*a*) $\sim q \to \sim p$ is contra positive of $p \to q$ (*b*) $q \to p$ is converse of $p \to q$
   (*c*) $\sim p \to \sim q$ is inverse of $p \to q$ (*d*) $\sim (q \to p)$ is inverse of $p \to q$.
4. The biconditional statement $p \leftrightarrow q$ is equivalent to
   (*a*) $(p \to q) \wedge (q \to p)$ (*b*) $(p \to q) \vee (q \to p)$
   (*c*) $(p \vee q) \wedge (q \to p)$ (*d*) $(p \vee q) \vee (q \to p)$.
5. Which of the following statements is true about the formula? $(p \to q) \leftrightarrow (\sim q \to \sim p)$?
   (*a*) It is a contradiction (*b*) It is contingency
   (*c*) It is tautology (*d*) None of these.

Answers and Explanations

1. (*d*) A proposition is either true or false.
2. (*b*) The statement is negation of the original one.
3. (*d*) $\sim p \to \sim q$ is inverse of $p \to q$
4. (*a*) Biconditional statement is equal to Anding of $(p \to q)$, $(q \to p)$.
5. (*c*) Draw the truth table for the statement and then check it.

14* MATRIX ALGEBRA

14.1. (a) MATRIX

A set of mn numbers (real or complex) arranged in a rectangular array having m rows (horizontal lines) and n columns (vertical lines), the numbers being enclosed by brackets [ ] or ( ), is called an $m \times n$ **matrix** (read as "m by n matrix").

An $m \times n$ matrix is also called a matrix of order $m \times n$. Each of the mn numbers is called an **element** of the matrix.

For example, $\begin{bmatrix} 2 & -1 & 5 \\ 3 & 0 & 4 \end{bmatrix}$ is a 2×3 matrix or matrix of order 2×3. It has two rows and three columns. The numbers 2, – 1, 5, 3, 0, 4 are its elements.

An $m \times n$ matrix is usually written as $\begin{bmatrix} a_{11} & a_{12} & a_{13} & \ldots & a_{1n} \\ a_{21} & a_{22} & a_{23} & \ldots & a_{2n} \\ a_{31} & a_{32} & a_{33} & \ldots & a_{3n} \\ \ldots & \ldots & \ldots & \ldots & \ldots \\ a_{m1} & a_{m2} & a_{m3} & \ldots & a_{mn} \end{bmatrix}$

Here each element has two suffixes. The first suffix indicates the row and the second suffix indicates the column in which the element lies. Thus, a_{23} is the element lying in the second row and third column, a_{ij} is the element lying in the ith row and jth column.

For brevity, a matrix is usually denoted by a single capital letter A or B or C etc.

Thus, an $m \times n$ matrix A may be written as

$$A = [a_{ij}]_{m \times n} \quad \text{or} \quad A = [a_{ij}], \quad \text{where } i = 1, 2, 3, \ldots, m\,;\ j = 1, 2, 3, \ldots, n.$$

14.1. (b) KINDS OF MATRICES

Real Matrix. A matrix is said to be real if all its elements are real numbers.

e.g., $\begin{bmatrix} \sqrt{5} & -3 & 1 \\ 0 & -\sqrt{2} & 7 \end{bmatrix}$ is a real matrix.

Square Matrix. A matrix in which the number of rows is equal to the number of columns is called a **square matrix,** otherwise, it is said to be a **rectangular matrix.**

Thus, a matrix $A = [a_{ij}]_{m \times n}$ is a square matrix if $m = n$ and a rectangular matrix if $m \neq n$.

* Not meant for P.T.U. B.Tech., "Discrete Structure" (BTCS–402) Course.

A square matrix having n rows and n columns is called "a square matrix of order n" or "an n-rowed square matrix",

e.g.,

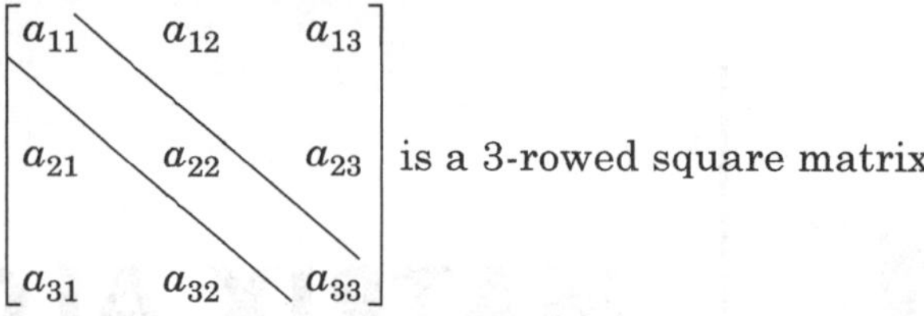

is a 3-rowed square matrix.

The elements a_{11}, a_{22}, a_{33} of a square matrix are called its **diagonal elements** and the diagonal along which these elements lie is called the **principal diagonal.**

In a square matrix $A = [a_{ij}]$,

(*i*) for elements along the principal diagonal, $i = j$

(*ii*) for elements above the principal diagonal, $i < j$

(*iii*) for elements below the principal diagonal, $i > j$

(*iv*) for non-diagonal elements, $i \neq j$

The sum of the diagonal elements of a square matrix is called its **trace.**

Thus, trace of the n rowed square matrix $A = [a_{ij}]$ is, $a_{11} + a_{22} + a_{33} + ... + a_{nn} = \sum_{i=1}^{n} a_{ij}$.

Row Matrix. A matrix having only one row and any number of columns *i.e.*, a matrix of order $1 \times n$ is called a **row matrix.** *e.g.*, $[2 \quad 5 \quad -3 \quad 0]$ is a row matrix.

Column Matrix. A matrix having only one column and any number of rows *i.e.*, a matrix of order $m \times 1$ is called a **column matrix.** *e.g.*, $\begin{bmatrix} \sqrt{2} \\ 0 \\ -1 \end{bmatrix}$ is a column matrix.

Null Matrix. A matrix in which each element is zero is called a **null matrix** or a **zero matrix.** A null matrix of order $m \times n$ is denoted by $O_{m \times n}$.

$$\begin{bmatrix} 0 & 0 \\ 0 & 0 \\ 0 & 0 \end{bmatrix} = O_{3 \times 2}, \begin{bmatrix} 0 & 0 & 0 & 0 \\ 0 & 0 & 0 & 0 \end{bmatrix} = O_{2 \times 4}$$

Sub-matrix. A matrix obtained from a given matrix A by deleting some of its rows or column or both is called a **sub-matrix of A.**

Thus, $B = \begin{bmatrix} 3 & 0 \\ 1 & 4 \end{bmatrix}$ is a sub-matrix of $A = \begin{bmatrix} 0 & -1 & 2 & 5 \\ 3 & 5 & 0 & 7 \\ 1 & 6 & 4 & -2 \end{bmatrix}$

obtained by deleting the first row, second and fourth columns of A.

Diagonal Matrix. A square matrix in which all non-diagonal elements are zero is called a **diagonal matrix.**

Thus, $A = [a_{ij}]_{n \times n}$ is a diagonal matrix if $a_{ij} = 0$ for $i \neq j$.

For example, $A = \begin{bmatrix} 2 & 0 & 0 \\ 0 & -1 & 0 \\ 0 & 0 & 0 \end{bmatrix}$ is a diagonal matrix.

An n-rowed diagonal matrix is briefly written as diag. $[d_1, d_2, \ldots\ldots, d_n]$, where $d_1, d_2, \ldots\ldots, d_n$ are the diagonal elements. Thus, the above diagonal matrix A can be written as diag. [2, – 1, 0].

Scalar Matrix. A diagonal matrix in which all the diagonal elements are equal to a scalar, say k, is called a **scalar matrix.**

Thus a scalar matrix is a square matrix in which all non-diagonal elements are zero and all diagonal elements are equal to a scalar, say k.

i.e., $A = [a_{ij}]_{n \times n}$ is a scalar matrix if $a_{ij} = \begin{cases} 0 & \text{when } i \neq j \\ k & \text{when } i = j \end{cases}$

For example, $\begin{bmatrix} 2 & 0 & 0 \\ 0 & 2 & 0 \\ 0 & 0 & 2 \end{bmatrix} \begin{bmatrix} -5 & 0 & 0 & 0 \\ 0 & -5 & 0 & 0 \\ 0 & 0 & -5 & 0 \\ 0 & 0 & 0 & -5 \end{bmatrix}$ are scalar matrices.

Unit Matrix or Identity Matrix. A scalar matrix in which each diagonal element is unity (*i.e.,* 1) is called a **unit matrix or identity matrix.**

Thus, a unit matrix is a square matrix in which all non-diagonal elements are zero and all diagonal elements are equal to 1.

i.e., $A = [a_{ij}]_{n \times n}$ is a unit matrix if $a_{ij} = \begin{cases} 0 & \text{when } i \neq j \\ 1 & \text{when } i = j \end{cases}$

A unit matrix of order n is denoted by I_n. If the order is evident, it may be simply denoted by I.

Thus, $\begin{bmatrix} 1 & 0 \\ 0 & 1 \end{bmatrix} = I_2, \begin{bmatrix} 1 & 0 & 0 \\ 0 & 1 & 0 \\ 0 & 0 & 1 \end{bmatrix} = I_3$

Upper Triangular Matrix. A square matrix in which all the elements *below* the principal diagonal are zero is called an **upper triangular matrix.**

Thus, $A = [a_{ij}]_{n \times n}$ is an upper triangular matrix if $a_{ij} = 0$ for $i > j$.

For example, $\begin{bmatrix} 2 & 3 & 4 \\ 0 & -1 & 5 \\ 0 & 0 & 3 \end{bmatrix}$ is an upper triangular matrix.

Lower Triangular Matrix. A square matrix in which all the elements *above* the principal diagonal are zero is called a **lower triangular matrix.**

Thus, $A = [a_{ij}]_{n \times n}$ is a lower triangular matrix if $a_{ij} = 0$ for $i < j$.

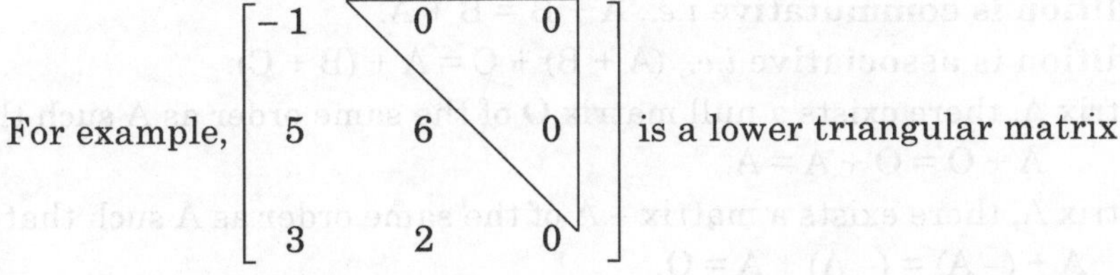

For example, $\begin{bmatrix} -1 & 0 & 0 \\ 5 & 6 & 0 \\ 3 & 2 & 0 \end{bmatrix}$ is a lower triangular matrix.

Triangular Matrix. A square matrix in which all the elements either below or above the principal diagonal are zero is called a *triangular matrix.* Thus, a triangular matrix is either upper triangular or lower triangular.

Equal Matrices. Two matrices A and B are said to be equal (written as A = B) if and only if they have the same order and their corresponding elements are equal.

Thus, if $A = [a_{ij}]_{m \times n}$ and $B = [b_{ij}]_{p \times q}$, then A = B if and only if

(*i*) $m = p$ and $n = q$ (*ii*) $a_{ij} = b_{ij}$ for all i and j.

14.2. ADDITION OF MATRICES

Two matrices are said to be conformable for addition if they have the same order.

If A and B are two matrices of the same order, then their **sum A + B** is a matrix each element of which is obtained by adding the corresponding elements of A and B.

In general, if $A = [a_{ij}]_{m \times n}$ and $B = [b_{ij}]_{m \times n}$, then $A + B = C = [c_{ij}]_{m \times n}$, where $c_{ij} = a_{ij} + b_{ij}$

Similarly, if A and B are two matrices of the same order, then their **difference A – B** is a matrix whose elements are obtained by subtracting the elements of B from the corresponding elements of A.

In general, if $A = [a_{ij}]_{m \times n}$ and $B = [b_{ij}]_{m \times n}$, then $A - B = C = [c_{ij}]_{m \times n}$, where $c_{ij} = a_{ij} - b_{ij}$

For example, if $$A = \begin{bmatrix} 2 & 5 & -1 \\ 3 & 0 & 4 \end{bmatrix} \text{ and } B = \begin{bmatrix} 1 & -6 & 2 \\ -2 & 5 & 7 \end{bmatrix}$$

then $$A + B = \begin{bmatrix} 2+1 & 5-6 & -1+2 \\ 3-2 & 0+5 & 4+7 \end{bmatrix} = \begin{bmatrix} 3 & -1 & 1 \\ 1 & 5 & 11 \end{bmatrix}$$

and $$A - B = \begin{bmatrix} 2-1 & 5-(-6) & -1-2 \\ 3-(-2) & 0-5 & 4-7 \end{bmatrix} = \begin{bmatrix} 1 & 11 & -3 \\ 5 & -5 & -3 \end{bmatrix}.$$

14.3. MULTIPLICATION OF A MATRIX BY A SCALAR

The product of a matrix $A = [a_{ij}]$ by a scalar k is denoted by kA and is obtained by multiplying every element of A by k.

Thus, $$kA = [ka_{ij}]$$

If $$A = \begin{bmatrix} a_1 & a_2 & a_3 \\ b_1 & b_2 & b_3 \end{bmatrix}, \text{ then } kA = \begin{bmatrix} ka_1 & ka_2 & ka_3 \\ kb_1 & kb_2 & kb_3 \end{bmatrix}$$

In particular, $$-A = (-1)\,A = \begin{bmatrix} -a_1 & -a_2 & -a_3 \\ -b_1 & -b_2 & -b_3 \end{bmatrix}.$$

14.4. PROPERTIES OF MATRIX ADDITION

(*i*) **Matrix addition is commutative** *i.e.*, A + B = B + A.

(*ii*) **Matrix addition is associative** *i.e.*, (A + B) + C = A + (B + C).

(*iii*) For any matrix A, there exists a null matrix O of the same order as A such that
A + O = O + A = A.

(*iv*) For any matrix A, there exists a matrix – A of the same order as A such that
A + (– A) = (– A) + A = O.

ILLUSTRATIVE EXAMPLES

Example 1. *Find x, y, z and w if* $3\begin{bmatrix} x & y \\ z & w \end{bmatrix} = \begin{bmatrix} x & 6 \\ -1 & 2w \end{bmatrix} + \begin{bmatrix} 4 & x+y \\ z+w & 3 \end{bmatrix}$.

Sol. The given equation is $\begin{bmatrix} 3x & 3y \\ 3z & 3w \end{bmatrix} = \begin{bmatrix} x+4 & 6+x+y \\ -1+z+w & 2w+3 \end{bmatrix}$

Equating the corresponding elements on the two sides, we get

$$3x = x + 4, \quad 3y = 6 + x + y, \quad 3z = -1 + z + w, \quad 3w = 2w + 3$$

$$\Rightarrow \quad 2x = 4, \quad 2y = 6 + x, \quad 2z = -1 + w, \quad w = 3$$

$$\Rightarrow \quad x = 2, \quad y = 4, \quad z = 1.$$

Example 2. *Express* $\begin{bmatrix} 2 & 5 & -7 \\ -9 & 12 & 4 \\ 15 & -13 & 6 \end{bmatrix}$ *as the sum of a lower triangular matrix and an upper triangular matrix with zero leading diagonal.*

Sol. Let $L = \begin{bmatrix} a & 0 & 0 \\ b & c & 0 \\ d & e & f \end{bmatrix}$ be the lower triangular matrix

and $U = \begin{bmatrix} 0 & p & q \\ 0 & 0 & r \\ 0 & 0 & 0 \end{bmatrix}$

be the upper triangular matrix with zero leading diagonal such that

$$\begin{bmatrix} 2 & 5 & -7 \\ -9 & 12 & 4 \\ 15 & -13 & 6 \end{bmatrix} = \begin{bmatrix} a & 0 & 0 \\ b & c & 0 \\ d & e & f \end{bmatrix} + \begin{bmatrix} 0 & p & q \\ 0 & 0 & r \\ 0 & 0 & 0 \end{bmatrix}$$

$$\Rightarrow \begin{bmatrix} 2 & 5 & -7 \\ -9 & 12 & 4 \\ 15 & -13 & 6 \end{bmatrix} = \begin{bmatrix} a+0 & 0+p & 0+q \\ b+0 & c+0 & 0+r \\ d+0 & e+0 & f+0 \end{bmatrix}$$

Equating the corresponding elements on the two sides

$$2 = a, \quad 5 = p, \quad -7 = q, \quad -9 = b, \quad 12 = c, \quad 4 = r, \quad 15 = d, \quad -13 = e, \quad 6 = f$$

$$\therefore \quad L = \begin{bmatrix} 2 & 0 & 0 \\ -9 & 12 & 0 \\ 15 & -13 & 6 \end{bmatrix} \text{ and } U = \begin{bmatrix} 0 & 5 & -7 \\ 0 & 0 & 4 \\ 0 & 0 & 0 \end{bmatrix}.$$

14.5. MATRIX MULTIPLICATION

(P.T.U., M.C.A. May 2007)

Two matrices A and B are said to be conformable for the product AB **(in this very order of A and B)** if the number of columns in A (called the pre-factor) is equal to the number of rows in B (called the post-factor).

Thus, if the orders of A and B are $m \times n$ and $p \times q$ respectively, then

(*i*) AB is defined if number of columns in A = number of rows in B, *i.e.*, if $n = p$.

(*ii*) BA is defined if number of columns in B = number of rows in A, *i.e.*, if $q = m$.

Let A = $[a_{ij}]_{m \times n}$ and B = $[b_{ij}]_{n \times p}$ be two matrices conformable for the product AB, then AB is defined as the matrix C = $[c_{ij}]_{m \times p}$ where $c_{ij} = a_{i1} b_{1j} + a_{i2}b_{2j} + \ldots\ldots + a_{in}b_{nj} = \sum_{k=1}^{n} a_{ik}b_{kj}$

i.e., (*i, j*)th element of AB = sum of the products of the elements of *i*th row of A with the corresponding elements of *j*th column of B.

The rule for multiplication of two conformable matrices is called **row-by-column method.**

Consider $A = \begin{bmatrix} a_{11} & a_{12} & a_{13} \\ a_{21} & a_{22} & a_{23} \\ a_{31} & a_{32} & a_{33} \end{bmatrix}$ and $B = \begin{bmatrix} b_{11} & b_{12} \\ b_{21} & b_{22} \\ b_{31} & b_{32} \end{bmatrix}$

Orders of A and B are 3 × 3 and 3 × 2 respectively. AB is defined and is of order 3 × 2.

Then, $AB = C = \begin{bmatrix} c_{11} & c_{12} \\ c_{21} & c_{22} \\ c_{31} & c_{32} \end{bmatrix}$

where c_{11} = Sum of products of elements of 1st row of A and 1st column of B

$$= [a_{11}\ a_{12}\ a_{13}] \begin{bmatrix} b_{11} \\ b_{21} \\ b_{31} \end{bmatrix} = a_{11}b_{11} + a_{12}b_{21} + a_{13}b_{31}$$

c_{12} = Sum of products of elements of 1st row of A and 2nd column of B

$$= [a_{11}\ a_{12}\ a_{13}] \begin{bmatrix} b_{12} \\ b_{22} \\ b_{32} \end{bmatrix} = a_{11}b_{12} + a_{12}b_{22} + a_{13}b_{32}$$

c_{21} = Sum of products of elements of 2nd row of A and 1st column of B

$$= [a_{21}\ a_{22}\ a_{23}] \begin{bmatrix} b_{11} \\ b_{21} \\ b_{31} \end{bmatrix} = a_{21}b_{11} + a_{22}b_{21} + a_{23}b_{31}$$

c_{22} = Sum of products of elements of 2nd row of A and 2nd column of B

$$= [a_{21}\ a_{22}\ a_{23}] \begin{bmatrix} b_{12} \\ b_{22} \\ b_{32} \end{bmatrix} = a_{21}b_{12} + a_{22}b_{22} + a_{23}b_{32}$$

c_{31} = Sum of products of elements of 3rd row of A and 1st column of B

$$= [a_{31}\ a_{32}\ a_{33}] \begin{bmatrix} b_{11} \\ b_{21} \\ b_{31} \end{bmatrix} = a_{31}b_{11} + a_{32}b_{21} + a_{33}b_{31}$$

c_{32} = Sum of products of elements of 3rd row of A and 2nd column of B

$$= [a_{31}\ a_{32}\ a_{33}] \begin{bmatrix} b_{12} \\ b_{22} \\ b_{32} \end{bmatrix} = a_{31}b_{12} + a_{32}b_{22} + a_{33}b_{32}$$

Thus, $$AB = \begin{bmatrix} a_{11}b_{11} + a_{12}b_{21} + a_{13}b_{31} & a_{11}b_{12} + a_{12}b_{22} + a_{13}b_{32} \\ a_{21}b_{11} + a_{22}b_{21} + a_{23}b_{31} & a_{21}b_{12} + a_{22}b_{22} + a_{23}b_{32} \\ a_{31}b_{11} + a_{32}b_{21} + a_{33}b_{31} & a_{31}b_{12} + a_{32}b_{22} + a_{33}b_{32} \end{bmatrix}$$

Note. Another useful notation to remember matrix multiplication.

$$A = \begin{bmatrix} a_{11} & a_{12} & a_{13} \\ a_{21} & a_{22} & a_{23} \\ a_{31} & a_{32} & a_{33} \end{bmatrix} = \begin{bmatrix} R_1 \\ R_2 \\ R_3 \end{bmatrix}, \quad \text{where } \begin{matrix} R_1 = \text{1st Row} \\ R_2 = \text{2nd Row} \\ R_3 = \text{3rd Row} \end{matrix}$$

$$B = \begin{bmatrix} b_{11} & b_{12} \\ b_{21} & b_{22} \\ b_{31} & b_{32} \end{bmatrix} = [C_1\ C_2], \quad \text{where } C_1 = \text{1st column, } C_2 = \text{2nd column}$$

$$AB = \begin{bmatrix} R_1 \\ R_2 \\ R_3 \end{bmatrix} [C_1\ C_2] = \begin{bmatrix} R_1C_1 & R_1C_2 \\ R_2C_1 & R_2C_2 \\ R_3C_1 & R_3C_2 \end{bmatrix}.$$

Example 3. *If* $A = \begin{bmatrix} 1 & -2 & 3 \\ 2 & 3 & -1 \\ -3 & 1 & 2 \end{bmatrix}$ *and* $B = \begin{bmatrix} 1 & 0 & 2 \\ 0 & 1 & 2 \\ 1 & 2 & 0 \end{bmatrix}$, *form the products AB and BA and show that* $AB \neq BA$.

Sol. $AB = \begin{bmatrix} 1 & -2 & 3 \\ 2 & 3 & -1 \\ -3 & 1 & 2 \end{bmatrix} \begin{bmatrix} 1 & 0 & 2 \\ 0 & 1 & 2 \\ 1 & 2 & 0 \end{bmatrix}$

$$= \begin{bmatrix} 1(1)-2(0)+3(1) & 1(0)-2(1)+3(2) & 1(2)-2(2)+3(0) \\ 2(1)+3(0)-1(1) & 2(0)+3(1)-1(2) & 2(2)+3(2)-1(0) \\ -3(1)+1(0)+2(1) & -3(0)+1(1)+2(2) & -3(2)+1(2)+2(0) \end{bmatrix} = \begin{bmatrix} 4 & 4 & -2 \\ 1 & 1 & 10 \\ -1 & 5 & -4 \end{bmatrix}$$

$$BA = \begin{bmatrix} 1 & 0 & 2 \\ 0 & 1 & 2 \\ 1 & 2 & 0 \end{bmatrix} \begin{bmatrix} 1 & -2 & 3 \\ 2 & 3 & -1 \\ -3 & 1 & 2 \end{bmatrix}$$

$$= \begin{bmatrix} 1(1)+0(2)+2(-3) & 1(-2)+0(3)+2(1) & 1(3)+0(-1)+2(2) \\ 0(1)+1(2)+2(-3) & 0(-2)+1(3)+2(1) & 0(3)+1(-1)+2(2) \\ 1(1)+2(2)+0(-3) & 1(-2)+2(3)+0(1) & 1(3)+2(-1)+0(2) \end{bmatrix} = \begin{bmatrix} -5 & 0 & 7 \\ -4 & 5 & 3 \\ 5 & 4 & 1 \end{bmatrix}$$

Orders of AB and BA are the same but their corresponding elements are not equal.

Hence $AB \neq BA$.

Example 4. *Illustrate with the help of examples that multiplication of matrices is not commutative in general, i.e.,* $AB \neq BA$. *Point out the various possibilities.*

Sol. Case I. AB is defined, but BA is not defined.

Let A be a matrix of order 2 × 3, and B be a matrix of order 3 × 4.

Then AB is defined and is a matrix of order 2 × 4, whereas BA is not defined.

Case II. AB and BA are both defined but their orders are different.

Let A be a matrix of order 2 × 3 and B be a matrix of order 3 × 2.

The product AB is defined and is a matrix of order 2 × 2.

The product BA is defined and is a matrix of order 3 × 3.

Since orders of AB and BA are different, $AB \neq BA$.

Case III. AB and BA are both defined and are matrices of the same order, yet $AB \neq BA$ (since corresponding elements are not equal).

Let $A = \begin{bmatrix} 1 & 2 \\ 3 & 4 \end{bmatrix}$ and $B = \begin{bmatrix} 2 & 1 \\ 4 & 0 \end{bmatrix}$

Here AB and BA are both defined and are of order 2×2.

$$AB = \begin{bmatrix} 1 & 2 \\ 3 & 4 \end{bmatrix}\begin{bmatrix} 2 & 1 \\ 4 & 0 \end{bmatrix} = \begin{bmatrix} 10 & 1 \\ 22 & 3 \end{bmatrix}$$

$$BA = \begin{bmatrix} 2 & 1 \\ 4 & 0 \end{bmatrix}\begin{bmatrix} 1 & 2 \\ 3 & 4 \end{bmatrix} = \begin{bmatrix} 5 & 8 \\ 4 & 8 \end{bmatrix}$$

Clearly, $AB \neq BA$.

Hence multiplication of matrices is not commutative in general, *i.e.*, in general $AB \neq BA$.

Note. If AB = BA, we say 'A and B commute'.

Example 5. *Show by means of an example that the matrix AB = O does not necessarily mean that either A = O or B = O, where O stands for the null matrix.*

Sol. Consider the matrices $A = \begin{bmatrix} 1 & 1 \\ 1 & 1 \end{bmatrix}$ and $B = \begin{bmatrix} 1 & 0 \\ -1 & 0 \end{bmatrix}$

Here $A \neq O, B \neq O$.

But $AB = \begin{bmatrix} 1 & 1 \\ 1 & 1 \end{bmatrix}\begin{bmatrix} 1 & 0 \\ -1 & 0 \end{bmatrix} = \begin{bmatrix} 1-1 & 0+0 \\ 1-1 & 0+0 \end{bmatrix} = \begin{bmatrix} 0 & 0 \\ 0 & 0 \end{bmatrix} = O.$

Example 6. *Prove by an example that AB = AC does not imply that B = C.*

Or

Give an example of matrices A, B, C such that B ≠ C, but AB = AC.

Sol. Let $A = \begin{bmatrix} 1 & 4 \\ 0 & 0 \end{bmatrix}, B = \begin{bmatrix} 1 & 5 \\ 2 & 1 \end{bmatrix}, C = \begin{bmatrix} 1 & 1 \\ 2 & 2 \end{bmatrix}$

Then $AB = \begin{bmatrix} 1 & 4 \\ 0 & 0 \end{bmatrix}\begin{bmatrix} 1 & 5 \\ 2 & 1 \end{bmatrix} = \begin{bmatrix} 1(1)+4(2) & 1(5)+4(1) \\ 0(1)+0(2) & 0(5)+0(1) \end{bmatrix} = \begin{bmatrix} 9 & 9 \\ 0 & 0 \end{bmatrix}$

$$AC = \begin{bmatrix} 1 & 4 \\ 0 & 0 \end{bmatrix}\begin{bmatrix} 1 & 1 \\ 2 & 2 \end{bmatrix} = \begin{bmatrix} 1(1)+4(2) & 1(1)+4(2) \\ 0(1)+0(2) & 0(1)+0(2) \end{bmatrix} = \begin{bmatrix} 9 & 9 \\ 0 & 0 \end{bmatrix}$$

Clearly, $B \neq C$ but $AB = AC$.

14.6. PROPERTIES OF MATRIX MULTIPLICATION

(*i*) Matrix multiplication is not commutative in general *i.e.*, $AB \neq BA$.

(*ii*) Matrix multiplication is associative *i.e.*, $(AB)C = A(BC)$.

(*iii*) Matrix multiplication is distributive with respect to matrix addition.

i.e., $$A(B + C) = AB + AC.$$

(*iv*) If A and I are square matrices of the same order, then $AI = IA = A$.

(*v*) If A is a square matrix of order *n*, then $A \times A = A^2$, $A \times A \times A = A^3$, $A \times A \times A$ ..... *m* times $= A^m$ are all square matrices of order *n*.

Also, we define $A^\circ = I$

(*vi*) For any positive integer *n*, $I^n = I$.

Note. (*i*) If A is a matrix such that $A^2 = I$, then A is called an **involutory matrix.**

(*ii*) If A is a matrix such that $A^2 = A$, then A is called **Idempotent matrix**.

(*P.T.U., M.C.A. May 2007*)

Example 7. *Evaluate $A^2 - 3A + 9I$, if I is the unit matrix of order 3 and* $A = \begin{bmatrix} 1 & -2 & 3 \\ 2 & 3 & -1 \\ -3 & 1 & 2 \end{bmatrix}$.

Sol. $A^2 = A \times A = \begin{bmatrix} 1 & -2 & 3 \\ 2 & 3 & -1 \\ -3 & 1 & 2 \end{bmatrix}\begin{bmatrix} 1 & -2 & 3 \\ 2 & 3 & -1 \\ -3 & 1 & 2 \end{bmatrix}$

$$= \begin{bmatrix} 1(1) - 2(2) + 3(-3) & 1(-2) - 2(3) + 3(1) & 1(3) - 2(-1) + 3(2) \\ 2(1) + 3(2) - 1(-3) & 2(-2) + 3(3) - 1(1) & 2(3) + 3(-1) - 1(2) \\ -3(1) + 1(2) + 2(-3) & -3(-2) + 1(3) + 2(1) & -3(3) + 1(-1) + 2(2) \end{bmatrix}$$

$$= \begin{bmatrix} -12 & -5 & 11 \\ 11 & 4 & 1 \\ -7 & 11 & -6 \end{bmatrix}$$

$$\therefore \quad A^2 - 3A + 9I = \begin{bmatrix} -12 & -5 & 11 \\ 11 & 4 & 1 \\ -7 & 11 & -6 \end{bmatrix} - 3\begin{bmatrix} 1 & -2 & 3 \\ 2 & 3 & -1 \\ -3 & 1 & 2 \end{bmatrix} + 9\begin{bmatrix} 1 & 0 & 0 \\ 0 & 1 & 0 \\ 0 & 0 & 1 \end{bmatrix}$$

$$= \begin{bmatrix} -12 & -5 & 11 \\ 11 & 4 & 1 \\ -7 & 11 & -6 \end{bmatrix} - \begin{bmatrix} 3 & -6 & 9 \\ 6 & 9 & -3 \\ -9 & 3 & 6 \end{bmatrix} + \begin{bmatrix} 9 & 0 & 0 \\ 0 & 9 & 0 \\ 0 & 0 & 9 \end{bmatrix}$$

$$= \begin{bmatrix} -12-3+9 & -5-(-6)+0 & 11-9+0 \\ 11-6+0 & 4-9+9 & 1-(-3)+0 \\ -7-(-9)+0 & 11-3+0 & -6-6+9 \end{bmatrix} = \begin{bmatrix} -6 & 1 & 2 \\ 5 & 4 & 4 \\ 2 & 8 & -3 \end{bmatrix}.$$

Note. Mathematical Induction

Mathematical induction is a very useful device for proving results for all positive integers. If the result to be proved involves n, where n is a positive integer, then the proof by mathematical induction consists of the following two steps :

Step 1. Verify the result for $n = 1$.

Step 2. Assume the result to be true for $n = k$ and **then prove** that it is true for $n = k + 1$.

Example 8. *By Mathematical Induction, prove that if*

$$A = \begin{bmatrix} 3 & -4 \\ 1 & -1 \end{bmatrix}, \text{ then } A^n = \begin{bmatrix} 1+2n & -4n \\ n & 1-2n \end{bmatrix} \text{ where } n \text{ is any positive integer.}$$

Sol. We prove the result by mathematical induction.

When n = 1, $A^n = \begin{bmatrix} 1+2n & -4n \\ n & 1-2n \end{bmatrix}$

$$\Rightarrow \quad A^1 = \begin{bmatrix} 1+2.1 & -4.1 \\ 1 & 1-2.1 \end{bmatrix} = \begin{bmatrix} 3 & -4 \\ 1 & -1 \end{bmatrix} = A$$

$\Rightarrow$ The result is true when $n = 1$

Let us assume that the result is true for any positive integer k.

i.e., let $A^k = \begin{bmatrix} 1+2k & -4k \\ k & 1-2k \end{bmatrix}$ ...(1)

Now $A^{k+1} = A^k . A = \begin{bmatrix} 1+2k & -4k \\ k & 1-2k \end{bmatrix}\begin{bmatrix} 3 & -4 \\ 1 & -1 \end{bmatrix}$ [Using (1)]

$$= \begin{bmatrix} 3(1+2k)-4k & -4(1+2k)+4k \\ 3k+1-2k & -4k-1+2k \end{bmatrix} = \begin{bmatrix} 3+2k & -4-4k \\ 1+k & -1-2k \end{bmatrix}$$

$$= \begin{bmatrix} 1+2(k+1) & -4(k+1) \\ k+1 & 1-2(k+1) \end{bmatrix}$$

$\Rightarrow$ The result is true for $n = k+1$

Hence, by mathematical induction, the result is true for all positive integers n.

TEST YOUR KNOWLEDGE 14.1

1. Given $\begin{pmatrix} x+y-1 & z-t+3 \\ x-y+1 & z+t-3 \end{pmatrix} = \begin{pmatrix} 2 & 4 \\ 3 & 5 \end{pmatrix}$, find x, y, z, t.

2. If $A = \begin{bmatrix} 3 & 2 & 0 \\ 4 & 1 & -1 \\ 1 & 2 & 2 \end{bmatrix}$ and $B = \begin{bmatrix} 5 & 1 & 3 \\ 2 & 1 & 1 \\ -1 & 5 & -3 \end{bmatrix}$, find 3A – 4B.

3. (*a*) If $A_{m \times n}$ and $B_{p \times q}$ are two matrices, state the conditions for the existence of

   (*i*) A + B (*ii*) AB (*iii*) BA (*iv*) AB + BA.

   (*b*) Matrix A has x rows and $x + 5$ columns. Matrix B has y rows and $11 - y$ columns. Both AB and BA exist. Find x and y.

4. If $A = \begin{bmatrix} 1 & 3 & 0 \\ -1 & 2 & 1 \\ 0 & 0 & 2 \end{bmatrix}$ and $B = \begin{bmatrix} 2 & 3 & 4 \\ 1 & 2 & 3 \\ -1 & 1 & 2 \end{bmatrix}$, form the products AB and BA and show that AB ≠ BA.

5. If $A = \begin{bmatrix} 0 & 1 & 2 \\ 1 & 2 & 3 \\ 2 & 3 & 4 \end{bmatrix}$ and $B = \begin{bmatrix} 1 & -2 \\ -1 & 0 \\ 2 & -1 \end{bmatrix}$, obtain the product AB and explain why BA is not defined.

6. If $A = \begin{bmatrix} -2 & 3 & -1 \\ -1 & 2 & -1 \\ -6 & 9 & -4 \end{bmatrix}$ and $B = \begin{bmatrix} 1 & 3 & -1 \\ 2 & 2 & -1 \\ 3 & 0 & -1 \end{bmatrix}$, verify that $AB = BA = I_3$.

7. Compute AB – BA, where $A = \begin{bmatrix} 2 & 9 \\ 4 & 3 \end{bmatrix}$ and $B = \begin{bmatrix} 1 & 5 \\ 7 & 2 \end{bmatrix}$.

8. If $A = \begin{bmatrix} 1 & 2 & 3 & 4 \\ 2 & 0 & 1 & 2 \\ 3 & 1 & 0 & 5 \end{bmatrix}$ and $B = \begin{bmatrix} 2 & 1 & 0 \\ 3 & 2 & 1 \\ 1 & 0 & 1 \end{bmatrix}$, find AB or BA, whichever exists.

9. If $A = \begin{bmatrix} 1 & 2 \\ -2 & 3 \end{bmatrix}$, $B = \begin{bmatrix} 2 & 1 \\ 2 & 3 \end{bmatrix}$ and $C = \begin{bmatrix} -3 & 1 \\ 2 & 0 \end{bmatrix}$, verify that (AB)C = A(BC) and A(B + C) = AB + AC.

10. If $A = \begin{bmatrix} 2 & 3 \\ 3 & 10 \end{bmatrix}$ and $I = \begin{bmatrix} 1 & 0 \\ 0 & 1 \end{bmatrix}$, show that (2I – A)(10I – A) = 9I.

11. (*a*) If $A + B = \begin{bmatrix} 1 & -1 \\ 3 & 0 \end{bmatrix}$ and $A - B = \begin{bmatrix} 3 & 1 \\ 1 & 4 \end{bmatrix}$, calculate the product AB.

    (*b*) If AB = A, BA = B where A and B are square matrices of same order, then show that $A^2 = A$ and $B^2 = B$.

12. If $A = \begin{bmatrix} 3 & 1 \\ -1 & 2 \end{bmatrix}$, show that $A^2 - 5A + 7I = O$, where I is a unit matrix of second order.

13. If $A = \begin{bmatrix} 0 & -\tan\frac{\alpha}{2} \\ \tan\frac{\alpha}{2} & 0 \end{bmatrix}$, show that $I + A = (I - A)\begin{bmatrix} \cos\alpha & -\sin\alpha \\ \sin\alpha & \cos\alpha \end{bmatrix}$.

14. Show that A is an involutory matrix where $A = \begin{bmatrix} 0 & 1 & -1 \\ 4 & -3 & 4 \\ 3 & -3 & 4 \end{bmatrix}$.

15. Given $A = \begin{bmatrix} 1 & -3 & 2 \\ 2 & 1 & -3 \\ 4 & -3 & -1 \end{bmatrix}$, $B = \begin{bmatrix} 1 & 4 & 1 & 0 \\ 2 & 1 & 1 & 1 \\ 1 & -2 & 1 & 2 \end{bmatrix}$ and $C = \begin{bmatrix} 2 & 1 & -1 & -2 \\ 3 & -2 & -1 & -1 \\ 2 & -5 & -1 & 0 \end{bmatrix}$, show that AB = AC.

16. Evaluate :

(*i*) $\begin{bmatrix} 1 & -1 & 1 \\ 2 & 0 & 1 \end{bmatrix} \times \begin{bmatrix} 1 & -1 & 0 \\ 0 & 1 & -1 \\ 1 & 1 & 1 \end{bmatrix} \times \begin{bmatrix} 1 & 0 \\ 0 & 1 \\ 1 & 1 \end{bmatrix}$ (*ii*) $[x\ y\ z]\begin{bmatrix} 7 & 3 & 0 \\ 3 & 7 & 4 \\ 0 & 4 & 7 \end{bmatrix}\begin{bmatrix} x \\ y \\ z \end{bmatrix}$.

17. If $A = \begin{bmatrix} 0 & 1 \\ 1 & 0 \end{bmatrix}$, $B = \begin{bmatrix} 0 & -i \\ i & 0 \end{bmatrix}$ where $i^2 = -1$, verify that $(A + B)^2 = A^2 + B^2$.

18. If $A = \begin{bmatrix} 1 & -1 \\ 0 & 2 \end{bmatrix}$, $B = \begin{bmatrix} 1 & 0 \\ 1 & 2 \end{bmatrix}$, verify that

(*i*) $(A + B)^2 \neq A^2 + 2AB + B^2$ (*ii*) $(A + B)(A - B) \neq A^2 - B^2$.

19. If $A = \begin{bmatrix} 1 & 1 & -1 \\ 2 & 0 & 3 \\ 3 & -1 & 2 \end{bmatrix}$, $B = \begin{bmatrix} 1 & 3 \\ 0 & 2 \\ -1 & 4 \end{bmatrix}$, $C = \begin{bmatrix} 1 & 2 & 3 & -4 \\ 2 & 0 & -2 & 1 \end{bmatrix}$, verify that A(BC) = (AB)C.

20. If $A = \begin{bmatrix} 1 & 2 \\ 3 & 4 \end{bmatrix}$, $B = \begin{bmatrix} 3 & 1 \\ 4 & 5 \end{bmatrix}$, $C = \begin{bmatrix} 1 & 1 \\ 2 & 2 \end{bmatrix}$, verify that

(*i*) A(B + C) = AB + AC (*ii*) (B + C)A = BA + CA.

21. If $A = \begin{bmatrix} 1 & 2 & -1 \\ 2 & 0 & 3 \\ 0 & 1 & 2 \end{bmatrix}$, $B = \begin{bmatrix} 3 & -1 & 1 \\ 0 & 0 & 2 \\ 4 & -3 & 2 \end{bmatrix}$, verify the result $(A + B)^2 = A^2 + AB + BA + B^2$.

22. If $A = \begin{bmatrix} 1 & 2 & 2 \\ 2 & 1 & 2 \\ 2 & 2 & 1 \end{bmatrix}$, compute $A^2 - 4A - 5I$.

23. (*a*) Show that $A = \begin{bmatrix} 1 & 2 & 3 \\ 2 & -1 & 4 \\ 3 & 1 & 1 \end{bmatrix}$ satisfies the matrix equation $A^3 - A^2 - 18A - 30I = O$.

(*b*) Prove that $A^3 - 4A^2 - 3A + 11\,I = O$, where $A = \begin{bmatrix} 1 & 3 & 2 \\ 2 & 0 & -1 \\ 1 & 2 & 3 \end{bmatrix}$.

24. If $f(x) = x^2 - 5x + 6$, find $f(A)$, where $A = \begin{bmatrix} 2 & 0 & 1 \\ 2 & 1 & 3 \\ 1 & -1 & 0 \end{bmatrix}$.

[Hint. $f(x) = x^2 - 5x + 6$

$\therefore$ $f(A) = A^2 - 5A + 6I = A^2 - 5A + 6I_3$]

25. If $A = \begin{bmatrix} \cos\alpha & \sin\alpha \\ -\sin\alpha & \cos\alpha \end{bmatrix}$, show that $A^n = \begin{bmatrix} \cos n\alpha & \sin n\alpha \\ -\sin n\alpha & \cos n\alpha \end{bmatrix}$, where n is a positive integer.

26. If $E = \begin{pmatrix} 0 & 1 & 0 \\ 0 & 0 & 1 \\ 0 & 0 & 0 \end{pmatrix}$ and $F = \begin{pmatrix} 0 & 0 & 0 \\ 1 & 0 & 0 \\ 0 & 1 & 0 \end{pmatrix}$, calculate EF and FE and show that $E^2F + FE^2 = E$.

27. If $A = \begin{bmatrix} k & 1 \\ 0 & k \end{bmatrix}$, prove that $A^n = \begin{bmatrix} k^n & nk^{n-1} \\ 0 & k^n \end{bmatrix}$ for all positive integers n.

28. If $A = \begin{bmatrix} 1 & 1 & 0 \\ 0 & 1 & 1 \\ 0 & 0 & 1 \end{bmatrix}$, then $A^n = \begin{bmatrix} 1 & n & \frac{1}{2}n(n-1) \\ 0 & 1 & n \\ 0 & 0 & 1 \end{bmatrix}$ for all positive integers n.

29. If $A = \begin{pmatrix} 0 & 1 \\ 0 & 0 \end{pmatrix}$, $I_2 = \begin{pmatrix} 1 & 0 \\ 0 & 1 \end{pmatrix}$ and n is a positive integer, then prove that $(aI_2 + bA)^n = a^n I_2 + na^{n-1} bA$.

30. Factorise the matrix $A = \begin{bmatrix} 5 & -2 & 1 \\ 7 & 1 & -5 \\ 3 & 7 & 4 \end{bmatrix}$ into the form LU, where L is lower triangular with each diagonal element one and U is upper triangular matrix.

Answers

1. $x = \frac{5}{2}, y = \frac{1}{2}, z = \frac{9}{2}, t = \frac{7}{2}$ **2.** $\begin{bmatrix} -11 & 2 & -12 \\ 4 & -1 & -7 \\ 7 & -14 & 18 \end{bmatrix}$

3. (*a*) (*i*) $m = p, n = q$ (*ii*) $n = p$ (*iii*) $q = m$

(*iv*) $m = n = p = q$ (*b*) $x = 3, y = 8$

4. $AB = \begin{bmatrix} 5 & 9 & 13 \\ -1 & 2 & 4 \\ -2 & 2 & 4 \end{bmatrix}$, $BA = \begin{bmatrix} -1 & 12 & 11 \\ -1 & 7 & 8 \\ -2 & -1 & 5 \end{bmatrix}$

5. $AB = \begin{bmatrix} 3 & -2 \\ 5 & -5 \\ 7 & -8 \end{bmatrix}$, no. of cols. in B ≠ no. of rows in A

7. $\begin{bmatrix} 43 & 4 \\ 3 & -43 \end{bmatrix}$ **8.** $BA = \begin{bmatrix} 4 & 4 & 7 & 10 \\ 10 & 7 & 11 & 21 \\ 4 & 3 & 3 & 9 \end{bmatrix}$ **11.** (*a*) $\begin{bmatrix} -2 & -2 \\ 0 & -6 \end{bmatrix}$

16. (*i*) $\begin{bmatrix} 4 & 1 \\ 4 & 0 \end{bmatrix}$ (*ii*) $[7(x^2 + y^2 + z^2) + 6xy + 8yz]$

22. $O_{3\times 3}$ **24.** $\begin{bmatrix} 1 & -1 & -3 \\ -1 & -1 & -10 \\ -5 & 4 & 4 \end{bmatrix}$

30. $\begin{bmatrix} 1 & 0 & 0 \\ 7/5 & 1 & 0 \\ 3/5 & 41/19 & 1 \end{bmatrix} \begin{bmatrix} 5 & -2 & 1 \\ 0 & 19/5 & -32/5 \\ 0 & 0 & 327/19 \end{bmatrix}$.

14.7. TRANSPOSE OF A MATRIX *(P.T.U., M.C.A. May 2008, May 2007)*

Given a matrix A, then the matrix obtained from A by changing its rows into columns and columns into rows is called the **transpose** of A and is denoted by A′ or A^t.

For example, if $A = \begin{bmatrix} 1 & 0 & 2 & 5 \\ 2 & -1 & 3 & 7 \end{bmatrix}$, then $A' = \begin{bmatrix} 1 & 2 \\ 0 & -1 \\ 2 & 3 \\ 5 & 7 \end{bmatrix}$

Clearly, (*i*) if the order of A is $m \times n$ then order of A′ is $n \times m$.

(*ii*) (*i, j*)th element of A′ = (*j, i*)th element of A.

In symbols, if $A = [a_{ij}]_{m \times n}$, then $A' = [b_{ij}]_{n \times m}$, where $b_{ij} = a_{ji}$.

14.8. PROPERTIES OF TRANSPOSE OF A MATRIX

If A′ and B′ denote the transposes of A and B respectively, then

(*a*) **(A′)′ = A *i.e.*, the transpose of the transpose of a matrix is the matrix itself.**

Let $A = [a_{ij}]_{m \times n}$ be any matrix,

Order of A′ is $n \times m$. ∴ Order of (A′)′ is $m \times n$

∴ The matrices (A′)′ and A are of the same order.

Also, (*i, j*)th element of (A′)′ = (*j, i*)th element of A′ = (*i, j*)th element of A.

Since (A′)′ and A are of the same order and their corresponding elements are equal

∴ $$(A')' = A.$$

(*b*) **(A + B)′ = A′ + B′ *i.e.*, the transpose of the sum of two matrices is equal to the sum of their transposes.**

Let $A = [a_{ij}]_{m \times n}$ and $B = [b_{ij}]_{m \times n}$ be two matrices.

Order of A + B is $m \times n$. ∴ Order of (A + B)′ is $n \times m$.

A′ and B′ are both $n \times m$ matrices. ∴ Order of A′ + B′ is $n \times m$.

Thus (A + B)′ and A′ + B′ are of the same order.

Also (*i, j*)th element of (A + B)′ = (*j, i*)th element of A + B
= (*j, i*)th element of A + (*j, i*)th element of B
= (*i, j*)th element of A′ + (*i, j*)th element of B′
= (*i, j*)th element of A′ + B′

Since (A + B)′ and A′ + B′ are of the same order and their corresponding elements are equal.

∴ $$(A + B)' = A' + B'.$$

(*c*) **(AB)′ = B′A′ *i.e.*, the transpose of the product of two matrices is equal to the product of their transposes taken in the reverse order.**

Let $A = [a_{ij}]_{m \times n}$ and $B = [b_{ij}]_{n \times p}$ be two matrices conformable for the product AB.

Order of AB is $m \times p$ ∴ Order of (AB)′ is $p \times m$.

Order of B′ is $p \times n$ and order of A′ is $n \times m$.

∴ Order of B′A′ is $p \times m$.

Thus, (AB)′ and B′A′ are of the same order.

Also (i, j)th element of $(AB)'$

$= (j, i)$th element of AB

= Sum of the products of the elements of jth row of A with the corresponding elements of ith column of B

= Sum of the products of elements of ith column of B with the corresponding elements of the jth row of A

= Sum of the products of elements of ith row of B′ with the corresponding elements of the jth column of A′

$= (i, j)$th element of $B'A'$.

Since $(AB)'$ and $B'A'$ are of the same order and their corresponding elements are equal.

$\therefore \qquad (AB)' = B'A'.$

14.9. SYMMETRIC MATRIX

A square matrix $A = [a_{ij}]$ is said to be symmetric if **A′ = A** *i.e.*, if the transpose of the matrix is equal to the matrix itself.

Thus, for a symmetric matrix $A = [a_{ij}]$, $a_{ij} = a_{ji}$

i.e., (i, j)th element of A = (j, i)th element of A

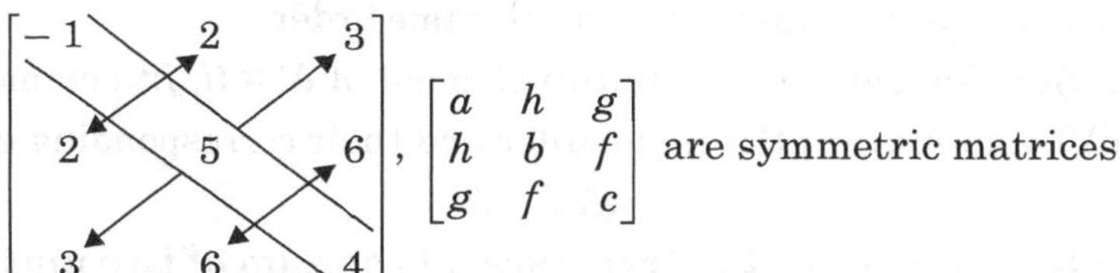

For example, $\begin{bmatrix} -1 & 2 & 3 \\ 2 & 5 & 6 \\ 3 & 6 & 4 \end{bmatrix}$, $\begin{bmatrix} a & h & g \\ h & b & f \\ g & f & c \end{bmatrix}$ are symmetric matrices.

14.10. SKEW-SYMMETRIC MATRIX (OR Anti-Symmetric Matrix)

A square matrix $A = [a_{ij}]$ is said to be skew-symmetric if **A′ = – A** *i.e.*, if the transpose of the matrix is equal to the negative of the matrix.

Thus, for a skew-symmetric matrix $A = [a_{ij}]$, $a_{ij} = -a_{ji}$

i.e., (i, j)th element of A = negative of (j, i)th element of A

Putting $j = i$, $a_{ii} = -a_{ii} \Rightarrow 2a_{ii} = 0$ or $a_{ii} = 0$ for all i.

Thus, all diagonal elements of a skew-symmetric matrix are zero.

For example, $\begin{bmatrix} 0 & 2 & -3 \\ -2 & 0 & 1 \\ 3 & -1 & 0 \end{bmatrix}$, $\begin{bmatrix} 0 & h & -g \\ -h & 0 & f \\ g & -f & 0 \end{bmatrix}$ are skew-symmetric matrices.

14.11. EVERY SQUARE MATRIX CAN UNIQUELY BE EXPRESSED AS THE SUM OF A SYMMETRIC MATRIX AND A SKEW-SYMMETRIC MATRIX

Proof. Let A be any square matrix.

Consider $\qquad B = \frac{1}{2}(A + A')$ and $C = \frac{1}{2}(A - A')$

$$B + C = \frac{1}{2}(A + A') + \frac{1}{2}(A - A') = A$$

Also $\qquad B' = \frac{1}{2}(A + A')' = \frac{1}{2}[A' + (A')'] = \frac{1}{2}(A' + A) = \frac{1}{2}(A + A') = B$

$\Rightarrow$ B is a symmetric matrix.

$$C' = \frac{1}{2}(A - A')' = \frac{1}{2}[A' - (A')'] = \frac{1}{2}(A' - A) = -\frac{1}{2}(A - A') = -C$$

$\Rightarrow$ C is a skew-symmetric matrix.

Hence every square matrix A can be expressed as $A = B + C$, where $B = \frac{1}{2}(A + A')$ is a symmetric matrix and $C = \frac{1}{2}(A - A')$ is a skew-symmetric matrix.

To prove that there is only one way in which A can be expressed as the sum of a symmetric matrix and a skew-symmetric matrix, suppose

$$A = P + Q \quad \text{...(1)}$$

is another such representation in which P is a symmetric matrix, and Q is a skew-symmetric matrix.

i.e., $P' = P$ and $Q' = -Q$

Then $A' = (P + Q)' = P' + Q' = P - Q$ ...(2)

From (1) and (2), $A + A' = 2P$ and $A - A' = 2Q$

$\therefore$ $P = \frac{1}{2}(A + A') = B$ and $Q = \frac{1}{2}(A - A') = C$ so that $P + Q = B + C$

Hence the result.

14.12. ORTHOGONAL MATRIX

A square matrix A is called an orthogonal matrix if $AA' = I = A'A$.

14.13. FOR ANY TWO ORTHOGONAL MATRICES A AND B, SHOW THAT AB IS AN ORTHOGONAL MATRIX

Proof. A and B are orthogonal matrices

$\Rightarrow$ $AA' = I = A'A$ and $BB' = I = B'B$ ...(1)

Now $(AB)(AB)' = (AB)(B'A')$

$= A(BB')\,A' = AIA'$ [Using (1)]

$= (AI)\,A' = AA' = I$ [Using (1)]

$\Rightarrow$ AB is an orthogonal matrix.

ILLUSTRATIVE EXAMPLES

Example 1. *If A and B are symmetric matrices, prove that AB – BA is a skew-symmetric matrix.*

Sol. A and B are symmetric matrices

$\Rightarrow$ $A' = A$ and $B' = B$ ...(1)

Now $(AB - BA)' = (AB)' - (BA)'$ [$\because$ $(A - B)' = A' - B'$]

$= B'A' - A'B'$ [$\because$ $(AB)' = B'A'$]

$= BA - AB$ [Using (1)]

$= -(AB - BA)$

$\therefore$ (AB – BA) is a skew-symmetric matrix.

Example 2. *Express the following matrix as the sum of a symmetric and a skew-symmetric matrix,*

$$\begin{bmatrix} -1 & 7 & 1 \\ 2 & 3 & 4 \\ 5 & 0 & 5 \end{bmatrix}.$$

Sol. Given matrix is $A = \begin{bmatrix} -1 & 7 & 1 \\ 2 & 3 & 4 \\ 5 & 0 & 5 \end{bmatrix} \quad \therefore \quad A' = \begin{bmatrix} -1 & 2 & 5 \\ 7 & 3 & 0 \\ 1 & 4 & 5 \end{bmatrix}$

$$A + A' = \begin{bmatrix} -1-1 & 7+2 & 1+5 \\ 2+7 & 3+3 & 4+0 \\ 5+1 & 0+4 & 5+5 \end{bmatrix} = \begin{bmatrix} -2 & 9 & 6 \\ 9 & 6 & 4 \\ 6 & 4 & 10 \end{bmatrix}$$

$$\therefore \qquad B = \tfrac{1}{2}(A + A') = \begin{bmatrix} -1 & \frac{9}{2} & 3 \\ \frac{9}{2} & 3 & 2 \\ 3 & 2 & 5 \end{bmatrix}$$

$$A - A' = \begin{bmatrix} -1+1 & 7-2 & 1-5 \\ 2-7 & 3-3 & 4-0 \\ 5-1 & 0-4 & 5-5 \end{bmatrix} = \begin{bmatrix} 0 & 5 & -4 \\ -5 & 0 & 4 \\ 4 & -4 & 0 \end{bmatrix}$$

$$\therefore \qquad C = \tfrac{1}{2}(A - A') = \begin{bmatrix} 0 & \frac{5}{2} & -2 \\ -\frac{5}{2} & 0 & 2 \\ 2 & -2 & 0 \end{bmatrix}$$

$$A = B + C = \begin{bmatrix} -1 & \frac{9}{2} & 3 \\ \frac{9}{2} & 3 & 2 \\ 3 & 2 & 5 \end{bmatrix} + \begin{bmatrix} 0 & \frac{5}{2} & -2 \\ -\frac{5}{2} & 0 & 2 \\ 2 & -2 & 0 \end{bmatrix}$$

where B is a symmetric and C is a skew-symmetric matrix.

Example 3. *Show that* $A = \begin{bmatrix} \cos\theta & 0 & \sin\theta \\ 0 & 1 & 0 \\ -\sin\theta & 0 & \cos\theta \end{bmatrix}$ *is orthogonal. Find the value of* $|A|$.

Sol. Here $\qquad A' = \begin{bmatrix} \cos\theta & 0 & -\sin\theta \\ 0 & 1 & 0 \\ \sin\theta & 0 & \cos\theta \end{bmatrix}$

$$AA' = \begin{bmatrix} \cos\theta & 0 & \sin\theta \\ 0 & 1 & 0 \\ -\sin\theta & 0 & \cos\theta \end{bmatrix} \begin{bmatrix} \cos\theta & 0 & -\sin\theta \\ 0 & 1 & 0 \\ \sin\theta & 0 & \cos\theta \end{bmatrix}$$

$$= \begin{bmatrix} \cos^2\theta + \sin^2\theta & 0 & 0 \\ 0 & 1 & 0 \\ 0 & 0 & \sin^2\theta + \cos^2\theta \end{bmatrix} = \begin{bmatrix} 1 & 0 & 0 \\ 0 & 1 & 0 \\ 0 & 0 & 1 \end{bmatrix} = I$$

Since AA′ = I, A is an orthogonal matrix.

$$|A| = \begin{vmatrix} \cos\theta & 0 & \sin\theta \\ 0 & 1 & 0 \\ -\sin\theta & 0 & \cos\theta \end{vmatrix}$$

Expanding by second column,

$$|A| = \begin{vmatrix} \cos\theta & \sin\theta \\ -\sin\theta & \cos\theta \end{vmatrix} = \cos^2\theta + \sin^2\theta = 1.$$

14.14. ADJOINT OF A SQUARE MATRIX

The adjoint of a square matrix is the transpose of the matrix obtained by replacing each element of A by its co-factor in | A |.

Adjoint of A is briefly written as adj A.

Thus, if $A = \begin{bmatrix} a_1 & b_1 & c_1 \\ a_2 & b_2 & c_2 \\ a_3 & b_3 & c_3 \end{bmatrix}$, then adj A = Transpose of $\begin{bmatrix} A_1 & B_1 & C_1 \\ A_2 & B_2 & C_2 \\ A_3 & B_3 & C_3 \end{bmatrix} = \begin{bmatrix} A_1 & A_2 & A_3 \\ B_1 & B_2 & B_3 \\ C_1 & C_2 & C_3 \end{bmatrix}$

where the capital letters denote the co-factors of corresponding small letters in

$$|A| = \begin{vmatrix} a_1 & b_1 & c_1 \\ a_2 & b_2 & c_2 \\ a_3 & b_3 & c_3 \end{vmatrix}.$$

14.15. AN IMPORTANT RELATION BETWEEN A SQUARE MATRIX A AND adj A

Theorem. *If A is an n-rowed square matrix, then A(adj A) = (adj A) A =* $|A|\, I_n$.

Proof. Let
$$A = \begin{bmatrix} a_{11} & a_{12} & \cdots\cdots & a_{1n} \\ a_{21} & a_{22} & \cdots\cdots & a_{2n} \\ \cdots & \cdots & \cdots\cdots & \cdots \\ a_{n1} & a_{n2} & \cdots\cdots & a_{nn} \end{bmatrix}$$

then
$$\text{adj } A = \text{Transpose of } \begin{bmatrix} A_{11} & A_{12} & \cdots\cdots & A_{1n} \\ A_{21} & A_{22} & \cdots\cdots & A_{2n} \\ \cdots & \cdots & \cdots\cdots & \cdots \\ A_{n1} & A_{n2} & \cdots\cdots & A_{nn} \end{bmatrix}$$

(where A_{ij} is co-factor of a_{ij} in | A |)

$$= \begin{bmatrix} A_{11} & A_{21} & \cdots\cdots & A_{n1} \\ A_{12} & A_{22} & \cdots\cdots & A_{n2} \\ \cdots & \cdots & \cdots\cdots & \cdots \\ A_{1n} & A_{2n} & \cdots\cdots & A_{nn} \end{bmatrix}$$

(i, j)th element in A (adj A) $= a_{i1}A_{j1} + a_{i2}A_{j2} + a_{i3}A_{j3} + \cdots\cdots + a_{in}A_{jn} = \begin{cases} |A| & \text{if } i = j \\ 0 & \text{if } i \neq j \end{cases}$

∴ In the product A (adj A), each diagonal element is | A | and each non-diagonal element is 0.

$$\therefore \quad A(\text{adj } A) = \begin{bmatrix} |A| & 0 & \ldots\ldots & 0 \\ 0 & |A| & \ldots\ldots & 0 \\ \ldots & \ldots & \ldots\ldots & \ldots \\ 0 & 0 & \ldots\ldots & |A| \end{bmatrix} = |A| \begin{bmatrix} 1 & 0 & \ldots\ldots & 0 \\ 0 & 1 & \ldots\ldots & 0 \\ \ldots & \ldots & \ldots\ldots & \ldots \\ 0 & 0 & \ldots\ldots & 1 \end{bmatrix} = |A|\, I_n$$

Similarly, $\quad (\text{adj } A)A = |A|\, I_n$

Hence $\quad A(\text{adj } A) = (\text{adj } A)A = |A|\, I_n.$

Example 4. *If* $A = \begin{bmatrix} 2 & 1 & 3 \\ 3 & 1 & 2 \\ 1 & 2 & 3 \end{bmatrix}$, *verify that* $A(\text{adj } A) = (\text{adj } A)A = |A|\, I.$

Sol.
$$A = \begin{bmatrix} 2 & 1 & 3 \\ 3 & 1 & 2 \\ 1 & 2 & 3 \end{bmatrix} = \begin{bmatrix} a_1 & b_1 & c_1 \\ a_2 & b_2 & c_2 \\ a_3 & b_3 & c_3 \end{bmatrix}$$

$$A_1 = \begin{vmatrix} 1 & 2 \\ 2 & 3 \end{vmatrix} = -1, \quad A_2 = -\begin{vmatrix} 1 & 3 \\ 2 & 3 \end{vmatrix} = 3, \quad A_3 = \begin{vmatrix} 1 & 3 \\ 1 & 2 \end{vmatrix} = -1$$

$$B_1 = -\begin{vmatrix} 3 & 2 \\ 1 & 3 \end{vmatrix} = -7, \quad B_2 = \begin{vmatrix} 2 & 3 \\ 1 & 3 \end{vmatrix} = 3, \quad B_3 = -\begin{vmatrix} 2 & 3 \\ 3 & 2 \end{vmatrix} = 5$$

$$C_1 = \begin{vmatrix} 3 & 1 \\ 1 & 2 \end{vmatrix} = 5, \quad C_2 = -\begin{vmatrix} 2 & 1 \\ 1 & 2 \end{vmatrix} = -3, \quad C_3 = \begin{vmatrix} 2 & 1 \\ 3 & 1 \end{vmatrix} = -1$$

$$|A| = a_1A_1 + b_1B_1 + c_1C_1 = 2(-1) + 1(-7) + 3(5) = 6$$

$$\text{adj } A = \text{Transpose of } \begin{bmatrix} A_1 & B_1 & C_1 \\ A_2 & B_2 & C_2 \\ A_3 & B_3 & C_3 \end{bmatrix} = \begin{bmatrix} A_1 & A_2 & A_3 \\ B_1 & B_2 & B_3 \\ C_1 & C_2 & C_3 \end{bmatrix} = \begin{bmatrix} -1 & 3 & -1 \\ -7 & 3 & 5 \\ 5 & -3 & -1 \end{bmatrix}$$

$$A(\text{adj } A) = \begin{bmatrix} 2 & 1 & 3 \\ 3 & 1 & 2 \\ 1 & 2 & 3 \end{bmatrix} \begin{bmatrix} -1 & 3 & -1 \\ -7 & 3 & 5 \\ 5 & -3 & -1 \end{bmatrix}$$

$$= \begin{bmatrix} 2(-1) + 1(-7) + 3(5) & 2(3) + 1(3) + 3(-3) & 2(-1) + 1(5) + 3(-1) \\ 3(-1) + 1(-7) + 2(5) & 3(3) + 1(3) + 2(-3) & 3(-1) + 1(5) + 2(-1) \\ 1(-1) + 2(-7) + 3(5) & 1(3) + 2(3) + 3(-3) & 1(-1) + 2(5) + 3(-1) \end{bmatrix}$$

$$= \begin{bmatrix} 6 & 0 & 0 \\ 0 & 6 & 0 \\ 0 & 0 & 6 \end{bmatrix} = 6 \begin{bmatrix} 1 & 0 & 0 \\ 0 & 1 & 0 \\ 0 & 0 & 1 \end{bmatrix} = |A|\, I$$

$$(\text{adj } A)A = \begin{bmatrix} -1 & 3 & -1 \\ -7 & 3 & 5 \\ 5 & -3 & -1 \end{bmatrix} \begin{bmatrix} 2 & 1 & 3 \\ 3 & 1 & 2 \\ 1 & 2 & 3 \end{bmatrix}$$

$$= \begin{bmatrix} -1(2) + 3(3) - 1(1) & -1(1) + 3(1) - 1(2) & -1(3) + 3(2) - 1(3) \\ -7(2) + 3(3) + 5(1) & -7(1) + 3(1) + 5(2) & -7(3) + 3(2) + 5(3) \\ 5(2) - 3(3) - 1(1) & 5(1) - 3(1) - 1(2) & 5(3) - 3(2) - 1(3) \end{bmatrix}$$

$$= \begin{bmatrix} 6 & 0 & 0 \\ 0 & 6 & 0 \\ 0 & 0 & 6 \end{bmatrix} = 6 \begin{bmatrix} 1 & 0 & 0 \\ 0 & 1 & 0 \\ 0 & 0 & 1 \end{bmatrix} = |A|\, I$$

Hence $A(\text{adj } A) = (\text{adj } A)A = |A|\, I.$

14.16. SINGULAR MATRICES AND NON-SINGULAR MATRICES

A square matrix A is said to be singular if $|A| = 0$ and non-singular if $|A| \neq 0$.

14.17. INVERSE (or Reciprocal) OF A SQUARE MATRIX

(P.T.U., M.C.A. May 2007, Dec. 2005)

Let A be an n-rowed square matrix. If there exists an n-rowed square matrix B such that

$$AB = BA = I$$

then the matrix **A is said to be invertible** and B is called the **inverse** (or **reciprocal**) of A.

Note 1. Only square matrices can be invertible.

2. From the definition, it is clear that if B is the inverse of A, then A is the inverse of B.

3. Inverse of A is denoted by A^{-1}, thus $B = A^{-1}$ and $AA^{-1} = A^{-1}A = I$.

14.18. THE INVERSE OF A SQUARE MATRIX, IF IT EXISTS, IS UNIQUE

Let A be an invertible square matrix. If possible, let B and C be two inverses of A. Then

$$AB = BA = I$$

$$AC = CA = I \qquad \text{(By definition of inverse)}$$

Now $\quad B = BI = B(AC)$

$= (BA)C \qquad$ [$\because$ Matrix multiplication is associative]

$= IC = C$

Hence the inverse of A is unique.

14.19. THE NECESSARY AND SUFFICIENT CONDITION FOR A SQUARE MATRIX A TO POSSESS INVERSE IS THAT $|A| \neq 0$ (i.e., A is Non-Singular)

Proof. (*a*) **The condition is necessary :**

i.e., Given that A has inverse, to show that $|A| \neq 0$

Let B be the inverse of A, then $AB = BA = I \Rightarrow |AB| = |BA| = |I|$

$\Rightarrow \quad |A|\,|B| = |B|\,|A| = 1 \quad \Rightarrow \quad |A| \neq 0.$

(*b*) **The condition is sufficient :**

i.e., Given that $|A| \neq 0$, to show that A has inverse.

$\because \quad |A| \neq 0$

$\therefore$ Consider $\quad B = \dfrac{\text{adj } A}{|A|}$

$$AB = A\left(\frac{\text{adj } A}{|A|}\right)\frac{1}{|A|}\,(A(\text{adj } A)) = \frac{1}{|A|}\,(|A|\,I)$$

$= I \qquad$ [$\because$ $A(\text{adj } A) = |A|\,I$]

$$BA = \left(\frac{\text{adj } A}{|A|}\right)A = \frac{1}{|A|}\,((\text{adj } A)A) = \frac{1}{|A|}\,(|A|\,I)$$

$= I \qquad$ [$\because$ $(\text{adj } A)A = |A|\,I$]

Thus $\quad AB = BA = I$

$\therefore$ The inverse of A exists and $A^{-1} = B = \dfrac{\text{adj } A}{|A|}$.

Note. From the above theorem, we conclude that A has inverse if and only if $|A| \neq 0$ and then $A^{-1} = \dfrac{\text{adj } A}{|A|}$.

Example 5. *If* $A = \begin{bmatrix} 3 & -3 & 4 \\ 2 & -3 & 4 \\ 0 & -1 & 1 \end{bmatrix}$, *(i) find* A^{-1} *(ii) show that* $A^3 = A^{-1}$.

Sol. (*i*)
$$A = \begin{bmatrix} 3 & -3 & 4 \\ 2 & -3 & 4 \\ 0 & -1 & 1 \end{bmatrix} = \begin{bmatrix} a_1 & b_1 & c_1 \\ a_2 & b_2 & c_2 \\ a_3 & b_3 & c_3 \end{bmatrix}$$

$$|A| = \begin{vmatrix} 3 & -3 & 4 \\ 2 & -3 & 4 \\ 0 & -1 & 1 \end{vmatrix} \quad \text{Operating } C_2 \to C_2 + C_3$$

$$= \begin{vmatrix} 3 & 1 & 4 \\ 2 & 1 & 4 \\ 0 & 0 & 1 \end{vmatrix} \quad \text{Expanding by third row}$$

$$= 1\begin{vmatrix} 3 & 1 \\ 2 & 1 \end{vmatrix} = 3 - 2 = 1$$

$\because$ $|A| = 1 \neq 0$ $\therefore$ A^{-1} exists.

$$A_1 = \begin{vmatrix} -3 & 4 \\ -1 & 1 \end{vmatrix} = 1, \quad A_2 = -\begin{vmatrix} -3 & 4 \\ -1 & 1 \end{vmatrix} = -1, \quad A_3 = \begin{vmatrix} -3 & 4 \\ -3 & 4 \end{vmatrix} = 0$$

$$B_1 = -\begin{vmatrix} 2 & 4 \\ 0 & 1 \end{vmatrix} = -2, \quad B_2 = \begin{vmatrix} 3 & 4 \\ 0 & 1 \end{vmatrix} = 3, \quad B_3 = -\begin{vmatrix} 3 & 4 \\ 2 & 4 \end{vmatrix} = -4$$

$$C_1 = \begin{vmatrix} 2 & -3 \\ 0 & -1 \end{vmatrix} = -2, \quad C_2 = -\begin{vmatrix} 3 & -3 \\ 0 & -1 \end{vmatrix} = 3, \quad C_3 = \begin{vmatrix} 3 & -3 \\ 2 & -3 \end{vmatrix} = -3$$

$$\therefore \quad \text{adj } A = \text{Transpose of} \begin{bmatrix} A_1 & B_1 & C_1 \\ A_2 & B_2 & C_2 \\ A_3 & B_3 & C_3 \end{bmatrix} = \begin{bmatrix} A_1 & A_2 & A_3 \\ B_1 & B_2 & B_3 \\ C_1 & C_2 & C_3 \end{bmatrix} = \begin{bmatrix} 1 & -1 & 0 \\ -2 & 3 & -4 \\ -2 & 3 & -3 \end{bmatrix}$$

$$A^{-1} = \frac{1}{|A|} \text{adj } A = \begin{bmatrix} 1 & -1 & 0 \\ -2 & 3 & -4 \\ -2 & 3 & -3 \end{bmatrix} \qquad (\because \; |A| = 1)$$

(*ii*)
$$A^2 = A \times A = \begin{bmatrix} 3 & -3 & 4 \\ 2 & -3 & 4 \\ 0 & -1 & 1 \end{bmatrix}\begin{bmatrix} 3 & -3 & 4 \\ 2 & -3 & 4 \\ 0 & -1 & 1 \end{bmatrix}$$

$$= \begin{bmatrix} 3(3) - 3(2) + 4(0) & 3(-3) - 3(-3) + 4(-1) & 3(4) - 3(4) + 4(1) \\ 2(3) - 3(2) + 4(0) & 2(-3) - 3(-3) + 4(-1) & 2(4) - 3(4) + 4(1) \\ 0(3) - 1(2) + 1(0) & 0(-3) - 1(-3) + 1(-1) & 0(4) - 1(4) + 1(1) \end{bmatrix}$$

$$= \begin{bmatrix} 3 & -4 & 4 \\ 0 & -1 & 0 \\ -2 & 2 & -3 \end{bmatrix}$$

$$A^4 = A^2 \times A^2 = \begin{bmatrix} 3 & -4 & 4 \\ 0 & -1 & 0 \\ -2 & 2 & -3 \end{bmatrix}\begin{bmatrix} 3 & -4 & 4 \\ 0 & -1 & 0 \\ -2 & 2 & -3 \end{bmatrix}$$

$$= \begin{bmatrix} 3(3)-4(0)+4(-2) & 3(-4)-4(-1)+4(2) & 3(4)-4(0)+4(-3) \\ 0(3)-1(0)+0(-2) & 0(-4)-1(-1)+0(2) & 0(4)-1(0)+0(-3) \\ -2(3)+2(0)-3(-2) & -2(-4)+2(-1)-3(2) & -2(4)+2(0)-3(-3) \end{bmatrix}$$

$$= \begin{bmatrix} 1 & 0 & 0 \\ 0 & 1 & 0 \\ 0 & 0 & 1 \end{bmatrix} = I$$

Now $A^4 = I$

$\Rightarrow$ $A.A^3 = A^3.\ A = I$

$\Rightarrow$ A^3 is the inverse of A

i.e., $A^3 = A^{-1}$.

14.20. IF A IS INVERTIBLE, THEN SO IS A^{-1} AND $(A^{-1})^{-1}$ = A

A is invertible $\Rightarrow$ A^{-1} exists and $AA^{-1} = A^{-1} A = I$

$\therefore$ By definition of inverse of a matrix, A^{-1} is invertible and inverse of A^{-1} is A.

i.e., $(A^{-1})^{-1} = A$.

14.21. IF A AND B BE TWO NON-SINGULAR SQUARE MATRICES OF THE SAME ORDER, THEN $(AB)^{-1} = B^{-1} A^{-1}$

Proof. A and B are non-singular square matrices

$\Rightarrow$ $|A| \neq 0,\ |B| \neq 0$

$\Rightarrow$ A^{-1} and B^{-1} both exist.

Also $|AB| = |A|\ |B| \neq 0$

$\Rightarrow$ AB is non-singular and $(AB)^{-1}$ exists.

Now $(AB)(B^{-1}A^{-1}) = A(BB^{-1})\ A^{-1} = AIA^{-1} = (AI)A^{-1} = AA^{-1} = I$ ...(1)

Again $(B^{-1}A^{-1})(AB) = B^{-1}(A^{-1}\ A)B = B^{-1}IB = B^{-1}(IB) = B^{-1}B = I$ ...(2)

From (1) and (2), we get $(AB)(B^{-1}A^{-1}) = (B^{-1}A^{-1})(AB) = I$

$\therefore$ Inverse of $AB = B^{-1}A^{-1}$ *i.e.,* $(AB)^{-1} = B^{-1}A^{-1}$.

Note. The result can be extended for more than two matrices.

i.e., $(A_1A_2A_3 \ldots\ldots A_{n-1}A_n)^{-1} = A_n^{-1}\ A_{n-1}^{-1} \ldots\ldots A_3^{-1}A_2^{-1}\ A_1^{-1}$

Putting $A_1 = A_2 = A_3 = \ldots\ldots = A_{n-1} = A_n = A$ we have (A.A.A..... n times)$^{-1} = A^{-1} . A^{-1}$ ..... n times

$\therefore$ $\mathbf{(A^n)^{-1} = (A^{-1})^n}$.

14.22. IF A IS A NON-SINGULAR SQUARE MATRIX, THEN SO IS A′ AND $(A')^{-1} = (A^{-1})'$

A is a non-singular square matrix

$\Rightarrow$ $|A| \neq 0$ and A^{-1} exists

$\therefore$ $|A'| = |A| \neq 0$

$\Rightarrow$ A′ is non-singular and $(A')^{-1}$ exists.

Now $AA^{-1} = A^{-1} A = I$

Taking transpose $= (AA^{-1})' = (A^{-1}A)' = I'$

$\Rightarrow$ $(A^{-1})'A' = A'(A^{-1})' = I$ [$\because$ $(AB)' = B'A'$]

$\Rightarrow$ Inverse of $A' = (A^{-1})'$ *i.e.*, $(A')^{-1} = (A^{-1})'$.

14.23. IF A AND B ARE TWO NON-SINGULAR SQUARE MATRICES OF THE SAME ORDER, THEN

adj (AB) = (adj B) (adj A)

Proof. A, B are non-singular $\Rightarrow$ $|A| \neq 0, |B| \neq 0$

$\therefore$ $|AB| = |A| |B| \neq 0$

$\Rightarrow$ AB is non-singular $\Rightarrow$ $(AB)^{-1}$ exists.

Now $(\text{adj } AB) AB = |AB| I$ ...(1)

Also, $(\text{adj } B)(\text{adj } A) AB = (\text{adj } B)(\text{adj } A . A) B = (\text{adj } B)(|A| I) B$

$= |A| (\text{adj } B) IB = |A| (\text{adj } B) B = |A| |B| I = |AB| I$ ...(2)

From (1) and (2), $(\text{adj } AB) AB = (\text{adj } B)(\text{adj } A) AB$

Post-multiplying by $(AB)^{-1}$, we have

$$(\text{adj } AB)(AB)(AB)^{-1} = (\text{adj } B)(\text{adj } A)(AB)(AB)^{-1}$$

$\Rightarrow$ $(\text{adj } AB) I = (\text{adj } B)(\text{adj } A) I$

$\therefore$ $\text{adj } (AB) = (\text{adj } B)(\text{adj } A)$.

Example 6. *If* $A = \begin{bmatrix} 1 & 0 & -2 \\ 2 & 2 & 4 \\ 0 & 0 & 2 \end{bmatrix}$, *verify* $A^2 - 3A + 2I = 0$ *and hence find* A^{-1}.

Sol.
$$A^2 = AA = \begin{bmatrix} 1 & 0 & -2 \\ 2 & 2 & 4 \\ 0 & 0 & 2 \end{bmatrix}\begin{bmatrix} 1 & 0 & -2 \\ 2 & 2 & 4 \\ 0 & 0 & 2 \end{bmatrix} = \begin{bmatrix} 1 & 0 & -6 \\ 6 & 4 & 12 \\ 0 & 0 & 4 \end{bmatrix}$$

$\therefore$
$$A^2 - 3A + 2I = \begin{bmatrix} 1 & 0 & -6 \\ 6 & 4 & 12 \\ 0 & 0 & 4 \end{bmatrix} - 3\begin{bmatrix} 1 & 0 & -2 \\ 2 & 2 & 4 \\ 0 & 0 & 2 \end{bmatrix} + 2\begin{bmatrix} 1 & 0 & 0 \\ 0 & 1 & 0 \\ 0 & 0 & 1 \end{bmatrix}$$

$$= \begin{bmatrix} 1-3+2 & 0 & -6+6+0 \\ 6-6+0 & 4-6+2 & 12-12+0 \\ 0 & 0 & 4-6+2 \end{bmatrix} = \begin{bmatrix} 0 & 0 & 0 \\ 0 & 0 & 0 \\ 0 & 0 & 0 \end{bmatrix}$$

$\Rightarrow$ $A^2 - 3A + 2I = O$ ...(1)

To find A^{-1}, multiplying (1) by A^{-1}, we get $A - 3I + 2A^{-1} = O$

$$\Rightarrow \quad 2A^{-1} = -A + 3I$$

$$= \begin{bmatrix} -1 & 0 & 2 \\ -2 & -2 & -4 \\ 0 & 0 & -2 \end{bmatrix} + 3\begin{bmatrix} 1 & 0 & 0 \\ 0 & 1 & 0 \\ 0 & 0 & 1 \end{bmatrix} = \begin{bmatrix} 2 & 0 & 2 \\ -2 & 1 & -4 \\ 0 & 0 & 1 \end{bmatrix}$$

$$\Rightarrow \quad A^{-1} = \begin{bmatrix} 1 & 0 & 1 \\ -1 & 1/2 & -2 \\ 0 & 0 & 1/2 \end{bmatrix}.$$

TEST YOUR KNOWLEDGE 14.2

1. If A is a square matrix, prove that
(*i*) A + A′ is symmetric (*ii*) A – A′ is skew-symmetric.

2. Prove that
(*i*) if A, B are symmetric, then so is A + B
(*ii*) if A, B are skew symmetric, then so is A + B
(*iii*) if A is a square matrix, then AA′ and A′A are both symmetric
(*iv*) if A is symmetric, then B′ AB is symmetric.
(*v*) if A and B are symmetric matrix of order n, show that AB + BA is also symmetric.

3. Express each of the following matrices as the sum of a symmetric and a skew symmetric matrix :

(*i*) $\begin{bmatrix} 4 & 2 & -3 \\ 1 & 3 & -6 \\ -5 & 0 & -7 \end{bmatrix}$ (*ii*) $\begin{bmatrix} a & a & b \\ c & b & b \\ c & a & c \end{bmatrix}$.

4. (*a*) If $A = \begin{bmatrix} 1 & 2 & -1 \\ 3 & 0 & 2 \\ 4 & 5 & 0 \end{bmatrix}$ and $B = \begin{bmatrix} 1 & 0 & 0 \\ 2 & 1 & 0 \\ 0 & 1 & 3 \end{bmatrix}$, verify that (AB)′ = B′A′.

(*b*) If $A = \begin{bmatrix} 2 & 3 \\ 4 & 5 \end{bmatrix}$, $B = \begin{bmatrix} 3 & 1 \\ 2 & 5 \end{bmatrix}$, find $(A + B)^T$.

5. If $A = \begin{bmatrix} 2 & 3 \\ 5 & -7 \end{bmatrix}$, verify that $(A^2)' = (A')^2$.

6. Show that the matrix $\begin{bmatrix} -4 & -3 & -3 \\ 1 & 0 & 1 \\ 4 & 4 & 3 \end{bmatrix}$ is its own adjoint.

7. Given the matrix $A = \begin{bmatrix} 1 & 2 & 3 \\ 1 & 3 & 5 \\ 1 & 5 & 12 \end{bmatrix}$. Compute adj A and verify that

$$A(\text{adj } A) = (\text{adj } A)A = |\ A\ |\ I.$$

8. Find the inverse of each of the following matrices :

(*i*) $\begin{bmatrix} 1 & 2 & 5 \\ 3 & 1 & 4 \\ 1 & 1 & 2 \end{bmatrix}$ (*ii*) $\begin{bmatrix} 2 & 5 & 3 \\ 3 & 1 & 2 \\ 1 & 2 & 1 \end{bmatrix}$ (*iii*) $\begin{bmatrix} 1 & 3 & 3 \\ 1 & 4 & 3 \\ 1 & 3 & 4 \end{bmatrix}$.

9. If D = diag. $[d_1, d_2, d_3]$, where $d_1 d_2 d_3 \neq 0$, then show that D^{-1} = diag. $[d_1^{-1}, d_2^{-1}, d_3^{-1}]$.

10. If $A = \frac{1}{3}\begin{bmatrix} 1 & 2 & 2 \\ 2 & 1 & -2 \\ -2 & 2 & -1 \end{bmatrix}$, prove that $A^{-1} = A'$.

11. If $A = \begin{bmatrix} 1 & 1 & 2 \\ 1 & 9 & 3 \\ 1 & 4 & 2 \end{bmatrix}, B = \begin{bmatrix} 1 & 2 & 0 \\ 2 & 3 & -1 \\ 1 & -1 & 3 \end{bmatrix}$, verify that $(AB)^{-1} = B^{-1}A^{-1}$.

12. Show that $\begin{bmatrix} \cos\theta & -\sin\theta \\ \sin\theta & \cos\theta \end{bmatrix} = \begin{bmatrix} 1 & -\tan\theta/2 \\ \tan\theta/2 & 1 \end{bmatrix} \begin{bmatrix} 1 & \tan\theta/2 \\ -\tan\theta/2 & 1 \end{bmatrix}^{-1}$.

13. Prove that $(ABC)^{-1} = C^{-1}B^{-1}A^{-1}$.

14. Prove that the matrix $\begin{bmatrix} -\frac{2}{3} & \frac{1}{3} & \frac{2}{3} \\ \frac{2}{3} & \frac{2}{3} & \frac{1}{3} \\ \frac{1}{3} & -\frac{2}{3} & \frac{2}{3} \end{bmatrix}$ is orthogonal.

Answers

3. (*i*) $\begin{bmatrix} 4 & \frac{3}{2} & -4 \\ \frac{3}{2} & 3 & -3 \\ -4 & -3 & -7 \end{bmatrix} + \begin{bmatrix} 0 & \frac{1}{2} & 1 \\ -\frac{1}{2} & 0 & -3 \\ -1 & 3 & 0 \end{bmatrix}$

(*ii*) $\begin{bmatrix} a & \frac{1}{2}(a+c) & \frac{1}{2}(b+c) \\ \frac{1}{2}(a+c) & b & \frac{1}{2}(a+b) \\ \frac{1}{2}(b+c) & \frac{1}{2}(a+b) & c \end{bmatrix} + \begin{bmatrix} 0 & \frac{1}{2}(a-c) & \frac{1}{2}(b-c) \\ \frac{1}{2}(c-a) & 0 & \frac{1}{2}(b-a) \\ \frac{1}{2}(c-b) & \frac{1}{2}(a-b) & 0 \end{bmatrix}$ **4.** (*b*) $\begin{bmatrix} 5 & 6 \\ 4 & 10 \end{bmatrix}$

8. (*i*) $\begin{bmatrix} -\frac{1}{2} & \frac{1}{4} & \frac{3}{4} \\ -\frac{1}{2} & -\frac{3}{4} & \frac{11}{4} \\ \frac{1}{2} & \frac{1}{4} & -\frac{5}{4} \end{bmatrix}$ (*ii*) $\begin{bmatrix} -\frac{3}{4} & \frac{1}{4} & \frac{7}{4} \\ -\frac{1}{4} & -\frac{1}{4} & \frac{5}{4} \\ \frac{5}{4} & \frac{1}{4} & -\frac{13}{4} \end{bmatrix}$ (*iii*) $\begin{bmatrix} 7 & -3 & -3 \\ -1 & 1 & 0 \\ -1 & 0 & 1 \end{bmatrix}$.

14.24. SOLUTION OF SIMULTANEOUS LINEAR EQUATIONS BY MATRIX INVERSION METHOD OR MATRIX METHOD

(*a*) **Non-homogeneous system of linear equations**

Let $AX = B$ be a given system of n linear equations in n-unknowns. Then

(*i*) if $|A| \neq 0$, we say that the system is consistent and has a unique solution given by $X = A^{-1}B$.

(*ii*) if $|A| = 0$ and $(\text{adj } A)B = 0$, we say that the system is consistent and has infinitely many solutions.

(*iii*) if $|A| = 0$ and $(\text{adj } A)B \neq 0$, we say that the system is inconsistent and has no solution.

(*b*) **Homogeneous system of linear equations**

Let $AX = 0$ be a given homogeneous system of n linear equations in n-unknowns. Then

(*i*) if $|A| \neq 0$, we say that the system is consistent and has a unique solution or trivial solution or $x = 0, y = 0, z = 0$.

(*ii*) if $|A| = 0$, we say that the system is consistent and has infinite solution or non-trivial solution.

Remark. A homogeneous system is always consistent.

14.25. IF A IS NON-SINGULAR MATRIX, THEN THE SYSTEM OF EQUATIONS AX = B HAS A UNIQUE SOLUTION GIVEN BY $X = A^{-1} B$

Proof. Let A is non-singular. $\therefore$ A^{-1} exists.

Now $AX = B$

$\Rightarrow$ $A^{-1}(AX) = A^{-1}B$ | Pre-multiplying by A^{-1}

$\Rightarrow$ $(A^{-1}A)X = A^{-1}B$

$\Rightarrow$ $IX = A^{-1}B$

$\Rightarrow$ $X = A^{-1}B$

Thus the system of equations AX = B has a solution given by $X = A^{-1} B$.

Uniqueness :

Let X_1 and X_2 be two solutions of AX = B, then

$AX_1 = B$

$AX_2 = B$

$\Rightarrow$ $AX_1 = AX_2$

$\Rightarrow$ $A^{-1}(AX_1) = A^{-1}(AX_2)$

$\Rightarrow$ $(A^{-1}A)X_1 = (A^{-1}A)X_2$

$\Rightarrow$ $IX_1 = IX_2$

$\Rightarrow$ $X_1 = X_2$ *i.e.,* the solution is unique.

ILLUSTRATIVE EXAMPLES

Example 1. *Solve the following equations*

$$x + y + z = 1$$
$$x + 2y + 3z = 6$$
$$x + 3y + 4z = 6, \text{ by matrix inversion method.}$$

Sol. In matrix notations, given system of equations can be written as AX = B, where

$$A = \begin{bmatrix} 1 & 1 & 1 \\ 1 & 2 & 3 \\ 1 & 3 & 4 \end{bmatrix}, X = \begin{bmatrix} x \\ y \\ z \end{bmatrix}, B = \begin{bmatrix} 1 \\ 6 \\ 6 \end{bmatrix}$$

$$|A| = \begin{vmatrix} 1 & 1 & 1 \\ 1 & 2 & 3 \\ 1 & 3 & 4 \end{vmatrix} = (8-9) - (4-3) + (3-2) = -1-1+1 = -1 \neq 0$$

$\therefore$ The given system has a unique solution given by $X = A^{-1}B$ ...(1)

To find A^{-1} : $A_{11} = \begin{vmatrix} 2 & 3 \\ 3 & 4 \end{vmatrix} = 8 - 9 = -1$; $A_{12} = -\begin{vmatrix} 1 & 3 \\ 1 & 4 \end{vmatrix} = -(4-3) = -1$

$$A_{13} = \begin{vmatrix} 1 & 2 \\ 1 & 3 \end{vmatrix} = 3 - 2 = 1 ; \quad A_{21} = -\begin{vmatrix} 1 & 1 \\ 3 & 4 \end{vmatrix} = -(4-3) = -1$$

$$A_{22} = \begin{vmatrix} 1 & 1 \\ 1 & 4 \end{vmatrix} = 4 - 1 = 3 ; \quad A_{23} = -\begin{vmatrix} 1 & 1 \\ 1 & 3 \end{vmatrix} = -(3-1) = -2$$

$$A_{31} = \begin{vmatrix} 1 & 1 \\ 2 & 3 \end{vmatrix} = 3 - 2 = 1 \; ; \quad A_{32} = -\begin{vmatrix} 1 & 1 \\ 1 & 3 \end{vmatrix} = -(3-1) = -2$$

$$A_{33} = \begin{vmatrix} 1 & 1 \\ 1 & 2 \end{vmatrix} = 2 - 1 = 1$$

$$\therefore \quad \text{adj } A = \begin{bmatrix} A_{11} & A_{12} & A_{13} \\ A_{21} & A_{22} & A_{23} \\ A_{31} & A_{32} & A_{33} \end{bmatrix}^T = \begin{bmatrix} -1 & -1 & 1 \\ -1 & 3 & -2 \\ 1 & -2 & 1 \end{bmatrix}$$

$$\therefore \quad A^{-1} = \frac{\text{adj } A}{|A|} = -\begin{bmatrix} -1 & -1 & 1 \\ -1 & 3 & -2 \\ 1 & -2 & 1 \end{bmatrix} = \begin{bmatrix} 1 & 1 & -1 \\ 1 & -3 & 2 \\ 1 & -2 & -1 \end{bmatrix}$$

Hence, from (1),
$$\begin{bmatrix} x \\ y \\ z \end{bmatrix} = \begin{bmatrix} 1 & 1 & -1 \\ 1 & -3 & 2 \\ -1 & 2 & -1 \end{bmatrix} \begin{bmatrix} 1 \\ 6 \\ 6 \end{bmatrix} = \begin{bmatrix} 1+6-6 \\ 1-18+12 \\ -1+12-6 \end{bmatrix} = \begin{bmatrix} 1 \\ -5 \\ 5 \end{bmatrix}$$

$\Rightarrow$ $x = 1, y = -5, z = 5$ is the required solution.

Example 2. *Solve the following system of simultaneous linear equations by matrix method.*

$$x + y + 2z = 4$$
$$2x - y + 3z = 9$$
$$3x - y - z = 2.$$

Sol. In matrix notations, given system of equations can be written as AX = B,

where
$$A = \begin{bmatrix} 1 & 1 & 2 \\ 2 & -1 & 3 \\ 3 & -1 & -1 \end{bmatrix}, X = \begin{bmatrix} x \\ y \\ z \end{bmatrix}, B = \begin{bmatrix} 4 \\ 9 \\ 2 \end{bmatrix}$$

$$|A| = \begin{vmatrix} 1 & 1 & 2 \\ 2 & -1 & 3 \\ 3 & -1 & -1 \end{vmatrix} = (1 + 3) - (-2 - 9) + 2(-2 + 3)$$
$$= 4 + 11 + 2 = 17 \neq 0$$

$\therefore$ The given system has a unique solution given by $X = A^{-1} B$ ...(1)

To find A^{-1}:
$$A_{11} = \begin{vmatrix} -1 & 3 \\ -1 & -1 \end{vmatrix} = 1 + 3 = 4 \; ; \quad A_{12} = -\begin{vmatrix} 2 & 3 \\ 3 & -1 \end{vmatrix} = -(-2-9) = 11$$

$$A_{13} = \begin{vmatrix} 2 & -1 \\ 3 & -1 \end{vmatrix} = -2 + 3 = 1 \; ; \quad A_{21} = -\begin{vmatrix} 1 & 2 \\ -1 & -1 \end{vmatrix} = -(-1+2) = -1$$

$$A_{22} = \begin{vmatrix} 1 & 2 \\ 3 & -1 \end{vmatrix} = -1 - 6 = -7 \; ; \quad A_{23} = -\begin{vmatrix} 1 & 1 \\ 3 & -1 \end{vmatrix} = -(-1-3) = 4$$

$$A_{31} = \begin{vmatrix} 1 & 2 \\ -1 & 3 \end{vmatrix} = 3 + 2 = 5 \; ; \quad A_{32} = -\begin{vmatrix} 1 & 2 \\ 2 & 3 \end{vmatrix} = -(3-4) = 1$$

$$A_{33} = \begin{vmatrix} 1 & 1 \\ 2 & -1 \end{vmatrix} = -1 - 2 = -3$$

$$\therefore \quad \text{adj } A = \begin{bmatrix} A_{11} & A_{12} & A_{13} \\ A_{21} & A_{22} & A_{23} \\ A_{31} & A_{32} & A_{33} \end{bmatrix}^T = \begin{bmatrix} 4 & -1 & 5 \\ 11 & -7 & 1 \\ 1 & 4 & -3 \end{bmatrix}$$

$$\therefore \quad A^{-1} = \frac{\text{adj } A}{|A|} = \frac{1}{17}\begin{bmatrix} 4 & -1 & 5 \\ 11 & -7 & 1 \\ 1 & 4 & -3 \end{bmatrix} = \begin{bmatrix} \frac{4}{17} & \frac{-1}{17} & \frac{5}{17} \\ \frac{11}{17} & \frac{-7}{17} & \frac{1}{17} \\ \frac{1}{17} & \frac{4}{17} & \frac{-3}{17} \end{bmatrix}$$

$\therefore$ From (1), $X = A^{-1} B$

$$\Rightarrow \quad \begin{bmatrix} x \\ y \\ z \end{bmatrix} = \begin{bmatrix} \frac{4}{17} & \frac{-1}{17} & \frac{5}{17} \\ \frac{11}{17} & \frac{-7}{17} & \frac{1}{17} \\ \frac{1}{17} & \frac{4}{17} & \frac{-3}{17} \end{bmatrix}\begin{bmatrix} 4 \\ 9 \\ 2 \end{bmatrix} = \begin{bmatrix} \frac{16-9+10}{17} \\ \frac{44-63+2}{17} \\ \frac{4+36-6}{17} \end{bmatrix} = \begin{bmatrix} \frac{17}{17} \\ -\frac{17}{17} \\ \frac{34}{17} \end{bmatrix} = \begin{bmatrix} 1 \\ -1 \\ 2 \end{bmatrix}$$

$\Rightarrow$ $x = 1, y = -1, z = 2$, is the required solution.

Example 3. *Solve, by matrix inversion method.*

$$x + y + z = 6$$
$$x + 2y + 3z = 10$$
$$x + 2y + 4z = 1.$$

(P.T.U., M.C.A. May 2007)

Sol. In matrix notations, given system of equations can be written as AX = B,

where
$$A = \begin{bmatrix} 1 & 1 & 1 \\ 1 & 2 & 3 \\ 1 & 2 & 4 \end{bmatrix}, X = \begin{bmatrix} x \\ y \\ z \end{bmatrix}, B = \begin{bmatrix} 6 \\ 10 \\ 1 \end{bmatrix}$$

Here
$$|A| = \begin{vmatrix} 1 & 1 & 1 \\ 1 & 2 & 3 \\ 1 & 2 & 4 \end{vmatrix} = (8-6) - (4-3) + (2-2) = 2 - 1 = 1 \neq 0$$

$\therefore$ The given system has a unique solution given by $X = A^{-1} B$ ...(1)

To find A^{-1}:

$$A_{11} = \begin{vmatrix} 2 & 3 \\ 2 & 4 \end{vmatrix} = 8 - 6 = 2\ ; \qquad A_{12} = -\begin{vmatrix} 1 & 3 \\ 1 & 4 \end{vmatrix} = -(4-3) = -1$$

$$A_{13} = \begin{vmatrix} 1 & 2 \\ 1 & 2 \end{vmatrix} = -2 - 2 = 0\ ; \qquad A_{21} = -\begin{vmatrix} 1 & 1 \\ 2 & 4 \end{vmatrix} = -(4-2) = -2$$

$$A_{22} = \begin{vmatrix} 1 & 1 \\ 1 & 4 \end{vmatrix} = 4 - 1 = 3\ ; \qquad A_{23} = -\begin{vmatrix} 1 & 1 \\ 1 & 2 \end{vmatrix} = -(2-1) = -1$$

$$A_{31} = \begin{vmatrix} 1 & 1 \\ 2 & 3 \end{vmatrix} = 3 - 2 = 1\ ; \qquad A_{32} = -\begin{vmatrix} 1 & 1 \\ 1 & 3 \end{vmatrix} = -(3-1) = -2$$

$$A_{33} = \begin{vmatrix} 1 & 1 \\ 1 & 2 \end{vmatrix} = 2 - 1 = -1$$

$\therefore$ $$\text{adj } A = \begin{bmatrix} A_{11} & A_{12} & A_{13} \\ A_{21} & A_{22} & A_{23} \\ A_{31} & A_{32} & A_{33} \end{bmatrix}^T = \begin{bmatrix} 2 & -2 & 1 \\ -1 & 3 & -2 \\ 0 & -1 & 1 \end{bmatrix}$$

$\therefore$ $$A^{-1} = \frac{\text{adj } A}{|A|} = \begin{bmatrix} 2 & -2 & 1 \\ -1 & 3 & -2 \\ 0 & -1 & 1 \end{bmatrix}$$

$\therefore$ From (1), $X = A^{-1} B$

$\Rightarrow$ $$\begin{bmatrix} x \\ y \\ z \end{bmatrix} = \begin{bmatrix} 2 & -2 & 1 \\ -1 & 3 & -2 \\ 0 & -1 & 1 \end{bmatrix} \begin{bmatrix} 6 \\ 10 \\ 1 \end{bmatrix} = \begin{bmatrix} 12 - 20 + 1 \\ -6 + 30 - 2 \\ 0 - 10 + 1 \end{bmatrix} = \begin{bmatrix} -7 \\ 22 \\ -9 \end{bmatrix}$$

$\Rightarrow$ $x = -7, y = 22, z = -9$, is the required solution.

TEST YOUR KNOWLEDGE 14.3

1. Apply Matrix method or matrix inversion method to solve the equations :
   (*i*) $x + y + z = 0,\ 2x + 5y + 3z = 1,\ -x + 2y + z = 2$
   (*ii*) $3x + y + 2z = 3,\ 2x - 3y - z = -3,\ x + 2y + z = 4$
   (*iii*) $x + y + z = 1,\ 2x + 3y + 2z = 2,\ 3x + 3y + 4z = 1$
   (*iv*) $x + 2y + 3z = 6,\ 2x + 4y + z = 7,\ 3x + 2y + 9z = 14$
   (*v*) $5x - 6y + 4z = 15,\ 7x + 4y - 3z = 19,\ 2x + y + 6z = 46$
   (*vi*) $7x + 10y + 5z = 42,\ 13x + 6y + 2z = 31,\ 11x + 14y + 8z = 63$
   (*vii*) $2x + y + 4z = 2,\ x + 3y = 2z + 7,\ 5z - 8 = 5x + 3y$
   (*viii*) $x + 3y + 6z = 2,\ 3x - y + 4z = 9,\ x - 4y + 2z = 7$
   (*ix*) $x + y + z = 6.6,\ x - y + z = 2.2,\ x + 2y + 3z = 15.2$.
   (*x*) $2yz - zx + xy = 3xyz,\ 3yz + 2zx + 4xy = 19xyz,\ 6yz + 7zx - xy = 17xyz$.

   $\left[\textbf{Hint.} \text{ Divide the equations by } xyz \text{ and put } \frac{1}{x} = u, \frac{1}{y} = v, \frac{1}{z} = w \right]$

   (*xi*) $x + y + z = 9,\ 2x + 5y + 72 = 5z,\ 2x + y - z = 0$. *(P.T.U., M.C.A. May 2008)*
   (*xii*) $10x + y + z = 12,\ x + 10y + z = 12,\ x + y + 10z = 12$.
2. Solve the following system of equations by matrix method

   $2x_1 - x_2 + 3x_3 = 9$

   $x_2 - x_3 = -1$

   $x_1 + x_2 - x_3 = 0$, by matrix method.
3. Solve by matrix method

   $x - y - z = 1 = y - z - x = z - x - y$

4. (a) Solve : $x + y - z = 2$

$$2x - y + z = 1$$

$$x - y - 2z = -3$$ by matrix method.

(b) Solve for x, y, z, the following equations

$$x + y + z = 3$$

$$2x - y + z = 2$$

$$x - 2y + 3z = 2$$

5. Does the system $x - 2y + 3z = 0, 2x + y - 4z = 0, x - y + z = 0$ have non-trivial solutions ? If so, find these solutions.

6. Determine the value of λ for which the equations $3x - 2y + z = 0, \lambda x - 14y + 15z = 0, x + 2y - 3z = 0$ have non-trivial solutions. Find the solutions.

7. Find the values of λ for which the equations

$$(\lambda - 1)x + (3\lambda + 1)y + 2\lambda z = 0,\ (\lambda - 1)x + (4\lambda - 2)y + (\lambda + 3)z = 0,\ 2x + (3\lambda + 1)y + 3(\lambda - 1)z = 0$$

are consistent, and find the ratios of x, y, z when λ has the smallest of these values. What happens when λ has the greatest of these values ?

8. Prove that if the equations $x = ay + z,\ y = z + ax,\ z = x + y,$ are consistent (have non-zero solution), then $a^3 + 1 = 0$.

9. Test the consistency of the following equations :

$$x + 2y + 3z + 14 = 0,\ 5x + 9y + 11z + 56 = 0,\ 3x + 5y + z + 16 = 0,\ 2x + 7y + 5z + 30 = 0.$$

10. Test for consistency the following system and find the solution, if it exists :

$$3x + 3y + 2z = 1,\ x + 2y = 4,\ 10y + 3z = -2,\ 2x - 3y - z = 5.$$

Answers

1. (i) $x = -1, y = 0, z = 1$ (ii) $x = 1, y = 2, z = -1$ (iii) $x = 3, y = 0, z = -2$
(iv) $x = y = z = 1$ (v) $x = 3, y = 4, z = 6$ (vi) $x = 1, y = 2, z = 3$
(vii) $x = -3, y = 4, z = 1$ (viii) $x = 2, y = -1, z = \frac{1}{2}$ (ix) $x = 1.2, y = 2.2, z = 3.2$
(x) $x = 1, y = \frac{1}{2}, z = \frac{1}{3}$ (xi) $x = 1, y = 3, z = 5$ (xii) $x = 1, y = 1, z = 1$.

2. $x = 1, y = 2, z = 3$ 3. $x = -1, y = -1, z = -1$

4. (a) $x = 1, y = 2, z = 1$ (b) $x = 1, y = 1, z = 1$

5. Yes ; $x = k, y = 2k, z = k$, where k is arbitrary.

6. $\lambda = 5$; $x = 2k, y = 5k, z = 4k$.

7. $\lambda = 0, 3$; when $\lambda = 0, x = y = z$; when $\lambda = 3$, equations become identical.

9. Inconsistent. 10. Consistent ; $x = 2, y = 1, z = -4$.

14.26. RANK OF A MATRIX (*P.T.U., M.C.A. May 2007*)

Let A be any $m \times n$ matrix. It has square sub-matrices of different orders. The determinants of these square sub-matrices are called minors of A. If **all** minors of order $(r + 1)$ are zero but there **is at least one** non-zero minor of order r, then r is called the rank of A. Symbolically, rank of A = r is written as $\rho(A) = r$ or $r(A) = r$.

From the definition of the rank of a matrix A, it follows that :

(*i*) If A is a null matrix, then $\rho(A) = 0$. [$\because$ every minor of A has zero value.]

(*ii*) If A is not a null matrix, then $\rho(A) \geq 1$.

(*iii*) If A is a non-singular $n \times n$ matrix, then $\rho(A) = n$

[$\because$ $|A| \neq 0$ is the largest minor of A.]

If I_n is the $n \times n$ unit matrix, then $|I_n| = 1 \neq 0 \Rightarrow \rho(I_n) = n$.

(*iv*) If A is an $m \times n$ matrix, then $\rho(A) \leq$ minimum of m and n.

(*v*) If all minors of order r are equal to zero, then $\rho(A) < r$.

14.27. TO DETERMINE THE RANK OF A MATRIX A

We adopt the following different methods :

Start with the highest order minor (or minors) of A. Let their order be r. **If any one of them** is non-zero, then $\rho(A) = r$.

If **all of them** are zero, start with minors of next lower order $(r - 1)$ and so on till you get a non-zero minor. The order of that minor is the rank of A.

This method **usually** involves a lot of computational work since we have to evaluate several determinants.

$\therefore$ We generally find the rank of a matrix by reducing it to echelon form.

14.28. ECHELON FORM OF A MATRIX

A matrix A is said to be in echelon form if it satisfies the following properties:

(*i*) All entries below the leading element are zero.

(*ii*) The number of zeros before the first non-zero element in a row is more than the number of such zeros in the previous row.

For example $A = \begin{bmatrix} 1 & 0 & 2 \\ 0 & 3 & 0 \\ 0 & 0 & 4 \end{bmatrix}$ is in echelon form

where as $B = \begin{bmatrix} 1 & 0 & 2 \\ 0 & 3 & 0 \\ 0 & 3 & 4 \end{bmatrix}$ is non-echelon form

14.29. RANK OF A MATRIX BY USING ECHELON FORM

The rank of a matrix is equal to number of non-zero rows in its echelon form.

For example $r(A) = 3$, where $A = \begin{bmatrix} 1 & 0 & 2 \\ 0 & 3 & 0 \\ 0 & 0 & 4 \end{bmatrix}$.

ILLUSTRATIVE EXAMPLES

Example 1. *Find the rank of the matrices*

(*i*) $\begin{bmatrix} 1 & 2 & 3 & 4 \\ -2 & 0 & 5 & 7 \end{bmatrix}$ (*ii*) $\begin{bmatrix} 1 & 2 & 3 \\ 1 & 4 & 2 \\ 2 & 6 & 5 \end{bmatrix}$.

Sol. (*i*) Here $A = \begin{bmatrix} 1 & 2 & 3 & 4 \\ -2 & 0 & 5 & 7 \end{bmatrix}$ is a 2 × 4 matrix.

$\therefore \quad r(A) \le 2$ (smaller of 2 and 4)

Operating $R_2 \to R_2 + 2R_1$, we get

$$A \sim \begin{bmatrix} 1 & 2 & 3 & 4 \\ 0 & 4 & 11 & 15 \end{bmatrix}, \text{ which is in echelon form}$$

$\therefore$ $r(A) = 2$ (No. of non-zero rows in echelon form)

(*ii*) Here $A = \begin{bmatrix} 1 & 2 & 3 \\ 1 & 4 & 2 \\ 2 & 6 & 5 \end{bmatrix}$ is a 3 × 3 matrix. $\therefore \quad r(A) \le 3$

Operating $R_2 \to R_2 - R_1$, $R_3 \to R_3 - 2R_1$, we get

$$A \sim \begin{bmatrix} 1 & 2 & 3 \\ 0 & 2 & -1 \\ 0 & 2 & -1 \end{bmatrix}, \text{ operating } R_3 \to R_3 - R_2$$

$$\sim \begin{bmatrix} 1 & 2 & 3 \\ 0 & 2 & -1 \\ 0 & 0 & 0 \end{bmatrix}, \text{ which is in echelon form.}$$

$\therefore$ $r(A) = 2$ (No. of non-zero rows in echelon form).

Example 2. *Find the rank of the matrices*

(*i*) $\begin{bmatrix} 2 & 3 & 4 & -1 \\ 5 & 2 & 0 & -1 \\ -4 & 5 & 12 & -1 \end{bmatrix}$. (*ii*) $\begin{bmatrix} 1 & 2 & 3 & 0 \\ 2 & 4 & 3 & 2 \\ 3 & 2 & 1 & 3 \\ 6 & 8 & 7 & 5 \end{bmatrix}$. (P.T.U., M.C.A. May 2008)

(*iii*) $\begin{bmatrix} 2 & 3 & -1 & -1 \\ 1 & -1 & -2 & -4 \\ 3 & 1 & 3 & -2 \\ 6 & 3 & 0 & -7 \end{bmatrix}$.

Sol. (*i*) Here $A = \begin{bmatrix} 2 & 3 & 4 & -1 \\ 5 & 2 & 0 & -1 \\ -4 & 5 & 12 & -1 \end{bmatrix}$ is a 3 × 4 matrix.

$\therefore$ $r(A) \le 3$ (smaller of 3 and 4)

Interchanging C_1 and C_4, we get

$$A \sim \begin{bmatrix} -1 & 3 & 4 & 2 \\ -1 & 2 & 0 & 5 \\ -1 & 5 & 12 & -4 \end{bmatrix}, \text{ operating } R_2 \to R_2 - R_1,\ R_3 \to R_3 - R_1$$

$$\sim \begin{bmatrix} -1 & 3 & 4 & 2 \\ 0 & -1 & -4 & 3 \\ 0 & 2 & 8 & -6 \end{bmatrix}, \text{ operating } R_3 \to R_3 + 2R_2$$

$$\sim \begin{bmatrix} -1 & 3 & 4 & 2 \\ 0 & -1 & -4 & 3 \\ 0 & 0 & 0 & 0 \end{bmatrix}, \text{ which is in echelon form.}$$

$\therefore$ $r(A) = 2$ (No. of non-zero rows in echelon form).

(*ii*) Here $A = \begin{bmatrix} 1 & 2 & 3 & 0 \\ 2 & 4 & 3 & 2 \\ 3 & 2 & 1 & 3 \\ 6 & 8 & 7 & 5 \end{bmatrix}$ is a 4×4 matrix.

$\therefore$ $r(A) \le 4$.

Operating $R_2 \to R_2 - 2R_1$, $R_3 \to R_3 - 3R_1$, $R_4 \to R_4 - 6R_1$, we get

$$A \sim \begin{bmatrix} 1 & 2 & 3 & 0 \\ 0 & 0 & -3 & 2 \\ 0 & -4 & -8 & 3 \\ 0 & -4 & -11 & 5 \end{bmatrix}, \text{ interchanging } R_2 \text{ and } R_3$$

$$\sim \begin{bmatrix} 1 & 2 & 3 & 0 \\ 0 & -4 & -8 & 3 \\ 0 & 0 & -3 & 2 \\ 0 & -4 & -11 & 5 \end{bmatrix}, \text{ operating } R_4 \to R_4 - R_2$$

$$\sim \begin{bmatrix} 1 & 2 & 3 & 0 \\ 0 & -4 & -8 & 3 \\ 0 & 0 & -3 & 2 \\ 0 & 0 & -3 & 2 \end{bmatrix}, \text{ operating } R_4 \to R_4 - R_3$$

$$\sim \begin{bmatrix} 1 & 2 & 3 & 0 \\ 0 & -4 & -8 & 3 \\ 0 & 0 & -3 & 2 \\ 0 & 0 & 0 & 0 \end{bmatrix}, \text{ which is in echelon form.}$$

$\therefore$ $r(A) = 3$ (No. of non-zero rows in echelon form).

(*iii*) Here $A = \begin{bmatrix} 2 & 3 & -1 & -1 \\ 1 & -1 & -2 & -4 \\ 3 & 1 & 3 & -2 \\ 6 & 3 & 0 & -7 \end{bmatrix}$, interchanging R_1 and R_2

$$\sim \begin{bmatrix} 1 & -1 & -2 & -4 \\ 2 & 3 & -1 & -1 \\ 3 & 1 & 3 & -2 \\ 6 & 3 & 0 & -7 \end{bmatrix}$$

Operating $R_2 \to R_2 - 2R_1$, $R_3 \to R_3 - 3R_1$, $R_4 \to R_4 - 6R_1$,

$$\sim \begin{bmatrix} 1 & -1 & -2 & -4 \\ 0 & 5 & 3 & 7 \\ 0 & 4 & 9 & 10 \\ 0 & 9 & 12 & 17 \end{bmatrix}, \text{ operating } R_3 \to R_3 + R_2$$

$$\sim \begin{bmatrix} 1 & -1 & -2 & -4 \\ 0 & 5 & 3 & 7 \\ 0 & 9 & 12 & 17 \\ 0 & 9 & 12 & 17 \end{bmatrix}, \text{ operating } R_4 \to R_4 - R_3$$

$$\sim \begin{bmatrix} 1 & -1 & -2 & -4 \\ 0 & 5 & 3 & 7 \\ 0 & 9 & 12 & 17 \\ 0 & 0 & 0 & 0 \end{bmatrix}, \text{ operating } R_3 \to R_3 - \frac{9}{5} R_2$$

$$\sim \begin{bmatrix} 1 & -1 & -2 & -4 \\ 0 & 5 & 3 & 7 \\ 0 & 0 & \frac{33}{5} & \frac{2}{5} \\ 0 & 0 & 0 & 0 \end{bmatrix}, \text{ which is in echelon form.}$$

$\therefore$ $r(A) = 3$ (No. of non-zero rows in echelon form).

TEST YOUR KNOWLEDGE 14.4

1. Find the rank of the matrices

(i) $\begin{bmatrix} 2 & -1 & 0 & 5 \\ 0 & 3 & 1 & 4 \end{bmatrix}$ (ii) $\begin{bmatrix} 1 & 2 & 3 \\ 2 & 4 & 7 \\ 3 & 6 & 10 \end{bmatrix}$ (iii) $\begin{bmatrix} 1 & 3 & 4 & 5 \\ 1 & 2 & 6 & 7 \\ 1 & 5 & 0 & 10 \end{bmatrix}$

(iv) $\begin{bmatrix} 1 & 0 & 0 \\ 0 & 1 & 0 \\ 0 & 0 & 1 \end{bmatrix}$ *(P.T.U., M.C.A. May 2007)* (v) $\begin{bmatrix} 1 & 2 & 3 \\ 3 & 4 & 5 \\ -1 & 1 & 1 \end{bmatrix}$. *(P.T.U., M.C.A. May 2007)* (vi) $\begin{pmatrix} 1 & 1 & -1 \\ 2 & -3 & 4 \\ 3 & -2 & 3 \end{pmatrix}$. *(P.T.U., M.C.A. Dec. 2008)*

2. Determine the rank of the matrix $A = \begin{bmatrix} 6 & 1 & 3 & 8 \\ 4 & 2 & 6 & -1 \\ 10 & 3 & 9 & 7 \\ 16 & 4 & 12 & 15 \end{bmatrix}$

3. Find the rank of

(i) $\begin{bmatrix} 3 & -1 & 2 \\ -6 & 2 & 4 \\ -3 & 1 & 2 \end{bmatrix}$ (ii) $\begin{bmatrix} 1 & 2 & 3 & 2 \\ 2 & 3 & 5 & 1 \\ 1 & 3 & 4 & 5 \end{bmatrix}$

(iii) $\begin{bmatrix} 1 & -1 & 2 & -3 \\ 4 & 1 & 0 & 2 \\ 0 & 3 & 1 & 4 \\ 0 & 1 & 0 & 2 \end{bmatrix}$ (iv) $\begin{bmatrix} 0 & 1 & -3 & -1 \\ 1 & 0 & 1 & 1 \\ 3 & 1 & 0 & 2 \\ 1 & 1 & -2 & 0 \end{bmatrix}$

(v) $\begin{bmatrix} 1 & 2 & -2 \\ 1 & 1 & 2 \\ 1 & 3 & 1 \end{bmatrix}$. *(P.T.U., M.C.A. Dec. 2006)*

4. Determine the rank of AB where

$$A = \begin{bmatrix} 1 & 1 & -1 \\ 2 & -3 & 4 \\ 3 & -2 & 3 \end{bmatrix}, \quad B = \begin{bmatrix} -1 & -2 & -1 \\ 6 & 12 & 6 \\ 5 & 10 & 5 \end{bmatrix}$$

Answers

1. (*i*) 2 (*ii*) 2 (*iii*) 3 (*iv*) 3 (*v*) 3 (*vi*) 2.

2. 2 **3.** (*i*) 2 (*ii*) 2 (*iii*) 4 (*iv*) 2 (*v*) 3.

4. does not defined.

14.30. SOLUTION OF A SYSTEM OF LINEAR EQUATIONS (Rank Method) or (Gauss Jordan Method)

Consider the system of equations $\left.\begin{aligned} a_1x + b_1y + c_1z &= d_1 \\ a_2x + b_2y + c_2z &= d_2 \\ a_3x + b_3y + c_3z &= d_3 \end{aligned}\right\}$ (3 equations in 3 unknowns)

In matrix notation, these equations can be written as

$$\begin{bmatrix} a_1x + b_1y + c_1z \\ a_2x + b_2y + c_2z \\ a_3x + b_3y + c_3z \end{bmatrix} = \begin{bmatrix} d_1 \\ d_2 \\ d_3 \end{bmatrix}$$

or

$$\begin{bmatrix} a_1 & b_1 & c_1 \\ a_2 & b_2 & c_2 \\ a_3 & b_3 & c_3 \end{bmatrix} \begin{bmatrix} x \\ y \\ z \end{bmatrix} = \begin{bmatrix} d_1 \\ d_2 \\ d_3 \end{bmatrix}$$

or

$$AX = B$$

where $A = \begin{bmatrix} a_1 & b_1 & c_1 \\ a_2 & b_2 & c_2 \\ a_3 & b_3 & c_3 \end{bmatrix}$ is called the co-efficient matrix, $X = \begin{bmatrix} x \\ y \\ z \end{bmatrix}$ is the column matrix of unknowns, $B = \begin{bmatrix} d_1 \\ d_2 \\ d_3 \end{bmatrix}$ is the column matrix of constants.

If $d_1 = d_2 = d_3 = 0$, then B = O and the matrix equation AX = B reduces to AX = O. Such a system of equations is called a system of **homogeneous linear equations.**

If at least one of d_1, d_2, d_3 is non-zero, then $B \neq O$.

Such a system of equations is called a system of **non-homogeneous linear equations.**

Solving the matrix equation AX = B means finding X, *i.e.*, finding a column matrix $\begin{bmatrix} \alpha \\ \beta \\ \gamma \end{bmatrix}$ such that

$$X = \begin{bmatrix} x \\ y \\ z \end{bmatrix} = \begin{bmatrix} \alpha \\ \beta \\ \gamma \end{bmatrix}. \text{ Then } x = \alpha,\ y = \beta,\ z = \gamma.$$

The matrix equation AX = B need not always have a solution. It may have no solution or a unique solution or an infinite number of solutions.

A system of equations having **no solution** is called an **inconsistent system** of equations.

A system of equations having **one or more solution** is called **a consistent system** of equations.

Thus for a system of non-homogeneous linear equations AX = B.

(*i*) if $\rho[A : B] \neq \rho(A)$, the system is inconsistent.

(*ii*) if $\rho[A : B] = \rho(A)$ = number of unknowns, the system has a unique solution.

(*iii*) $\rho[A : B] = \rho(A)$ < number of unknowns, the system has an infinite number of solutions.

The matrix [A : B] in which the elements of A and B are written side by side is called the **augmented matrix.**

For a system of homogeneous linear equations AX = O

(*i*) X = O is always a solution. This solution in which each unknown has the value zero is called the **Null Solution** or the **Trivial Solution.** Thus a homogeneous system is always consistent.

A system of homogeneous linear equations has either the trivial solution or an infinite number of solutions.

(*ii*) if ρ (A) = number of unknowns, the system has only the trivial solution.

(*iii*) if ρ (A) < number of unknowns, the system has an infinite number of non-trivial solutions.

14.31. THEOREM : IF A IS A NON-SINGULAR MATRIX, THEN THE MATRIX EQUATION AX = B HAS A UNIQUE SOLUTION

Proof. The given equation is $\quad AX = B \quad$ ...(1)

$\because$ A is a non-singular matrix, $\therefore$ A^{-1} exists.

Pre-multiplying both sides of (1) by A^{-1}, we get

$$A^{-1}AX = A^{-1}B \quad \text{or} \quad (A^{-1}A)X = A^{-1}B$$

or $$IX = A^{-1}B \quad \text{or} \quad X = A^{-1}B$$

which is the required unique solution (since A^{-1} is unique).

ILLUSTRATIVE EXAMPLES

Example 1. *Solve, with the help of rank method or by Gauss Jordan method, the simultaneous equations :*

$$x + y + z = 3, \quad x + 2y + 3z = 4, x + 4y + 9z = 6.$$

Sol. Augmented matrix $[A : B] = \begin{bmatrix} 1 & 1 & 1 & : & 3 \\ 1 & 2 & 3 & : & 4 \\ 1 & 4 & 9 & : & 6 \end{bmatrix}$,

Operating $R_2 \to R_2 - R_1$, $R_3 \to R_3 - R_1$, we get

$$[A : B] \sim \begin{bmatrix} 1 & 1 & 1 & : & 3 \\ 0 & 1 & 2 & : & 1 \\ 0 & 3 & 8 & : & 3 \end{bmatrix}, \text{ operating } R_3 \to R_3 - 3R_2$$

$$\sim \begin{bmatrix} 1 & 1 & 1 & : & 3 \\ 0 & 1 & 2 & : & 1 \\ 0 & 0 & 2 & : & 0 \end{bmatrix}, \text{ which is in echelon form}$$

$\therefore$ $r(A : B) = 3$, $r(A) = 3$, no. of unknowns = 3

$\therefore$ The system has a unique solution.

From 3rd row, $2z = 0 \Rightarrow z = 0$

From 2nd row, $y + 2z = 1 \Rightarrow y = 1$

From 1st row, $x + y + z = 3 \Rightarrow x = 2$

$\therefore$ $x = 2, \quad y = 1, \quad z = 0$

Example 2. *Solve the system of equations by using Gauss Jordan method*

$$2x_1 + x_2 + 2x_3 + x_4 = 6$$
$$4x_1 + 3x_2 + 3x_3 - 3x_4 = -1$$
$$6x_1 - 6x_2 + 6x_3 + 12x_4 = 36$$
$$2x_1 + 2x_2 - x_3 + x_4 = 10.$$

Sol. In matrix notation, the given system of equations is AX = B, where

$$A = \begin{bmatrix} 2 & 1 & 2 & 1 \\ 4 & 3 & 3 & -3 \\ 6 & -6 & 6 & 12 \\ 2 & 2 & -1 & 1 \end{bmatrix}, X = \begin{bmatrix} x_1 \\ x_2 \\ x_3 \\ x_4 \end{bmatrix}, B = \begin{bmatrix} 6 \\ -1 \\ 36 \\ 10 \end{bmatrix}$$

$$\text{Augmented matrix } [A : B] = \begin{bmatrix} 2 & 1 & 2 & 1 & : & 6 \\ 4 & 3 & 3 & -3 & : & -1 \\ 6 & -6 & 6 & 12 & : & 36 \\ 2 & 2 & -1 & 1 & : & 10 \end{bmatrix}$$

Operating $R_2 \to R_2 - 2R_1$, $R_3 \to R_3 - 3R_1$, $R_4 \to R_4 - R_2$, we get

$$[A : B] \sim \begin{bmatrix} 2 & 1 & 2 & 1 & : & 6 \\ 0 & 1 & -1 & -5 & : & -13 \\ 0 & -9 & 0 & 9 & : & 18 \\ 0 & 1 & -3 & 0 & : & 4 \end{bmatrix}, \text{ operating } R_3 \to R_3 + 9R_2, R_4 \to R_4 - R_2$$

$$\sim \begin{bmatrix} 2 & 1 & 2 & 1 & : & 6 \\ 0 & 1 & -1 & -5 & : & -13 \\ 0 & 0 & -9 & -36 & : & -99 \\ 0 & 0 & -2 & 5 & : & 17 \end{bmatrix}, \text{ operating } R_3\left(-\frac{1}{9}\right)$$

$$\sim \begin{bmatrix} 2 & 1 & 2 & 1 & : & 6 \\ 0 & 1 & -1 & -5 & : & -13 \\ 0 & 0 & 1 & 4 & : & 11 \\ 0 & 0 & -2 & 5 & : & 17 \end{bmatrix}, \text{ operating } R_4 \to R_4 + 2R_3$$

$$\sim \begin{bmatrix} 2 & 1 & 2 & 1 & : & 6 \\ 0 & 1 & -1 & -5 & : & -13 \\ 0 & 0 & 1 & 4 & : & 11 \\ 0 & 0 & 0 & 13 & : & 39 \end{bmatrix}, \text{ which is in echelon form.}$$

$\therefore \quad r\,[A : B] = 4,\ r(A) = 4$, number of unknowns = 4

$\therefore$ The system has a unique solution.

From 4th row, $13x_4 = 39 \Rightarrow x_4 = 3$

From 3rd row, $x_3 + 4x_4 = 11 \Rightarrow x_3 + 12 = 11 \Rightarrow x_3 = 11 - 12 = -1$

From 2nd row, $x_2 - x_3 - 5x_4 = -13$

$\Rightarrow \quad x_2 = -13 - 1 + 15 = 1$

From Ist row, $2x_1 + x_2 + 2x_3 + x_4 = 6$

$\Rightarrow \quad 2x_1 + 1 - 2 + 3 = 6$

$\Rightarrow \quad 2x_1 = 4 \Rightarrow x_1 = 2.$

Hence $x_1 = 2, x_2 = 1, x_3 = -1, x_4 = 3$ is the required solution

Example 3. *Show that the equations*

$3x + 3y + 2z = 1,\ 10y + 3z = -2,\ x + 2y = 4,\ 2x - 3y - z = 5,$

are consistent and hence obtain the solutions for x, y, z by using Rank method.

Sol. In matrix notation, the given system of equations can be written as AX = B, where

$$A = \begin{bmatrix} 3 & 3 & 2 \\ 0 & 10 & 3 \\ 1 & 2 & 0 \\ 2 & -3 & -1 \end{bmatrix},\ X = \begin{bmatrix} x \\ y \\ z \end{bmatrix},\ B = \begin{bmatrix} 1 \\ -2 \\ 4 \\ 5 \end{bmatrix}$$

$$\text{Augmented matrix } [A : B] = \begin{bmatrix} 3 & 3 & 2 & : & 1 \\ 0 & 10 & 3 & : & -2 \\ 1 & 2 & 0 & : & 4 \\ 2 & -3 & -1 & : & 5 \end{bmatrix}, \text{ interchanging } R_1 \text{ and } R_3, \text{ we get}$$

$$[A : B] \sim \begin{bmatrix} 1 & 2 & 0 & : & 4 \\ 0 & 10 & 3 & : & -2 \\ 3 & 3 & 2 & : & 1 \\ 2 & -3 & -1 & : & 5 \end{bmatrix}, \text{ operating } R_3 \to R_3 - 3R_1, R_4 \to R_4 - 2R_1$$

$$\sim \begin{bmatrix} 1 & 2 & 0 & : & 4 \\ 0 & 10 & 3 & : & -2 \\ 0 & -3 & 2 & : & -11 \\ 0 & -7 & -1 & : & -3 \end{bmatrix}, \text{ operating } R_3 \to R_3 + R_4$$

$$\sim \begin{bmatrix} 1 & 2 & 0 & : & 4 \\ 0 & 10 & 3 & : & -2 \\ 0 & -10 & 1 & : & -14 \\ 0 & -7 & -1 & : & -3 \end{bmatrix}, \text{ operating } R_3 \to R_3 + R_2$$

$$\sim \begin{bmatrix} 1 & 2 & 0 & : & 4 \\ 0 & 10 & 3 & : & -2 \\ 0 & 0 & 4 & : & -16 \\ 0 & -7 & -1 & : & -3 \end{bmatrix}, \text{ operating } R_4 \to R_4 + \frac{7}{10} R_2$$

$$\sim \begin{bmatrix} 1 & 2 & 0 & : & 4 \\ 0 & 10 & 3 & : & -2 \\ 0 & 0 & 4 & : & -16 \\ 0 & 0 & \frac{11}{10} & : & -\frac{44}{10} \end{bmatrix}, \text{ operating } R_4 \left[\frac{10}{11}\right]$$

$$\sim \begin{bmatrix} 1 & 2 & 0 & : & 4 \\ 0 & 10 & 3 & : & -2 \\ 0 & 0 & 4 & : & -16 \\ 0 & 0 & 1 & : & -4 \end{bmatrix}, \text{ operating } R_4 \to R_4 - \frac{1}{4} R_3$$

$$\sim \begin{bmatrix} 1 & 2 & 0 & : & 4 \\ 0 & 10 & 3 & : & -2 \\ 0 & 0 & 4 & : & -16 \\ 0 & 0 & 0 & : & 0 \end{bmatrix}, \text{ which is in echelon form.}$$

r [A : B] = 3, r (A) = 3, number of unknowns = 3.

∴ The system has a unique solution.

From 3rd row, $4z = -16 \Rightarrow z = -4$

From 2nd row, $10y + 3z = -2 \Rightarrow 10y - 12 = -2 \Rightarrow y = 1$

From Ist row, $x + 2y = 4 \Rightarrow x = 2.$

Hence $x = 2$, $y = 1$, $z = -4$, is the required solution

Example 4. *Test for consistency the equations*

$$x - 3y - 8z = -10 ; \quad 3x + y - 4z = 0 ; 2x + 5y + 6z = 13 \text{ and hence solve it.}$$

Sol. In matrix notation, the given system of equations can be written as AX = B,

where
$$A = \begin{bmatrix} 1 & -3 & -8 \\ 3 & 1 & -4 \\ 2 & 5 & 6 \end{bmatrix}, X = \begin{bmatrix} x \\ y \\ z \end{bmatrix}, B = \begin{bmatrix} -10 \\ 0 \\ 13 \end{bmatrix}$$

$$\text{Augmented matrix } [A : B] = \begin{bmatrix} 1 & -3 & -8 & : & -10 \\ 3 & 1 & -4 & : & 0 \\ 2 & 5 & 6 & : & 13 \end{bmatrix}$$

Operating $R_2 \to R_2 - 3R_1$, $R_3 \to R_3 - 2R_1$, we get

$$[A : B] \sim \begin{bmatrix} 1 & -3 & -8 & : & -10 \\ 0 & 10 & 20 & : & 30 \\ 0 & 11 & 22 & : & 33 \end{bmatrix}, \text{ operating } R_2 \left(\frac{1}{10}\right), R_3 \left(\frac{1}{11}\right)$$

$$\sim \begin{bmatrix} 1 & -3 & -8 & : & -10 \\ 0 & 1 & 2 & : & 3 \\ 0 & 1 & 2 & : & 3 \end{bmatrix}, \text{ operating } R_3 \to R_3 - R_2$$

$$\sim \begin{bmatrix} 1 & -3 & -8 & : & -10 \\ 0 & 1 & 2 & : & 3 \\ 0 & 0 & 0 & : & 0 \end{bmatrix}$$

⇒ r [A : B] = r(A) = 2 < 3 = number of unknowns.

∴ The system has infinite solution. Take $z = k$

From 2nd row, $y + 2z = 3 \quad \Rightarrow \quad y = 3 - 2k$

From 1st row, $x - 3y - 8z = -10 \Rightarrow \quad x - 9 + 6k - 8k = -10$

$$\Rightarrow \quad x = -10 + 9 + 2k = -1 + 2k$$

∴ The required solution is

$$x = -1 + 2k, \quad y = 3 - 2k, \quad z = k.$$

Example 5. *For what values of λ and μ do the system of equations :*

$$x + y + z = 6, \quad x + 2y + 3z = 10, \quad x + 2y + \lambda z = \mu$$

have (i) no solution (ii) unique solution (iii) more than one solutions ?

Sol. In matrix notation, the given system of equations can be written as

$$AX = B$$

where $A = \begin{bmatrix} 1 & 1 & 1 \\ 1 & 2 & 3 \\ 1 & 2 & \lambda \end{bmatrix}, X = \begin{bmatrix} x \\ y \\ z \end{bmatrix}, B = \begin{bmatrix} 6 \\ 10 \\ \mu \end{bmatrix}$

Augmented matrix $[A \vdots B]$

$$= \left[\begin{array}{ccc:c} 1 & 1 & 1 & 6 \\ 1 & 2 & 3 & 10 \\ 1 & 2 & \lambda & \mu \end{array}\right], \text{ operating } R_2 \to R_2 - R_1, R_3 \to R_3 - R_1$$

$$\sim \left[\begin{array}{ccc:c} 1 & 1 & 1 & 6 \\ 0 & 1 & 2 & 4 \\ 0 & 1 & \lambda - 1 & \mu - 6 \end{array}\right], \text{ operating } R_3 \to R_3 - R_2$$

$$\sim \left[\begin{array}{ccc:c} 1 & 1 & 1 & 6 \\ 0 & 1 & 2 & 4 \\ 0 & 0 & \lambda - 3 & \mu - 10 \end{array}\right]$$

Case I. If $\lambda = 3, \mu \neq 10$, then

$$\rho(A) = 2, \rho(A \vdots B) = 3 \quad \Rightarrow \quad \rho(A) \neq \rho(A \vdots B)$$

∴ The system has no solution.

Case II. If $\lambda \neq 3$, μ may have any value, then

$$\rho(A) = \rho(A \vdots B) = 3 = \text{number of unknowns}$$

∴ The system has unique solution.

Case III. If $\lambda = 3, \mu = 10$, then $\rho(A) = \rho(A \vdots B) = 2 <$ number of unknowns

∴ The system has an infinite number of solutions.

Example 6. *Solve the equations by using Guass Jordan method*

$$x_1 + 3x_2 + 2x_3 = 0, \quad 2x_1 - x_2 + 3x_3 = 0, \quad 3x_1 - 5x_2 + 4x_3 = 0, \quad x_1 + 17x_2 + 4x_3 = 0.$$

Sol. In matrix notation, the given system of equations can be written as $AX = O$

where $A = \begin{bmatrix} 1 & 3 & 2 \\ 2 & -1 & 3 \\ 3 & -5 & 4 \\ 1 & 17 & 4 \end{bmatrix}, X = \begin{bmatrix} x_1 \\ x_2 \\ x_3 \end{bmatrix}, O = \begin{bmatrix} 0 \\ 0 \\ 0 \end{bmatrix}$

Operating $R_2 - 2R_1$, $R_3 - 3R_1$, $R_4 - R_1$, we get

$$A \sim \begin{bmatrix} 1 & 3 & 2 \\ 0 & -7 & -1 \\ 0 & -14 & -2 \\ 0 & 14 & 2 \end{bmatrix}, \text{ operating } R_3 \to R_3 - 2R_2,\ R_4 \to R_4 + 2R_2$$

$$\sim \begin{bmatrix} 1 & 3 & 2 \\ 0 & -7 & -1 \\ 0 & 0 & 0 \\ 0 & 0 & 0 \end{bmatrix}, \text{ which is in Echelon form}$$

$\rho(A) = 2 < 3$ = number of unknowns

$\Rightarrow$ The system has an infinite number of non-trivial solutions given by (take $x_2 = k$)

$x_1 + 3x_2 + 2x_3 = 0,\quad -7x_2 - x_3 = 0$ or $x_3 = -7k$

i.e., $x_1 = 11k,\quad x_2 = k,\ x_3 = -7k$, where k is any number.

Different values of k give different solutions.

Example 7. *Solve the following system of homogeneous equations*

$$2x + 3y - z = 0$$
$$x - y - 2z = 0$$
$$3x + y + 3z = 0.$$

Sol. In matrix notation, the given system of equations can be written as $AX = O$,

where $A = \begin{pmatrix} 2 & 3 & -1 \\ 1 & -1 & -2 \\ 3 & 1 & 3 \end{pmatrix}, X = \begin{pmatrix} x \\ y \\ z \end{pmatrix}, O = \begin{pmatrix} 0 \\ 0 \\ 0 \end{pmatrix}$

Here $|A| = \begin{vmatrix} 2 & 3 & -1 \\ 1 & -1 & -2 \\ 3 & 1 & 3 \end{vmatrix} = 2(-3 + 2) - 3(3 + 6) - 1(1 + 3)$

$$= -2 - 27 - 4 = -33 \neq 0$$

$\therefore$ The system has a unique solution or trivial solution given by $x = 0, y = 0, z = 0$.

TEST YOUR KNOWLEDGE 14.5

1. Write the following equations in matrix form $AX = B$ and solve for X by finding A^{-1} :

(*i*) $2x - 2y + z = 1$
$x + 2y + 2z = 2$
$2x + y - 2z = 7$

(*ii*) $2x_1 - x_2 + x_3 = 4$
$x_1 + x_2 + x_3 = 1$
$x_1 - 3x_2 - 2x_3 = 2.$

2. Using the loop current method on a circuit, the following equations were obtained :

$$7i_1 - 4i_2 = 12, - 4i_1 + 12i_2 - 6i_3 = 0, - 6i_2 + 14i_3 = 0.$$

By matrix method, solve for i_1, i_2 and i_3.

3. Solve the following systems of equations by matrix method :

(i) $x + y + z = 8, \quad x - y + 2z = 6, \quad 3x + 5y - 7z = 14$

(ii) $x + y + z = 6, \quad x - y + 2z = 5, \quad 3x + y + z = 8$

(iii) $x + 2y + 3z = 1, \quad 2x + 3y + 2z = 2, \quad 3x + 3y + 4z = 1.$

4. Show that the equations $x + 2y - z = 3$, $3x - y + 2z = 1$, $2x - 2y + 3z = 2$, $x - y + z = -1$ are consistent and solve them.

5. Test for consistency the equations $2x - 3y + 7z = 5$, $3x + y - 3z = 13$, $2x + 19y - 47z = 32$.

6. Solve the equations $x + 3y + 2z = 0$, $2x - y + 3z = 0$, $3x - 5y + 4z = 0$, $x + 17y + 4z = 0$.

7. For what values of a and b do the equations $x + 2y + 3z = 6$, $x + 3y + 5z = 9$, $2x + 5y + az = b$ have (i) no solution (ii) a unique solution (iii) more than one solution ?

8. Find the value of k so that the equations $x + y + 3z = 0$, $4x + 3y + kz = 0$, $2x + y + 2z = 0$ have a non-trivial solution.

9. Show that the equations $3x + 4y + 5z = a$, $4x + 5y + 6z = b$, $5x + 6y + 7z = c$ do not have a solution unless $a + c = 2b$.

10. Investigate the values of λ and μ so that the equations $2x + 3y + 5z = 9$, $7x + 3y - 2z = 8$, $2x + 3y + \lambda z = \mu$ have (i) no solution, (ii) a unique solution and (iii) an infinite number of solutions.

11. Determine the values of λ for which the following set of equations may possess non-trivial solution. $3x_1 + x_2 - \lambda x_3 = 0$, $4x_1 - 2x_2 - 3x_3 = 0$, $2\lambda x_1 + 4x_2 + \lambda x_3 = 0$. For each permissible value of λ, determine the general solution.

12. Investigate for consistency of the following equations and if possible find the solutions :

$$4x - 2y + 6z = 8, \quad x + y - 3z = -1, \quad 15x - 3y + 9z = 21.$$

13. For what values of k the equations $x + y + z = 1$, $2x + y + 4z = k$, $4x + y + 10z = k^2$ have a solution and solve them completely in each case.

14. Show that if $\lambda \neq -5$, the system of equations $3x - y + 4z = 3$, $x + 2y - 3z = -2$, $6x + 5y + \lambda z = -3$ have a unique solution. If $\lambda = -5$, show that the equations are consistent. Determine the solutions in each case.

15. Find the values of a and b for which the equations $x + ay + z = 3$, $x + 2y + 2z = b$, $x + 5y + 3z = 9$ are consistent. When will these equations have a unique solution ?

16. Show that the simultaneous linear equations $-2x + y + z = a$; $x - 2y + z = b$; $x + y - 2z = c$, have no solution unless $a + b + c = 0$. In which case, the system has infinitely many solutions ? Find the solution when $a = b = 1$ and $c = -2$. *(P.T.U., M.C.A. Dec. 2006)*

Answers

1. (i) $x = 2, y = 1, z = -1$ (ii) $x_1 = 1, x_2 = -1, x_3 = 1$ **2.** $i_1 = \frac{396}{175}, i_2 = \frac{24}{25}, i_3 = \frac{72}{175}$

3. (i) $x = 5, y = \frac{5}{3}, z = \frac{4}{3}$ (ii) $x = 1, y = 2, z = 3$ (iii) $x = -\frac{3}{7}, y = \frac{8}{7}, z = -\frac{2}{7}$

4. $x = -1, y = 4, z = 4$ **5.** Inconsistent

6. $x = 11k, y = k, z = -7k$, where k is arbitrary

7. (i) $a = 8, b \neq 15$ (ii) $a \neq 8$, b may have any value (iii) $a = 8, b = 15$

8. $k = 8$ **10.** (i) $\lambda = 5, \mu \neq 9$ (ii) $\lambda \neq 5$, μ arbitrary (iii) $\lambda = 5, \mu = 9$

11. $\lambda = 1, -9$. For $\lambda = 1$, solution is $x = k, y = -k, z = 2k$. For $\lambda = -9$, solution is $x = 3k, y = 9k, z = -2k$.
12. Consistent ; $x = 1, y = 3k - 2, z = k$ where k is arbitrary.
13. $k = 1, x = -3\lambda, y = 1 + 2\lambda, z = \lambda$; $k = 2, x = 1 - 3\lambda, y = 2\lambda, z = \lambda$ where λ is arbitrary.
14. $\lambda \neq -5, x = \frac{4}{7}, y = -\frac{9}{7}, z = 0$; $\lambda = -5, x = \frac{4-5k}{7}, y = \frac{13k-9}{7}, z = k$ where k is arbitrary.
15. If $a = -1, b = 6$, equations will be consistent and have infinitely many solutions
If $a = -1, b \neq 6$, equations will be inconsistent.
If $a \neq -1$, b is arbitrary, equations will be consistent and have a unique solution.
16. Infinite solution when $a + b + c = 0$. When $a = 1, b = 1, c = -2$, the solution is $x = k - 1, y = k - 1, z = k$, where k is arbitrary.

14.32. GAUSS ELIMINATION METHOD

In this method the unknowns are eliminated successively by transforming the given system into an equivalent system with upper triangular co-efficient matrix (*i.e.*, a matrix in which all the elements below the principal diagonal are zero) by means of elementary row operations, from which the unknowns are found by back substitution. Here we shall explain it by considering a system of three equations in three unknowns. Consider the system

$$\left.\begin{aligned} a_1x + b_1y + c_1z = d_1 \\ a_2x + b_2y + c_2z = d_2 \\ a_3x + b_3y + c_3z = d_3 \end{aligned}\right\} \quad \ldots(1)$$

where x, y, z are unknowns. The system in matrix form is AX = B, where

$$A = \begin{bmatrix} a_1 & b_1 & c_1 \\ a_2 & b_2 & c_2 \\ a_3 & b_3 & c_3 \end{bmatrix}, \; X = \begin{bmatrix} x \\ y \\ z \end{bmatrix} \text{ and } B = \begin{bmatrix} d_1 \\ d_2 \\ d_3 \end{bmatrix}$$

Consider the augmented matrix [A : B]

$$[A : B] = \left[\begin{array}{ccc|c} a_1 & b_1 & c_1 & d_1 \\ a_2 & b_2 & c_2 & d_2 \\ a_3 & b_3 & c_3 & d_3 \end{array}\right] \quad \ldots(2)$$

Now, our aim is to reduce (2) to an upper triangular matrix. Let $a_1 \neq 0$. Then

$$\begin{aligned} R_2 &\to R_2 - \frac{a_2}{a_1}R_1 \\ R_3 &\to R_3 - \frac{a_3}{a_1}R_1 \end{aligned} \sim \left[\begin{array}{ccc|c} a_1 & b_1 & c_1 & d_1 \\ 0 & b_2' & c_2' & d_2' \\ 0 & b_3' & c_3' & d_3' \end{array}\right] \quad \ldots(3)$$

Here a_1 is called the first pivot and $b_2', c_2', d_2', b_3', c_3', d_3'$ are transformed elements. Now take b_2' as the pivot ($b_2' \neq 0$). Then

$$R_3 \to R_3 - \frac{b_3'}{b_2'}R_2 \sim \left[\begin{array}{ccc|c} a_1 & b_1 & c_1 & d_1 \\ 0 & b_2' & c_2' & d_2' \\ 0 & 0 & c_3'' & d_3'' \end{array}\right] \quad \ldots(4)$$

Now if $c_3'' \neq 0$, from (4), the given system of linear equations is equivalent to

$$a_1x + b_1y + c_1z = d_1$$
$$b_2'y + c_2'y = d_2'$$
$$c_3''z = d_3''$$

Using back substitution $\qquad z = d_3''/c_3''$

$$y = \frac{1}{b_2'c_3''}\{d_2'\,c_3'' - c_2'\,d_3''\}$$

and $\qquad x = \dfrac{1}{a_1b_2'c_3''}\{d_1b_2'c_3'' - b_1d_2'c_3'' + b_1c_2'd_3'' - b_2'c_1d_3''\}.$

Notes : 1. This method fails if any one of the pivots a_1, b_2' or c_3'' becomes zero. In such cases by interchanging the rows we can get the non-zero pivots.

2. Partial pivoting. From the first column of (2) [called the pivot column if $a_i \neq 0$, $i = 1, 2, 3$] select the component with the largest absolute value. This is called the pivot. Then at the second stage *i.e.*, from the second column of (3) select once again the component with the largest absolute value as the pivot. Continue this process. This procedure is called partial pivoting. [Refer example (1)]

3. Complete pivoting. If we are not interested in the elimination of x, y, z in a particular order, then we can choose at each stage the numerically largest co-efficient of the entire co-efficient matrix. This requires an interchange of equations and also an interchange of the position of the variables.

Example 1. *Solve the system of equations* $3x + y - z = 3$, $2x - 8y + z = -5$, $x - 2y + 9z = 8$ *using Gauss-elimination method.*

Sol. The given system is equivalent to

$$\underset{A}{\begin{bmatrix} 3 & 1 & -1 \\ 2 & -8 & 1 \\ 1 & -2 & 9 \end{bmatrix}} \underset{X}{\begin{bmatrix} x \\ y \\ z \end{bmatrix}} = \underset{B}{\begin{bmatrix} 3 \\ -5 \\ 8 \end{bmatrix}}$$

$\therefore$ The augmented matrix is $\quad [A : B] = \left[\begin{array}{ccc|c} 3 & 1 & -1 & 3 \\ 2 & -8 & 1 & -5 \\ 1 & -2 & 9 & 8 \end{array}\right]$

Now we will make A as upper triangular choosing 3 as pivot

$$\begin{array}{l} R_2 \to R_2 - \frac{2}{3}R_1 \\ R_3 \to R_3 - \frac{1}{3}R_1 \end{array} \sim \left[\begin{array}{ccc|c} 3 & 1 & -1 & 3 \\ 0 & -\frac{26}{3} & \frac{5}{3} & -7 \\ 0 & -\frac{7}{3} & \frac{28}{3} & 7 \end{array}\right]$$

Now choosing $-\dfrac{26}{3}$ as the pivot from the 2nd column

$$R_3 \to R_3 - \frac{7}{26}R_2 \sim \left[\begin{array}{ccc|c} 3 & 1 & -1 & 3 \\ 0 & -\frac{26}{3} & \frac{5}{3} & -7 \\ 0 & 0 & \frac{693}{78} & \frac{231}{26} \end{array}\right]$$

From this we get $3x + y - z = 3$;

$$-\frac{26}{3}y + \frac{5}{3}z = -7 \quad \text{and} \quad \frac{693}{78}z = \frac{231}{26}.$$

Now by back substitution $z = 1$

$$-\frac{26}{3}y = -7 - \frac{5}{3}z = -7 - \frac{5}{3} = -\frac{26}{3} \quad \therefore \; y = 1$$

and
$$x = \frac{1}{3}[3 - y + z] = \frac{1}{3}[3 - 1 + 1] = 1$$

$$\therefore \quad x = 1,\; y = 1,\; z = 1.$$

Example 2. *Solve by Gauss-elimination method :*

$$2x + 3y + z = 2,$$
$$x + 2y - z = 6,$$
$$3x + z = 2,$$

Sol. The given equations in matrix notation are

$$\begin{bmatrix} 2 & 3 & 1 \\ 1 & 2 & -1 \\ 3 & 0 & 1 \end{bmatrix}\begin{bmatrix} x \\ y \\ z \end{bmatrix} = \begin{bmatrix} 2 \\ 6 \\ 2 \end{bmatrix} \quad \text{or} \quad AX = B$$

$\therefore$ The augmented matrix is $[A : B] = \left[\begin{array}{ccc:c} 2 & 3 & 1 & 2 \\ 1 & 2 & -1 & 6 \\ 3 & 0 & 1 & 2 \end{array}\right]$

In First column, the component with largest absolute value is 3. So interchange R_1 and R_3, we get

$$[A : B] = \left[\begin{array}{ccc:c} 3 & 0 & 1 & 2 \\ 1 & 2 & -1 & 6 \\ 2 & 3 & 1 & 2 \end{array}\right], \text{ operating } R_2 \to R_2 - \frac{1}{3}R_1,\; R_3 \to R_3 - \frac{2}{3}R_1$$

$$\sim \left[\begin{array}{ccc:c} 3 & 0 & 1 & 2 \\ 0 & 2 & -\frac{4}{3} & \frac{16}{3} \\ 0 & 3 & \frac{1}{3} & \frac{2}{3} \end{array}\right]$$

In second column, the component with largest absolute value is 3. So interchanging R_2 and R_3, we get

$$[A : B] \sim \left[\begin{array}{ccc:c} 3 & 0 & 1 & 2 \\ 0 & 3 & \frac{1}{3} & \frac{2}{3} \\ 0 & 2 & -\frac{4}{3} & \frac{16}{9} \end{array}\right], \text{ operating } R_3 \to R_3 - \frac{2}{3}R_2$$

$$\sim \left[\begin{array}{ccc:c} 3 & 0 & 1 & 2 \\ 0 & 3 & \frac{1}{3} & \frac{2}{3} \\ 0 & 0 & -\frac{14}{9} & \frac{44}{9} \end{array}\right]$$

From third row, $-\frac{14}{9}z = \frac{44}{9} \quad \Rightarrow \quad z = -\frac{22}{7}$

From second row, $3y + \frac{z}{3} = \frac{2}{3}$

$$\Rightarrow \qquad 3y = \frac{2}{3} - \frac{z}{3} = \frac{2}{3} - \frac{1}{3}\left(-\frac{22}{7}\right)$$

$$= \frac{2}{3} + \frac{22}{21} = \frac{14+22}{21} = \frac{36}{21} = \frac{12}{7} \quad \Rightarrow \quad y = \frac{4}{7}$$

From first row, $3x + z = 2$

$$\Rightarrow \qquad 3x = 2 - z = 2 + \frac{22}{7} = \frac{14+22}{7} = \frac{36}{7} \quad \Rightarrow \quad x = \frac{12}{7}$$

Hence, the required solution is

$$x = \frac{12}{7},\ y = \frac{4}{7},\ z = -\frac{22}{7}.$$

Example 3. *Solve*

$2x + 3y + z = 9$,
$x + 2y + 3z = 6$,
$3x + y + 2z = 8$, by Gauss-elimination method.

Sol. The given equations in matrix notations are

$$\begin{bmatrix} 2 & 3 & 1 \\ 1 & 2 & 3 \\ 3 & 1 & 2 \end{bmatrix} \begin{bmatrix} x \\ y \\ z \end{bmatrix} = \begin{bmatrix} 9 \\ 6 \\ 8 \end{bmatrix} \quad \text{or } AX = B$$

$\therefore$ The augmented matrix is $[A : B] = \left[\begin{array}{ccc:c} 2 & 3 & 1 & 9 \\ 1 & 2 & 3 & 6 \\ 3 & 1 & 2 & 8 \end{array}\right]$

In first column, the component with largest absolute value is 3. So interchanging R_1 and R_3, we get

$$[A : B] = \left[\begin{array}{ccc:c} 3 & 1 & 2 & 8 \\ 1 & 2 & 3 & 6 \\ 2 & 3 & 1 & 9 \end{array}\right], \text{ operating } R_2 \to R_2 - \frac{1}{3}R_1,\ R_3 \to R_3 - \frac{2}{3}R_1$$

$$\sim \left[\begin{array}{ccc:c} 3 & 1 & 2 & 8 \\ 0 & \frac{5}{3} & \frac{7}{3} & \frac{10}{3} \\ 0 & \frac{7}{3} & -\frac{1}{3} & \frac{11}{3} \end{array}\right]$$

In second column, the component with largest absolute value is $\frac{7}{3}$. So interchaning R_2 and R_3, we get

$$\sim \left[\begin{array}{ccc:c} 3 & 1 & 2 & 8 \\ 0 & \frac{7}{3} & -\frac{1}{3} & \frac{11}{3} \\ 0 & \frac{5}{3} & \frac{7}{3} & \frac{10}{3} \end{array}\right] \text{ operating } R_3 \to R_3 - \frac{5}{7}R_2,$$

$$\sim \left[\begin{array}{ccc:c} 3 & 1 & 2 & 8 \\ 0 & \frac{7}{3} & -\frac{1}{3} & \frac{11}{3} \\ 0 & 0 & \frac{54}{21} & \frac{15}{21} \end{array}\right]$$

From third row, $\frac{54}{21}z = \frac{15}{21} \Rightarrow z = \frac{15}{54} = \frac{5}{18}$

From second row, $\frac{7}{3}y - \frac{1}{3}z = \frac{11}{3} \Rightarrow \frac{7}{3}y = \frac{11}{3} + \frac{z}{3} = \frac{11}{3} + \frac{5}{54}$

$\Rightarrow \qquad \frac{7}{3}y = \frac{203}{54} \Rightarrow y = \frac{203}{54} \times \frac{3}{7} = \frac{29}{18}$

From first row, $2x + 3y + z = 9 \Rightarrow 2x = 9 - 3y - z$

or $\qquad 2x = 9 - \frac{87}{18} - \frac{5}{18} = \frac{162 - 87 - 5}{18} = \frac{70}{18} \Rightarrow x = \frac{35}{18}$

Hence, the required solution is

$$x = \frac{35}{18},\ y = \frac{29}{18},\ z = \frac{5}{18}.$$

TEST YOUR KNOWLEDGE 14.6

Solve the following system of linear equations by (i) Gauss-elimination method (ii) Gauss Jordan method.

1. $3x + 4y - z = 8$; $-2x + y + z = 3$; $x + 2y - z = 2$
2. $10x + y + z = 12$; $x + 10y + z = 12$; $x + y + 10z = 12$
3. $3x + y + 2z = 3$; $2x - 3y - z = -3$; $x - 2y + z = -4$
4. $2x - 6y + 8z = 24$; $5x + 4y - 3z = 2$; $3x + y + 2z = 16$
5. $x + 2y - z = 3$; $3x - y + 2z = 1$, $2x - 2y + 3z = 2$
6. $2x - 3y + z = -1$, $x + 4y + 5 = 25$, $3x - 4y + z = 2$
7. $2x + y + 4z = 12$, $8x - 3y + 2z = 20$; $4x + 11y - z = 33$
8. $2x + y + z = 10$, $3x + 2y + 3z = 18$, $x + 4y + 9z = 16$
9. $x + 2y + z = 8$; $2x + 3y + 4z = 20$, $4x + 3y + 2z = 16$
10. $5x_1 + 2x_2 + x_3 = 12$; $-x_1 + 4x_2 + 2x_3 = 2$, $2x_1 - 3x_2 + 10x_3 = -45$

11. $x_1 - x_2 - x_3 - x_4 = 2$; $2x_1 + 4x_2 - 3x_3 = 6$, $3x_2 - 4x_3 - 2x_4 = -1$; $-2x_1 + 4x_3 + 3x_4 = -3$

12. $x_1 + 2x_2 + 3x_3 + 4x_4 = 20$; $3x_1 - 2x_2 + 8x_3 + 4x_4 = 26$, $2x_1 + x_2 - 4x_3 + 7x_4 = 10$; $4x_1 + 2x_2 - 8x_3 - 4x_4 = 2$

13. $5x_1 + x_2 + x_3 + x_4 = 4$; $x_1 + 7x_2 + x_3 + x_4 = 12$, $x_1 + x_2 + 6x_3 + x_4 = -5$; $x_1 + x_2 + x_3 + 4x_4 = -6$

14. $2x_1 + x_2 + 5x_3 + x_4 = 5$; $x_1 + x_2 - 3x_3 + 4x_4 = -1$, $3x_1 + 6x_2 - 2x_3 + x_4 = 8$; $2x_1 + 2x_2 + 2x_3 - 3x_4 = 2$

15. $x + y + 2z - w = 5$; $x + 3y + 2z + w = 17$, $x + y + 3z + 2w = 20$; $x + 3y + 4z + 2w = 27$.

Answers

1. 1, 2, 3	**2.** 1, 1, 1	**3.** 1, 2, – 1	**4.** 1, 3, 5
5. – 1, 4, 4	**6.** 8.7, 5.7, – 1.3	**7.** 3, 2, 1	**8.** 7, – 9, 5
9. 1, 2, 3	**10.** 2, 3, – 4	**11.** 2, – 1, – 2, 3	**12.** 4, 3, 2, 1
13. 1, 2, – 1, – 2	**14.** 2, $\frac{1}{5}$, 0, $\frac{4}{5}$	**15.** 1, 2, 3, 4.	

15A ARITHMETIC PROGRESSION

15.1. SEQUENCE

Consider the following arrangement 3, 7, 11, 15, ...

Each term of the above arrangement is four more than the proceeding term. We call it **sequence** *i.e.,*

"A succession of numbers formed according to a certain rule, arranged in a certain order, is called a **sequence.**"

The nth term of a sequence is denoted by T_n, t_n, a_n, $a(n)$ etc.

15.2. FINITE AND INFINITE SEQUENCE

A sequence containing finite number of terms is called a **finite sequence.** A sequence having infinite number of terms is called an **infinite sequence.**

For *e.g.,* the sequence 1, 3, 5, 7, 9 is a finite sequence and the sequence 2, 4, 6, 8, 10, ... , is an infinite sequence. Consider some more examples given below :

Example 1. *Write the first three terms of the sequence* $<a_n>$, *where* $a_n = \dfrac{n(n+1)(2n+1)}{6}$.

Sol. For $n = 1, 2, 3$, we have

$$a_1 = \frac{1 \times 2 \times 3}{6} = 1,\ a_2 = \frac{2 \times 3 \times 5}{6} = 5,\ a_3 = \frac{3 \times 4 \times 7}{6} = 14.$$

Example 2. *Find* T_{10} *for the sequence* $< T_n >$ *where* $T_1 = 1$, $T_2 = 6$, $T_n = \dfrac{T_{n-1}}{T_{n-2}}$, $n \geq 3$.

Sol. Given $T_1 = 1$, $T_2 = 6$, $T_n = \dfrac{T_{n-1}}{T_{n-2}}$, $n \geq 3$. Put $n = 3, 4, 5, \ldots\ldots$, so that

$$T_3 = \frac{T_2}{T_1} = \frac{6}{1} = 6,\ T_4 = \frac{T_3}{T_2} = \frac{6}{6} = 1,\ T_5 = \frac{T_4}{T_3} = \frac{1}{6} = \frac{1}{6}$$

$$T_6 = \frac{T_5}{T_4} = \frac{\frac{1}{6}}{1} = \frac{1}{6},\ T_7 = \frac{T_6}{T_5} = \frac{\frac{1}{6}}{\frac{1}{6}} = 1,\ T_8 = \frac{T_7}{T_6} = \frac{1}{\frac{1}{6}} = 6$$

Similarly, $$T_9 = \frac{T_8}{T_7} = \frac{6}{1} = 6,\ T_{10} = \frac{T_9}{T_8} = \frac{6}{6} = 1.$$

Example 3. *Find the second term of the sequence where* $T_n = (-1)^{n-1}\, 2^{n-1},\ n \in N$

(P.T.U., M.B.A. May 2003)

Sol. Given $\qquad T_n = (-1)^{n-1}\, 2^{n-1}$ ...(1)

Put $n = 2$ in (1), $\qquad T_2 = (-1)^1\, 2^1 = -2.$

15.3. ARITHMETIC PROGRESSION (A.P.)

A sequence $< a_n >$ is said to be an arithmetic progression (written as A.P.) if there exists a number, say, **d** such that $a_{n+1} - a_n = d \;\forall\; n \geq 1$ *i.e.,*

If the difference of each term of a sequence except the first one, from its proceeding term is always same, then the sequence is called **Arithmetic Progression.**

For *e.g.*, the sequence 4, 8, 12, 16, ..., is an A.P. since 8 – 4 = 12 – 8 = 16 – 12 = ... The constant number **d** is called the **common difference** of the concerned **A.P.** Also the first term of an A.P. is denoted by a.

Theorem I. *If 'a' and 'd' be the first term and common difference of the A.P. given by* $< T_n >$, *then* $T_n = a + (n-1)\, d,\ n \in N$

Proof. We apply principle of mathematical induction to prove

$$T_n = a + (n-1)\, d,\ n \in \text{N}$$

Step I. For $n = 1$, L.H.S. $= T_1 =$ First term $= a$

$$\text{R.H.S.} = a + (1-1)d = a$$

$\therefore$ The result holds for $n = 1$

Step II. Assume the result holds for $n = m$ *i.e.,* $T_m = a + (m-1)\, d$ (1)

Step III. Take $n = m + 1$. By definition, $T_{m+1} - T_m = d$

$\Rightarrow \qquad T_{m+1} = T_m + d = a + (m-1)d + d$ | Using (1)

$$= a + md = a + (m + 1 - 1)d$$

i.e, the result holds for $n = m + 1$, which completes the induction. Therefore by the principle of mathematical induction, the given result holds for each n *i.e.,*

$$T_n = a + (n-1)d,\ n \in \text{N}.$$

ILLUSTRATIVE EXAMPLES

Example 1. *Find the 20th term and nth term of the sequence 4, 9, 14, 19 ...*

Sol. Given sequence 4, 9, 14, 19, ..., is an A.P. Here $a = 4$, d = common difference = 9 – 4 = 5. Using $T_n = a + (n-1)d$, we have

$$T_{20} = 4 + (20-1)\,(5) = 4 + 95 = 99$$

Also $\qquad T_n = 4 + (n-1)\,(5) = 5n - 5 + 4 = 5n - 1.$

Example 2. *Show that the sequence* $\log a,\ \log \dfrac{a^2}{b},\ \log \dfrac{a^3}{b^2},\ \ldots$ *is an A.P.*

Sol. Consider $\quad T_2 - T_1 = \log \dfrac{a^2}{b} - \log a = \log \dfrac{a^2/b}{a} = \log \dfrac{a}{b}$ $\quad\Big|\ \log m - \log n = \log \dfrac{m}{n}$

$$T_3 - T_2 = \log \frac{a^3}{b^2} - \log \frac{a^2}{b} = \log \frac{a^3}{b^2} \times \frac{b}{a^2} = \log \frac{a}{b}$$

Thus $\qquad T_2 - T_1 = T_3 - T_2 = \ldots = \log \frac{a}{b}$

$\therefore$ The given sequence is an A.P.

Example 3. *Which term of A.P. 3, 15, 27, 39, ... will be 132 more than its 54th term ?*

(P.T.U., M.B.A. May 2006)

Sol. Here $a = 3$, $d = 12$, Let T_r be the required term such that

$$T_r + 132 = T_{54}$$

$\Rightarrow \quad a + (r-1)d + 132 = a + (54-1)d$

$\Rightarrow \quad (r-1)d + 132 = 53d \qquad \ldots(1)$

$\Rightarrow \quad 12(r-1) + 132 = 53 \times 12 = 636$

$\Rightarrow \quad 12(r-1) = 636 - 132 = 504$

$\Rightarrow \quad r - 1 = \frac{504}{12} = 42$

$\Rightarrow \quad r = 43$ *i.e.*, the required term $= T_{43}$.

Example 4. *If $\log_{10} 2$, $\log_{10} (2^x - 1)$, $\log_{10} (2^x + 3)$ form an A.P., find the value of x.*

Sol. Given $\log_{10} 2$, $\log_{10} (2^x - 1)$, $\log_{10} (2^x + 3)$ are in A.P.

$\Rightarrow \quad \log_{10} (2^x - 1) - \log_{10} 2 = \log_{10} (2^x + 3) - \log_{10} (2^x - 1)$

$\Rightarrow \quad \log_{10} \frac{2^x - 1}{2} = \log_{10} \frac{2^x + 3}{2^x - 1} \Rightarrow \frac{2^x - 1}{2} = \frac{2^x + 3}{2^x - 1}$

$\Rightarrow \quad \frac{t-1}{2} = \frac{t+3}{t-1}, t = 2^x \Rightarrow (t-1)^2 = 2t + 6$

$\Rightarrow \quad t^2 + 1 - 2t = 2t + 6 \Rightarrow t^2 - 4t - 5 = 0$

$\Rightarrow \quad t^2 - 5t + t - 5 = 0 \Rightarrow t(t-5) + (t-5) = 0$

$\Rightarrow \quad (t+1)(t-5) = 0 \Rightarrow t = -1, t = 5$

When $t = -1$, we have $2^x = -1$, not possible

When $t = 5$, we have $2^x = 5 \Rightarrow x = \log_2 5$.

Example 5. *Show that the sequence $\{n^2 + 4\}$ is not an A.P.*

Sol. Let $\qquad T_n = n^2 + 4, n \in \mathbf{N}$.

$\therefore \qquad T_1 = (1)^2 + 4 = 5, \quad T_2 = (2)^2 + 4 = 8,$

$\qquad T_3 = (3)^2 + 4 = 13, \quad T_4 = (4)^2 + 4 = 20,$

....

$\therefore$ The sequence is 5, 8, 13, 20, ...

Here $\qquad T_2 - T_1 = 8 - 5 = 3$

$\qquad T_3 - T_2 = 13 - 8 = 5$

....

$\therefore$ The given sequence is not an A.P., because the differences $T_2 - T_1$, $T_3 - T_2$, ... are not equal.

Example 6. *For what value of n, the nth term of the series :*
"3 + 10 + 17 + ..." and "63 + 65 + 67 + ..." are equal.

Sol. Ist series. We have $3 + 10 + 17 + \ldots\ldots$

Here $a_1 = 3$ and $d_1 = 10 - 3 = 17 - 10 = 7$

$\therefore \quad T_n = a_1 + (n-1)d_1 = 3 + (n-1)\,7 = 7n - 4.$

IInd series. We have $63 + 65 + 67 + \ldots\ldots$

Here $a_2 = 63$ and $d_2 = 65 - 63 = 67 - 65 = 2$

$\therefore \quad T_n = a_2 + (n-1)\,d_2 = 63 + (n-1)\,2 = 2n + 61.$

Let nth term of given series be equal.

$\therefore \quad 7n - 4 = 2n + 61 \quad$ or $\quad 5n = 65$ *i.e.*, $n = 13$.

Example 7. *Determine the Ist term and the 40th term of the A.P. whose 7th term is 34 and 15th term is 74.*

Sol. Let a be the first term and d, the common difference of the A.P.

$\therefore$ We have $\quad T_7 = a + 6d = 34 \quad \ldots(1)$

and $\quad T_{15} = a + 14d = 74 \quad \ldots(2)$

$(2) - (1) \Rightarrow \quad 8d = 40 \Rightarrow d = 5.$

Putting $d = 5$ in (1), we get $a + 6(5) = 34$ or $a = 34 - 30 = 4$.

$\therefore$ First term $= a = 4$ and 40th term $= T_{40} = a + 39d = 4 + 39\,(5) = 199.$

Example 8. *The 2nd, 31st and last terms of an A.P. are* $7\frac{3}{4}, \frac{1}{2}$ *and* $-6\frac{1}{2}$ *respectively. Find the first term and the number of terms.*

Sol. Let a be the first term and d, the common difference of the A.P.

$\therefore$ We have $\quad T_2 = 7\frac{3}{4} = \frac{31}{4}, \quad T_{31} = \frac{1}{2}$ and last term, $l = -6\frac{1}{2} = -\frac{13}{2}$.

Now $\quad T_2 = a + d$ and $T_{31} = a + 30d$.

$$\therefore \quad a + d = \frac{31}{4} \quad \ldots(1)$$

$$\text{and} \quad a + 30d = \frac{1}{2} \quad \ldots(2)$$

$$(2) - (1) \Rightarrow \quad 29d = \frac{1}{2} - \frac{31}{4} = \frac{2 - 31}{4} = -\frac{29}{4} \quad \text{or} \quad d = -\frac{29}{4} \times \frac{1}{29} = -\frac{1}{4}$$

Putting $d = -\frac{1}{4}$ in (1), we get $a + \left(-\frac{1}{4}\right) = \frac{31}{4}$ or $a = \frac{31}{4} + \frac{1}{4} = \mathbf{8}.$

Let number of terms be n.

$$\therefore \quad l = T_n = a + (n-1)d = 8 + (n-1)\left(-\frac{1}{4}\right) = \frac{32 - n + 1}{4} = \frac{33 - n}{4}.$$

$$\therefore \quad \frac{33 - n}{4} = -\frac{13}{2} \quad \left(\because \ l = -\frac{13}{2}\right)$$

$$\Rightarrow \quad 33 - n = -26 \quad \textit{i.e.,} \quad n = 59.$$

Example 9. *If (p + 1)th term of an A.P. is twice the (q + 1)th term, prove that the (3p + 1)th term is twice the (p + q + 1)th term.*

Sol. Let a be the first term and d, the common difference of the A.P.

We have $T_{p+1} = 2T_{q+1}.$

$\Rightarrow \quad a + (p + 1 - 1)d = 2[a + (q + 1 - 1)d]$

$\Rightarrow \quad a + pd = 2a + 2qd \Rightarrow a = (p - 2q)d$ ...(1)

Now $T_{3p+1} = a + (3p + 1 - 1)d = (p - 2q)d + 3pd = 2(2p - q)d$ (Using (1))

and $T_{p+q+1} = a + (p + q + 1 - 1)d = (p - 2q)d + (p + q)d = (2p - q)d$

$\therefore \quad T_{3p+1} = 2T_{p+q+1}.$

Example 10. *The fourth term of an A.P. is equal to three times the first term and the seventh term exceeds twice the third term by one. Find the first term and the common difference.*

Sol. Let a be the first term and d, the common difference of the A.P.

$\therefore \quad T_n = a + (n - 1)d, n \in \mathbf{N}$

By the given conditions, $T_4 = 3T_1$ ...(1) and $T_7 = 2T_3 + 1$ ...(2)

(1) $\Rightarrow \quad a + 3d = 3a$ or $a = \frac{3d}{2}$ ...(3)

(2) $\Rightarrow \quad a + 6d = 2(a + 2d) + 1$ or $2d = a + 1$

or $\quad 2d = \frac{3d}{2} + 1$ or $\frac{d}{2} = 1$ or $d = 2$ [by (3)]

$\therefore$ (3) $\Rightarrow \quad a = \frac{3}{2}(2) = 3.$

Example 11. *If m times the mth term of an A.P. is equal to n times its nth term, then show that (m + n)th term of the A.P. is zero.*

Sol. Let a be the first term and d, the common difference of the A. P.

$\therefore \quad T_m = a + (m - 1)d$ and $T_n = a + (n - 1)d$

By the given condition, $mT_m = nT_n$

$\Rightarrow \quad m\{a + (m - 1)d\} = n\{a + (n - 1)d\} \Rightarrow ma - na + m(m - 1)d - n(n - 1)d = 0$

$\Rightarrow \quad (m - n)a + (m^2 - m - n^2 + n)d = 0 \Rightarrow (m - n)a + (m - n)(m + n - 1)d = 0$

$\Rightarrow \quad (m - n)\{a + (m + n - 1)d\} = 0 \Rightarrow a + (m + n - 1)d = 0$ $(\because m \neq n)$

$\therefore \quad T_{m+n} = 0.$

Example 12. *If pth, qth, rth terms of an A.P. be a, b, c respectively, then prove that*

$$p(b - c) + q(c - a) + r(a - b) = 0.$$

Sol. Let A be the first term and d, the common difference of the A.P.

We have $T_p = a, \quad T_q = b$ and $T_r = c$

$\therefore \quad A + (p - 1)d = a, \quad A + (q - 1)d = b$ and $A + (r - 1)d = c.$

L.H.S. $= p(b - c) + q(c - a) + r(a - b)$

$= p[(A + (q - 1)d) - (A + (r - 1)d] + q[(A + (r - 1)d) - (A + (p - 1)d)] + r[(A + (p - 1)d) - (A + (q - 1)d)]$

$= p[d(q - 1 - r + 1)] + q[d(r - 1 - p + 1)] + r[d(p - 1 - q + 1)]$

$= pd\,(q - r) + qd\,(r - p) + rd\,(p - q) = d\,[(pq - pr) + (qr - qp) + (rp - rq)]$

$= d(0) = 0 =$ R.H.S.

Theorem II. *If a, b, c, are in A.P., then*

(i) $a + k, b + k, c + k$ *are also in A.P.* *(ii)* $a - k, b - k, c - k$ *are also in A.P.*

(iii) $ka, kb, kc, k \neq 0$ *are also in A.P.* *(iv)* $\frac{a}{k}, \frac{b}{k}, \frac{c}{k}, k \neq 0$ *are also in A.P.*

Proof. Given a, b, c are in A.P.

$\therefore \quad b - a = c - b$

(i) $a + k, b + k = c + k$ will be in A.P. if $b + k - (a + k) = c + k - (b + k)$ *i.e.*, if $b - a = c - b$, which is true

(ii) $a - k, b - k, c - k$ will be in A.P. if $b - k - (a - k) = c - k - (b - k)$ *i.e.*, if $b - a = c - b$, which is true

(iii) ka, kb, kc are in A.P. if $kb - ka = kc - kb$ *i.e.*, if $k(b - a) = k(c - b)$ *i.e.*, if $b - a = c - b$, which is true

(iv) $\frac{a}{k}, \frac{b}{k}, \frac{c}{k}$ will be in A.P. if $\frac{b}{k} - \frac{a}{k} = \frac{c}{k} - \frac{b}{k}$ *i.e*, if $\frac{b-a}{k} = \frac{c-b}{k}$ *i.e.*, if $b - a = c - b$, which is true

Theorem III. *If each term of an A.P. is*

(i) increased by the same number *(ii) decreased by the same number*

(iii) multiplied by the same number *(iv) divided by the same number,*

then the resulting sequence is also an A.P.

Proof. Let the given A.P. is $a, a + d, a + 2d$ ...(1)

(i) Let each term is increased by "k", then the new sequence will be

$$a + k, a + k + d, a + k + 2d, \ldots$$

This is again an A.P. with common difference d and first term $a + k$.

(ii) Let each term of (1) is decreased by k. The new sequence will be

$$a - k, a - k + d, a - k + 2d, \ldots$$

It is again an A.P. with common difference d and first term $a - k$.

(iii) Let each term of (1) is multiplied by $k \neq 0$ the new sequence will be

$$ka, k(a + d), k\,(a + 2d), \ldots$$

It is again an A.P. with common difference kd and first term ka.

(iv) Let each term is divided by k, $k \neq 0$ the new sequence will be

$$\frac{a}{k}, \frac{a+d}{k}, \frac{a+2d}{k}, \ldots \quad \text{or} \quad \frac{a}{k}, \frac{a}{k} + \frac{d}{k}, \frac{a}{k} + \frac{2d}{k}, \ldots$$

It is again an A.P with common difference $\frac{d}{k}$, $k \neq 0$ and first term $\frac{a}{k}$, $k \neq 0$.

15.4. IMPORTANT OBSERVATIONS

(i) If sum of three numbers in A.P. is given, the numbers are $a - d, a, a + d$.

(ii) If sum of four numbers in A.P. is given, the numbers are $a - 3d, a - d, a + d, a + 3d$.

(iii) If sum of five numbers in A.P. is given, the numbers are $a - 2d, a - d, a, a + d, a + 2d$.

Note. If sum of numbers in A.P. is not given, then the numbers are $a, a + d, a + 2d, \ldots$

Example 13. *If a, b, c are in A. P., show that :*

(i) $\frac{1}{bc}, \frac{1}{ca}, \frac{1}{ab}$ *are in A.P.* *(ii)* $a^2(b+c), b^2(c+a), c^2(c+b)$ *are in A.P.*

(iii) $\frac{a(b+c)}{bc}, \frac{b(c+a)}{ca}, \frac{c(a+b)}{ab}$ *are in A.P.* *(iv)* $b+c-a, c+a-b, a+b-c$ *are in A.P.*

(v) $\frac{1}{\sqrt{b}+\sqrt{c}}, \frac{1}{\sqrt{c}+\sqrt{a}}, \frac{1}{\sqrt{a}+\sqrt{b}}$ *are in A.P.*

(vi) $(b+c)^2 - a^2, (c+a)^2 - b^2, (a+b)^2 - c^2$ *are in A.P.*

Sol. *(i)* $\frac{1}{bc}, \frac{1}{ca}, \frac{1}{ab}$ are in A.P. if $\frac{abc}{bc}, \frac{abc}{ca}, \frac{abc}{ab}$ are in A.P.

(Multiplying each term by *abc*)

if *a, b, c* are in A.P., which is given to be true.

$\therefore$ $\frac{1}{bc}, \frac{1}{ca}, \frac{1}{ab}$ are in A.P.

Alternative method

$\frac{1}{bc}, \frac{1}{ca}, \frac{1}{ab}$ are in A.P. if $\frac{1}{ca} - \frac{1}{bc} = \frac{1}{ab} - \frac{1}{ca}$, if $\frac{b-a}{abc} = \frac{c-b}{abc}$,

if $b - a = c - b$, if *a, b, c* are in A.P., which is given to be true.

$\therefore$ $\frac{1}{bc}, \frac{1}{ca}, \frac{1}{ab}$ are in A.P.

(ii) $a^2(b+c), b^2(c+a), c^2(a+b)$ are in A.P.

if $a^2b + a^2c, b^2c + b^2a, c^2a + c^2b$ are in A.P.

if $a^2b + a^2c + abc, b^2c + b^2a + abc, c^2a + c^2b + abc$ are in A.P.

(Adding *abc* to each term)

if $a(ab + ac + bc), b(bc + ba + ac), c(ca + cb + ab)$ are in A.P.

if *a, b, c* are in A.P. which is given to be true.

(Dividing each term by $ab + bc + ca$)

$\therefore$ $a^2(b+c), b^2(c+a), c^2(a+b)$ are in A.P.

(iii) $\frac{a(b+c)}{bc}, \frac{b(c+a)}{ca}, \frac{c(a+b)}{ab}$ are in A.P.

if $\frac{ab+ac}{bc}, \frac{bc+ba}{ca}, \frac{ca+cb}{ab}$ are in A.P.

if $\frac{ab+ac}{bc} + 1, \frac{bc+ba}{ca} + 1, \frac{ca+cb}{ab} + 1$ are in A.P.

(Adding '1' to each term)

if $\frac{ab+ac+bc}{bc}, \frac{bc+ba+ca}{ca}, \frac{ca+cb+ab}{ab}$ are in A.P.

if $\frac{1}{bc}, \frac{1}{ca}, \frac{1}{ab}$ are in A.P. (Dividing each term by $ab + bc + ca$)

if $\frac{abc}{bc}, \frac{abc}{ca}, \frac{abc}{ab}$ are in A.P. (Multiplying each term by *abc*)

if *a, b, c* are in A.P., which is given to be true.

$\therefore$ $\frac{a(b+c)}{bc}, \frac{b(c+a)}{ca}, \frac{c(a+b)}{ab}$ are in A.P.

(*iv*) $b + c - a, c + a - b, a + b - c$ are in A.P.

if $(b + c - a) - (a + b + c), (c + a - b) - (a + b + c), (a + b - c) - (a + b + c)$ are in A.P.

(Subtracting $a + b + c$ from each term)

if $-2a, -2b, -2c$ are in A.P.

if *a, b, c* are in A.P. which is given to be true.

$\therefore$ $b + c - a, c + a - b, a + b - c$ are in A.P.

(*v*) $\frac{1}{\sqrt{b}+\sqrt{c}}, \frac{1}{\sqrt{c}+\sqrt{a}}, \frac{1}{\sqrt{a}+\sqrt{b}}$ are in A.P.

if $\frac{1}{\sqrt{c}+\sqrt{a}} - \frac{1}{\sqrt{b}+\sqrt{c}} = \frac{1}{\sqrt{a}+\sqrt{b}} - \frac{1}{\sqrt{c}+\sqrt{a}}$

if $\frac{\sqrt{b}+\sqrt{c}-\sqrt{c}-\sqrt{a}}{(\sqrt{c}+\sqrt{a})(\sqrt{b}+\sqrt{c})} = \frac{\sqrt{c}+\sqrt{a}-\sqrt{a}-\sqrt{b}}{(\sqrt{a}+\sqrt{b})(\sqrt{c}+\sqrt{a})}$ if $\frac{\sqrt{b}-\sqrt{a}}{\sqrt{b}+\sqrt{c}} = \frac{\sqrt{c}-\sqrt{b}}{\sqrt{a}+\sqrt{b}}$

if $b - a = c - b$ if *a, b, c* are in A.P., which is given to be true.

$\therefore$ $\frac{1}{\sqrt{b}+\sqrt{c}}, \frac{1}{\sqrt{c}+\sqrt{a}}, \frac{1}{\sqrt{a}+\sqrt{b}}$ are in A.P.

(*vi*) $(b+c)^2 - a^2, (c+a)^2 - b^2, (a+b)^2 - c^2$ are in A.P.

if $(b + c + a)(b + c - a), (c + a + b)(c + a - b), (a + b + c)(a + b - c)$ are in A.P.

if $b + c - a, c + a - b, a + b - c$ are in A.P. (Dividing each term by $a + b + c$)

if $(b + c - a) - (a + b + c), (c + a - b) - (a + b + c), (a + b - c) - (a + b + c)$ are in A. P.

if $-2a, -2b, -2c$ are in A.P.

if *a, b, c* are in A.P., which is given to be true.

$\therefore$ $(b+c)^2 - a^2, (c+a)^2 - b^2, (a+b)^2 - c^2$ are in A.P.

Example 14. *The sum of the first three consecutive terms of an A.P. is 9 and the sum of their squares is 35. Find* T_n.

Sol. Let the first three consecutive terms of the A.P. be $a - d, a, a + d$.

$\therefore$ Sum $= (a - d) + a + (a + d) = 9$ (Given)

$\therefore$ $3a = 9$ *i.e.,* $a = 3$.

$\therefore$ The numbers are $3 - d, 3, 3 + d$.

Also, sum of squares $= (3-d)^2 + (3)^2 + (3+d)^2 = 35$ (Given)

$\therefore \quad (9 + d^2 - 6d) + 9 + (9 + d^2 + 6d) = 35$

$\Rightarrow \quad 2d^2 = 35 - 27 = 8 \quad$ *i.e.,* $\quad d^2 = 4$ or $d = \pm 2$.

Case I. d = 2. $\therefore$ The numbers are $3 - 2 = 1, 3, 3 + 2 = 5$.

$\therefore$ The A.P. is 1, 3, 5, ... and $T_n = a + (n-1)d = 1 + (n-1)2 = 2n - 1$.

Case II. d = – 2. $\therefore$ The numbers are $3 - (-2) = 5, 3, 3 + (-2) = 1$.

$\therefore$ The A.P. is 5, 3, 1, ... and $T_n = a + (n-1)d = 5 + (n-1)(-2) = -2n + 7$.

Example 15. *Find four numbers in A.P. whose sum is 20 and the sum of whose squares is 120.*

Sol. Let the numbers be $a - 3d, a - d, a + d, a + 3d$.

$\therefore$ Sum $= (a - 3d) + (a - d) + (a + d) + (a + 3d) = 20$ (Given)

$\therefore \quad 4a = 20 \quad$ *i.e.,* $\quad a = 5$

$\therefore$ The numbers are $5 - 3d, 5 - d, 5 + d, 5 + 3d$.

Also, sum of squares $= (5-3d)^2 + (5-d)^2 + (5+d)^2 + (5+3d)^2 = 120$ (Given)

$\therefore \quad (25 + 9d^2 - 30d) + (25 + d^2 - 10d) + (25 + d^2 + 10d) + (25 + 9d^2 + 30d) = 120$

$\Rightarrow \quad 20d^2 = 20 \quad$ *i.e.,* $\quad d^2 = 1$ or $d = \pm 1$.

Case I. d = 1. The numbers are $5 - 3(1), 5 - (1), 5 + (1), 5 + 3(1)$ or **2, 4, 6, 8.**

Case II. d = – 1. The numbers are $5 - 3(-1), 5 - (-1), 5 + (-1), 5 + 3(-1)$ or **8, 6, 4, 2.**

Example 16. *If $x^{18} = y^{21} = z^{28}$, prove that 3, $3 \log_y x$, $3 \log_z y$ and $7 \log_x z$ are in A.P.*

Sol. Let $\quad x^{18} = y^{21} = z^{28} = k$.

Taking common logarithms, we get

$$\log x^{18} = \log y^{21} = \log z^{28} = \log k$$

$$\Rightarrow \quad 18 \log x = 21 \log y = 28 \log z = \log k.$$

$$\therefore \quad 3 \log_y x = 3\left(\frac{\log x}{\log y}\right) = 3\left(\frac{(\log k)/18}{(\log k)/21}\right) = \frac{3 \times 21}{18} = \frac{7}{2}$$

$$3 \log_z y = 3\left(\frac{\log y}{\log z}\right) = 3\left(\frac{(\log k)/21}{(\log k)/28}\right) = \frac{3 \times 28}{21} = 4$$

and
$$7 \log_x z = 7\left(\frac{\log z}{\log x}\right) = 7\left(\frac{(\log k)/28}{(\log k)/18}\right) = \frac{7 \times 18}{28} = \frac{9}{2}.$$

Now the numbers $3, \frac{7}{2}, 4, \frac{9}{2}$ are in A.P. with *c.d.* $= \frac{1}{2}$. $\therefore$ Given numbers are in A.P.

TEST YOUR KNOWLEDGE 15.1

1. Show that 4, 10, 16, 22, ... is an A.P. Find its 7th and 9th terms.
2. Show that $6, 5\frac{1}{3}, 4\frac{2}{3}, 4, \ldots\ldots$ is an A.P. Find its 10th and kth terms.
3. Find the 20th, 25th and nth terms of the A.P. given by 21, 16, 11, 6, ...
4. Show that $\log a, \log ab, \log ab^2, \ldots\ldots$ is an A.P. Find its 7th and nth terms.
5. If $\frac{2}{3}, k, \frac{5}{8}k$ are in A.P., find the value of k.
6. Each term of an A.P. is doubled. Is the resulting sequence also an A.P. ? If it is, write its first term, common difference and the nth term.
7. Determine the number of terms in the sequence $17, 14\frac{1}{2}, 12, \ldots\ldots, -38$.
8. Which term of the A.P. 1, 6, 11, 16, ... is 301 ?
9. Which term of the series 20 + 16 + 12 + ... is – 96 ?
10. Is 310 a term of the sequence 3, 8, 13, ... ?
11. If the 9th term of an A.P. be zero, show that its 29th term is twice its 19th term.
12. In a certain A.P., the 24th term is twice the 10th term. Show that its 72nd term is twice the 34th term.
13. Determine the 25th term of the A.P. whose 9th term is – 6 and common difference is 5/4.
14. The 3rd term of an A.P. is 1 and the 6th term is – 11. Determine its 15th and rth terms.
15. Determine the 2nd and rth terms of an A.P. whose 6th term is 12 and the 8th term is 22.
16. Determine x so that $2x + 1$, $x^2 + x + 1$ and $3x^2 - 3x + 3$ are consecutive terms of an A.P.
17. If 5 times the 5th term of an A.P. is equal to the 10 times the 10th term, find the 15th term of the A.P.
18. Which term of the A.P. $8 - 6i, 7 - 4i, 6 - 2i, \ldots\ldots$ is (*i*) purely real (*ii*) purely imaginary.
19. If a, b, c are in A.P., show that :

    (*i*) $5a, 5b, 5c$ are in A.P.
    (*ii*) $\frac{2}{bc}, \frac{2}{ca}, \frac{2}{ba}$ are in A.P.

    (*iii*) $b + c, c + a, a + b$ are in A.P.
    (*iv*) $\frac{b+c-4a}{3}, \frac{c+a-4b}{3}, \frac{a+b-4c}{3}$ are in A.P.

20. (*i*) If $a + b + c \neq 0$ and $\frac{b+c}{a}, \frac{c+a}{b}, \frac{a+b}{c}$ are in A.P., show that $\frac{1}{a}, \frac{1}{b}, \frac{1}{c}$ are also in A.P.

    (*ii*) If the numbers a^2, b^2, c^2 are in A.P., show that $\frac{1}{b+c}, \frac{1}{c+a}, \frac{1}{a+b}$ are also in A.P.

    (*iii*) If the numbers a^2, b^2, c^2 are in A.P., show that $\frac{a}{b+c}, \frac{b}{c+a}, \frac{c}{a+b}$ are in A.P.

21. If $\log_5 2, \log_5 (2^x - 3), \log_5 \left(\frac{17}{2} + 2^{x-1}\right)$ are in A.P., then find the value of x.
22. The sides of a right angled triangle are in A.P., show that these are in the ratio 3 : 4 : 5.

23. The pth term of an A.P. is q and the qth term is p, show that its rth term is $p + q - r$.

24. If pth, qth, rth terms of an A.P. be a, b, c respectively, then show that $a(q - r) + b(r - p) + c(p - q) = 0$.

25. Find the 8th and nth terms of the series $\frac{1}{n} + \frac{2n+1}{n} + \frac{4n+1}{n} + \ldots\ldots$.

Can the 8th term be deduced from the nth term. If not, why ?

26. In any A.P. show that $T_{m+n} + T_{m-n} = 2T_m$.

27. How many numbers of two digits are divisible by 7 ?

28. For the A.P., $a_1, a_2, a_3, \ldots\ldots$, if $\frac{a_4}{a_7} = \frac{2}{3}$, find $\frac{a_6}{a_8}$.

29. If a, b, c are in A.P., show that $bc - a^2, ca - b^2, ab - c^2$ are in A.P.

30. If the roots of the equation $(b - c)x^2 + (c - a)x + (a - b) = 0$ are equal, then show that a, b, c are in A.P.

31. Find three numbers in A.P. whose sum is –3 and product is 8.

32. (*i*) The sum of three numbers in A.P. is 9 and the sum of their squares is 59.Find the numbers.

(*ii*) The sum of three numbers in A.P. is 15 and the sum of the squares of the two extremes is 58. Find the numbers.

33. Split 69 into 3 parts in A.P. such that the product of two smaller parts is 483.

34. Four numbers are in A.P., their sum is 16 and sum of their squares is 84. Find the numbers.

35. Divide 32 into four parts which are in A.P. such that the product of the extremes is to the product of means as 7 : 15.

Answers

1. 40, 52 **2.** 0, $\frac{20-2k}{3}$ **3.** – 74, – 99, $26 - 5n$ **4.** $\log ab^6, \log ab^{n-1}$

5. 16/33 **6.** yes **7.** 23 **8.** 61

9. 30 **10.** No **13.** 14 **14.** – 47, $13 - 4r$

15. – 8, $5r - 18$ **16.** 1, 2 **17.** 0 **18.** 4, 9

21. 3 **25.** $\frac{14n+1}{n}, \frac{2n^2-2n+1}{n}$, No **27.** 13

28. 4/5 **31.** – 4, – 1, 2 or 2, – 1, – 4

32. (*i*) – 1, 3, 7, or 7, 3, – 1 (*ii*) 3, 5, 7, or 7, 5, 3

33. 21, 23, 25 **34.** 1, 3, 5, 7 or 7, 5, 3, 1 **35.** 2, 6, 10, 14 or 14, 10, 6, 2

Hints

6. Let the A.P. be $a, a + d, a + 2d, \ldots\ldots$ $\therefore$ New sequence is $2a, 2a + 2d, 2a + 4d, \ldots\ldots$.
This is also an A.P. with first term $2a$, common difference $2d$ and nth term $= 2a + (n - 1)2d$.

10. $T_n = 310$ implies $n = 62.4$, which is not in **N**.

17. $5T_5 = 10T_{10} \Rightarrow 5(a + 4d) = 10(a + 9d) \Rightarrow 5a + 70d = 0 \Rightarrow a = -14d$

$\therefore T_{15} = a + 14d = -14d + 14d = 0$.

18. $T_n = 8 - 6i + (n - 1)(-1 + 2i) = 9 - n + i(2n - 8)$

(*i*) *If* $T_n = 0 \Rightarrow 2n - 8 = 0$.

21. $\log_5(2^x - 3) - \log_5 2 = \log_5\left(\frac{17}{2} + 2^{x-1}\right) - \log_5(2^x - 3)$

$\Rightarrow \quad \frac{2^x - 3}{2} = \frac{17 + 2^x}{2\,(2^x - 3)} \quad \Rightarrow \quad 2^x = 8, -1.$

22. $(a + d)^2 = a^2 + (a - d)^2 \quad \Rightarrow \quad a = 4d$

$\therefore$ Sides are $4d - d$, $4d$, $4d + d$.

25. T_8 cannot be deduced from T_n because '*c.d.*' is in terms of n.

27. Required numbers are 14, 21, 28, ..., 98.

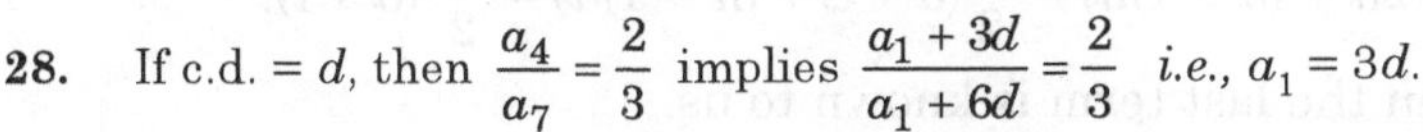

28. If c.d. = d, then $\frac{a_4}{a_7} = \frac{2}{3}$ implies $\frac{a_1 + 3d}{a_1 + 6d} = \frac{2}{3}$ *i.e.,* $a_1 = 3d$.

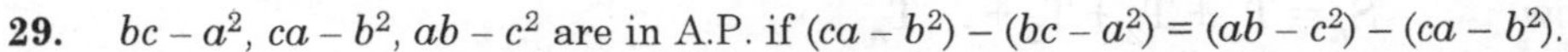

29. $bc - a^2$, $ca - b^2$, $ab - c^2$ are in A.P. if $(ca - b^2) - (bc - a^2) = (ab - c^2) - (ca - b^2)$.

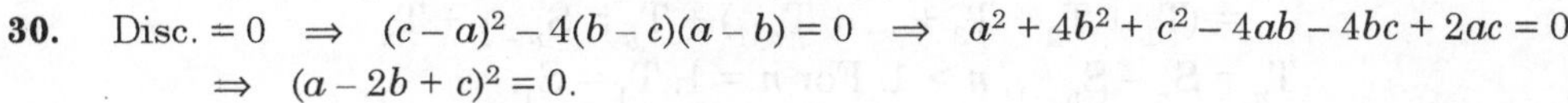

30. Disc. = 0 $\Rightarrow$ $(c - a)^2 - 4(b - c)(a - b) = 0$ $\Rightarrow$ $a^2 + 4b^2 + c^2 - 4ab - 4bc + 2ac = 0$

$\Rightarrow \quad (a - 2b + c)^2 = 0.$

33. Let the parts be $a - d, a, a + d$ where $d > 0$. $\therefore$ $(a - d) + a + (a + d) = 69$ implies $(23 - d)23 = 483$.

15.5. SUM OF FIRST n TERMS OF AN A.P.

If $<T_n>$ denotes the nth term of an A.P. then the sum of first n terms is denoted by S_n, where $S_n = T_1 + T_2 + T_3 + \ldots + T_n$.

Theorem IV. *If 'a' and 'd' be the first term and common difference of an A.P., then show that the sum of first n terms is given by* $S_n = \frac{n}{2}\,[2a + (n - 1)d],\ n \in N$

(P.T.U., M.B.A. Dec. 2001)

Proof. We apply principle of mathematical induction to prove

$$S_n = \frac{n}{2}\,[2a + (n - 1)d] \qquad \ldots(1)$$

Step I. For $n = 1$ in (1),

$$\text{L.H.S.} = S_1 = T_1 = a,\ \text{R.H.S.} = \frac{1}{2}\,(2a + 0) = a \ \Rightarrow\ \text{L.H.S.} = \text{R.H.S.}$$

$\therefore$ The result holds for $n = 1$.

Step II. Assume the result holds for $n = m$ *i.e.,* $S_m = \frac{m}{2}\,(2a + (m - 1)d)$ ...(2)

Step III. Consider

$$S_{m+1} = T_1 + T_2 + T_3 + \ldots T_m + T_{m+1} = S_m + T_{m+1}$$

$$= \frac{m}{2}(2a + (m - 1)d) + a + (m + 1 - 1)\,d \qquad \text{| Using (2)}$$

$$= ma + \frac{m}{2}\,(m - 1)d + a + md = (m + 1)a + md\left(\frac{m - 1}{2} + 1\right)$$

$$= (m + 1)a + \frac{md}{2}\,(m - 1 + 2) = (m + 1)a + \frac{md}{2}\,(m + 1)$$

$$= \frac{m+1}{2}(2a + md) = \frac{m+1}{2}(2a + (m + 1 - 1)d)$$

i.e., the result holds for $n = m + 1$, which completes the induction. Therefore by the principle of mathematical induction, the given result is true for each n *i.e.*,

$$S_n = \frac{n}{2}(2a + (n - 1)d).$$

Particular Cases

(*a*) We know $\quad S_n = \frac{n}{2}(2a + (n - 1)d) = \frac{n}{2}(a + a + (n - 1)d) = \frac{n}{2}(a + l).$

The above form is used when the last term is known to us.

(*b*) We know $\quad S_n = T_1 + T_2 + T_3 + \dots T_n$

$$= (T_1 + T_2 + T_3 + \dots + T_{n-1}) + T_n = S_{n-1} + T_n$$

$\Rightarrow \quad T_n = S_n - S_{n-1},\ n > 1.$ For $n = 1$, $T_1 = S_1$.

ILLUSTRATIVE EXAMPLES

Example 1. *Evaluate*

(*i*) *2 + 4 + 6 + ... 50 terms* (*ii*) *45 + 47 + 49 + ... + 99*

(*iii*) $\frac{1}{9} + \frac{2}{9} + \frac{1}{3} + \dots$ *25 terms.*

Sol. (*i*) Here $a = 2$, $d = 2$, $n = 50$. Using $S_n = \frac{n}{2}[2a + (n - 1)d]$, we have

$$S_{50} = \frac{50}{2}(2 \times 2 + 49 \times 2) = 25\ (4 + 98) = 25 \times 102 = 2550.$$

(*ii*) Given series is an A.P. with $a = 45$, $d = 2$, $l = 99$

$\therefore$ To find n, $\quad l = 99$

$\Rightarrow \quad a + (n - 1)d = 99$

$\Rightarrow \quad 45 + (n - 1) \times 2 = 99$

$\Rightarrow \quad 2(n - 1) = 99 - 45 = 54 \Rightarrow n - 1 = 27 \Rightarrow n = 28.$

Using $S_n = \frac{n}{2}(a + l)$, we have

$$S_{28} = \frac{28}{2}(45 + 99) = 14\ (144) = 2016.$$

(*iii*) Here $\quad T_2 - T_1 = \frac{2}{9} - \frac{1}{9} = \frac{1}{9},\ T_3 - T_2 = \frac{1}{3} - \frac{2}{9} = \frac{3-2}{9} = \frac{1}{9}$

$\therefore$ Given series is an A.P. with $a = \frac{1}{9}$, $d = \frac{1}{9}$, $n = 25$. $\therefore$ Using $S_n = \frac{n}{2}(2a + (n - 1)d)$, we have

$$S_{25} = \frac{25}{2}\left(2 \times \frac{1}{9} + 24 \times \frac{1}{9}\right) = \frac{25}{2} \times \frac{26}{9} = \frac{325}{9}.$$

Example 2. *Find the A.P. in which the sum of any number of terms is always three times the square of the number of these terms.*

Sol. Let the A.P. be a, $a + d$, $a + 2d$, ...

$\therefore$ By the given condition, $S_n = 3n^2$.

Using $T_n = S_n - S_{n-1}$, we have

$$\therefore \quad T_n = 3n^2 - 3(n-1)^2 = 3n^2 - 3(n^2 - 2n + 1) = 6n - 3$$

$$\therefore \quad T_1 = 6(1) - 3 = 3, \quad T_2 = 6(2) - 3 = 9, \quad T_3 = 6(3) - 3 = 15, \; ......$$

$\therefore$ The A.P. is 3, 9, 15, ...

Example 3. *Find the sum of all integers which are divisible by 7 and lying between 50 and 500.*

Sol. Integers lying between 50 and 500 and divisible by 7 are 56, 63, 70, ..., 497.

$\therefore$ Required sum $= 56 + 63 + 70 + ... + 497$

This is an A.P. with $a = 56$ and $d = 7$

Let n be the number of terms. $\therefore$ $497 = T_n$

$\Rightarrow$ $56 + (n - 1)7 = 497$ Solving, we get $n = 64$

$$\therefore \quad \text{Required sum} = S_{64} = \frac{64}{2}(56 + 497) = 32(553) = 17696. \qquad \left(S_n = \frac{n}{2}(a + l)\right)$$

Example 4. *How many terms of the sequence – 12, – 9, – 6, – 3, ... must be taken to make the sum 54 ?*

Sol. The given sequence is – 12, – 9, – 6, –3, ...

$$\text{Here} \quad T_2 - T_1 = -9 - (-12) = 3$$

$$T_3 - T_2 = -6 - (-9) = 3$$

$$T_4 - T_3 = -3 - (-6) = 3$$

....

$\therefore$ The given sequence is an A.P. with $a = -12$ and $d = 3$.

Let 54 be the sum of the first n terms. $\therefore$ $S_n = 54$

$$\Rightarrow \quad \frac{n}{2}[2a + (n-1)d] = 54 \quad \Rightarrow \quad n[2(-12) + (n-1)3] = 108$$

$$\Rightarrow \quad n(3n - 27) = 108 \quad \Rightarrow \quad n^2 - 9n - 36 = 0 \quad \Rightarrow \quad n = -3, 12$$

$$\therefore \quad n = 12, \text{ because } n = -3 \text{ is impossible.}$$

Example 5. *Determine the sum of the first 35 terms of an A.P. if its second term is 2 and seventh term is 22.*

Sol. Let the A.P. be a, $a + d$, $a + 2d$, ...

$$\therefore \quad T_2 = a + d = 2 \qquad ...(1)$$

$$\text{and} \quad T_7 = a + 6d = 22 \qquad ...(2)$$

(2) – (1) $\Rightarrow$ $5d = 20$ *i.e.,* $d = 4$

Putting $d = 4$ in (1), we get $a + 4 = 2$ *i.e.,* $a = 2 - 4 = -2$.

$$\therefore \quad S_{35} = \frac{35}{2}[2(-2) + (35 - 1)4] = \frac{35}{2} \times 132 = 2310. \qquad \left(\because S_n = \frac{n}{2}[2a + (n-1)d]\right)$$

Example 6. *If the first term of an A.P. is 2 and the sum of first five terms is equal to one-fourth of the sum of the next five terms, then :*

(*i*) *show that* $T_{20} = -112$ (*ii*) *find the sum of first 30 terms.*

Sol. Let the A.P. be $a, a + d, a + 2d, \ldots\ldots$

We have $T_1 + T_2 + T_3 + T_4 + T_5 = \frac{1}{4}[T_6 + T_7 + T_8 + T_9 + T_{10}]$

$$\therefore \quad \frac{5}{2}(T_1 + T_5) = \frac{1}{4} \times \frac{5}{2}(T_6 + T_{10}) \qquad \left[\text{Sum} = \frac{n}{2}(\text{first term + last term})\right]$$

$$\Rightarrow \quad \frac{5}{2}[a + (a + 4d)] = \frac{5}{8}[(a + 5d) + (a + 9d)] \quad \Rightarrow \quad 2a + 4d = \frac{1}{4}[2a + 14d]$$

$$\Rightarrow \quad 4a + 8d = a + 7d \quad \Rightarrow \quad d = -3a \quad \Rightarrow \quad d = -3(2) = -6$$

($\because$ $a = 2$ is given)

$$\therefore \quad a = 2 \text{ and } d = -6.$$

(*i*) $T_{20} = a + (20 - 1)d = 2 + 19(-6) = 2 - 114 = -112.$

(*ii*) $S_{30} = \frac{30}{2}[2(2) + (30 - 1)(-6)] = 15(-170) = -2550.$ $\left[S_n = \frac{n}{2}[2a + (n - 1)d]\right]$

Example 7. *If the sum of the first n terms of a sequence is of the form $An^2 + Bn$, where A and B are constants, independent of n, show that the sequence is an A.P. Is the converse true ? Justify your answer.*

Sol. We have $S_n = An^2 + Bn.$

Replacing n by $n - 1$, we get $S_{n-1} = A(n - 1)^2 + B(n - 1) = An^2 - 2An + A + Bn - B$

$$\therefore \quad T_n = S_n - S_{n-1} = (An^2 + Bn) - (An^2 - 2An + A + Bn - B) = 2An - A + B.$$

Replacing n by $n - 1$, we get $T_{n-1} = 2A(n - 1) - A + B$

$$\therefore \quad T_n - T_{n-1} = (2An - A + B) - (2An - 2A - A + B)$$

$$= 2A, \text{ a constant independent of } n.$$

$\therefore$ The sequence is an A.P.

Conversely, let $a, a + d, a + 2d, \ldots\ldots$ be an A.P.

$$\therefore \quad S_n = \frac{n}{2}[2a + (n - 1)d] = \frac{n}{2}[nd + (2a - d)] = \frac{d}{2}n^2 + \left(a - \frac{d}{2}\right)n$$

$$= An^2 + Bn, \quad \text{where} \quad A = \frac{d}{2} \quad \text{and} \quad B = a - \frac{d}{2}.$$

$\therefore$ $S_n = An^2 + Bn$, where A and B are constants, independent of n.

$\therefore$ Converse is also true.

Example 8. *If the sum of m terms of an A.P. be n and the sum of n terms be m, show that the sum of m + n terms is $-(m + n)$.*

Sol. Let a be the first term and d, the common difference of the A.P.

$$\therefore \quad S_m = n \Rightarrow \frac{m}{2}[2a + (m - 1)d] = n \Rightarrow 2am + m(m - 1)d = 2n \qquad \ldots(1)$$

and

$$S_n = m \Rightarrow \frac{n}{2}[2a + (n - 1)d] = m \Rightarrow 2an + n(n - 1)d = 2m \qquad \ldots(2)$$

$$(1) - (2) \quad \Rightarrow \quad 2a(m-n) + d(m^2 - m - n^2 + n) = 2n - 2m$$

$$\Rightarrow \quad 2a(m-n) + d((m^2 - n^2) - (m-n)) = -2(m-n)$$

Dividing by $m - n$, we get $\quad 2a + d(m + n - 1) = -2 \quad$ ...(3)

$$\therefore \quad S_{m+n} = \frac{m+n}{2}[2a + (m+n-1)d]$$

$$= \frac{m+n}{2}(-2) = -(m+n). \quad \text{(Using (3))}$$

Example 9. *If the sum of first n, 2n, 3n terms of an A.P. are S_1, S_2, S_3 respectively show that $S_3 = 3(S_2 - S_1)$.*

Sol. Let a be the first term and d, the common difference of the A.P.

$$\therefore \quad S_1 = \frac{n}{2}[2a + (n-1)d], \quad S_2 = \frac{2n}{2}[2a + (2n-1)d] \quad \text{and} \quad S_3 = \frac{3n}{2}[2a + (3n-1)d]$$

$$\text{R.H.S.} = 3(S_2 - S_1) = 3\left[\frac{2n}{2}[2a + (2n-1)d] - \frac{n}{2}[2a + (n-1)d]\right]$$

$$= \frac{3n}{2}[2[2a + (2n-1)d] - [2a + (n-1)d]]$$

$$= \frac{3n}{2}[4a + 4nd - 2d - 2a - nd + d] = \frac{3n}{2}[2a + (3n-1)d] = S_3 = \text{L.H.S.}$$

Example 10. *If the sum of m terms of an A.P. is to the sum of n terms of the same A.P. is as $m^2 : n^2$; prove that the ratio of its mth and nth terms is $2m - 1 : 2n - 1$.*

Sol. Let a and d be respectively the first term and common difference of the A.P.

By the given condition, $\quad \dfrac{S_m}{S_n} = \dfrac{m^2}{n^2}$

$$\Rightarrow \quad \frac{\frac{m}{2}[2a + (m-1)d]}{\frac{n}{2}[2a + (n-1)d]} = \frac{m^2}{n^2} \quad \Rightarrow \quad \frac{2a + (m-1)d}{2a + (n-1)d} = \frac{m}{n}$$

$$\Rightarrow \quad 2an + (m-1)nd = 2am + (n-1)md \quad \Rightarrow \quad 2a(n-m) = (nm - m - mn + n)d$$

$$\Rightarrow \quad 2a = d \quad \text{...(1)}$$

$$\text{Required ratio} \quad = \frac{T_m}{T_n} = \frac{a + (m-1)d}{a + (n-1)d} = \frac{a + (m-1).2a}{a + (n-1).2a} \quad \text{[by (1)]}$$

$$= \frac{1 + 2m - 2}{1 + 2n - 2} = \frac{2m - 1}{2n - 1}.$$

($\because$ $a \neq 0$. For if $a = 0$, then d would also be zero.)

TEST YOUR KNOWLEDGE 15.2

1. (*i*) Find the sum of first 50 natural numbers
   (*ii*) Find the sum of first 35 even natural numbers.
   (*iii*) Find the sum of first 65 odd natural numbers.

2. Find the sum of indicated number of terms of each of the following A.P.

(i) 5, 2, –1, ... n terms. (ii) 0.9, 0.91, 0.92, ... 100 terms.

3. Find the following sums :

(i) $2 + 5 + 8 + \ldots\ldots + 44$ (ii) $6 + 5\frac{2}{3} + 5\frac{1}{3} + \ldots\ldots + \frac{2}{3}$.

4. If S_n denotes the sum of n terms of an A.P. whose common difference is d, show that $d = S_n - 2S_{n-1} + S_{n-2}$, $n > 2$.

5. (i) Find the sum of 10 terms of the sequence $(x + y)^2, x^2 + y^2, (x - y)^2, \ldots\ldots$

(ii) Find the sum of 27 terms of the sequence $\frac{x - y}{x + y}, \frac{3x - 2y}{x + y}, \frac{5x - 3y}{x + y}, \ldots\ldots$

6. (i) Find the sum of all odd numbers between 100 and 200.

(ii) Find the sum of first 100 even natural numbers divisible by 5.

(iii) Find the sum of all natural numbers between 200 and 400 and divisible by 7.

(iv) Find the sum of all positive integers less than 500 and divisible by both 3 and 7.

7. (i) Find the rth term of the A.P., the sum of whose first n terms is $2n + 3n^2$.

(ii) Find the sum of first n terms of the A.P. whose kth term is given by $5k + 1$.

(iii) Find the sum of first n terms of an A.P. whose nth term is $5 - 6n$, $n \in \mathbf{N}$.

8. (i) How many terms of the A.P. 1, 4, 7, ... are needed to give sum 715 ?

(ii) How many terms of the A.P. – 6, – 11/2, – 5, ... are needed to give sum – 25 ?

9. (i) How many terms of the A.P. 18, 16, 14, ... are needed to give sum 78 ? Explain the double answer.

(ii) If the first term of an A.P. is 22, the common difference is – 4 and the sum to n terms is 64, find n. Explain the double answer.

10. Solve the equation : $1 + 6 + 11 + \ldots\ldots + x = 148$.

11. If S_1, S_2, S_3 are the sums of n terms of three A.P.s, the first term of each being unity and the respective common differences being 1, 2, 3, show that $S_1 + S_3 = 2S_2$.

12. (i) Find the sum of first n terms of an A.P. whose 7th term is 30 and 13th term is 54.

(ii) If the 5th and the 12th terms of an A.P. are 30 and 65 respectively, what is the sum of the first 20 terms ?

13. If S_n denotes the sum of n terms of an A.P. show that $S_{30} = 3(S_{20} - S_{10})$.

14. Find the common difference of an A.P. whose first term is 100 and the sum of whose first six terms is five times the sum of the next six terms.

15. Show that the sum of first n even natural numbers is equal to $\left(1 + \frac{1}{n}\right)$ times the sum of first n odd natural numbers.

16. If the 12th term of an A.P. is – 13 and the sum of the first four terms is 24, what is the sum of the first 10 terms ?

17. The sum of first n terms of an A.P. is zero, show that the sum of next m terms is $-\frac{am(m + n)}{n - 1}$, a being the first term.

18. The sum of n terms of two A.P. are in the ratio $5n + 4 : 9n + 6$. Find the ratio of their 18th terms.

19. The first, second and the last terms of a finite A.P. are a, b, c respectively. Show that the sum of all terms of the A.P. is $\dfrac{(a+c)(b+c-2a)}{2(b-a)}$.

20. If the sums of the first p, q, r terms of an A.P. be a, b, c respectively, show that

$$\frac{a}{p}(q-r)+\frac{b}{q}(r-p)+\frac{c}{r}(p-q)=0.$$

21. If the sum of first p terms of an A.P. is equal to the sum of first q terms, show that the sum of first $p+q$ terms is zero.

22. If the pth term of an A.P. is $\dfrac{1}{q}$ and qth term is $\dfrac{1}{p}$, show that the sum of pq terms is $\dfrac{1}{2}(pq+1)$.

23. If S_n denotes the sum of n terms of an A.P. and if $S_1 = 6$, $S_7 = 105$, show that $\dfrac{S_n}{S_{n-3}}=\dfrac{n+3}{n-3}$.

Answers

1. (*i*) 1275 (*ii*) 1260 (*iii*) 4225

2. (*i*) $n(13-3n)/2$ (*ii*) 139.5 **3.** (*i*) 345 (*ii*) $56\frac{2}{3}$

5. (*i*) $10(x^2+y^2-7xy)$ (*ii*) $27\left(\dfrac{27x-14y}{x+y}\right)$ **6.** (*i*) 7500 (*ii*) 50500

(*iii*) 8729 (*iv*) 5796 **7.** (*i*) $6r-1$ (*ii*) $n(5n+7)/2$

(*iii*) $n(2-3n)$ **8.** (*i*) 22 (*ii*) 5, 20 **9.** (*i*) 6, 13

(*ii*) 4, 8 **10.** 36 **12.** (*i*) $2n(n+2)$ (*ii*) 1150

14. -10 **16.** 0 **18.** 179 : 321.

Hints

4. $S_n - 2S_{n-1} + S_{n-2} = (S_n - S_{n-1}) - (S_{n-1} - S_{n-2}) = T_n - T_{n-1} = d.$

7. (*i*) $T_r = S_r - S_{r-1}$.

(*ii*) The A.P. is 5(1) + 1, 5(2) + 1, 5(3) + 1, ... *i.e.*, 6, 11, 16, ...

10. We have $x = 1 + (n-1)5$ and $148 = \dfrac{n}{2}(1+x)$.

14. $S_6 = 5(S_{12} - S_6) \Rightarrow 6S_6 = 5S_{12} \Rightarrow 6\left[\dfrac{6}{2}(200+5d)\right] = 5\left[\dfrac{12}{2}(200+11d)\right]$

16. Solve $a + 11d = -13$ and $\dfrac{4}{2}[2a+3d] = 24$.

17. If the A.P. be a, $a+d$, $a+2d$, ... then $\dfrac{n}{2}[2a+(n-1)d] = 0$ implies $d = \dfrac{2a}{n-1}$.

$\therefore$ Required sum $= S_{m+n} - S_n = S_{m+n} - 0 = S_{m+n}$.

18. Let a_1, a_2 and d_1, d_2 be the first terms and common differences of the two A.P.s respectively.

$$\Rightarrow \frac{\frac{n}{2}[2a_1+(n-1)d_1]}{\frac{n}{2}[2a_2+(n-1)d_2]}=\frac{5n+4}{9n+6} \Rightarrow \frac{a_1+\frac{n-1}{2}d_1}{a_2+\frac{n-1}{2}d_2}=\frac{5n+4}{9n+6}$$

Now put $\dfrac{n-1}{2} = 18 - 1$ *i.e.*, $n = 35$.

19. $d = T_2 - T_1 = b - a, \quad c = T_n = a + (n-1)d \Rightarrow \dfrac{c-a}{b-a} = n - 1 \Rightarrow n = \dfrac{b+c-2a}{b-a}$

$\therefore$ Sum $= \dfrac{n}{2}(a+c)$.

20. $a = \dfrac{p}{2}(2A + (p-1)D]$ implies $\dfrac{a}{p} = A + (p-1)\dfrac{D}{2}$ etc.

21. $S_p = S_q$ implies $\dfrac{p}{2}[2a + (p-1)d] = \dfrac{q}{2}[2a + (q-1)d]$

$\Rightarrow \quad 2a(p-q) + (p-q)(p+q-1)d = 0$ *i.e.*, $2a + (p+q-1)d = 0$ etc.

23. $S_1 = 6 \Rightarrow a = 6, \quad S_7 = 105 \Rightarrow \dfrac{7}{2}[2(6) + (7-1)d] = 105 \Rightarrow d = 3.$

15.6. ARITHMETIC MEANS

If three or more than three numbers are in A.P., then all the numbers lying between the first and the last numbers are called the **arithmetic means (A.M.s)** between them.

Equivalently, if $a, A_1, A_2, ..., A_n, b$ are in A.P., then $A_1, A_2, ..., A_n$ are called the n arithmetic means between a and b.

For example,

(*i*) 5, 10, 15 are in A.P. $\therefore$ 10 is the single A.M. between 5 and 15.

(*ii*) 5, 10, 15, 20, 25, 30 are in A.P. $\therefore$ 10, 15, 20, 25 are the four A.M.s between 5 and 30.

15.7. SINGLE A.M. BETWEEN ANY TWO GIVEN NUMBERS

Let a, b be any two numbers. Let A be the single A.M. between a and b.

$\therefore$ By definition, a, A, b are in A.P.

$\therefore \quad A - a = b - A$ (each = common difference)

$\Rightarrow \quad 2A = a + b \quad$ or $\quad A = \dfrac{a+b}{2}$.

Remark. The single A.M. between any two numbers is simply referred as the A.M. between the numbers. Thus, the A.M. between given two numbers is equal to half their sum.

For example, the A.M. between 7 and 29 is $\dfrac{7+29}{2} = \dfrac{36}{2} = 18$.

15.8. n A.M.s BETWEEN ANY TWO GIVEN NUMBERS

Let a, b be any two numbers. Let $A_1, A_2, ..., A_n$ be the n A.M.s between a and b.

$\therefore$ By definition, $a, A_1, A_2, ..., A_n, b$ are in A.P. Let d be the common difference of this A.P.

Now $\quad b = T_{n+2} = a + (n+1)d. \quad \therefore \quad d = \dfrac{b-a}{n+1}$

$\therefore \quad A_1 = a + d = a + \dfrac{b-a}{n+1}$

$$A_2 = a + 2d = a + 2\left(\frac{b-a}{n+1}\right)$$

....

$$A_n = a + nd = a + n\left(\frac{b-a}{n+1}\right).$$

$\therefore$ **The n A.M.s between a and b are** $\mathbf{a + \frac{b-a}{n+1}, a + 2\left(\frac{b-a}{n+1}\right), ..., a + n\left(\frac{b-a}{n+1}\right)}$.

Theorem V. *The sum of n A.M.s between any two numbers is equal to n times the A.M. between them.*

Proof. Let $A_1, A_2, ..., A_n$ be the n A.M.s between numbers a and b.

$\therefore$ Sum of n A.M.s between a and b

$$= A_1 + A_2 + ... + A_n = (a + A_1 + A_2 + ... + A_n + b) - (a + b)$$

$$= \frac{n+2}{2}(a+b) - (a+b) \qquad (\because \; a, A_1, A_2, ..., A_n, b \text{ is an A.P. of } n+2 \text{ terms})$$

$$= (a+b)\left[\frac{n+1}{2} - 1\right] = n\left(\frac{a+b}{2}\right) = n \text{ times the A.M. between } a \text{ and } b.$$

$\therefore$ **Sum of n A.M.s between a and b = n (A.M. between a and b).**

Note. The five A.M.s between a and b are given by $a + b$, $a + 2d$, $a + 3d$, $a + 4d$, and $a + 5d$, respectively. *(P.T.U. M.B.A. Dec. 2002, May 2002)*

ILLUSTRATIVE EXAMPLES

Example 1. *Insert three A.M.s between 11 and 14.*

Sol. Let A_1, A_2, A_3 be the three A.M.s between 11 and 14.

$\therefore$ 11, A_1, A_2, A_3, 14 are in A.P. Let d be the common difference of this A.P.

Now $\quad 14 = T_5 = 11 + 4d. \quad \therefore \; d = 3/4$

$$\therefore \quad A_1 = 11 + d = 11 + \frac{3}{4} = \frac{47}{4}, \quad A_2 = A_1 + d = \frac{47}{7} + \frac{3}{4} = \frac{25}{2}$$

and $$A_3 = A_2 + d = \frac{50}{4} + \frac{3}{4} = \frac{53}{4}.$$

Example 2. *Prove that if the number of terms of an A.P. is odd, then the middle term is the A.M. between the first and the last terms.*

Sol. Let the A.P. be $a, a + d, a + 2d, ...$ $(2n + 1)$ terms. (Note that $2n + 1$ is odd)

$\therefore$ Middle term $= T_{n+1} = a + (n + 1 - 1)d = a + nd.$

A.M. between first and last term

$$= \frac{T_1 + T_{2n+1}}{2} = \frac{a + (a + (2n + 1 - 1)d)}{2} = \frac{2a + 2nd}{2} = a + nd = T_{n+1}$$

$\therefore$ The result holds.

Example 3. *The sum of two numbers is 13/6. An even number of A.M.s are being inserted between them and their sum exceeds their number by 1. Find the number of A.M.s inserted.*

Sol. Let the numbers be a and b. $\therefore \quad a+b=13/6$.

Let $A_1, A_2, \ldots\ldots, A_{2n}$ be the $2n$ (even number) A.M.s between a and b.

$\therefore \quad A_1 + A_2 + \ldots\ldots + A_{2n} = 2n$ (A.M. between a and b)

$$=2n\left(\frac{a+b}{2}\right) = n(a+b) = n \times \frac{13}{6} = \frac{13}{6}n$$

Also $A_1 + A_2 + \ldots\ldots + A_{2n} = 2n+1$ (Given)

$$\therefore \quad \frac{13}{6}n = 2n+1 \quad \text{or} \quad \frac{n}{6} = 1 \text{ i.e., } n = 6.$$

$\therefore \quad$ No. of A.M.s inserted $= 2n = 2(6) = 12$.

Example 4. *Find n such that $\frac{a^n + b^n}{a^{n-1} + b^{n-1}}$ may be the A.M. between distinct numbers a and b.*

Sol. The A.M. between a and b is $\frac{a+b}{2}$. Let $\frac{a^n + b^n}{a^{n-1} + b^{n-1}} = \frac{a+b}{2}$

$$\Rightarrow \quad 2a^n + 2b^n = a^n + ab^{n-1} + a^{n-1}b + b^n \quad \Rightarrow \quad a^n + b^n = ab^{n-1} + a^{n-1}b$$

$$\Rightarrow \quad a^n - a^{n-1}b = ab^{n-1} + b^n \quad \Rightarrow \quad a^{n-1}(a-b) = b^{n-1}(a-b)$$

$$\Rightarrow \quad a^{n-1} = b^{n-1} \qquad (\because \ a \neq b \Rightarrow a - b \neq 0)$$

$$\therefore \quad \frac{a^{n-1}}{b^{n-1}} = 1 \Rightarrow \left(\frac{a}{b}\right)^{n-1} = \left(\frac{a}{b}\right)^0 \Rightarrow n-1=0 \text{ i.e., } n=1.$$

Example 5. *There are n A.M.s between 7 and 85 such that (n – 3)th mean : nth mean is as 11 : 24. Find n.*

Sol. Let $A_1, A_2, \ldots\ldots, A_n$ be the n A.M.s between 7 and 85.

$\therefore \quad 7, A_1, A_2, \ldots\ldots, A_n, 85$ is an A.P. of $n+2$ terms. Let d be the common difference of this A.P.

Now $$85 = T_{n+2} = 7 + (n+1)d \quad \therefore \quad d = \frac{78}{n+1}$$

We are given $\frac{(n-3)\text{th A.M.}}{n\text{th A.M.}} = \frac{11}{24}$.

$$\Rightarrow \quad \frac{T_{(n-3)+1}}{T_{n+1}} = \frac{11}{24} \quad \Rightarrow \quad \frac{7+(n-2-1)d}{7+(n+1-1)d} = \frac{11}{24} \quad \Rightarrow \quad \frac{7+(n-3)\frac{78}{n+1}}{7+n\frac{78}{n+1}} = \frac{11}{24}$$

$$\Rightarrow \quad \frac{7n+7+78n-234}{7n+7+78n} = \frac{11}{24} \quad \Rightarrow \quad \frac{85n-227}{85n+7} = \frac{11}{24} \quad \Rightarrow \quad 2040n - 5448 = 935n + 77$$

$$\Rightarrow \quad 1105n = 5525 \quad \Rightarrow \quad n = 5.$$

Example 6. *If the A.M. between pth and qth terms of an A.P. be equal to the A.M. between rth and sth terms of the A.P., show that p + q = r + s.*

Sol. Let the A.P. be $a, a + d, a + 2d, ...$

$\therefore \quad T_p = a + (p-1)d,\ T_q = a + (q-1)d,\ T_r = a + (r-1)d,\ T_s = a + (s-1)d.$

We are given that :

A.M. between T_p and T_q = A.M. between T_r and T_s

$$\Rightarrow \quad \frac{T_p + T_q}{2} = \frac{T_r + T_s}{2} \quad \Rightarrow \quad T_p + T_q = T_r + T_s$$

$$\Rightarrow \quad [a + (p-1)d] + [a + (q-1)d] = [a + (r-1)d] + [a + (s-1)d]$$

$$\Rightarrow \quad (p - 1 + q - 1)d = (r - 1 + s - 1)d \quad \Rightarrow \quad p + q - 2 = r + s - 2$$

$\Rightarrow \quad p + q = r + s. \quad \therefore$ The result holds.

TEST YOUR KNOWLEDGE 15.3

1. Find the A.M. between 5 and 9.
2. Find the A.M. between $(x-y)^2$ and $(x+y)^2$.
3. Find the sum of 500 A.M.s between 2 and 3.
4. If a, b, c are in A.P., show that $(a-c)^2 = 4(b^2 - ac)$.
5. If x, y, z are in A.P. show that $(x + 2y - z)(2y + z - x)(z + x - y) = 4xyz$.
6. Find the ratio of the sum of m A.M.s between any two numbers to the sum of n A.M.s between the same numbers.
7. Insert 3 A.M.s between 3 and 19.
8. Insert 5 A.M.s between 8 and 26.
9. Insert 6 A.M.s between 3 and 24.
10. Insert p A.M.s between 1 and p^2.
11. Insert 8 A.M.s between 2 and 29. Also verify that the sum of these 8 A.M.s is equal to 8 times the A.M. between 2 and 29.
12. Find n such that $\dfrac{a^{n+1} + b^{n+1}}{a^n + b^n}$ may be the A.M. between a and b.
13. n A.M.s are inserted between 5 and 86 such that the ratio of the first and the last mean is as 2 : 11. Find n.
14. There are n A.M.s between 3 and 17. The ratio of the last mean to the first mean is 3 : 1. Find the value of n.

Answers

1. 7 **2.** $x^2 + y^2$ **3.** 1250 **6.** $m : n$

7. 7, 11, 15 **8.** 11, 14, 17, 20, 23 **9.** 6, 9, 12, 15, 18, 21

10. $p, 2p - 1, 3p - 2, ..., p^2 - p + 1$ **11.** 5, 8, 11, 14, 17, 20, 23, 26

12. 0 **13.** 8 **14.** 6.

Hint

4. Use $b = \dfrac{a+c}{2}$ in the R.H.S.

15B GEOMETRIC PROGRESSION

15.9. GEOMETRICAL PROGRESSION

A sequence of non-zero numbers is said to form a **geometric progression** (G.P.) if the ratio of each term (except the first one) to its proceeding term is always same.

For example, $1, \frac{1}{2}, \frac{1}{4}, \frac{1}{8}, \ldots$, is a G.P. or A sequence $< T_n >$, with $T_n \neq 0$ is said to be a *geometric progression* if there exists a number r, $r \neq 0$, such that $\frac{T_{n+1}}{T_n} = r \; \forall \; n \geq 1$. The number '$r$' is called the **common ratio** of the concerned G.P.

Remark. 1. The first term of a G.P. is generally denoted by 'a'.

2. In case of a G.P., $a \neq 0$, $r \neq 0$.

Theorem VI. *If 'a' and 'r' be the first term and common ratio of the G.P.* $<T_n>$, *then show that* $T_n = ar^{n-1}$, $n \in N$.

Proof. We apply principle of mathematical induction to prove $T_n = ar^{n-1}$ ...(1)

Step I. For $n = 1$ in (1), L.H.S. $= T_1 = a$; R.H.S. $= ar^{1-1} = a$

$\therefore$ The result holds for $n = 1$.

Step II. Assume the result holds for $n = m$ *i.e.*, $T_m = ar^{m-1}$ ...(2)

Step III. Take $n = m + 1$ and consider

$$\frac{T_{m+1}}{T_m} = r \Rightarrow T_{m+1} = r\,T_m = r.ar^{m-1} = ar^m = ar^{m+1-1} \qquad | \text{ By definition}$$

i.e., the result holds for $n = m + 1$. It completes the induction. Therefore by the principle of mathematical induction, the given result holds for each n

i.e., $T_n = ar^{n-1}$, $n \in N$.

15.10. IMPORTANT OBSERVATIONS

(*i*) If product of three numbers in a G.P. is given, then the numbers are $\frac{a}{r}, a, ar$.

(*ii*) If product of four numbers in a G.P. is given, the numbers are $\frac{a}{r^3}, \frac{a}{r}, ar, ar^3$.

(*iii*) If product of five numbers in a G.P. is given, the numbers are $\frac{a}{r^2}, \frac{a}{r}, a, ar, ar^2$.

Remark. If product of numbers in G.P. is not given, the numbers to be taken as $a, ar, ar^2, \ldots\ldots$

ILLUSTRATIVE EXAMPLES

Example 1. *Find the 12th term and general term of the sequence 3, 6, 12, 24,...*

Sol. Given sequence 3, 6, 12, 24, ... is a G.P. with $a = 3$, $r = 2$

Using $T_n = ar^{n-1}$, we have

$$T_{12} = ar^{11} = 3 \times 2^{11} = 6144$$

Also $\quad T_n = ar^{n-1} = 3 \times 2^{n-1}$.

Example 2. *Which term of the series* $\frac{1}{4} - \frac{1}{2} + 1 + ...$, *is 256 ?*

Sol. Given series is $\frac{1}{4} - \frac{1}{2} + 1 + ...$

Here $\quad \frac{T_2}{T_1} = \frac{-1/2}{1/4} = -2, \frac{T_3}{T_2} = \frac{1}{-1/2} = -2$ and so on

$\therefore$ Given series is a G.P. with $a = \frac{1}{4}$, $r = -2$.

Let $T_n = 256 \Rightarrow ar^{n-1} = 256 \Rightarrow \frac{1}{4}(-2)^{n-1} = 256 \Rightarrow (-2)^{n-1} = 1024 = 2^{10} = (-2)^{10}$

$\Rightarrow \quad n - 1 = 10 \Rightarrow n = 11$ *i.e.*, $T_{11} = 256$.

Example 3. *For what value of n, the nth term of the series* $5 + 10 + 20 + ...$ *and* $1280 + 640 + 320 + ...$ *are equal ?*

Sol. In first series, $a = 5$, $\quad r = 2 \quad \therefore \quad T_n = ar^{n-1} = 5 \times 2^{n-1}$

In second series, $a = 1280$, $\quad r = \frac{1}{2} \quad \therefore \quad T_n = ar^{n-1} = 1280 \times \frac{1}{2^{n-1}}$

If nth term of first series and second series are equal, then we must have

$$5 \times 2^{n-1} = 1280 \times \frac{1}{2^{n-1}} \Rightarrow 2^{n-1} \times 2^{n-1} = \frac{1280}{5} = 256$$

$$\Rightarrow \quad 2^{n-1+n-1} = 256 \Rightarrow 2^{2n-2} = 2^8$$

$$\Rightarrow \quad 2n - 2 = 8 \Rightarrow 2n = 10 \Rightarrow n = 5$$

i.e., 5th term of the two series are equal.

Example 4. *The 6th, 12th and 18th terms of a G.P. are a, b and c respectively. Show that* $b^2 = ac$.

Sol. Let A be the first term and R, the common ratio of the G.P.

$\therefore$ The G.P. is $A, AR, AR^2, ...$

$$\therefore \quad a = T_6 = AR^5, \quad b = T_{12} = AR^{11} \quad \text{and} \quad c = T_{18} = AR^{17}.$$

$$\therefore \quad b^2 = (AR^{11})^2 = A^2R^{22} \quad \text{and} \quad ac = (AR^5)(AR^{17}) = A^2R^{22}$$

$$\therefore \quad b^2 = ac.$$

Example 5. *The first term of a G.P. is 1. The sum of third and fifth terms is 90. Find the common ratio of the G.P.*

Sol. Let r be the common ratio of the G.P. Here $a = 1$. We are given $T_3 + T_5 = 90$.

$\Rightarrow \quad ar^2 + ar^4 = 90 \quad \Rightarrow \quad r^2 + r^4 = 90 \qquad (\because \; a = 1)$

$$\Rightarrow \quad r^4 + r^2 - 90 = 0 \quad \Rightarrow \quad r^2 = \frac{-1 \pm \sqrt{1+360}}{2} = \frac{-1 \pm 19}{2} = 9, -10$$

$r^2 = 9$ implies $r = -3, 3$ $\qquad r^2 = -10$ is impossible.

$\therefore$ Common ratio $= -3$ or 3.

Example 6. *If the 4th, 10th and 16th terms of a G.P. are x, y, z respectively, prove that x, y, z are in G.P. Generalise the result.*

Sol. Let A be the first term and R, the common ratio of the G.P.

$\therefore$ The G.P. is $A, AR, AR^2, \ldots\ldots$

$$x = T_4 = AR^3, \quad y = T_{10} = AR^9 \quad \text{and} \quad z = T_{16} = AR^{15}.$$

Now
$$\frac{y}{x} = \frac{AR^9}{AR^3} = R^{9-3} = R^6 \quad \text{and} \quad \frac{z}{y} = \frac{AR^{15}}{AR^9} = R^{15-9} = R^6$$

$$\therefore \quad \frac{y}{x} = \frac{z}{y} \quad \text{i.e., } x, y, z \text{ are in G.P.}$$

Generalisation. Let $x = T_p = AR^{p-1}, \quad y = T_q = AR^{q-1}$ and $z = T_r = AR^{r-1}$.

$$\therefore \quad \frac{y}{x} = \frac{AR^{q-1}}{AR^{p-1}} = R^{q-1-p+1} = R^{q-p} \quad \text{and} \quad \frac{z}{y} = \frac{AR^{r-1}}{AR^{q-1}} = R^{r-1-q+1} = R^{r-q}.$$

$\therefore$ x, y, z are in G.P. if $\frac{y}{x} = \frac{z}{y}$ i.e., if $R^{q-p} = R^{r-q}$ or if $q - p = r - p$

or if p, q, r are in A.P.

$\therefore$ **In general, if p, q, r are in A.P., then T_p, T_q, T_r of any G.P. are also in G.P.**

Example 7. *If the 5th term of a G.P. is 16 and the 10th term is $\frac{1}{2}$, find the G.P. Also find its 15th term.*

Sol. Let a be the first term and r, the common ratio of the G.P.

$$\therefore \quad T_n = ar^{n-1}, \; n \in \mathbf{N}.$$

We have
$$T_5 = 16 \quad \text{and} \quad T_{10} = \frac{1}{2}.$$

$$\Rightarrow \quad ar^4 = 16 \quad \text{and} \quad ar^9 = \frac{1}{2} \quad \Rightarrow \quad \frac{ar^9}{ar^4} = \frac{\frac{1}{2}}{16}$$

$$\Rightarrow \quad r^5 = \frac{1}{32} = \left(\frac{1}{2}\right)^5 \quad \Rightarrow \quad r = \frac{1}{2}$$

$$\therefore \quad ar^4 = 16 \quad \Rightarrow \quad a\left(\frac{1}{2}\right)^4 = 16 \quad \Rightarrow \quad a = 256$$

$\therefore$ The G.P. is $256, 256\left(\frac{1}{2}\right), 256\left(\frac{1}{2}\right)^2, \ldots\ldots$ or $256, 128, 64, \ldots\ldots$

Also, $T_{15} = ar^{14} = 256\left(\frac{1}{2}\right)^{14} = \frac{1}{64}.$

Example 8. *If p, q, r are in A.P. and x, y, z are in G.P., prove that* $x^{q-r} \cdot y^{r-p} \cdot z^{p-q} = 1$.

Sol. p, q, r are in A.P. Let d be the common difference

$\therefore$ $q = p + d$ and $z = p + 2d$

Also, x, y, z are in G.P. Let R be the common ratio

$\therefore$ $y = xR$ and $z = xR^2$.

Putting the values of q, r, y, z in L.H.S. we get

$$\text{L.H.S.} = x^{q-r}y^{r-p}z^{p-q} = x^{(p+d)-(p+2d)}\,(xR)^{(p+2d)-p}(xR^2)^{p-(p+d)}$$
$$= x^{-d}(xR)^{2d}\,(xR^2)^{-d} = x^{-d} \cdot x^{2d}R^{2d} \cdot x^{-d}R^{-2d}$$
$$= x^{-d+2d-d} \cdot R^{2d-2d} = x^0R^0 = 1 \times 1 = 1 = \text{R.H.S.}$$

Example 9. *Find all the sequences which are simultaneously arithmetic and geometric progressions.*

Sol. Let $T_1, T_2, T_3, \ldots\ldots$ be a sequence which is A.P. as well as G.P.

Let $T_n = a + (n-1)d \;\; \forall \;\; n \in \mathbf{N}$

$\therefore$ The sequence is $a, a + d, a + 2d, \ldots\ldots$

This is also a G.P. $\therefore$ $\frac{T_{n+1}}{T_n} = \frac{T_{n+2}}{T_{n+1}} \;\; \forall \;\; n \in \mathbf{N}.$

$$\Rightarrow \quad \frac{a+nd}{a+(n-1)d} = \frac{a+(n+1)d}{a+nd} \quad \Rightarrow \quad (a+nd)^2 = [(a+nd)+d]\,[(a+nd)-d]$$

$$\Rightarrow \quad (a+nd)^2 = (a+nd)^2 - d^2 \quad \Rightarrow \quad d^2 = 0 \quad i.e., d = 0$$

$\therefore$ The sequence is $a, a + 0, a + 2(0), \ldots\ldots$ *i.e.,* $a, a, a, \ldots\ldots$

$\therefore$ **Only a constant sequence can be both A.P. and G.P.**

Example 10. *If a, b, c, d in G.P., show that the following are also in G.P.*

(*i*) $a^2 - b^2, b^2 - c^2, c^2 - d^2$ (*ii*) $a^2 + b^2 + c^2, ab + bc + cd, b^2 + c^2 + d^2$.

Sol. a, b, c, d are in G.P.

Let r be the common ratio of this G.P. $\therefore$ $b = ar, c = ar^2, d = ar^3$.

(*i*) $a^2 - b^2, b^2 - c^2, c^2 - d^2$ are in G.P.

if $\frac{b^2 - c^2}{a^2 - b^2} = \frac{c^2 - d^2}{b^2 - c^2}$. if $\frac{(ar)^2 - (ar^2)^2}{a^2 - (ar)^2} = \frac{(ar^2)^2 - (ar^3)^2}{(ar)^2 - (ar^2)^2}$

if $\frac{a^2r^2\,(1 - r^2)}{a^2\,(1 - r^2)} = \frac{a^2r^4\,(1 - r^2)}{a^2r^2\,(1 - r^2)}$ if $r^2 = r^2$, which is true.

(Since a, b, c, d are in G.P. therefore, $a \neq 0$. We also assume that $r \neq 1$.)

$\therefore$ $a^2 - b^2, b^2 - c^2, c^2 - d^2$ are in G.P.

(*ii*) $a^2 + b^2 + c^2, ab + bc + cd, b^2 + c^2 + d^2$ are in G.P.

if $\frac{ab + bc + cd}{a^2 + b^2 + c^2} = \frac{b^2 + c^2 + d^2}{ab + bc + cd}$

if $$\frac{(a)(ar)+(ar)(ar^2)+(ar^2)(ar^3)}{a^2+(ar^2)+(ar^2)^2}=\frac{(ar)^2+(ar^2)^2+(ar^3)^2}{(a)(ar)+(ar)(ar^2)+(ar^2)(ar^3)}$$

if $$\frac{a^2r\,(1+r^2+r^4)}{a^2\,(1+r^2+r^4)}=\frac{a^2r^2\,(1+r^2+r^4)}{a^2\,(1+r^2+r^4)}$$ if $r = r$, which is true.

$\therefore$ $a^2 + b^2 + c^2$, $ab + bc + cd$, $b^2 + c^2 + d^2$ are in G.P.

Example 11. *Find four numbers forming a G.P. where the third term is greater than the first by 9 and the second term is greater than the fourth by 18.*

Sol. Let the four numbers in G.P. be a, ar, ar^2, and ar^3 .

$\therefore$ By the given conditions : $ar^2 = a + 9$ ...(1) and $ar = ar^3 + 18$ ...(2)

(1) $\Rightarrow$ $a(r^2 - 1) = 9$ ...(3)

(2) $\Rightarrow$ $ar\,(1 - r^2) = 18$ $\Rightarrow$ $-ar(r^2 - 1) = 18$ ...(4)

Dividing (4) by (3), we get $-r = 2$ *i.e.,* $r = -2$

Putting $r = -2$ in (3), we get $a\,((-2)^2 - 1) = 9$ *i.e.,* $3a = 9$ or $a = 3$.

$\therefore$ The numbers are $3, 3(-2), 3(-2)^2, 3(-2)^3$ *i.e.,* $3, -6, 12, -24$.

Example 12. *The sum of three numbers in G.P. is 13/12 and their product is – 1. Find the numbers.*

Sol. Let the numbers be $\frac{a}{r}$, a and ar.

$\therefore$ Product $=\frac{a}{r}.\,a\,.\,ar = -1$ *(Given)*

$\Rightarrow$ $a^3 = -1 \Rightarrow a^3 = (-1)^3$ or $a = -1$.

$\therefore$ The numbers are $\frac{-1}{r}, -1$ and $-1(r) = -r$.

Also, sum $=\left(-\frac{1}{r}\right)+(-1)+(-r)=\frac{13}{12}$ *(Given)*

$\Rightarrow$ $-\left(\frac{1}{r}+1+r\right)=\frac{13}{12}$ $\Rightarrow$ $-12 - 12r - 12r^2 = 13r$

$\Rightarrow$ $12r^2 + 25r + 12 = 0$ $\Rightarrow$ $r = -3/4, -4/3$

$\therefore$ $r = -3/4$	and $r = -4/3$
The numbers are $-\frac{1}{-3/4}, -1, -(-3/4)$	The numbers are $-\frac{1}{(-4/3)}, -1, -(-4/3)$
or $4/3, -1, 3/4$	or $3/4, -1, 4/3$.

Example 13. *The product of three numbers in G.P. is 1000. If we add 6 to its second number and 7 to its third number, the resulting three numbers form an A.P. Find the numbers in G.P.* (P.T.U., M.B.A. March 1999)

Sol. Let the numbers in G.P. be $\frac{a}{r}$, a and ar.

$\therefore$ Product $=\frac{a}{r}\,.\,a\,.\,ar = 1000$ *(Given)*

$\Rightarrow$ $a^3 = 1000$ *i.e.,* $a^3 = (10)^3$ or $a = 10$

$\therefore$ The numbers are $\frac{10}{r}$, 10 and $10r$.

New numbers are $\frac{10}{r}$, 10 + 6, $10r + 7$. These are in A.P.

$$\Rightarrow \quad 16 - \frac{10}{r} = (10r + 7) - 16 \quad \Rightarrow \quad 16 - \frac{10}{r} = 10r - 9$$

$$\Rightarrow \quad 16r - 10 = 10r^2 - 9r \quad \Rightarrow \quad 2r^2 - 5r + 2 = 0 \quad \Rightarrow \quad r = 1/2, 2.$$

Case I. $r = \frac{1}{2}$. The numbers in G.P. are $\frac{10}{1/2}$, 10, 10 (1/2) *i.e.,* 20, 10, 5.

Case II. r = 2. The numbers in G.P. are $\frac{10}{2}$, 10, 10(2) *i.e.,* 5, 10, 20.

Example 14. *The sum of the first three terms of a G.P. is 7 and sum of their squares is 21. Determinate first five terms of the G.P.*

Sol. Let the G.P. be $a, ar, ar^2, \ldots\ldots$

By the given conditions, $a + ar + ar^2 = 7$ ...(1) and $(a)^2 + (ar)^2 + (ar^2)^2 = 21$ ...(2)

(1) $\Rightarrow$ $a(1 + r + r^2) = 7$ ...(3) (2) $\Rightarrow$ $a^2 (1 + r^2 + r^4) = 21$ ...(4)

(3) and (4) implies $\dfrac{a^2 (1+r^2+r^4)}{[a (1+r+r^2)]^2} = \dfrac{21}{(7)^2}$.

$$\Rightarrow \quad \frac{(1+2r^2+r^4)-r^2}{(1+r+r^2)^2} = \frac{3}{7} \quad \Rightarrow \quad \frac{(1+r^2+r)(1+r^2-r)}{(1+r+r^2)^2} = \frac{3}{7} \quad \Rightarrow \quad \frac{1+r^2-r}{1+r+r^2} = \frac{3}{7}$$

$$\Rightarrow \quad 7 + 7r^2 - 7r = 3 + 3r + 3r^2 \quad \Rightarrow \quad 2r^2 - 5r + 2 = 0 \quad \Rightarrow \quad r = 2, \frac{1}{2}.$$

Case I. r = 2. In this case, (3) $\Rightarrow$ $a(1 + 2 + 4) = 7$ $\Rightarrow$ $a = 1$

$\therefore$ The G.P. is 1, 1(2), 1(2)2, ... *i.e.,* 1, 2, 4 ......

The first five terms are 1, 2, 4, 8, 16.

Case II. $r = \frac{1}{2}$. In this case, (3) $\Rightarrow$ $a\left(1+\frac{1}{2}+\frac{1}{4}\right) = 7$ $\Rightarrow$ $a = 4$

The G.P. is 4, $4\left(\frac{1}{2}\right), 4\left(\frac{1}{2}\right)^2$, ... *i.e.,* 4, 2, 1, ...

The first five terms are 4, 2, 1, $\frac{1}{2}, \frac{1}{4}$.

Example 15. *The sum of three numbers in G.P. is 56. If we subtract 1, 7, 21 from these numbers in this order, we obtain an A.P. Find the numbers.*

Sol. Let the numbers in G.P. be a, ar, ar^2.

By the given conditions : $a + ar + ar^2 = 56$ ...(1)

and $a - 1, ar - 7, ar^2 - 21$ are in A.P. ...(2)

(1) $\Rightarrow$ $a(1 + r + r^2) = 56$ ...(3)

(2) $\Rightarrow$ $(ar - 7) - (a - 1) = (ar^2 - 21) - (ar - 7)$

$\Rightarrow$ $ar - a - 6 = ar^2 - ar - 14 = a - 2ar + ar^2 = 8$

$\Rightarrow \qquad a(1-2r+r^2)=8 \qquad ...(4)$

Dividing (4) by (3), we get $\dfrac{a(1+r+r^2)}{a(1-2r+r^2)}=\dfrac{56}{8}$ *i.e.,* $\dfrac{1+r+r^2}{1-2r+r^2}=7$

$\Rightarrow \quad 1+r+r^2=7-14r+7r^2 \quad \Rightarrow 6r^2-15r+6=0 \Rightarrow 2r^2-5r+2=0$

$\therefore \quad r=1/2,\ 2.$

Case I. r = $\frac{1}{2}$. (1) $\Rightarrow a\left(1+\frac{1}{2}+\frac{1}{4}\right)=56 \quad \Rightarrow \frac{7a}{4}=56$ *i.e.,* $a=32$

$\therefore$ The numbers are 32, 32(1/2), 32(1/2)^2 *i.e.,* 32, 16, 8.

Case II. r = 2. (1) $\Rightarrow a(1+2+4)=56 \quad \Rightarrow 7a=56$ *i.e.,* $a=8$

$\therefore$ The numbers are 8, 8(2), 8(2)^2 *i.e.,* 8, 16, 32.

Example 16. *There are four numbers such that the first three of these form an A.P. and the last three form a G.P. The sum of the first and the third numbers is 2 and that of the second and fourth is 26, what are these numbers ?*

Sol. Let the numbers be a, b, c, d. By the given conditions :

$a+c=2 \quad ...(1) \qquad b+d=26 \quad ...(2)$

(1) $\Rightarrow c=2-a$ and (2) $\Rightarrow d=26-b.$

$\therefore$ The numbers are $a, b, 2-a, 26-b$.

Also, $a, b, 2-a$ are in A.P. $\therefore \ b-a=(2-a)-b$ *i.e.,* $2b=2$ or $b=1$.

$\therefore$ The numbers become $a, 1, 2-a, 26-1=25$.

Also, last three numbers are in G.P. $\therefore \ \dfrac{2-a}{1}=\dfrac{25}{2-a}$.

$\Rightarrow \quad (2-a)^2=25 \quad \Rightarrow \quad a^2-4a-21=0 \Rightarrow a=-3, 7$

Case I. a = – 3. In this case, the numbers are – 3, 1, 2 – (– 3), 25 *i.e.,* – 3, 1, 5, 25.

Case II. a = 7. In this case, the numbers are 7, 1, 2 – 7, 25 *i.e.,* 7, 1, – 5, 25.

TEST YOUR KNOWLEDGE 15.4

1. Find the common ratio of the following G.P. :

(*i*) 0.01, 0.0001, 0.000001, ... (*ii*) $1, -a, a^2$, ...

(*iii*) $\sqrt{2}, \dfrac{1}{\sqrt{2}}, \dfrac{1}{2\sqrt{2}}$, ... (*iv*) a^{m-n}, a^m, a^{m+n}, ...

2. Show that each of the following is a G.P. Also find nth term in each case :

(*i*) 128, 64, 32, ... (*ii*) 5/2, 5/4, 5/8, ...

(*iii*) $\sqrt{3}$, 3, $3\sqrt{3}$, ... (*iv*) ak, ak^3, ak^5, ...

3. Each term of a G.P. is doubled. Is the resulting sequence also a G.P. ?

4. If a, b, c are in G.P. and $k\ (\neq 0) \in \mathbf{R}$, then show that :

(*i*) $a+k, b+k, c+k$ may not be in G.P. (*ii*) $a-k, b-k, c-k$ may not be in G.P.

(*iii*) ka, kb, kc are in G.P. (*iv*) $a/k, b/k, c/k$ are in G.P.

5. If each term of a G.P. is :

(*i*) multiplied by the same non-zero number

(*ii*) divided by the same non-zero number, then show that the resulting sequence is also a G.P.

6. (*i*) Find the sequence whose *n*th term is 3^{n-1}. What type of sequence is it ? What is the 5th term ?

(*ii*) Find the sequence whose *n*th term is $3^n - 1$. Is it a G.P. ? What is its 6th term ?

7. Determine the number of terms in the sequence 5/2, 5, 10, ..., 640.

8. (*i*) Which term of the G.P. 2, $2\sqrt{2}$, 4,...... is 128 ?

(*ii*) Which term of the series 5 + 10 + 20 + ... is 2560 ?

9. Is 512 a term of the sequence 1/256, 1/64, 1/16, ... ?

10. (*i*) The 5th, 8th and 11th terms of a G.P. are *p*, *q* and *r* respectively. Show that $q^2 = pr$.

(*ii*) In any G.P. prove that $T_{n-r} \times T_{n+r} = (T_n)^2$.

11. (*i*) The 3rd term of a G.P. is 24 and 6th term is 192. Find the 10th term.

(*ii*) The 5th term of a G.P. is 16 and 10th term is 1/2. Find the G.P. Also find the 15th term.

12. (*i*) The fourth term of a G.P. is square of its second term and the first term is – 3. Determine the seventh term of the G.P.

(*ii*) Find the G.P. for which the sum of the first two terms is – 4 and the fifth term is 4 times the third term.

13. If *a*, *b*, *c* are in A.P., then show that for any non-zero real number *k*, the numbers k^a, k^b, k^c are in G.P.

14. If $\frac{1}{x+y}, \frac{1}{2y}, \frac{1}{y+z}$ are in A.P., show that *x*, *y*, *z* are in G.P.

15. (*i*) Find the value of *x* if – 2/7, *x*, – 7/2 are in G.P.

(*ii*) For what value(s) of *k*, the numbers $1 + k, \frac{5}{6} + k, \frac{13}{18} + k$ are in G.P. ?

16. If *a*, *b*, *c* are in G.P., show that $\frac{1}{\log_a m}, \frac{1}{\log_b m}, \frac{1}{\log_c m}$ are in A.P.

17. If $a^x = b^y = c^z$ and *x*, *y*, *z* are in G.P., show that $\log_b a = \log_c b$.

18. If the *m*th, *n*th and *p*th terms of a G.P. are in G.P., show that *m*, *n* and *p* are in A.P.

19. If the *p*th, *q*th and *r*th terms of a G.P. are *x*, *y* and *z* respectively, show that :

(*i*) $x^{q-r} y^{r-p} z^{p-q} = 1$ (*ii*) $(q - r) \log x + (r - p) \log y + (p - q) \log z = 0$.

20. Find the value of *x* for which $x + 9$, $x - 6$ and 4 are the first three terms of a G.P. and calculate the fourth term of the progression.

21. If the *p*th, *q*th, *r*th and *s*th terms of an A.P. are in G.P., show that $p - q$, $q - r$, $r - s$ are also in G.P.

22. If *a*, *b*, *c* are in G.P., show that :

(*i*) a^2, b^2, c^2 are in G.P. (*ii*) $\frac{1}{a^2}, \frac{1}{b^2}, \frac{1}{c^2}$ are in G.P.

(*iii*) b^2c^2, c^2a^2, a^2b^2 are in G.P., (*iv*) $a^2 + b^2, ab + bc, b^2 + c^2$ are in G.P.

23. If *a*, *b*, *c*, *d* are in G.P., show that :

(*i*) $a + b, b + c, c + d$ are in G.P. (*ii*) $a^2 + b^2, b^2 + c^2, c^2 + d^2$ are in G.P.

(*iii*) $(a - c)^2 + (b - c)^2 + (b - d)^2 = (a - d)^2$ (*iv*) $(a^2 + b^2 + c^2)(b^2 + c^2 + d^2) = (ab + bc + cd)^2$.

24. If *a* be the first term of a G.P., *l* the *n*th term and *P*, the product of first *n* terms, show that $P = (al)^{n/2}$.

25. If the $(m + n)$th term of a G.P. is *p* and $(m - n)$th term is *q*, show that *m*th and *n*th terms are $\sqrt{pq}$ and $p(q/p)^{m/2n}$ respectively.

26. The *p*th, *q*th and *r*th terms of an A.P. as well as of a G.P. are *a*, *b*, *c* respectively. Prove that $a^{b-c} b^{c-a} c^{a-b} = 1$.

27. If p, q, r are in G.P. and the equations $px^2 + 2qx + r = 0$ and $dx^2 + 2ex + f = 0$ have a common root, then show that

$$\frac{d}{p}, \frac{e}{q}, \frac{f}{r} \text{ are in A.P.}$$

28. If a, b, c are in A.P., b, c, d are in G.P. and $\frac{1}{c}, \frac{1}{d}, \frac{1}{e}$ are in A.P. ; prove that a, c, e are in G.P.

29. (*i*) The sum of first three terms of a G.P. is $\frac{39}{10}$ and their product is 1. Find the three terms.

(*ii*) Obtain three consecutive terms a, b, c of a G.P. if $a + b + c = 93$ and $abc = 3375$.

30. (*i*) The lengths of the sides of a triangle form a G.S. If the perimeter of the triangle is 37 cm, and the shortest side is of length 9 cm ; find the lengths of the other two sides.

(*ii*) Find three numbers in G.P., whose sum is 19 and the sum of their squares is 133.

31. (*i*) The sum of three numbers in G.P. is 56. If we subtract 1, 7, 21 from these numbers in this order, we obtain an A.P. Find these numbers.

(*ii*) The sum of three numbers which are consecutive terms of an A.P. is 21. If the second term is reduced by 1 while the third is increased by 1, three consecutive terms of a G.P. are formed. Find these numbers.

(*iii*) The sum of three numbers in A.P. is 15. If 1, 4 and 19 are added to these numbers respectively, the resulting numbers are in G.P. Find the numbers.

32. (*i*) In a set of four numbers, the first three are in G.P. and the last three are in A.P. with common difference 6. If the first number is the same as the fourth, find the four numbers.

(P.T.U., M.B.A. Dec. 2000)

(*ii*) The sum of four numbers in G.P. is 60 and the A.M. between the first and the last is 18. Find the numbers.

Answers

1. (*i*) 0.01 (*ii*) $-a$ (*iii*) 1/2 (*iv*) a^n

2. (*i*) 2^{8-n} (*ii*) $5/2^n$ (*iii*) $3^{n/2}$ (*iv*) ak^{2n-1}

3. yes **6.** (*i*) 1, 3, 9, 27, ... ; G.P. ; 81

(*ii*) 2, 8, 26, 80, ... ; This is not a G.P. ; 728 **7.** 9. **8.** (*i*) 13th

(*ii*) 10th **9.** No

11. (*i*) 3072 (*ii*) 256, 128, 64, ... ; $\frac{1}{64}$

12. (*i*) -2187 (*ii*) 4, -8, 16, ... or $-\frac{4}{3}, -\frac{8}{3}, -\frac{16}{3}$, ... **15.** (*i*) ± 1

(*ii*) $-1/2$ **20.** $x = 0, T_4 = -8/3$; $x = 16, T_4 = 8/5$ **29.** (*i*) $\frac{2}{5}, 1, \frac{5}{2}$ or $\frac{5}{2}, 1, \frac{2}{5}$

(*ii*) 3, 15, 75, or 75, 15, 3 **30.** (*i*) 12 cm, 16 cm (*ii*) 4, 6, 9 or 9, 6, 4

31. (*i*) 8, 16, 32 or 32, 16, 8 (*ii*) 3, 7, 11 or 12, 7, 2 (*iii*) 2, 5, 8 or 26, 5, -16

32. (*i*) 8, -4, 2, 8. (*ii*) 4, 8, 16, 32, or 32, 16, 8, 4.

Hints

4. (*i*) Let $a = 1, b = 2, c = 4, k = 2$. $\therefore$ a, b, c are in G.P. but $a + k = 3, b + k = 4, c + k = 6$ are not in G.P.

12. (*i*) If the G.P. be a, ar, ar^2, ... then $a = -3$ and $ar^3 = (ar)^2$.

13. $\frac{k^b}{k^a} = \frac{k^c}{k^b}$ implies $b - a = c - b$.

16. $\dfrac{1}{\log_a m}, \dfrac{1}{\log_b m}, \dfrac{1}{\log_c m}$ are in A.P. if $\log_m a, \log_m b, \log_m c$ are in A.P. if $\log_m \dfrac{b}{a} = \log_m \dfrac{c}{b}$.

17. Let $a^x = b^y = c^z = k$. $\quad \therefore \quad x = \dfrac{\log_e k}{\log_e a} = \log_a k, y = \log_b k, z = \log_c k$

x, y, z in G.P. $\Rightarrow \dfrac{\log_b k}{\log_a k} = \dfrac{\log_c k}{\log_b k} \Rightarrow \dfrac{\log_k a}{\log_k b} = \dfrac{\log_k b}{\log_k c} \Rightarrow \log_b a = \log_c b$.

19. (*i*) Let $x = AR^{p-1}$ etc.

21. Let the A.P. be $a, a+d, a+2d, ...$ $\therefore \dfrac{a+(q-1)d}{a+(p-1)d} = \dfrac{a+(r-1)d}{a+(q-1)d} = \dfrac{a+(s-1)d}{a+(r-1)d}$

This implies $\dfrac{[a+(q-1)d]-[a+(r-1)d]}{[a+(p-1)d]-[a+(q-1)d]} = \dfrac{[a+(r-1)d]-[a+(s-1)d]}{[a+(q-1)d]-[a+(r-1)d]}$ *i.e.*, $\dfrac{q-r}{p-q} = \dfrac{r-s}{q-r}$.

22. Let $b = ar, \quad c = ar^2$

23. Let $b = ar, \quad c = ar^2, \quad d = ar^3$.

24. If the G.P. be $a, ar, ar^2, ...$, then $l = ar^{n-1}$

and $P = a \,.\, ar \,.\, ar^2 ...... ar^{n-1} = a^n r^{1+2+......+(n-1)} = a^n r^{\frac{(n-1)n}{2}} = a^n (r^{n-1})^{n/2} = a^n (l/a)^{n/2}$.

25. $p = ar^{m+n-1}$ and $q = ar^{m-n-1}$ implies $pq = a^2 r^{2m-2} = (ar^{m-1})^2 = T_m^{\ 2}$.

Also, $\dfrac{p}{q} = r^{m+n-1-m+n+1} = r^{2n} \quad \therefore \quad r = (p/q)^{1/2n}$

Now $\quad T_n = ar^{n-1} = ar^{m+n-1}.r^{-m} = p[(p/q)^{1/2n}]^{-m}$

26. Let the A.P. and G.P. be A, A + D, A + 2D, ... and B, BR, BR^2,......

$\therefore \quad a = A + (p-1)D = BR^{p-1}$

$b = A + (q-1)D = BR^{q-1}$

and $\quad C = A + (r-1)D = BR^{r-1}$

$\therefore \quad b - c = (q-r)D, \; c - a = D(r-p), \; a - b = D(p-q)$

$\therefore \quad a^{b-c}\, b^{c-a}\, c^{a-b} = (BR^{p-1})^{(q-r)D} \,.\, (BR^{q-1})^{(r-p)D} \,.\, (BR^{r-1})^{(p-q)D}$.

27. We have $q^2 = pr$.

Roots of $px^2 + 2qx + r = 0$ are $\dfrac{-2q \pm \sqrt{4q^2 - 4pr}}{2p}$ or $\dfrac{-q \pm \sqrt{q^2 - pr}}{p}$ or $\dfrac{-q \pm 0}{p}$ or $-\dfrac{q}{p}, -\dfrac{q}{p}$.

$\therefore \quad -\dfrac{q}{p}$ is a root of $dx^2 + 2ex + f = 0$. $\therefore \quad d\left(-\dfrac{q}{p}\right)^2 + 2e\left(-\dfrac{q}{p}\right) + f = 0$ or $dq^2 - 2epq + p^2 f = 0$

Dividing by pq^2, we get $\dfrac{d}{p} - \dfrac{2e}{q} + \dfrac{pf}{q^2} = 0$ or $\dfrac{d}{p} - \dfrac{2e}{q} + \dfrac{pf}{pr} = 0$ or $\dfrac{d}{p} + \dfrac{f}{r} = 2\left(\dfrac{e}{q}\right)$.

28. We have $\quad b = \dfrac{a+c}{2}, \quad c^2 = bd, \quad \dfrac{1}{d} = \dfrac{1}{2}\left(\dfrac{1}{c} + \dfrac{1}{e}\right)$ or $d = \dfrac{2ce}{c+e}$.

$\therefore \quad c^2 = \left(\dfrac{a+c}{2}\right)\left(\dfrac{2ce}{c+e}\right)$.

32. (*ii*) Take numbers as a, ar, ar^2, ar^3 and use $1 + r + r^2 + r^3 = (1+r)(1+r^2)$.

15.11. SUM OF *n* TERMS OF A G.P.

If $<T_n>$ be the nth term of a G.P., then the sum of first n terms of the G.P. is given by S_n, where $S_n = T_1 + T_2 + T_3 + ... + T_n$, $n \in N$

Theorem. *If 'a' and 'r' be the first term and common ratio of a G.P. respectively, then show that*

$$S_n = \begin{cases} \dfrac{a(1-r^n)}{1-r}, & r \neq 1 \\ na, & r = 1 \end{cases}.$$

Proof. Case I. $r \neq 1$: Let T_n = nth term of the given G.P. $= ar^{n-1}$, $n \in N$

$$\therefore \quad S_n = T_1 + T_2 + T_3 + ... + T_{n-1} + T_n$$

$$= a + ar + ar^2 + ... + ar^{n-2} + ar^{n-1} \quad ...(1)$$

$$\Rightarrow \quad r S_n = ar + ar^2 + ar^3 + ... + ar^{n-1} + ar^n \quad ...(2)$$

Subtracting (1) and (2), we get

$$(1-r) S_n = a + 0 + 0 + ... + 0 - ar^n = a(1-r^n)$$

$$\Rightarrow \quad S_n = \frac{a(1-r^n)}{1-r}.$$

Case II. When $r = 1$, then from (1)

$$S_n = a + a + a + ... + a(n \text{ times}) = na$$

15.12. PARTICULAR CASES

(*a*) We have $S_n = \dfrac{a(1-r^n)}{1-r}$, $r < 1$.

If $r > 1$, then $\quad S_n = -\dfrac{a(r^n-1)}{-(r-1)} = \dfrac{a(r^n-1)}{r-1}.$

(*b*) Using $l = T_n = ar^{n-1}$, we have

$$S_n = \begin{cases} \dfrac{a(1-r^n)}{1-r}, & r < 1 \\ \dfrac{a(r^n-1)}{r-1}, & r > 1 \end{cases}$$

$$= \begin{cases} \dfrac{a - ar^{n-1}.r}{1-r}, & r < 1 \\ \dfrac{ar^{n-1}.r - 1}{r-1}, & r > 1 \end{cases}$$

$$= \begin{cases} \dfrac{a - lr}{1-r}, & r < 1 \\ \dfrac{lr - 1}{r-1}, & r > 1 \end{cases}.$$

ILLUSTRATIVE EXAMPLES

Example 1. *Find the sum of the following series upto 50 terms.*

$$1 - \frac{1}{3} + \frac{1}{3^2} - \frac{1}{3^3} + \frac{1}{3^4} \ldots$$ (P.T.U., M.B.A. May 2003)

Sol. Given series is a G.P. with $a = 1$, $r = -\frac{1}{3}$, $n = 50$.

Using $S_n = a\left(\frac{1-r^n}{1-r}\right)$, we have $S_{50} = \frac{1-\left(-\frac{1}{3}\right)^{50}}{1+\frac{1}{3}} = \frac{1-\frac{1}{3^{50}}}{\frac{4}{3}} = \frac{3}{4}\left(1-\frac{1}{3^{50}}\right)$.

Example 2. *Evaluate* $\frac{2}{625} + \frac{2}{125} + \frac{2}{25} + \ldots 50.$

Sol. Given series is a G.P. with $a = \frac{2}{625}$, $r = \frac{\frac{2}{125}}{\frac{2}{625}} = 5$, $l = 50$.

$\therefore$ Required sum $= \frac{lr-a}{r-1}$, $r > 1 = \frac{50 \times 5 - \frac{2}{625}}{5-1} = \frac{156248}{625 \times 4} = \frac{39062}{625}$.

Example 3. *Determine the number of terms in the G.P. $\{T_n\}$ if $T_1 = 3$, $T_n = 96$ and $S_n = 189$.*

Sol. Let the G.P. be $a, ar, ar^2, \ldots\ldots$ We are given : $T_1 = 3$, $T_n = 96$, $S_n = 189$.

$$\therefore \quad a = 3, \quad ar^{n-1} = 96, \quad \frac{a(1-r^n)}{1-r} = 189$$

$$\Rightarrow \quad 189 = \frac{a - ar^{n-1}.r}{1-r} = \frac{3-96r}{1-r}$$

Cross-multiplying, we get $189 - 189r = 3 - 96r$. Solving, we get $r = 2$.

$\therefore$ $ar^{n-1} = 96$ implies $3(2)^{n-1} = 96$ or $2^{n-1} = 32 = 2^5 \Rightarrow n - 1 = 5$ *i.e.*, $n = 6$.

Example 4. *How many terms of the G.P. $\sqrt{3}, 3, 3\sqrt{3}, \ldots\ldots$ sum up to $39 + 13\sqrt{3}$?*

Sol. The given G.P. is $\sqrt{3}, 3, 3\sqrt{3}, \ldots\ldots$

Here $a = \sqrt{3}$, $r = \frac{3}{\sqrt{3}} = \sqrt{3}$.

Let $39 + 13\sqrt{3}$ be the sum of n terms of the G.P. $\therefore$ $S_n = 39 + 13\sqrt{3}$

$$\Rightarrow \quad \frac{\sqrt{3}\,(1-(\sqrt{3})^n)}{1-\sqrt{3}} = 39 + 13\sqrt{3}$$

$$\Rightarrow \quad \sqrt{3}\,(1-(\sqrt{3})^n) = 39 - 39\sqrt{3} + 13\sqrt{3} - 39 \Rightarrow \sqrt{3}\,(1-(\sqrt{3})^n) = -26\sqrt{3}$$

$$\Rightarrow \quad 1 - 3^{n/2} = -26 \Rightarrow 3^{n/2} = 27 = 3^3 \Rightarrow \frac{n}{2} = 3 \Rightarrow n = 6.$$

Example 5. *The second and fifth terms of a G.P. are – 1/2 and 1/16 respectively. Find the sum of the G.P. upto 8 terms.*

Sol. Let a be the first term and r, the common ratio of the G.P.

$$\therefore \quad T_2 = ar^{2-1} = ar \quad \text{and} \quad T_5 = ar^{5-1} = ar^4.$$

$$\therefore \quad ar = -\frac{1}{2} \quad \text{...(1)} \qquad \text{and} \qquad ar^4 = \frac{1}{16} \quad \text{...(2)}$$

Dividing (2) by (1), we get $\dfrac{ar^4}{ar} = \dfrac{1/16}{-1/2}$ *i.e.,* $r^3 = -\dfrac{1}{8}$ or $r = -\dfrac{1}{2}$.

Putting $r = -\dfrac{1}{2}$ in (1), we get $a\left(-\dfrac{1}{2}\right) = -\dfrac{1}{2}$ *i.e.,* $a = 1$.

$$\therefore \quad S_8 = \frac{1\left(1-\left(-\frac{1}{2}\right)^8\right)}{1-\left(-\frac{1}{2}\right)} = \frac{1-\frac{1}{256}}{\frac{3}{2}} = \frac{255}{256} \times \frac{2}{3} = \frac{85}{128}. \qquad \left(S_n = \frac{a(1-r^n)}{1-r} \text{ for } r < 1\right)$$

Example 6. *Find the sum of 50 terms of the sequence 7, 7.7, 7.77, 7.777, ...*

Sol. The given sequence is 7, 7.7, 7.77, 7.777, ...

Required sum $= S_{50}$

$= 7 + 7.7 + 7.77 +$ ...... 50 terms $= 7\,(1 + 1.1 + 1.11 +$ ...... 50 terms)

$= \dfrac{7}{9}\,(9 + 9.9 + 9.99 +$ ...... 50 terms) $= \dfrac{7}{9}\,[(10 - 1) + (10 - 0.1) + (10 - 0.01)$ + ... 50 terms]

$= \dfrac{7}{9}\,[(10 + 10 + 10 +$ ...... 50 terms) – (1 + 0.1 + 0.01 + ... 50 terms)]

$= \dfrac{7}{9}\,[500 -$ (Sum of G.P. of 50 terms with $a = 1$, $r = 0.1$)]

$$= \frac{7}{9}\left[500 - \frac{1(1-(0.1)^{50})}{1-0.1}\right] = \frac{7}{9}\left[500 - \frac{10}{9}\left(1 - \frac{1}{10^{50}}\right)\right]$$

$$= \frac{7}{81}\,(4500 - 10 + 10^{-49}) = \frac{7}{81}\,(4490 + 10^{-49}).$$

Example 7. *Find the sum of n terms of the sequence* $x + y,\ x^2 + xy + y^2,\ x^3 + x^2y + xy^2 + y^3$, ...

Sol. The given sequence is $x + y,\ x^2 + xy + y^2,\ x^3 + x^2y + xy^2 + y^3$, ...

Required sum $= (x + y) + (x^2 + xy + y^2) + (x^3 + x^2y + xy^2 + y^3) +$ ...... n terms

$$= \frac{x^2 - y^2}{x - y} + \frac{x^3 - y^3}{x - y} + \frac{x^4 - y^4}{x - y} + \text{...... } n \text{ terms}$$

$$= \frac{1}{x-y} [(x^2 - y^2) + (x^3 - y^3) + (x^4 - y^4) \text{ ...... } n \text{ terms}]$$

$$= \frac{1}{x-y} [(x^2 + x^3 + x^4 + \text{ ...... } n \text{ terms}) - (y^2 + y^3 + y^4 + \text{ ...... } n \text{ terms})]$$

$$= \frac{1}{x-y} [(\text{Sum of G.P. of } n \text{ terms with } a = x^2, r = x)$$

$$- (\text{Sum of G.P. of } n \text{ terms with } a = y^2, r = y)]$$

$$= \frac{1}{x-y}\left[\frac{x^2(1-x^n)}{1-x} - \frac{y^2(1-y^n)}{1-y}\right].$$

Example 8. *If S be the sum, P be the product, and R be the sum of reciprocals of n terms of a G.P., show that* $P^2 = \left(\frac{S}{R}\right)^n$. (P.T.U., M.B.A. Dec. 2002)

Sol. Let the G.P. is $a, ar, ar^2, \ldots$

Given $S = a + ar + ar^2 + \ldots n \text{ terms} = \frac{a(1-r^n)}{1-r}$ ...(1)

$$P = a \cdot ar \cdot ar^2 \ldots n \text{ terms}$$

$$= a^n \cdot r^{1+2+3+\ldots(n-1)}$$

$$= a^n \cdot r^{\frac{(n-1)n}{2}} \quad \ldots(2)$$

$$R = \frac{1}{a} + \frac{1}{ar} + \frac{1}{ar^2} + \ldots n \text{ terms}$$

$$= \frac{1}{a}\left(\frac{1-\frac{1}{r^n}}{1-\frac{1}{r}}\right) = \frac{r^n - 1}{a(r-1)r^{n-1}} \quad \ldots(3)$$

From (2), $$P^2 = \left(a^n \cdot r^{\frac{(n-1)n}{2}}\right)^2 = a^{2n} r^{(n-1)n} \quad \ldots(4)$$

From (1) and (3), $$\frac{S}{R} = \frac{a(1-r^n)}{1-r} \times \frac{a(r-1)r^{n-1}}{r^n - 1} = a^2 r^{n-1}$$

$$\Rightarrow \quad \left(\frac{S}{R}\right)^n = a^{2n} r^{(n-1)n} \quad \ldots(5)$$

From (5) and (4), $P^2 = \left(\frac{S}{R}\right)^n$

Example 9. *What is the difference between geometrical progression and arithmetical progression ?* (P.T.U., M.B.A. May 2004, Dec. 2003)

Sol. **Difference between A.P. and G.P.**

G.P.	*A.P.*
• The common ratio remains same. • In G.P., first term cannot be zero. • In G.P., product of kth term from the beginning and kth term from the end is independent of k and equals the product of first and last terms.	• The common difference remains same. • In A.P., first term may be zero • In A.P., sum of kth term from the beginning and kth term from the end is independent of k and equals the sum of first and last terms.

TEST YOUR KNOWLEDGE 15.5

1. Find the sum of the series : $\sqrt{2} + 2 + \sqrt{8} + ...$ 12 terms
2. Find the sum of the series $\frac{1}{4} + \frac{1}{2} + 1 + ...$ n terms.
3. Find the following sums :

   (*i*) $\sqrt{3} + 3 + 3\sqrt{3} + ... + 729$ (*ii*) $\frac{2}{9} - \frac{1}{3} + \frac{1}{2} - ...... + \frac{81}{32}$.
4. Given a G.P. with first term = 729, $T_7 = 64$; determine S_7.
5. (*i*) How many terms of the sequence 3, 3^2, 3^3, ... are needed to give the sum 120 ?

   (*ii*) How many terms of the G.P. 3, 3/2, 3/4, ... are needed to give the sum 3069/512 ?
6. Evaluate the following :

   (*i*) $\sum_{j=1}^{11} (2 + 3^j)$ (*ii*) $\sum_{j=1}^{8} (2^j + 3^{j-1})$
7. If $a + b + ... + l$ is a geometric series, show that its sum is $\frac{bl - a^2}{b - a}$.
8. (*i*) The fourth and seventh terms of a G.P. are 1/27 and 1/729 respectively. Find the sum of n terms of the G.P.

   (*ii*) The first term of a G.P. is 27 and its 8th term is 1/81. Find the sum of its 10 terms.
9. If $v = \frac{1}{1+\lambda}$, show that $v + v^2 + v^3 + ... + v^n = \frac{1 - v^n}{\lambda}$.
10. The sum of first three terms of a G.P. to the sum of first six terms is as 125 : 152. Find the common ratio of the G.P.
11. Determine the third term of a G.P. whose common ratio is 3 and the sum to first seven terms is 2186.
12. Find the following sums :

    (*i*) 9 + 99 + 999 + ... n terms (*ii*) 5 + 55 + 555 + ... n terms.

    (*iii*) 0.6 + 0.66 + 0.666 + ... n terms (*iv*) 4 + 0.4 + 0.44 + 0.444 + ... n terms.
13. (*i*) Find the sum of the first n terms of the sequence 7, 77, 777, 7777, ...

    (*ii*) Find the sum of the first 50 terms of the sequence 7, 7.7, 7.77, 7.777, ...

14. Find the following sums :

(*i*) $(a + b) + (a^2 + 2b) + (a^3 + 3b) + \ldots\ldots$ *n* terms

(*ii*) $a + br + ar^2 + br^3 + ar^4 + br^5 + \ldots\ldots$ $2n$ terms.

15. Find the following sums :

(*i*) $x (x + y) + x^2 (x^2 + y^2) + x^3 (x^3 + y^3) + \ldots\ldots$ *n* terms

(*ii*) $1 + (1 + x) + (1 + x + x^2) + (1 + x + x^2 + x^3) + \ldots\ldots$ *n* terms

(*iii*) $1 + (1 + a)\, x + (1 + a + a^2)\, x^2 + \ldots\ldots$ *n* terms.

16. If S_1, S_2, S_3 be the sums of $n, 2n, 3n$ terms of a G.P., show that $S_1(S_3 - S_2) = (S_2 - S_1)^2$.

Answers

1. $63(\sqrt{2} + 2)$ **2.** $\dfrac{2^n - 1}{4}$ **3.** (*i*) $364\,(3 + \sqrt{3})$ (*ii*) $\dfrac{463}{288}$.

4. 463, 2059 **5.** (*i*) 4 (*ii*) 10

6. (*i*) 265741 (*ii*) 3790 **8.** (*i*) $\dfrac{3}{2}\left(1 - \dfrac{1}{3^n}\right)$ (*ii*) $\dfrac{81}{2}\left(1 - \dfrac{1}{3^{10}}\right)$

10. $\dfrac{3}{5}$ **11.** 18

12. (*i*) $\dfrac{1}{9}(10^{n+1} - 9n - 10)$ (*ii*) $\dfrac{5}{81}(10^{n+1} - 9n - 10)$ (*iii*) $\dfrac{2}{27}\left(9n - 1 + \dfrac{1}{10^n}\right)$

(*iv*) $4 + \dfrac{4}{81}\left(9n - 10 + \dfrac{1}{10^{n-1}}\right)$ **13.** (*i*) $\dfrac{7}{9}\left[\dfrac{10}{9}(10^n - 1) - n\right]$ (*ii*) $\dfrac{7}{81}(4490 + 10^{-49})$

14. (*i*) $\dfrac{a\,(1 - a^n)}{1 - a} + \dfrac{bn\,(n + 1)}{2}$ (*ii*) $\dfrac{(a + br)\,(1 - r^{2n})}{1 - r^2}$

15. (*i*) $\dfrac{x^2\,(1 - x^{2n})}{1 - x^2} + \dfrac{xy\,(1 - x^n y^n)}{1 - xy}$ (*ii*) $\dfrac{1}{1 - x}\left[n - \dfrac{x\,(1 - x^n)}{1 - x}\right]$

(*iii*) $\dfrac{1}{1 - a}\left[\dfrac{1 - x^n}{1 - x} - \dfrac{a\,(1 - a^n x^n)}{1 - ax}\right]$

Hints

3. (*i*) Use sum $= \dfrac{lr - a}{r - 1}$ (*ii*) Use sum $= \dfrac{a - lr}{1 - r}$

6. (*ii*) $\displaystyle\sum_{j=1}^{8} (2^j + 3^{j-1}) = (2^1 + 2^2 + \ldots\ldots + 2^8) + (1 + 3 + \ldots\ldots + 3^7)$

7. Sum $= \dfrac{a - lr}{1 - r} = \dfrac{a - l\,(b/a)}{1 - (b/a)}$.

8. (*i*) Let the G.P. be $a, ar, ar^2, \ldots\ldots$ $\therefore$ $ar^3 = 1/27$ and $ar^6 = 1/729$.

$$\therefore \quad \frac{ar^6}{ar^3} = \frac{27}{729} \quad \text{i.e.,} \quad r = \frac{1}{3} \quad \therefore \quad a\left(\frac{1}{3}\right)^3 = \frac{1}{27} \quad \text{i.e.,} \quad a = 1.$$

10. $\dfrac{\dfrac{a(1-r^3)}{1-r}}{\dfrac{a(1-r^6)}{1-r}} = \dfrac{125}{152} \Rightarrow \dfrac{1}{1+r^3} = \dfrac{125}{152} \Rightarrow r^3 = \dfrac{152}{125} - 1 \Rightarrow r^3 = \left(\dfrac{3}{5}\right)^3.$

12. (*iv*) $4 + 0.4 + 0.44 + 0.444 + \ldots\ldots\ n$ terms

$$= 4 + \frac{4}{9}\left[\left(1-\frac{1}{10}\right)+\left(1-\frac{1}{100}\right)+\left(1-\frac{1}{1000}\right)+\ldots\ldots (n-1) \text{ terms}\right]$$

$$= 4 + \frac{4}{9}\left[(n-1)-\left(\frac{1}{10}+\frac{1}{100}+\frac{1}{1000}+\ldots\ldots (n-1) \text{ terms}\right)\right]$$

13. (*i*) Reqd. sum $= 7 + 77 + 777 + \ldots\ldots\ n$ terms

$$= \frac{7}{9}[(10-1)+(100-1)+(1000-1)+\ldots\ldots\ n \text{ terms}]$$

(*ii*) Reqd. sum $= 7 + 7.7 + 7.77 + \ldots\ldots$ 50 terms

$$= 7 + \frac{7}{9}[9.9 + 9.99 + \ldots\ldots 49 \text{ terms}]$$

15. (*i*) Reqd. sum $= (x^2 + x^4 + x^6 + \ldots\ldots\ n \text{ terms} + (xy + x^2y^2 + x^3y^3 + \ldots\ldots\ n \text{ terms})$

(*iii*) Reqd. sum $= \dfrac{1}{1-a}\,[(1-a) + (x - a^2x) + (x^2 - a^3x^2) + \ldots\ldots\ n \text{ terms}]$

$$= \frac{1}{1-a}\,[(1 + x + x^2 + \ldots\ldots\ n \text{ terms}) - a(1 + ax + a^2x^2 + \ldots\ldots\ n \text{ terms})]$$

15.13. SUM TO INFINITY OF A G.P.

We know that for the G.P. $a, ar, ar^2, \ldots\ldots$, the sum of first n terms is given by

$$S_n = \begin{cases} \dfrac{a(1-r^n)}{1-r}, & \text{if } r \neq 1 \\ na, & \text{if } r = 1. \end{cases}$$

This sum is defined for any natural number n.

Now we shall explore the possibility of finding the sum to infinity of a G.P.

Consider the G.P. 1, 2, 4, 8, ...

For this G.P., $a = 1$ and $r = 2$.

$\therefore$ $$S_n = \frac{1(2^n - 1)}{2-1} = 2^n - 1.$$

As n increases, $2^n - 1$ increases very rapidly.

$\therefore$ S_n keep on increasing as n increases. In this case, we do not expect to have a number which may be equal to the sum to infinity of the G.P.

In the next theorem, we shall derive a formula to find the sum to infinity of a G.P., provided the common ratio of the G.P. is within certain limits.

Theorem VII. *If a and r be respectively the first term and common ratio of a G.P. such that* $|r| < 1$, *then prove that the sum to infinity (S) of this G.P. is given by*

$$S = \frac{a}{1-r}.$$ (P.T.U., M.B.A. Dec. 2000)

Proof. Let S_n denotes the sum of the first n terms of the given G.P.

$$\therefore \quad S_n = \frac{a(1-r^n)}{1-r} \qquad (\because \ |r| < 1 \text{ implies } r \neq 1)$$

$$\therefore \quad \lim_{n\to\infty} S_n = \lim_{n\to\infty} \frac{a(1-r^n)}{1-r} = \frac{a - a\lim_{n\to\infty} r^n}{1-r} = \frac{a-a(0)}{1-r} = \frac{a}{1-r}.$$

$$(\because \ |r| < 1 \Rightarrow r^n \to 0 \text{ as } n \to \infty)$$

$$\therefore \quad S = \frac{a}{1-r}.$$

Remark. The sum up to infinity of a G.P. is also denoted by S_∞.

ILLUSTRATIVE EXAMPLES

Example 1. *Find the sum of the infinite series,* $1 + \frac{1}{2} + \frac{1}{4} + \frac{1}{8} + \ldots$

(P.T.U., M.B.A. May 2000)

Sol. Given series is an infinite G.P. with $a = 1$, $r = \frac{1}{2}$. Using $S_\infty = \frac{a}{1-r}$, we have

$$\text{Sum of the given series} = \frac{1}{1-\frac{1}{2}} = 2.$$

Example 2. *Prove that* $9^{1/3} \,.\, 9^{1/9} \,.\, 9^{1/27} \ldots\ldots$ *to* $\infty = 3$.

Sol. $9^{1/3} \,.\, 9^{1/9} \,.\, 9^{1/27} \ldots\ldots$ to ∞

$$= 9^{\frac{1}{3}+\frac{1}{9}+\frac{1}{27}+\ldots\ldots\infty} = 9^S \text{ where } S \text{ is } \frac{1}{3}+\frac{1}{9}+\frac{1}{27}+\ldots\ldots \text{ to } \infty.$$

We have $$S = \frac{1}{3}+\frac{1}{9}+\frac{1}{27}+\ldots\ldots \text{ to } \infty$$

This is an infinite G.S. with $a = \frac{1}{3}$, $r = \frac{1/9}{1/3} = \frac{1}{3}$ and $|r| = \left|\frac{1}{3}\right| = \frac{1}{3} < 1$.

$$\therefore \quad S = \frac{1/3}{1-\frac{1}{3}} = \frac{1/3}{2/3} = \frac{1}{2}. \qquad \left(S = \frac{a}{1-r}\right)$$

$\therefore$ Required product $= 9^S = 9^{1/2} = 3.$

Example 3. *Find* S_∞ *of the G.P. whose first term is 28 and the fourth term is* $\frac{4}{49}$.

Sol. Let a and r be respectively the first term and common ratio of the G.P.

By the given conditions $\quad T_1 = a = 28$ and $T_4 = ar^3 = \frac{4}{49}$.

$$\therefore \quad 28r^3 = \frac{4}{49} \Rightarrow r^3 = \frac{1}{343} \Rightarrow r = \frac{1}{7}.$$

$$| r | = \left|\frac{1}{7}\right| = \frac{1}{7} < 1.$$

$\therefore$ S_∞ is defined and is given by, $S_\infty = \dfrac{a}{1-r} = \dfrac{28}{1-\frac{1}{7}} = 28\left(\dfrac{7}{6}\right) = \dfrac{98}{3}.$

Example 4. *The first term of a G.P. exceeds the second term by 2 and the sum to infinity is 50, find the G.P.*

Sol. Let a be the first term and r ($| r | < 1$), the common ratio of the G.P.

$\therefore$ The G.P. is $a, ar, ar^2, ...$

By the given conditions, $T_1 = T_2 + 2$ ...(1) and $S_\infty = 50$ ...(2)

(1) $\Rightarrow$ $a = ar + 2$ $\Rightarrow$ $a - ar = 2 \Rightarrow a(1-r) = 2$ ...(3)

(2) $\Rightarrow$ $\dfrac{a}{1-r} = 50$ $\Rightarrow$ $a = 50\,(1-r)$ ...(4)

Putting the value of a in (3), we get $[50\,(1-r)]\,(1-r) = 2$

$\Rightarrow$ $(1-r)^2 = \dfrac{2}{50} = \dfrac{1}{25}$ $\Rightarrow$ $1 - r = \pm\dfrac{1}{5}$

Case I. $\mathbf{1 - r = \frac{1}{5}}$**.** In this case, $r = 1 - \dfrac{1}{5} = \dfrac{4}{5}$ and $| r | = \left|\dfrac{4}{5}\right| = \dfrac{4}{5} < 1.$ $\therefore$ $r = \dfrac{4}{5}$ is valid.

(4) $\Rightarrow$ $a = 50\left(1 - \dfrac{4}{5}\right) = 10.$

$\therefore$ The G.P. is 10, 10 (4/5), $10(4/5)^2$, ... *i.e.,* 10, 8, 32/5, ...

Case II. $\mathbf{1 - r = -\frac{1}{5}}$**.** In this case, $r = 1 + \dfrac{1}{5} = \dfrac{6}{5}$ and $| r | = \left|\dfrac{6}{5}\right| = \dfrac{6}{5} > 1.$ This is impossible.

Example 5. *The sum of infinite number of terms of a G.P. is 23 and the sum of their squares is 69. Find the G.P.*

Sol. Let the required G.P. is $a, ar, ar^2, ...$

Given $a + ar + ar^2 + ...\infty = 23$

$\Rightarrow$ $\dfrac{a}{1-r} = 23$ ...(1)

Also $a^2 + (ar)^2 + (ar^2)^2 + ...\infty = 69$

$\Rightarrow$ $a^2 + a^2r^2 + a^2r^4 + ...\infty = 69$

$\Rightarrow$ $a^2\,(1 + r^2 + r^4 + ...\infty) = 69$

$\Rightarrow$ $a^2 \cdot \dfrac{1}{1-r^2} = 69$

$\Rightarrow$ $(23)^2\,(1-r)^2 \times \dfrac{1}{1-r^2} = 69$ | Using (1)

$\Rightarrow$ $\dfrac{529\,(1-r)^2}{(1-r)(1+r)} = 69$

$\Rightarrow$ $\dfrac{1-r}{1+r} = \dfrac{69}{529} = \dfrac{3}{23}$

$$\Rightarrow \quad 23(1-r) = 3 + 3r$$
$$\Rightarrow \quad 23 - 3 = 26r$$
$$\Rightarrow \quad 26r = 20$$
$$\Rightarrow \quad r = \frac{20}{26} = \frac{10}{13}$$

From (1), $= 23(1-r) = 23\left(1-\frac{10}{13}\right) = 23 \times \frac{3}{13} = \frac{69}{13}$

$\therefore$ The required G.P. is $\frac{69}{13}, \frac{69}{13} \times \frac{10}{13}, \frac{69}{13} \times \left(\frac{10}{13}\right)^2, \ldots\ldots$

Example 6. *Find the sum the following infinite series 0.7 + 0.07 + 0.007 + ...*

(P.T.U., M.B.A. Dec. 2005)

Sol. Given series is

$$0.7 + 0.07 + 0.007 + \ldots \infty = 7(0.1 + 0.01 + 0.001 + \ldots \infty)$$
$$= 7\left(\frac{1}{10} + \frac{1}{100} + \frac{1}{1000} + \ldots \infty\right)$$
$$= 7\left(\frac{\frac{1}{10}}{1 - \frac{1}{10}}\right) = 7\left(\frac{1}{10} \times \frac{10}{9}\right) = \frac{7}{9} = 0.77.$$

TEST YOUR KNOWLEDGE 15.6

1. Find the sum of the following series :

   (*i*) $1 - \frac{1}{3} + \frac{1}{9} - \frac{1}{27} + \ldots\ldots$ (*ii*) $0.3 + 0.18 + 0.108 + 0.0648 + \ldots\ldots$
2. Find the sum of the following sequences :

   (*i*) $2, -1, 1/2, -1/4, \ldots\ldots$ (*ii*) $\sqrt{2}+1, 1, \sqrt{2}-1, \ldots\ldots$
3. Find the sum of the following sequences :

   (*i*) $a, br, ar^2, br^3, ar^4, br^5, \ldots\ldots \mid r \mid < 1$ (*ii*) $1, \frac{1}{3}, \frac{1}{2}, \frac{1}{3^2}, \frac{1}{2^2}, \frac{1}{3^3}, \ldots\ldots$
4. Show that :

   (*i*) $x^{1/2} \,.\, x^{1/4} \,.\, x^{1/8} \ldots\ldots \infty = x$ (*ii*) $3^{1/2} \,.\, 3^{1/4} \,.\, 3^{1/8}, \ldots\ldots \infty = 3$
5. If $\mid x \mid < 1$ and $y = 1 + x + x^2 + \ldots\ldots \infty$, show that $x = \frac{y-1}{y}$.
6. If $\mid a \mid < 1, \mid b \mid < 1$ and $x = 1 + a + a^2 + \ldots\ldots \infty, y = 1 + b + b^2 + \ldots\ldots \infty$, show that

   $1 + ab + a^2b^2 + \ldots\ldots \infty = \frac{xy}{x+y-1}$.
7. (*i*) The sum of an infinite G.P. is 6 and its first term is 2. Find the series.

   (*ii*) The common ratio of a G.P. is – 4/5 and the sum to infinity is 80/9. Find the first term.
8. If the first term of a G.P. is 729 and the seventh term is 64, find S_∞.
9. Find the value(s) of p if S_∞ for the G.P. $p, 1, 1/p, \ldots\ldots$ is 25/4.
10. The sum of an infinite G.P. is 6 and the sum of its first two terms is 9/2. Find the first term and the common ratio.
11. The sum of infinite number of terms in G.P. is 15 and the sum of their squares is 45. Find the series.

12. The sum of an infinite G.P. is 57 and the sum of their cubes is 9747, find the G.P.

13. (*i*) The first term of an infinite G.P. is unity and any term is equal to the sum of all the terms that follow it. Find the series.

(*ii*) If each term of an infinity G.P. is twice the sum of the terms following it, then find the common ratio of the G.P.

Answers

1. (*i*) $\frac{3}{4}$ (*ii*) $\frac{3}{4}$ **2.** (*i*) $\frac{4}{3}$ (*ii*) $\frac{4+3\sqrt{2}}{2}$

3. (*i*) $\frac{a+br}{1-r^2}$ (*ii*) $\frac{5}{2}$ **7.** (*i*) $2+\frac{4}{3}+\frac{8}{9}+......$ (*ii*) 16

8. 2187 or $\frac{2187}{5}$ **9.** 5, $\frac{5}{4}$ **10.** $a=3, r=\frac{1}{2}$; $a=9, r=-\frac{1}{2}$

11. $5+\frac{10}{3}+\frac{20}{9}+......$ **12.** 19, $\frac{38}{3}, \frac{76}{9}, ...$

13. (*i*) $1+\frac{1}{2}+\frac{1}{4}+......$ (*ii*) $\frac{1}{3}$.

Hints

3. (*i*) $a+br+ar^2+br^3+ar^4+br^5+......=(a+ar^2+ar^4+......)+(br+br^3+br^5+......)$

$$=\frac{a}{1-r^2}+\frac{br}{1-r^2}.$$

9. Use $\frac{p}{1-1/p}=\frac{25}{4}$

10. Eliminate a from $\frac{a}{1-r}=6,\quad a+ar=\frac{9}{2}$.

11. Eliminate a from $\frac{a}{1-r}=15,\quad \frac{a^2}{1-r^2}=45$.

13. (*ii*) $T_n=2[T_{n+1}+T_{n+2}+......]$

$$\Rightarrow \quad ar^{n-1}=2ar^n[1+r+......]=\frac{2ar^n}{1-r}.$$

15.14. GEOMETRIC MEAN

If three or more than three numbers are in G.P., then all the numbers lying between the first and the last numbers are called *geometric means* between them. *i.e.*

If a, G_1, G_2, ... G_n, b, be a G.P., then, G_1, G_2, ... G_n are called geometric means (G.P.s) between a and b.

Theorem VIII. *To find single G.M. between two given positive numbers.*

Proof. Let a and b be two positive numbers and G be the single G.M. between them. Then, by definition, a, G, b are in G.P.

$$\Rightarrow \quad \frac{G}{a}=\frac{b}{G} \quad \Rightarrow \quad G^2=ab \quad \Rightarrow \quad G=\sqrt{ab}.$$

Theorem IX. *To find n G.M.s between two given positive numbers.*

Proof. Let a and b be any two positive numbers and $G_1, G_2, \ldots G_n$, be n G.M.s between them. Then, by definition, $a, G_1, G_2, \ldots G_n, b$ are in G.P.

Now $$b = T_{n+2} = ar^{n+1}$$

$$\Rightarrow \quad r^{n+1} = \frac{b}{a} \quad \Rightarrow \quad r = \left(\frac{b}{a}\right)^{\frac{1}{n+1}}$$

$$\therefore \quad G_1 = T_2 = ar = a\left(\frac{b}{a}\right)^{\frac{1}{n+1}}$$

$$G_2 = T_3 = ar^2 = a\left(\frac{b}{a}\right)^{\frac{2}{n+1}}$$

....

$$G_n = T_{n+1} = ar^n = a\left(\frac{b}{a}\right)^{\frac{n}{n+1}}.$$

Theorem X. *The product of n G.M.s between any two positive numbers is equal to the nth power of the G.M. between them.*

Proof. Proceeding in above theorem, we have, product of n G.M.s between a and b

$$= G_1 . G_2 . G_3 . \ldots\ldots G_n$$

$$= \left(a . \left(\frac{b}{a}\right)^{\frac{1}{n+1}}\right)\left(a\left(\frac{b}{a}\right)^{\frac{2}{n+1}}\right) \ldots\ldots \left(a\left(\frac{b}{a}\right)^{\frac{n}{n+1}}\right)$$

$$= a^n \left[\left(\frac{b}{a}\right)^{\frac{1}{n+1}}\right]^{1+2+3+\ldots+n}$$

$$= a^n \left[\left(\frac{b}{a}\right)^{\frac{1}{n+1}}\right]^{\frac{n(n+1)}{2}} = a^n\left(\frac{b}{a}\right)^{\frac{n}{2}} = a^{n-\frac{n}{2}} b^{\frac{n}{2}} = a^{\frac{n}{2}} b^{\frac{n}{2}}$$

$$= \left(\sqrt{ab}\right)^n = (\text{G.M. between } a \text{ and } b)^n.$$

ILLUSTRATIVE EXAMPLES

Example 1. *Insert 5 Geometric means between 320 and 5.* (P.T.U., M.B.A. May 2003)

Sol. Let G_1, G_2, G_3, G_4, G_5 be the required 5 G.M.s between 320 and 5. Then, by definition, 320, G_1, G_2, G_3, G_4, G_5, 5 are in G.P.

Now $$5 = T_7 = ar^6 = 320\, r^6$$

$$\Rightarrow \quad r^6 = \frac{5}{320} = \frac{1}{64} = \left(\frac{1}{2}\right)^5 \quad \Rightarrow \quad r = \frac{1}{2}$$

$$\therefore \quad G_1 = T_2 = ar = 320 \times \frac{1}{2} = 160$$

$$G_2 = T_3 = ar^2 = 320 \times \frac{1}{4} = 80$$

$$G_3 = T_4 = ar^3 = 320 \times \frac{1}{8} = 40$$

$$G_4 = T_5 = ar^4 = 320 \times \frac{1}{16} = 20$$

$$G_5 = T_6 = ar^5 = 320 \times \frac{1}{32} = 10.$$

Example 2. *The A.M. between two positive numbers is 34 and their G.M.s is 16. Find the numbers.*

Sol. Let the positive numbers be a and b.

$\therefore$ A.M. $= 34 \Rightarrow \frac{a+b}{2} = 34 \Rightarrow a + b = 68$ ...(1)

and G.M. $= 16 \Rightarrow \sqrt{ab} = 16 \Rightarrow ab = 256$ ...(2)

Now $(a+b)^2 - (a-b)^2 = 4ab \Rightarrow (68)^2 - (a-b)^2 = 4(256)$

$\Rightarrow (a-b)^2 = 3600 \Rightarrow a - b = 60$ ...(3)

(We assume $a > b$)

$(1) + (3) \Rightarrow 2a = 68 + 60 \Rightarrow a = 64$

$\therefore$ (1) $\Rightarrow 64 + b = 68 \Rightarrow b = 4$

$\therefore$ The numbers are 64, 4.

Example 3. *Insert 5 G.M.s between 1/3 and 9 and verify that their product is the 5th power of the G.M. between 1/3 and 9.*

Sol. Let G_1, G_2, G_3, G_4, G_5, be the 5 G.M.s between 1/3 and 9.

$\therefore \frac{1}{3}, G_1, G_2, G_3, G_4, G_5, 9$ are in G.P.

Let r be the common ratio of this G.P.

$$\therefore \quad 9 = T_7 = \frac{1}{3} r^6 \Rightarrow r^6 = 27 = (\sqrt{3})^6 \Rightarrow r = \sqrt{3}$$

$$\therefore \quad G_1 = \frac{1}{3} r = \frac{\sqrt{3}}{3} = \frac{1}{\sqrt{3}}, \quad G_2 = G_1 r = \frac{1}{\sqrt{3}} \sqrt{3} = 1, \quad G_3 = G_2 r = 1 . \sqrt{3} = \sqrt{3},$$

$$G_4 = G_3 r = \sqrt{3} . \sqrt{3} = 3 \quad \text{and} \quad G_5 = G_4 r = 3 . \sqrt{3} = 3\sqrt{3}.$$

Product of 5 G.M.s $= G_1 G_2 G_3 G_4 G_5 = \left(\frac{1}{\sqrt{3}}\right) . (1) . (\sqrt{3}) . (3) . (3\sqrt{3}) = 9\sqrt{3}.$

Also, 5th power of the G.M. between $\frac{1}{3}$ and $9 = \left(\sqrt{\frac{1}{3} . 9}\right)^5 = (\sqrt{3})^5 = (\sqrt{3})^4 \sqrt{3} = 9\sqrt{3}.$

Hence the result holds.

Example 4. *If two G.M.s g_1 and g_2 and one A.M. A be inserted between two numbers, then show that* $2A = \frac{g_1^2}{g_2} + \frac{g_2^2}{g_1}$.

Sol. Let the numbers be a, b.

$\therefore$ a, g_1, g_2, b are in G.P. ($\because$ g_1, g_2 are G.M.s between a, b)

$$\therefore \quad \frac{g_1}{a} = \frac{g_2}{g_1} = \frac{b}{g_2} \qquad \therefore \quad g_1^2 = ag_2 \text{ and } g_2^2 = bg_1$$

$$\Rightarrow \quad \frac{g_1^2}{g_2} = a \text{ and } \frac{g_2^2}{g_1} = b \quad \Rightarrow \quad \frac{g_1^2}{g_2} + \frac{g_2^2}{g_1} = a + b \qquad \text{...(1)}$$

Also a, A, b are in A.P. ($\because$ A is A.M. between a, b)

$$\therefore \quad A - a = b - A \qquad \therefore \quad A = \frac{a+b}{2} \quad \text{or} \quad 2A = a + b$$

$$\therefore \quad (1) \Rightarrow \quad 2A = \frac{g_1^2}{g_2} + \frac{g_2^2}{g_1}.$$

Example 5. *If A and G be respectively the A.M. and G.M. between two positive numbers, prove that the numbers are*

$$A \pm \sqrt{A^2 - G^2}.$$

Sol. Let the positive numbers be a, b. Let $a > b$.

$$\therefore \quad A = \frac{a+b}{2} \quad \text{or} \quad a + b = 2A \qquad \text{...(1)}$$

$$\text{and} \quad G = \sqrt{ab} \quad \text{or} \quad ab = G^2 \qquad \text{...(2)}$$

$$\text{Now} \quad (a + b)^2 = (a - b)^2 + 4ab$$

$$\Rightarrow \quad 4A^2 = (a - b)^2 + 4G^2 \quad \Rightarrow \quad (a - b)^2 = 4\,(A^2 - G^2)$$

$$\Rightarrow \quad a - b = 2\sqrt{A^2 - G^2} \qquad \text{...(3)}$$

$$(1) + (3) \Rightarrow \quad 2a = 2A + 2\sqrt{A^2 - G^2} \quad \Rightarrow \quad a = A + \sqrt{A^2 - G^2}.$$

$$(1) - (3) \Rightarrow \quad 2b = 2A - 2\sqrt{A^2 - G^2} \quad \Rightarrow \quad b = A - \sqrt{A^2 - G^2}.$$

$\therefore$ The numbers are $A \pm \sqrt{A^2 - G^2}$.

Example 6. *Find n such that* $\frac{a^n + b^n}{a^{n-1} + b^{n-1}}$ *may be the G.M. between distinct positive numbers a and b.*

Sol. The G.M. between a and b is $\sqrt{ab}$.

$$\text{Let} \quad \frac{a^n + b^n}{a^{n-1} + b^{n-1}} = \sqrt{ab}$$

$$\Rightarrow \quad a^n + b^n = a^{1/2}\, b^{1/2}\, (a^{n-1} + b^{n-1})$$

$$\Rightarrow \quad a^n + b^n = a^{n-\frac{1}{2}}\, b^{\frac{1}{2}} + a^{\frac{1}{2}}\, b^{n-\frac{1}{2}}$$

$$\Rightarrow \quad a^n - a^{n-\frac{1}{2}}\, b^{\frac{1}{2}} = a^{\frac{1}{2}}\, b^{n-\frac{1}{2}} - b^n$$

$$\Rightarrow \quad a^{n-\frac{1}{2}}(a^{1/2} - b^{1/2}) = b^{n-\frac{1}{2}}(a^{1/2} - b^{1/2})$$

$$\Rightarrow \quad a^{n-\frac{1}{2}} = b^{n-\frac{1}{2}} \qquad (\because \;\; a \neq b \;\Rightarrow\; a^{1/2} - b^{1/2} \neq 0)$$

$$\therefore \quad \frac{a^{n-\frac{1}{2}}}{b^{n-\frac{1}{2}}} = 1 \;\Rightarrow\; \left(\frac{a}{b}\right)^{n-\frac{1}{2}} = \left(\frac{a}{b}\right)^0 \;\Rightarrow\; n - \frac{1}{2} = 0 \;\; i.e., \;\; n = \frac{1}{2}.$$

Example 7. *If the A.M. between positive numbers a and b be equal to n times their G.M., find the ratio of a to b.*

Sol. We have $\quad \text{A.M.} = \dfrac{a+b}{2} \quad$ and $\quad \text{G.M.} = \sqrt{ab}$.

By the given condition, $\quad \text{A.M.} = n\,(\text{G.M.})$

$$\Rightarrow \quad \frac{a+b}{2} = n\sqrt{ab} \quad \text{or} \quad \frac{a+b}{2\sqrt{ab}} = \frac{n}{1} \qquad \textbf{(Note this step)}$$

$$\text{or} \quad \frac{a+b+2\sqrt{ab}}{a+b-2\sqrt{ab}} = \frac{n+1}{n-1} \qquad \text{| Componendo and dividendo}$$

$$\text{or} \quad \frac{(\sqrt{a}+\sqrt{b})^2}{(\sqrt{a}-\sqrt{b})^2} = \frac{n+1}{n-1} \quad \text{or} \quad \frac{\sqrt{a}+\sqrt{b}}{\sqrt{a}-\sqrt{b}} = \frac{\sqrt{n+1}}{\sqrt{n-1}} \qquad (\text{Assuming } a > b)$$

$$\text{or} \quad \frac{(\sqrt{a}+\sqrt{b})+(\sqrt{a}-\sqrt{b})}{(\sqrt{a}+\sqrt{b})-(\sqrt{a}-\sqrt{b})} = \frac{\sqrt{n+1}+\sqrt{n-1}}{\sqrt{n+1}-\sqrt{n-1}} \qquad \text{| Componendo and dividendo}$$

$$\text{or} \quad \frac{2\sqrt{a}}{2\sqrt{b}} = \frac{\sqrt{n+1}+\sqrt{n-1}}{\sqrt{n+1}-\sqrt{n-1}}$$

$$\text{or} \quad \frac{a}{b} = \frac{(n+1)+(n-1)+2\sqrt{(n+1)(n-1)}}{(n+1)+(n-1)-2\sqrt{(n+1)(n-1)}} \qquad (\text{On squaring})$$

$$= \frac{2n+2\sqrt{n^2-1}}{2n-2\sqrt{n^2-1}} = \frac{n+\sqrt{n^2-1}}{n-\sqrt{n^2-1}}.$$

$$\therefore \quad a : b = n + \sqrt{n^2-1} : n - \sqrt{n^2-1}.$$

TEST YOUR KNOWLEDGE 15.7

1. (*i*) Find the G.M. between $\frac{8}{9}$ and $\frac{49}{50}$. (*ii*) Find the G.M. between 0.008 and 0.2.
2. The G.M. between two positive numbers is 16. If one number is 32, find the other number.
3. If $k - 1$ is the G.M. between $k - 2$ and $k + 1$, then find the value of k.
4. If A and G are the A.M. and G.M. between distinct positive numbers a and b respectively, then show that $A > G$.
5. If A and G be the A.M. and G.M. between positive numbers a and b respectivly, then show that a and b are the roots of the equation $x^2 - 2Ax + G^2 = 0$.
6. If a, b, c are in A.P., x is the G.M. between a and b, y is the G.M. between b are c, show that b^2 is the A.M. between x^2 and y^2.
7. Find two positive numbers whose difference is 2 and whose A.M. exceeds the G.M. by 1/2.
8. If the A.M. between two positive numbers exceeds their G.M. by 2 and the ratio of the numbers be 1 : 4, find the numbers.
9. The A.M. between two positive numbers, whose sum is 100, is to their G.M. as 5 : 4, find the numbers.
10. Insert 4 G.M.s between 1/2 and 512. **11.** Insert 6 G.M.s between 1 and 2187.
12. Insert 4 G.M.s between 1/9 and 27 and verify that their product is equal to the fourth power of the G.M. between 1/9 and 27.
13. If G_1 is the first of n G.M.s between positive numbers a and b, then show that $G_1^{n+1} = a^n b$.
14. If $\frac{a^{n+1} + b^{n+1}}{a^n + b^n}$ is the G.M. between positive numbers a and b, then find the value of n.
15. If the A.M. between positive numbers a and b is twice their G.M., then show that

    $a : b = 2 + \sqrt{3} : 2 - \sqrt{3}$.
16. If a, b, c be in G.P. and x, y be the A.M. between a, b and b, c respectively, show that :

    (*i*) $\frac{1}{x} + \frac{1}{y} = \frac{2}{b}$ (*ii*) $\frac{a}{x} + \frac{c}{y} = 2$.
17. If one geometric mean G and two arithmetic means p and q be inserted between two positive quantities, show that $G^2 = (2p - q)(2q - p)$.
18. If G_1 and G_2 are the two geometric means between b and c and a is the arithmetic mean, then show that $G_1^3 + G_2^3 = 2abc$.

Answers

1. (*i*) $\frac{14}{15}$ (*ii*) 0.04 **2.** 8 **3.** 3

7. $\frac{1}{4}, \frac{9}{4}$ **8.** 4, 16 **9.** 20, 80 **10.** 2, 8, 32, 128

11. 3, 9, 27, 81, 243, 729 **12.** $\frac{1}{3}$, 1, 3, 9 **14.** $-\frac{1}{2}$.

Hints

4. Establish : $A - G = \dfrac{(\sqrt{a} - \sqrt{b})^2}{2}$.

5. Reqd. equation is $x^2 - (a + b)x + ab = 0$

6. We have $b = \dfrac{a + c}{2}$, $x = \sqrt{ab}$, $y = \sqrt{bc}$. $\therefore$ $\dfrac{x^2 + y^2}{2} = \dfrac{ab + bc}{2} = \dfrac{b(a + c)}{2} = b^2$.

7. If the numbers are a, $a + 2$, then $\dfrac{a + (a + 2)}{2} = \sqrt{a(a + 2)} + \dfrac{1}{2}$ *i.e.*, $a + \dfrac{1}{2} = \sqrt{a^2 + 2a}$.

8. If the numbers are a, $4a$, then $\dfrac{a + 4a}{2} = \sqrt{a \,.\, 4a} + 2$.

9. If the numbers are a, $100 - a$, then $\dfrac{\dfrac{a + (100 - a)}{2}}{\sqrt{a\,(100 - a)}} = \dfrac{5}{4}$.

16. $\dfrac{1}{x} + \dfrac{1}{y} = \dfrac{1}{\dfrac{a + b}{2}} + \dfrac{1}{\dfrac{b + c}{2}} = \dfrac{2}{a + b} + \dfrac{2}{b + c} = 2\left[\dfrac{b + c + a + b}{ab + ac + b^2 + bc}\right] = \dfrac{2\,[a + 2b + c]}{ab + b^2 + b^2 + bc}$.

17. Let the numbers be c, b. $\therefore$ $p - a = q - p = b - q$, $\dfrac{G}{a} = \dfrac{b}{G}$ $\therefore$ $2p - q = a$, $2q - p = b$, $G^2 = ab$.

18. Let r be *c.r.* of b, G_1, G_2, c

$\therefore$ $G_1 = br$, $G_2 = br^2$, $c = br^3$.

$\therefore$ $G_1^{\,3} + G_2^{\,3} = b^3r^3 + b^3r^6 = b^3r^3(1 + r^3)$

and $2abc = 2\left(\dfrac{b + c}{2}\right)bc = (b + br^3)b \,.\, br^3 = b^3r^3(1 + r^3)$.

16 SEQUENCES

16.1. SEQUENCE OF NATURAL NUMBERS : 1, 2, 3, ...

The sum of first n terms of the sequence 1, 2, 3, ... is denoted by Σn.

$\therefore \quad \Sigma n = 1 + 2 + 3 + \ldots\ldots + n$. This is an arithmetic series of n terms with $a = 1$ and $d = 1$.

$$\therefore \quad \Sigma n = \frac{n}{2}[2(1) + (n-1)1] \qquad \left(\text{Using } S_n = \frac{n}{2}[2a + (n-1)d]\right)$$

$$\therefore \quad \mathbf{\Sigma n = \frac{n(n+1)}{2}, \quad n \in N.}$$

Under the notation Σn, we write $\sum_{k=1}^{8} k$ as $(\Sigma n)_{n=8}$.

Remark. Σn *i.e.,* $1 + 2 + 3 + \ldots\ldots + n$ can also be written as $\sum_{k=1}^{n} k$.

16.2. SEQUENCE OF SQUARES OF NATURAL NUMBERS : $1^2, 2^2, 3^2$, ...

The sum of first n terms of the sequence $1^2, 2^2, 3^2$, ... is denoted by Σn^2.

$$\therefore \quad \Sigma n^2 = 1^2 + 2^2 + 3^2 + \ldots\ldots + n^2.$$

In the next theorem, we establish a formula to find the value of Σn^2.

Theorem. Prove that $\mathbf{\Sigma n^2 = 1^2 + 2^2 + 3^2 + \ldots\ldots + n^2 = \frac{n(n+1)(2n+1)}{6}, \; n \in N.}$

Proof. For any x, we have $(x+1)^3 = x^3 + 3x^2 + 3x + 1$.

$$\Rightarrow \quad (x+1)^3 - x^3 = 3x^2 + 3x + 1. \qquad \ldots(1)$$

Putting $x = 1, 2, 3, \ldots\ldots, n$ in (1), we get

$$2^3 - 1^3 = 3.1^2 + 3.1 + 1$$
$$3^3 - 2^3 = 3.2^2 + 3.2 + 1$$
$$4^3 - 3^3 = 3.3^2 + 3.3 + 1$$
$$\ldots\ldots\ldots\ldots\ldots\ldots\ldots\ldots\ldots\ldots$$
$$(n+1)^3 - n^3 = 3.n^2 + 3.n + 1$$

Adding these vertically, we get

$$(n+1)^3 - 1^3 = 3(1^2 + 2^2 + 3^2 + \ldots\ldots + n^2) + 3(1 + 2 + 3 + \ldots\ldots + n) + n$$

$$\Rightarrow \quad (n+1)^3 - 1 = 3\Sigma n^2 + 3\Sigma n + n$$

$$\Rightarrow \quad 3\Sigma n^2 = (n+1)^3 - 3\Sigma n - n - 1 = (n+1)^3 - 3\,\frac{n(n+1)}{2} - (n+1)$$

$$= \frac{(n+1)}{2}\,[2(n+1)^2 - 3n - 2] = \frac{n+1}{2}\,[2n^2 + n] = \frac{(n+1)\,n\,(2n+1)}{2}.$$

$$\therefore \quad \mathbf{\Sigma n^2 = \frac{n(n+1)(2n+1)}{6}, \quad n \in N.}$$

Remark. Σn^2 *i.e.,* $1^2 + 2^2 + 3^2 + \ldots\ldots + n^2$ can also be written as $\sum_{k=1}^{n} k^2$.

Illustration. $\sum_{k=1}^{10} k^2 = 1^2 + 2^2 + 3^2 + \ldots\ldots + 10^2 = \frac{10\,(10+1)\,(20+1)}{6} = 385.$

Under the notation Σn^2, we write $\sum_{k=1}^{10} k^2$ as $(\Sigma n^2)_{n=10}$.

16.3. SEQUENCE OF CUBES OF NATURAL NUMBERS : $1^3, 2^3, 3^3, \ldots\ldots$

The sum of first n terms of the sequence $1^3, 2^3, 3^3, \ldots\ldots$ is denoted by Σn^3.

$$\therefore \quad \Sigma n^3 = 1^3 + 2^3 + 3^3 + \ldots\ldots + n^3.$$

Now, we prove a result regarding the value of Σn^3.

Theorem I. Prove that : $\mathbf{\Sigma n^3 = 1^3 + 2^3 + 3^3 + \ldots\ldots + n^3 = \left[\frac{n(n+1)}{2}\right]^2, \quad n \in N.}$

Proof. For any x, we have $(x+1)^4 = x^4 + 4x^3 + 6x^2 + 4x + 1$.

$$\Rightarrow \quad (x+1)^4 - x^4 = 4x^3 + 6x^2 + 4x + 1 \qquad \ldots(1)$$

Putting $x = 1, 2, 3, \ldots\ldots, n$ in (1), we get

$$2^4 - 1^4 = 4.1^3 + 6.1^2 + 4.1 + 1$$

$$3^4 - 2^4 = 4.2^3 + 6.2^2 + 4.2 + 1$$

$$4^4 - 3^4 = 4.3^3 + 6.3^2 + 4.3 + 1$$

$$\ldots\ldots\ldots\ldots\ldots\ldots\ldots\ldots\ldots\ldots$$

$$(n+1)^4 - n^4 = 4.n^3 + 6.n^2 + 4.n + 1$$

Adding these vertically, we get

$$(n+1)^4 - 1^4 = 4(1^3 + 2^3 + 3^3 + \ldots\ldots + n^3) + 6(1^2 + 2^2 + 3^2 + \ldots\ldots + n^2) + 4(1 + 2 + 3 + \ldots\ldots + n) + n$$

$$\Rightarrow \quad (n+1)^4 - 1 = 4\Sigma n^3 + 6\Sigma n^2 + 4\Sigma n + n$$

$$\Rightarrow \quad 4\Sigma n^3 = (n+1)^4 - 6\Sigma n^2 - 4\Sigma n - n - 1$$

$$= (n+1)^4 - 6\,\frac{n(n+1)(2n+1)}{6} - 4\,\frac{n(n+1)}{2} - (n+1)$$

$$= (n+1)\,[(n+1)^3 - n(2n+1) - 2n - 1]$$

$$= (n + 1)\,[n^3 + 3n^2 + 3n + 1 - 2n^2 - n - 2n - 1]$$
$$= (n + 1)\,(n^3 + n^2) = (n + 1)n^2\,(n + 1) = [n(n + 1)]^2.$$

$$\therefore \qquad \mathbf{\Sigma n^3 = \left[\frac{n\,(n+1)}{2}\right]^2.}$$

Remark. Σn^3 *i.e.,* $1^3 + 2^3 + 3^3 + \ldots\ldots + n^3$ can also be written as $\sum_{k=1}^{n} k^3$.

Illustration. $\sum_{k=1}^{7} k^3 = 1^3 + 2^3 + 3^3 + \ldots\ldots + 7^3 = \left[\frac{7(7+1)}{2}\right]^2 = 784.$

Under the notation Σn^3, we write $\sum_{k=1}^{7} k^3$ as $(\Sigma n^3)_{n=7}$.

Theorem II. If for a sequence, $T_n = an^3 + bn^2 + cn + d$, then prove that $S_n = a\Sigma n^3 + b\Sigma n^2 + c\Sigma n + nd$.

Proof. We have $\qquad T_n = an^3 + bn^2 + cn + d.$

$\therefore \qquad T_1 = a.1^3 + b.1^2 + c.1 + d.$

$T_2 = a.2^3 + b.2^2 + c.2 + d.$

$T_3 = a.3^3 + b.3^2 + c.3 + d.$

....

$T_n = an^3 + bn^2 + cn + d.$

Adding vertically, we get $\quad S_n = T_1 + T_2 + T_3 + \ldots\ldots + T_n$

$= a(1^3 + 2^3 + 3^3 + \ldots\ldots + n^3) + b(1^2 + 2^2 + 3^2 + \ldots\ldots + n^2) + c\,(1 + 2 + 3 + \ldots\ldots + n) + nd$

$\therefore \qquad \mathbf{S_n = a\Sigma n^3 + b\Sigma n^2 + c\Sigma n + nd.}$

$\therefore$ The result holds.

WORKING RULES TO FIND THE SUM OF SERIES

Step I. *Write the nth term T_n of the given series.*

Step II. *Simplify T_n and write it as a polynomial in n. Let $T_n = an^3 + bn^2 + cn + d$.*

Step III. *Write the sum S_n of the first n terms of the series as $a\Sigma n^3 + b\Sigma n^2 + c\Sigma n + nd$.*

Step IV. *Put* $\Sigma n = \frac{n(n+1)}{2}$, $\Sigma n^2 = \frac{n\,(n+1)\,(2n+1)}{6}$ *and* $\Sigma n^3 = \left[\frac{n\,(n+1)}{2}\right]^2$ *in the expression :* $S_n = a\Sigma n^3 + b\Sigma n^2 + c\Sigma n + nd$ *and simplify.*

ILLUSTRATIVE EXAMPLES

Example 1. *Find the following sums :*

(i) $12 + 13 + 14 + \ldots\ldots + 37$ *(ii)* $1^2 + 2^2 + 3^2 + \ldots\ldots + 20^2$

(iii) $18^2 + 19^2 + 20^2 + \ldots\ldots + 47^2$ *(iv)* $11^3 + 12^3 + 13^3 + \ldots\ldots + 25^3$

(v) $2^3 + 4^3 + 6^3 + \ldots\ldots + 18^3$.

Sol. (*i*) $12 + 13 + 14 + ... + 37 = (1 + 2 + ... + 11 + 12 + 13 + 14 + ... + 37)$
$- (1 + 2 + ... + 11)$

$$= (\Sigma n)_{n=37} - (\Sigma n)_{n=11} = \frac{37(37+1)}{2} - \frac{11(11+1)}{2} = 703 - 66 = \mathbf{637.}$$

(*ii*) $1^2 + 2^2 + 3^2 + ... + 20^2 = (\Sigma n^2)_{n=20}$

$$= \frac{20(20+1)(2(20)+1)}{6} = \frac{20(21)(41)}{6} = \mathbf{2870.}$$

(*iii*) $18^2 + 19^2 + 20^2 + ... + 47^2 = (1^2 + 2^2 + 3^2 + ... + 47^2) - (1^2 + 2^2 + 3^2 + ... + 17^2)$

$$= (\Sigma n^2)_{n=47} - (\Sigma n^2)_{n=17}$$

$$= \frac{47(47+1)(2(47)+1)}{6} - \frac{17(17+1)(2(17)+1)}{6}$$

$$= \frac{47(48)(95)}{6} - \frac{17(18)(35)}{6} = 35720 - 1785 = \mathbf{33935.}$$

(*iv*) $11^3 + 12^3 + 13^3 + ... + 25^3 = (1^3 + 2^3 + 3^3 + ... + 25^3) - (1^3 + 2^3 + 3^3 + ... + 10^3)$

$$= (\Sigma n^3)_{n=25} - (\Sigma n^3)_{n=10}$$

$$= \left[\frac{25(25+1)}{2}\right]^2 - \left[\frac{10(10+1)}{2}\right]^2 = 105625 - 3025 = \mathbf{102600.}$$

(*v*) $2^3 + 4^3 + 6^3 + ... + 18^3 = (2 \times 1)^3 + (2 \times 2)^3 + (2 \times 3)^3 + ... (2 \times 9)^3$

$$= 2^3 [1^3 + 2^3 + 3^3 + ... + 9^3] = 8(\Sigma n^3)_{n=9} = 8\left[\frac{9(9+1)}{2}\right]^2 = \mathbf{16200.}$$

Example 2. *Find the sum of first n terms of the series : $1^2 + 3^2 + 5^2 + ...$*

Sol. The given series is $1^2 + 3^2 + 5^2 + ...$

Let T_n be the nth term.

$\therefore \quad T_n = (n\text{th term of } 1, 3, 5, ...)^2$
$= (1 + (n-1)2)^2 = (2n-1)^2 = 4n^2 - 4n + 1.$

$\therefore$ Sum of n terms $= S_n = \Sigma T_n{}^* = \Sigma(4n^2 - 4n + 1) = 4\Sigma n^2 - 4\Sigma n + n$

$$= 4\left(\frac{n(n+1)(2n+1)}{6}\right) - 4\left(\frac{n(n+1)}{2}\right) + n$$

$$= \frac{n}{3}[2(n+1)(2n+1) - 6(n+1) + 3] = \mathbf{\frac{n(4n^2-1)}{3}}.$$

Alternative method. $1^2 + 3^2 + 5^2 + ...$ n terms.

$= (1^2 + 2^2 + 3^2 + 4^2 + 5^2 + 6^2 + ...\ 2n \text{ terms}) - (2^2 + 4^2 + 6^2 + ...\ n \text{ terms})$
$= (1^2 + 2^2 + 3^2 + ...\ 2n \text{ terms}) - 2^2(1^2 + 2^2 + 3^2 + ...\ n \text{ terms})$

$$= \frac{2n(2n+1)(2(2n)+1)}{6} - 4\,\frac{n(n+1)(2n+1)}{6}$$

$$= \frac{2n(2n+1)}{6}[4n+1-2(n+1)] = \frac{n(2n+1)}{3}[2n-1] = \mathbf{\frac{n(4n^2-1)}{3}}.$$

* ΣT_n stands for $T_1 + T_2 + T_3 + ... + T_n$.

Example 3. *Find the sum of the series : 1.2.3 + 2.3.4 + 3.4.5 + ... to n terms.*

Sol. The given series is 1.2.3 + 2.3.4 + 3.4.5 + ... Let T_n be the nth term.

$$\therefore \quad T_n = \begin{pmatrix} n\text{th term of} \\ 1, 2, 3, \ldots\ldots \end{pmatrix} \begin{pmatrix} n\text{th term of} \\ 2, 3, 4, \ldots\ldots \end{pmatrix} \begin{pmatrix} n\text{th term of} \\ 3, 4, 5, \ldots\ldots \end{pmatrix}$$

$$= n(n + 1)(n + 2) = n^3 + 3n^2 + 2n.$$

$$\therefore \quad \text{Sum of } n \text{ terms} \quad = S_n = \Sigma T_n = \Sigma n^3 + 3\Sigma n^2 + 2\Sigma n$$

$$= \left[\frac{n(n+1)}{2}\right]^2 + 3\frac{n(n+1)(2n+1)}{6} + 2\frac{n(n+1)}{2}$$

$$= \frac{n(n+1)}{4}[n(n + 1) + 2(2n + 1) + 4] = \frac{\mathbf{n(n+1)(n^2+5n+6)}}{\mathbf{4}}.$$

Example 4. *Sum the series : $1.3^2 + 2.5^2 + 3.7^2 + \ldots\ldots$ to 20 terms.*

Sol. The given series is $1.3^2 + 2.5^2 + 3.7^2 + \ldots\ldots$

Let T_n be the nth term

$$\therefore \quad T_n = \begin{pmatrix} n\text{th term of} \\ 1, 2, 3, \ldots\ldots \end{pmatrix} \begin{pmatrix} n\text{th term of} \\ 3, 5, 7, \ldots\ldots \end{pmatrix}^2 = n\,(3 + (n - 1)\,2)^2 = n(2n + 1)^2 = 4n^3 + 4n^2 + n.$$

$$\therefore \quad \text{Sum of 20 terms} = S_{20} = (\Sigma T_n)_{n=20} = 4(\Sigma n^3)_{n=20} + 4(\Sigma n^2)_{n=20} + (\Sigma n)_{n=20}.$$

$$= 4\left[\frac{20(20+1)}{2}\right]^2 + 4\left[\frac{20(20+1)(2(20)+1)}{6}\right] + \frac{20(20+1)}{2}$$

$$= 176400 + 11480 + 210 = \mathbf{188090.}$$

Example 5. *If S_1, S_2, S_3 are the sums of the first n natural numbers, their squares and their cubes respectively, show that* $\quad S_3\,(1 + 8S_1) = 9S_2^2.$

Sol. We have $\quad S_1 = \dfrac{n(n+1)}{2}, \quad S_2 = \dfrac{n(n+1)(2n+1)}{6}, \quad S_3 = \left(\dfrac{n(n+1)}{2}\right)^2$

$$\therefore \quad S_3(1 + 8S_1) = \left[\frac{n(n+1)}{2}\right]^2 \left[1 + 8 \cdot \frac{n(n+1)}{2}\right] = \frac{n^2(n+1)^2}{4}[1 + 4n^2 + 4n]$$

$$= \frac{n^2(n+1)^2(2n+1)^2}{4}$$

and $$9S_2^2 = 9\left[\frac{n(n+1)(2n+1)}{6}\right]^2 = \frac{n^2(n+1)^2(2n+1)^2}{4}.$$

$$\therefore \quad \mathbf{S_3(1 + 8S_1) = 9S_2^2.}$$

Example 6. *Find the sum of all possible products of the first n natural numbers taken two at a time.*

Sol. Let S = Sum of all possible products of the first n natural numbers taken two at a time.

We have $\quad (1 + 2 + 3 + \ldots\ldots + n)^2 = (1^2 + 2^2 + 3^2 + \ldots\ldots + n^2) + 2S \qquad$ **(Note this step)**

$$\Rightarrow \quad (\Sigma n)^2 = \Sigma n^2 + 2S \;\Rightarrow\; 2S = (\Sigma n)^2 - \Sigma n^2$$

$$\therefore \quad S = \frac{1}{2}[(\Sigma n)^2 - (\Sigma n^2)] = \frac{1}{2}\left[\left(\frac{n(n+1)}{2}\right)^2 - \frac{n(n+1)(2n+1)}{6}\right]$$

$$= \frac{n(n+1)}{4}\left[\frac{n(n+1)}{2} - \frac{2n+1}{3}\right]$$

$$= \frac{n(n+1)}{24}[3n^2 + 3n - 4n - 2] = \frac{\mathbf{n(n+1)(3n^2 - n - 2)}}{\mathbf{24}}.$$

Example 7. *Sum of the series :* $\frac{1^3}{1} + \frac{1^3 + 2^3}{1+3} + \frac{1^3 + 2^3 + 3^3}{1+3+5} + ...$ *to n terms.*

Sol. Let T_n be the nth term of the given series.

$$\therefore \quad T_n = \frac{1^3 + 2^3 + 3^3 + ... n \text{ terms}}{1 + 3 + 5 + ... n \text{ terms}} = \frac{\Sigma n^3}{\frac{n}{2}[2(1) + (n-1)2]} = \frac{\left[\frac{n(n+1)}{2}\right]^2}{\frac{n}{2}(2n)} = \frac{(n+1)^2}{4}$$

$$= \frac{1}{4}(n^2 + 2n + 1).$$

$$\therefore \quad \text{Sum of } n \text{ terms} \quad = S_n = \Sigma T_n = \frac{1}{4}\Sigma(n^2 + 2n + 1)$$

$$= \frac{1}{4}(\Sigma n^2 + 2\Sigma n + n] = \frac{1}{4}\left[\frac{n(n+1)(2n+1)}{6} + 2\left(\frac{n(n+1)}{2}\right) + n\right]$$

$$= \frac{n}{24}[(n+1)(2n+1) + 6(n+1) + 6] = \frac{\mathbf{n(2n^2 + 9n + 13)}}{\mathbf{24}}.$$

TEST YOUR KNOWLEDGE

Find the following sums (Q. No. 1–3) :

1. (*i*) $1 + 2 + 3 + ... + 30$ (*ii*) $1^2 + 2^2 + 3^2 + ... + 45^2$
(*iii*) $1^3 + 2^3 + 3^3 + ... + 22^3$.
2. (*i*) $5 + 6 + 7 + ... + 100$ (*ii*) $5^2 + 6^2 + 7^2 + ... + 20^2$
(*iii*) $15^3 + 16^3 + 17^3 + ... + 30^3$.
3. (*i*) $15 + 30 + 45 + ... + 390$ (*ii*) $3^2 + 6^2 + 9^2 + ... + 42^2$
(*iii*) $4^3 + 8^3 + 12^3 + ... + 40^3$.

Find the sum of the first n terms of the series whose nth term is given by (Q. No. 4—9) :

4. $n(n + 2)$ 5. $n(n + 3)$ 6. $n^2 + n + 1$
7. $n(n + 1)(n + 4)$ 8. $n^2 + 2^n$ 9. $n^3 - 4^n$
10. Find the sum of the series $1^2 + 4^2 + 7^2 + ...$ to n terms.
11. Find the sum of the series $3.8 + 6.11 + 9.14 + ...$ to n terms.
12. Find the sum of the first n terms of the series $3.1^2 + 5.2^2 + 7.3^2 + ...$
13. Find the sum of the first 25 terms of the series $5.6 + 6.7 + 7.8 + ...$
14. Find the sum of the first n terms of the series $1^2 + (1^2 + 2^2) + (1^2 + 2^2 + 3^2) + ...$
15. Evaluate : $1^3 + 3^3 + 5^3 + ... + (2n - 1)^3$.
16. Show that $\frac{1.2^2 + 2.3^2 + ... + n(n+1)^2}{1^2.2 + 2^2.3 + ... + n^2(n+1)} = \frac{3n+5}{3n+1}$.
17. Show that the sum of the cubes of any number of consecutive positive integers is divisible by the sum of those integers.

Answers

1. (i) 465 (ii) 31395 (iii) 64009
2. (i) 5040 (ii) 2840 (iii) 205200
3. (i) 5265 (ii) 9135 (iii) 193600

4. $\dfrac{n(n+1)(2n+7)}{6}$ 5. $\dfrac{n(n+1)(n+5)}{3}$ 6. $\dfrac{n^3+3n^2+5n}{3}$ 7. $\dfrac{n(n+1)(3n^2+23n+34)}{12}$

8. $\dfrac{n(n+1)(2n+1)}{6} + 2(2^n - 1)$ 9. $\dfrac{n^2(n+1)^2}{4} - \dfrac{4(4^n-1)}{3}$

10. $\dfrac{n(6n^2-3n-1)}{2}$ 11. $3n(n+1)(n+3)$ 12. $\dfrac{n(n+1)(3n^2+5n+1)}{6}$

13. 8950 14. $\dfrac{n(n+1)^2(n+2)}{12}$ 15. $n^2(2n^2-1)$.

Hints

14. $T_n = 1^2 + 2^2 + 3^2 + \ldots\ldots + n^2 = \Sigma n^2 = \dfrac{n(n+1)(2n+1)}{6}$.

16. $\text{L.H.S.} = \dfrac{\sum\limits_{k=1}^{n} k(k+1)^2}{\sum\limits_{k=1}^{n} k^2(k+1)} = \dfrac{\sum\limits_{k=1}^{n} k^3 + 2\sum\limits_{k=1}^{n} k^2 + \sum\limits_{k=1}^{n} k}{\sum\limits_{k=1}^{n} k^3 + \sum\limits_{k=1}^{n} k^2} = \dfrac{\Sigma n^3 + 2\Sigma n^2 + \Sigma n}{\Sigma n^3 + \Sigma n^2}$

17. Let the positive numbers be $m+1, m+2, \ldots\ldots, m+n$.

$\therefore$ Sum of cubes $= (m+1)^3 + (m+2)^3 + \ldots\ldots + (m+n)^3$

$= [1^3 + 2^3 + \ldots\ldots + (m+n)^3] - [1^3 + 2^3 + \ldots\ldots + m^3]$

$= \Sigma(m+n)^3 - \Sigma m^3 = \left[\dfrac{(m+n)(m+n+1)}{2}\right]^2 - \left[\dfrac{m(m+1)}{2}\right]^2$

Also, sum of these numbers $= (m+1) + (m+2) + \ldots\ldots + (m+n)$

$= [1 + 2 + \ldots\ldots + (m+n)] - [1 + 2 + \ldots.. + m]$

$= \Sigma(m+n) - \Sigma m = \dfrac{(m+n)(m+n+1)}{2} - \dfrac{m(m+1)}{2}$

$\therefore \dfrac{\text{Sum of cubes of numbers}}{\text{Sum of numbers}} = \dfrac{(m+n)(m+n+1)}{2} + \dfrac{m(m+1)}{2}$

$[\because\ x^2 - y^2 = (x-y)(x+y)]$

$= [1 + 2 + \ldots\ldots + (m+n)] + [1 + 2 + \ldots\ldots + m] \in \mathbf{N}$.

$\therefore$ Sum of cubes of numbers is divisible by sum of numbers.

17 PARTIAL FRACTIONS

17.1. INTRODUCTION

We know that the quotient of any two polynomials is called a rational function. In the present chapter, we shall learn the techniques of resolving a fraction into partial fractions. This method will help us to integrate any given rational function.

17.2. RESOLUTION OF A FRACTION INTO PARTIAL FRACTIONS

We know the method of finding the sum of two or more algebraic fractions by reducing the denominators of fractions to a common denominator, which is their L.C.M. For example :

$$\frac{2}{x-3}+\frac{5}{x+2}=\frac{2x+4+5x-15}{(x-3)(x+2)}=\frac{7x-11}{x^2-x-6}.$$

The reverse process of breaking up a single fraction into simpler fractions whose denominators are the factors of the denominator of the given fraction is called the **resolution of a fraction into its partial fractions.**

For example, $\dfrac{2x}{x^2-1}=\dfrac{1}{x-1}+\dfrac{1}{x+1}$.

Here $\dfrac{1}{x-1}$ and $\dfrac{1}{x+1}$ are the partial fractions of the fraction $\dfrac{2x}{x^2-1}$.

17.3. METHOD OF RESOLUTION INTO PARTIAL FRACTIONS

If $f(x)$ and $g(x)$ are polynomials, then $\dfrac{f(x)}{g(x)}$ is a rational function or a **rational fraction.**

If deg. $f(x)$ < deg. $f(x)$, then the rational fraction $\dfrac{f(x)}{g(x)}$ is called **proper,** otherwise $\dfrac{f(x)}{g(x)}$ is called **improper.**

For example, the rational fraction $\dfrac{x^2+x+3}{(x+2)(x^2+7)}$ is *proper* and the rational fraction $\dfrac{x^3+x+9}{x^2+5x-6}$ is *improper.*

If $\dfrac{f(x)}{g(x)}$ is an improper rational fraction, then we can divide $f(x)$ by $g(x)$ so as to write $\dfrac{f(x)}{g(x)}$ as the sum of a polynomial and a proper rational fraction.

∴ **Any improper fraction can be expressed as the sum of a polynomial and a proper fraction.**

It can be proved mathematically that any proper fraction may be resolved into partial fractions and :

(*i*) if $ax + b$ is any linear non-repeated factor in the denominator, then there corresponds a partial fraction of the form $\frac{A}{ax+b}$.

(*ii*) if $ax + b$ is any linear factor repeated r ($\in$ N) times in the denominator, then there corresponds partial fractions of the form $\frac{A}{ax+b}, \frac{B}{(ax+b)^2}, \frac{C}{(ax+b)^3}, \ldots\ldots$ r terms.

(*iii*) if $ax^2 + bx + c$ is any irreducible quadratic factor in the denominator, then there corresponds a partial fraction of the form $\frac{Ax+B}{ax^2+bx+c}$, and

(*iv*) if $ax^2 + bx + c$ is any irreducible factor repeated r ($\in$ N) times in the denominator, then there corresponds partial fractions of the form $\frac{Ax+B}{ax^2+bx+c}, \frac{Cx+D}{(ax^2+bx+c)^2}, \ldots\ldots$ r terms.

The quantities A, B, C, D, ... are all constants independent of x.

$\therefore$ The given proper fraction can be expressed as the sum of its partial fractions. **The constants A, B, C, D, ... occurring in the numerators of the partial fractions are determined by simplifying the sum of partial fractions and then giving various values to x, to obtain equations involving unknown constants or by comparing the coefficients of like powers of x.**

Illustrations. (*i*) The partial fractions of the proper fraction $\frac{x-1}{(x+1)(x-2)(x+3)^3}$ are of the types $\frac{A}{x+1}, \frac{B}{x-2}, \frac{C}{x+3}, \frac{D}{(x+3)^2}$ and $\frac{E}{(x+3)^3}$.

(*ii*) The partial fractions of the proper fraction $\frac{x^2+4}{(x-7)(x+2)^2(x^2+2x+3)}$ are of the types $\frac{A}{x-7}, \frac{B}{x+2}, \frac{C}{(x+2)^2}$ and $\frac{Dx+E}{x^2+2x+3}$.

(*iii*) The partial fractions of the proper fraction $\frac{x^3+x+7}{(x-1)(x+2)^2(x^2+5)^2}$ are of the types $\frac{A}{x-1}, \frac{B}{x+2}, \frac{C}{(x+2)^2}, \frac{Dx+E}{x^2+5}$ and $\frac{Fx+G}{(x^2+5)^2}$.

ILLUSTRATIVE EXAMPLES

Example 1. *Resolve the following fractions into partial fractions :*

(*i*) $\frac{3x-1}{x^2-1}$ (*ii*) $\frac{6x^3+5x^2-7}{3x^2-2x-1}$ (*iii*) $\frac{2x+1}{(x+2)(x-3)^2}$

(*iv*) $\frac{2x+3}{(x^2+1)(x+1)}$ (*v*) $\frac{x^2+5}{(x^2+x+1)(x^2-x+2)}$ (*vi*) $\frac{x}{(x-1)(x^2+1)^2}$.

Sol. (*i*) $\frac{3x-1}{x^2-1}$ is a proper fraction and the denominator has linear non-repeated factors.

Let $\frac{3x-1}{x^2-1} = \frac{3x-1}{(x-1)(x+1)} = \frac{A}{x-1} + \frac{B}{x+1}$. $\therefore$ $\frac{3x-1}{(x-1)(x+1)} = \frac{A(x+1)+B(x-1)}{(x-1)(x+1)}$

Multiplying by $(x-1)(x+1)$, we get, $3x-1 = A(x+1) + B(x-1)$ ...(1)

Now we find the values of A and B.

Method I. By comparing the coefficients of like powers of x.

(1) $\Rightarrow$ $3x - 1 = (A + B)x + (A - B)$.

$\therefore$ $A + B = 3$ and $A - B = -1$. Solving these equations, we get $A = 1$ and $B = 2$.

Method II. By giving specific values to x.

Let us put $x = 1$ and $x = 2$ in (1).

$x = 1 \Rightarrow 3(1) - 1 = A(1 + 1) + B(1 - 1) \Rightarrow 2A = 2 \Rightarrow A = 1$

$x = 2 \Rightarrow 3(2) - 1 = A(2 + 1) + B(2 - 1) \Rightarrow 3A + B = 5 \Rightarrow B = 5 - 3A = 5 - 3(1) = 2.$

$\therefore$ $A = 1$ and $B = 2$.

Method III. In this method, it is always convenient to use those values of x which make linear factors in the denominator zero.

Here $x - 1 = 0 \Rightarrow x = 1$ and $x + 1 = 0 \Rightarrow x = -1$. We put $x = 1$ and $x = -1$ in (1).

$x = 1 \Rightarrow 3(1) - 1 = A(1 + 1) + B(0) \Rightarrow A = 2/2 = 1$

$x = -1 \Rightarrow 3(-1) - 1 = A(0) + B(-1 - 1) \Rightarrow B = \frac{-4}{-2} = 2.$

$\therefore$ $A = 1$ and $B = 2$.

$$\therefore \quad \frac{3x-1}{(x-1)(x+1)} = \frac{\mathbf{1}}{\mathbf{x-1}} + \frac{\mathbf{2}}{\mathbf{x+1}}.$$

Remark. In the second method, we observe that

$A =$ Value of $\frac{3x-1}{x+1}$ when $x = 1$ and $B =$ Value of $\frac{3x-1}{x-1}$ when $x = -1$.

$\therefore$ The value of A can be found out by putting $x = 1$ in the given proper fraction, after omitting $x - 1$ from the denominator. Similarly, the value of B can be found out by putting $x = -1$ in the given proper fraction, after omitting $x + 1$ from the denominator.

Thus, when the factors in the denominator are *linear* and *non-repeated,* we can decompose the given proper fraction into partial fractions as follows :

$$\frac{3x-1}{(x-1)(x+1)} = \frac{3(1)-1}{(x-1)(1+1)} + \frac{3(-1)-1}{(-1-1)(x+1)} = \frac{2}{(x-1)2} + \frac{-4}{(-2)(x+1)} = \frac{1}{x-1} + \frac{2}{x+1}.$$

This is called the **short cut method** of finding the values of A, B etc. It is very important to bear in mind that this short cut method is to be used only when the denominator has *only linear and non-repeated factors.*

(*ii*) Given fraction is not a proper fraction. Dividing $6x^3 + 5x^2 - 7$ by $3x^2 - 2x - 1$, we get

$6x^3 + 5x^2 - 7 = (2x + 3)(3x^2 - 2x - 1) + (8x - 4)$.

$$\therefore \quad \frac{6x^3+5x^2-7}{3x^2-2x-1} = 2x + 3 + \frac{8x-4}{(x-1)(3x+1)} \qquad (\because \; 3x^2 - 2x - 1 = (x-1)(3x+1))$$

$$= 2x + 3 + \left(\frac{8(1) - 4}{(x - 1)(3(1) + 1)} + \frac{8\left(-\frac{1}{3}\right) - 4}{\left(-\frac{1}{3} - 1\right)(3x + 1)} \right)$$

(Using short-cut method)

$$= 2x + 3 + \frac{4}{(x - 1)4} + \frac{-20/3}{\left(-\frac{4}{3}\right)(3x + 1)} = \mathbf{2x + 3 + \frac{1}{x - 1} + \frac{5}{3x + 1}}.$$

(*iii*) $\frac{2x + 1}{(x + 2)(x - 3)^2}$ is a proper fraction. Let the partial fraction corresponding to the factor $x + 2$ be $\frac{A}{x + 2}$ and the partial fractions, corresponding to the factor $(x - 3)^2$ be $\frac{B}{x - 3}$ and $\frac{C}{(x - 3)^2}$.

$$\therefore \qquad \frac{2x + 1}{(x + 2)(x - 3)^2} = \frac{A}{x + 2} + \frac{B}{x - 3} + \frac{C}{(x - 3)^2}.$$

Multiplying both sides by $(x + 2)(x - 3)^2$, we get

$$2x + 1 = A(x - 3)^2 + B(x + 2)(x - 3) + C(x + 2) \qquad ...(1)$$

Now $x + 2 = 0 \Rightarrow x = -2$ and $x - 3 = 0 \Rightarrow x = 3$.

$$x = -2 \text{ in (1) implies } -3 = A(-5)^2 + B(0) + C(0) \Rightarrow A = \frac{-3}{25}$$

$$x = 3 \text{ in (1) implies } 7 = A(0) + B(0) + C(5) \Rightarrow C = \frac{7}{5}$$

Comparing the coefficients of x^2 in (1), we get $0 = A + B$.

$$\therefore \qquad B = -A = -\left(\frac{-3}{25}\right) = \frac{3}{25}.$$

$$\therefore \qquad \frac{2x + 1}{(x + 2)(x - 3)^2} = \frac{\frac{-3}{25}}{x + 2} + \frac{\frac{3}{25}}{x - 3} + \frac{\frac{7}{5}}{(x - 3)^2} = \mathbf{-\frac{3}{25(x + 2)} + \frac{3}{25(x - 3)} + \frac{7}{5(x - 3)^2}}.$$

(*iv*) $\frac{2x + 3}{(x + 1)(x^2 + 1)}$ is a proper fraction. Let the partial fraction corresponding to the factor $x + 1$ be $\frac{A}{x + 1}$ and the partial fraction corresponding to the factor $x^2 + 1$ be $\frac{Bx + C}{x^2 + 1}$.

$$\therefore \qquad \frac{2x + 3}{(x + 1)(x^2 + 1)} = \frac{A}{x + 1} + \frac{Bx + C}{x^2 + 1}.$$

Multiplying both sides by $(x + 1)(x^2 + 1)$, we get

$$2x + 3 = A(x^2 + 1) + (Bx + C)(x + 1) \qquad ...(1)$$

$$x = -1 \text{ in (1)} \Rightarrow 1 = A(2) + (-B + C)(0) \Rightarrow A = \frac{1}{2}$$

Comparing the coefficients of x^2 in (1), we get $0 = A + B$. $\therefore$ $B = -A = -\frac{1}{2}$.

Comparing the coefficients of x, we get $2 = B + C$. $\therefore$ $C = 2 - B = 2 + \frac{1}{2} = \frac{5}{2}$.

$$\therefore \quad \frac{2x+3}{(x+1)(x^2+1)} = \frac{1/2}{x+1} + \frac{(-1/2)x + 5/2}{x^2+1} = \frac{\mathbf{1}}{\mathbf{2(x+1)}} + \frac{\mathbf{5-x}}{\mathbf{2(x^2+1)}}.$$

(*v*) $\frac{x^2+5}{(x^2+x+1)(x^2-x+2)}$ is a proper fraction with non-repeated irreducible quadratic factors $x^2 + x + 1$ and $x^2 - x + 2$.

Let $\frac{x^2+5}{(x^2+x+1)(x^2-x+2)} = \frac{Ax+B}{x^2+x+1} + \frac{Cx+D}{x^2-x+2}$.

Multiplying both sides by $(x^2 + x + 1)(x^2 - x + 2)$, we get

$$x^2 + 5 = (Ax + B)(x^2 - x + 2) + (Cx + D)(x^2 + x + 1) \quad ...(1)$$

Comparing the coefficients of x^3, x^2, x and the constant terms, we get

$0 = A + C$ ...(1) $\quad 1 = -A + B + C + D$ ...(2)

$0 = 2A - B + C + D$ ...(3) $\quad 5 = 2B + D$ ...(4)

(1) $\Rightarrow$ $C = -A$. $\therefore$ (2) $\Rightarrow$ $-A + B - A + D = 1$ *i.e.*, $-2A + B + D = 1$ ...(5)

(3) $\Rightarrow$ $2A - B - A + D = 0$ *i.e.*, $A - B + D = 0$ ...(6)

(5) + 2(6) $\Rightarrow$ $-B + 3D = 1$ ...(7)

Solving (4) and (7), we get $B = 2$ and $D = 1$.

(5) $\Rightarrow$ $-2A + 2 + 1 = 1$ or $A = 1$. (1) $\Rightarrow$ $C = -A = -1$.

$\therefore$ $A = 1, B = 2, C = -1, D = 1$.

$$\therefore \quad \frac{x^2+5}{(x^2+x+1)(x^2-x+2)} = \frac{1.x+2}{x^2+x+1} + \frac{-1.x+1}{x^2-x+2} = \frac{\mathbf{x+2}}{\mathbf{x^2+x+1}} - \frac{\mathbf{x-1}}{\mathbf{x^2-x+2}}.$$

(*vi*) $\frac{x}{(x-1)(x^2+1)^2}$ is a proper fraction. Let $\frac{x}{(x-1)(x^2+1)^2} = \frac{A}{x-1} + \frac{Bx+C}{x^2+1} + \frac{Dx+E}{(x^2+1)^2}$.

Multiplying by $(x - 1)(x^2 + 1)^2$, we get

$$x = A(x^2 + 1)^2 + (Bx + C)(x - 1)(x^2 + 1) + (Dx + E)(x - 1) \quad ...(1)$$

$x = 1$ in (1) $\Rightarrow$ $1 = A(4) + (B + C)(0)(2) + (D + E)(0)$ $\Rightarrow$ $A = \frac{1}{4}$

Comparing the coefficients x^4, x^3, x^2, x, in (1), we get

$0 = A + B$ ...(2) $\quad 0 = -B + C$ ...(3)

$0 = 2A + B - C + D$ ...(4) $\quad 1 = -B + C - D + E$ ...(5)

(2) $\Rightarrow$ $B = -A = -\frac{1}{4}$, (3) $\Rightarrow$ $C = B = -\frac{1}{4}$ (4) $\Rightarrow$ $D = -2A - B + C = -\frac{1}{2} + \frac{1}{4} - \frac{1}{4} = -\frac{1}{2}$

(5) $\Rightarrow$ $E = 1 + B - C + D = 1 - \frac{1}{4} + \frac{1}{4} - \frac{1}{2} = \frac{1}{2}$

$$\therefore \quad A = \frac{1}{4}, B = -\frac{1}{4}, C = -\frac{1}{4}, D = -\frac{1}{2}, E = \frac{1}{2}.$$

$$\therefore \quad \frac{x}{(x-1)(x^2+1)^2} = \frac{1/4}{x-1} + \frac{(-1/4)x + (-1/4)}{x^2+1} + \frac{(-1/2)x + 1/2}{(x^2+1)^2}$$

$$= \frac{\mathbf{1}}{\mathbf{4(x-1)}} - \frac{\mathbf{x+1}}{\mathbf{4(x^2+1)}} - \frac{\mathbf{x-1}}{\mathbf{2(x^2+1)^2}}.$$

Example 2. *Resolve the following fractions into partial fractions*

(*i*) $\dfrac{1}{(x+1)(2x+1)}$ (*ii*) $\dfrac{2x+1}{(x+1)(x-2)}$ (*iii*) $\dfrac{x^2}{(x-1)(x-2)}$

(*iv*) $\dfrac{2x-1}{(x-1)(x+2)(x-3)}$ (*v*) $\dfrac{x^2+2}{(x^2+1)(x^2+4)}$ (*vi*) $\dfrac{3x+4}{x^2-5x+6}$.

Sol. (*i*) Let $\dfrac{1}{(x+1)(2x+1)} = \dfrac{A}{x+1} + \dfrac{B}{2x+1}$.

Multiplying by $(x + 1)(2x + 1)$, we get $1 = A(2x + 1) + B(x + 1)$ ...(1)

$x + 1 = 0 \Rightarrow x = -1$ and $2x + 1 = 0 \Rightarrow x = -1/2$.

In (1), $x = -1 \Rightarrow 1 = A(-2 + 1) + B(0) \Rightarrow A = -1$

$x = -1/2 \Rightarrow 1 = A(0) + B(-\frac{1}{2} + 1) \Rightarrow B = 2$

$$\therefore \quad \frac{1}{(x+1)(2x+1)} = -\frac{1}{x+1} + \frac{2}{2x+1}.$$

(*ii*) The factors in the denominator are *linear* and *non-repeated*. We use the *short cut method* to break up the given fraction

$$\frac{2x+1}{(x+1)(x-2)} = \frac{2(-1)+1}{(x+1)(-1-2)} + \frac{2(2)+1}{(2+1)(x-2)} = \frac{-1}{-3(x+1)} + \frac{5}{3(x-2)} = \frac{1}{3}\left[\frac{1}{x+1} + \frac{5}{x-2}\right].$$

(*iii*) Since the given fraction is not a *proper* fraction, we have by long division,

$$\frac{x^2}{(x-1)(x-2)} = \frac{x^2}{x^2-3x+2} = 1 + \frac{3x-2}{x^2-3x+2} = 1 + \frac{3x-2}{(x-1)(x-2)} \quad ...(1)$$

By *short cut method* :

$$\frac{3x-2}{(x-1)(x-2)} = \frac{3(1)-2}{(x-1)(1-2)} + \frac{3(2)-2}{(2-1)(x-2)} = \frac{1}{(x-1)(-1)} + \frac{4}{1(x-2)} = -\frac{1}{x-1} + \frac{4}{x-2}$$

$$\therefore \quad (1) \Rightarrow \quad \frac{x^2}{(x-1)(x-2)} = 1 - \frac{1}{x-1} + \frac{4}{x-2}.$$

(*iv*) The factors in the denominator are *linear* and *non-repeated*.

$$\therefore \quad \frac{2x-1}{(x-1)(x+2)(x-3)} = \frac{2(1)-1}{(x-1)(1+2)(1-3)} + \frac{2(-2)-1}{(-2-1)(x+2)(-2-3)} + \frac{2(3)-1}{(3-1)(3+2)(x-3)}$$

$$= \frac{1}{-6(x-1)} + \frac{-5}{15(x+2)} + \frac{5}{10(x-3)}.$$

(*v*) Let $z = x^2$. $\therefore \quad \dfrac{x^2+2}{(x^2+1)(x^2+4)} = \dfrac{z+2}{(z+1)(z+4)}$

Let $\dfrac{z+2}{(z+1)(z+4)} = \dfrac{A}{z+1} + \dfrac{B}{z+4}$. $\therefore \quad z + 2 = A(z + 4) + B(z + 1)$

$z = -1 \Rightarrow 1 = 3A + B(0) \Rightarrow A = 1/3$

$z = -4 \Rightarrow -2 = A(0) + (-3)B \Rightarrow B = 2/3$

$\therefore \quad \dfrac{z+2}{(z+1)(z+4)} = \dfrac{1/3}{z+1} + \dfrac{2/3}{z+4}. \quad \therefore \quad \dfrac{x^2+2}{(x^2+1)(x^2+4)} = \dfrac{1}{3(x^2+1)} + \dfrac{2}{3(x^2+4)}.$

(*vi*) Let $\dfrac{3x+4}{x^2-5x+6} = \dfrac{3x+4}{(x-2)(x-3)} = \dfrac{A}{x-2} + \dfrac{B}{x-3}$.

Multiplying by $(x-2)(x-3)$, we get $3x + 4 = A(x-3) + B(x-2)$ ...(1)

$x - 2 = 0 \Rightarrow x = 2$ and $x - 3 = 0 \Rightarrow x = 3$.

We put $x = 2$ and $x = 3$ in (1)

$x = 2 \Rightarrow 3(2) + 4 = A(2-3) + B(2-2) \Rightarrow 10 = -A + 0 \Rightarrow A = -10$

$x = 3 \Rightarrow 3(3) + 4 = A(3-3) + B(3-2) \Rightarrow 13 = 0 + B \Rightarrow B = 13$

$\therefore \quad \dfrac{3x+4}{(x-2)(x-3)} = \dfrac{-10}{x-2} + \dfrac{13}{x-3}.$

Example 3. *Resolve the following fractions into partial fractions*

(*i*) $\dfrac{12x^2-2x-9}{(4x^2-1)(x+3)}$ (*ii*) $\dfrac{x^2+8x+4}{x^3-4x}$

(*iii*) $\dfrac{2x}{(x^2+2)(x^2+3)}$.

Sol. (*i*) The factors of $(4x^2 - 1)(x + 3)$ are *linear* and *non-repeated.*

Let $\dfrac{12x^2-2x-9}{(2x-1)(2x+1)(x+3)} = \dfrac{A}{2x-1} + \dfrac{B}{2x+1} + \dfrac{C}{x+3}$

$$= \frac{A(2x+1)(x+3) + B(2x-1)(x+3) + C(2x-1)(2x+1)}{(2x-1)(2x+1)(x+3)}$$

$\therefore \quad 12x^2 - 2x - 9 = A(2x+1)(x+3) + B(2x-1)(x+3) + C(2x-1)(2x+1)$ ...(1)

Now, $2x - 1 = 0 \Rightarrow x = \dfrac{1}{2}$, $2x + 1 = 0 \Rightarrow x = -\dfrac{1}{2}$ and $x + 3 = 0 \Rightarrow x = -3$.

$x = \dfrac{1}{2} \Rightarrow 3 - 1 - 9 = A(2)\left(\dfrac{7}{2}\right) + 0 + 0 \Rightarrow -7 = 7A \Rightarrow A = -1$

$x = -\dfrac{1}{2} \Rightarrow 3 + 1 - 9 = 0 + B(-2)\left(\dfrac{5}{2}\right) + 0 \Rightarrow -5 = -5B \Rightarrow B = 1$

and $x = -3 \Rightarrow 108 + 6 - 9 = 0 + 0 + C(-7)(-5) \Rightarrow 105 = 35C \Rightarrow C = 3$.

$\therefore \quad \dfrac{12x^2-2x-9}{(2x-1)(2x+1)(x+3)} = \dfrac{-1}{2x-1} + \dfrac{1}{2x+1} + \dfrac{3}{x+3}.$

(*ii*) $\dfrac{x^2+8x+4}{x^3-4x} = \dfrac{x^2+8x+4}{x(x+2)(x-2)} = \dfrac{A}{x} + \dfrac{B}{x+2} + \dfrac{C}{x-2}$ ...(1)

$$= \frac{A(x+2)(x-2) + Bx(x-2) + Cx(x+2)}{x(x+2)(x-2)}$$

$\therefore \quad x^2 + 8x + 4 = A(x + 2)(x - 2) + Bx(x - 2) + Cx(x + 2)$

$x = 0 \quad \Rightarrow \quad 0 + 0 + 4 = -4A \quad \text{or} \quad A = -1$

$x = -2 \Rightarrow \quad 4 - 16 + 4 = 8B \quad \text{or} \quad B = -1$

$x = 2 \quad \Rightarrow \quad 4 + 16 + 4 = 8C \quad \text{or} \quad C = 3.$

Hence (1) can be written as $\dfrac{x^2 + 8x + 4}{x^3 - 4x} = \dfrac{-1}{x} - \dfrac{1}{x+2} + \dfrac{3}{x-2}$.

(*iii*) $\dfrac{2x}{(x^2+2)(x^2+3)} = \dfrac{Ax + B}{x^2 + 2} + \dfrac{Cx + D}{x^2 + 3}$

$$= \frac{(Ax + B)(x^2 + 3) + (Cx + D)(x^2 + 2)}{(x^2 + 2)(x^2 + 3)}$$

Multiplying by $(x^2 + 2)(x^2 + 3)$, we get

$$2x = (Ax + B)(x^2 + 3) + (Cx + D)(x^2 + 2)$$

Equating the coefficient of x^3, both sides, we get

$$0 = A + C \qquad ...(1)$$

Equating the coefficient of x^2, both sides, we get

$$0 = B + D \qquad ...(2)$$

Equating the coefficient of x, both sides, we get

$$2 = 3A + 2C \qquad ...(3)$$

Equating the constant term, we get

$$0 = 3B + 2D \qquad ...(4)$$

From (1), $C = -A$ and from (3), $2 = 3A - 2A \Rightarrow A = 2$

$\therefore \quad C = -2$

From (2) $B = -D$ and from (4) $-3D + 2D = 0$

$\Rightarrow \quad D = 0, B = 0$

$$\therefore \quad \frac{2x}{(x^2+2)(x^2+3)} = \frac{2x}{x^2+2} - \frac{2x}{x^2+3}.$$

Example 4. *Resolve the following fractions into partial fractions :*

(*i*) $\dfrac{5x^3 + 18x^2 - 10x - 6}{x(x+3)(5x-2)}$ (*ii*) $\dfrac{x^4 + x^3 + 2x^2 + 4x + 1}{x(x+1)}$

Sol. (*i*) Since the given fraction is *not a proper* fraction, we have by long division,

$$\frac{5x^3 + 18x^2 - 10x - 6}{x(x+3)(5x-2)} = 1 + \frac{5x^2 - 4x - 6}{5x^3 + 13x^2 - 6x} \qquad ...(1)$$

We resolve $\dfrac{5x^2 - 4x - 6}{5x^2 + 13x^2 - 6x}$ into partial fractions.

Let $\dfrac{5x^2 - 4x - 6}{x(x+3)(5x-2)} = \dfrac{A}{x} + \dfrac{B}{x+3} + \dfrac{C}{5x-2}$ ...(2)

$$= \frac{A(x+3)(5x-2) + Bx(5x-2) + Cx(x+3)}{x(x+3)(5x-2)}.$$

$\therefore \quad 5x^2 - 4x - 6 = A(x+3)(5x-2) + Bx(5x-2) + Cx(x+3)$ ...(3)

When $x = 0$, (3) $\Rightarrow -6 = -6A$ $\Rightarrow A = 1$

When $x = -3$, (3) $\Rightarrow 45 + 12 - 6 = B(-3)(-15-2)$ $\Rightarrow B = 1$

When $x = \frac{2}{5}$, (3) $\Rightarrow 5\left(\frac{2}{5}\right)^2 - 4\left(\frac{2}{5}\right) - 6 = C\left(\frac{2}{5}\right)\left(\frac{2}{5}+3\right)$ $\Rightarrow C = -5.$

$$\therefore \quad (3) \Rightarrow \frac{5x^2 - 4x - 6}{x(x+3)(5x-2)} = \frac{1}{x} + \frac{1}{x+3} - \frac{5}{5x-2}$$

(*ii*) Since the given fraction is *not a proper* fraction, we use long division to get

$$\frac{x^4 + x^3 + 2x^2 + 4x + 1}{x(x+1)} = x^2 + 2 + \frac{2x+1}{x^2+x} \quad ...(1)$$

We resolve $\frac{2x+1}{x^2+x}$ into partial fractions.

Let $\frac{2x+1}{x(x+1)} = \frac{A}{x} + \frac{B}{x+1} = \frac{A(x+1)+Bx}{x(x+1)}$ ...(2)

$\therefore \quad 2x + 1 = A(x+1) + Bx.$ ...(3)

When $x = 0$, (3) $\Rightarrow 1 = A$

When $x = -1$, (3) $\Rightarrow -2 + 1 = -B$ or $B = 1.$

$$\therefore \quad (2) \Rightarrow \frac{2x+1}{x(x+1)} = \frac{1}{x} + \frac{1}{x+1}.$$

Example 5. *Resolve the following fractions into partial fractions :*

(*i*) $\frac{x^2+1}{(x-1)^2(x+3)}$ (*ii*) $\frac{3x+1}{(x-2)^2(x+2)}$.

Sol. (*i*) Let $\frac{x^2+1}{(x-1)^2(x+3)} = \frac{A}{x-1} + \frac{B}{(x-1)^2} + \frac{C}{x+3}$.

$\therefore \quad x^2 + 1 = A(x-1)(x+3) + B(x+3) + C(x-1)^2$ ...(1)

$x = 1 \Rightarrow 1 + 1 = A(0)(4) + B(4) + C(0) \Rightarrow B = 1/2$

$x = -3 \Rightarrow 9 + 1 = A(-4)(0) + B(0) + C(16) \Rightarrow C = 5/8$

Let us put $x = 0$ in (1).

$\therefore \quad 0 + 1 = A(-1)(3) + B(3) + C(1)$

$$\Rightarrow \quad 3A = 3B + C - 1 = 3\left(\frac{1}{2}\right) + \frac{5}{8} - 1 = \frac{9}{8} \quad \Rightarrow \quad A = \frac{3}{8}.$$

$$\therefore \quad \frac{x^2+1}{(x-1)^2(x+3)} = \frac{3/8}{x-1} + \frac{1/2}{(x-1)^2} + \frac{5/8}{x+3}$$

(*ii*) Let $\frac{3x+1}{(x-2)^2(x+2)} = \frac{A}{x-2} + \frac{B}{(x-2)^2} + \frac{C}{x+2}$.

$\therefore \quad 3x + 1 = A(x-2)(x+2) + B(x+2) + C(x-2)^2$ ...(1)

$x = 2 \Rightarrow 7 = A(0)(4) + B(4) + C(0) \Rightarrow B = 7/4$

$x = -2 \Rightarrow -5 = A(-4)(0) + B(0) + C(16) \Rightarrow C = -5/16$

Let us put $x = 0$ in (1).

$\therefore \quad 1 = A(-2)(2) + B(2) + C(4)$

$\Rightarrow \quad 1 = -4A + 2\left(\frac{7}{4}\right) + 4\left(-\frac{5}{16}\right) \quad \Rightarrow \quad 4A = \frac{7}{2} - \frac{5}{4} - 1 = \frac{5}{4} \quad \Rightarrow \quad A = \frac{5}{16}$

$$\therefore \quad \frac{3x+1}{(x-2)^2(x+2)} = \frac{5/16}{x-2} + \frac{7/4}{(x-2)^2} + \frac{-5/16}{x+2}$$

Example 6. *Resolve the following fractions into partial fractions:*

(*i*) $\frac{x^4}{x^4 - 1}$ (*ii*) $\frac{x^4}{x^4 - 16}$

(*iii*) $\frac{x}{x^3 + x^2 + x + 1}$

Sol. (*i*) By long division, $\frac{x^4}{x^4-1} = 1 + \frac{1}{x^4-1}$

Let $\quad \frac{1}{x^4-1} = \frac{1}{(x^2-1)(x^2+1)} = \frac{1}{(x-1)(x+1)(x^2+1)} = \frac{A}{x-1} + \frac{B}{x+1} + \frac{Cx+D}{x^2+1}$ ...(1)

Multiplying by $(x-1)(x+1)(x^2+1)$, we get

$\therefore \quad 1 = A(x+1)(x^2+1) + B(x-1)(x^2+1) + (Cx+D)(x^2-1)$

$x = 1 \quad \Rightarrow \quad 1 = A(2)(2) + B(0)(2) + (C + D)(0) \quad \Rightarrow \quad 1 = 4A \quad \Rightarrow \quad A = \frac{1}{4}$

$x = -1 \quad \Rightarrow \quad 1 = A(0)(2) + B(-2)(2) + (-C + D)(0) \quad \Rightarrow \quad 1 = -4B \quad \Rightarrow \quad B = -\frac{1}{4}$

Comparing the coefficient of x^3, we get $0 = A + B + C$.

$\therefore \quad C = -A - B = -\frac{1}{4} + \frac{1}{4} = 0.$

Comparing the coefficient of x^2, we get $0 = A - B + D$.

$\therefore \quad D = B - A = -\frac{1}{4} - \frac{1}{4} = -\frac{1}{2}.$

$\therefore \quad A = \frac{1}{4}, B = -\frac{1}{4}, C = 0, D = -\frac{1}{2}.$

$$\therefore \quad \frac{1}{x^4-1} = \frac{1/4}{x-1} + \frac{-1/4}{x+1} + \frac{0.x + (-1/2)}{x^2+1} = \frac{1}{4(x-1)} - \frac{1}{4(x+1)} - \frac{1}{2(x^2+1)}$$

$$\therefore \quad (1) \Rightarrow \quad \frac{x^4}{x^4-1} = 1 + \frac{1}{4(x-1)} - \frac{1}{4(x+1)} - \frac{1}{2(x^2+1)}$$

(*ii*) By long division, $\frac{x^4}{x^4-16} = 1 + \frac{16}{x^4-16}$ ...(1)

Consider $\frac{16}{x^4-16} = \frac{16}{(x^2-4)(x^2+4)} = \frac{16}{(y-4)(y+4)}$ where $y = x^2$.

$$= \frac{16}{(y-4)(y+4)} + \frac{16}{(-4+4)(y+4)} = \frac{2}{y-4} - \frac{2}{y+4} = \frac{2}{x^2-4} - \frac{2}{x^2+4}$$

$\therefore$ (1) $\Rightarrow$ $\dfrac{16}{x^4-16} = 1 + \dfrac{2}{x^2-4} - \dfrac{2}{x^2+4}$.

(*iii*) Let $\dfrac{x}{(x^2+1)(x+1)} = \dfrac{Ax+B}{x^2+1} + \dfrac{C}{x+1}$.

$\therefore$ $x = (Ax + B)(x + 1) + C(x^2 + 1)$ ...(1)

$x = -1 \Rightarrow -1 = (-A + B)(0) + C(2) \Rightarrow C = -1/2$

Comparing the coefficients of like powers of x in (1), we get

$0 = A + C$ ...(2) $1 = A + B$ ...(3)

(2) $\Rightarrow$ $A = -C = -(-1/2) = 1/2$

(3) $\Rightarrow$ $B = 1 - A = 1 - (1/2) = 1/2$

$$\therefore \quad \frac{x}{(x^2+1)(x+1)} = \frac{\frac{1}{2}x + \frac{1}{2}}{x^2+1} + \frac{-\frac{1}{2}}{x+1} = \frac{x+1}{2(x^2+1)} - \frac{1}{2(x+1)}.$$

Example 7. *Resolve the following fractions into partial fractions :*

(*i*) $\dfrac{2x^3}{(x^2+1)^2}$ (*ii*) $\dfrac{x^5 - x^4 + 4x^3 - 4x^2 + 8x - 4}{(x^2+2)^3}$.

Sol. (*i*) Let $\dfrac{2x^3}{(x^2+1)^2} = \dfrac{Ax+B}{x^2+1} + \dfrac{Cx+D}{(x^2+1)^2}$

$\therefore$ $2x^3 = (Ax + B)(x^2 + 1) + (Cx + D)$

$\Rightarrow$ $2x^3 = Ax^3 + Bx^2 + (A + C)x + B + D$

Comparing the coefficients, we get $A = 2, B = 0, A + C = 0, B + D = 0$

Solving, we get $A = 2, B = 0, C = -2, D = 0$.

$$\therefore \quad \frac{2x^3}{(x^2+1)^2} = \frac{2x+0}{x^2+1} + \frac{-2x+0}{(x^2+1)^2} = \frac{2x}{x^2+1} - \frac{2x}{(x^2+1)^2}$$

(*ii*) Let $\dfrac{x^5 - x^4 + 4x^3 - 4x^2 + 8x - 4}{(x^2+2)^3} = \dfrac{Ax+B}{x^2+2} + \dfrac{Cx+D}{(x^2+2)^2} + \dfrac{Ex+F}{(x^2+2)^3}$

$\therefore$ $x^5 - x^4 + 4x^3 - 4x^2 + 8x - 4 = (Ax + B)(x^2 + 2)^2 + (Cx + D)(x^2 + 2) + (Ex + F)$

$= Ax^5 + Bx^4 + (4A + C)x^3 + (4B + D)x^2 + (4A + 2C + E)x + (4B + 2D + F)$

Comparing the coefficients, we get

$A = 1, B = -1, 4A + C = 4, 4B + D = -4, 4A + 2C + E = 8, 4B + 2D + F = -4$.

$\Rightarrow$ $A = 1, B = -1, C = 0, D = 0, E = 4, F = 0$.

$$\therefore \quad \frac{x^5 - x^4 + 4x^3 - 4x^2 + 8x - 4}{(x^2+2)^3} = \frac{x-1}{x^2+2} + \frac{4x}{(x^2+2)^3}.$$

TEST YOUR KNOWLEDGE

1. Resolve the following fractions into partial fractions :

(*i*) $\dfrac{3x+2}{(x-1)(2x+3)}$ (*ii*) $\dfrac{x}{(x+2)(3-2x)}$ (*iii*) $\dfrac{x-1}{(x+1)(x-2)}$

(*iv*) $\dfrac{2x-5}{(x-1)^2}$ (*v*) $\dfrac{x^3-5x^2+6x-2}{x(x-2)^2(x-1)}$.

2. Resolve the following fractions into partial fractions :

(*i*) $\dfrac{x^2+6x+6}{(x+1)^2(x+3)}$ (*ii*) $\dfrac{x^3+x^2+1}{x^2(x-1)(x+2)}$ (*iii*) $\dfrac{2x^2+9x-6}{(x-1)(x+3)(x+2)}$

(*iv*) $\dfrac{x}{(x^2+4)(x^2+9)}$.

3. Resolve the following fractions into partial fractions :

(*i*) $\dfrac{3x^2-x^3-4x+52}{(x^2+4)(x+1)(x-2)}$ (*ii*) $\dfrac{17}{(x^2+4)(x^2+2x+5)}$ (*iii*) $\dfrac{3x-33}{(x-1)(x+2)(x^2+9)}$

(*iv*) $\dfrac{1}{x^2(x^2+1)}$.

4. Resolve the following fractions into partial fractions :

(*i*) $\dfrac{1}{(x-1)(x+1)(x^2+1)}$ (*ii*) $\dfrac{x^2-3x+1}{(x-1)^3}$ (*iii*) $\dfrac{5x}{(1-x)(x+1)(x^2+4)}$

(*iv*) $\dfrac{x^2+1}{x^2(x+1)^2}$ (*v*) $\dfrac{(x^2+1)(x^2+2)}{(x^2+3)(x^2+4)}$.

Answers

1. (*i*) $\dfrac{1}{x-1}+\dfrac{1}{2x+3}$ (*ii*) $\dfrac{2}{7(x+2)}+\dfrac{3}{7(3-2x)}$ (*iii*) $\dfrac{2}{x-1}-\dfrac{3}{(x-1)^2}$

(*iv*) $\dfrac{2}{x-1}-\dfrac{3}{(x-1)^2}$ (*v*) $\dfrac{1}{2x}+\dfrac{1}{2(x-2)}-\dfrac{1}{(x-2)^2}$.

2. (*i*) $\dfrac{7}{4(x+1)}+\dfrac{1}{2(x+1)^2}-\dfrac{3}{4(x+1)^3}$ (*ii*) $-\dfrac{1}{4x}-\dfrac{1}{2x^2}+\dfrac{1}{x-1}+\dfrac{1}{4(x+2)}$

(*iii*) $\dfrac{5}{12(x-1)}-\dfrac{15}{4(x+3)}+\dfrac{16}{3(x+2)}$ (*iv*) $\dfrac{1}{5(x^2+4)}-\dfrac{1}{5(x^2+9)}$

3. (*i*) $\frac{-4}{x+1}+\frac{2}{x-1}+\frac{x-6}{x^2+4}$ (*ii*) $\frac{-2x+1}{x^2+4}+\frac{2x+3}{x^2+2x+5}$

(*iii*) $\frac{-1}{x-1}+\frac{1}{x+2}+\frac{3}{x^2+9}$ (*iv*) $\frac{1}{x^2}-\frac{1}{x^2+1}$

4. (*i*) $\frac{1}{4(x-1)}-\frac{1}{4(x+1)}-\frac{1}{2(x^2+1)}$ (*ii*) $\frac{1}{x-1}-\frac{1}{(x-1)^2}-\frac{1}{(x-1)^3}$

(*iii*) $-\frac{1}{2(x-1)}-\frac{1}{2(x+1)}+\frac{x}{(x^2+4)}$ (*iv*) $-\frac{2}{x}+\frac{1}{x^2}+\frac{2}{x+1}+\frac{2}{(x+1)^3}$

(*v*) $1+\frac{2}{x^2+3}-\frac{6}{x^2+4}$.

INDEX

A

B

C

D

E

F

G

H

I

J

K

L

Q

R

S

T

U

V

W

Z